Infrared Thermography and Thermal Nondestructive Testing

Vladimir Vavilov • Douglas Burleigh

Infrared Thermography and Thermal Nondestructive Testing

Springer

Vladimir Vavilov
Institute of Nondestructive Testing
Tomsk Polytechnic University
Tomsk, Russia

Douglas Burleigh
La Jolla Cove Consulting
San Diego, CA, USA

ISBN 978-3-030-48001-1 ISBN 978-3-030-48002-8 (eBook)
https://doi.org/10.1007/978-3-030-48002-8

This Springer imprint is published by the registered company Springer Nature Switzerland AG
The registered company address is: Gewerbestrasse 11, 6330 Cham, Switzerland

Acknowledgment

Many experimental illustrations in this book were obtained in the framework of the Tomsk Polytechnic University Competitiveness Enhancement Program.

Introduction

Subtle is the Lord but malicious He is not.
– A. Einstein

Temperature is a quantification of the internal energy in a physical system or object. In any system, thermal processes, such as energy (heat) generation, conversion, transport, storage, and usage, take place continuously.

Analyzing the thermal processes reveals useful information. In nondestructive testing (NDT), temperature evaluation is typically accomplished by analyzing the thermal, or infrared (IR), radiation emitted by test objects.

While there is some controversy on the early history of IR (infrared) science, the first proposals of practical applications of IR can be traced to the end of the nineteenth century, and the first implementations of IR technology took place in the first half of the twentieth century. However, as noted by Busse, the first investigations on the dynamics of heat propagation can be traced to very early works by Fourier and Ångstrom [1–3]. Historic milestones in both IR technology and thermal nondestructive testing (TNDT) are presented in Table 1 [4–6]. An exhaustive review of the history of IR engineering was published by Hudson [7]. In 1935, Nichols suggested the use of an IR radiometer for analyzing hot rolled metals [8]. In 1937 Vernotte suggested an active IR method for the study of the thermal properties of materials [9], and similar methods are still in use today. An active TNDT process that is similar to contemporary techniques was proposed in 1965 by Beller for the inspection of Polaris rocket motor cases [10]. In that same time period, Green performed basic research on the active IR testing of nuclear reactor fuel elements, where he described a notorious emissivity problem [11]. In the 1970s, during the "Space Race," a number of aerospace researchers got involved in TNDT and the American Society for Nondestructive Testing (ASNT) established an IR committee, which even published its transactions. In that time period, most active TNDT was performed using either thermal wave or pulse techniques. The discussion of which technique is the most effective has continued for decades [12–15], and in recent years this discussion has included the Lock-in method.

By the end of the 1970s, IR thermography applications were still rather qualitative, thus preventing an accurate comparison of TNDT with other inspection techniques. Improvements to TNDT were achieved by the use of heat conduction theory, the

Table 1 IR technology and TNDT in short: historical milestones. (Adapted from [4–6])[a]

BCE. The existence of thermal rays was hypothesized by Titus Lukretius Carus (99 – 55 BCE), a Roman poet and the author of the epic "De Rerum Natura" ("On the Nature of the Universe").	
1770.	Pictet, a French scientist, described his famous experiment on focusing heat and cold.
1800.	Sir William Herschel was credited with the discovery of IR radiation.
1840.	John Herschel, William's son, proposed a prototype of an evaporograph and introduced the term "thermogram."
1900	*s.* Einstein, Golitzyn, Kirchhoff, Planck, Wien et al. discovered the laws of thermal radiation.
1927.	Pokrovsky (in Russia) performed experiments on the remote detection of war ships.
1934.	Holst invented the image converter tube (wavelength range up to 1.5 μm).
1935.	Nichols performed IR radiometric analysis of hot rolled metals [8].
1937.	Vernotte: Determination of material thermal properties [9].
1940	*s.* Night vision equipment was developed. Infrared photodetectors.
1948.	Parker and Marshall: Analyzed the temperature distribution in brake shoes [16].
1949.	Gorrill: Inspection of soldered seams on a tin can [17]. Leslie and Wait: power transmission line surveys [18].
1954.	Barnes airborne scanner served as a prototype for IR imagers.
1960.	IR scanners can produce images in 5 minutes.
1961.	Detection of overheated components on circuit boards [19].
1965.	AGA (later AGEMA Infrared, now FLIR Systems) introduced the first radiometric IR camera.
1965.	Beller: Active TNDT of Polaris rockets [8]. Green: Active TNDT of nuclear fuel elements [11].
1970	*s.* Balageas, Karpelson[b], MacLaughlin, Popov[b], Vavilov et al: 1D, 2D and 3D Thermal NDT models, using classical heat transfer physics.
1973.	AGA introduced the model 750, the first battery-operated portable industrial IR imager.
1980	*s.* Almond, Berardi, Busse, Carlomagno, Cielo, Favro, Hartikainen, Mandelis, Milne, Reynolds, Rosencwaig et al.: Thermal wave theory, pulsed TNDT, thermal property measurements. Degiovanni, Maillet et al. introduced "thermal quadrupoles."
1987.	AGEMA introduced the model 470, a thermoelectrically cooled IR imager with onboard digital image storage.
1990	*s.* Almond, Balageas, Bison, Bremond, Budadin[b], Burleigh, Busse, Cramer, Degiovanni, Grinzato, Krapez, Luong, MacLachlan Spicer, Maldague, Marinetti, Rantala, Vavilov, Walther, Winfree et al.: Practical implementation of TNDT. Development of defect characterization algorithms. IR computerized systems. Study of fatigue phenomena. Various industrial applications were developed. IR thermographic measurements of thermal diffusivity and effusivity.

(continued)

Table 1 continued

1990	*s*. Uncooled microbolometric IR imagers. Quantum well arrays. Dramatic drop in imager prices. A camcorder-style FPA camera from Inframetrics (1995). A low-cost snapshot IR camera from Infrared Solutions (1997).
2000.	FLIR Systems ThermaCAM PM 695 incorporated thermal/visual/voice and text data logging in an IR imager.
2000	*s*. Avdelidis, Ayvasyan, Balageas, Batsale, Bendada, Busse, Cernuschi, Cramer, Dillenz, Shepard, Maillet, Maldague, Mulaveesala, Nowakowski, Oswald-Tranta, Sakagami, Tuli, Więcek, Wu et al.: Commercial Thermal NDT equipment. Novel stimulation techniques. Sophisticated data processing algorithms.
2010	*s*. Maierhofer, Oswald-Tranta, Rajic, Safai, Schlichting, Zalameda et al.: Eddy current IR thermography. Portable equipment for active TNDT.

[a]The concept for this table and some references have been borrowed from [4–6]. Some milestone studies are also presented in references [3, 12, 20–32]. The emphasis is on active TNDT. Due to space limitations in the table, many researchers who contributed to passive IR thermographic inspection are unfortunately not mentioned

[b]Early Russian publications, not easily available to Western readers [33–35]

basics of which were summarized in the well-known books by Carslaw and Jaeger [36] and Luikov [37]. A "thermophysical" approach to TNDT has been separately and cooperatively developed by Carlomagno and Berardi [13], Vavilov and Taylor [21], MacLaughlin and Mirchandani [22], Mandelis [23], Breitenstein and Langekamp [24], Balageas et al. [25], Popov and Karpelson [33], Degiovanni, Maillet et al. [26] who introduced one- (1D), two- (2D), and three-dimensional (3D) TNDT models.

In condition monitoring and predictive maintenance, IR diagnostics is considered to be a reliable and practical tool that provides significant economic benefit. The appearance of second generation IR imagers in the last decade, as well as an accompanying dramatic increase in computer processing, resulted in significant improvements to TNDT.

Nowadays, IR thermographic diagnostics and TNDT represent a mature technology field that combines achievements in the understanding of heat conduction, material science, IR technology, and computer data processing. Interest in TNDT increased due to its wide variety of applications, high inspection rate, and low relative cost.

Most major aerospace companies have tried the use of TNDT for the inspection of composite materials and some are using it as a primary inspection method.

The use of TNDT may be complimented by other NDT methods, such as ultrasonic, eddy current, and laser methods (these techniques may involve similar physical principles and hardware). The weaknesses of one method may be offset by the strength of another.

There is a strong relationship between the measurement of thermal properties and the detection of hidden defects because a defect, such as a delamination, will cause a variation in the local material properties, while a more widespread concentration of micro-defects, such as porosity, can cause a variation in the bulk thermal properties. Defects in composite materials that can be detected by TNDT include delaminations, skin-to-core (honeycomb or other) disbonds, voids, porosity, foreign material inclusions, trapped water, variations in thickness and other geometric details, and variations in thermal or physical properties.

The first reviews of the pioneering research in the area of active TNDT were written by Willburn [6] and McGonnagle [38]. Pulsed TNDT of composites was summarized by Milne and Reynolds in 1984 [39]. Periodic thermal waves in application to TNDT were described by Almond and Patel [14]. Later, a literature survey and bibliography on TNDT of composite materials was compiled by Burleigh [40]. General features of active thermal NDT of composites were described by Maldague [41] and Vavilov [4, 5, 21] and summarized in the ASNT IR handbook [20]. A recent review of TNDT was prepared by Vavilov and Burleigh [42].

The authors of this book have been discussing and collaborating on TNDT research and applications since they first met at the Society of Photo-optical Instrumentation Engineers (SPIE) conference in 1992. Others contributed to this book by providing their opinions, suggestions, and constructive criticism. These, include Xavier Maldague (Canada), Shepard (USA), Greg Stockton (USA), Daniel Balageas (France), Gerd Busse (Germany), Giovanni Carlomagno (Italy), Timo Kauppinen (Finland), Takahide Sakagami (Japan), Sharon Semanovich (USA), Grinzato, Bison and Marinetti (Italy), and many others. The chapter devoted to IR applications in electrical power systems was edited by Bob Madding, who is one of the world pioneers in applications of IR. Bob Madding also started Thermosense, the first conference dedicated solely to IR. This conference is now in its 42nd year and is managed by SPIE.

References to Introduction

1. Fourier, J.: Théory du mouvement de la chaleur dans les corps solides, 1er partie. Mémoires de l'Académie des Sciences. **4**, 185–555 (1824). **5**, 153–246 (1826)
2. Ångstrom, M.A.J.: New method of determining the thermal conductivity of bodies. Philos. Mag. **25**, 130–142 (1863)
3. Busse, G.: Imaging with optically generated thermal waves. In: Physical Acoustics: Principles and Methods, vol. 18, (1985)
4. Vavilov, V.P.: Thermal NDT: historical milestones, state-of-the-art and trends. QIRT. J. **11**(1), 66–83 (2014)
5. Vavilov, V.P.: Pulsed thermal NDT of materials: Back to basics. Nondestruct. Test. Eval. **22**(2–3), 177–197 (2007)
6. Willburn, D.K.: Survey of infrared inspection and measurement techniques. Mater. Res. Stand. **1**, 528 (1961)
7. Hudson, R.D.: Infrared System Engineering. Wiley-Interscience, New York (1969)
8. Nichols, J.T.: Temperature measuring. US Patent 2,008,793 (1935)

9. Vernotte, P.: Mesure de la conductibilite thermique des isolants. methode de toushau. Chaleur Ind. **208**, 331–337 (1937). (in French)
10. Beller, W.S.: Navy sees promise in infrared thermography for solid case checking. Missiles Rockets. **22**, 1234–1241 (1965)
11. Green, D.R.: Emissivity-independent infrared thermal testing method. Mater. Eval. **23**(2), 79–85 (1965)
12. Rosencwaig, A., Gersho, A.: Thermal-wave imaging. Science. **218**, 223–228 (1982)
13. Carlomagno, G.M., Berardi, P.G.: Unsteady thermophototopography in nondestructive testing. Proc. 3rd biannual exchange, St. Louis, USA, pp. 33–39 (1976)
14. Almond, D., Patel, P.: Photothermal Science and Techniques. Chapman & Hall, London (1996)
15. Meierhofer, C., Myrach, P., Röllig, M., Steinfurth, H.: Validation of active thermography technique for the characterization of CFRP structures. Proc. 11th Europ. Conf. on NDT, Prague, Czech Republic: 1–10 (available on CD) (2014, 6–10 Oct)
16. Parker, R.C., Marshall, P.R.: The measurement of the temperature of sliding surfaces with particular references in railway brake blocks and shoes. Proc. Inst. Mech. Eng. **158**, 209 (1948)
17. Gorril, W.S.: Industrial high-speed infrared pyrometer to measure the temperature of a soldered seam on a tin can. Electronics. **22**, 112 (1949)
18. Leslie, J.R., Wait, J.R.: Detection of overheated transmission line joints by means of a bolometer. Trans. Am. Inst. Electr. Eng. **68**, 64 (1949)
19. Infrared camera spots malfunctions. Electron. Des. **12**, 9 (1961) https://www.electronicdesign.com/technologies/boards/article/21772563/infrared-camera-spots-malfunctions
20. Infrared and thermal testing. In: Nondestructive Testing Handbook, Techn. ed. Xavier P.V. Maldague, Ed. Patrick O. Moore, vol. 3. A.S.N.T, Bellingham (2001)
21. Vavilov, V., Taylor, R.: Theoretical and practical aspects of the thermal NDT of bonded structures. In: Sharpe, R. (ed.) Res. Techn. in NDT, vol. 5, pp. 239–280. Academic, London (1982)
22. MacLaughlin, P.V., Mirchandani, H.G.: Aerostructure NDT evaluation by thermal field detection (Phase II), Final Rep., AIRTASK, Naval Air System Command AIR-310G. Washington, D.C., U.S.A (1984)
23. Mandelis, A.: Chap. 9. In: Diffusion-Wave Fields: Mathematical Methods and Green Functions. Springer, New York (2001)
24. Breitenstein, O., Warta, W., Langekamp, M.: Lock-in Thermography.- Springer Series in Advanced Microelectronics, vol. 10. Springer, Germany, (2010). 250 p
25. Balageas, D.L., Krapez, J.-C., Cielo, P.: Pulsed photo-thermal modeling of layered materials. J. Appl. Phys. **59**(2), 348–357 (1986)
26. Maillet, D., Andre, S., Batsale, J.-C., et al.: Thermal Quadrupoles: Solving the Heat Equation Through Integral Transforms. Wiley, Chichester (2000)
27. Lyon, Jr. B.R., Orlove, G.L.: A brief history of 25 years (or more) of infrared imaging radiometry. Proc. SPIE "Thermosense-XXV", Vol. 5078, pp. 17–19 (2003)

28. Krapez, J.-C., Maldague, X., Cielo, P.: Thermographic NDE: data inversion procedure (part II: 2D analysis and experimental results). Res. in NDE. **2**, 101–124 (1991)
29. Luong, M.P.: Infrared thermography of fatigue in metals. Proc. SPIE "Thermosense-XIV", Vol. 1682, pp. 222–232 (1992)
30. Aamodt, L.C., Maclachlan Spicer, J.W., Murphy, J.C.: Analysis of characteristic thermal transit times for time-resolved infrared radiometry studies of multilayered coatings. J. Appl. Phys. **68**(12), 6087–6098 (1990)
31. Prabhu, D.R., Howell, P.A., Syed, H.I., Winfree, W.P.: Application of artificial neural networks to thermal detection of disbonds. Rev. Prog. Quant. NDE. **11**, 1331–1338 (1992)
32. Shepard, S.: Advances in pulsed thermography. Proc. SPIE "Thermosense-XXIII", Vol. 4360, pp. 511–515 (2001)
33. Popov Yu, A., Karpelson, A.E., Strokov, V.A.: Thermal NDT of multi-layer structures. Defectoscopiya (Soviet J. NDT). **3**, 76–81 (1976)
34. Kush, D.V., Rapoport, D.A., Budadin, O.N.: Inverse problem of automated thermal NDT. Defectoscopiya (Soviet J. NDT). **5**, 64–68 (1988)
35. Bekeshko, N.A., Popov Yu, A., Upadyshev, A.B.: A plasmatron as a source of localized heating in thermal nondestructive testing of materials and products. Defectoscopia (Sov. J. NDT). (5), 88–91 (1972). (in Russian)
36. Carslow, H.S., Jaeger, T.S.: Conduction of Heat in Solids, 580 P. Oxford University Press, Oxford, U.K. (1959)
37. Luikov, A.V.: In: Hartnett, J.P. (ed.) Analytical Heat Conduction Theory, 685 P. Academic Press, New York, U.S.A. (1969)
38. McGonnagle, W., Park, F.: Nondestructive testing. Int. Sci. Technol. **7**, 14 (1964)
39. Milne, J.M., Reynolds, W.N.: The non-destructive evaluation of composites and other materials by thermal pulse video thermography. "Thermosense-VII", Vol. 520, pp. 119–122 (1984)
40. Burleigh, D.: A bibliography of nondestructive testing (NDT) of composite materials performed with infrared thermography and liquid crystals. Proc. SPIE "Thermosense-IX", Vol. 780, pp. 250–255 (1987)
41. Maldague, X.: Theory and Practice of Infrared Technology for Nondestructive Testing, Wiley Series in Microwave and Optical Engineering. Wiley, New York (2001)
42. Vavilov, V.P., Burleigh, D.D.: Review of pulsed thermal NDT: physical principles, theory and data processing. NDT E Int. **73**, 28–52 (2015)

Contents

Chapter 1
Physical Models of TNDT

Abstract This chapter is an introduction to TNDT. It describes the terminology, general physics and common heating methods used in the TNDT inspection method.

1.1 Terminology

With continuing research and expanding computer capabilities, as many new TNDT terms and concepts have appeared in recent years (see Table 1.1). Most of these terms are related to new computer processing algorithms.

1.2 Basic Inspection Procedures

TNDT is typically classified as passive or active, and also as steady-state (stationary) or transient (non-stationary or dynamic). In a passive mode, tested objects are characterized by temperature distributions that appear naturally due to the function of an object or because of other technical reasons, while active inspection requires the application of external thermal stimulation. Additionally, defects can be active or passive. Active defects generate or absorb thermal energy and therefore they can be detected in a passive regime. Passive testing samples are at the same temperature as the environment prior to inspection. Therefore a passive test object must be heated or cooled to produce detectable temperature signals in defective areas. In passive TNDT, the testing scheme is determined by access to the sample surface and how subsurface defects might be detected most effectively.

In transient TNDT, the material is generally heated by a pulse from a Xenon flash lamp or other source. In this case, the temperature difference between a defect and the surrounding "good" material changes as a function of the time after heating. As a result, active TNDT requires using advanced data processing algorithms. The dynamic character of active TNDT is illustrated in Fig. 1.1 where a the evolution of time versus the differential temperature signal, ΔT, over the defect in a turbine blade thermal protection coating, is shown. The maximum variation in the temperature

V. Vavilov, D. Burleigh, *Infrared Thermography and Thermal Nondestructive Testing*, https://doi.org/10.1007/978-3-030-48002-8_1

Table 1.1 Pulsed TNDT terminology (in alphabetic order, adapted from [1])

Term	Description
Active IR thermography (active TNDT)*	Infrared thermographic examination of materials and objects that require the addition of external thermal stimulation
Ampligram (image of magnitude)	Image in which each pixel represents the magnitude of Fourier values at a particular frequency
Characteristic heat transit time	A particular time $\tau*$ in a temperature versus time (temporal) evolution that is used for defect detection and characterization; this is considered a temporal informative parameter
Chronological thermogram	Temperature versus time evolution at a particular time
Cross-sectional timegram	Image that represents the evolution of a surface temperature profile at one time
«crawling-spot» technique	See *«Flying spot» technique* (low scanning speed)
Defect characterization, (identification, parameters retrieving)	Determining defect parameters by applying inverse algorithms to experimental data
Depth profiling	Retrieving the structure of an object by depth (see *Depthgram*)
Depthgram	Image that contains pixel-based values of defect depth
Difference temperature signal	$\Delta T(x, y, \tau) = T(x, y, \tau) - T_{ref}(x, y, \tau)$, where $T(x, y, \tau)$ is the sample temperature, $T_{ref}(x, y, \tau)$ is the reference temperature.
Differential hyperbolic equation of heat conduction	$a\nabla^2 T = (\partial T/\partial\tau) + \tau_r(\partial^2 T/\partial\tau^2)$, τ_r is relaxation time
Differential parabolic equation of heat conduction	$a\,\nabla^2 T = \partial T/\partial\tau$, a is thermal diffusivity
Differentiated contrast	$Con_{dif}\left(x,y,\tau^{(n)}\right) = T\left(x,y,\tau\right) - \sqrt{\tau^{(n)}/\tau}\;\; T\left(x,y,\tau^{(n)}\right)$, where $\tau^{(n)}$ is a normalization time. This term is based on the classic solution for the adiabatic heating of a semi-infinite body [2]*
Direct TNDT mathematical problem	The temperature evolution in space and time is calculated by using the sample geometry and the thermal properties of the material
Dynamic thermal tomography	Tomographic representation created by an analysis of the time versus temperature evolution in a one-sided (front-surface) test
Early detection time	A time $\tau*$ when a temperature signal $\Delta T(\tau)$ first exceeds the noise level Typically, $\tau*$ is much shorter than the optimum observation time τ_m
Filtered contrast (*FC*)	$FC(x, y) = T(x, y) - Filter\,T(x, y)$, where $Filter\,T(x, y)$ is the "smoothed" temperature (*FC* is similar to *differential temperature signal*) [3]
Flash technique	Inspection with a thermal pulse whose duration is significantly shorter than the observation time; typically, samples are heated with a pulsed laser or flash tubes (see *Pulsed TNDT*)

(continued)

Table 1.1 (continued)

Term	Description
«Flying spot» technique	TNDT by heating an object with a localized moving heat source while monitoring the surface temperature at a fixed distance (and time) from the heated point (high scanning speed)
Forced diffusion thermography	IR thermography of a moving object that is thermally stimulated by the use of a slit mask
Frequency-domain analysis	Analysis of a temperature frequency spectrum
Image of a polynomial coefficient	Image that contains pixel-based values of polynomial coefficients (see *Polynomial fitting*).
Induction (eddy-current) thermography	Thermography that stimulates objects by electromagnetic radiation (radio waves)
Infrared thermographic testing (thermographic testing)*	Inspection of materials and products using *infrared thermography*
Infrared thermography (IR thermography) *	Imaging an object by sensing the infrared (thermal) radiation emitted by it
Infrared (IR) thermogram*	A thermal map or image of a target where the grey tones or color hues represent the distribution of infrared thermal radiant energy emitted by the surface of the target
Infrared tomography	Tomographic presentation of semi-transparent objects by their IR radiation using the principles of computer X ray tomography
Inverse TNDT problem	Sample/defect thermal properties and geometrical parameters are determined by experimentally measured temperature evolution in space and time
Lock-in IR thermography	Phase-sensitive IR thermography that uses low-frequency thermal stimulation of tested objects
Maxigram	Image that contains pixel-based maximum values of a selected parameter (differential temperature signal) independent of the time of their appearance
«Mirage» technique	A *Photothermal* method where a sample is stimulated by modulated laser radiation and temperature anomalies are detected by the deflection of a laser beam that sweeps across the area of interest
Normalized temperature contrast	$$Con_{norm}(x,y,\tau)=\frac{T(x,y,\tau)-T_{in}(x,y)}{T^{(n)}\left(x,y,\tau^{(n)}\right)-T_{in}(x,y)}-\frac{T_{ref}(x,y,\tau)-T_{in}(x,y)}{T_{ref}^{(n)}\left(x,y,\tau^{(n)}\right)-T_{in}(x,y)},$$ where $T^{(n)}(x,y,\tau^{(n)})$is the temperature of a current point at time $\tau^{(n)}$, which is assumed to be normalizing (frequently, $\tau^{(n)}$corresponds to the end of heating τ_h), $T_{in}(x,y)$ - sample initial (ambient) temperature

(continued)

Table 1.1 (continued)

Term	Description
One-dimensional (1D) defect	A defect whose largest dimension is along a particular coordinate Such defects produce temperature signals independent of their location along this coordinate
Optimum observation time	A time τ_m when a maximum value of a chosen critical parameter, e.g. signal-to-noise ratio, appears
Passive thermography (passive TNDT)*	Thermographic technique for inspecting objects or structures by monitoring their emitted thermal radiation, without using an additional energy source for thermal stimulation
Phasegram	Image that contains pixel-based values of Fourier phase at a particular frequency
Photoacoustic method	An ultrasonic NDT method in which acoustic signals are produced by the optical illumination of samples (this can be regarded as a combination of an ultrasonic and thermal method)
Photothermal (optothermal) method	The thermal stimulation of a sample is provided by optical radiation (typically, in a small area thus requiring scanning the heat over the sample
Polynomial coefficient image	Image that contains pixel-based values of polynomial coefficients (see *Polynomial fitting*).
Polynomial fitting	Fitting "noisy" temporal evolution functions with polynomial functions (a source sequence of arbitrary length can be replaced with a few *Images made up of polynomial coefficients*)
Principal component analysis (PCA)	A type of *Singular value decomposition*
Pulse-echo thermal wave technique	An approach in which surface temperature signals are treated as the superposition of thermal waves reflected from layer boundaries (see *Pulsed TNDT*)
Pulse phase thermography (PPT)*	Processing technique used in *pulsed thermography* in which data are analyzed in the frequency domain rather than in the time domain Note: Phase data is often of particular interest
Pulsed thermography*	Active infrared thermographic inspection technique in which a test sample is stimulated with a pulse of energy and the recorded infrared image sequences are analyzed to enhance defect 'visibility' and to characterize defect parameters
Reflection mode (front-surface, one-sided test)	Thermal stimulation and temperature monitoring occur on the same surface of a sample
Running temperature contrast	$Con_{run}(x,y,\tau) = \dfrac{\Delta T(x,y,\tau)}{T(x,y,\tau) - T_{in}(x,y)}$
Significant parameter	Parameter used to make a decision on object quality (see *Space, Time* or *Frequency domain analysis*)

(continued)

Table 1.1 (continued)

Term	Description
Singular value decomposition (SVD)	Calculation of a covariation matrix and the eigenvectors of its input data Eigenvectors are arranged in the order of their magnitude, thus providing the components of an analyzed statistical set By neglecting low-order components, it is possible to improve results, By reducing the effect of such things as uneven heating [4]
Space-domain analysis	Analysis of temperature amplitude parameters
Spatio-temporal filtering (STF)	The algorithm involves simultaneous spatial and temporal filtering to enhance the signal-to-noise ratio. The input parameters are thermal diffusivity and the sample thickness or the maximum depth of detection [3]
Space/time mapping (STM) JPEG-based data compression	To compress raw thermographic data, the algorithm consists of 3 steps: (1) extraction of the dynamically-changing part of the data; (2) space/time mapping; (3) JPEG compression [5] This enables a high compression ratio while maintaining a high reconstruction quality.
Steady-state, stationary	Temperature is independent of time
Stimulated (forced, active, transient) TNDT	Additional thermal stimulation is applied
Synchronous (vector lock-in) technique	Filtering temperature signals in a narrow frequency band by a reference signal (suppressing non-coherent noise)
Synthetic signal processing	A data processing technique patented by Shepard [6, 7] that involves: (1) polynomial fitting; (2) restoring a source function by polynomial coefficients; (3) analysis of temperature derivatives in time
Synthetic thermal time-of-flight technique (STTOF)	This data processing technique was proposed by Ringermacher et al. [8]. It requires the determination of inflection points in $\Delta T(\tau)$ curves that allows the conversion of temperature images into depthgrams
Thermal contrast*	The degree of detectable temperature difference between adjacent areas that have unequal temperatures at a particular time Note: Thermal contrast is a processing technique used to enhance defect visibility. In its simplest form, the thermal contrast is computed by calculating the difference between the temperature of the target area and the temperature of a (sound) reference area
Thermal/infrared NDT	See *Infrared nondestructive testing*
Thermal resistance image	Image that contains pixel-based values of thermal resistance (defect thermal resistance)
Thermal tomography*	A data processing technique used in pulsed thermography in which data is analyzed in comparison with a particular time of interest, such as the time of maximum thermal contrast

(continued)

Table 1.1 (continued)

Term	Description
Thermal tomogram	Image that contains pixel-based values of thermal properties (defect indications) within a material layer at a chosen depth
Thermal wave imaging*	An active infrared thermographic inspection technique, in which a test sample is heated with periodic pulses of thermal energy [9–12], also - the name of a U.S. company that manufactures TNDT systems
Thermoelastic stress image	IR thermogram image that shows the thermoelastic stresses
Thermographic signal reconstruction (TSR)*	A signal-processing technique for reconstructing and improving time-resolved thermal images, used in pulsed thermography testing and based on polynomial fitting of temperature decay Note: This is a type of *Synthetic signal processing* (patented by Shepard [6, 7])
Thermography	Analysis of spatial and temporal distribution of thermal parameters (temperature) in objects, typically accomplished by creating corresponding infrared images (*thermograms*)
Thermovision	Term introduced by AGEMA infrared systems (sometimes used as *Infrared thermography*
Thicknessgram	Image that contains pixel-based values of defect thickness
Time-domain analysis	Analysis of temperature temporal parameters
Timegram	Image that contains pixel-based values of a characteristic heat transit time, e.g. *Optimum observation time*
Time-resolved infrared radiometry (TRIR)	Analyzing temporal evolution of temperature signals (see *Synchronous technique*)
Transient, dynamic	Temperature is dependent on the time
Transmission mode (rear surface, two-sided) test	The thermal stimulation and temperature monitoring occur on opposite sides of a sample
Ultrasonic lock-in thermography	Thermography in which objects are stimulating by periodic ultrasonic waves
Ultrasonic burst excitation, sonic IR imaging, thermosonic method	Thermography in which objects are stimulated by pulsed ultrasonic waves
Vibrothermography*	Thermographic technique in which temperature differences are created by mechanical vibrations
Videothermography	Real-time IR thermography (obsolete term)
Wavelet thermography	A type of *Pulsed phase thermography* in which a source temperature function is represented by wavelets

*These terms are given in accordance with the ISO # 10878 international standard (first edition 11 January 2013)

**References are given for some recently proposed terms

over the delamination, $\Delta T = 3.03$ °C, occurs at 65 ms after flash heating, and at 2 seconds this has dropped to 0.004 °C, nearly back to equilibrium.

Models of active TNDT can be classified by: (1) the type of thermal stimulation (Fig. 1.2), (2) the arrangement of the sample and the thermal stimulation source (Fig. 1.3), and (3) the size and shape of stimulated area (Fig. 1.4).

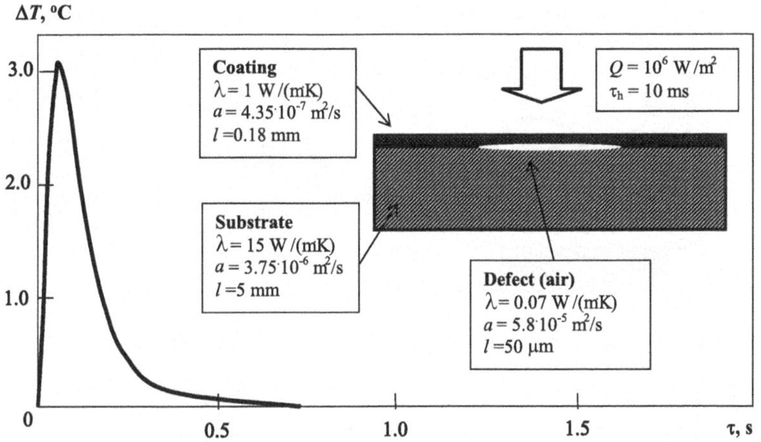

Fig. 1.1 Differential temperature signal *vs.* time in the detection of delaminations between the coating and the substrate in turbine blades (parameter specification: a-thermal diffusivity, λ-thermal conductivity, l-coating or substrate thickness)

Thermal stimulation can be either positive or negative; both heating and cooling are acceptable. But heating is preferable as it generally provides a higher power density and the heating parameters are easy to control. The highest optical radiation power density is provided by lasers and certain Xenon flash tubes (Fig. 1.2a), while common electrical quartz lamps allow more 'mild' stimulation. However, the heating rate provided by lasers and flash tubes can be high enough to damage the material being inspected. This happens because lasers usually provide a small intense spot that is scanned across the surface and Xenon flash lamps provide an intense pulse over a very short period (milliseconds) of time. Quartz lamps can provide a large amount of energy over a relatively large area at a lower rate (seconds) that is less likely to damage materials. And while some test cases require a very fast (short) heating pulse, in others, slower heating provides satisfactory results.

Materials that are electrically conductive can be heated by induction. The absorbed power density is rather low, but the inductive heating produces low noise. And by this method, metals may be heated through non-metallic layers (Fig. 1.2b). A particular example of this method is *induction* (*eddy current*) *IR thermography* (Fig. 1.2c) which involves the use of high current frequency (~100 kHz). Note that eddy current devices can also be used as near-surface temperature sensors.

In some cases, metallic samples can be resistively heated by the application of electrical current (Fig. 1.2d). This technique also produces low noise. And this method can be used to detect cracks that are perpendicular to the direction of the current. However, establishing a good electrical contact capable of carrying a powerful current through a sample can be challenging, due to the contact problem. Note that electrical resistive heating is the basis of the infrared testing of the electrical connections in electrical power distribution systems.

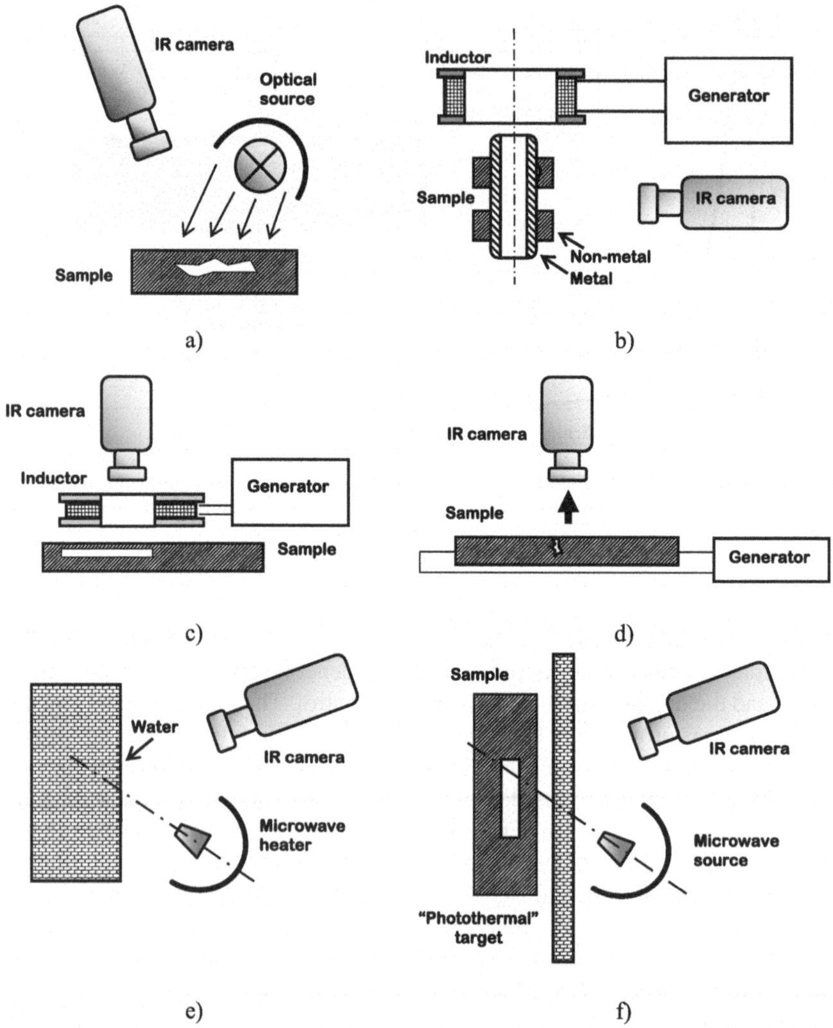

Fig. 1.2 Thermal NDT procedures: (**a**) – optical heating, (**b**) – eddy current heating (low frequency), (**c**) - eddy current heating (high frequency), (**d**) – electric current heating, (**e**) – microwave heating, (**f**) – electromagnetic IR thermography, (**g**) – internal heating with gas/liquid, (**h**) – external blowing with air, (**i**) – mechanical heating, (**k**) – natural heating, (**l**) – "mirage' technique. (**m**) – photodeformation technique

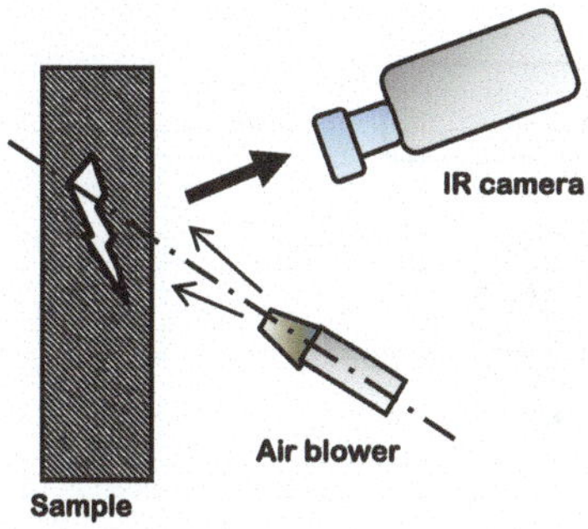

g)

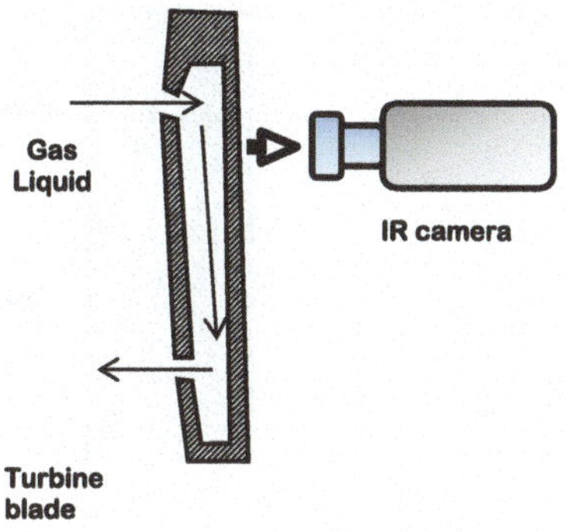

h)

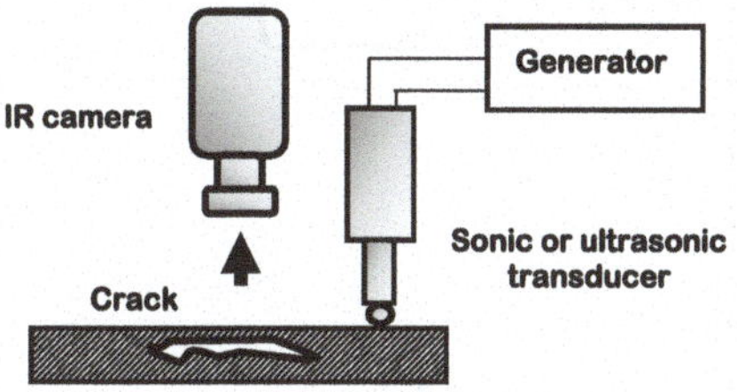

i)

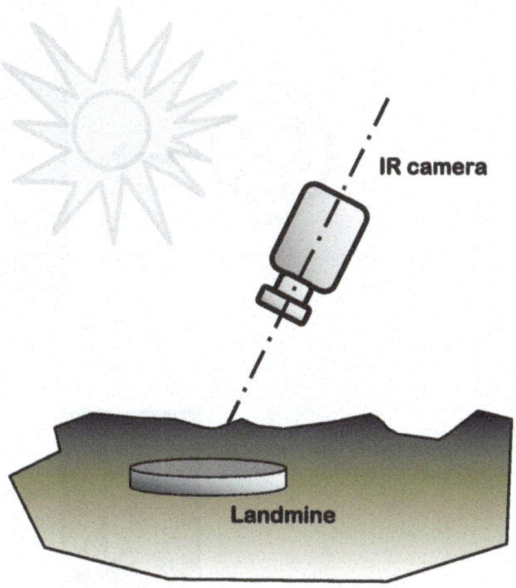

k)

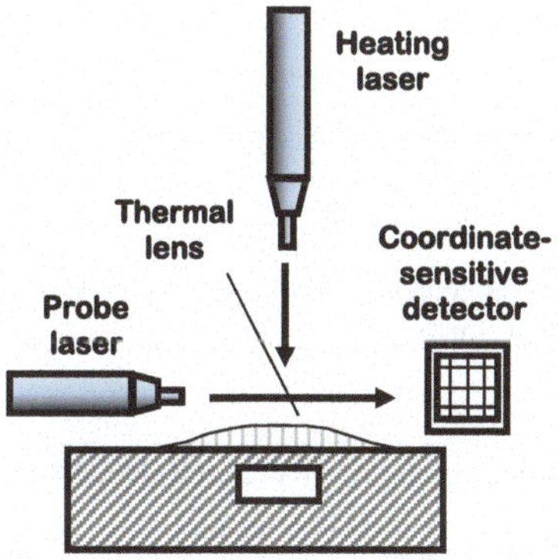

l)

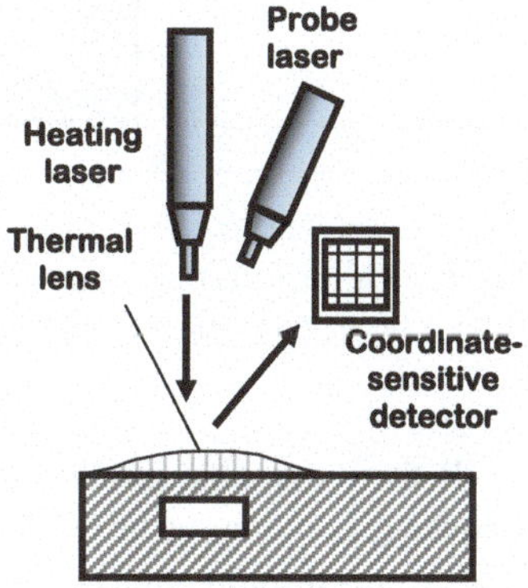

m)

Fig. 1.2 (continued)

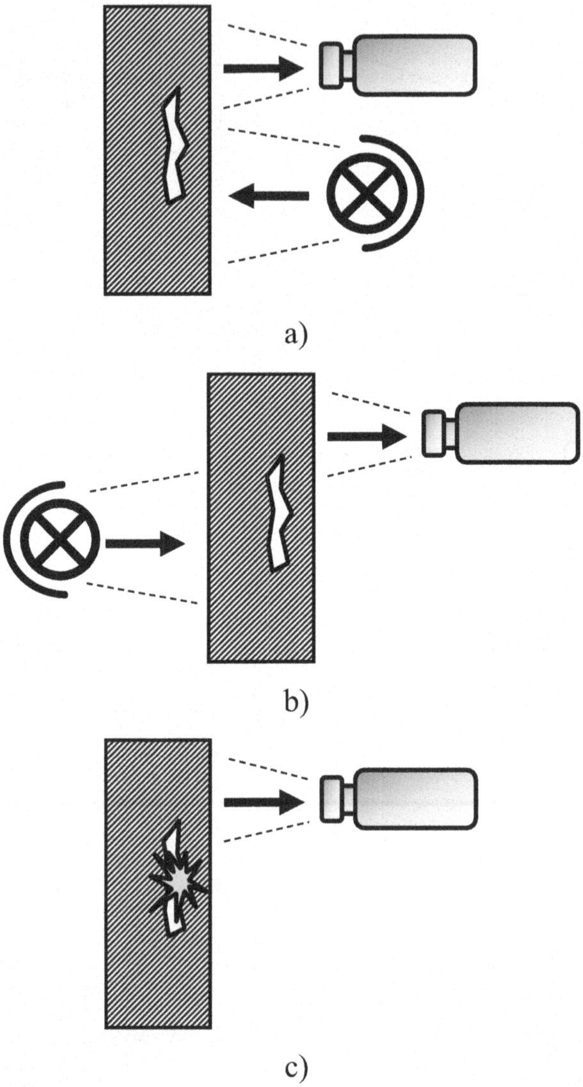

Fig. 1.3 Thermal NDT procedures (heater *vs.* IR camera arrangement): (**a**) one-sided procedure, (**b**) two-sided procedure, (**c**) stimulation with an internal heat source

IR thermographic monitoring of materials subjected to microwave (ultra-high frequency) heating can be used to detect moist areas in porous non-metals (Fig. 1.2e). Another testing approach combining a microwave source and an IR camera is shown in Fig. 1.2f. The *electromagnetic IR thermography* technique involves a photothermal target that is placed close to a test object. An IR imager monitors a target to observe a temperature rise that appears due to the superposition of incident and

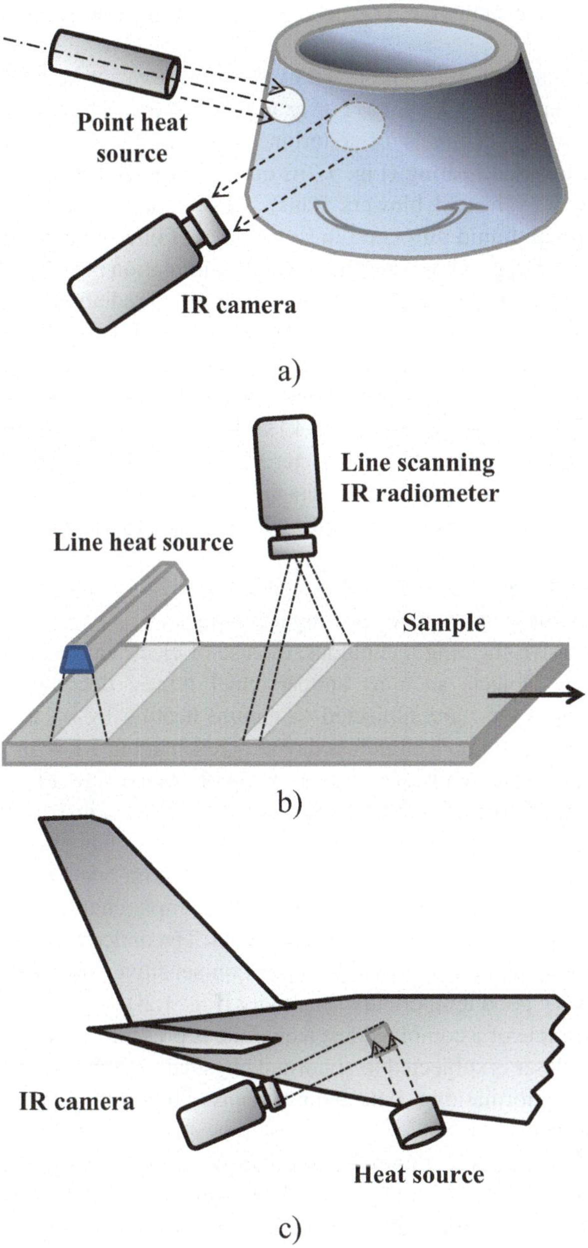

Fig. 1.4 Thermal NDT procedures (heated/monitored area): (**a**) – point-by-point scanning, (**b**) – line scanning, (**c**) – IR thermographic NDT

reflected microwave radiation. This technique has been used for visualizing terahertz radiation.

Convective heating can be employed by spraying a gas or liquid on or near the surface of a sample (Fig. 1.2g, h). For example, hot or cold water can be passed through the cooling channels of a turbine blade to thermally stimulate it by (Fig. 1.2g). One-sided heating (Fig. 1.2h) may be provided by hot air from "hair dryers" or other by heated blowers (fans). Also, samples have been cooled by spraying them with liquid nitrogen.. In some cases of one-sided TNDT, convective air heating may be as good or better than optical stimulation because of lower noise, a higher heating efficiency, and the absence of reflected radiation.

Mechanical stimulation, or vibrothermography, uses internal friction to heat an area of interest, usually a crack. The main advantage of mechanical stimulation, other than low noise, is the fact that temperature anomalies appear only in defect areas due to friction between defect (crack) surfaces or another phenomena, such as plastic deformation and mechanical hysteresis (Fig. 1.2i). This heating approach has been very efficient in detecting defects in fiber reinforced composites excited by ultrasonic transducers. However it is difficult to apply to some parts and has found little use in real applications.

The types of thermal stimulation depicted in Fig. 1.2a–i require the use of electrical equipment. These are "active" heating techniques.

In other "passive' heating techniques, the use of electrical heating devices may be unnecessary. Objects such as antipersonnel mines, decorative coatings on building envelopes etc., are subjected to natural heating by solar radiation. The power of such radiation on a cloudless day may reach about 1 kW/m^2 in the middle latitudes, varying periodically during a day-night (diurnal) cycle (Fig. 1.2k). Heating methods such as this are similar to those shown in Fig. 1.2a, just slower.

Some methods of thermal stimulation are only used in laboratory research. For example, a "mirage" technique which utilizes two lasers, one for heating and one for sensing temperature. The heated air over the sample changes density and this cause a slight deflection of the sensing laser beam, The deflection is proportional to the air temperature and it is detected by a position-sensitive photodetector; this system provides very good temperature sensitivity (Fig. 1.2l).

Another example of a combined test technique is presented in Fig. 1.2m: a material heated by a laser is subjected to abnormal physical deformation over subsurface defects, and the deformation is detected by means of a probing laser and a position-sensitive photodetector.

The relative displacement between a heater and a temperature recording device influences the defect detectability. The advantages of transient TNDT are clearly demonstrated in a one-sided technique (Fig. 1.3a) which is also called a 'reflection' technique due to the analogy with the ultrasonic pulse echo method. A two-sided or transmission technique requires through-the-sample heating and it cannot be applied to thick or thermally-insulating materials (Fig. 1.3b). However, a two-sided technique is the most efficient to detect defects through the entire sample thickness.

One method of applying internal heating is to pass an electric current through a sample. Concurrently, a temperature-recording device should monitor the sample areas when defects produce the maximum temperature anomalies (Fig. 1.3c).

Another classification of TNDT procedures is shown in Fig. 1.4. In this case, a point heating device scans or "rasters" the heating spot across a sample to heat it. Concurrently, a point temperature detector is scanned across the sample at a distance or time delay behind the heating spot. The depth of the defects determines the magnitude of the delay, deeper defects requiring larger delays (Fig. 1.4a). Some early studies used this technique and it was later revived as a *flying-spot* or *crawling-spot* method. This technique may also be used to detect vertical cracks in a surface. The main drawback of this technique is its low speed. A somewhat similar, but faster method is a *line-scanning* technique. In this technique, a sample is stimulated with a strip heater and the temperature is recorded by a line-scanning IR radiometer (Fig. 1.4b). Strip heating can be applied uniformly, and the speed of a line scanning technique is higher than a point scanning technique by a factor of N, where N is the number of pixels in the scanned line. Because a heated line sweeps across a sample surface, vertical cracks can also be also detected by line scanning. IR thermographic testing involves the heating of a large area monitored by an IR imager (Fig. 1.4c). In recent years, this technique became very popular due to the availability of small IR imagers and powerful flash heaters. The efficiency of IR thermographic NDT can be significantly enhanced by applying advanced algorithms of IR image processing.

Many test procedures can be regarded as variants of a generic, pulsed TNDT procedure, in which a sample is heated with a heat pulse while recording the surface temperature pattern either during the *heating period* or after it, i.e. within a *cooling period*. Temperature can be monitored on the front (heated) or rear surface of the sample, as well as along the edges of the sample. Another widely-used procedure of transient TNDT is a *thermal wave* method (lock-in IR thermography) in which a sample is heated with a periodically varying heat flux. The test methods described in Table 1.1 are variations in either the heating method or in data processing algorithms. The test methods also vary in the method used to extract qualitative or quantitative information from a basic temperature function $T(x, y, \tau)$ which describes the evolution of sample excess temperature as a function of time at each surface point (x, y). Note that the sample excess temperature is calculated from an initial sample temperature, which is often equal to the ambient temperature. In TNDT, the basic function is defined for a sequence of recorded IR images (Fig. 1.5a), and the coordinates (x, y) are replaced with the pixels (i,j).

A typical temperature evolution $T(i, j, \tau)$ is shown in Fig. 1.5b, for both defect (d) and non-defect (nd) areas. On the front surface of the sample, the *excess* temperature T increases from a zero level up to the maximum that occurs at the end of a heat pulse of duration τ_h. On the rear surface, the maximum of the excess temperature is 'shifted' in time relative to the cessation of heating by an amount that depends on the material thickness and the thermal diffusivity. The process ends when the excess temperature drops to the initial value, e.g. to the ambient temperature.

The fundamental parameter in TNDT is the "difference" (delta) temperature signal, which is defined as the temperature difference between a reference point,

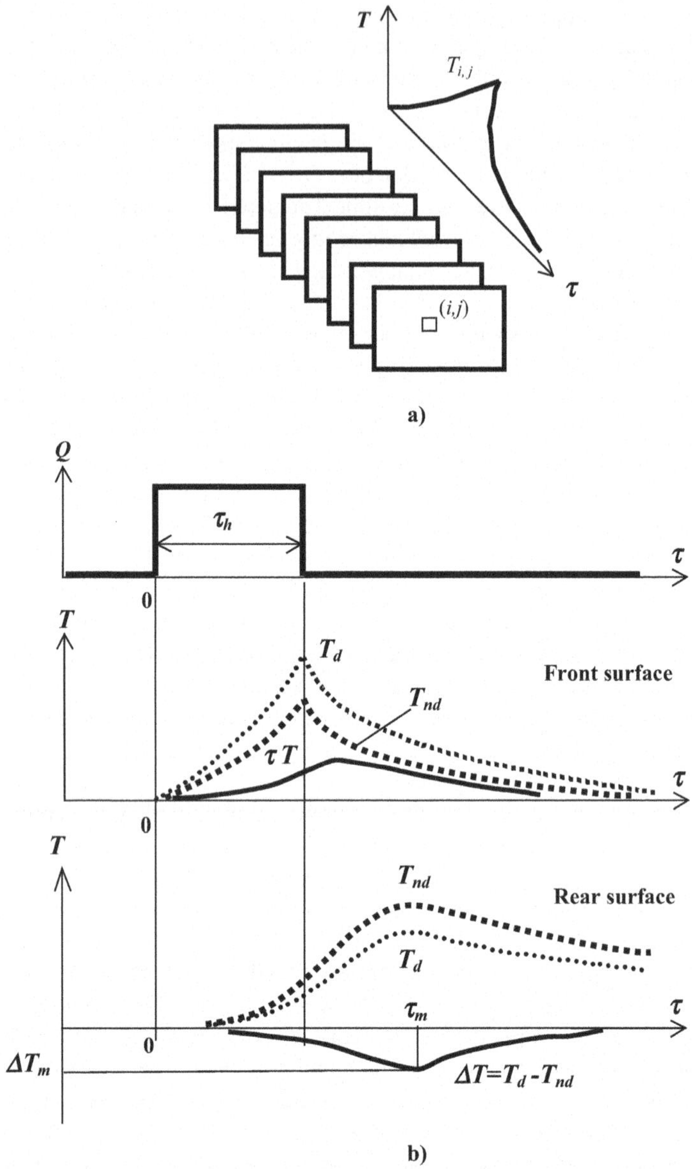

Fig. 1.5 Basic temperature functions in TNDT: (**a**) – producing IR image sequence and $T(i,j,\tau)$ function, (**b**) – heating pulse and producing ΔT differential temperature signal

assumed to be over a "good" (non-defect) area and a point over another point that may be over a defect: $\Delta T(x, y, \tau) = T(x, y, \tau) - T_{nd}(x, y, \tau)$. Sometimes this parameter is called the 'temperature contrast', although the term "contrast" is normally applied to dimensionless values. Under this definition of the difference signal, $\Delta T > 0$ on the heated (front) surface of a sample with defects whose thermal conductivity is lower than that of the bulk material. On the rear surface, such defects produce$\Delta T < 0$. Conversely, defects which are more conductive than a host material produce $\Delta T < 0$ on the front surface and $\Delta T > 0$ on the rear surface. Typical defects in composites include delaminations, voids and porosity, all of which have lower thermal conductivity than the composite itself.

While a sample is heated by a pulse and then allowed to cool, the difference temperature signals reach their maximum values ΔT_m at time τ_m (see Fig. 1.1). By using the $T(x, y, \tau)$ and $\Delta T(x, y, \tau)$ notation, many popular procedures of transient TNDT can be identified (Table 1.1). These procedures will be discussed in detail in Chap. 3.

The statistical reliability of TNDT increases with a higher signal-to-noise ratio. In transient tests, the maximum signal-to-noise ratio occurs at an *optimum observation time*. As a first approximation, we can assume that the time $\tau_m(\Delta T_m)$ at which the maximum difference signal occurs is the *optimum observation time*. However, it is worth noting that the maximum signal-to-noise ratio does not always appear at the same time as ΔT_m. It follows that a TNDT test might be accomplished by recording a single IR image at a τ_m time instead of capturing a whole sequence of images.

Heating may be performed with a short heat pulse (flash) and temperature may be recorded during the cooling period. Such a procedure is typically called *pulsed* TNDT. If the heating period is long enough, the optimum observation time may occur during the heating period. This fact is used in the technique of *Temperature-Resolved Infrared Radiometry* (TRIR). At first glance, the *Early Detection* technique seems to be paradoxical because it records temperature at very early observation times, i.e. $\tau < \tau_m$ before the heating period is completed. Since the full effect of heating has not yet been achieved, the signal-to-noise ratio is low and therefore, only large defects can be detected with this technique. The main advantage of early detection is the lack of significant heat diffusion in lateral directions, which provides sharper images of hidden defects than those achieved at later times after lateral diffusion has occurred. These early images can appear similar to those obtained by ultrasonic methods.

In the *averaging technique*, N IR thermograms are averaged around a τ_m time, reducing the random noise by $\sqrt{N}$ times.

Signal processing in the amplitude domain is often complemented by additional treatment in the time or phase domain. This achieves a higher signal-to-noise ratio.

Dynamic thermal tomography is based on the analysis of ΔT_m and τ_m. This method can be considered a particular implementation of the above-mentioned approach, as well as a method of *Pulse Phase Thermography* (PPT) which uses the Fourier transformation or wavelet functions.

The *Synthetic Signal Processing* (SSP) technique combines polynomial fitting and analysis of derivatives. This technique is used in commercial TNDT equipment.

The TNDT techniques discussed above were developed for pulse heating test procedures. However, most of them also apply to thermal wave procedures, even if the final goal is producing phase images (phasegrams).

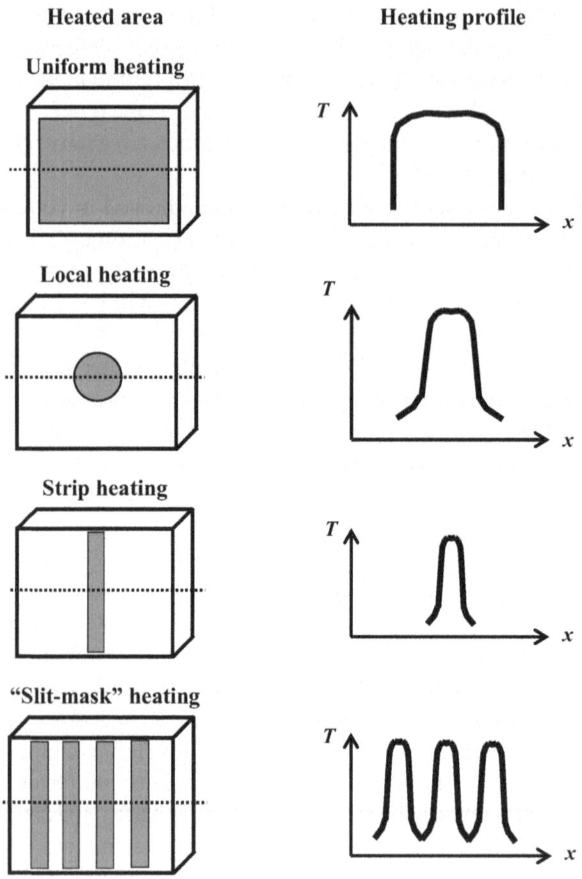

Fig. 1.6 Heating spatial profiles

1.2.1 Heating Methods in TNDT

The following types of heating are common in TNDT: (1) radiation (optical) (2) convection (using hot gas or liquid) (3) microwave (4) mechanical (sonic) (5) electrical resistive and (6) electrical induction. Typical heating profiles are shown in Fig. 1.6. According to TNDT theory, the highest probability of detection of subsurface defects is provided by short and powerful heating, which can be achieved by all of the above-mentioned methods except mechanical and convection (gas/liquid). However, some of these techniques produce enough thermal noise that some low-power or slower heating techniques are preferable. And in some cases, the magnitude of the heating pulse that is needed to achieve the best theoretical result can actually damage the material being tested. Or the heating pulse that theory recommends cannot be achieved. So other means of heating must be evaluated.

Heating (resistive) with a transient pulse of electric current is mainly used in laboratory tests but IR thermographic diagnostic testing is widely used in electronics and electrical engineering where components are self-heated when functioning. In aviation, flexible heaters called "heat blankets" may be used to warm up large aircraft components in order to detect water hidden in honeycomb panels. An Airbus procedure uses this method.

Inductive heating, which is commonly used in the processing of metals, is rarely used in TNDT. Steel may be efficiently heated by using industrial current frequencies (50 or 60 Hz). However, there are some special cases where inductive heating may be effective. One example is the heating of a metal located behind a non-metal such as locating reinforcing bars behind concrete. And the high-frequency inductive excitation of eddy currents in metals and composites (low electrical conductivity) has been shown to be effective in detecting both surface and subsurface cracks.

Microwaves in the kHz and MHz range will heat water that is trapped in porous or hollow materials such as honeycombs and cavities in aircraft structures. This allows the water to be detected, but the use of microwave heaters requires special safety measures.

While very high power densities can be achieved by *carbon arc* (*plasmotron) heaters* (up to $2{\cdot}10^{10}$ W/m^2) and *cathode-ray guns* (up to 10^{13} W/m^2 in welding), these are cumbersome and dangerous to use and the high cost of equipment limits their TNDT applications to laboratory research. .

In aerospace, optical radiation heating remains the most widely used. It can be achieved with: (1) Xenon flash tubes, (2), halogen lamps (3), lasers and (3) common electrical light bulbs.

In commercial TNDT equipment, such as that supplied by Thermal Wave Imaging, USA, powerful *Xenon flash tubes (lamps)* are typically used. Such tubes are used for pumping lasers and studio photography. Photographic sets commercialized by Balcar, Calumet, Hensei and other manufacturers typically include two lamps with the energy of about 3 kJ each and one power supply unit; flash duration is about 5 ms (Fig. 1.7). In specialized TNDT systems tubular lamps with the energy of 2–5 kJ may be used (Fig. 1.8a). The radiation spectrum of flash tubes is from 0.2 to 1.4 μm with some noticeable spikes at particular wavelengths. Some world TNDT teams use several flash tubes in parallel to achieve energy over 20 kJ but such devices are not very safe and practical.

a) b)

Fig. 1.7 Xenon flash tubes: (**a**) – optical head with reflector, (**b**) - heating element (lamp)

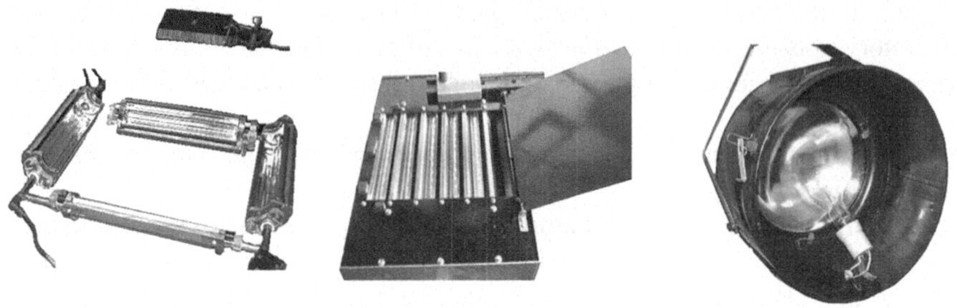

a) b) c)

Fig. 1.8 Optical heaters: (**a**) – linear Xenon flash tubes (energy density up to 25 kJ/m^2, pulse duration 1 ms), (**b**) – halogen lamps (power density up to 15 kW/m^2, heating duration over 1 s), (**c**) – incandescent lamp with reflector (power density up to 2 kW/m^2, heating duration over 1 s)

Table 1.2 Heating temporal profiles

Dirac pulse	Square pulse	Flash	Thermal waves (periodic)
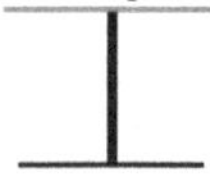	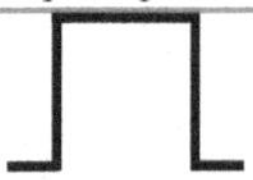		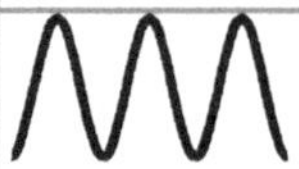

The advantages of *lasers* are monochromatic radiation, short pulse duration (up to few ns) and high power density (up to 10^{21} W/m^2). However, lasers are bulky and expensive. And the laser beam must be scanned or rastered across the material. Or the laser beam must be expanded, by the use of a lens, to project the laser energy across the entire surface; and this results in a non-uniform heat pattern. In the measurement of material properties, which is important for TNDT modeling and calculations, lasers are often used in the Parker technique for determining thermal properties.

Recent evaluation of Light Emitting Diodes (LED) has shown that these devices may eventually become practical optical heaters for TNDT. Commercial LED panels can provide several hundred watts of heating power and they operate in the red part of the visual spectrum, which is outside the detection range of IR cameras. The latter prevents the reflections that occur with radiation from sources that produce radiation that is in the detection range of the IR camera. However, the radiation from current LED heaters is not efficiently absorbed by light colored surfaces, resulting in low efficiency heating.

Since metals, especially aluminum, reflect optical radiation very well, they are usually painted to enhance their emissivity. But very short, powerful optical pulses may overheat and damage a sample or a coating. In such case, pulse duration must be increased to 10–100 ms.

IR heaters may be classified as: (1) high-temperature (1500°C, filament bulbs, quartz halogen lamps), (2) middle-temperature (400-1000°C, quartz, ceramic or glass), and low-temperature (<450°C, quartz, ceramic or glass).

Quartz halogen lamps have power from 0.5 to 2.5 kW and are manufactured in different sizes and shapes (Fig. 1.8b). Heating duration is usually up to 10–30 s, which is optimum for the TNDT of thick composites. Sometimes standard halogen lamps with reflectors (projector-type) can be used (Fig. 1.8c). Such optical heaters can provide power density up to 500 W/m^2.

Optical radiation can be easily modulated to provide the heat pulse shapes that are required for some types of TNDT. Some of these are shown in Table 1.2.

Line-scanning radiometers can be combined with linear heaters to inspect cylindrical or flat areas. The linear heater is passed closely over the surface and the line scanner is in parallel with the heater but at some distance behind it. This distance must be established, by experiment or calculation, to allow detection of the defects of interest. In the case of cylindrical test articles, the heater and line scanner may be stationary and the test article is rotated on a rotating platform. Linear electrical heaters may be ceramic bars ("globars"), or metal elements. The latter may be either a single linear element or a wire filament wound on a bar. Such heaters can provide a power density up to 100 kW/m^2.

Convective heating with hot air is very simple, and in spite of a relatively low power output (~ 500 W/m^2), it has been widely used for the TNDT inspection of composite structures (Atlas rocket and others). However, it is generally not effective in the TNDT inspection of metallic structures.

It is worth noting that in the 1960's rocket motor cases were tested simply by moving them from a cold to a hot location and monitoring temperature thermographically.

As mentioned above, test samples can be also cooled down by placing them into a refrigerator or spraying them with liquid nitrogen or evaporating solvents.

Without active heating, water can easily be detected in honeycomb-reinforced composite aircraft panels. Water has a very high heat capacity and there is a significant difference between the temperature at high altitude (−50° C or −60 F) and on the ground. Water is frozen or chilled during flight and it stays cold for quite a while after the plane lands. The water is easily detected as long as the inspection is done fairly soon after the plane lands and before the water temperature approaches ambient.

Cyclic mechanical loading of test samples has been used in laboratory experiments on thermoelasticity and this loading generates heat. Sometimes this heating is increased at the location of a defect, allowing the defect to be detected. The heat generated by low-frequency mechanical loading is generally not sufficient to detect defects but high-frequency (ultrasonic) excitation has been successfully used in the inspection of composites (*ultrasonic, sonic, lock-in* or *burst IR thermography*).

In general the following factors should be taken into account when choosing an appropriate heating method: (1) power density required (energy emitted and absorbed must be considered), (2) required heat pulse duration, (3) contact or non-contact method, (4) modulation requirements, (5) spectral range (optical sources), (6) noise, and (7) safety requirements.

References

1. Vavilov, V.P., Burleigh, D.D.: Review of pulsed thermal NDT: physical principles, theory and data processing.-NDT & E International. **73**, 28–52 (2015)
2. Burleigh, D.: A bibliography of nondestructive testing (NDT) of composite materials performed with infrared thermography and liquid crystals. Proc. SPIE "Thermosense-IX". **780**, 250–255 (1987)
3. Alsaadawi, Z., Netzelmann, U.: Spatio-temporal filtering of active thermography data for noise reduction and data compression. Proc. 10 th Intern. Conf. on Quant. Infr. Therm. (2010). Québec, Canada:8 (available on CD)
4. Rajic, N.: Principal component thermography for flaw contrast enhancement and flaw depth characterisation in composite structures. Compos. Struct. **58**(4), 521–528 (2002)
5. Lugin, S., Netzelmann, U.: Effective compression algorithms for pulsed thermography data. Quant. Infr. Therm. J. **3**(2), 201 (2006)
6. Shepard, S.: Advances in pulsed thermography. Proc. SPIE "Thermosense-XXIII". **4360**, 511–515 (2001)
7. Shepard S. Temporal noise reduction, compression and analysis of thermographic image data sequences. U.S. Patent No. 6516084, (2003)
8. Ringermacher H.I., Howard D.R., Knight B. Thermal imaging NDT at General Electric. 18th WCNDT, 16–20 April 2012 Durban, South Africa: 4 (available on CD)
9. Rosencwaig, A., Gersho, A.: Thermal-wave imaging. Science. **218**, 223–228 (1982)
10. Almond, D., Patel, P.: Photothermal Science and Techniques. Chapman & Hall, London (1996)
11. Mandelis, A.: Diffusion-Wave Fields: Mathematical Methods and Green Functions. Chap. 9. Springer, New York (2001)
12. Breitenstein, O., Warta, W., Langekamp, M.: Lock-in Thermography.- Springer Series in Advanced Microelectronics, vol. 10, p. 250. Springer (2010)

Chapter 2
Heat Transfer in Solid Bodies

Abstract This chapter presents the basics of classical heat conduction in solids. A mathematical formulation of typical TNDT problems is given, and direct and inverse inspection problems are discussed. Tables of solutions of heat conduction equations for various types of heating and for both a semi-infinite body and a plate are provided. Both amplitude and time (frequency or phase) TNDT parameters are introduced.

2.1 Heat Transfer Mechanisms

2.1.1 Heat Conduction, Convection and Radiation

If a temperature gradient exists between two points, thermal energy can be transported by three mechanisms: *heat conduction* (mostly in solids), *convection* (in liquids and gases) and radiation (through gasses, vacuum, and transparent solids). The thermal flux density Q between two points with different temperatures T_1 and T_2 is governed by the Fourier law (Fig. 2.1):

$$Q = -\lambda(T_2 - T_1)/\Delta x \text{ (algebraic presentation)},$$

$$Q = -\lambda\left(\partial T / \partial x\right)\left(\text{differential presentation}\right). \quad (2.1)$$

Hence, the *thermal conduction coefficient* λ in W/(m·K) determines the density of the heat flux in a solid if the temperature gradient is unity. This parameter applies only to a steady-state condition because its defining equation does not contain time.

Thermal energy will flow between two points within a solid body by conduction. Heat will flow between two solid bodies if they are in contact.

Convection involves the movement of gas or liquid molecules.

The surface of a body is heated or cooled by convection according to Newton's law (Fig. 2.1):

$$Q = \alpha_{cv}\ \left(T_s - T_{amb}\right), \quad (2.2)$$

V. Vavilov, D. Burleigh, *Infrared Thermography and Thermal Nondestructive Testing*, https://doi.org/10.1007/978-3-030-48002-8_2

where α_{cv} is the convective heat exchange coefficient, T_s and T_{amb} are the temperatures of the body and the convective medium (fluid) respectively. The value of α_{cv} varies, as this parameter describes the interaction of two media rather than a particular material; therefore, its value is dependent on temperature gradient, object geometry and the velocity of the moving media.

For vertical surfaces, the value of α_{cv} can be calculated by the following expression:

$$\alpha_{cv} = 1.7 \cdot \left(T_s - T_{amb}\right)^{1/3}. \tag{2.3}$$

Outdoors, as in the inspection of buildings, chimneys etc., convection is strongly dependent on wind velocity V, or altitude H. An example of an available empirical formula is [1]:

$$\alpha_{cv} = 6.3\left(KV\right)^{0,66}, \tag{2.4}$$

where the constant K depends on H:

H, m	25	50	100	150	200	>350
K	1.2	1.6	2.1	2.3	2.6	3.1

Radiation is an energy transfer process. All bodies having a temperature above absolute zero (0 K) emit an energy flux. Radiant heat exchange between two bodies is described by the Stephan-Boltzmann law (Fig. 2.1):

$$Q_{12} = \sigma F_g F_\varepsilon \left(T_1^4 - T_2^4\right), \tag{2.5}$$

where Q_{12} is the heat flux, σ is the Stephan-Boltzmann constant ($\sigma = 5.67 \cdot 10^{-8}$W/m^2·K^4), F_g is a dimensionless geometric factor and F_ε is a reduced emissivity. The latter reduces the energy emitted.

When radiation heat exchange occurs between two plates ($T_1 > T_2$):

$$Q_{rd} \approx \varepsilon_{red} \sigma \left(T_1^4 - T_2^4\right), \tag{2.6}$$

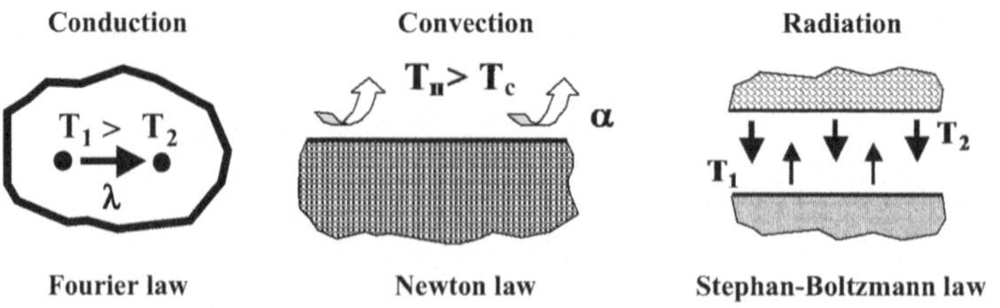

Fig. 2.1 Heat transfer mechanisms

where $\varepsilon_{red} = 1/\left(\frac{1}{\varepsilon_1} + \frac{1}{\varepsilon_1} - 1\right)$ is the reduced emissivity.

The heat exchange between a warm solid body (T_s) and a gas or liquid surrounding (T_{amb}) occurs due to both convection and radiation:

$$Q = Q_{cv} + Q_{rd} = \alpha_{cv}\left(T_s - T_{amb}\right) + \sigma F_g\left(\varepsilon T_s^4 - \varepsilon_{amb} T_{amb}^4\right). \tag{2.7}$$

In TNDT, the temperature difference $T_s - T_{amb}$ is typically small, and emissivities are typically high ($\varepsilon \cong 1$). Therefore:

$$\sigma F_g\left(\varepsilon T_s^4 - \varepsilon_{amb} T_{amb}^4\right) \approx 4\sigma F_g\left(\frac{T_s + T_a}{2}\right)^3\left(T_s - T_{amb}\right) = \alpha_{rd}\left(T_s - T_{amb}\right), \tag{2.8}$$

where α_{rd} is the heat exchange coefficient for the radiation component.

Expressions (2.2) and (2.7) become identical with the introduction of the combined heat exchange coefficient $\alpha = \alpha_{cv} + \alpha_{rd}$. The value of α depends on the shape of an object and its orientation in space, as well as on the temperature difference $T_s - T_a$.

Some recommended values of α are given in Table 2.1.

It is often assumed that $\alpha = 10$ W/(m$^2\cdot$K) at ambient temperatures. When there is no surface heat exchange ($\alpha_{cv} = \alpha_{rd} = 0$), this constitutes *adiabatic* boundary conditions. These conditions appear in the TNDT inspection of metals and thin nonmetals, especially at short observation times.

Table 2.1 Recommended values of combined coefficient of heat exchange. ($\varepsilon = 0, 9$; $T_{amb} = 20^oC$; α_{cv} is determined by (2.3); α_{rd} is determined by (2.8)

$T_s - T_a$, ^oC	α_{cv}, W/(m$^2\cdot$K)	α_{rd}, W/(m$^2\cdot$K)	α, W/(m$^2\cdot$K)
1	1.7	5.2	6.9
5	2.9	5.3	8.2
10	3.7	54	9.1
20	4.6	5.7	10.3
30	5.3	6.0	11.3
40	5.8	6.3	12.1
50	6.3	6.6	12.9
60	6.7	6.9	13.6
70	7.0	7.3	14.3
80	7.3	7.6	14.9
90	7.6	8.0	15.6
100	7.9	8.4	16.3

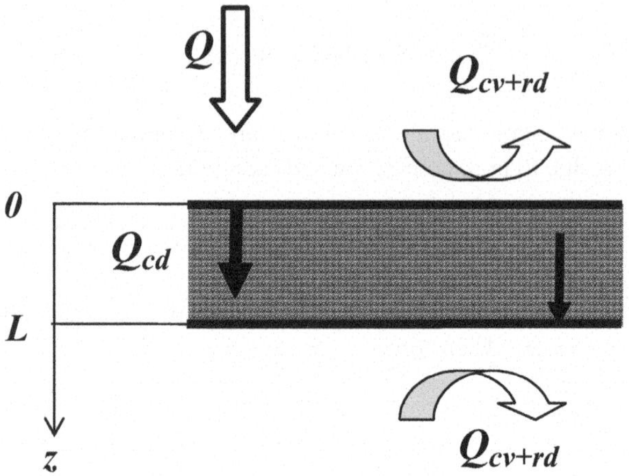

Fig. 2.2 Boundary conditions on the surface of a test object

2.1.2 Boundary Conditions

On a heated surface, the boundary condition consists of the external heating of the sample and its cooling by convection and radiation (Fig. 2.2). The corresponding expression represents the relationship between three heat fluxes: the heating flux Q, the convective/radiant flux Q_{cv+rd} and the conductive flux Q_{cd}, which passes through the sample:

$$Q = Q_{cv+rd} + Q_{cd}. \tag{2.9}$$

The typical presentation of (2.8) is as follows:

$$-\lambda \frac{\partial T(z=0)}{\partial z} = Q - \alpha\left[T(z=0) - T_{amb}\right], \tag{2.10}$$

where $z = 0$ indicates the front surface and $z = L$ indicates the rear surface. L is the sample thickness, Since there is no heating flux at the rear surface, only two heat fluxes are involved:

$$-\lambda \frac{\partial T(z=L)}{\partial z} = \alpha\left[T(z=L) - T_{amb}\right], \tag{2.11}$$

where the term Q_{cv+rd} has a negative sign due to the reversed direction of the heat flux with respect to the front surface (Fig. 2.2).

Non-adiabatic heat exchange involves all three of the heat transfer mechanisms mentioned above. Convection cannot be evaluated with high accuracy. Therefore, solving inverse TNDT problems encounters difficulties in such cases, e.g. in buildings. In active TNDT, the added heat flux is generally much more powerful than "ambient" heat fluxes caused by convection and radiation. Therefore, in this case, heat exchange becomes adiabatic, and the corresponding heat transfer solutions become simple.

2.1.3 Heat Transfer in Defects and on Layer Boundaries

Most typical TNDT defects can be considered as thin air gaps located perpendicularly to the main heat flux. There are conduction, convection and radiation in such defects. Each heat flux component is determined by the temperature gradient $\Delta T = T_1 - T_2$ between defect walls:

$$Q_{cd} = \lambda_d (T_2 - T_1)/d \text{ (conduction)},$$

$$Q_{cv} = \lambda_d^c (T_2 - T_1)/d \,(convection), \tag{2.12}$$

$Q_{rd} \approx \varepsilon_{red}\sigma\left(T_1^4 - T_2^4\right) \approx 4\varepsilon_{red}\sigma T_{0.5}^3 \left(T_1 - T_2\right)$ (radiation). Here λ_d^c is the effective thermal conductivity determined by convection, and $T_{0.5} = \dfrac{T_1 + T_2}{2}$, K.

The convective component of thermal conductivity λ_d^c is determined by the product of two criteria – Grashof and Prandtl (Gr Pr):

$$\lambda_d^c = (\zeta - 1)\lambda_d. \tag{2.13}$$

If 10^3<Gr Pr<10^6,

$$\zeta = 0{,}105(\mathrm{Gr\,Pr})^{0.3}, \tag{2.14}$$

where $\mathrm{Gr} = \dfrac{gd^3}{\nu^2}\beta\left(T_1 - T_2\right)$, $\mathrm{Pr} = \nu/a$, g is acceleration of free gravity, m²/s, $\nu = \mu/\rho$ is the coefficient of air kinematic viscosity, m²/s, μ is the coefficient of air dynamic viscosity, N·s/m², $a = \dfrac{\lambda}{C\rho}$ is thermal diffusivity, m²/s, C is specific heat, J/(kg·K), $\rho$is density, kg/m³, $\beta = \dfrac{1}{t_{0.5} + 273}$, and $t_{0.5} = (t_1 + t_2)/2$, °C.

The ratio between the conduction and radiation components is:

$$\frac{Q_{cd}}{Q_{rd}} \approx \frac{\lambda_d}{4d\varepsilon_{red}\sigma T^3}. \tag{2.15}$$

Respectively, the ratio between the conduction and convection components is:

$$\frac{Q_{cd}}{Q_{cv}} = \frac{1}{\zeta - 1}. \tag{2.16}$$

Evaluate the Q_{cd}/Q_{rd} and Q_{cd}/Q_{cv} ratios for thin air-filled defects. Let the sample excess temperature be under 50 °C that is typical in TNDT. Since the temperature gradient across the defect will not exceed this value, assume $t_1 - t_2 = 50$ °C, $T_{0.5} = \frac{T_1 + T_2}{2} = \frac{293 + 343}{2} = 318$ K and $t_{0.5} = \frac{t_1 + t_2}{2} = \frac{20 + 50}{2} = 35\,^{\circ}$C. Other parameters: $d = 250$ μm, $g = 9.8$ m²/s, $\nu = 1.51 \cdot 10^{-5}$ m²/s, $a = 57.8 \cdot 10^{-6}$ m²/s. Then:

$$Q_{cd} / Q_{rd} = 57{,}6, \tag{2.17}$$

$$Q_{cd} / Q_{cv} \to \infty.$$

Hence, in thin, air-filled defects, the conductive heat flux greatly exceeds the heat fluxes provided by convection and radiation. For example, if the thickness of the air gap (defect) is less than 250 μm, the increase in the value of the effective thermal conductivity of air is less than 1.5% more than the handbook value for air.

The situation changes if the thickness of the air gap (defect) exceeds 5–20 mm, as it may in honeycomb cells. In such gaps, a significant convection component may occur, and the effective thermal conductivity of air can be written as follows:

$$\lambda_d^{eq} = \lambda_d + (1 - \zeta)\lambda_d + \psi d, \tag{2.18}$$

$$\psi = \varepsilon_{red}\sigma\left(T_1^4 - T_1^4\right)/\left(T_1 - T_4\right).$$

By applying Eq. (2.18) to the TNDT of 25 mm-thick honeycomb cells, one can show that the efficient thermal conductivity of air in the honeycomb cells increases from 0.07 to 0.28 W/(м·K).

By assuming pure heat conduction through defects, the following two types of boundary conditions can be introduced for contacting layers.

In general, at the boundary between layers i and $i + 1$, both heat fluxes and temperatures change in the same way:

$$\lambda_i \frac{\partial T_i}{\partial z} = \lambda_{i+1} \frac{\partial T_{i+1}}{\partial z}, \quad T_i = T_{i+1}. \tag{2.19}$$

If Eq. 2.19 is true, a differential heat conduction equation must be solved for both a host (bulk) material and for each defect by matching the solutions at the interfaces. In this case, the temperature distribution in a sample will depend on the thermal conduction and thermal diffusivity (heat capacity) of each constituent. Therefore, defects may be described as *thermally-capacitive*.

If the heat capacity of a $i + 1$ layer can be neglected, its most significant thermal parameter is *thermal resistance* $R_{i+1} = d_{i+1}/\lambda_{i+1}$; and the corresponding defects may be called *thermally-resistive*. At the boundary of a resistive layer, the temperature drops and the heat flux remains constant:

$$-\lambda_i \frac{\partial T_i}{\partial z} = -\lambda_{i+1} \frac{\partial T_{i+1}}{\partial z}; \quad T_{i+1} - T_i = R_{i+1} \frac{\partial T_{i+1}}{\partial z}. \tag{2.20}$$

Simulating defects according to the models shown in Eqs. (2.19) and (2.20) is described in Sect. 4.1.

2.2 Differential Equation of Heat Conduction

2.2.1 Parabolic Equation of Heat Conduction in Cartesian Coordinates

Transient (non-stationary) temperature distributions in solids with internal heat sources are governed by the *parabolic* differential equation of heat conduction:

$$\frac{\partial}{\partial x}\left(\lambda_x \frac{\partial T}{\partial x}\right) + \frac{\partial}{\partial y}\left(\lambda_y \frac{\partial T}{\partial y}\right) + \frac{\partial}{\partial z}\left(\lambda_z \frac{\partial T}{\partial z}\right) + w(x,y,z) = C\rho \frac{\partial T}{\partial \tau}, \tag{2.21}$$

where $w(x, y, z)$ is the rate of energy generation per unit volume (W/m^3) in the medium, and C and ρ are *specific heat capacity* (J/(kg·K) and *material density* (kg/m^3) respectively. Equation 2.16 reflects the principle of energy conservation in media where heat energy is generated and transported by diffusion.

The anisotropic nature of heat diffusion in Eq. 2.21 is expressed by the thermal conduction coefficients λ_x, λ_y, λ_z.

In the case of an isotropic material:

$$\frac{\partial^2 T}{\partial x^2} + \frac{\partial^2 T}{\partial y^2} + \frac{\partial^2 T}{\partial z^2} + \frac{w}{\lambda} = \frac{1}{a}\frac{\partial T}{\partial \tau}, \tag{2.22}$$

where $a = \lambda/(C\rho)$ is thermal diffusivity (m^2/s).

When there is no internal heat source ($w = 0$), as is true in active TNDT, this equation becomes the well-known presentation of Eq. 2.21:

$$\frac{\partial^2 T}{\partial x^2} + \frac{\partial^2 T}{\partial y^2} + \frac{\partial^2 T}{\partial z^2} = \frac{1}{a}\frac{\partial T}{\partial \tau}. \tag{2.23}$$

In a steady-state regime with internal heat sources:

$$\frac{\partial^2 T}{\partial x^2}+\frac{\partial^2 T}{\partial y^2}+\frac{\partial^2 T}{\partial z^2}+\frac{w}{\lambda}=0. \tag{2.24}$$

A steady-state regime without internal heat sources is described by the Laplace equation:

$$\frac{\partial^2 T}{\partial x^2}+\frac{\partial^2 T}{\partial y^2}+\frac{\partial^2 T}{\partial z^2}=0. \tag{2.25}$$

In an adiabatic situation, i.e. in bodies which are thermally-insulated from the ambient, the temperature signals produced by hidden defects are fully depleted due to heat diffusion in the material. In practice, such temperature signals can survive due to heat exchange with the ambient, but their amplitude is typically much lower than the maximum temperature signals that appear at optimum observation times (see Sect. 1.2). Hence, subsurface defects are detected by applying active TNDT procedures that are described with Eqs. 2.21, 2.22, and 2.23.

2.2.2 Parabolic Equation of Heat Conduction in Cylindrical and Spherical Coordinates

Equation (2.23) is given in Cartesian coordinates. Its general presentation is:

$$\nabla^2 T=\frac{1}{a}\frac{\partial T}{\partial \tau}, \tag{2.26}$$

which is a parabolic equation.

In cylindrical coordinates r, φ, z:

$$\nabla^2 T=\frac{\partial^2 T}{\partial r^2}+\frac{1}{r}\frac{\partial T}{\partial r}+\frac{1}{r^2}\frac{\partial^2 T}{\partial \phi^2}+\frac{\partial^2 T}{\partial z^2}. \tag{2.27}$$

And in spherical coordinates r, Θ, φ:

$$\nabla^2 T=\frac{\partial^2 T}{\partial r^2}+\frac{2}{r}\frac{\partial T}{\partial r}+\frac{1}{r^2}\frac{\partial}{\partial \mu}\left[\left(1-\mu^2\right)\frac{\partial T}{\partial \mu}\right]+\frac{1}{r^2\left(1-\mu^2\right)}\frac{\partial^2 T}{\partial \phi^2}, \tag{2.28}$$

where $\mu = \cos\ \Theta$.

2.2.3 *Hyperbolic Equation of Heat Conduction in Cartesian Coordinates*

The Fourier law in the form of (2.1) excludes the cause-and-effect relationship between a temperature gradient and the heat flux that caused it. In other words, all equations in Sect. 2.2.1, describing conduction heat transfer, imply an infinite velocity of heat transportation. Physically, heat transfer occurs due to a transfer of heat energy between material particles. In dielectric materials, heat energy is transferred through oscillations in a lattice, while in metals heat energy is carried by free electrons. Obviously, both processes require time; they are not instantaneous. To describe a finite velocity of heat transfer, one introduces a *thermal relaxation time* τ_r and the parabolic equation (2.26) will be replaced with a *hyperbolic* one [2]:

$$\nabla^2 T = \frac{1}{a}\frac{\partial T}{\partial \tau} + \frac{\tau_r}{a}\frac{\partial^2 T}{\partial \tau^2}. \tag{2.29}$$

Heat transfer velocity is given by:

$$V_t = \sqrt{\lambda / (C\rho\tau_r)}. \tag{2.30}$$

In metals, the thermal relaxation time is about 10^{-11} seconds and in aluminum $V_f \approx 3$ km/s.

The validity of the hyperbolic equation (2.29) is limited by the material thickness $L_{\min}$ [2]:

$$L_{\min} < 2\pi a / V_t. \tag{2.31}$$

For metals, $L_{\min}$~10^{-6}, while for ceramics $L_{\min}$~$10^{-4} - 10^{-3}$ m.

The differential eq. (2.24) is of a rather academic interest in the analysis of *phonon* heat conduction in pure crystals under super-low temperatures, as well as of explosive processes. However, the increasing use of very thin films and the ongoing increase in IR imager speed (up to several kilohertz now) can make the study of high-speed thermal processes practical.

2.3 Thermal Properties of Materials

The basic thermal properties of materials are thermal conductivity λ, heat capacity C and density ρ (see Table 2.2).

Thermal **conductivity** is the steady-state material heat transport property that is responsible for the attenuation of heat flux.

Two dynamic thermal parameters are thermal diffusivity $a = \lambda/(C\rho)$, m²/s, and *thermal effusivity* $e = \sqrt{\lambda C\rho}$ (W·s$^{1/2}$/(m^2·K).

Table 2.2 Thermal properties of materials (in order of increasing thermal conductivity)

Material**	Density ρ, kg/m^3	Heat capacity C, J/(kg·K)	Thermal conductivity λ, W/(m·K)	Thermal diffusivity a, 10^{-6} m^2/s	Thermal effusivity e, W·s$^{1/2}$/(m^2·K)
Air (in thin gaps)	1.2	1005	0.070	58.0	9.19
Polystirol foamed	20	1500	0.034	1.13	31.9
Rubber (soft)	1100	2010	0.130	0.0588	536
Polyisopren	913	1905	0.134	0.0770	483
Polyvinylchloride	–	–	0.140	0.11	422
Polyaramide	1330	–1047	0.142 (⊥)	0.102 (⊥)**	445 (⊥)
			1.69 (∥)	0.121 (∥)	485 (∥)
Epoxy resin	1100…1300	800…1200	0.17…0.19	0.15	465
Pine (parallel to	550	2512	0.174	0.126	490
fibers)	1200	1850	0.2	0.09	666
Formaldehyde resin	–	–	0.3 (⊥)	0.13 (⊥)	832 (⊥)
Glass fiber			0.38 (∥)	0.17 (∥)	922 (∥)
reinforced plastic*					
(GFRP)					
Plaster	1100	1005	0.233	0.211	507
Water	1000	4193	0.586	0.140	1570
Teflon	2210	1050	0.23	0.099	731
Trinitrotoluene	–	–	0.23	0.0925	–
Sand (river dry)	1520	800	0.34	0.28	643
Caprolon B	–	–	0.37	0.21	–
Carbon fiber	–	–	0.64 (⊥)	0.52 (⊥)	888 (⊥)
reinforced plastic*			1.28 (∥)	1.04 (∥)	1260 (∥)
(CFRP)					
Adhesive (epoxy	1985	2461	0.66	13.5	1795
50%,	1700	879	0.755	0.505	1060
Aluminum 50%)	2442	837	0.879	0.430	1340
Brick (red)	1650	2090	1.13	0.328	1974
Glass	2200	745	1.38	0.842	1504
Sand (river moist)	2400	837	1.51	0.752	1740
Silicon dioxide	–	–	2.6	0.5	–
Concrete*	8300	440	9.5	2.60	5890
Soil	2000	1000	11 (⊥)	5,5 (⊥)	–
Nickel alloy			16 (∥)	8,0 (∥)	
Boro-epoxy					
composite*					

(continued)

Table 2.2 (continued)

Material**	Density ρ, kg/m^3	Heat capacity C, J/(kg·K)	Thermal conductivity λ, W/(m·K)	Thermal diffusivity a, 10^{-6} m^2/s	Thermal effusivity e, W·s$^{1/2}$/(m^2·K)
Steel (AISI 316)	8240	468	13.4	3.47	7190
Silicon nitride*	3000	700	15–30	7.14–14.28	5612–7937
Titanium	4500	522	21.9	9.32	7170
Zirconia	6570	278	22.7	12.4	6440
Boron	2340	600	27,4	20,0	–
Aluminum oxide	3970	765	46.0	15.2	11,800
Brass	8800	420	52.0	14.1	13,900
Steel (AISI 1010)	7830	434	63.9	18.8	14,700
Silicon	2330	800	84–126	45.1–67.6	12,513–
Graphite	2300	670	116	75.3	15,325
Tungsten	19,300	132	174	68.3	13,400
Aluminum alloy	2770	875	177	73.0	21,100
2024-T6	19,450	134	313	120	20,700
Gold	9000	406	365	100	28,600
Copper	3160	675	490	230.0	36,500
Silicon carbide	3516	502	2200–3200	1247–1814	32,300
Diamond	–	–	4840–5300	–	62,305–
Graphen***					75,143
					–

*Approximate values (depend on manufacturer and manufacturing process)
**(⊥) – perpendicular to fibers; (||) – parallel to fibers. In composites, three orthogonal components of thermal conductivity can be defined as a function of fiber direction.
***Nano-structured carbon-based material (Nobel prize 2010, A. Geim and K. Novoselov)

Thermal **diffusivity** is the rate at which heat diffuses through a material. Generally, materials that have high thermal conductivityλ also have high thermal diffusivity a.

Thermal **effusivity** is a measure of the ability of a material to increase its temperature as a response to a given energy (heat) input. It is often referred to as *thermal inertia* and is used in calculating a *thermal mismatch factor*, or *reflection coefficient*, Γ that characterizes a thermal contact between two bodies, 1 and 2:

$$\Gamma = \frac{e_1 - e_2}{e_1 + e_2}. \tag{2.32}$$

$\Gamma = 0$ means there is no thermal contact between two bodies, $\Gamma = 1$ means the second material is a perfect thermal conductor, and $\Gamma = -1$ means the second material is a perfect insulator.

Another dynamic thermal parameter is the *thermal diffusion length* μ, m:

$$\mu = \sqrt{0/(\pi f)}, \tag{2.33}$$

which is related to the frequency f (in Hz) of a periodic (time-varying) heat flux. In the theory of thermal waves, this parameter helps to explain that low-frequency

Table 2.3 Classical heat conduction solutions for a semi-infinite body heated with a uniform heat flux*

Body	Heating function	Surface heat exchange	Surface solutions
Semi-infinite body 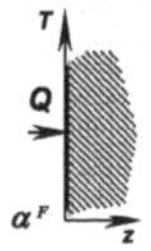		$\alpha^F = 0$	$\frac{T}{Q/\lambda} = 2\left(\frac{a\tau}{\pi}\right)^{1/2}$
		$\alpha^F \neq 0$	$\frac{T}{Q/\alpha} = 1 - e^{H^2 a\tau}\ erfc\ (H\sqrt{a\tau});\ \ H = \alpha/\lambda$
	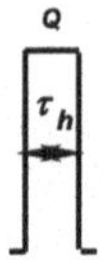	$\alpha^F = 0$	$\frac{T}{Q/\lambda} = 2\left(\frac{a\tau}{\pi}\right)^{1/2} - 2\left[\frac{a(\tau-\tau_h)}{\pi}\right]^{1/2}$
		$\alpha^F \neq 0$	$\frac{T}{Q/\alpha} = e^{H^2 a(\tau-\tau_h)}\ erfc\ [H\sqrt{a(\tau-\tau_h)}] - e^{H^2 a\tau}\ erfc\ (H\sqrt{a\tau})$
		$\alpha^F = 0$	$\frac{T}{W} = \frac{1}{e\sqrt{\pi\tau}};\ \ e = \sqrt{C\rho\lambda}$
		$\alpha^F \neq 0$	$\frac{T}{W} = \frac{1}{e\sqrt{\pi\ \tau}} - \frac{\alpha}{e^2}\ [e^{H^2 a\tau}\ erfc\ (H\sqrt{a\tau})]$

thermal oscillations penetrate deeper into materials than higher frequency oscillations. In other words, it can also be stated that a thermal wave of a given frequency will penetrate deeper into a material of a higher diffusivity. In one-sided TNDT, the thermal diffusion length provides an estimate of the defect detection limit when a material is subjected to thermal waves of various frequencies.

2.4 Classical Solutions of Heat Conduction

An exhaustive analysis of transient heat conduction was performed by Carslaw and Jaeger in their fundamental monograph [3]. In Russia, a similar book was published by Luikov [4]. In these books one can find many 1D analytical solutions to thermal heat conduction problems, and these are used to calculate the temperature of sound areas of materials under thermal loading. The basic TNDT solutions are shown systematically in Tables 2.3 and 2.4 for a semi-infinite body and a plate respectively. The solutions are given in the general form of:

$$T = Q'\ \theta, \tag{2.34}$$

where T is the surface excess temperature, Q' (or W') is the term that characterizes the heating efficiency, θ defines the normalized excess surface temperature. In the discussion that follows, θ will be used with the corresponding indices; for example, θ^R_{SP} represents the normalized excess temperature on the rear (R) surface of a plate

Table 2.4 Classical heat conduction solutions for a plate heated with a uniform heat flux

Body	Heating function	Surface heat exchange	Front (heated) surface solution	Rear surface solution
Plate 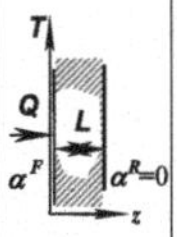		$\alpha^F = 0$ $\alpha^R = 0$	$\frac{T}{Q'} = Fo + \frac{1}{3} - \frac{2}{\pi^2}\sum_{n=1}^{\infty}\frac{1}{n^2}e^{-n^2\pi^2 Fo} = \theta_{ST}^F(Fo)$ $Q' = \frac{QL}{\lambda};\ \frac{T}{Q'} = \theta_{ST}^F(Fo)$ $\frac{\partial T/Q'}{\partial Fo} = 1 + 2\sum_{n=1}^{\infty} e^{-n^2\pi^2 Fo}$ $\frac{\partial T/T}{\partial Fo/Fo} = Fo\,\frac{1 + 2\sum_{n=1}^{\infty} e^{-n^2\pi^2 Fo}}{\theta_{ST}^F(Fo)}$	$\frac{T}{Q'} = Fo - \frac{1}{6} - \frac{2}{\pi^2}\sum_{n=1}^{\infty}\frac{-1^n}{n^2}e^{-n^2\pi^2 Fo} = \theta_{ST}^R(Fo)$ $\frac{\partial T}{\partial Q'} = \theta_{ST}^R(Fo)$ $\frac{\partial T/Q'}{\partial Fo} = 1 + 2\sum_{n=1}^{\infty}(-1)^n e^{-n^2\pi^2 Fo}$ $\frac{\partial T/T}{\partial Fo/Fo} = Fo\,\frac{1 + 2\sum_{n=1}^{\infty}(-1)^n e^{-n^2\pi^2 Fo}}{\theta_{ST}^R(Fo)}$
		$\alpha^F \neq 0$ $\alpha^R = 0$	$\frac{T}{Q'} = 1 - \sum_{n=1}^{\infty}\frac{2\,Bi}{Bi(Bi+1)+\mu_n^2}e^{-\mu_n^2 Fo} = \theta_{ST}^F(Bi, Fo)$; $Q' = Q/\alpha;\ \frac{\partial T}{\partial Q'} = \theta_{ST}^F(Bi, Fo)$ $\frac{\partial T/Q'}{Bi} = \sum_{n=1}^{\infty}\frac{Bi(6Bi+4)}{[Bi(Bi+1)+\mu_n^2]^2}e^{-\mu_n^2 Fo}$ $\frac{\partial T/Q'}{Fo} = \sum_{n=1}^{\infty}\mu_n^2\frac{2\,Bi}{Bi(Bi+1)+\mu_n^2}e^{-\mu_n^2 Fo}$ μ_n are positive roots of the transcendent equation: $\mu\, tg\, \mu = Bi$	$\frac{T}{Q'} = 1 - \sum_{n=1}^{\infty}\frac{1}{\cos\mu_n}\frac{2\,Bi}{Bi(Bi+1)+\mu_n^2}e^{-\mu_n^2 Fo} = \theta_{ST}^R(Bi, Fo)$ $\frac{\partial T}{\partial Q'} = \theta_{ST}^R(Bi, Fo)$ $\frac{\partial T/Q'}{Bi} = \sum_{n=1}^{\infty}\frac{1}{\cos\mu_n}\frac{Bi(6Bi+4)}{[Bi(Bi+1)+\mu_n^2]^2}e^{-\mu_n^2 Fo}$ $\frac{\partial T/Q'}{Fo} = \sum_{n=1}^{\infty}\frac{\mu_n^2}{\cos\mu_n}\frac{2\,Bi}{Bi(Bi+1)+\mu_n^2}e^{-\mu_n^2 Fo}$
	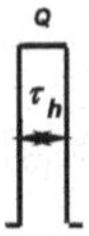	$\alpha^F = 0$ $\alpha^R = 0$	$\frac{T}{Q'} = Fo_h + \frac{2}{\pi^2}\sum_{n=1}^{\infty}\frac{1}{n^2}e^{-n^2\pi^2 Fo}(e^{n^2\pi^2 Fo_h} - 1);\ Q' = QL/\lambda$ $\frac{\partial T}{\partial Q'} = Fo_h + \frac{2}{\pi^2}\sum_{n=1}^{\infty}\frac{1}{n^2}e^{-n^2\pi^2 Fo}(e^{n^2\pi^2 Fo_h} - 1)$ $\frac{\partial T/T}{\partial Fo_h/Fo_h} = Fo_h\,\frac{1 + 2\sum_{n=1}^{\infty} e^{-n^2\pi^2(Fo - Fo_h)}}{Fo_h + \frac{2}{\pi^2}\sum_{n=1}^{\infty}\frac{1}{n^2}e^{-n^2\pi^2 Fo}(e^{n^2\pi^2 Fo_h} - 1)}$ $\frac{\partial T/Q'}{\partial Fo} = -Fo\,\frac{2\sum_{n=1}^{\infty} e^{-n^2\pi^2 Fo}(e^{n^2\pi^2 Fo} - 1)}{Fo_h + \frac{2}{\pi^2}\sum_{n=1}^{\infty}\frac{1}{n^2}e^{-n^2\pi^2 Fo}(e^{n^2\pi^2 Fo_h} - 1)}$	$\frac{T}{Q'} = Fo_h + \frac{2}{\pi^2}\sum_{n=1}^{\infty}\frac{-1^n}{n^2}e^{-n^2\pi^2 Fo}(e^{n^2\pi^2 Fo_h} - 1)$ $\frac{\partial T}{\partial Q'} = Fo_h + \frac{2}{\pi^2}\sum_{n=1}^{\infty}\frac{-1^n}{n^2}e^{-n^2\pi^2 Fo}(e^{n^2\pi^2 Fo_h} - 1)$ $\frac{\partial T/T}{\partial Fo_h/Fo_h} = Fo_h\,\frac{1 + 2\sum_{n=1}^{\infty}(-1)^n e^{-n^2\pi^2(Fo - Fo_h)}}{Fo_h + \frac{2}{\pi^2}\sum_{n=1}^{\infty}\frac{(-1)^n}{n^2}e^{-n^2\pi^2 Fo}(e^{n^2\pi^2 Fo_h} - 1)}$ $\frac{\partial T/Q'}{\partial Fo} = -Fo\,\frac{2\sum_{n=1}^{\infty}(-1)^n e^{-n^2\pi^2 Fo}(e^{n^2\pi^2 Fo} - 1)}{Fo_h + \frac{2}{\pi^2}\sum_{n=1}^{\infty}\frac{(-1)^n}{n^2}e^{-n^2\pi^2 Fo}(e^{n^2\pi^2 Fo_h} - 1)}$
		$\alpha^F \neq 0$ $\alpha^R = 0$	$\frac{T}{Q'} = \sum_{n=1}^{\infty}\frac{2\,Bi}{Bi(Bi+1)+\mu_n^2}e^{-\mu_n^2 Fo}(e^{\mu_n^2 Fo_h} - 1) =$ $= \theta_{SP}^F(Bi, Fo, Fo_h)$ $Q' = Q/\alpha;\ \frac{\partial T}{\partial Q'} = \theta_{SP}^F(Bi, Fo, Fo_h)$ $\frac{\partial T/Q'}{Bi} = \sum_{n=1}^{\infty}\frac{Bi(6Bi+4)}{Bi(Bi+1)+\mu_n^2}e^{-\mu_n^2 Fo}(e^{\mu_n^2 Fo_h} - 1)$ $\frac{\partial T/Q'}{Fo_h} = \sum_{n=1}^{\infty}\mu_n^2\frac{2\,Bi}{Bi(Bi+1)+\mu_n^2}e^{-\mu_n^2 Fo}e^{\mu_n^2 Fo_h}$ $\frac{\partial T/Q'}{Fo} = -\sum_{n=1}^{\infty}\mu_n^2\frac{2\,Bi}{Bi(Bi+1)+\mu_n^2}e^{-\mu_n^2 Fo}(e^{\mu_n^2 Fo_h} - 1)$	$\frac{T}{Q'} = \sum_{n=1}^{\infty}\frac{1}{\cos\mu_n}\frac{2\,Bi}{Bi(Bi+1)+\mu_n^2}e^{-\mu_n^2 Fo}(e^{\mu_n^2 Fo_h} - 1) =$ $= \theta_{SP}^R(Bi, Fo, Fo_h)$ $Q' = Q/\alpha;\ \frac{\partial T}{\partial Q'} = \theta_{SP}^R(Bi, Fo, Fo_h)$ $\frac{\partial T/Q'}{Bi} = \sum_{n=1}^{\infty}\frac{1}{\cos\mu_n}\frac{Bi(6Bi+4)}{Bi(Bi+1)+\mu_n^2}e^{-\mu_n^2 Fo}(e^{\mu_n^2 Fo_h} - 1)$ $\frac{\partial T/Q'}{Fo_h} = \sum_{n=1}^{\infty}\frac{\mu_n^2}{\cos\mu_n}\frac{2\,Bi}{Bi(Bi+1)+\mu_n^2}e^{-\mu_n^2 Fo}e^{\mu_n^2 Fo_h}$ $\frac{\partial T/Q'}{Fo} = -\sum_{n=1}^{\infty}\frac{\mu_n^2}{\cos\mu_n}\frac{2\,Bi}{Bi(Bi+1)+\mu_n^2}e^{-\mu_n^2 Fo}(e^{\mu_n^2 Fo_h} - 1)$
		$\alpha^F = 0$ $\alpha^R = 0$	$\frac{T}{W'} = 1 + 2\sum_{n=1}^{\infty} e^{-n^2\pi^2 Fo} = \theta_D^F;\ W' = \frac{W\,a}{\lambda L}$ $\frac{\partial T}{\partial W'} = \theta_D^F$ $\frac{\partial T/T}{\partial Fo/Fo} = \frac{-2\pi^2 Fo\sum_{n=1}^{\infty} n^2 e^{-n^2\pi^2 Fo}}{1 + 2\sum_{n=1}^{\infty} e^{-n^2\pi^2 Fo}}$	$\frac{T}{W'} = 1 + 2\sum_{n=1}^{\infty}(-1)^n e^{-n^2\pi^2 Fo} = \theta_D^R$ $\frac{\partial T}{\partial W'} = \theta_D^R$ $\frac{\partial T/T}{\partial Fo/Fo} = \frac{-2\pi^2 Fo\sum_{n=1}^{\infty}(-1)^n n^2 e^{-n^2\pi^2 Fo}}{1 + 2\sum_{n=1}^{\infty}(-1)^n e^{-n^2\pi^2 Fo}}$
		$\alpha^F \neq 0$ $\alpha^R = 0$	$\frac{T}{W'} = \sum_{n=1}^{\infty}\frac{2\mu_n^2}{Bi(Bi+1)+\mu_n^2}e^{-\mu_n^2 Fo} = \theta_D^F(Bi, Fo)$ $W' = \frac{W\,a}{\lambda L};\ \frac{\partial T}{\partial W'} = \theta_D^F(Bi, Fo)$ $\frac{\partial T/W'}{\partial Fo} = -\sum_{n=1}^{\infty}\frac{2\mu_n^4}{Bi(Bi+1)+\mu_n^2}e^{-\mu_n^2 Fo}$	$\frac{T}{W'} = \sum_{n=1}^{\infty}\frac{1}{\cos\mu_n}\frac{2\mu_n^2}{Bi(Bi+1)+\mu_n^2}e^{-\mu_n^2 Fo} = \theta_D^R(Bi, Fo)$ $\frac{\partial T}{\partial W'} = \theta_D^R(Bi, Fo)$ $\frac{\partial T/W'}{\partial Fo} = -\sum_{n=1}^{\infty}\frac{1}{\cos\mu_n}\frac{2\mu_n^4}{Bi(Bi+1)+\mu_n^2}e^{-\mu_n^2 Fo}$

Table 2.5 Classical heat conduction solutions for harmonic (periodic) thermal waves (uniform heating)

		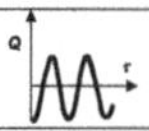 Heating function $Q = Q_m \cos(\omega\tau)$	
Body	**Surface heat exchange**	**General solution**	**Front (heated) surface solution**
Semi-infinite body 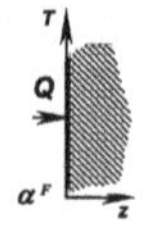	$\alpha^F = 0$	$\frac{T(\tau)}{Q_m} = \frac{1}{\lambda\sqrt{\omega/a}} e^{-z\sqrt{\omega/2a}} \cos[\omega\tau - (z\sqrt{\omega/2a}) - \Theta]$; $\Theta = 45^0$	$\frac{T(z=0,\tau)}{Q_m} = \frac{1}{\lambda\sqrt{\omega/a}} \cos[\omega\tau - \Theta]$
	$\alpha^F \neq 0$	$\frac{T(z,\tau)}{Q_m/\alpha} = A_o e^{-z\sqrt{\omega/2a}} \cos[\omega\tau - (z\sqrt{\omega/2a}) - \Omega]$; $A_o = (1 + 2\sqrt{\omega/2a}/H + \omega/aH^2)^{-1/2}$; $\Omega = Arctg\ (1 + H\sqrt{2a/\omega})^{-1}$; $H = \alpha/\lambda$.	$\frac{T(z=0,\tau)}{Q_m/\alpha} = A_o \cos[\omega\tau - (z\sqrt{\omega/2a}) - \Omega]$; $A_o = (1 + 2\sqrt{\omega/2a}/H + \omega/aH^2)^{-1/2}$; $\Omega = Arctg\ (1 + H\sqrt{2a/\omega})^{-1}$; $H = \alpha/\lambda$
Plate 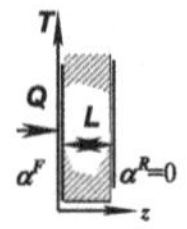	$\alpha^F \neq 0$ $\alpha^R = 0$	$\frac{T(Z,\tau)}{Q_m/\alpha} = A_0 \cos(Pd\ Fo - \Psi)$; $Z = z/L$; $A_0 = (N_i N_{-i})^{1/2}$; $\Psi = Arctg\ (i\frac{N_i - N_{-i}}{N_i + N_{-i}})$; $N_i = \frac{ch\sqrt{iPd}\,(1-Z)}{ch\sqrt{iPd} + (1/Bi)\sqrt{iPd}\ sh\sqrt{iPd}}$; $N_{-i} = \frac{ch\sqrt{-iPd}\,(1-Z)}{ch\sqrt{-iPd} + (1/Bi)\sqrt{-iPd}\ sh\sqrt{-iPd}}$; $Bi = \alpha L/\lambda$; $Pd = \omega L^2/a$	$\frac{T(z=0,\tau)}{Q_m/\alpha} = A_0\, c(Pd\ Fo - \Psi)$; $A_0 = (N_i N_{-i})^{1/2}$; $\Psi = Arctg\ (i\frac{N_i - N_{-i}}{N_i + N_{-i}})$; $N_i = \frac{ch\sqrt{iPd}}{ch\sqrt{iPd} + (1/Bi)\sqrt{iPd}\ sh\sqrt{iPd}}$; $N_{-i} = \frac{ch\sqrt{-iPd}}{ch\sqrt{-iPd} + (1/Bi)\sqrt{-iPd}\ sh\sqrt{-iPd}}$; $Bi = \alpha L/\lambda$; $Pd = \omega L^2/a$

Table 2.6 Classical heat conduction solutions for an immovable point-like heat source

Body	**Surface heat exchange**	**General solution**	**Front (heated) surface**
Semi-infinite body 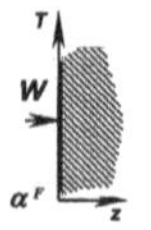	$\alpha^F = 0$	$\frac{T}{W'} = \frac{1}{4(\pi a\tau)^{3/2}} e^{-\frac{r^2}{4a\tau}}$; $W' = W/\rho C$; $r^2 = x^2 + y^2 + z^2$	$\frac{T}{W'} = \frac{1}{4(\pi a\tau)^{3/2}} e^{-\frac{r^2}{4a\tau}}$ $r^2 = x^2 + y^2$
	$\alpha^F \neq 0$	$\frac{T}{W'} = \frac{1}{4(\pi a\tau)^{3/2}} e^{-\frac{r^2}{4a\tau}} - \frac{H}{4\pi a\tau} erfc(\frac{z}{2\sqrt{a\tau}} + H\sqrt{a\tau}) e^{Hz + aH^2\tau - r^2/4a\tau}$; $W' = W/\rho C$; $H = \alpha/\lambda$	$\frac{T}{W'} = \frac{1}{4(\pi a\tau)^{3/2}} e^{-\frac{r^2}{4a\tau}} - \frac{H}{4\pi a\tau} erfc(H\sqrt{a\tau}) e^{aH^2\tau - r^2/4a\tau}$
Plate 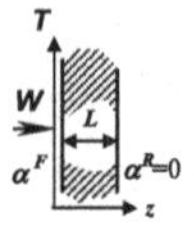	$\alpha^F = 0$ $\alpha^R = 0$	$\frac{T}{W'} = \frac{1}{4\pi aL\tau} e^{-\frac{r^2}{4a\tau}} [1 + 2\sum_{n=1}^{\infty} e^{-n^2\pi^2 Fo} \cos(n\pi Z)]$; $W' = W/\rho C$; $r^2 = x^2 + y^2 + z^2$; $Z = z/L$; $Fo = a\tau/L^2$	$\frac{T}{W'} = \frac{1}{4\pi aL\tau} e^{-\frac{r^2}{4a\tau}} (1 + 2\sum_{n=1}^{\infty} e^{-n^2\pi^2 Fo})$; $r^2 = x^2 + y^2$
	$\alpha^F \neq 0$ $\alpha^R \neq 0$	$\frac{T}{W'} = \frac{1}{2\pi aL\tau} e^{-\frac{r^2}{4a\tau}} \sum_{n=1}^{\infty} \frac{\mu_n^2 \cos\mu_n Z + \mu_n Bi \sin\mu_n Z}{\mu_n^2 + 2Bi + Bi^2} e^{-\mu_n^2 Fo}$; $W' = W/\rho C$; $r^2 = x^2 + y^2 + z^2$; $Bi = \alpha L/\lambda$; $Fo = a\tau/L^2$; $Z = z/L$; μ_n are roots of the transcendent equation: $tg\ \mu = 2\mu Bi/(\mu^2 - Bi^2)$	$\frac{T}{W'} = \frac{1}{2\pi aL\tau} e^{-\frac{r^2}{4a\tau}} \sum_{n=1}^{\infty} \frac{\mu_n^2}{\mu_n^2 + 2Bi + Bi^2} e^{-\mu_n^2 Fo}$; ; $r^2 = x^2 + y^2$

heated with a square pulse (SP). Table 2.4 also contains some temperature derivatives that are used in Sect. 3.7 in the analysis of *sensitivity functions*.

It is worth noting that, in many practical cases, the solutions for heating bodies of cylindrical and spherical shape can be approximated by using planar geometry, i.e. using the solutions in Table 2.3 and 2.4. The analysis using classical heat conduction solutions in TNDT is described in Sect. 2.7 [3, 4].

Three classical solutions for heating objects with periodic thermal waves are shown in Table 2.5.

Table 2.7 Classical heat conduction solutions for an immovable Gaussian heat source

Body	Surface heat exchange	General solution	Front (heated) surface
Semi-infinite body 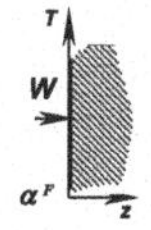	$\alpha^F = 0$	$T(r,z,\tau) = \frac{W}{2\pi e\sqrt{a\tau_h}} \int_0^{+\infty} J_0(\sigma r) e^{-\frac{\sigma^2 R^2}{8}} [I(\sigma,z,\tau) -$ $- \Theta(\tau - \tau_h) I(\sigma, z, \tau - \tau_h)] d\sigma;$ $r^2 = x^2 + y^2 + z^2;$ $I(\sigma,z,\tau) = [e^{-z\sigma} erfc(-\sigma\sqrt{a\tau} + \frac{1}{2}\frac{z}{\sqrt{\alpha\tau}}) - e^{z\sigma} erfc(\sigma\sqrt{a\tau} + \frac{1}{2}\frac{z}{\sqrt{\alpha\tau}})];$ J_0 – Bessel function; Heat energy distribution in space: $e^{-(x^2+y^2)/2R^2}$; $\Theta(\tau - \tau_h)$ - step-function (Heviside function)	$T(r,z=0,\tau) = \frac{W}{2\pi e\sqrt{a\tau_h}} \int_0^{+\infty} J_0(\sigma r) e^{-\frac{\sigma^2 R^2}{8}} x$ $x[erf(\sigma\sqrt{\alpha\tau}) - \Theta(\tau - \tau_h) erf(\sigma\sqrt{\alpha}\sqrt{\tau - \tau_h})] d\sigma;$ $r^2 = x^2 + y^2$
Plate 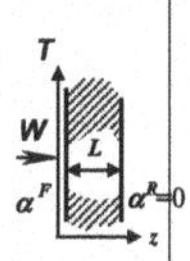	$\alpha^F = 0$ $\alpha^R = 0$	$T(r,z,\tau) = \frac{2W}{\pi\rho CL} \frac{e^{-\frac{2r^2}{R^2+8a\tau}}}{(R^2 + 8a\tau)} [1 + 2\sum_{n=1}^{+\infty} e^{-\frac{n^2\pi^2 a\tau}{L^2}} \cos(n\pi Z)];$ $r^2 = x^2 + y^2 + z^2$ Heat energy distribution in space: $e^{-(x^2+y^2)/2R^2}$; $e = \sqrt{\rho C\lambda}$	$T(r,z=0,\tau) = \frac{2W}{\pi\rho CL} \frac{e^{-\frac{2r^2}{R^2+8a\tau}}}{(R^2 + 8a\tau)} (1 + 2\sum_{n=1}^{+\infty} e^{-\frac{n^2\pi^2 a\tau}{L^2}});$ $r^2 = x^2 + y^2$

Table 2.8 Heat conduction solution for a plate heated with a circular uniform heat source [6]

Body	Surface heat exchange	General solution
Plate 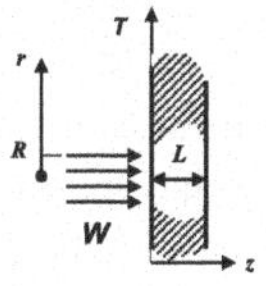	$\alpha^F = 0$ $\alpha^R = 0$	$T(r,z,\tau) = \frac{W}{\rho CL} e^{-\frac{r^2}{4a\tau}} [1 + 2\sum_{n=1}^{\infty} e^{-\frac{n^2\pi^2 a\tau}{L^2}} \cos(n\pi Z)] \times$ $\times \{1 - e^{-\frac{R^2}{4a\tau}} + \sum_{k=1}^{\infty} \frac{(-1)^{k+1}}{(k!)^2} (\frac{r^2}{4a\tau})^k [(-\frac{R^2}{4a\tau})^k e^{-\frac{R^2}{4a\tau}} - k \int_0^{-R^2/4a\tau} \mu^{k-1} e^{\mu} d\mu]\};$ $Z = z/L$; R – heat source radius

A number of solutions for heating bodies with localized, Gaussian and circular heat sources are presented in Tables 2.6, 2.7, and 2.8. These solutions involve all spatial coordinates and are often used in thermal property measurements.

In Table 2.9, there are solutions for 3D problems of heating a semi-infinite body with a localized and Gaussian moving heat source. These have been used to examine welds and they might be used in the detection of hidden defects by scanning a sample surface with a localized source (the "Flying spot" and "Crawling spot" techniques, see Table 1.1) because, as will be shown below, best detectability of defects is achieved by heating a sample with a heat flux that follows the Dirac behavior in both time and space. Another benefit is that a moving heat source technique also enables the detection of defects (cracks) located perpendicular to the front surface of a sample.

In 2002, Maillet et al. summarized the "thermal quadrupole" method applied to 1D and 2D multilayer heat conduction problems [5]. This book contains numerous examples related to TNDT, as well as several computer programs implemented on the Matlab platform.

Table 2.9 Classical heat conduction solutions for heating a semi-infinite body with a moving heat source [7]

Body	General solution	Solution for a line crossing a heat source
Semi-infinite body **Point-like heat source** 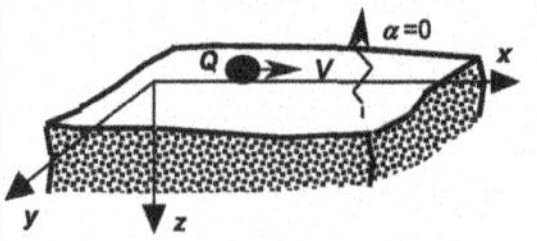 	$T = \frac{Q}{\lambda}\frac{1}{\pi\sqrt{\pi}}\frac{e^{-\frac{V\cdot x}{2a}}}{R}\int_{\frac{R}{2\sqrt{a\tau}}}^{\infty} e^{-\mu^2-\frac{V^2\cdot R^2}{16a^2\cdot\mu^2}}d\mu;$ $R = x^2 + y^2 + z^2;$ Q in [W]	$T = \frac{Q}{\lambda}\frac{1}{\pi\sqrt{\pi}}\frac{e^{-\frac{V x}{2a}}}{x^2}\int_{\frac{x^2}{2\sqrt{a\tau}}}^{\infty} e^{-\mu^2-\frac{V^2 x^2}{16a^2\mu^2}}d\mu$
Semi-infinite body **Gaussian heat source** 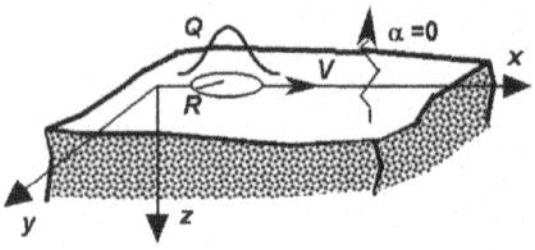 	$T(\tilde{x},\tilde{y},\tilde{z},Fo) = \frac{Q}{\lambda}\frac{1}{\pi\sqrt{\pi}}\frac{1}{R}x$ $x\int_0^{Fo}\frac{\exp[-\frac{(\tilde{x}-\tilde{v}\cdot Fo')^2+\tilde{y}^2+\tilde{z}^2}{1+4(Fo-Fo')}]}{\sqrt{Fo-Fo'}[1+4(Fo-Fo')]}\cdot dFo';$ $Fo = \frac{a\tau}{R^2};\quad \tilde{v} = \frac{V R}{a};$ $\tilde{x} = x/R;\ \tilde{y} = y/R;\ \tilde{z} = z/R;$ energy distribution across the heat source: $F(x,y,z=0) = \frac{Q a}{C\rho}\frac{1}{\pi R^2}\exp[-\frac{x^2+y^2}{R^2}]$ Q in [W]	$T(\tilde{x},Fo) = \frac{Q}{\lambda}\frac{1}{\pi\sqrt{\pi}}\frac{1}{R}x$ $x\int_0^{Fo}\frac{\exp[-\frac{(\tilde{x}-\tilde{v}\cdot Fo')^2}{1+4(Fo-Fo')}]}{\sqrt{Fo-Fo'}[1+4(Fo-Fo')]}dFo'$

Mathematically, the "thermal quadrupole" method belongs to a class of analytical methods used for solving linear differential equations in simple geometries. It utilizes such analytical tools as the integral Laplace transformation (in time), as well as the spatial Fourier and Hankel transformations based on the separation of variables. The corresponding solutions are expressed as linear matrix relationships between vectors of the temperature and heat fluxes on boundaries of a multilayer system. This allows us to obtain solutions that are independent of boundary conditions.

Solving differential equations with integral transformations is a well-known traditional mathematical procedure [3, 4], but presently this method is being complemented by novel mathematical approaches, such as computer-based numerical solutions for the *inverse* Laplace and Fourier transformations. This gave birth to a hybrid of both analytical and numerical methods.

In short, the "thermal quadrupole" method allows us to derive *algebraic* solutions for heating a multilayer plate in the Laplace domain. Therefore, the analysis of dependencies between defect parameters (defect depth and thickness) and the "Laplace temperature" can be fulfilled in Laplace space, and the return to real temperatures can be achieved by applying numerical algorithms.

This approach is of a particular interest in solving multilayer thermal problems.

2.5 TNDT Parameters of Interest

TNDT *parameters* of interest are *quantitative characteristics of dynamic heat transfer* in defective structures which are used in decision making (Table 1.1). When analyzing the surface temperature $T(x, y, \tau)$, *amplitude* and *temporal* parameters are introduced. After applying the 1D Fourier transformation in time ($F[T(x, y, \tau)]$), *magnitude* and *phase* are used to distinguish between defective and sound areas. The corresponding parameters appear after applying other signal transformations, e.g. by analyzing *wavelet functions*. Those parameters, which are defined either by testing or by a computer model, can be also called *parameters of detection* for hidden defects. An optimum detection parameter is characterized by a maximum signal-to-noise ratio.

2.5.1 Amplitude and Temporal Parameters of Interest

A dynamic surface temperature distribution is described by a $T(x, y, \tau)$ function. As a first approximation, it can be assumed that the temperature over a sound (non-defect) area is either constant or known. Therefore, a reference temperature $T_{nd}(x_{nd}, y_{nd}, \tau) = T_{nd}(\tau)$ can be introduced. In TNDT, one typically deals with *excess* temperature, which is an increase in temperature relative to the initial temperature of the sample, often assumed to be the same as the ambient temperature T_{amb}. In a defective area, the temperature distribution becomes distorted and the so-called local *differential temperature signals* appear:

$$\Delta T(x,y,\tau) = T_d(x,y,\tau) - T_{nd}(x_{nd},y_{nd},\tau). \tag{2.35}$$

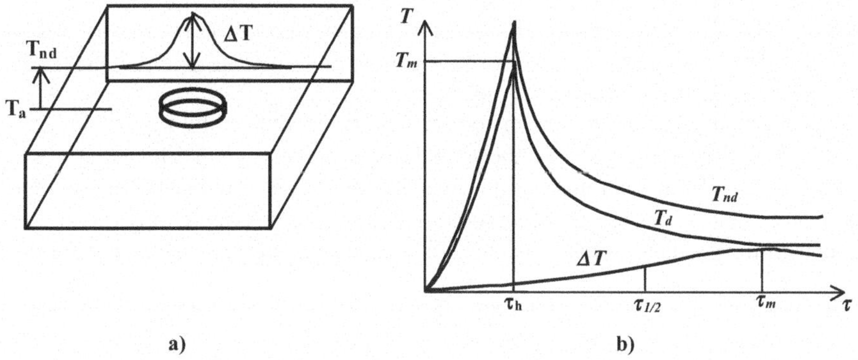

Fig. 2.3 Spatial and temporal profiles $\Delta T(x, y, \tau)$ in pulsed TNDT. (**a**) – spatial temperature profile over a defect, (**b**) – temperature evolution in time

In one-sided TNDT, the spatial temperature profile over small defects resembles a Gaussian bell-shaped curve (Fig. 2.3a). Frequently, one deals only with the maximum temperature signals $\Delta T_m (x, y, \tau) = \Delta T_m(\tau)$. It is important to understand that temperature-dependent parameters, such as the excess temperature $T_{nd}(x_{nd}, y_{nd}, \tau)$, $T_d(x, y, \tau)$ and the signal $\Delta T (x, y, \tau)$, are directly proportional to the absorbed power Q due to the linearity of the heat conduction behavior of materials. A parameter that is independent of Q is the *running temperature contrast*:

$$C^{run}\left(x,y,\tau\right) = \Delta T\left(x,y,\tau\right) / T_{nd}\left(\tau\right), \tag{2.36}$$

which describes a particular testing procedure, a particular material and a particular type a defect. Note that normalization by absorbed energy according to Eq. (2.36) makes sense only if Q is constant across the material.

In practice, absorbed energy may vary even in sound areas because of variations in absorptivity or emissivity. In this case, the so-called *normalized temperature contrast* is more efficient. It is illustrated with temperature profiles $T(\tau)$ and signals $\Delta T (x, y, \tau)$ depicted in Fig. 2.3b. Within a heating period τ_h, the temperature increases from an initial (zero) level to a maximum value $T_m(\tau_h)$, after which it begins to decrease. The differential signal reaches its maximum value ΔT_m at an *optimum observation time* τ_m. In sound areas, the temperature is determined by the absorbed energy. Therefore, $T(x_i, y_i, \tau)$ curves at different points are similar and differ by a constant coefficient if we neglect some subtle phenomena. In defect areas, the presence of defects at a certain depth distorts the surface temperature distribution and creates the differential signal $\Delta T(x_i, y_i)$.

The normalized temperature contrast is defined by the following expression:

$$C^{norm} = \frac{T\left(x_i, y_i, \tau\right)}{T_m\left(x_i, y_i, \tau_h\right)} - \frac{T\left(x_{nd}, y_{nd}, \tau\right)}{T_m\left(x_{nd}, y_{nd}, \tau_h\right)}. \tag{2.37}$$

Each surface point will be ‘normalized’, and this eliminates the effect of uneven heating. The use of normalized temperature contrast is common in procedures of active TNDT. It is worth noting that normalization can be done not only with respect to the end of heating (τ_h), but also with respect to other times, e.g. at the beginning or end of the dynamic heat transfer process.

A drawback of using the dimensionless parameters defined by Eqs. (2.35, 2.36, and 2.37) is the requirement to choose a reference point (x_{nd}, y_{nd}). This point can be easily selected by an operator, but it is difficult for an automated inspection system. Assuming that subsurface defects modify the surface temperature only after a particular time, Maldague et al. introduced the *differentiated contrast* in the following form:

$$C^{dif}\left(x,y,\tau^{(n)}\right) = T\left(x,y,\tau\right) - \sqrt{\frac{\tau^{(n)}}{\tau}}\, T\left(x,y,\tau^{(n)}\right), \tag{2.38}$$

where $\tau^{(n)}$ is the time that corresponds to a 'non-defect' behavior of an analyzed point (note that C^{dif} has the dimension of temperature, therefore, the use of the term 'contrast' is not absolutely correct). Eq. (2.38) follows from the solution for heating an adiabatic semi-infinite body with a Dirac pulse and it requires normalization of each surface point separately.

Defining temperature contrast by Eqs. (2.36, 2.37, and 2.38) can be considered as *1D filtering* (3D heat diffusion is not taken into account). In high-conductivity materials, 3D heat diffusion may seriously distort spatial temperature profiles and thus complicate the detection of small defects.

In order to take 3D heat diffusion into consideration, Vavilov proposed using an algorithm of *3D filtering*. Experimental temperature functions $T^{\exp}$ are normalized by the reference temperature function T_{nd}^{theor}, which is obtained by solving the corresponding 3D TNDT problem:

$$C^{3D} = \frac{T^{\exp}(x,y,\tau)}{T_{nd}^{theor}(x,y,\tau)} = \frac{T_{nd}^{\exp}(x,y,\tau) + \Delta T(x,y,\tau)}{T_{nd}^{theor}(x,y,\tau)}. \tag{2.39}$$

If experimental and theoretical temperature functions are identical in sound areas, the *3D contrast* defined by

$$C^{3D} = 1 + \frac{\Delta T(x,y,\tau)}{T_{nd}^{theor}(x,y,\tau)}, \tag{2.40}$$

deviates from unity (or another constant value) only over subsurface defects.

The introduction of temporal (time) parameters of interest takes advantage of the fact that higher signal-to-noise ratios can be provided in the time domain than in the amplitude domain. This statement can be illustrated by the plot of $\Delta T(\tau)$ in Fig. 2.3b, where the optimum observation time τ_m can be determined with a higher accuracy than the maximum signal ΔT_m. In this way, τ_m serves as a temporal parameter of interest. We shall see later that, in a one-sided test, deeper defects are characterized by longer τ_m observation times. This feature of active one-sided TNDT serves as a basis for performing dynamic thermal tomography, which assumes that defects characterized by similar values of τ_m are located at a similar depth in a sample. Instead of using τ_m, some TNDT procedures also involve an analysis of the time $\tau_{1/2}$ when the signal $\Delta T(\tau)$ reaches half its maximum value $\Delta T_m/2$ (see Fig. 2.3b), as well as the time of the maximum temperature derivative $\tau_{m.d.}$.

In the *Early Detection* method, very short observation times τ_{early} are used when signal amplitudes $\Delta T(\tau)$ are low, but also the 3D heat diffusion is weak. The advantages of using temporal detection parameters become clear if one remembers that the shape of the temperature evolution of sound material $T(\tau)$ is independent of absorbed energy. Hence, all such functions are characterized by identical temporal parameters. Keep in mind that the difference between two sound temperatures may be significant, reaching its maximum value at the end of heating. Obviously, the presence of noise and heat exchange with the ambient also alters temporal

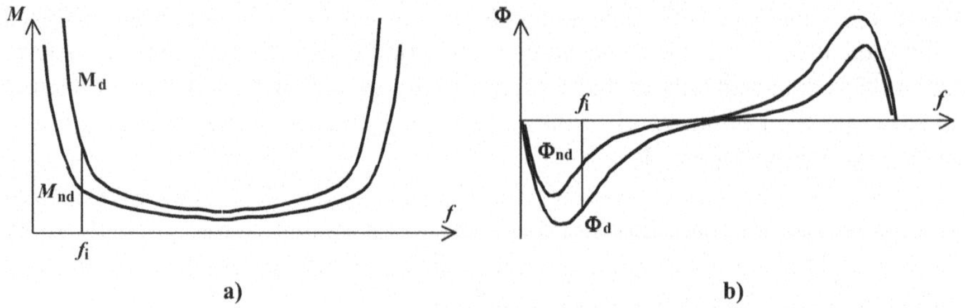

Fig. 2.4 Magnitude and phase parameters (using the Fourier transform in pulsed TNDT). (**a**) – magnitude spectrum, (**b**) – phase spectrum

parameters. Therefore, in TNDT, both temporal and amplitude parameters are combined to provide the highest detection efficiency.

Following the concept of amplitude and temporal parameters, Vavilov and Shiryaev introduced "*maxigrams*" and "*timegrams*" (see Table 1.1). A maxigram is a synthetic (artificial) image that is produced by analyzing a sequence $T(i, j, \tau)$. Each pixel of this image represents the maximum differential signal $\Delta T_m(i, j)$ with respect to a chosen reference point. It is important to understand that maxigram pixel values will occur at *different* observation times. A timegram is a synthetic image in which each pixel represents the time, $\tau_m(i, j)$, when the differential temperature signal at that point reaches a maximum value. The timegram concept can be also applied to $\tau_{1/2}$, $\tau_{m.\,d.}$and τ_{early} times.

2.5.2 Magnitude and Phase Parameters of Interest

If one applies the temporal 1D Fourier transformation to a "normal" temperature function $T(x, y, \tau)$, the amplitude parameters of interest become '*magnitude*' and the temporal "*frequency*", or "*phase*" parameters. The examples of both magnitude and phase Fourier spectrums are presented in Fig. 2.4 for the functions T_{nd} and T_d shown in Fig. 2.4b. Note that Fourier parameters are determined at each particular Fourier frequencyf_i.

The temporal Fourier transformation allows an analysis of the penetration depths of thermal waves at different frequencies. It is known that material layers behave as low-frequency filters; in other words, the penetration depth increases as the frequency decreases. Therefore, high-frequency thermal waves are effective for detecting near-surface defects, while low-frequency waves are better for detecting deeper defects.

In the technique of Pulse Phase Thermography (PPT), defects are detected by phase anomalies. The basic idea of this technique is to use temporal parameters of interest. At each frequencyf_i, sound areas are characterized by identical or approximately equal values of phase, while defect areas are characterized by significant

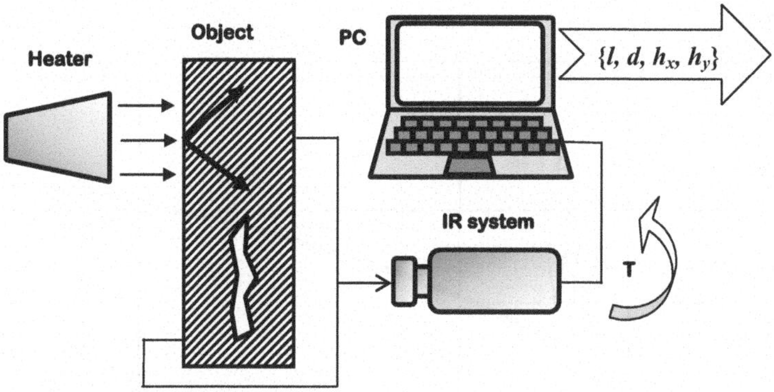

Fig. 2.5 The scheme of active TNDT

deviations in phase. This technique is becoming increasingly popular and is used in some commercial equipment.

The IR technique using heating by harmonic thermal waves is also called *synchronous* or *lock-in* thermography. This is also a version of the PPT technique.

2.6 Direct and Inverse TNDT Problems

The mathematical analysis of TNDT problems involves solving *direct* (modeling) and *inverse* (processing experimental data) heat conduction problems. In a broad sense, direct and inverse problems involve solving differential equations that describe particular inspection procedures. Direct problems are preferred to inverse problems because the latter are *poorly-posed*, i.e. they have no exact solutions. Therefore, the solutions to inverse problems, e.g. those used in defect characterization, are always approximate, and one should choose an "optimum" or best solution from among those available.

Consider a general TNDT scheme as shown in Fig. 2.5, where an object to be tested is regarded as a "black box" and is thermally stimulated on its surface or throughout its volume. In the course of the inspection, the temperature distribution is monitored on the two principle surfaces (front and rear) of the sample. The output signal of the monitoring device can be written as follows:

$$U(x,y,\tau) = E(x,y,\tau)T(\tau,[L],[D],[Q]) + N_s(x,y,\tau) + N_d, \qquad (2.41)$$

where E (x, y, τ) is the expression for the emission-absorption phenomena on the sample surface. T $(x, y, \tau, [L], [D], [Q])$ is the surface temperature, which depends on sample geometry $[L]$, defect parameters $[D]$, and heat source parameters $[Q]$. N_s (x, y, τ) is the expression for additive noise, which appears because of spurious reflections from the sample surface. And N_d is the expression for noise generated by

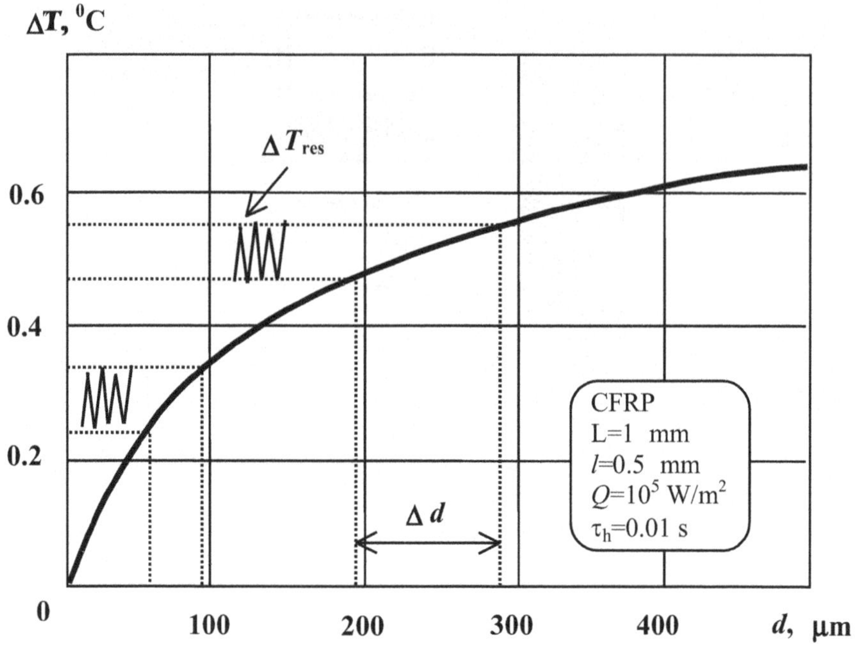

Fig. 2.6 Inverse TNDT problem: evaluating defect thickness in Carbon Fiber Reinforced Plastic (CFRP)

the IR detector and electronics. However, the N_d term which limits the temperature resolution of an IR system can be neglected. Then Eq. 2.40 becomes a well-known equation of IR thermography where the total energy recorded by an IR system is composed of the sample's thermal radiation and the radiation from surrounding objects reflected from the sample surface. The structure of Eq. 2.40 illustrates well the multi-parametric nature of TNDT procedures where the temperature signal caused by a hidden defect is accompanied by noise caused by emissivity-absorptivity phenomena.

Defect parameters can be characterized by inverting Eq. (2.40) with respect to the parameters sought:

$$\{D\} = \{l, d, h_x, h_C\} = \{U(x, y, \tau)\}^{-1}, \tag{2.42}$$

where l is defect depth, d is defect thickness and h_x, h_y are the lateral dimensions of the defect.

Since temperature signal values are functions of time and space, they serve as input data for solving inverse TNDT problems. The accuracy of the defect characterization depends on the accuracy of the measured input parameters. Small variations in input data may cause a significant dispersion of retrieved values and sometimes the variations can fully destroy an inverse algorithm. For example, a

Table 2.10 Direct and inverse TNDT problems

Direct problems	Inverse problems
Theoretical input data:	*Experimental input data:*
Material thermal properties Heat exchange parameters Defect lateral size Defect depth Defect thickness	Material thermal properties Spatial temperature distribution Temperature dynamics
Output parameters:	*Output parameters:*
Spatial temperature distribution Temperature dynamics Differential temperature signal over defect Optimum observation time	Defect lateral size Defect depth Defect thickness
Features:	*Features:*
Correct and accurate solution	Poorly posed problem may have many possible solutions

certain combination of input parameters may lead to calculated defect depths that are greater than the sample thickness. The defect depth is determined by the ΔT signal (Fig. 2.6) and the sensitivity of temperature measurement is limited only by the temperature resolution $\Delta T_{res} = 0.1^{o}$C. The calculated function $\Delta T(d)$ shows that the accuracy of evaluating d is 30 μm for thin materials and for thicker materials it can increase to as much as 100 μm. Since the $\Delta T(d)$ curves experience saturation, thick defects cannot be characterized with high accuracy.

Since time is typically measured more accurately than temperature, temporal, or phase, parameters are increasingly being used for TNDT.

As previously stated, inverse problems are used for characterizing defects, and direct problems are used in mathematical modeling. The popularity of modeling has increased as computers improved. Its advantages are: (1) deriving 'pure', i.e. noise-free, relationships between inspection parameters, (2) saving inspection time, in particular, while optimizing TNDT procedures, and (3) developing inverse algorithms for calculating defect parameters.

Direct and inverse TNDT problems are compared in Table 2.10.

2.7 Analysis of Classical Heat Conduction Solutions

Solutions of transient heat conduction problems require material thermal properties, sample geometry (typically, 1D) and heating parameters. According to the *similarity theory*, the above-mentioned parameters can be united as *dimensionless criteria* and *numbers,* thus allowing the calculation of solutions in a general form. The common dimensionless criteria are as follows:

- Biot criterion $Bi = \dfrac{\alpha\ L}{\lambda}$ (L is the sample thickness) which specifies the generalized sample thickness,
- Fourier similarity number $Fo = \dfrac{a\ \tau}{L^2}$ which characterizes the generalized time,
- Predvoditelev criterion $Pd = \dfrac{\omega L^2}{a}$ (ω is the frequency of thermal waves) which

  In TNDT problems involving heating or cooling by heat pulses or thermal waves, non-stationary temperatures are fully determined by the complexes $\{Bi,\ Fo\}$ or $\{Bi,\ Fo,\ Pd\}$.

The consideration of hidden defects complicates a general analysis. The following dimensionless defect parameters have been proposed in TNDT (see Sects. 3.5 and 4.1):

- $h_x^* = \dfrac{h_x}{L}\left(\dfrac{\lambda_z}{\lambda_x}\right)^{1/2}$ is the generalized lateral defect size of a rectangular defect at location x in an anisotropic material,
- $R_d^* = \dfrac{R_d}{L/\lambda_z}$ is the generalized thermal resistance of a defect,
- $Fo_d = a\ \tau/l^2$ is the Fourier number that represents the flowing time in a defect area (one-sided TNDT), I don't know what this means…time to reach max temperature, temperature and a function of time? I don't think it means those.
- $Bi_d = l/(\lambda\ R_d)$ is the Biot criterion which reflects the ratio of thermal resistance of the host material and the defect (one-sided TNDT).

When determining TNDT parameters, i.e. solving the corresponding inverse problem, one should take into account the following:

1. in dynamic thermal processes, time serves as a specific measured parameter
2. in certain sample geometries and TNDT procedures, some parameters may be complex;, therefore their separate determination is impossible
3. multi-parametric identification problems are typically solved by minimizing the functional, which is defined as a difference between experimental and theoretical data as they evolve in time; with this technique, the parameters that more strongly affect the measured parameters of interest, are determined with higher accuracy.
4. 1D classical heat conduction solutions allow the determination of defect depth and thickness but not lateral size. Consequently, from the point of view of TNDT, 2D and 3D heat conduction problems differ from 1D only by taking into account heat diffusion, which becomes significant around small defects.

References

1. Volkov, E.P., Gavrilov, E.I., Duzhikh, F.P.: Gas-Collecting Chimneys at Thermal and Nuclear Power Plants, p. 340. Energoatomizdat Publ, Moscow, Russia (1987)
2. Tivanov, G.G.: Analysis of mathematical models used in thermal NDT of composite materials. Defectoscopiya (Rus. J. NDT). **5**, 83–85 (1987)
3. Carslow, H.S., Jaeger, T.S.: Conduction of Heat in Solids, p. 580. Oxford Univ Press, Oxford, U.K. (1959)
4. Luikov, A.V.: In: Hartnett, J.P. (ed.) Analytical Heat Conduction Theory, p. 685. Academic (1969)
5. Maillet, D., Andre, S., Batsale, J.-C., et al.: Thermal Quadrupoles: Solving the Heat Equation through Integral Transforms. John Wiley & Sons Publisher, England (2000)
6. Troitsky, O., Reiss, H.: Estimation of thermal diffusivity of thin film materials by a one-level, two-point 'diverging'-thermal-wave technique. Part I: Theory.-High Temp.-High Pressures. **35/36**, 265–280 (2003/2004)
7. Bison, P.G., Grinzato, E., Bragiotti, A., Bressan, C., Mazzoldi, A., Vavilov, V.: Nondestructive Evaluation of Cracks in Porous Building Materials by Use of Crawling Spot Thermal Technique, vol. 15, pp. 653–658. Rev. of Progress in Quant. NDE, Seattle

Chapter 3
Determining Thermal Properties of Materials

3.1 Temperature Evolution in Different Coordinates

TNDT deals with the sample surface temperature distribution $T(\tau)$ evolving in time. In IR thermographic NDT, it is represented by the dynamic pixel function $T(i,j,\tau)$. Two examples of this function are shown in Fig. 3.1a for heating a steel plate with a Dirac and square pulse of duration 0.01 s. The temperature curves are practically identical, thus demonstrating that the calculation results are independent of the duration of heat pulse.

A $T(\tau)$ function can be analyzed in different coordinates that allows: (1) reducing the signal dynamic range, (2) discovering characteristic points, such as extreme, inflections etc., and (3) the easier use of some processing algorithms, e.g. polynomial fitting or derivative analysis.

For example, the difference between the two above-mentioned cases becomes noticeable in logarithmic coordinates $Ln\ (T)$ – $Ln\ (\tau)$ as shown in Fig. 3.1b. Originally, the logarithmic presentation is related to the adiabatic heating of a semi-infinite body with a Dirac pulse (see the solution in Sect. 3.2). The Dirac-pulse solution is expressed by a straight line while the use of square pulse heating causes deviation from the straight-line behavior in the beginning of the process. Note that at the end of the process, the finite thickness of the plate starts to affect the front-surface temperature thus underlying the difference between a semi-infinite body and a plate.

These two processes look even more different in the coordinates $T\sqrt{\tau}-\sqrt{\tau}$ (Fig. 3.1c). Another variant of the temperature function presentation in the coordinates $T\tau^{1/3}$ – τ will be shown in Sect. 3.7.6.

In Fig. 3.2, the logarithmic presentation of $T(\tau)$ illustrates the particularly strong difference between the defect and non-defect evolutions in TNDT of a turbine blade coating.

Polynomial fitting of the non-defect curve from Fig. 3.2 is illustrated in Fig. 3.3. The source temperature function $T(\tau)$ is corrupted by random noise (Fig. 3.2a). To

V. Vavilov, D. Burleigh, *Infrared Thermography and Thermal Nondestructive Testing*, https://doi.org/10.1007/978-3-030-48002-8_3

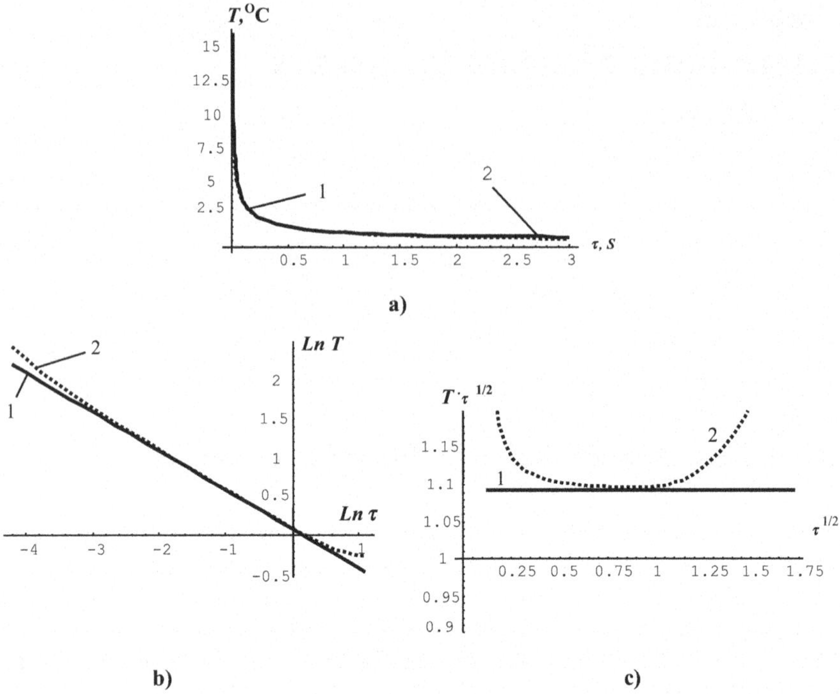

Fig. 3.1 Mathematical model of adiabatic heating of an AISI-304 steel sample ($a = 3.75\cdot10^{-6}$ m²/s, $\lambda = 15$ W/(m·K) with a Dirac ($W = 1.5\cdot10^{4}$ J/m², curve (1) and square pulse ($Q = 1.5\cdot10^{6}$ W/m², $\tau_h = 0.01$ s, curve (2): (**a**) – temperature vs. time ($T - \tau$ presentation), (**b**) – $Ln(T) - Ln(\tau)$; (**c**) – $T\sqrt{\tau} - \sqrt{\tau}$

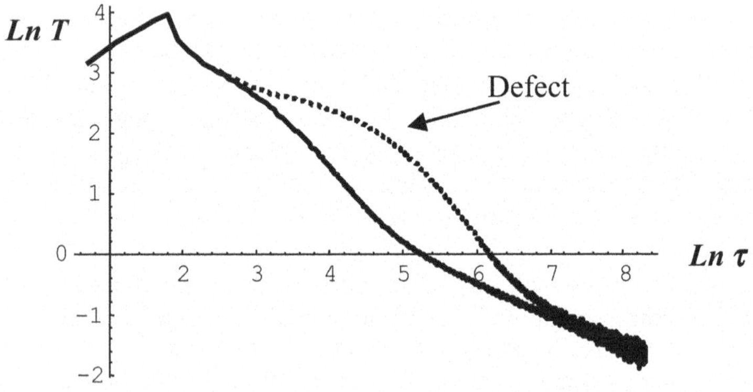

Fig. 3.2 Mathematical model of adiabatic heating of a turbine blade coating (5 ms square pulse, coating thickness 150 μm, delamination thickness 20 μm)

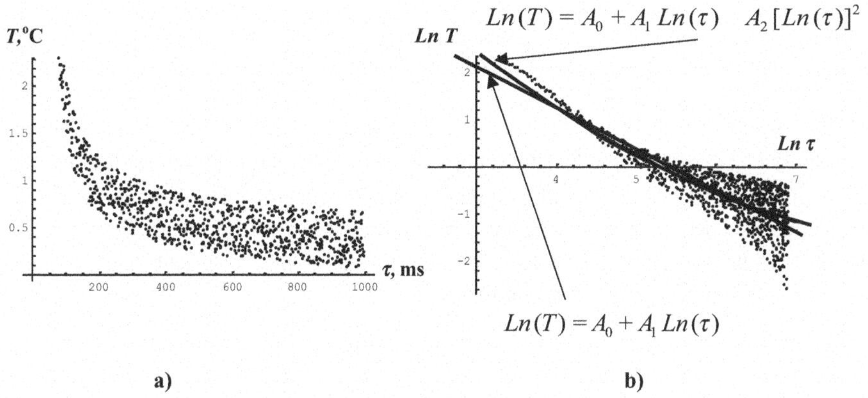

Fig. 3.3 Polynomial fitting of temperature functions (adiabatic heating of a turbine blade coating from Fig. 3.2, no defect, imposed white noise with amplitude from −0.3 to +0.3 °C): (**a**) – $T - \tau$ presentation, (**b**) – $Ln(T) - Ln(\tau)$

better reveal the peculiarities of temperature evolution, coordinate system is changed from natural to logarithmic (Fig. 3.2b).

In logarithmic coordinates, it is expedient to approximate noisy curves with polynomial functions of a particular order. The linear approximation is good enough in non-defect zones where temperature evolutions resemble straight lines. The presence of defects leads to the appearance of significant coefficients at non-linear members of the polynomial presentation (in Fig. 3.3b, only two polynomial members are presented). The technique of polynomial fitting will be discussed in details in Sect. 6.10.1.

3.2 Determining Material Thermal Effusivity (Front Surface of an Adiabatic Semi-Infinite Body Heated with a Dirac Pulse)

The solution for heating a semi-infinite body with a Dirac pulse is as follows (see Table 2.3):

$$T = \frac{W}{e\sqrt{\pi}}\frac{1}{\sqrt{\tau}}. \tag{3.1}$$

It is convenient to present Eq. (3.1) as a straight line in the $Ln\ (T)$ - $Ln\ (\tau)$ coordinates (see Sect. 3.1). In the ideal case, at any time $\tau*$, a material effusivity $e$can be evaluated using the surface temperature and absorbed energy:

$$e = \frac{W}{T(\tau *)} \frac{1}{\sqrt{\pi \tau *}}. \tag{3.2}$$

Equation (3.2) illustrates the fact that, by using the semi-infinite body solution, effusivity can be determined as the combination of three thermophysical parameters ($e = \sqrt{C\rho\lambda}$) but not as an independent parameter. In any case, it is necessary to measure absorbed energy very accurately and that is not always easy. The latter difficulty can be overcome by performing a comparative measurement using a *reference material* whose effusivity is known.

$$\frac{e}{e_{ref}} = \frac{T_{ref}(\tau *)}{T(\tau *)}, \tag{3.3}$$

where the subscript "*ref*" is related to a reference. Below, this method will also be applied to evaluating subsurface defects.

3.3 Determining Material Thermal Diffusivity (Rear Surface of an Adiabatic Plate Heated with a Dirac Pulse)

In their classical work [1], Parker et al. proposed to determine the thermal properties of materials, in particular, thermal diffusivity, by using the response for the rear surface of an adiabatic plate heated on the front surface with a short (Dirac) heat pulse, see Table 2.4:

$$\frac{T^R}{(Wa/\lambda L)} = \theta_D^R = 1 + 2\sum_{n=1}^{\infty}(-1)^n e^{-n^2\pi^2 Fo}. \tag{3.4}$$

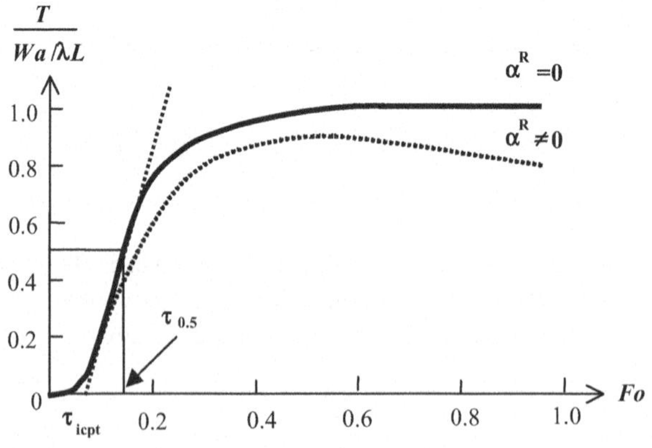

Fig. 3.4 Mathematical model of temperature evolution on the rear surface of a plate heated with a Dirac pulse

The plot of θ_D^R shown in Fig. 3.4 starts from zero and increases to a constant value equal to $T \to Wa/\lambda L$. In the case of a real (non-adiabatic) plate, the rear-surface temperature reaches a maximum value and then slowly goes down, due to heat losses to the environment (see the dashed line in Fig. 3.4).

The rising rear-surface temperature signal allows one to choose some characteristic points with the corresponding specific heat transit times $\tau*$, in particular:

- the time when the temperature reaches a half of its maximum value $\tau_{1/2}$ (*time-at-half-maximum*),
- the time when the temperature derivative $\tau_{m.\,d.}$ reaches a maximum,
- the intercept time τ_{icpt} when a straight line drawn across the straight section of the θ_D^R plot intersects the time axis.

Using the definition for the Fourier number, it is obvious that, at any experimental time $\tau*$, thermal diffusivity can be calculated by:

$$a = Fo*\frac{L^2}{\tau *}, \tag{3.5}$$

where the coefficient $Fo*$ can be easily found from Eq. (3.4), for example:

$Fo* = 0.1388$ for $\tau_{1/2}$;

$$Fo* = 0.0918 \,\text{for}\, \tau_{m.d.}; \tag{3.6}$$

$Fo* = 0.04863$for τ_{icpt}.

The advantage of using the $\tau_{m.\,d.}$ parameter is that thermal diffusivity can be evaluated in real time, i.e. without waiting until the end of the thermal process, which is necessary if the $\tau_{1/2}$ parameter is used. Also, in the case of $\tau_{m.\,d.}$, diffusivity estimates are less-dependent on heat losses due to lower sample temperature. However, determining temperature derivatives $\partial T/\partial \tau$ for noisy functions inevitably increases high-frequency noise, thus reducing the accuracy of using the $\tau_{m.\,d.}$ parameter. Therefore, the practical implementation of the Parker technique is usually based on the analysis of $\tau_{1/2}$.

The Parker, or flash, method of measuring thermal diffusivity is highly accurate and used worldwide. There are several modifications of this technique where the accuracy can be enhanced by: (1) taking into consideration heat losses and finite sample size, (2) using a more convenient $\tau*$ parameter (different form the introduced above), etc.

In addition to the reliable evaluation of thermal diffusivity, the Parker method also allows the determination of thermal conductivity, if the absorbed energy W is measured. If a, λ and ρ are known, it is also possible to determine heat capacity: $C = \lambda/(\rho a)$.

3.4 Determining Material Thermal Diffusivity (Front Surface of an Adiabatic Plate Heated with a Dirac Pulse)

The curve that represents the solution of front-surface pulsed heating of an adiabatic plate (see Table 2.4) reveals no characteristic points and the determination of material thermal properties is more difficult than by analyzing the rear surface data. Moreover, the front-surface solution is less sensitive to diffusivity variations than the rear surface solution. A more correct comparison between two solutions will be given in Sect. 3.7. Here, we will see how the technique of *non-linear fitting* (see also Sect. 3.7.5 and Chap. 5) can be used to suppress experimental noise in the evaluation of thermal diffusivity.

A general approach to determining the thermal diffusivity in a one-sided test was described in [2]. It is based on comparing the experimental data points calculated by using the corresponding 1D theoretical model:

$$T^F(\tau) = T_\infty \frac{1+2\exp\left[\dfrac{-L^2}{a\left(\tau-\tau_{shift}\right)}\right]}{\sqrt{\dfrac{\pi a\left(\tau-\tau_{shift}\right)}{L^2}}}, \text{if } \tau \le 0,62\frac{L^2}{a}; \tag{3.7}$$

$$T^F(\tau) = T_\infty\left[1+\exp\left(-\frac{a\pi^2\left(\tau-\tau_{shift}\right)}{L^2}\right)\right], \text{if } \tau > 0,62\frac{L^2}{a},$$

where the fitting parameters are the temperature at the end of the process T_∞ and the term a/L^2, while the parameter τ_{shift} takes into account a pulse finite duration (in fact, this parameter can be also evaluated as a result of fitting). For a known sample thickness L, a value of thermal diffusivity a can be easily calculated.

If heating is performed with a Dirac pulse, the same authors recommend a simpler model [3]:

$$T^F = W\sqrt{a}\left(\frac{1+2e^{-L^2/(a\tau)}}{\lambda\sqrt{\pi\tau}}\right) \text{for } \tau < 0.062L^2/a, \tag{3.8}$$

$$T^F = \frac{W}{C\rho L}\left(1+2e^{-a\pi^2\tau/L^2}\right)$$

for

$$\tau > 0.062L^2/a.$$

Non-linear fitting is performed by using the parameters $p_1 = a/L^2$ and $T_\infty = W/(C\rho L)$ when presenting Eq. (3.8) in the following form:

$$T = T_\infty \left(\frac{1 + 2e^{-1/(p_1\tau)}}{\sqrt{\pi\tau p_1}} \right), \tag{3.9}$$

$$T = T_\infty \left(1 + 2e^{-p_1\pi^2\tau} \right).$$

3.5 Determining Sample Thickness and Corrosion Material Loss (Front Surface of an Adiabatic Plate Heated with a Dirac Pulse)

In recent years, a certain number of TNDT research was done on determining material thickness, for example, in order to evaluate corrosive degradation and material loss in metals. This is accomplished by analyzing the sample front-surface temperature response after pulsed thermal stimulation. The method of non-linear fitting described above can be applied to evaluating L, if the thermal diffusivity of the material is known.

The amount of corrosion is expressed as the relative *material loss* $\Delta L/L$, rather than the absolute thickness value L. The basic theory has been developed for thin metallic plates, in which heating is adiabatic due to small values of the Biot criterion ($Bi < 0.1$). Therefore, at the end of the thermal process, large sample areas tend to keep the temperature close to the stationary value $T = Wa/\lambda L$. If the sample thickness in a corroded area decreases to the $l \leq L$ value, the temperature difference between corroded and sound areas appears at the end of the process:

$$\Delta T = T_d - T_{nd} = \frac{Wa}{\lambda l} - \frac{Wa}{\lambda L}, \tag{3.10}$$

and from this, the material loss $\Delta L = L - l$ can be expressed as:

$$\frac{\Delta L}{L} = 1 - \frac{T_{nd}}{T_d}. \tag{3.11}$$

To determine the residual thickness of a sample, the non-linear fitting technique by Eq. (3.7) can be also used. In this case, thermal diffusivity must be known and the fitting parameter is a/l^2.

Below, Eq. (3.11) will be modified by taking into account 3D heat diffusion around small defects.

3.6 Determining Optimum Observation Time for Subsurface Defects (an Adiabatic Semi-Infinite Body Heated with a Dirac Pulse)

The temperature of an adiabatic semi-infinite body at the depth z is determined by the corresponding general solution (see Table 2.3):

$$\frac{T}{Wa/\lambda z} = \frac{1}{\sqrt{\pi Fo_z}} e^{-1/4Fo_z}, \tag{3.12}$$

where $Fo_z = a\tau/z^2$. It can be assumed that a discontinuity (defect) located at depth $z = l$ will be detected by its surface temperature 'footprint' if a *packet of thermal energy* released on surface will reach the point z and, after having reflected from the defect surface, returns to the sample surface. By analyzing the temperature derivative $\partial T/\partial\tau$ in Eq. (3.12), it can be seen that the maximum temperature at depth l appears at $\tau_l = l^2/2a$. Knowing this, it becomes possible to estimate the optimum time for detecting a subsurface defect, this is as follows:

$$\tau_m = 2\tau_l = l^2/a. \tag{3.13}$$

Equation (3.13) has been mentioned by several authors as a rule-of-thumb for evaluating defect depth. Its limits will be discussed below.

3.7 Sensitivity Functions

Sensitivity functions following from classical solutions of heat conduction (Tables 2.3 and 2.4) may serve as an effective tool for evaluating the efficiency of determining particular solution parameters (Tables 2.3 and 2.4).

Let us to introduce a sensitivity function in the dimensionless form:

$$S_p = \frac{\partial T/T}{\partial p_i/p_i}, \tag{3.14}$$

where p_i is a particular solution parameter, e.g. thermal diffusivity a or sample thickness L. Notice that the time τcan be regarded as a specific parameter which is being measured in an experiment, e.g. while determining diffusivity by the Parker technique. A relationship between two parameters can be derived by combining the corresponding sensitivity functions. For instance, the error in measuring the process time is related to the error in determining diffusivity by the expression:

$$\frac{\partial\tau/\tau}{\partial\alpha/\alpha} = S_\alpha/S_\tau. \tag{3.15}$$

Sensitivity functions can be defined for either physical parameters ($\{\tau, \tau_h, Q\ (W), \alpha, \lambda, a\}$ for a semi-infinite body and $\{\tau, \tau_h, Q\ (W), \alpha, \lambda, a, L\}$ for a plate) or complexes $\{Q'\ (W'), \theta\}$.

For each sensitivity function, the following expression is true:

$$S_p = \frac{\partial T / T}{\partial Q' / Q'} \frac{\partial Q' / Q'}{\partial p_i / p_i} + \frac{\partial T / T}{\partial \theta / \theta} \frac{\partial \theta / \theta}{\partial p_i / p_i}. \tag{3.16}$$

Using the complexes $Q'\ (W')$ and θ makes the analysis more general but in practice some particular physical parameters are of special interest. both approaches will be discussed below

3.7.1 Semi-Infinite Body

There are 12 solutions in Table 2.3 valid for adiabatic and non-adiabatic semi-infinite bodies heated with step-wise, square and Dirac pulses. Seven parameters of these solutions, namely $\{\tau, \tau_h, Q\ (W), \alpha, \lambda, a\}$, allow the introduction of 36 sensitivity functions which are described in Table A1.1, ***Appendix 1***. The table proves the advantage of adiabatic solutions which are characterized by constant values of sensitivity functions under both step and Dirac heating. In square pulse heating, only two sensitivity functions (S_τ and S_{τ_h}) vary in time. Hence, conclusions on the accuracy of the corresponding inverse solutions are very simple in adiabatic cases.

The sensitivity functions for a semi-infinite non-adiabatic body are presented in Table A2.2 in the evident form and the corresponding graphs are shown in Fig. A2.1. The graph numbers correspond to the numbers of the solutions in Tables A1.1 and A1.2, ***Appendix 1***. The initial points of the curves correspond to the adiabatic case ($\alpha^F = 0$).

3.7.2 Plate

Plate solutions additionally include the sample thickness L. In this case, we will limit ourselves to the generalized derivatives reported in Table 2.4 along with the original solutions. The graphs of the generalized sensitivity functions are presented in Fig. A1.2, ***Appendix 1***.

3.7.3 Evaluating Material Loss

Equation (3.11) for determining the material loss $\Delta L/L$ is valid if heat exchange is adiabatic. The influence of observation time on the efficiency of evaluating $\Delta L/L$ can be illustrated with the corresponding sensitivity function. For an adiabatic plate heated with a Dirac pulse it is as follows:

$$\frac{\partial T/T}{\partial L/L} = -1 - 2\frac{\partial T/T}{\partial Fo/Fo}. \tag{3.17}$$

The plot of this function shown in Fig. 3.5 illustrates that procedures of one-sided TNDT are insensitive to thickness variations until a particular time $Fo^{(0)}$ ($Fo^{(0)} = 0.1238$ for the 1% threshold).

The maximum sensitivity occurs at the end of a thermal process ($Fo \approx 1$), i.e. in the quasi-stationary regime where Eq. (3.11) becomes viable. As shown in [4], this equation produces reasonable results in quasi-adiabatic cases, e.g. in the inspection of metals, where observation times are within the $Fo = 0.7$–1 interval. If surface heat losses cannot be neglected, the accuracy of applying Eq. (3.11) declines, as illustrated by the 'non-adiabatic' sensitivity functions presented in Fig. 3.6 for 4 different values of the Biot criterion $Bi = \alpha L/\lambda$ (the corresponding solution from Table 2.4 has been used). The highest accuracy takes place at $Fo \approx 1$ and at longer times a value of $(\partial T/T)/(\partial L/L)$ diminishes with greater Bi.

Marinetti et al. proposed taking into account the 3D heat diffusion phenomena by introducing the coefficient κ shown in Fig. 3.7 as a function of D_d/L, where D_d is the diameter of a corrosion area [5].

In the range $D/L = 1$–6, this correction coefficient was analytically expressed in the following form:

$$k(D/L) = 0.8871 + 14.573e^{-0.4D/L} - 35.149e^{-0.5D/L} + 19.690e^{-0.6D/L}. \tag{3.18}$$

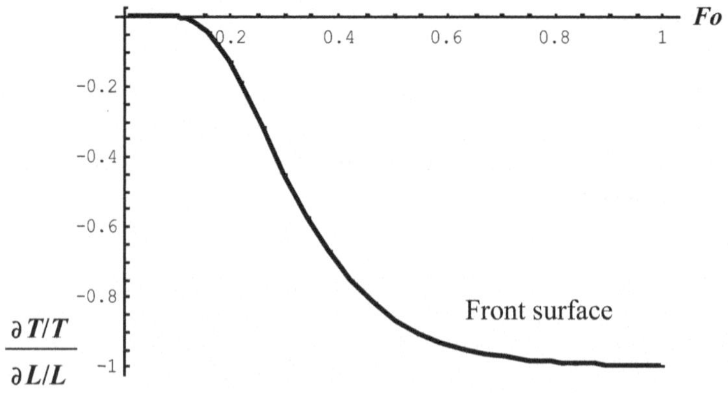

Fig. 3.5 Front-surface sensitivity to plate thickness (adiabatic case)

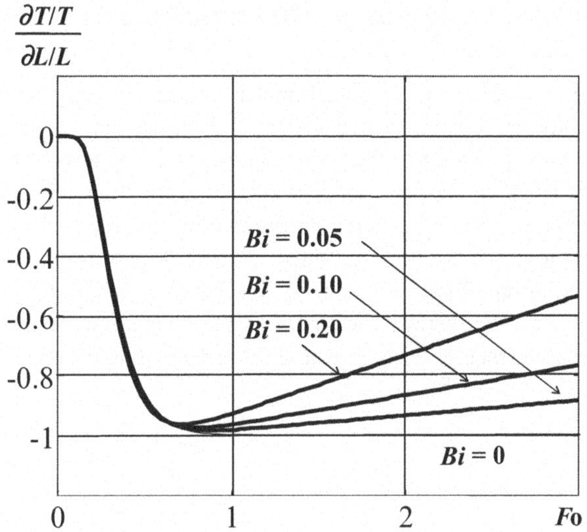

Fig. 3.6 Front-surface sensitivity to plate thickness (non-adiabatic case)

Originally, the coefficient κ was introduced for the fixed observation time $Fo = 0.68$. Therefore, the corrosion inversion formula in this case is:

$$\Delta L / L = \left[1 - T_{nd}\left(Fo = 0.68\right) / T_d\left(Fo = 0.68\right)\right] / k\left(D / L\right). \tag{3.19}$$

A more general approach to the analysis of lateral heat diffusion will be discussed in Chap. 4.

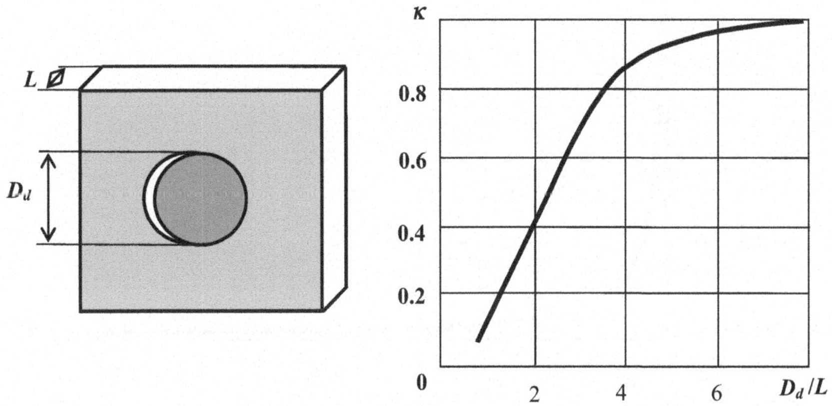

Fig. 3.7 Compensating 3D heat diffusion in corrosion evaluation

3.7.4 Determining Thermal Diffusivity

A two-sided test to determine thermal diffusivity by the rear-surface temperature response was described in Sec. 3.3. As mentioned earlier, one-sided tests are less convenient for this purpose because of the following factors: (1) weaker sensitivity of the front-surface temperature toward diffusivity variations, (2) absence of characteristic points in temperature evolutions that makes difficult diffusivity determination, (3) higher influence of heat losses, (4) presence of spurious reflected radiation which is present in both heating and cooling stages.

Let us illustrate the first statement by comparing the corresponding front- and rear-surface sensitivity functions for an adiabatic plate heated with a Dirac pulse:

$$S_a^F = 1 - \frac{2\pi^2 Fo \sum_{n=1}^{\infty} n^2 e^{-n^2\pi^2 Fo}}{1 + 2\sum_{n=1}^{\infty} e^{-n^2\pi^2 Fo}}, \tag{3.20}$$

$$S_a^R = 1 - \frac{2\pi^2 Fo \sum_{n=1}^{\infty} (-1)^n n^2 e^{-n^2\pi^2 Fo}}{1 + 2\sum_{n=1}^{\infty} (-1)^n e^{-n^2\pi^2 Fo}}.$$

The evolution of the S_a^R / S_a^F ratio in time is shown in Fig. 3.8.

Within the time range of interest ($Fo < 0.6$), the rear-surface temperature response is more sensitive to variations of a compared to the front-surface one. For example,

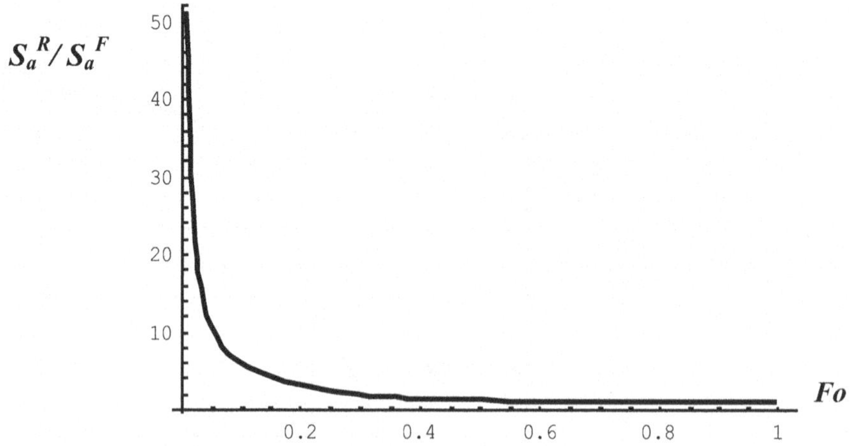

Fig. 3.8 Sensitivity functions in diffusivity measurement (adiabatic heating of a plate with a Dirac pulse)

$S_a^R / S_a^F = 4.5$ at $Fo = 0.139$. The ratio S_a^R / S_a^F tends to unity at the end of a thermal process when the sample temperature becomes $T \approx (Wa)/(\lambda L)$.

If heating is performed with a square pulse, some characteristic heat transit times, such as $\tau_{1/2}$, may appear in the front-surface response. As an example, consider a heat pulse of the duration $Fo_h = 0.05$ (Fig. 3.9).

On the front surface, the time $\tau_{1/2}$can be determined within both the heating ($Fo* = 0.0125$) and cooling ($Fo* = 0.1048$) stages, while the rear-surface response produces only a single $\tau_{1/2}$ value ($Fo* = 0.1646$). The corresponding sensitivity functions are characterized by the following expressions:

$$\left(\frac{\partial T / T}{\partial a / a}\right)^F = 0.5$$

at

$$Fo* = 0.0125,$$

$$\left(\frac{\partial T / T}{\partial a / a}\right)^F = -0.691 \text{at } Fo* = 0.1048, \tag{3.21}$$

$$\left(\frac{\partial T / T}{\partial a / a}\right)^R = 1.5325$$

at

$$Fo* = 0.1646.$$

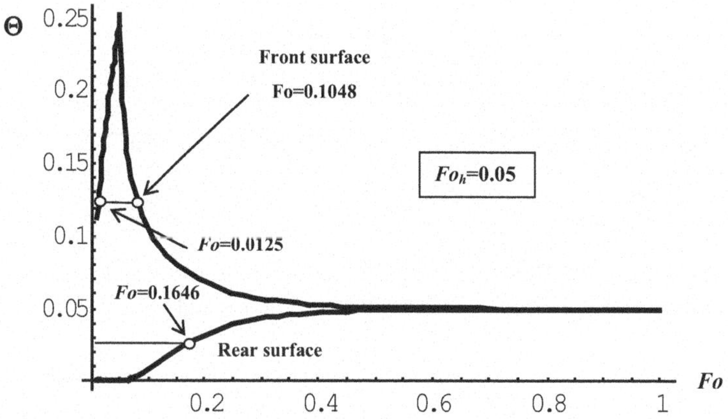

Fig. 3.9 Temperature response on the surface of an adiabatic plate (square-pulse heating, $Fo_h = 0,05$)

The examples above prove that two-sided TNDT, along with some technical advantages, provides at least the two-fold improvement in sensitivity when determining thermal diffusivity to compare to one-sided TNDT.

3.7.5 Separating Information Between Thermal Diffusivity and Heat Exchange Coefficient

While determining material thermal properties in a one-sided test, it is difficult to evaluate the surface heat losses (described by the heat exchange coefficient α) separately from other thermal parameters because of noise. For instance, when determining thermal diffusivity by the Parker method, we deal with three unknown parameters (W, a and α) composing the solution for heating a plate of known thickness L with a Dirac pulse (see Table 2.4). In mathematical physics, such problems are typically solved by *minimizing* (using non-linear fitting) the following function:

$$\chi^2(p) = \sum_{n=1}^{N} \left(\frac{T_n^{\exp} - T_n^{theor}}{\sigma_n} \right)^2, \tag{3.22}$$

constituted by the square of the differences between the experimental (index "*exp*") and theoretical (index "*theor*") temperatures at each nth time point (N time points in total). A value of p in Eq. (3.22) describes the desired complex $\{W, \ a, \ \alpha\}$. Examples of using this approach in TNDT can be found in [6, 7], see also Chap. 5.

The application of least-square fitting for solving the problem by Eq. (3.22) is reduced to finding a set of values of p which correspond to a minimum of the χ^2 function in the hyper-space of the features $W, \ a, \ \alpha$. Confidence intervals for p can be obtained by approximating χ^2 with a parabolic function and calculating a *curvature matrix* A of which elements are defined as follows:

$$A_{k,l} = \sum_{n=1}^{N} \frac{1}{\sigma_n} \left[\frac{\partial T(\tau_n, p)}{\partial p_k} \frac{\partial T(\tau_n, p)}{\partial p_l} \right]. \tag{3.23}$$

The covariation matrix Cov for an approximating function is:

$$\mathrm{cov} = A^{-1}, \tag{3.24}$$

where matrix diagonal elements represent variations (quadratic deviations) of the desired parameters p. The *correlation matrix* can be found by:

$$Corr(i,j) = \frac{\mathrm{cov}(i,j)}{\sqrt{\mathrm{cov}(i,j)\mathrm{cov}(j,i)}}. \tag{3.25}$$

Consider determining thermal diffusivity of a non-adiabatic plate heated with a Dirac pulse. Diffusivity values a are calculated by applying non-linear best fitting to the front- or rear-surface temperature responses. Let the front-surface heat losses be the main noise factor; assume that on the rear surface $\alpha_R = 0$. Assume also that the standard deviation of noise is $\sigma_n = 3\%$ in the maximum rear-surface excess temperature, and the noise affects temperature at each time point in the same way. In Fig. 3.10, the coefficient of correlation between a and $Bi = (\alpha L)/\lambda$, calculated by Eq. (3.25) at $Bi = 0.0025$, is presented as the function of time.

On the front surface, the correlation coefficient is close to unity, thus making the independent estimation of a and α impossible, except at an initial time period where surface heat losses are negligible. On the rear surface, the efficiency of determining diffusivity is weakly dependent on front surface heat losses. Moreover, there is a particular time, given by the intersection with the zero axis, when the accuracy of the determination of a is optimal.

3.7.6 *Manipulating the Temperature Response on the Plate Front Surface*

It was mentioned above that a difficulty in the analysis of the front-surface temperature response is a lack of characteristic heat transit points in the corresponding functions. However, such points can be produced artificially by appropriately treating these functions. Moreover, in Chap. 4, we will see that characteristic points will appear in the front-surface profile in the case of multilayer-structures being produced by layer interfaces, including defect boundaries. Therefore, by forming appropriate complex respectiveheat transit times, e.g., for locating layer depths.

In this paragraph, we introduce an artificial function $T \times Fo^n$ with n varying from 0.01 to 1.0 in both non-defect and defect areas [8]. Figure 3.11 shows the family of curves $T \times Fo^n$ *vs.* Fo. One should analyze n in the range from 0.01 to 0.49 where temporal minimums appear, as shown in Fig. 3.11.

It has been found that each n value is related to a particular minimum time point $Fo_{\min}$. The relationship $Fo_{\min}$ *vs.* n and its best fit is shown in Fig. 3.12.

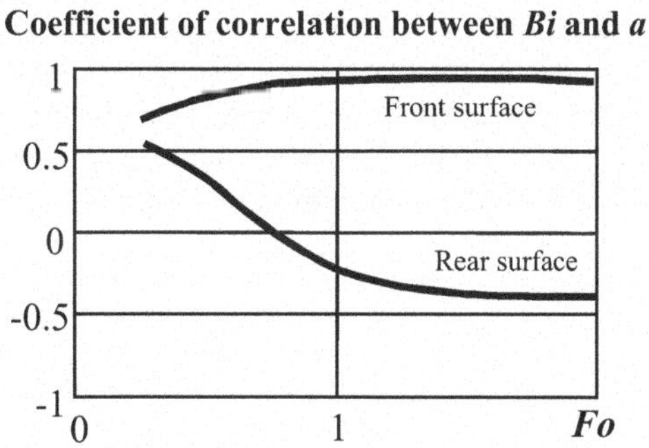

Fig. 3.10 Coefficient of correlation between Bi (α) and a for a non-adiabatic plate ($Bi = 0.0025$; $\sigma_n = 3\%$; courtesy S. Marinetti)

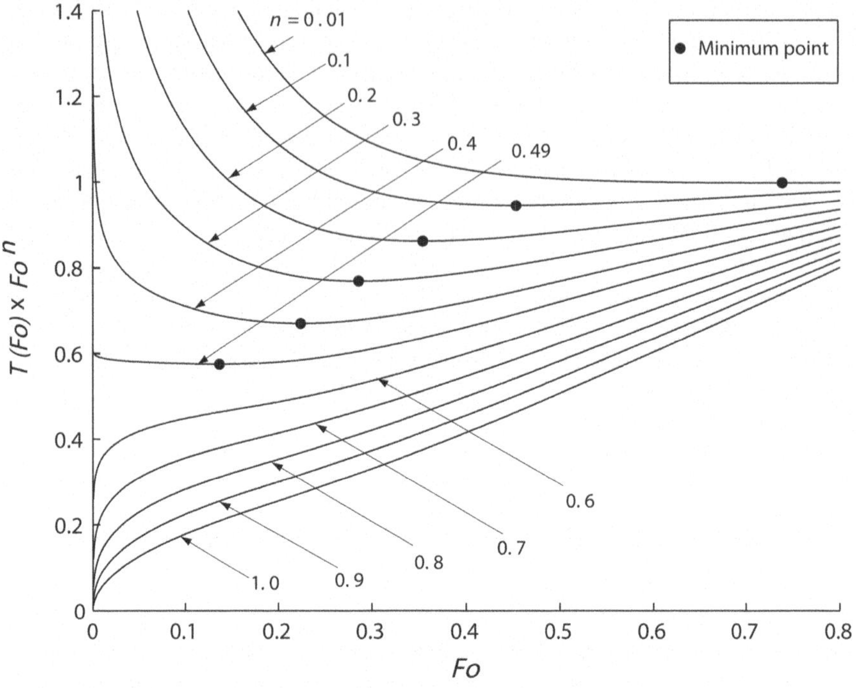

Fig. 3.11 Artificial function $T \times Fo^n$ *vs.* Fo for $n = 0.01$–1.0 and minimums for $n = 0.01$–0.49

We assume that an optimal n ensures the steepest derivative of the $T(\tau) \times \tau^n$ near a minimum point. The following function has been considered as the optimization parameter:

$$OP(n) = \frac{\left|\left[T(Fo) \times Fo^n\right]'_{-\min}\right| + \left|\left[T(Fo) \times Fo^n\right]'_{+\min}\right|}{2\left[T(Fo_{\min}) \times Fo^n_{\min}\right]}, \tag{3.26}$$

where the quotation mark (') means the derivative by Fo, and (−min) and (+min) correspond to the derivatives values at time points $\pm\Delta Fo$ in regard to the minimum time point $Fo_{\min}$. By using (3.26), one may obtain the $OP(n,\ Fo_{\min})$function which specifies the relative mean derivative of the analyzed function to the left and to the right from a minimum point. The plot of this function shown in Fig. 3.13 for $\Delta Fo = 0.001$ reaches its maximum value at $n = 0.4$ that is supposed to be the optimal value when defining minimums of the $T(\tau) \times \tau^n$ function.

For $n = 0.4$, the corresponding $Fo_{\min} = 0.2237$, hence, material diffusivity can be calculated in this case as:

$$a = 0.2237L^2 / \tau_{\min}. \tag{3.27}$$

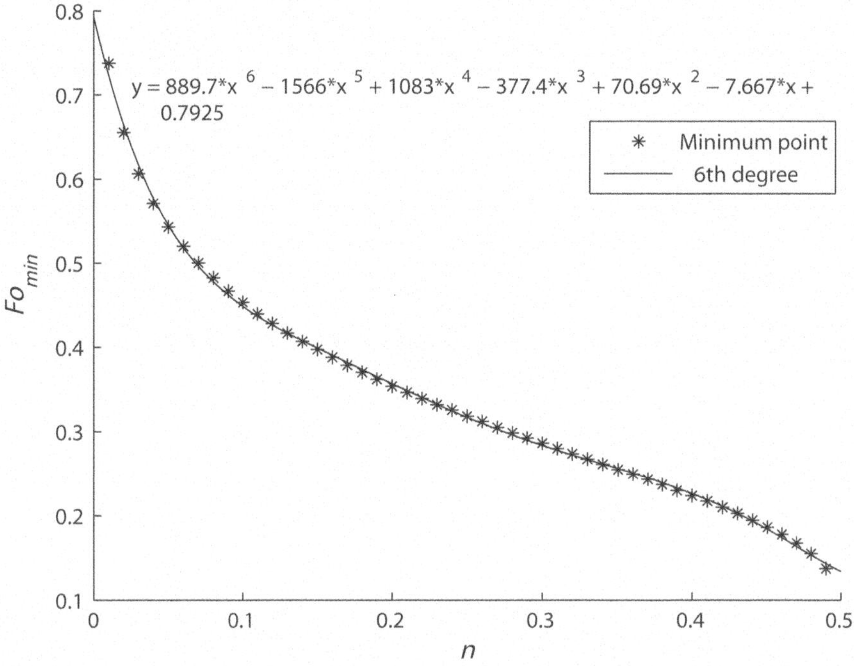

Fig. 3.12 $Fo_{\min}$ *vs.* n discrete function

Another approach to manipulating with the front-surface temperature response was elaborated by Troyitsky who introduced the function called an *index of sample thermal uniformity* [9]:

$$K = \frac{T(\tau)}{\tau\left[\partial T / \partial \tau\right]}. \tag{3.28}$$

Its physical sense is related to the so-called regular stage of heating where the *heating rate,* or the temperature derivative in time, remains constant.

In [10], the evolution of K has been analyzed in the following cases.

Heating a semi-infinite body with a Dirac pulse of energy W:

$$T = \frac{W}{e\sqrt{\pi\tau}}. \tag{3.29}$$

Heating a semi-infinite body with a square pulse of energy W:

$$T = \frac{W}{\tau_h e\sqrt{\pi}}\sqrt{\tau}, \text{if } \tau \le \tau_h, \text{and} \tag{3.30}$$

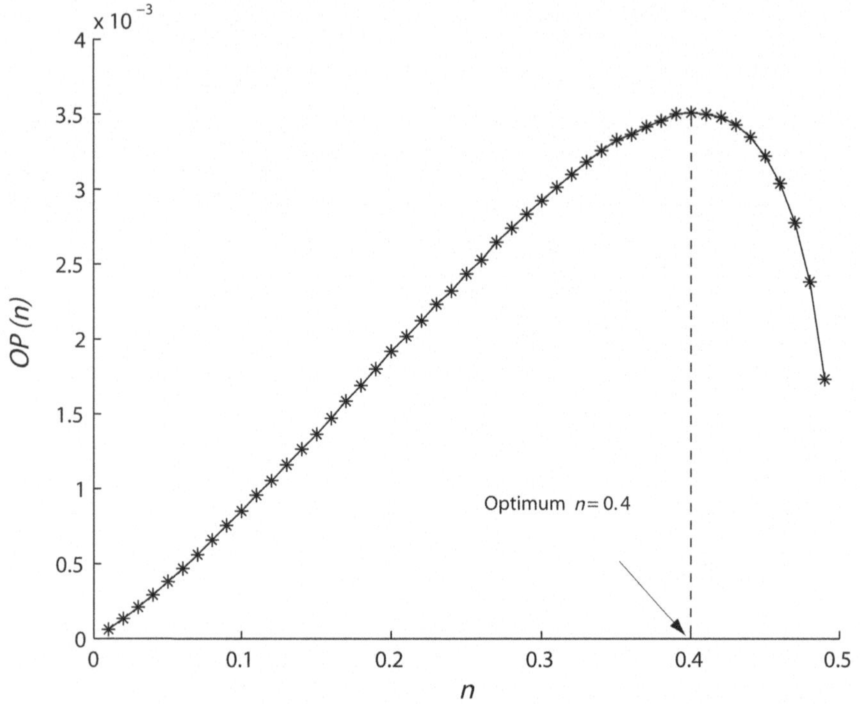

Fig. 3.13 Optimizing n power ($\Delta Fo = 0.001$)

$$T = \frac{W}{\tau_h e\sqrt{\pi}}\left(\sqrt{\tau} - \sqrt{\tau - \tau_h}\right), \text{ if } \tau > \tau_h.$$

If a coating on a semi-infinite substrate is heated with a Dirac pulse of energy W:

$$T = \frac{W}{e\sqrt{\pi\tau}}\left[1 + 2\sum_{n=1}^{\infty}\Gamma^n \exp\left(-n^2\frac{L_1^2}{a_1\tau}\right)\right]. \tag{3.31}$$

Equations (3.29 and 3.30) directly follow from Table 2.3, and Eq. (3.31) will be discussed in Sect. 4.2.4. Here $\Gamma = (e_1 - e_2)/(e_1 + e_2)$ is the *coefficient of reflection* of a pulsed thermal wave from an interface between two layers, the subscript "1" is related to the first layer, and $L_1$is the thickness of the first layer.

The evolution of $K$is shown in Fig. 3.14 for a two layer structure including a 3 mm thick steel layer and a thermally-thick aluminum substrate. The coordinate origin is at the end of a 90 μs long heat pulse; Eqs. (3.29, 3.30, and 3.31) are used with $Fo_1 = a_1\tau / L_1^2$.

For orientation, the inflection point E ($Fo_1 = 0.13$) can be used for the analysis, however choosing the time Fo_1 which corresponds to the start of the regular regime,

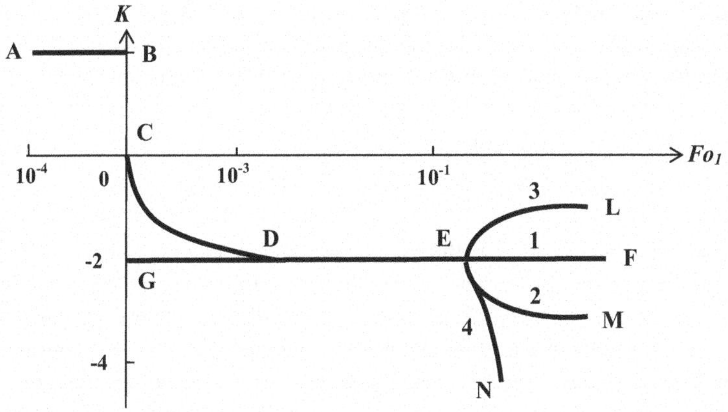

Fig. 3.14 Index of sample thermal uniformity vs. time for a steel-aluminum structure heated with a 90 μs pulse (adapted from [9]): curve 1 GDEF - semi-infinite steel body, Dirac pulse; curve ABCDEF - semi-infinite steel body, square pulse; curve AB – within square pulse; curve CD - influence of square pulse rear front; curve 2 GDEM – thermal inertia of the first layer higher than that of the second layer (Dirac pulse); curve 3 GDEL - thermal inertia of the first layer lower than that of the second layer (Dirac pulse); curve 4 GDEN – first layer adiabatic (Dirac pulse).

is more correct. In this regime, the temperature can be described with the first member of the infinite series only:

$$T = \frac{W}{C_1 \rho_1 L_1}\left[1 + 2A_1 \exp\left(-\pi^2 Fo_1\right)\right]. \tag{3.32}$$

A coefficient A_1 in Eq. (3.32) can be excluded after some transformations, thus allowing the determination of the thermal properties of a particular layer. During the time period of $Fo_1 < 0.25$ the heat conduction is essential only in the first layer, therefore, applying regression analysis for $0.16<Fo_1<0.25$ has lead to the following expressions for determining the thermal properties of the first layer [9]:

$$a_1 = -\frac{L_1 \ln F}{\Delta\tau\pi^2};\quad \gamma_1 = C_1\rho_1 = \frac{W(1-F)}{L_1 D};\lambda_1 = a_1\gamma_1;$$
$$F = \frac{\sum_{i=1}^{N} T_i \sum_{i=1}^{N} T_{i-1} - N\sum_{i=1}^{N} T_i T_{i-1}}{\left(\sum_{i=1}^{N} T_{i-1}\right)^2 - N\sum_{i=1}^{N} T_{i-1}^2};D = \frac{\sum_{i=1}^{N} T_{i-1}\sum_{i=1}^{N} T_i T_{i-1} - \sum_{i=1}^{N} T_i \sum_{i=1}^{N} T_{i-1}^2}{\left(\sum_{i=1}^{N} T_{i-1}\right)^2 - N\sum_{i=1}^{N} T_{i-1}^2}., \tag{3.33}$$

As the start point for analyzing the data ($Fo_1 = 0.16$), one should choose the time $\tau = \tau_E + 0, 23\tau_E$ on the K evolution curve, while the duration of the time interval should meet the condition: $\Delta\tau = 0.7\tau_E/N$.

Similarly, the analysis of the temperature dynamics at $Fo_1 > 0.25$ allows the determination of the 'effective' thermal properties of both layers $a_{12}, \gamma_{12}, \lambda_{12}$, whereupon thermal properties of the subsequent layers will be calculated by:

$$\gamma_2 = \frac{\gamma_{12} - (1-m)\gamma_1}{m}, \lambda_2 = \frac{m}{1/\lambda_{12} - (1-m)/\lambda_1}, a_2 = \frac{\lambda_2}{\gamma_2}, \quad (3.34)$$

where $m = L_2/(L_1 + L_2)$ and L_2 is thickness of the second layer.

The feature of time derivatives to emphasize evolution characteristic points was used in the *Synthetic Signal Processing* (SSP), or *Thermographic Signal Reconstruction* (TSR), method patented (U.S. patent 6,516,084) by Thermal Wave Imaging, USA, see Sect. 6.10.3. It is stated that the second time derivative of the temperature $T^F(\tau)$ on the front surface of an adiabatic plate heated with a Dirac pulse is characterized by the maximum $\tau_{2nd\ der}$ which meets the following condition:

$$\tau_{2nd der} = \frac{L^2}{\pi a}. \quad (3.35)$$

In other words, the maximum value of the second derivative corresponds to the Fourier number $Fo_{2ndder} = 1/\pi = 0.31831$ [11]. A graphical illustration of the first and second derivative technique is shown in Fig. 3.15. The time derivatives of the logarithmic time history are particularly useful in discriminating between defect and non-defect points.

Experimental implementation of the above-described approaches may be complicated by the fact that both division and taking derivatives of noisy functions tends to enhance noise, therefore noisy temperature evolutions $T(\tau)$ should be smoothed prior conducting further analysis, e.g. by applying polynomial fitting.

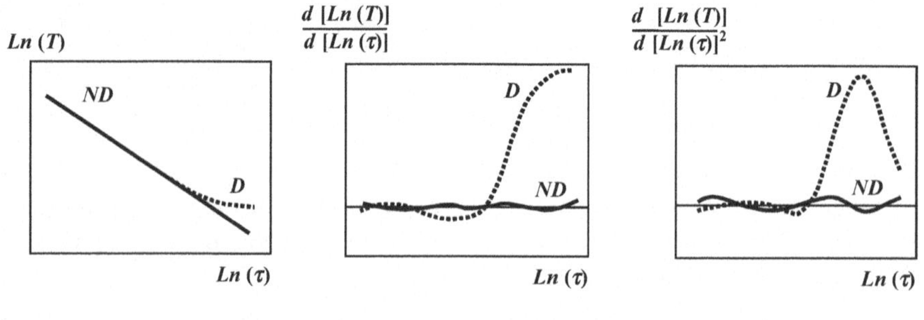

Fig. 3.15 Logarithmic temperature and its derivatives *vs.* logarithmic time (adapted from [11])

3.7.7 Evaluating TNDT Limits in Determining Material Thermal Properties

Sensitivity functions can be effectively used for evaluating minimal variations in thermal properties which can be reliably sensed by IR thermography. For example, the rear-surface sensitivity toward diffusivity variations (adiabatic plate, Dirac pulse heating) at $Fo_{1/2} = 0.139$ is:

$$S_a^R = \frac{\partial T / T}{\partial a / a}(Fo = 0.139) = 2.3. \tag{3.36}$$

Following the methodology proposed in [12], let us determine a TNDT sensitivity limit for a signal-to-noise ratio equal to unity. IR imaging radiometers intended for measuring temperature are characterized by the temperature resolution ΔT_{res}, which is typically from 0.05 to 0.1 °C for most commercial units. Research IR imagers may provide ΔT_{res} up to 0.001 °C, although working at such a level of sensitivity requires keeping a tested object in a thermostatically controlled environment. Therefore, the value ΔT_{res}~ 0.01 °C may be considered as a practical lower limit of temperature resolution. The minimum relative variation of thermal diffusivity that can be sensed with such equipment is as follows:

$$\frac{\Delta a}{a} = \frac{\Delta T_{res} / T}{2.3}, \tag{3.37}$$

where$T$is the sample excess temperature at $\tau_{1/2}$, which is linearly proportional to the energy $W(Q)$ applied to a sample, and is inversely proportional to the sample thickness.

For instance, in many TNDT applications, typical values of the above-mentioned parameters are: $\Delta T_{res} = 0.1^0$C, $T = 3 - 30^0$C. Hence, the limit of diffusivity measurement is $\Delta a/a$ = 1.4 – 0.14% being limited by equipment temperature resolution only; the sensitivity can be enhanced by increasing the absorbed energy or bytesting thinner samples.

In the Sect. 4.5 devoted to noise, we will demonstrate that, in practice, the TNDT sensitivity limit should be estimated by using the concept of noise running contrast $C_n = (\Delta T/T)_{noise}$. For example, carbon fiber reinforced plastic (CFRP) is characterized by C_n~ 6%, hence, the realistic sensitivity limit in this case will be about $\Delta a/a$~2.6%.

3.7.8 Determining Anisotropic Thermal Diffusivity

Mathematically, thermal diffusivity, like thermal conductivity, is a tensor value. In practice, one typically deals with *orthotropic* materials characterized by three tensor components of which one (a_z) is directed in-depth and two others (a_x and a_y) are

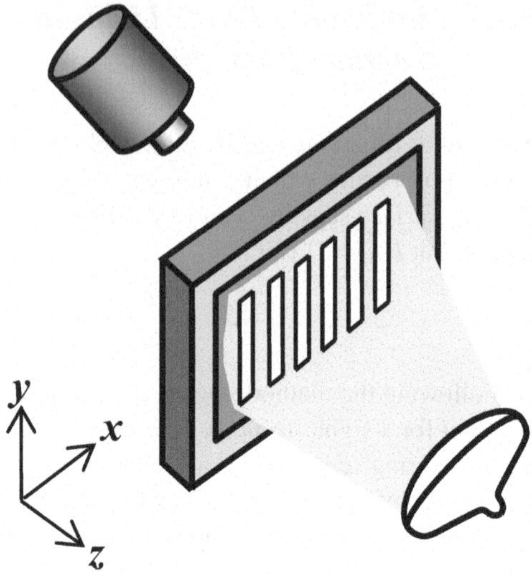

Fig. 3.16 Slit-mask IR thermography

planar. Classical procedures, such as Parker's method, usually deal with determining a_z, however, the increasing use of anisotropic composite materials makes it necessary to evaluate lateral diffusivities, and this requires applying a heat flux which would propagate in lateral directions. For this purpose, point, line and so-called "random" heating are used, while test results are evaluated by applying a 2D spatial Fourier transform.

For instance, a line heating technique for determining a_x and a_y was developed by Krapez et al. [13]. The experiment is schematically presented in Fig. 3.16.

A planar sample is heated with a pulsed heat source (flash tube) in front of which a slit-mask is placed to produce a line heating profile. Sample temperature is monitored on the rear surface with an IR imager. The formulation of the corresponding non-adiabatic boundary problem supposes that the heating function can be represented as $Q(x,y)Q_z(z)$:

$$\begin{aligned}
&\rho C\frac{\partial T}{\partial \tau}=\lambda_x\frac{\partial^2 T}{\partial x^2}+\lambda_y\frac{\partial^2 T}{\partial y^2}+\frac{\partial}{\partial z}\left[\lambda_z(z)\frac{\partial T}{\partial x}\right],\\
&T(x,y,z,\tau=0)=Q(x,y)Q_z(z)/\rho C,\\
&-\lambda_z\frac{\partial T(z=0)}{\partial z}=-\alpha^F T(x,y,z=0,\tau),\\
&-\lambda_z\frac{\partial T(z=L)}{\partial z}=\alpha^R T(x,y,z=L,\tau).
\end{aligned} \tag{3.38}$$

The problem above has been solved by applying the Laplace transform to time and the Fourier transform to lateral coordinates x and y:

$$\bar{T}(\omega_x,\omega_y,z,p)=\int\limits_{\tau'=0}^{\tau'=\infty}\int\limits_{x'=-\infty}^{x'=\infty}\int\limits_{y'=-\infty}^{y'=\infty}T(x',y',z,\tau)e^{i\omega_x x'}e^{i\omega_y y'}e^{-p\tau}d\tau' dy' dx', \quad (3.39)$$

that leads to the expression:

$$\left(\rho Cp+\lambda_x\omega_x^2+\lambda_y\omega_y^2\right)\bar{\tilde{T}}-\tilde{Q}Q_z(z)=\frac{\partial}{\partial z}\left[\lambda_z(z)\frac{\partial \bar{\tilde{T}}}{\partial z}\right]. \quad (3.40)$$

Let $T_z(z,\tau)$ be the solution for the problem where a 1D plate with thermal conductivity of $\lambda_z(z)$ is heated *uniformly* by heat flux $Q_z(z)$. By using the shift feature of the Laplace transform, we obtain:

$$\tilde{T}(\omega_x,\omega_y,z,\tau)=\tilde{Q}(\omega_x,\omega_y)T_z(z,\tau)e^{-a_x\omega_x^2\tau-a_y\omega_y^2\tau}.. \quad (3.41)$$

From Eq. (3.41) it follows that, for zero spatial frequencies $\omega_x=\omega_y=0$ and any coordinate z, the dual spatial Fourier transform which involves spatial integration by x and y experiences the same evolution in time as the temperature conditioned by uniform heating by heat flux $Q_z(z)$. This conclusion is also true for a finite-size plate with thermally-insulated edges.

Note that, if $a_z(z)$ is constant and $Q_z(z)$ describes the Dirac pulse, the above-mentioned tests for determining thermal diffusivity, e.g. Parker's method, can be applied to the rear-surface function $\tilde{T}(0,0,\tau)$.

The basic expression for determining lateral components of diffusivity can be obtained from Eq. (3.41) by excluding T_z:

$$\tilde{T}(\omega_x,\omega_y,z,\tau)/\tilde{T}(0,0,z,\tau)=\tilde{Q}(\omega_x,\omega_y)/\tilde{Q}(0,0)e^{-a_x\omega_x^2\tau-a_y\omega_y^2\tau}. \quad (3.42)$$

The important feature of Eq. (3.42) is its validity for any z coordinate even if heat losses are present on both the rear and front surfaces. Moreover, the condition of providing surface heating is not obligatory, therefore, Eq. (3.42) can be also applied to semi-transparent materials.

Line heating, as shown in Fig. 3.16, implies periodic variation of temperature along a lateral coordinate, e.g. by x, with the period Δ. The temperature is to be integrated by the coordinate y from $-Y$ to $+Y$ and the Fourier transform is to be applied by the coordinate x for a particular number of periods $n\Delta$ (the origin of the coordinate axis is placed in such way that the periodic heating function becomes *even* and achieves a maximum at $x=0$):

$$\tilde{T}(\omega_x,0,z,\tau) = \frac{1}{n\Delta}\frac{1}{2Y}\int_{x'=0}^{x'=n\Delta}\int_{y'=-Y}^{y'=+Y} T(x',y',z,\tau)\cos(\omega_x x')\,dy'dx'. \quad (3.43)$$

The a_x component can be determined by the following linear equation:

$$Ln\left[\tilde{T}(\omega_x,0,z,\tau)/\tilde{T}(0,0,z,\tau)\right] = Ln\left[\tilde{Q}(\omega_x,0)/\tilde{Q}(0,0)\right] - a_x\omega_x^2\tau. \quad (3.44)$$

The last equation is also valid on the rear surface ($z = 0$). Besides, due to the periodic variation of temperature along the coordinate x, the cyclic frequency $\omega_x = 2\pi/\Delta$ is of the most interest. Performing linear regression on the function $Ln\left[\tilde{\tilde{T}}(\omega_x,0,z,\tau)/\tilde{\tilde{T}}(0,0,z,\tau)\right]$ allows the determination of a_x directly by calculating the slope of the corresponding straight line. To determine the component a_y, one has to turn a slit-mask by 90° and repeat the experiment.

It was found that the accuracy of evaluating diffusivity components increases with a greater *thermal aspect ratio*:

$$A = \Delta / L\sqrt{a_z / a_x}. \quad (3.45)$$

However, for A higher than 4–5, measurement improvement becomes insignificant [13].

Philippi et al. analyzed a case of random heating [14] with the emphasis on a two-sided test, same as in the work by Krapez et al. [13].

Bison et al. extended the algorithm above onto a one-sided test [15] by having solved the problem for heating a non-adiabatic semi-infinite body with a Dirac pulse $\delta(\tau)$ characterized by the energy W and arbitrary spatial distribution of energy$f(x, y)$:

$$\begin{aligned}
&\frac{\partial^2 T}{\partial x^2} + \frac{\partial^2 T}{\partial x^2} + \frac{\partial^2 T}{\partial x^2} = \frac{1}{a}\frac{\partial T}{\partial \tau}, \\
&-\lambda\frac{\partial T(z=0)}{\partial z} = W\delta(\tau)f(x,y) - \alpha T(x,y,z=0), \\
&\int_{surf} f(x,y)\,dxdy = 1.
\end{aligned} \quad (3.46)$$

By applying the Laplace transform to time and the Fourier transform to coordinates x and y, we can convert the problem formulas as follows:

$$\frac{\partial^2 \tilde{\bar{\bar{T}}}}{\partial z^2} + \sigma^2 \tilde{\bar{\bar{T}}} = 0, \sigma = \sqrt{\omega_x^2 + \omega_y^2 + p/a},$$
$$-\lambda \frac{\partial \tilde{\bar{\bar{T}}}(z=0)}{\partial z} = W\Phi(\omega_x, \omega_y) - \alpha \tilde{\bar{\bar{T}}}(z=0), \tag{3.47}$$

where $\Phi(\omega_x, \omega_y)$ is the Fourier-image of the function $f(x, y)$.

Equation (3.46) has the solution:

$$\tilde{\bar{\bar{T}}} = \frac{W\Phi}{\lambda\sigma + \alpha} e^{-\sigma z}. \tag{3.48}$$

Having returned from the Laplace domain, we obtain the solution for the surface Fourier temperature as a function of time and Fourier frequencies:

$$\bar{\bar{T}}(\omega_x, \omega_y, z = 0, \tau) = \frac{W\Phi(\omega_x, \omega_y)}{\lambda} e^{-a(\omega_x^2 + \omega_y^2)\tau} \left[\sqrt{\frac{a}{\pi\tau}} - \frac{\alpha a}{\lambda} e^{\left(\frac{\alpha}{\lambda}\right)^2 a\tau} erf\left(\frac{\alpha}{\lambda}\sqrt{a\tau}\right) \right]. \tag{3.49}$$

The solution of Eq. (3.49) is quite general and represents the Fourier image of the surface temperature, which decays exponentially in time and diminishes with increasing diffusivity and square of a Fourier frequency. The solution member, which is dependent on time and coordinate z, can be excluded by using the same approach as when deriving Eq. (3.42); in this way, the surface heat loss member can also be excluded. For $z = 0$:

$$\frac{\bar{\bar{T}}(\omega_x, \omega_y, z = 0, \tau)}{\bar{\bar{T}}(\omega_x = 0, \omega_y = 0, z = 0, \tau)} = \frac{\Phi(\omega_x, \omega_y)}{\Phi(\omega_x = 0, \omega_y = 0)} e^{-a(\omega_x^2 + \omega_y^2)\tau}. \tag{3.50}$$

The identification procedure starts with integrating the temperature by the coordinate which is orthogonal to an analyzed coordinate direction. For example, while determining a_x, the temperature values in the y direction are to be summed at each time point to produce a temporal profile of the temperature along the coordinate x. Then, one has to reduce Eq. (3.50) to the logarithmic form and approximate the experimental function in the left section of Eq. (3.50) with a straight line whose slope represents a sought diffusivity component multiplied by the squared Fourier frequency. A value of the Fourier frequency can be determined if the true pixel size is known. Thus, the procedure for identifying a_x is expressed with the formula:

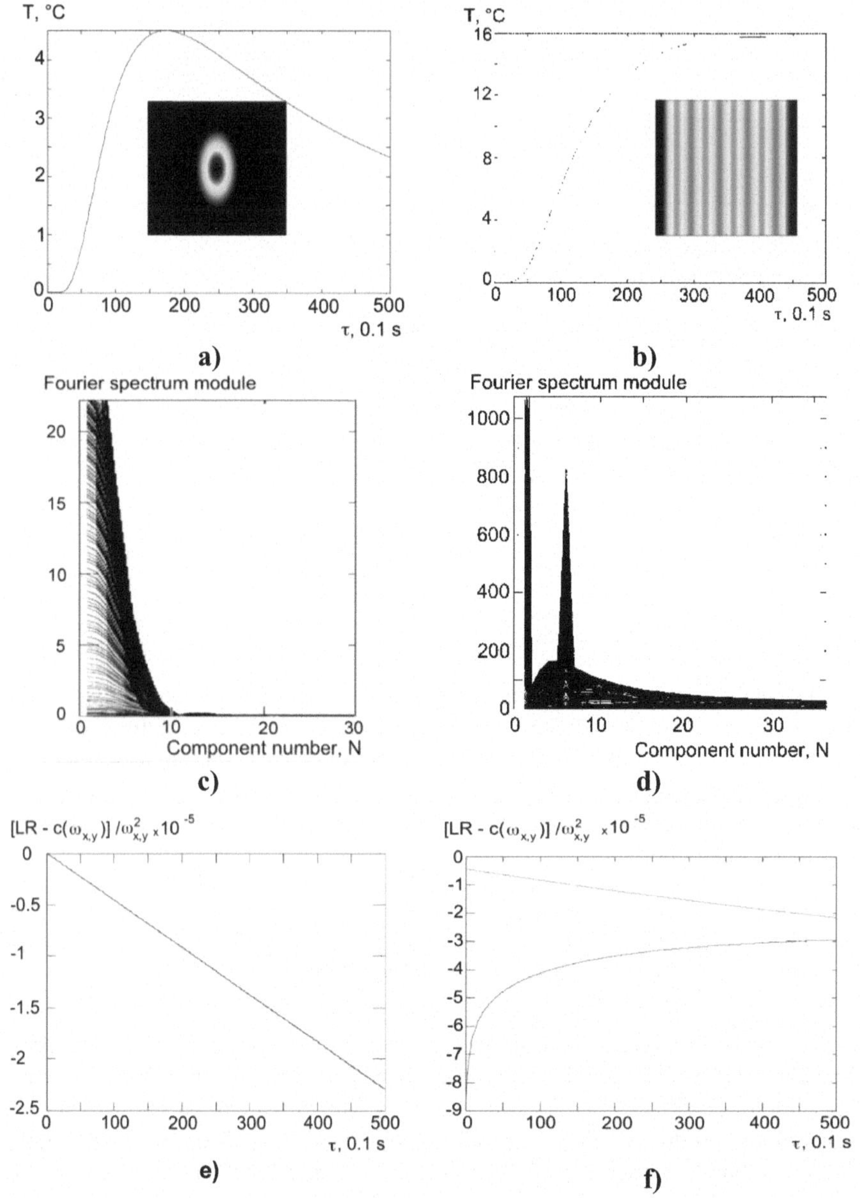

Fig. 3.17 Determining lateral diffusivity (CFRP sample 100x100 mm; rear surface; heating duration 0.1 s): (**a**, **c**, **e**) – square-mask 10×10 mm heating; (**b**, **d**, **f**) – slit-mask heating, 5 slits): (**a**, **b**) – temperature distribution and central point temperature vs. time. (**c**, **d**) – Fourier spectrum, (**e**, **f**) – *logarithmic ratio vs. time*

$$Ln\left[\frac{\overline{\overline{T}}(\omega_n,\tau)}{\overline{\overline{T}}(0,\tau)}\right]=c(\omega_n)-a\omega_n^2\tau;\omega_n=\frac{2\pi n}{N\Delta x};n=0\ldots N-1, \tag{3.51}$$

$$a\tau=-\left[LR-c(\omega_n)\right]/\omega_n^2,LR=Ln\left[\frac{\overline{\overline{T}}(\omega_n,\tau)}{\overline{\overline{T}}(0,\tau)}\right],$$

where N is the number of pixels along a chosen line, and $\Delta x$is pixel size.

The steps of the above-described method are illustrated by the theoretical data in Fig. 3.17.

A 5 mm-thick CFRP sample with a unidirectional fiber layout is heated through a 10x10 mm square or slit mask. Both heating cases clearly reveal the classical behavior of temperature signals and their Fourier transforms. Note that the square-mask profile of heating converts into the elliptical pattern of the temperature distribution on the rear surface due to anisotropic diffusivity (Fig. 3.17a). It is important to note that the Fourier spectrum shown in Fig. 3.17c reveals a zero component at the 11th frequency (the first frequency is assumed to be zero), therefore, thermal diffusivity values evaluated by the logarithmic curves, such as in Fig. 3.17e, are constant at all Fourier frequencies except the 11th frequency. In the case of slit-mask heating (Fig. 3.17b), the Fourier spectrum (Fig. 3.17d) reveals a prominent spike which corresponds to a carrier frequency. Subsequently, only at this frequency is the dependence of the logarithmic ratio of time linear (see the straight line in Fig. 3.17f), while other frequencies supply incorrect data (see the second curve in Fig. 3.17f).

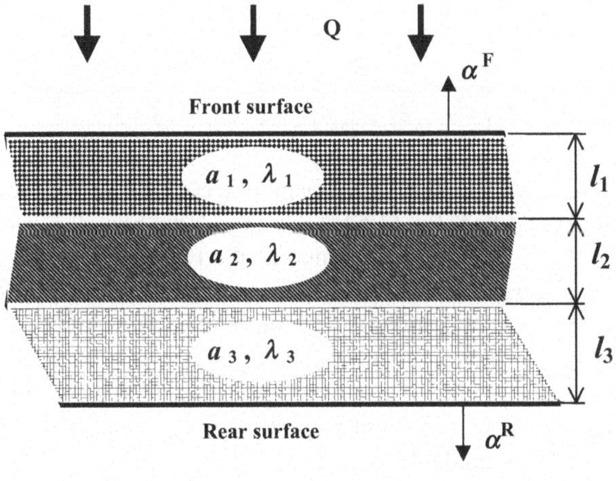

Fig. 3.18 Three-layer TNDT model (thermally-capacitive defect)

3.8 Non-stationary Heating of a Multi-Layer Plate

In the framework of the analytical approach to solving TNDT problems, multi-layer models are most convenient. For example, bonding two materials with an adhesive may be simulated by a three-layer plate where the middle layer represents either a bonding material (adhesive) or a defect. Then, the difference between non-stationary solutions, of which one is related to a defect area and another to a sound area, will serve as a 1D differential temperature signal.

3.8.1 *Three-Layer Non-adiabatic Plate with an Ideal Contact Between Layers*

The mathematical formulation of this model which involves capacitive defects is identical in both sound and defect areas (Fig. 3.18):

$$a_i \cdot \frac{\partial^2 T}{\partial z^2} = \frac{\partial T}{\partial \tau}; i = 1 \div 3; z = 0 - L; \tau > 0; \tag{3.52}$$

$$T(z, \tau = 0) = 0; \tag{3.53}$$

$$-\lambda_1 \cdot \frac{\partial T_1(z = 0, \tau)}{\partial z} = Q(\tau) - \alpha_F T_1(z = 0, \tau); \tag{3.54}$$

$$\lambda_3 \cdot \frac{\partial T_3(z = L, \tau)}{\partial z} = -\alpha_R T_3(z = L, \tau); \tag{3.55}$$

$$\lambda_j \frac{\partial T_j(z = l_j, \tau)}{\partial z} = \lambda_{j+1} \frac{\partial T_{J+1}(z = l_j, \tau)}{\partial z}; j = 1 - 2; \tag{3.56}$$

$$T_j(z = l_j, \tau) = T_{j+1}(z = l_j, \tau); j = 1 - 2. \tag{3.57}$$

Here: Eq. (3.52) is the 1D differential parabolic equation of heat conduction, Eq. (3.53) is the initial condition, Eq. (3.54) accounts for the boundary condition on the front (heated) surface, Eq. (3.55) accounts for the boundary condition on the rear surface, and Eqs. (3.56 and 3.57) account for the boundary conditions on the layer interface which express the continuity of both temperatures and heat fluxes while passing from one layer into another. Depending on the definition of $Q(\tau)$, various types of heating can be modeled. It is worth pointing out that solutions for heating with a Dirac (T_D) and square (T_{sp}) pulses will be easily obtained if the solution for step-wise heating (T_{sf}) is available:

$$T_D(\tau) = \frac{\partial T_{sf}(\tau)}{\partial \tau}; T_{sp}(\tau, \tau_h) = T_{sf}(\tau) - T_{sf}(\tau - \tau_h). \quad (3.58)$$

The solution $T_D(\tau)$ for heating a body with a Dirac pulse is often regarded as basic. Then, a case of arbitrary heating $\phi(\tau)$ can be analyzed by means of the Duhamel integral:

$$T_\phi(\tau) = \int_0^\tau T_D(\tau - \tau')\phi(\tau')d\tau', \int_0^\infty \phi(\tau')d\tau' = 1. \quad (3.59)$$

In numerical calculus, square-pulse solutions are typically used, with an easy transition to ether a Dirac pulse or continuous (step) heating.

The three-layer model expressed by Eqs. (3.52) is fairly general and allows the simulation of several test cases, such as a coating on a substrate, multilayer and composite structures etc. The model involves most of TNDT parameters that affect the temperature signal $\Delta T(\tau)$, except defect lateral size, which is absent in all 1D defect geometries. However, this model is useless if heating is performed with a localized (motionless or moving) heat source, although some realistic estimates can be obtained even in this case if a heat spot diameter is much larger than the sample thickness.

Solving 'multi-layer' problems requires performing tedious calculus, especially in finding roots of bulky transcendent equations. Since this can be achieved only by using computer-based numerical methods, advantages of analytical solutions become minimal to compare to very flexible numerical solutions. However, multi-layer analytical solutions are recommended for testing the accuracy of numerical solutions in defect zones that are difficult to analyze by using only classical expressions.

If the number of layers exceeds three, solving non-adiabatic heat conduction problems analytically is possible but using calculus is not practical as this results in extremely complex mathematical expressions. Also, it is worth stating that, if there are several layer interfaces in a body, a one-sided test allows the detection of only 2–3 interfaces which are close to the front surface. Deeper defects can be detected by applying a two-sided test.

Heating of adiabatic multi-layer plates can be comfortably analyzed by using the method of thermal quadrupoles. The theory and relating computer programs are given in [16]. In this approach, the analysis is often limited by the Laplace domain without returning to the space of original temperature functions. Using the thermal quadrupole technique for characterizing defects will be described in Sect. 5.5.

To illustrate how cumbersome 'multi-layer' solutions are, here is the solution for the step heating of a three-layer non-adiabatic plate, using Eqs. (3.52) and Fig. 3.18:

$$T(z = 0, \tau) = A_0 - \sum_{n=1}^{\infty} A_n e^{-\mu_n^2 a_1 \tau / l_1^2}; \quad (3.60)$$

$$T(z=L,\tau)=D_0-\sum_{n=1}^{\infty}D_n e^{-\mu_n^2 a_1\tau/l_1^2};$$

$$A_0=\Theta\frac{R}{R_1};D_0=\frac{\Theta}{R_1};R=1+\alpha_R\left(\frac{l_1}{\lambda_1}+\frac{l_2}{\lambda_2}+\frac{l_3}{\lambda_3}\right);R_1=R+\frac{\alpha_R}{\alpha_F};\Theta=\frac{Q}{\alpha_F};\Theta_s=\Theta\gamma_1;$$

$$A_n=D_n\{B_1\left[\cos(\mu_n\eta_1)+\frac{\gamma_2}{\mu_n}\sin(\mu_n\eta_1)\right]+B_2\left[\cos(\mu_n\eta_2)-\frac{\gamma_2}{\mu_n}\sin(\mu_n\eta_2)\right]+$$
$$+B_3\left[\cos(\mu_n\eta_3)-\frac{\gamma_2}{\mu_n}\sin(\mu_n\eta_3)\right]+B_4\left[\cos(\mu_n\eta_4)+\frac{\gamma_2}{\mu_n}\sin(\mu_n\eta_4)\right]\};$$

$$D_n=\frac{2\Theta_s}{\mu_n\{C_1[(1+\mu_0+\varsigma_1\eta_1)\sin(\mu_n\eta_1)+\mu_n\eta_1(1-\mu_0)\cos(\mu_n\eta_1)+}$$
$$\overline{+C_2[(1-\mu_0-\varsigma_1\eta_2)\sin(\mu_n\eta_2)+\mu_n\eta_2(1+\mu_0)\cos(\mu_n\eta_2)-}$$
$$\overline{-C_3[(1-\mu_0-\varsigma_3\eta_3)\sin(\mu_n\eta_3)+\mu_n\eta_3(1+\mu_0)\cos(\mu_n\eta_3)-}$$
$$\overline{-C_4\left[(1+\mu_0+\varsigma_4\eta_4)\sin(\mu_n\eta_4)+\mu_n\eta_4(1-\mu_0)\cos(\mu_n\eta_4)\right]\}};$$

μ_n are the roots of the transcendent equation:

$$C_1\left[\mu(1-\mu_0)\sin(\mu\eta_1)-\varsigma_1\cos(\mu\eta_1)\right]+C_2\left[\mu(1+\mu_0)\sin(\mu\eta_2)+\varsigma_2\cos(\mu\eta_2)\right]-$$
$$-C_3\left[\mu(1+\mu_0)\sin(\mu\eta_3)+\varsigma_2\cos(\mu\eta_3)\right]-C_4\left[\mu(1-\mu_0)\sin(\mu\eta_4)-\varsigma_1\cos(\mu\eta_4)\right]=0;$$

$$p_1=\frac{\lambda_2}{\lambda_1}\sqrt{\frac{a_1}{a_2}};p_2=\frac{\lambda_3}{\lambda_2}\sqrt{\frac{a_2}{a_3}};\gamma_1=\frac{\alpha_R l_1}{\lambda_1};\gamma_2=\frac{\alpha_R l_1}{\lambda_3}\sqrt{\frac{a_3}{a_1}};$$

$$\varsigma_1=\gamma_1+\gamma_2;\varsigma_2=\gamma_2-\gamma_1;\mu_0=\frac{\gamma_1\gamma_2}{\mu^2};$$

$$C_1=B_1=\frac{(1+p_1)(1+p_2)}{4};C_2=B_2=\frac{(1-p_1)(1+p_2)}{4};$$

$$-C_3=B_3=\frac{(1+p_1)(1-p_2)}{4};-C_4=B_4=\frac{(1-p_1)(1-p_2)}{4};$$

$$\eta_1 = \frac{l_1 + l_2\sqrt{\frac{a_1}{a_2}} + l_3\sqrt{\frac{a_1}{a_3}}}{l_1};\eta_2 = \frac{l_1 - l_2\sqrt{\frac{a_1}{a_2}} - l_3\sqrt{\frac{a_1}{a_3}}}{l_1};$$

$$\eta_3 = \frac{l_1 + l_2\sqrt{\frac{a_1}{a_2}} - l_3\sqrt{\frac{a_1}{a_3}}}{l_1};\eta_4 = \frac{l_1 - l_2\sqrt{\frac{a_1}{a_2}} + l_3\sqrt{\frac{a_1}{a_3}}}{l_1};$$

Q is the heat power density, W/m^2.

The solution expressed by Eqs. (3.60) was derived by applying the Laplace transform and implementing in the Layer-3 Analytic software from Innovation, Ltd., Russia. It has been used for testing 2D and 3D numerical solutions obtained with the ThermoCalc-2D and ThermoCalc-3D computer programs (Tomsk Polytechnic University.

3.8.2 *Three-Layer Adiabatic Plate with an Ideal Contact Between Layers*

Multi-layer solutions become simpler if surface heat losses can be neglected. Adiabatic solutions are also more convenient for characterizing subsurface defects.

As an example, let us give the solution for a three-layer plate adiabatically heated with a Dirac pulse (model by Fig. 3.18 if $\alpha^F = \alpha^R = 0$, solution obtained by Balageas et al. [17]):

$$\frac{T_D(\tau, z=0)}{T_m} = 1 + 2\frac{\sum_{i=1}^{4}\chi_i v_i}{\sum_{i=1}^{4}\chi_i}\sum_{n=1}^{\infty}\frac{\sum_{i=1}^{4}\chi_i\cos(v_i\gamma_n)}{\sum_{i=1}^{4}\chi_i v_i\cos(v_i\gamma_n)}e^{-\gamma_n^2\tau/\eta_3^2}, \tag{3.61}$$

where γ_n are the roots of the transcendent equation:

$$\sum_{i=1}^{4}\chi_i\sin(v_i\gamma) = 0;$$

$$T_m = W / \sum_{i=1}^{3}\rho_i C_i l_i;$$

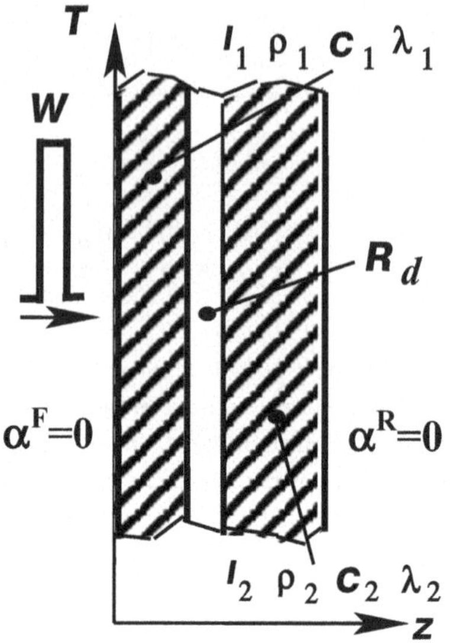

Fig. 3.19 Two-layer TNDT model (thermally-resistive defect)

$$\chi_1 = e_{13} + e_{12} + e_{23} + 1; \omega_1 = \eta_{13} + \eta_{23} + 1;$$
$$\chi_2 = e_{13} - e_{12} + e_{23} - 1; \omega_2 = \eta_{13} + \eta_{23} - 1;$$
$$\chi_3 = e_{13} + e_{12} - e_{23} + 1; \omega_3 = \eta_{13} - \eta_{23} + 1;$$
$$\chi_4 = e_{13} + e_{12} - e_{23} - 1; \omega_4 = \eta_{13} - \eta_{23} - 1;$$
$$e_{ij} = e_i / e_j; e_i = \sqrt{\lambda_i C_i \rho_i}; \eta_{ij} = \eta_i / \eta_j; \eta_i = l_i / \sqrt{a_i};$$

W is the heat energy density, J/m^2.

3.8.3 Two-Layer Adiabatic Plate with a Thermal Resistance Between (Square-Pulse Heating)

The three-layer model of Eqs. (3.59, 3.60) can be further simplified by replacing a thermally-capacitive middle layer with a thermally-resistive one. Let a test body be represented by two adiabatic layers with the thickness l_1 and l_2, separated by a layer with thermal resistance $R_d = d/\lambda_d$ (Fig. 3.19).

The front-surface temperature response to heating with a square pulse of duration τ_h was given by Delpeche et al. [18]:

$$T^F(\tau > \tau_h) = \sum_{n=1}^{\infty} T_\infty \frac{4L\mu_n}{D(\mu_n)} \frac{\exp(a\mu_n\tau_h) - 1}{a\mu_n\tau_h} \exp[-a\mu_n\tau]; \qquad (3.62)$$

$$T_{\infty} = \frac{Q\tau_h a}{\lambda L} = \frac{W}{C\rho L}; W = Q\tau_h;$$

$$D(\mu_n) = \left[1 + \left(\frac{\alpha}{\lambda\mu_n}\right)^2\right] 2\mu_n l_1 + \left[1 - \left(\frac{\alpha}{\lambda\mu_n}\right)^2 + 2\frac{\alpha}{\lambda\mu_n} tg(\mu_n l_1)\right] \sin(2\mu_n l_1) +$$

$$+ \left(\frac{\frac{\alpha}{\lambda\mu_n}\cos(\mu_n l_1) - \sin(\mu_n l_1)}{\sin(\mu_n l_2) - \frac{\alpha}{\lambda\mu_n}\cos(\mu_n l_2)}\right)^2 x$$

$$x\left\{\left[1 + \left(\frac{\alpha}{\lambda\mu_n}\right)^2\right] 2\mu_n l_2 + \left[1 - \left(\frac{\alpha}{\lambda\mu_n}\right)^2 + 2\frac{\alpha}{\lambda\mu_n} tg(\mu_n l_2)\right] \sin(2\mu_n l_2)\right\},$$

where T_{∞} is the stationary temperature of the adiabatic ($\alpha = 0$) plate at longer times and μ_n are roots of the transcendent equation:

$$\frac{\lambda\mu}{1/R_d} = ctg\left[\mu l_1 - arctg\left(\frac{\alpha}{\lambda\mu}\right)\right] + ctg\left[\mu l_2 - arctg\left(\frac{\alpha}{\lambda\mu}\right)\right], \tag{3.63}$$

which may be found by using any of several mathematical techniques.

3.8.4 Two-Layer Adiabatic Plate with the Thermal Resistance Between (Dirac-Pulse Heating)

An even simpler solution was derived by Balageas et al. for heating a two-layer adiabatic plate with a Dirac pulse [17] (Fig. 3.19):

$$\frac{T_D(z=0,\tau)}{T_m} = 1 + 2\frac{\sum_{i=1}^{2}\chi_i\gamma_i}{\sum_{i=1}^{2}\chi_i} x \tag{3.64}$$

$$x\sum_{k=1}^{\infty}\frac{\sum_{i=1}^{2}\chi_i\cos(\gamma_i v_k) + (-1)^i R*v_k\sin(\gamma_i v_k)}{\sum_{i=1}^{2}\chi_i\gamma_i\cos(\gamma_i v_k) - (-1)^i R*\cos(\gamma_i v_k) + (-1)^i R*\gamma_i v_k\sin(\gamma_i v_k)} e^{-v_k^2\tau/\eta_2^2};$$

$$\eta_{12} = \eta_1/\eta_2; R* = e_1 R_d/\eta_2; e_i = \sqrt{C_i\rho_i\lambda_i};$$

$$e_{12} = e_1 / e_2; \eta_i = l_i / \sqrt{a_i};$$

ν_k are the roots of the transcendent equation:

$$\sum_{i=1}^{2} \chi_i \sin(\gamma_i \nu) - (-1)^i R * \nu \cos(\gamma_i \nu) = 0.$$

Note that $R*$ is the dimensionless thermal resistance of a defect which is dependent on the thermal properties of both layers.

In the 1D approach, Eqs. (3.64) can be successfully used in the TNDT of metals and composites containing thin defects, such as delaminations, disbonds, etc.

3.9 1D Thermal Waves

As mentioned above, thermal waves appear as a result of stimulating a sample with modulated heat energy. The concept of thermal waves and the analysis of their penetration depth can be traced to the earlier works by Fourier [19] and Angström [20].

3.9.1 *Semi-Infinite Adiabatic Body*

Below we will consider some general features of harmonic thermal waves which are propagating in a semi-infinite adiabatic body as the result of stimulation with a heat flux described by:

$$Q(\tau) = Q_0 \left[1 + \cos(\omega\tau)\right], \tag{3.65}$$

where ω is the cyclic frequency. At $z = 0$, the adiabatic boundary condition takes place:

$$-\lambda \frac{\partial T}{\partial x} = Q_0 \left[1 + \cos(\omega\tau)\right] = \mathrm{Re}\left\{Q_0 \left[1 + \exp(j\omega\tau)\right]\right\}, \tag{3.66}$$

where Re specifies the real part of a complex expression. The heating function includes two members of which one is constant (Q_0) and another is oscillating ($\exp(j\omega\tau)$). Below, the constant member will be omitted. The oscillating solution of the 1D differential Eq. (2.12) is as follows:

$$T(z,\tau)=\frac{Q_0}{\lambda\sigma}\exp(-\sigma x+j\omega\tau)$$
$$=\frac{Q_0}{\sqrt{C\rho\lambda\omega}}\exp\left(-z\sqrt{\frac{\omega}{2a}}\right)\exp\left[j\left(\omega\tau-z\sqrt{\frac{\omega}{2a}}-\frac{\pi}{4}\right)\right], \quad (3.67)$$

where $\sigma=(1+j)\sqrt{\omega/(2a)}$. Eq. (3.67) is analogous to that for the amplitude of an electromagnetic wave in a skin-layer of electrically-conductive materials (metals). The analogy with optical radiation allows the consideration of some phenomena which are characteristic for waves in general, such as reflection, refraction, dispersion, etc.

Like any propagating wave, thermal wave is characterized by the *wave vector* $\mathrm{Re}(\sigma)=1/\mu=\sqrt{\omega/(2a)}$, where μ is the thermal diffusion length. With μ taken into account, Eq. (3.67) acquires the following form:

$$T(z,\tau)=\frac{Q_0}{\sqrt{C\rho\lambda\omega}}\exp(-z/\mu)\exp\left[j\left(\omega\tau-z/\mu-\frac{\pi}{4}\right)\right], \quad (3.68)$$

Thermal waves are damping exponentially with increasing material depth. Therefore, it can be assumed that they do not penetrate deeper than three thermal diffusion lengths ($z = 3\mu$). Thermal wave penetration depth increases with higher thermal diffusivity and/or lower frequency. For example, there are two types of thermal waves appearing on the Earth's surface. One has a period of 24 hours (daily) and one has a period of 1 year (yearly). It is obvious that yearly waves, which have a super-low frequency penetrate soil up to maximum depths of 2–3 m. Daily waves, which propagate in soil up to 10–20 cm, may be used in IR thermographic detection of landmines (see Sect. 4.10.2).

Thermal waves are subject to substantial dispersion because high-frequency waves propagate faster than low-frequency. The *phase velocity* of thermal waves is determined as follows:

$$v=\omega\mu=\sqrt{2a\omega}. \quad (3.69)$$

The phase shift between a *stimulating* thermal wave and the *surface temperature* wave ($z = 0$) is equal to -45^0, increasing with depth by the angle $\phi = -z/\mu - \pi/4$.

The thermal wave *impedance*

$$Z=\frac{1}{\lambda\sigma}=\frac{1}{\sqrt{j\omega C\rho\lambda}} \quad (3.70)$$

determines the surface wave amplitude which increases with decreasing material effusivity $e=\sqrt{C\rho\lambda}$. Like thermal diffusivity, thermal effusivity is a dynamic characteristic of a material. As a rule, materials with high diffusivity also have high

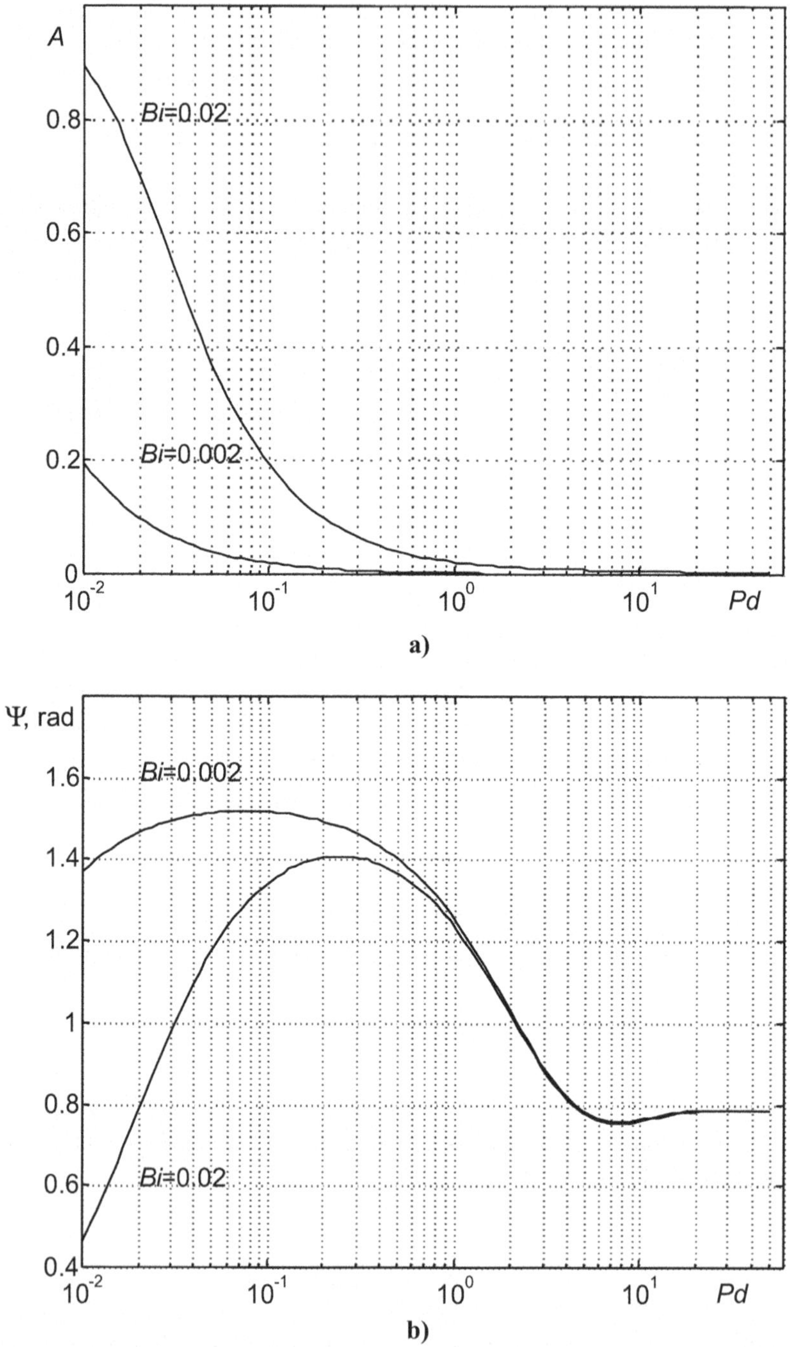

Fig. 3.20 Thermal wave amplitude (**a**) and phase (**b**) defined by Eqs. (3.60)

effusivity. However, there are exceptions, most notably air which, due to its low density, has a high diffusivity close to that of metals, but very low effusivity.

3.9.2 Semi-Infinite Non-adiabatic Body

The solution for heating a semi-infinite non-adiabatic body with a harmonic heat flux presented in Table 2.5 is more general than that by Eq. (3.68) but it contains no sample thickness. Below a deeper analysis of thermal wave amplitude and phase will be done for a non-adiabatic plate [21].

3.9.3 Non-adiabatic Plate

The solution for a plate which experiences heat losses on the front surface with the coefficient α and is adiabatic on the rear surface is as follows [22]:

$$\begin{aligned}
&\frac{T(Z,\tau)}{Q_m/\alpha} = A\cos(PdFo - \Psi); Z = z/L;. \\
&A = (N_i N_{-i})^{1/2}; \\
&\Psi = Arctg\left(i\frac{N_i - N_{-i}}{N_i + N_{-i}}\right); \\
&N_i = \frac{ch\sqrt{iPd}(1-Z)}{ch\sqrt{iPd} + (1/Bi)\sqrt{iPd}sh\sqrt{iPd}}; \\
&N_{-i} = \frac{ch\sqrt{-iPd}(1-Z)}{ch\sqrt{-iPd} + (1/Bi)\sqrt{-iPd}\,sh\sqrt{-iPd}}; \\
&Bi = \alpha L/\lambda; Pd = \omega L^2/a.
\end{aligned} \tag{3.71}$$

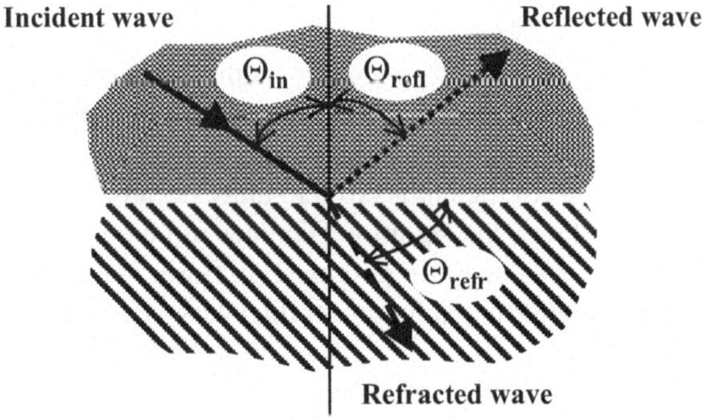

Fig. 3.21 Reflection and refraction of thermal waves at the interface between two media

Here z is depth, Q_m is the maximum absorbed heat power density, A is the thermal wave amplitude, Ψ is the thermal wave phase, is the sample thickness, λ is the thermal conductivity, Bi and Pd are the Biot and Predvoditelev criteria.

By assuming $L \to \infty$ in Eq. (3.71), the expressions for the amplitude and phase on the plate surface become identical to those available for a semi-infinite body.

In the theory of thermal waves, amplitude damping and phase shifting with depth are typically analyzed, while TNDT mainly deals with surface solutions ($z = 0$) where depth serves as a parameter. In further discussion we will use Eq. (3.71) where the surface temperature is a function of three variables: . The thermal wave amplitude and phase depend on two parameters: . The plots of the corresponding functions are shown in Fig. 3.20 for values of and typical in TNDT of composite materials.

With a greater value of the Predvoditelev criterion, the thermal wave amplitude drops asymptotically to zero. In accordance with the conclusions above, this means that high-frequency waves produce no significant temperature signals on the surface of thermally-thick samples because they are characterized by a high thermal inertia.

3.9.4 Temperature Waves at the Interface of Two Media

Let us consider reflection and refraction (transmission) of thermal waves at the interface between two media (Fig. 3.21). The coefficients of reflection Γ and transmission T are [23]:

$$\Gamma = \frac{\cos(\Theta_{in}) - b\cos(\Theta_{refr})}{\cos(\Theta_{in}) + b\cos(\Theta_{refr})}; T = \frac{2\cos(\Theta_{in})}{\cos(\Theta_{in}) + b\cos(\Theta_{refr})}, \tag{3.72}$$

where $b = e_2/e_1$ characterizes thermal match of two media and indices *in* and *refr* specify incident and refraction angles respectively.

In the case of 1D (planar) waves:

$$\Gamma = \frac{1-b}{1+b}; T = \frac{2}{1+b}. \tag{3.73}$$

At the interface between two media, thermal waves reveal the same phenomena as electromagnetic and acoustic waves; their transmission and reflection is affected by a thermal mismatch between media. If a thermal wave leaves a solid to enter air, a reflection coefficient is close to unity because of very low air effusivity. However, in thin defects, the situation becomes more complicated due to the interference of thermal waves in a thin air gap. Therefore, in this case, the coefficient of reflection is mainly determined by defect thickness (see Sect. 4.2.3).

3.10 The Relationship Between Pulsed and Harmonic Thermal Waves

Generation and propagation of both pulsed and harmonic thermal waves in materials are governed by the same physics. Mathematically, solutions for modulated (harmonic) processes are obtained in the *frequency domain*, while pulsed solutions appear in the *time domain*. It is known that the relationship between such solutions is realized through the Laplace transform. A solution in the time domain $T(\tau)$ can be derived by representing the corresponding frequency-domain solution with the Fourier series:

$$T(z,\tau)=\sum_{n=1}^{\infty}A(\omega_n)T(z,\omega_n,\tau), \tag{3.74}$$

where $T(z, \omega_n, \tau)$is a planar thermal wave with the cyclic frequency ω_n propagating along the z coordinate (e.g., see Eqs. (3.68) and (3.71), $A(\omega_n)$ characterizes the contribution of a wave of a particular frequency in the general pulsed response.

Since the spectrum of a square pulse is continuous, the Fourier series can be substituted with an integral transform.

Let us consider the heating of a semi-infinite body with a Dirac pulse. Dirac-like heat energy generation at the surface $z = 0$ is equivalent to heating with thermal waves which are characterized by a continuous spectrum and the same energy being in-phase at $z = 0$ and $\tau = 0$. In-depth propagation of a primary heat pulse can be treated as propagation of a packet of thermal waves which are of different frequencies, and which experience damping by amplitude and retardation by phase depending on frequency. The temperature response at depth z can be obtained by summing amplitudes of all thermal waves at (z, τ), and this can be done by using Eq. (3.68):

$$T(z,\tau)=\sum_{n=0}^{\infty}\frac{Q_0}{\lambda\sigma_n}\exp(-\sigma_n z)\exp(j\omega_n\tau). \tag{3.75}$$

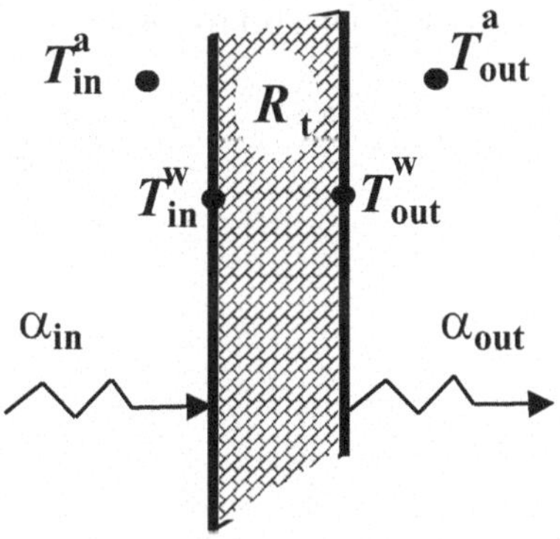

Fig. 3.22 Heat conduction through a planar wall

Here the dimensions of Q_0 are W/m^2. By substituting the series with the integral and introducing $j\omega = s$, we have

$$T(z,\tau) = \frac{1}{\pi j}\int_0^\infty \frac{Q_0}{\lambda\,\sigma(s)}\exp\left[-\sigma(s)z\right]\exp(s\tau)ds$$
$$= W_0 L^{-1}\left[\frac{\exp(-\sigma z)}{\lambda\sigma}\right] \quad (3.76)$$

where L^{-1} specifies the inverse Laplace transform, W_0 is the total pulse energy in J/m^2. By using the tables of inverse Laplace transforms and assuming $\sigma = \sqrt{s/a}$, we come to the final solution:

$$T(z,\tau) = \frac{Q_0}{\sqrt{\pi C\rho\lambda\tau}}\exp\left(-z^2/4a\tau\right), \quad (3.77)$$

which is identical to that in Table 2.3.

3.11 Steady-State Heat Conduction through a Planar Wall and Determination of Wall Thermal Resistance

This book is mainly devoted to active TNDT, therefore, all heat conduction problems analyzed above describe transient thermal processes. As an exception, we willl briefly consider steady-state heat conduction through a planar wall because problems arise in the IR thermographic inspection of building envelopes, chimneys and pipes. In particular, the determination of wall thermal resistance will be of interest.

Let us consider a planar wall which separates two media with different temperatures T_{in}^a and T_{out}^a (Fig. 3.22).

The surface heat exchange is characterized by the coefficients α_{in} and α_{out}, also called surface conductance coefficients. A uniform wall has thermal resistance $R_t = L/\lambda$ and a multi-layer wall - $R_t = \sum_{i=1}^{N} L_i/\lambda_i$. The heat flux through the wall is:

$$Q = \frac{T_{in}^w - T_{out}^w}{R_t}$$

or

$$Q = \frac{T_{in}^a - T_{out}^a}{R}, \quad (3.78)$$

where $R = 1/\alpha_{in} + R_t + 1/\alpha_{out}$is the total thermal resistance of the wall which takes into account both the insulation abilities of a wall material and the thermal resistance of the interface between the wall and the surrounding environment, with the latter being defined by the heat exchange coefficients. Note that the reverse quantity $U = 1/R_t$ is called the *overall (total) coefficient of heat transmission*, or simply *U-value*.

In practical applications, the values of R_t, α_{in}, α_{out} are determined by measuring both the density of the heat flux through a wall and the temperatures of the wall surface and the ambient air:

$$R_t = \frac{T_{in}^w - T_{out}^w}{Q}; \alpha_{in} = \frac{Q}{T_{in}^a - T_{in}^w}; \alpha_{out} = \frac{Q}{T_{out}^w - T_{out}^a} \tag{3.79}$$

The scheme shown in Fig. 3.22 appears in the IR thermographic inspection of the thermal insulation quality in building envelopes and chimneys. Using an IR imaging radiometer, the surface temperature T_{out}^w of a tested object is monitored, and by evaluating the variations in the temperature field and other temperature anomalies one makes decision about the presence/absence of structural defects (see Chap. 11).

The expression which connects the surface temperature and thermal insulation parameters results from solving the inverse problem of steady-state heat conduction:

$$T_{out}^w = \frac{T_{in}^a + \alpha_{out} T_{out}^a \left(\dfrac{1}{\alpha_{in}} + R_t \right)}{1 + \dfrac{\alpha_{out}}{\alpha_{in}} + \alpha_{out} R_t}. \tag{3.80}$$

Determining insulation thermal resistance represents a typical inverse problem of technical diagnostics. Hypothetically, it can be solved with two-fold approaches: (1) measuring both the front/rear-surface temperatures (with an IR imager) and the

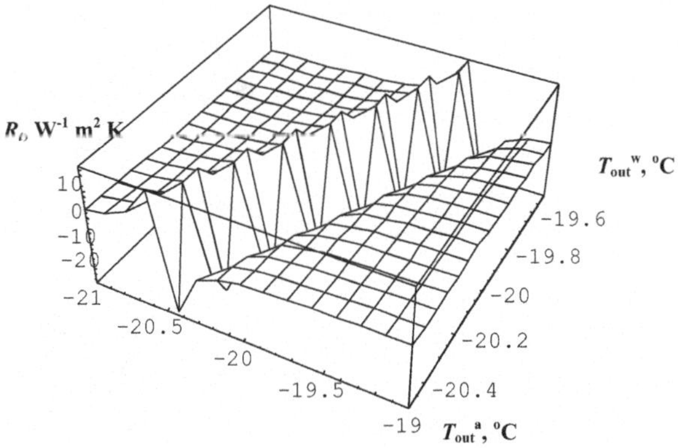

Fig. 3.23 Determining thermal resistance of a wall by Eq. (3.75); true value $R_t = 2$ W^{-1} m^2 K at $T_{out}^a = -20$ °C and $T_{out}^w = -19.19$ °C

heat flux density through an object (with a heat fluxmeter), (2) measuring only surface temperatures along with the ambient temperature.

The first approach is typical in building and approved by many national standards (in Russia - GOST 26254–84). Since on-site measurements are complicated by natural dynamics of tested objects, mainly caused by variations in solar radiation and ambient temperature, heat flux and temperature measurements must be conducted over a rather long period of time (up to 15 days). However, quasi-stationary periods from 1 to 3 days long depending on object thermal inertia and thermal resistance may be evaluated by Eq. (3.78).

The second approach requires using coefficients of heat exchange according to the following formula:

$$R_t = \frac{T_{in}^a - \frac{\alpha_{out}}{\alpha_{in}}\left(T_{out}^w - T_{out}^a\right) - T_{out}^w}{\alpha_{out}\left(T_{out}^w - T_{out}^a\right)}. \quad (3.81)$$

Let us illustrate the "instability" of the inverse solution using Eq. (3.81) for a typical building envelope ($R_t = 2\ \mathrm{W^{-1} \cdot m^2 \cdot K}$). Let the indoor and outdoor temperatures be: $T_{out}^a = -20$ °C and $T_{in}^a = +20$ °C. According to the Russian building norms, assume α_{in} = 7.8 W/(m²·K) and α_{out} = 23 W/(m²·K). Using the 'direct' Eq. (3.79) yields in $T_{out}^w = -19.19$ °C. Now assume that the temperatures of the wall and air are measured with an IR imaging radiometer that inevitably leads to some errors caused by varying emissivity, the presence of solar radiation, wall temperature pre-history etc. Let the measured values be in the following ranges: $T_{out}^a = -19\ldots-21$ °C and $T_{out}^w = -19.5\ldots-20.5$ °C. The plot of the function $R_t\left(T_{out}^a, T_{out}^w\right)$ calculated by Eq. (3.80) is presented in Fig. 3.23. It is seen that some combinations of input data result in unrealistic values of thermal resistance, including negative (the true value is

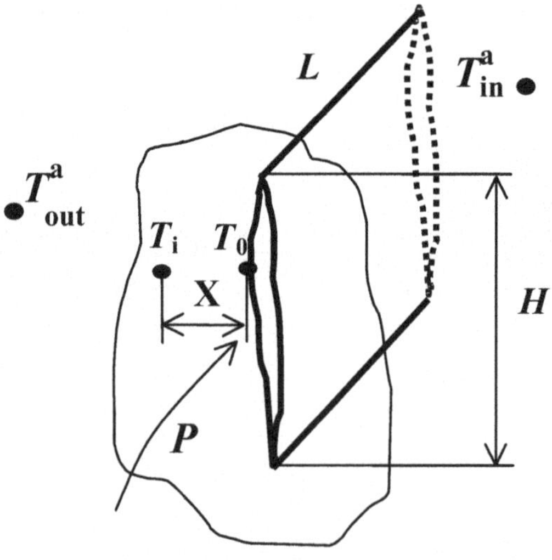

Fig. 3.24 Modeling air leakage by a local heat sink

$R_t = 2\ \mathrm{W^{-1}{\cdot}m^{2}{\cdot}K}$). The valley in Fig. 3.23 is a good graphical illustration of the instability of the inverse problem (3.81).

The example above explains why in the testing of buildings practice thermal resistance is always determined by directly measuring heat fluxes and temperatures as required by Eq. (3.79) but not by solving Eq. (3.81), where recommended values of heat exchange coefficients should be used.

It has been concluded that IR thermography is more convenient for evaluating local variations in thermal resistance rather than for measuring its absolute value [24]. Indeed, if the differential temperature signal over an expected defect is ΔT_{out}^{w}, the resulting variation in R_t is:

$$\Delta R_t = \Delta T_{out}^{w} / \frac{\partial T_{out}^{w}}{\partial R_t} = \frac{\left(1 + \frac{\alpha_{out}}{\alpha_{in}} + \alpha_{out} R_t\right)^2}{\alpha_{out}\left(T_{out}^{a} - T_{in}^{a}\right)} \Delta T_{out}^{w}. \tag{3.82}$$

For example, in the inspection of a typical chimney ($R_t = 1.37\ \mathrm{W^{-1}{\cdot}m^{2}{\cdot}K}$), incorrect input data to Eq. (3.81) cause the error in thermal resistance to be $\Delta R_t = 0.26\ \mathrm{W^{-1}{\cdot}m^{2}{\cdot}K}$, which constitutes 19% of the total thermal resistance. Such a great variation in thermal resistance may occur only as a result of total destruction of a chimney refractory layer. However, if an R_t value is known *a priori*, e.g. from the technical documentation, the sensitivity of IR thermographic evaluation using Eq. (3.81), may reach $\Delta R_t = 0.032\ \mathrm{W^{-1}{\cdot}m^{2}{\cdot}K}$, or 2.3% of the total thermal resistance.

3.12 Evaluating Air Leaks by Using IR Thermography

In this section, we will present a method which enables the simple evaluation of an air leak occurring in thermally-insulated objects under a pressure gradient.

Assume that a leak exists through a thin long crack in a building wall, as often occurs in inter-panel seams. The crack length and thickness are *H* and *L* respectively (see Fig. 3.24).

If indoor pressure is lower than outdoor and the outside temperature is colder than the indoor temperature, cold air will penetrate into a building through the crack by carrying the energy *q* (W):

$$q = C^{air} \rho^{air} P\left(T_{in}^{a} - T_{out}^{a}\right), \tag{3.83}$$

where C^{air}, ρ^{air} are air heat capacity and density and *P* is leak rate in $\mathrm{m^3/s}$. In the IR thermogram, both the leak and neighboring areas look colder than the main wall. To some extent, it can be assumed that there is a heat source (sink) of power *q* in the wall. Adjacent to both sides of the crack, according to the Fourier law, a temperature gradient appears being directed toward the crack:

$$T_x - T_o = \frac{q}{2HL}\frac{X}{\lambda}, \tag{3.84}$$

where T_o is the temperature on the lateral surface of the crack, T_x is the temperature at the distance $X$from the crack side, and λ is thermal conductivity of a wall material. The leak rate is:

$$P = \frac{2HL\lambda\left(T_x - T_o\right)}{C\rho X\left(T_{in}^a - T_{out}^a\right)}. \tag{3.85}$$

For example, if on the chimney surface, the temperature gradient at distance X= 1.2 m from a 2.5 m-long crack is $T_i - T_o$ = 1.8°C, the leak rate will be 43 liters an hour [24].

The model (3.85) has been derived for a very simple uniform structure and includes no surface heat exchange. Moreover, quantitative evaluation of air leaks by using IR thermography is an indirect technique which probably would not be competitive with more traditional methods.

References

1. Parker, W.J., Jenkins, R.J., Butler, C.P., Abbot, G.L.: Flash method of determining thermal diffusivity, heat capacity and thermal conductivity. J. Appl. Phys. **32**, 1679–1686 (1961)
2. Zalameda, J.N., Winfree, W.P.: Improved sampling of thermal transients using focal plane Array infrared imagers. Proc. SPIE "Thermosense-XXVI". **5405**, 374–381 (2004)
3. Winfree, W.P., Zalameda, J.N.: Single sided thermal diffusivity imaging in composites with a shuttered thermographic inspection system. Proc. SPIE "Thermosense-XXIV". **4710**, 536–544 (2002)
4. Vavilov, V., Grinzato, E., Bison, P.G., et al.: Inversion for hidden corrosion characterization: theory and applications. Int. J. Heat Mass Transf. **39**, 355–371 (1996)
5. Marinetti, S., Bison, P.G., Grinzato, E.: 3D heat flux effects in the experimental evaluation of corrosion by IR thermography. Proc. European Seminar "Quant. IR Thermography – QIRT'02", Dubrovnik, Croatia. P. 92–97 (2002)
6. Krapez, J.-C., Maldague, X., Cielo, P.: Thermographic NDE: data inversion procedure (part II: 2D analysis and experimental results). Res. in NDE. (2), 101–124 (1991)
7. Bison, P.G., Grinzato, E., Martinetti, S.: Moisture evaluation by dynamic thermography data modeling. Proc. SPIE "Thermosense-XVI". **2245**, 176–182 (1994)
8. Vavilov, V.P., Pawar, S.S.: A novel approach for one-sided thermal nondestructive testing of composites by using infrared thermography. Polym. Test. **44**, 224–233 (2015)
9. Yu, T.O.: Pulsed thermal nondestructive testing of layered materials. SPIE Proc. V. "Thermosense XV". 309–312 (1933)
10. Yu, T.O., Reiss, H.: Remote nondestructive monitoring of coatings and materials by the flash technique. High Temperatures-High Pressures. **32**, 391–395 (2000)
11. Shepard, S., Hou, Y., Lhota, J.: Quantitative characterization of thermographic sequence data. CD-ROM Proc. 16th World Conf. on NDT. Montreal, Canada. 30 Aug.-3 Sept. 392–398 (2004)
12. Vavilov, V.P., Ivanov, A.I.: Pulsed thermal NDT of layered structures. Defectoscopiya (Rus. J. NDT). **6**, 39–47 (1984). (in Russian)

13. Krapez, J.-C., Spagnolo, L., Frieß, M.: Measurement of in-plane diffusivity in non-homogeneous slabs by applying flash thermography. Intern. J. of Thermal Sciences. **43**, 967–977 (2004)
14. Philippi, I., Batsale, J.-C., Maillet, D., Degiovanni, A.: Measurements of thermal diffusivities through processing of infrared images. Rev. Sci. Instrum. **66**(1), 223–229 (1995)
15. Bison, P.G., Grinzato, E., Marinetti, S.: Local thermal diffusivity measurement. J. Quant. Infr. Thermogr. **1**(2), 241–250 (2004)
16. Maillet, D., Andre, S., Batsale, J.-C., et al.: Thermal Quadrupoles: Solving the Heat Equation through Integral Transforms. John Wiley & Sons Publisher, England (2000)
17. Balageas, D.L., Krapez, J.-C., Cielo, P.: Pulsed photo-thermal modeling of layered materials. J. Appl. Phys. **59**(2), 348–357 (1986)
18. Delpesh, P., Krapez, J.C., Balageas, D.L.: Thermal defectometry using the temperature decay rate method. Proc. "Quant. Infr. Thermography QIRT-94", Eurotherm Seminar #42, Sorrento, Italy, August 23–26. 220–225 (1994)
19. Fourier, J.: Théory du mouvement de la chaleur dans les corps solides, 1er partie. Mémoires de l'Académie des Sciences. **4**, 185–555 (1824). 1826;5:53–246
20. Ångstrom, M.A.J.: New method of determining the thermal conductivity of bodies. Phil. Mag. **25**, 130–142 (1863)
21. Vavilov, V.P., Marinetti, S.: Pulsed phase thermography and tomography by using the Fourier transformation. Defectoscopiya (Rus. J. NDT). **2**, 58–72 (1999). (in Russian)
22. Luikov, A.V.: In: Hartnett, J.P. (ed.) Analytical Heat Conduction Theory, p. 685. Academic (1969)
23. Almond, D., Patel, P.: Photothermal Science and Techniques, p. 242. Chapman & Hall, London (1996)
24. Vavilov, V., Demin, V., Shiryaev, V.: Inspecting smokestacks by IR thermographic surveying and heat conduction modeling. Proc. SPIE "Thermosense-XXIII". **4360**, 324–332 (2001)

Chapter 4
Heat Conduction in Structures Containing Defects and the Optimization of TNDT Procedures

Abstract This chapter contains a detailed description of the mathematical methods used for solving TNDT problems. Practical two and three dimensional mathematical models are used to derive the basic mathematical relationships between the physical parameters of defects and the surface temperature. The optimization of pulsed and thermal wave TNDT procedures is discussed along with numerous examples of optimal detection parameters. Also, some advanced TNDT models that allow the analysis of subtle testing phenomena are considered.

4.1 Methods for Solving TNDT Problems

The methods for solving classical heat conduction problems discussed in Chaps. 2 and 3 are well known [1, 2]. 1D multilayer problems can be solved analytically by using the *operation* or *Green's function method*, while multi-dimensional problems should be analyzed numerically. In this Chapter, we will discuss peculiarities of using both analytical and numerical methods for analyzing heat conduction in solids that contain subsurface discontinuities (defects).

4.1.1 Thermophysical Description of Defects

A structural inhomogeneity of any type distorts the uniform flow of heat in materials and thus causes a local temperature anomaly which may be observed on the surface of the material as a differential (delta) temperature signal $\Delta T(x, y, z = 0, \tau)$.

In classical mathematical models, an internal defect is considered to be a material layer in which heat transfer takes place by conduction only. In this case, Eqs. (2.14) or (2.15) is valid:

$$\lambda_{nd} \frac{\partial T_{nd}}{\partial z} = \lambda_d \frac{\partial T_d}{\partial z}; \quad T_{nd} = T_d, \text{or} \tag{4.1}$$

V. Vavilov, D. Burleigh, *Infrared Thermography and Thermal Nondestructive Testing*, https://doi.org/10.1007/978-3-030-48002-8_4

$$\lambda_{nd}\frac{\partial T_{nd}}{\partial z} = \lambda_d \frac{\partial T_d}{\partial z}; \quad T_d - T_{nd} = R_d \frac{\partial T_d}{\partial z}. \tag{4.2}$$

When using Eqs. (4.1 and 4.2), the differential equation for heat conduction should be solved for both the host material and for defects. Therefore, temperature versus time profiles of "defective" areas will depend on both the thermal conductivity and diffusivity (heat capacity) of defects. If the heat capacity of a defect is comparable to that of the sample being tested (in a two-sided test), or of the material layer above the defect (in a one-sided test), it can significantly affect the amplitude and temporal evolution of temperature signals during heating/cooling. As mentioned above, such defects are called thermally-capacitive. In such defects, heat conduction boundary conditions are described by Eq. (4.1). Landmines buried in soil at a depth up to 10 cm and water trapped in honeycombs or porous materials are examples of thermally-capacitive objects. This "capacitive" nature may also be noticeable in the case of relatively large shallow delaminations in composite materials. The heat capacity of a defect may cause the so-called *temperature inversion* of temperature signals when the material above a low-conductivity defect becomes colder ($\Delta T(\tau) < 0$) than the surrounding area, even if this appears to violate classical TNDT theory; see Sect. 4.8.1.

If the heat capacity of a defect can be neglected, thermal resistance $R_d = d/\lambda_d$ is the basic defect characteristic. Defects described by Eq. (4.2) are called thermally-resistive. The differential equation of heat conduction cannot be solved within such defects because their heat capacity is not involved into consideration.

When two materials are in contact but not bonded, there is "*contact thermal resistance*" between them. Defects in fusion welds may consist of contact between two materials without fusion and these have contact resistance. Similar defects, sometimes called "kissing disbonds", may appear in composites. And disbonds between coatings and a substrate have similar characteristics.

TNDT models with thermally-resistive defects are relatively simple, and are often used in the development of defect characterization algorithms.

4.1.2 Laplace Transform and Method of Thermal Quadrupoles

In classical heat conduction theory, most of the solutions for bodies of simple geometry have been obtained by using the integral Laplace transform.

The Laplace transform $F(p)$ of the function $f(\tau)$ is given by:

$$F(p) = \int_0^\infty f(\tau)\, e^{-p\tau}\, d\tau = L\left[f(\tau)\right]. \tag{4.3}$$

Similarly, the inverse Laplace transform is accomplished by the following expression:

$$f(\tau) = L^{-1}\left[F(p)\right] = \frac{1}{2\pi j}\int_{\sigma - j\infty}^{\sigma + j\infty} F(p)e^{p\tau}\,dp, \qquad (4.4)$$

where the integration is performed in the complex plane along a straight line $\sigma = const$ parallel to the imaginary axis. Another form of Eq. (4.4) is:

$$f(\tau) = \lim_{n\to\infty}\left[\frac{(-1)^n}{n!}\left(\frac{n}{\tau}\right)^{n+1} F^{(n)}\left(\frac{n}{\tau}\right)\right]. \qquad (4.5)$$

If the image $F(p)$ is a fractional function of p, which is a typical case in TNDT:

$$F(p) = \frac{\phi(p)}{\eta(p)} = \frac{A_0 + A_1 p + A_2 p^2 + \ldots}{B_1 p + B_2 p^2 + \ldots}, \qquad (4.6)$$

the original function can be obtained by using the expansion theorem:

$$f(\tau) = L^{-1}\left[\frac{\phi'(p)}{\eta(p)}\right] = \sum_{n=1}^{\infty}\frac{\phi'(p_n)}{\eta(p_n)}\,e^{p_n\tau}, \qquad (4.7)$$

where p_n are simple roots of $\eta\ (p)$ with the denominator $\eta\ (p)$ having no free member.

If $F(p)$ is the ratio of two polynomial functions and the repetition factor of the $\eta\ (p)$ roots is k in the points p_m (the degree of the polynomial function $\phi(p)$ is to be lower than that of $\eta\ (p)$):

$$f(\tau) = L^{-1}\left[\frac{\phi(p)}{\eta(p)}\right] = \sum_m \frac{1}{(k-1)!}\lim_{p\to p_m}\left\{\frac{d^{k-1}}{dp^{k-1}}\left[\frac{\phi(p)(p-p_m)^k}{\eta(p)}\,e^{p\tau}\right]\right\}, \qquad (4.8)$$

where summing is done by all roots of $F(p)$.

In simple cases, solutions can be found in Laplace transform tables, while in more complex cases, as in TNDT, numerical methods, such as the Stefest algorithm, are often used. Important relationships between thermal properties of sample layers (including defects) and surface temperature signals can be derived within the Laplace domain. The main advantage of Laplace domain analysis is that respective expressions are algebraic and are easy to manipulate further.

TNDT theory provides an understanding of the mechanism that governs the generation of temperature anomalies in defective areas, and also develops identification algorithms intended for characterizing defects by their surface temperature "footprints". Unfortunately, analytical solutions, particularly the classical ones, are so cumbersome that their inversion is generally impossible. In 1987, Maillet et al. proposed a technique called the "thermal quadrupoles method" [3] for analyzing 1D multilayer heat conduction problems. This method has been successfully used in obtaining relatively simple solutions for direct and inverse TNDT problems in the Laplace domain [3, 4] and thus deserves additional consideration.

Let an extended homogeneous plate of thickness L be heated by with a heat flux Q. A 1D temperature distribution in such plate is described by the function $T(z,\tau)$. As assumed earlier, the sample initial temperature is zero and, therefore all results will be expressed in terms of excess temperature. This function is a solution of the following differential equation:

$$\frac{\partial T}{\partial \tau} = a\frac{\partial^2 T}{\partial z^2}, \tag{4.9}$$

with $T(\tau = 0) = 0$. The heat flux density in the plate is given by the Fourier law:

$$Q(z,\tau) = -\lambda\frac{\partial T(z,\tau)}{\partial z}. \tag{4.10}$$

The Laplace transforms for $T(z,\tau)$ and $q(z,\tau)$ are:

$$\Theta(z,p) = \int_0^\infty T(z,\tau)e^{-p\tau}d\tau; \qquad \Phi(z,p) = \int_0^\infty Q(z,\tau)e^{-p\tau}d\tau. \tag{4.11}$$

The method of thermal quadrupoles is most descriptive when dealing with Laplace transforms of *surface* (F-front and R-rear) temperatures and heat fluxes: Θ^F, Θ^R, Φ^F, Φ^R.

These four functions are interrelated by the simple matrix equation:

$$\begin{bmatrix}\Theta^F\\ \Phi^F\end{bmatrix} = \begin{bmatrix}A & B\\ C & D\end{bmatrix}\begin{bmatrix}\Theta^R\\ \Phi^R\end{bmatrix} = M\begin{bmatrix}\Theta^R\\ \Phi^R\end{bmatrix};$$
$$A = D = \mathrm{ch}(k\ L); \quad B = \frac{1}{\lambda k}\mathrm{sh}(k\ L); \tag{4.12}$$
$$C = \lambda k \mathrm{sh}(k\ L); \quad k = \sqrt{p/a}.$$

Coefficients $A,\ B,\ C,\ D$ depend on the Laplace transform parameter p, plate thickness L, and the thermal properties of the plate material. It is important to note that Eq. (4.12) is valid for any boundary conditions.

The matrix M and the coefficients $A,\ B,\ C,\ D$ characterize the analyzed system 'externally' because functions Θ^F, Θ^R, Φ^F, Φ^R are defined only for system inputs and outputs (see the corresponding graphical presentation in Fig. 4.1).

The solution of Eq. (4.12) is particularly effective in the analysis of multi-layer structures where one has to multiply matrices M_i defined for each layer.

Let us consider a two-layer plate heated by a Dirac pulse. Between the two plates is an additional layer that simulates a resistive $R_d = d/\lambda_d$ defect. The two plates are

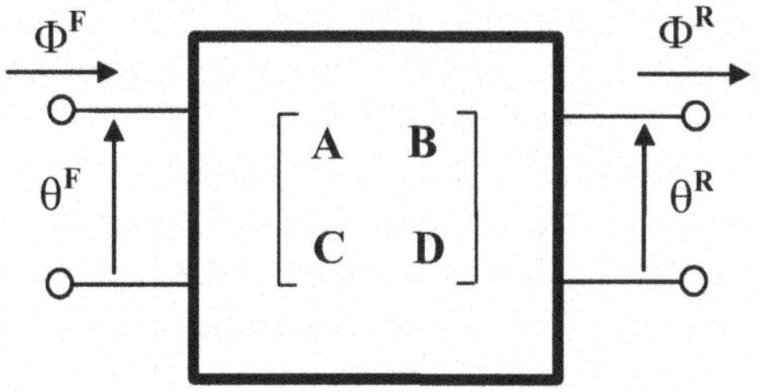

Fig. 4.1 Heating a plate: matrix presentation

made of the same material and have thicknesses l_1 and l_2 and l is the thickness of the resistive layer.

The problem is solved in the matrix form:

$$\left|\frac{\theta^F}{W}\right| = M_1 M_d M_3 \left|\frac{\theta^R}{0}\right|, \tag{4.13}$$

where $\theta^F(z,p)$, $\theta^R(z,p)$ are the Laplace transforms on the front and rear surfaces, M_1, M_2 are the matrices characterizing plate layers (except the defect):

$$M = \left|\frac{A}{E} - \frac{B}{D}\right|, \tag{4.14}$$

where

$$\begin{aligned} &A = D = \mathrm{ch}(kl);\ E = \lambda k \mathrm{sh}(kl); \\ &B = \mathrm{sh}(kl)/(\lambda k);\ k = \sqrt{p/a} \end{aligned} \tag{4.15}$$

Respectively, for the matrix M_d describing the defect layer:

$$A_d = D_d = 1;\ B_d = R_d;\ E_d = 0. \tag{4.16}$$

In the Laplace domain, the front and rear-surface solutions are algebraic, thus allowing the calculation of defect depth l_1 and thermal resistance $R_d = d/\lambda_d$.

Winfree and Zalameda obtained these solutions in the following form [5]:

$$\begin{aligned} &\theta^F(p) = \frac{W}{\lambda k} \frac{\mathrm{sh}(kl_1)\mathrm{sh}(kl_2) + \mathrm{ch}(kl_1)\left[\mathrm{ch}(kl_2) + k\lambda R_d \mathrm{sh}(kl_2)\right]}{\mathrm{ch}(kl_2)\mathrm{sh}(kl_1) + \left[\mathrm{ch}(kl_1) + k\lambda R_d \mathrm{sh}(kl_1)\right]\mathrm{sh}(kl_2)}; \\ &\theta^R(p) = \frac{W}{\lambda k} \frac{\left[\mathrm{ch}^2(kl_1) - \mathrm{sh}^2(kl_1)\right]\left[\mathrm{ch}^2(kl_2) - \mathrm{sh}^2(kl_2)\right]}{\mathrm{ch}(kl_2)\mathrm{sh}(kl_1) + \left[\mathrm{ch}(kl_1) + k\lambda R_d \mathrm{sh}(kl_1)\right]\mathrm{sh}(kl_2)}, \end{aligned} \tag{4.17}$$

where W is the absorbed energy in J/m^3 and $l_1 + l_2 = L$ is the total plate thickness. In the case of a layer of zero resistance ($R_d = 0$), Eq. (4.17) becomes the solution for a homogeneous plate of thickness L. The important feature of the rear-surface solution θ^R is its symmetry about the center of the plate; the solutions at $l_1 = x$ and $l_2 = L - x$ are identical.

Usually, the solutions are normalized by the constant temperature $T_\infty = W/(C\rho\lambda)$ that appears at the end of the thermal process. Then temperature signals that represent the temperature difference ΔT between solutions for the defective and sound plate ($R_d = 0$ for a sound plate) are analyzed.

In the Laplace domain:

$$\Delta\Theta^F = \frac{R^* \mathrm{sh}^2\left[\alpha(1-x*)\right]}{\mathrm{sh}(\alpha)\left\{\mathrm{sh}(\alpha)+R^*\alpha\cdot\mathrm{sh}(\alpha x*)\cdot\mathrm{sh}\left[\alpha(1-x*)\right]\right\}};$$
$$\Delta\Theta^R = -\frac{R^*\mathrm{sh}(\alpha x*)\cdot\mathrm{sh}^2\left[\alpha(1-x*)\right]}{\mathrm{sh}(\alpha)\left\{\mathrm{sh}(\alpha)+R^*\alpha\cdot\mathrm{sh}(\alpha x*)\cdot\mathrm{sh}\left[\alpha(1-x*)\right]\right\}}, \tag{4.18}$$

where $\Delta\Theta$ is the Laplace transform of the normalized temperature $\Delta T/T_\infty$; $R* = R_d/R$; $R = L/\lambda$; $\alpha = \sqrt{p*}$; $p* = L^2 p/a$; $x* = l/L$.

By analyzing Eqs. (4.17) and (4.18), some important TNDT features can be obtained for both one and two-sided tests. In particular, expressions for defect depth l and thickness R_d can be derived (see Sect. 5.5). An important step in applying the thermal quadrupoles technique is using the numerical Stefest algorithm for performing the inverse Laplace transform, i.e. returning from the Laplace domain $F(p)$ into the time-domain of original functions $f(\tau)$:

$$f(\tau) \cong \frac{\ln(2)}{\tau}\sum_{i=1}^{n} v_i\, F\left[i\ln(2)/\tau\right]. \tag{4.19}$$

We reiterate that the solution for the above-described problem was given in the one form in Sect. 3.9.4.

4.1.3 Numerical Methods

As mentioned above, analytical methods provide nice graphs but do not provide accurate solutions, especially when modeling finite defects in objects of complicated shape.

Numerical methods have become ubiquitous due to the exponential growth of computer capabilities. For example, 3D TNDT problems that required hours of processing time on mainframe computers in the 1970's, can now be solved on laptops in minutes or even seconds.

There are two primary methods of numerical analysis: the *finite-element method* and the *finite-difference* method. The finite-element method appeared in the 1950's in the theory of elasticity and stress engineering in the aerospace insdustry. This has become a general method for solving differential equations using "*residuals*". Specifically, one is looking for an optimum solution that satisfies a differential equation (set of equations) together with specific boundary conditions. The method of finite elements is very efficient in the analysis of objects of complicated shape.

Next we will discuss a simple *finite-difference* method that can easily produce accurate solutions for many practical TNDT problems.

Depending on the geometry of a test, the differential Eq. (2.18) of transient heat conduction can be written in Cartesian, cylindrical or spherical coordinates in the following general form:

$$\frac{1}{a}\frac{\partial T}{\partial \tau} = \sum_{i=1}^{N} L_i T, \tag{4.20}$$

where $L_i = \partial^2 / \partial q_i^2$ is the differential operator, and N is the problem metrics.

While formulating a problem numerically, the second partial derivatives with respect to dimensions (x, y, z) and the first partial derivative with respect to time will be replaced with their finite-difference substitutes. The general finite-difference presentation of Eq. (4.20) is:

$$\frac{1}{a}\frac{T^{p+1} - T^p}{\Delta\tau} = \Theta\left(\Lambda T\right)^{p+1} + \left(1-\Theta\right)\Lambda T)^p, \tag{4.21}$$

where Θ is a constant, and $\Lambda = \sum_{i=1}^{N}\Lambda_i$ (Λ_i is the finite-difference analog of L_i). The derivatives are expressed in the following way:

$$\begin{aligned} \frac{\partial^2 T}{\partial q_i^2} &\approx \frac{1}{\left(\Delta h\right)^2}\left(T_{i+1} + T_{i-1} - 2T_i\right); \\ \frac{\partial T}{\partial \tau} &\approx \frac{T_i^{p+1} - T_i^p}{\Delta\tau}. \end{aligned} \tag{4.22}$$

Here "i" specifies nodes of the numerical grid, "p" denotes time steps, Δh is the step by a particular coordinate, and $\Delta\tau$ is the time step. Case $\Theta = 0$ leads to the *explicit* numerical scheme, Θ = 0, 5- (Use. not), corresponds to the *Crank-Nicholson* scheme, and $\Theta - 1$ specifies the *implicit* scheme. The explicit scheme for a 1D heat conduction equation is as follows:

$$T_i^{p+1} = \frac{a\Delta\tau}{\left(\Delta x\right)^2}\left(T_{i+1}^p + T_{i-1}^p\right) + \left[1 - 2\frac{a\Delta\tau}{\left(\Delta x\right)^2}\right]T_i^p. \tag{4.23}$$

Equation (4.23) allows the calculation of temperature T_i in the i-th node at time $\tau + \Delta\tau$, if temperatures of neighboring nodes at a previous time τ are known.

Analogous numerical schemes appear when solving 2D and 3D problems. An illustration of 3D numerical modeling with a uniform spatial/temporal mesh is shown in Fig. 4.2. Obviously, computation time in solving numerical problems is proportional to the number of nodes in a spatial/temporal grid. In most practical cases, the computation time is just a few minutes for a mesh that includes millions of nodes.

An explicit scheme is stable if $\frac{(\Delta x)^2}{a\Delta\tau} \geq 2$ (1D case). This condition limits the choice of a time step if the spatial steps are fixed. When solving multi-dimensional problems, this could lead to lengthy calculations because many time steps are needed. Explicit solutions are pictorial and are easily programmed, but in practice, implicit schemes free of such drawbacks are typically used [1, 2]. It is quite efficient to use the implicit 1D local scheme, which is calculated by applying a method of *alternating directions*, i.e. by splitting each time step into three sub-steps and calculating temperature for each sub-step along a particular coordinate. Such a scheme developed for solving TNDT problems in cylindrical geometry was described in detail in [6]. As mentioned previously, in the practical computation of 3D TNDT problems, several million grid nodes may need to be analyzed.

Both spatial and temporal grids may be non-uniform. Typically, a spatial grid should be dense in the areas that include defects, and a time step should be shorter than the characteristic times when significant temperature variations occur, such as, when switching a heater on or off. But non-uniform grids are more difficult to program and their accuracy has not been evaluated for TNDT analysis.

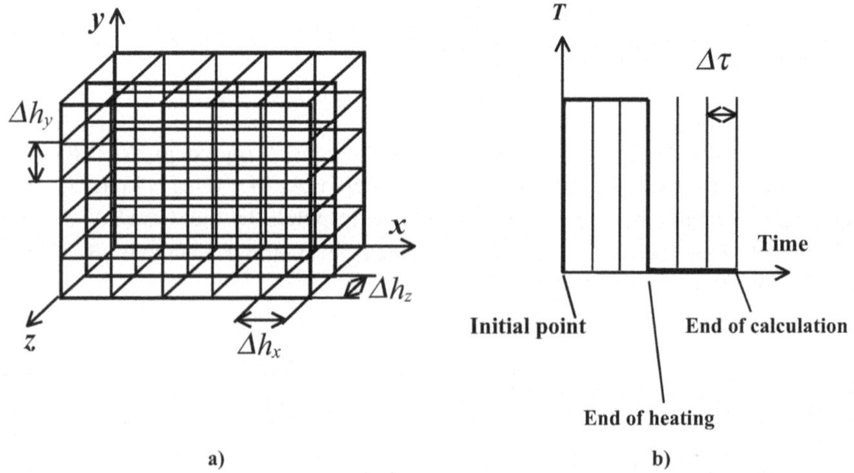

Fig. 4.2 Spatial (**a**) and temporal (**b**) mesh in the numerical modeling of TNDT problems

Table 4.1 Recommended spatial/temporal mesh steps in solving TNDT problems with ThermoCalc-2D software (*L*-material thickness, τ_h-heating duration, *Q*-heating power, defect depth l = 10%*L*; defect thickness d = 10%*L*, material properties: Al – $\lambda = 210$ $W/m \cdot K$, $a = 8.6 \cdot 10^{-5}$ m^2/s; steel – $\lambda = 32$ $W/m \cdot K$, $a = 7.3 \cdot 10^{-6}$ m^2/s; CFRP: $\lambda = 0.64$ $W/m \cdot K$, $a = 5.2 \cdot 10^{-7} m^2/s$; concrete: $\lambda = 1.5$ $W/m \cdot K$, $a = 7.5 \cdot 10^{-7}$ m^2/s; air defect – $\lambda = 0.07$ $W/m \cdot K$, $a = 5.8 \cdot 10^{-5}$ m^2/s; heat exchange coefficients on both surfaces $\alpha = 10$ $W/(m^2 \cdot K)$; accuracy in brackets is given with regard to the corresponding 1D case calculated with Layer-3 Analytic software)

L, mm	Heating conditions	Recommended spatial/temporal mesh steps upper line is $R \times Z$ mesh size, second line is time step, s; and the bottom line, is accuracy in % (in brackets)			
		Al	Steel	Concrete	CFRP
1	$\tau_h = 0.01\ s$, $Q = 10^5\ W/m^2$	100 × 200 0.001 (1.6)	50 × 300 0.01 (3)	50 × 200 0.01 (2.5)	50 × 200 0.005 (2.6)
	$\tau_h = 100\ s$, $Q = 10^4\ W/m^2$	50 × 100 0.01 (0.4)	50 × 200 0.01 (0.5)	50 × 200 0.05 (3.1)	50 × 200 0.05 (3.6)
5	$\tau_h = 0.01\ s$, $Q = 10^5\ W/m^2$	200 × 400 0.01 (0.3)	50 × 400 0.01 (2.5)	200 × 400 0.01 (5)	200 × 400 0.01 (3)
	$\tau_h = 100\ s$, $Q = 10^4\ W/m^2$	100 × 100 1 (1.6)	100 × 200 1 (3)	100 × 200 1 (3)	100 × 100 1 (0.8)
50	$\tau_h = 0.01\ s$, $Q = 10^5\ W/m^2$	100 × 800 0.01 (5)	50 × 2000* 0.01 (10)	100 × 2000* 0.01 (10)	100 × 2000* 0.01 (9)
	$\tau_h = 100\ s$, $Q = 10^4\ W/m^2$	50 × 800 1 (1.5)	50 × 300 5 (2)	50 × 300 5 (2.5)	50 × 800 5 (1.5)

*These cases require a very small spatial depth step

4.1.4 Accuracy of Numerical Solutions

The accuracy of numerical calculations is evaluated with regard to the corresponding analytical solutions. This issue is well explored in classical heat conduction theory if one deals with non-defect solutions. A typical recommendation is that spatial/temporal grid steps should be as small as possible to ensure the best accuracy. However, evaluating accuracy in defect areas seems to be complicated due to a lack of corresponding analytical solutions. The authors have analyzed the accuracy of 2D and 3D numerical algorithms and solutions provided by the ThermoCalc-2D and ThermoCalc-3D software (Tomsk Polytechnic University, Russia). This software is commercially available, it is designed to perform the analysis described above, and it is optimized for TNDT analysis. However, the reader may perform these analyses by other methods and software, see Sect. 4.1.5,

ThermoCalc-2D allows the simulation of circular defects in three-layer disk samples (see Sect. 4.3). If the defect diameter tends to infinity, the 2D cylindrical model becomes a 1D three-layer model, as discussed in Chap. 3. Table 4.1 contains recommendations on choosing steps of a uniform spatial/temporal numerical mesh when solving some typical TNDT problems.

Table 4.1 shows that errors in calculating temperatures in defect areas (values in brackets) are typically from 0.5% to 5%. An analytical solution for the corresponding 1D three-layer problem realized in the Layer-3 Analytic software (Tomsk Polytechnic University, Russia) was used as a reference. The calculation of non-defect temperatures usually have smaller errors (<1%). Note that the maximum errors occur in cases where a thermally-thick body is stimulated with a short heat pulse. Fortunately, such test cases are not practical because thick or low thermal conductivity samples are typically heated with longer and lower power heat pulses.

4.1.5 Commercial Software for Numerical Solutions of TNDT Problems

4.1.5.1 General Purpose Software

Commercial mathematical software packages designed for providing numerical solutions for heat transfer problems are available. Some of these are capable of modeling objects of complicated shapes, as well as for visualizing results. Many of these can be used to simulate and evaluate TNDT processes.

Models may be constructed in general purpose Computer-Aided Design (CAD) and Computer-Aided Modeling (CAM) programs that are widely used in aerospace and other industries. Some of the tools in these programs may be applied to solving 2D and 3D problems. Existing models (geometry, time, and other components) can be transferred among several programs. This "model portability" may lead to a reduction in the time necessary for analyzing data. And it allows analysts to use software tools that they already possess.

In commercial computer programs, the designing an appropriate numerical grid is particularly important when analyzing objects having a complicated shape. The numerical mesh density can be increased in space or time depending on where and when temperature anomalies on defect areas are expected. For example, while analyzing shallow defects in thick samples, the spatial grid should be denser over defects and the time steps should be shorter at the optimum observation time and longer at the end of the process. This allows a reduction in the computation time without reducing the accuracy of the calculation. For modeling special shapes and conditions, such as a rough surface, specially shaped grid elements may be used. Material anisotropy and the variation of certain parameters as a function of time can be also simulated by available models. Calculation schemes can be explicit or implicit by manually or automatically changing grid step sizes in time. In the case of a complex test object, finite-element models are typically used, while simpler objects are often analyzed by means of finite difference models which are efficient when using very small spatial/temporal grid steps.

An example of modeling corrosion in a boiler is shown in Fig. 4.3 (Comsol Multiphysics software, finite-element method).

Some specialized programs, such as those from ThermoAnalytics (USA), can simulate heat transfer characteristics and can also predict IR signatures in various spectral wavelength bands. Such programs are used in aerospace, military, architecture, and other applications.

A short review of some commercial software packages which can be used in TNDT analysis is presented in Table 4.2. The features of particular programs can be found on Web sites of the corresponding manufacturers.

Commercial software packages require powerful computers. Some difficulties may arise when modeling 3D non-linear and pulsed problems. At this time, the best results are typically obtained in the UNIX medium, but the use of (Windows-based) PCs is also possible. The license cost for using these programs is reasonable for universities but can be very high for industries.

4.1.5.2 Specialized Programs

Specialized TNDT software has been developed by some research teams, such as NASA (USA), ONERA (France), Wayne State University (USA), ITC-CNR (Italy), IZfP, IKP and BAM (Germany), University Laval (Canada) and others. In Table 4.3, some programs commercialized by Tomsk Polytechnic University (Russia) are shortly described.

4.2 1D TNDT Models

4.2.1 Temperature Signals in 1D Models

Within 1D models, it is assumed that heat conduction occurs independently and differently in defective and sound areas, i.e. lateral (3D) heat diffusion is neglected and, at any observation time, a spatial profile ΔT looks as shown in Fig. 4.4.

In 1D models, temperature signals depend only on time and the depth coordinate z (Fig. 4.4):

$$\Delta T(z,\tau) = T_d(z,\tau) - T_{nd}(z,\tau). \tag{4.24}$$

A differential temperature signal ΔT is defined as a difference between two solutions obtained in defective and sound areas independently. Such an approach allows the analysis of relationships between ΔT and the thermal properties of a host material, as well as the defect depth, thickness and thermal properties. The limits of 1D models, where ΔT is not affected by lateral heat diffusion, will be considered below. In practice, the 3D heat diffusion that takes place at the boundaries of defects

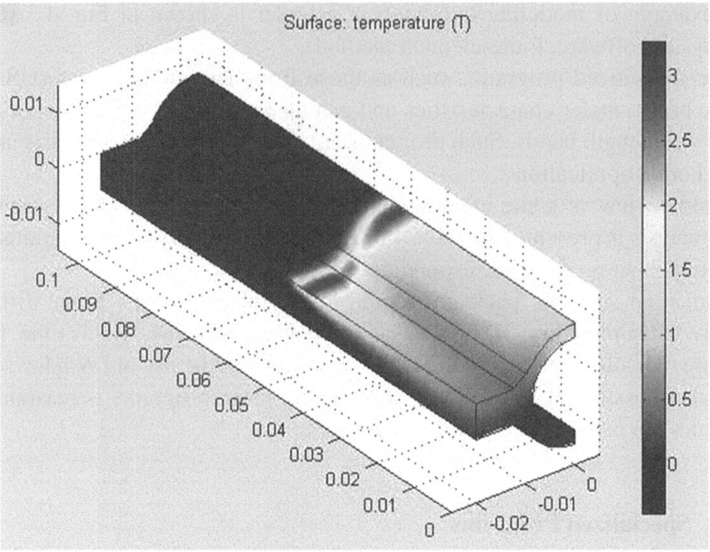

Fig. 4.3 Modeling corrosion detection in a boiler by applying line heating (courtesy S. Marinetti)

smoothes spatial temperature profiles and may significantly diminish temperature signals ΔT (see dashed line in Fig. 4.4).

4.2.2 1D TNDT Model with a Thermally-Insulating Defect

4.2.2.1 Heating a Non-adiabatic Plate (Classical Solution)

The 1D defect model proposed for one-sided TNDT is shown in Fig. 4.5, being a further simplification of the 1D model from Fig. 4.4. Since the defect is a thermal insulator, the boundary condition is $\partial T/\partial z = 0$. The defect area is considered to be a material layer of thickness l, with l also considered to be the defect depth. Obviously, the lower the thermal conductivity of a defect is in comparison to that of the host material, the more valid this model is. Below, we shall see that such a model is more suitable for determining plate thickness (see the corrosion detection model in Sect. 3.7.3) than for detecting hidden voids. Moreover, the model in Fig. 4.5, unlike the more general model in Fig. 4.4, cannot be used for simulating two-sided TNDT. The model in Fig. 4.5 will produce exaggerated ΔT values under the defect, which are to be assumed equal to the non-defect temperature T_{nd} (the excess temperature under a defect is zero at any time due to a lack of heat conduction through the defect).

Below we will analyze the one-sided TNDT process.

Consider heating an adiabatic plate as shown in Fig. 4.5 with a square heat pulse of duration τ_h. In the sound area, the mathematical formula is trivial:

Table 4.2 General-purpose commercial software packages for thermophysical (and TNDT) calculations

Program/Thermophysical module	Web site	Characteristic*
Abaqus/Standard	www.simuleon.com/simulia-abaqus	
Adina-Adina-T	www.adina.com	
Algor/Multiphysics	www.algor.com/products	
Ansys/Multiphysics	www.ansys.com	Finite element method Graphical data processor
CosmosM/HSTAR	www.brs.kyushu-u.ac.jp	Controlled numerical grid 2D & 3D geometry
Comsol Multiphysics (Femlab)	www.comsol.ru/comsol-multiphysics	Transient heat transfer Thermal properties depend on temperature
Marc	www.mscsoftware.com/product/marc	Anisotropic materials Heat exchange by convection and radiation
Matlab/PDE	www.matlab.com	Boundary conditions depend on time Layers separated by a thermally-resistive gap
MSC-Nastran	www.mscsoftware.com/produc/msc-nastran	Signatures of IR radiation in typical wavelength bands Calculation of R (or U)-values for building construction parts Heating/cooling performance Numerical analysis of thermal bridges in building Operating in MS Windows & UNIX
LMS Samsef Solver Suite	www.plm.automation.simens.com	
Mathematica/Industrial Thermics applications	www.unisoftwareplus.com	
CYPECAD MEP	www.cype.com	
HEAT2-Buildingphysics	www.buildingphysics.com	
NumeriCAL	www.jbtfoodtech.com	
MuSES, TAIThermIR	www.thermoanalytics.com	

*Some features apply to particular programs only

Table 4.3 Specialized TNDT software from Tomsk Polytechnic University (Russia)

Program	Characteristic
Layer-3 Analytic	*Analytical* solution for heating a 1D three-layer non-adiabatic plate with a square pulse
ThermoCalc-2D	*Numerical* solution for heating a 2D three-layer disk body with a concentric defect (cylindrical geometry) Heating with a square or periodic pulse Thermal waves Adaptive convection & radiation
ThermoCalc-3D	*Numerical* solution for heating a 3D six-layer parallelepiped-like body up to 40 parallelepiped-like defects (Cartesian geometry) Heating with a square or periodic pulse Arbitrary heating function Thermal properties can be functions of time and coordinates Spatial heating mask (synthetic or experimental) Producing IR image sequences
ThermoCalc-30L	Similar to ThermoCalc-3D, but up to 30 layers with an arbitrary orientation of the thermal property tensor
ThermoSource	Similar to ThermoCalc-3D, but with internal heat sources

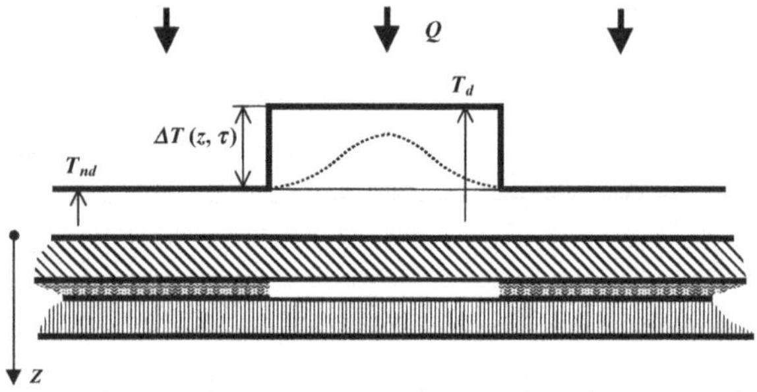

Fig. 4.4 1D TNDT model (dashed line: 3D heat diffusion case)

$$\frac{\partial T}{\partial \tau} = a\frac{\partial^2 T}{\partial z^2}; \tag{4.25}$$

$$T\left(z, \tau = 0\right) = 0; \tag{4.26}$$

$$-\lambda\frac{\partial T\left(z = 0\right)}{\partial z} = Q - \alpha_F T\left(z = 0\right); \tag{4.27}$$

$$-\lambda \frac{\partial T(z=L)}{\partial z} = \alpha_R T(z=L), \tag{4.28}$$

where (4.25) is the 1D differential equation of heat conduction, (4.26) is the initial condition, and Eqs. (4.27) and (4.28) represent the boundary conditions on plate surface. According to this model, before heating the plate, all temperatures are equal to the ambient temperature. Therefore, after heating, we will want to mode the "excess temperatures", which are the temperature differences ($\Delta T T\Delta T$'s) with respect to the original, ambient temperatures.

Since in the defective area $\partial T(z = l)/\partial z = 0$, to obtain a comparable solution, one should assume the same adiabatic boundary condition on the rear surface, i.e. $\alpha^R = 0$, then the heating model becomes identical in both the defective and sound areas, with either l (defect depth) or L (sample thickness) being the model parameter.

Note that the model expressed by Eqs. (4.25, 4.26, 4.27, and 4.28) also describes the heating (cooling) of a plate in a medium with temperature T_a. Indeed, if heating is accomplished by a medium, Eq. (4.27) can be written as:

$$-\lambda \frac{\partial T(z=0)}{\partial z} = \alpha^F \left[T_a - T(z=0)\right], \tag{4.29}$$

which is identical to (4.27) if we assume $Q = \alpha^F T_a$. In the case of cooling, one has to change the sign in the right side of Eq. (4.29).

The classical heat conduction theory deals with a plate of thickness $2L$ heated on *both* sides with a medium of temperature T_a, with the first temperature derivative being zero in the centre of the plate due to temperature symmetry [2]. It is clear that such model is equivalent to that described by Eqs. (4.25, 4.26, 4.27, and 4.28). Then, the front-surface solution is (see Table 2.4):

$$\frac{T^F}{Q'} = 1 - \sum_{n=1}^{\infty} \frac{2Bi}{Bi(Bi+1) + \mu_n^2} e^{-\mu_n^2 Fo}, \tag{4.30}$$

where $Q' = Q/\alpha$ and μ_n are positive roots of the transcendent equation $\mu_n tg(\mu_n) = Bi$.

Some values of μ_n are given in Table 4.4.

If $\tau \to \infty$, $T^F \to Q/\alpha$. While heating a plate in a fluid medium of temperature $T_a = Q/\alpha$, the surface temperature tends to T_a (if the rear surface of the plate is adia-

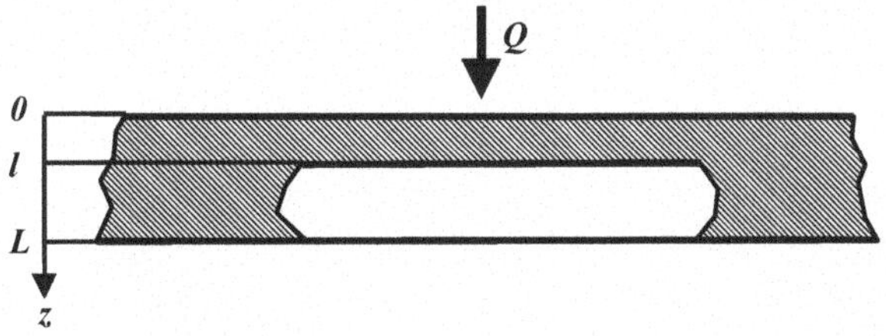

Fig. 4.5 One-sided 1D TNDT model (defect as absolute insulator)

batic). At shorter times, the number of significant roots and respective series members increases. For samples that meet the condition $Bi \leq 0.1$, only the first series member $\mu_1 = \sqrt{Bi}$ can be taken into account [2].

Then, the 'non-defect' solution will be:

$$T_{nd}^{F}(\tau) = \frac{Q}{\alpha}\left(1 - \cos\sqrt{Bi}\, e^{-BiFo}\right) = \frac{Q}{\alpha}\left(1 - \cos\sqrt{Bi}\, e^{-\frac{\alpha a}{\lambda L}\tau}\right). \tag{4.31}$$

In spite of the fact that modern computers allow the calculation of any number of roots of the transcendent equation above, the use of the solution (4.30) is cumbersome, particularly, if its derivatives are to be found. Marinetti and Vavilov proposed the following analytical expressions for determining μ_n [7]:

$$\mu_1 = a_1 Bi^2 + b_1 \sqrt[3]{Bi} + c_1 Bi + d_1 \sqrt{Bi} + e_1;$$

$$\mu_n = a_n Bi^2 + \frac{1}{\pi(n-1)} Bi + \pi(n-1); \quad n = 2,3,\ldots; \tag{4.32}$$

$$a_n = \frac{1}{\alpha_a n^3 + \beta_a n^2 + \gamma_a n + \delta_a};$$

$$a_1 = 5.77 \cdot 10^{-2}; \quad \alpha_a = -13.5;$$
$$b_1 = -2.12 \cdot 10^{-1}; \quad \beta_a = -345.0;$$
$$c_1 = 1.68 \cdot 10^{-2}; \quad \gamma_a = 6184.4;$$
$$d_1 = 9.97 \cdot 10^{-1}; \quad \delta_a = -22718.0;$$
$$e_1 = 1.70 \cdot 10^{-4}.$$

When using Eq. (4.30) with approximate roots calculated by Eq. (4.32), errors are less than 2% for and $Bi = 0.05 - 1$, but become significant for $Bi > 2$.

Table 4.4 Positive roots of the transcendent equation $\mu_n tg(\mu_n) = Bi$ [2]

Bi	μ_1	μ_2	μ_3
0	0.0000	3.1416	6.2832
0.001	0.0316	3.1419	6.2833
0.01	0.0998	3.1448	6.2848
0.1	0.3111	3.1731	6.2991
0.2	0.4328	3.2039	6.3148
0.3	0.5218	3.2341	6.3305
0.4	0.5932	3.2636	6.3461
0.5	0.6533	3.2923	6.3616

4.2.2.2 Evaluating Temperature Signals in Pulsed TNDT of an Adiabatic Plate

At short times (less than 0.01–0.1 second for thin samples), calculations made by using Eq. (4.31) with only one member of the series do not produce accurate results, while using many series members may be tedious. For $Bi \le 0.1$, which is a common case in TNDT, it is possible to use the solution for heating an adiabatic plate with a square pulse (Table 2.4):

$$T_{nd}^{F} = \frac{QL}{\lambda}\left\{Fo_h + \frac{2}{\pi^2}\sum_{n=1}^{\infty}\frac{1}{n^2}e^{-n^2\pi^2 Fo}\left(e^{n^2\pi^2 Fo_h} - 1\right)\right\}, \tag{4.33}$$

which becomes the Dirac-pulse solution for $\tau \gg \tau_h$ (Table 2.4):

$$T_{nd}^{F} = \frac{Wa}{\lambda L}\left(1 + 2\sum_{n=1}^{\infty}e^{-n^2\pi^2 Fo}\right). \tag{4.34}$$

Eq. (4.34) will be analyzed below to illustrate the limits of the simplest 1D model in Fig. 4.5.

In a defective area, the sample thickness L will be replaced by the defect depth l:

$$T_{d}^{F} = \frac{Wa}{\lambda l}\left(1 + 2\sum_{n=1}^{\infty}e^{-n^2\pi^2 Fo\, L^2/l^2}\right). \tag{4.35}$$

The energy of a Dirac heat pulse is related to the corresponding square pulse by the following equation: $W = Q\tau_h$. Some features of the solutions (4.34 and 4.35) for evaluating sample thickness were considered in Sect. 3.7.3. In Fig. 4.6, evolutions of the normalized temperature signal $\Delta T/(Wa/\lambda L)$ and running contrast C^{run} are presented for the defect located at depth l. At $\tau \to \infty$, both parameters tend to $L/l - 1$, which means they are independent of the thermal properties of the material.

To demonstrate how such a 1D model represents practical TNDT cases, Fig. 4.7 shows the evolution of the running contrast C^{run} for a circular air-filled defect of radius $r_d = 5$ mm and thickness $d = 0.2$ mm located in several materials. The sample thickness is 5 mm and the heating is provided by a pulse of duration $\tau_h = 0.01$ s. Contrast values were calculated with the ThermoCalc-2D program. According to the 1D model in Fig. 4.7, the maximum running contrast should be: $C_m^{run} = L/l - 1 = 4$. Note that a close approximation is provided by an air-filled defect in steel. This may be due to the differences between the thermal properties of the defect and the host material. The accuracy of the simplest 1D model improves with greater defect thickness and/or greater thermal conductivity of the host material. It is worth noting that this model predicts very long observation times which are close to the equilibrium stage of the thermal process.

4.2.3 Periodic Thermal Waves

4.2.3.1 Interference in Adiabatic Thin Layers and Defects

In Sect. 3.9.4, we analyzed the reflection of thermal waves from a boundary between two media. Following the work of Almond and Patel [8], consider the behavior of thermal waves in thin air gaps (defects) where interference between incident and reflected waves may significantly modify the $\Delta T(\tau)$ amplitude compared to that of a semi-infinite body.

Examine the important practical case of a coating of thickness L_c on a substrate (Fig. 4.8). The substrate is regarded as a semi-infinite body since its thickness is much greater than the thermal diffusion length. Let the total heat energy be fully absorbed by the outer surface, although, in a more general case, one can take into account the in-depth absorption of heat energy in a semi-transparent coating.

The surface temperature is the sum of incident and reflected temperature waves which propagate in the coating and experience reflection at the coating-air and coating-substrate interfaces (Fig. 4.8a).

Bennett and Patty proposed the following expression for $T(z = 0)$ [9]:

$$T(z=0) = A\theta_{c-a}\left[\frac{1+R_{c-s}\exp(-2\sigma_c L_c)}{1-R_{c-s}R_{c-a}\exp(-2\sigma_c L_c)}\right], \tag{4.36}$$

where A is the thermal wave amplitude, θ_{c-a} is the transmittance of the coating-air interface, R_{c-s} is the reflectance of the coating-substrate interface, and the subscripts "c", "s" and "a" specify the coating, substrate and air respectively. At the coating-air interface, $R_{c-a} \approx 1$ and $\theta_{c-a} \approx 2$ because b~0 due to the low effusivity of air. The normalized module and phase of the surface temperature wave $T_{norm}(z = 0)$ are shown in Fig. 4.9, having been calculated by the following expression:

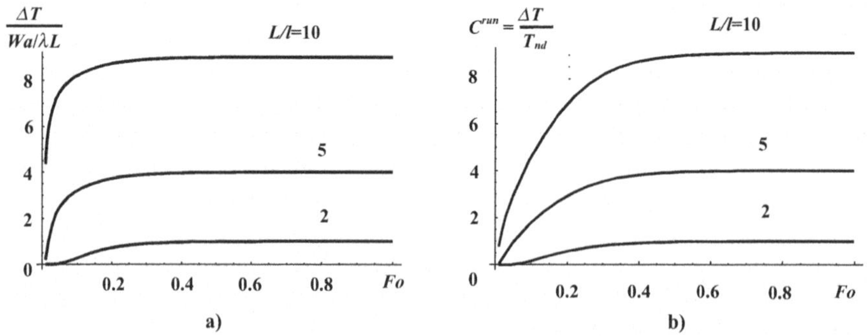

Fig. 4.6 Evolution of temperature signal (**a**) and running contrast (**b**) in adiabatic model by Eqs. (4.34) and (4.35)

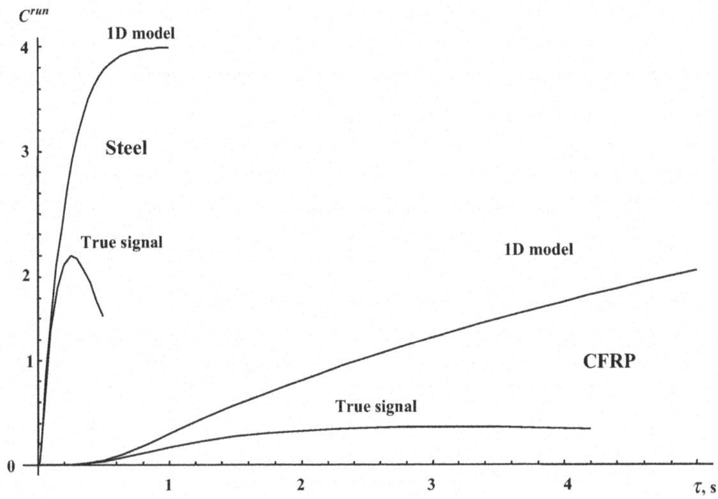

Fig. 4.7 Validating TNDT adiabatic 1D model for 0.2 mm-thick air-filled defect by 5 mm radius at depth of 1 mm in AISI 1010 steel and CFRP (heating with 0.01 s pulse, see material thermal properties in Table 2.2)

Fig. 4.8 Thermal wave interference in coating (**a**) and defect (**b**)

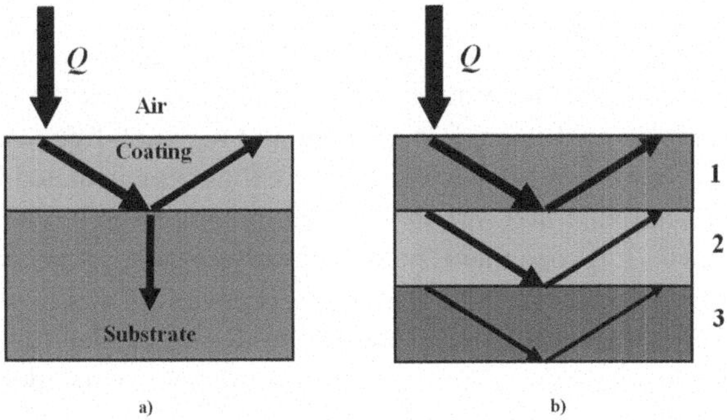

$$T_{norm}(z=0)=\frac{1+R_{c-s}\exp(-2\sigma_c L_c)}{1-R_{c-s}R_{c-a}\exp(-2\sigma_c L_c)};$$

$$\text{Module}=\sqrt{\left[\text{Re}\{T_{norm}\}\right]^2+\left[Im\{T_{norm}\}\right]^2};\tag{4.37}$$

$$\text{Phase}=\text{Arctg}\frac{Im\{T_{norm}\}}{\text{Re}\{T_{norm}\}},,$$

where Re and Im respectively specify the real and imaginary parts of a complex number.

The graphs in Fig. 4.9 clearly show that the interference of thermal waves is substantial if the coating thickness is less than one thermal diffusion length. A point of interest on the curves is the intersection with the zero axis; for amplitudes, it occurs at $L_c/\mu = 0.785$ and for phases at $L_c/\mu = 1.571$. This fact, which will be considered in details below, may indicate that the thermal wave phase penetrates deeper into solids than the amplitude.

The practical meaning of Eq. (4.37) is that it is possible to determine either the coating thickness or its thermal properties by recording the surface temperature of the coating. Note that, due to the complex nature of σ_c, the temperature function is also complex. Therefore, the temperature wave and the stimulating heat wave differ in phase at all points.

Interference also occurs in defects such as cracks, delaminations etc. Therefore, the reflection of thermal waves from defects is frequency-dependent. A thin inclusion (defect) can be characterized by the effective coefficient of reflection (Fig. 4.8b) [8]:

$$\Gamma = \frac{\Gamma_1 + \Gamma_2 \exp(-2\sigma_2 L_2)}{1 + \Gamma_1 \Gamma_2 \exp(-2\sigma_2 L_2)}, \tag{4.38}$$

where $\Gamma_1 = \frac{1-b_{21}}{1+b_{21}}$; $\Gamma_2 = \frac{1-b_{32}}{1+b_{32}}$; $b_{i+1/i} = \frac{\lambda_{i+1}\sigma_{i+1}}{\lambda_i \sigma_i}$. Here indices 1, 2 and 3 are related to the sample layers (Fig. 4.8b), Γ_1 and Γ_2 are the coefficients of reflection at the interfaces between layers 1–2 and 2–3, L_2 is the thickness the 2nd layer. The complex coefficient Γ is a function of material, frequency and defect thickness.

Figure 4.10 shows reflection coefficients calculated by Eq. (4.38) for air defects in aluminum and carbon fiber reinforced plastic (CFRP) at two frequencies: 1 and 10 Hz, It is seen that Γ increases with frequency, material thermal conductivity and defect thickness. Note that the respective function experiences saturation. In other words, defect thickness, starting from a particular value, does not affect the wave reflectance. The same conclusion applies to the dependence of ΔT on defect thickness.

Equation (4.38) is valid for thermally-capacitive defects produced by layers with well-defined boundaries. For thermally-resistive defects, the reflectance is given by [8]:

$$\Gamma = \frac{1 - b_{21} + R_d \lambda_2 \sigma_2}{1 + b_{21} + R_d \lambda_2 \sigma_2}, \tag{4.39}$$

where, unlike Eq. (4.38), indices 1 and 2 are both related to the layers of the host material separated by a layer of thermal resistance R_d.

A comparison of Eqs. (4.38) and (4.39) for defects in aluminum and CFRP reveal identical reflection coefficients (see Fig. 4.10); thus, in many solids, thin air filled defects can be modeled by the contact thermal resistance R_d, thus simplifying the corresponding analytical solutions.

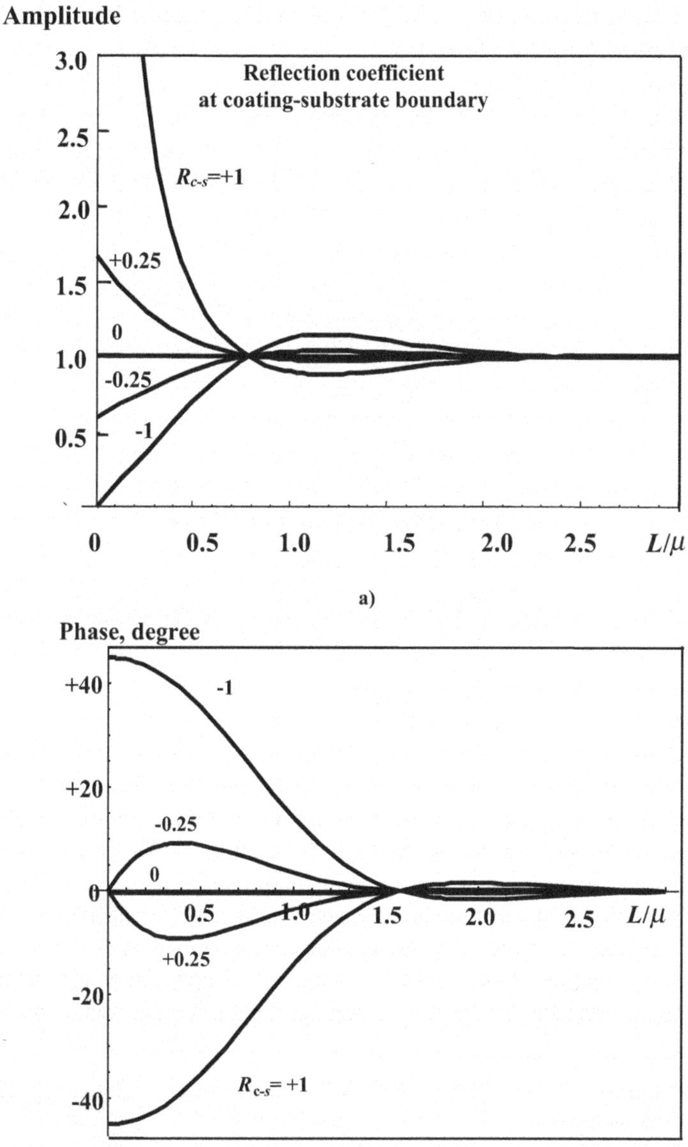

Fig. 4.9 Thermal wave surface amplitude (**a**) and phase (**b**) for different reflection coefficients at coating-substrate boundary (substrate is modeled as semi-infinite body, $R_{c-a}=1$; $T_{c-a}=2$; $L=L_c$)

4.2.3.2 A Simple Evaluation of TNDT Defect Signals for a Non-adiabatic Plate of Varying Thickness

The solution for the heating of a non-adiabatic plate with a periodic heat flux is given in Table 2.5.

By taking $\sqrt{\pm j} = \pm\sqrt{0.5}\,(1 \pm j)$, the front-surface solution can be written as:

$$
\begin{aligned}
&A = \left(\frac{B}{B + \frac{2}{Bi}\sqrt{\frac{Pd}{2}}S + \frac{Pd}{Bi^2}V} \right)^{1/2}; \\
&\Psi = \frac{W}{S + B/\left(\frac{1}{Bi}\sqrt{\frac{Pd}{2}} \right)}; \\
&B = ch^2(u)\cos^2(u) + sh^2(u)\sin^2(u); \\
&S = ch(u)sh(u) - \cos(u)\sin(u); \\
&V = sh^2(u)\cos^2(u) + ch^2(u)\sin^2(u); \\
&W = ch(u)sh(u) + \cos(u)\sin(u); \\
&u = \sqrt{\frac{Pd}{2}}.
\end{aligned}
\tag{4.40}
$$

The Predvoditelev criterion $Pd = \omega L^2/a$ is the main parameter in the solution above. At $L \to \infty$, Eq. (4.40) become those for a semi-infinite body. Note that, for simplicity, the "Periodic" solutions in Table 2.5 contain neither a constant heating member, nor an expression that shows initial conditions.

The plots in Fig. 4.11 illustrate how thermal waves of different frequencies interact with a CFRP plate as the thickness changes. In agreement with the conclusions above, it is seen that low-frequency waves penetrate deeper into a test material and result in a higher surface temperature. In this case, there is no reason to introduce an

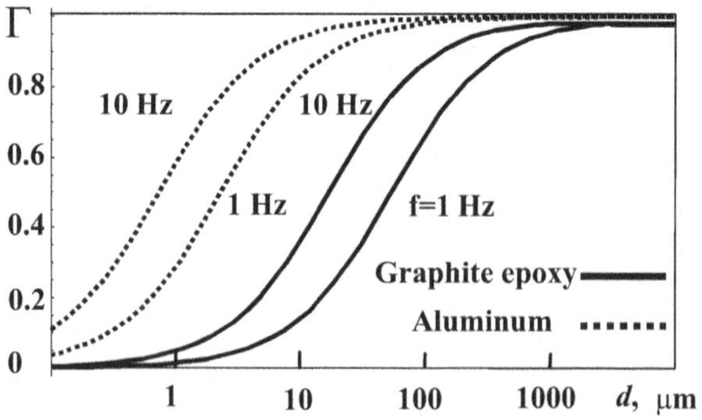

Fig. 4.10 Reflection coefficient Γ for thermal waves of different frequencies (thin air-filled defects in aluminum and CFRP, see material thermal properties in Table 2.2)

optimum stimulating frequency, at least, in a sound area. In fact, a maximum surface temperature will be achieved in the case of a "zero frequency" wave, i.e. by applying continuous heating. In Fig. 4.11, the two graphs reveal large differences in temperature and phase at particular sample thicknesseses (in the previous section, this was treated as an interference of waves in a material layer). Some specific points in the curves of Fig. 4.11 corresponding to the minimums of $A(L)$ and $\Psi(L)$ are of interest. We assume that these points are a result of thermal wave interference.

By comparing the amplitude and phase curves in Fig. 4.9, it can be proposed that, at the same frequency, the thermal wave phase penetrates approximately *twice as* deep as the amplitude. Consider this phenomenon in the example of CFRP composite using the data in Fig. 4.11. A maximum penetration depth for the phase ($L_{\max}^{Phase}$) and amplitude ($L_{\max}^{Ampl}$) parameters can be found at the minimums of the functions $A(L)$ and $\Psi(L)$ which, to some extent, represent the transition to the semi-infinite body case. Then, by analyzing graphs in Fig. 4.11, one can obtain $L_{\max}^{Phase} / L_{\max}^{Ampl} \oplus 1.5$ for $f = 0.2$ Hz and $L_{\max}^{Phase} / L_{\max}^{Ampl} \oplus 1.7$ for $f = 1$ Hz. The analysis of penetration depths of thermal wave phase and amplitude will be continued in Sect. 6.5 by using the Fourier transform.

As shown earlier, by assuming a 100%-reflective (thermally-insulating) defect (Sect. 4.2.2, model by Fig. 4.5), the solution (4.40) can be used for analyzing the heating in both the sound and defective areas; in the latter case, $L = l$.

Consider modeling the active detection of landmines by solar heating. Solar heating is periodic with a maximum generally occurring at midday. The landmine is modeled as a disk of 10 cm thickness and 20 cm diameter buried in soil at a depth of 5 cm. In Table 4.5, the amplitude and phase of surface temperature waves are presented for both the sound and 'defective' areas and Eq. (4.40) are used for the calculations. Also, the data for the corresponding 3D geometry is supplied for comparison. It is seen that the variations in both phase and amplitude due to the presence of the defect are significant in the 1D model, while in the real 3D case the signals have lower amplitude. Landmine detection is discussed further in Sect. 4.10.2.

In Fig. 4.12, three periods of the temperature signal $\Delta T(\tau)$ are shown in both models. Note that the influence of the initial conditions vanishes within the first period. When the material in the defect is changed from air to trinitrotoluene (TNT), the maximum signal is reduced from $\Delta T_m = 4.5$ °C to $\Delta T_m = 3.8$ °C. In soils having low thermal conductivity (dry versus wet, etc.), the temperature signals will be even lower. The necessity to change the probing frequency for the dection of defects at different depths is an unpleasant feature of the thermal wave technique. Mulaveesala and Tuli proposed a technique of frequency-modulated thermal waves thus improving inspection productivity [10].

4.2.4 *Pulsed Thermal Waves*

4.2.4.1 Phenomena at Material-Defect Boundaries

In Sect. 4.2.3, we discussed the propagation of periodic thermal waves through a coating-substrate boundary. In the same way, one can treat pulsed thermal waves penetrating a non-homogeneous media. As indicated in Sect. 3.7.6, the reflection

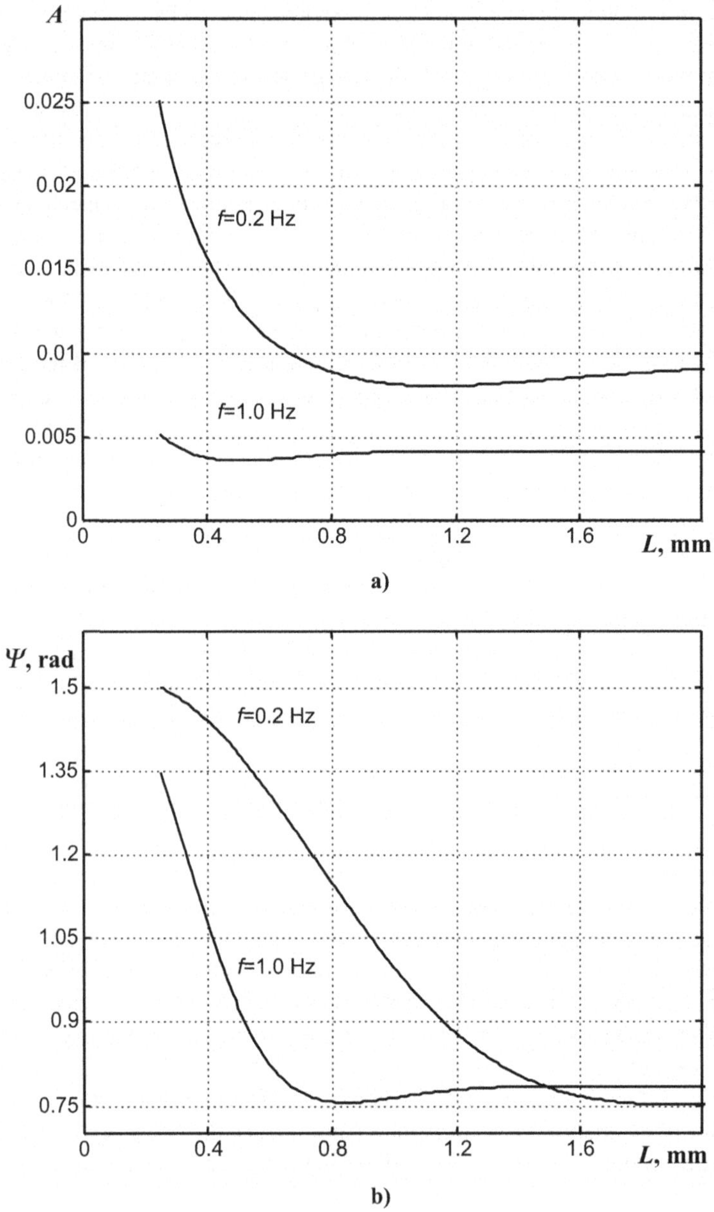

Fig. 4.11 Thermal wave amplitude (**a**) and phase (**b**) *vs.* CFRP thickness

Table 4.5 Comparing 1D and 3D thermal wave models in the detection of disk objects in soil (simulating landmine detection, object diameter 20 cm, thickness 10 cm, depth 5 cm, see soil thermal properties in Table 2.2, heating function $Q(\tau) = Q_m \cos\left(\frac{2\pi}{T}\tau\right)$; T = 24 *hrs*; $Q_m = 500\ W/m^2$; $\alpha^F = 10\ W/(m^2 \cdot K)$)

TNDT model	Surface thermal wave parameters	
	Amplitude	Phase, radians
1D model		
Defect area	0.256	0.603
Non-defect area	0.449	0.988
3D model		
Defect area	0.254	0.590∗
Non-defect area (defect – air)	0.349	0.654∗
Non-defect area (defect – TNT)	0.322	0.654∗

∗With one time step accuracy (0.065 rad, or 900 s)

coefficient at the boundary between two materials 1 and 2 (material-defect and defect-material) is the real number:

$$\Gamma = \frac{e_2 - e_1}{e_2 + e_1}. \tag{4.41}$$

Analytical solutions for multi-layer thermally-capacitive structures are complicated. Therefore, thermally-resistive layers characterized by the thermal resistance R_d are typically analyzed in defective areas. Along with purely analytical solutions, the method of thermal quadrupoles has been used to analyze such phenomena using an arbitrary heating profile, surface heat exchange, in-depth penetration of stimulating heat energy, etc.

4.2.4.2 Adiabatic Coating on Substrate

Consider a coating of thickness L on a substrate that can be considered as a semi-infinite body. Under adiabatic conditions, the surface temperature is described by the following expressions.

Heating with a Dirac pulse of the energy W:

$$T^F(\tau) = \frac{W}{e_c\sqrt{\pi\tau}}\left[1 + 2\sum_{n=1}^{\infty}(-\Gamma)^n e^{-\tau*/\tau}\right]. \tag{4.42}$$

Heating with a square pulse of the power Q:

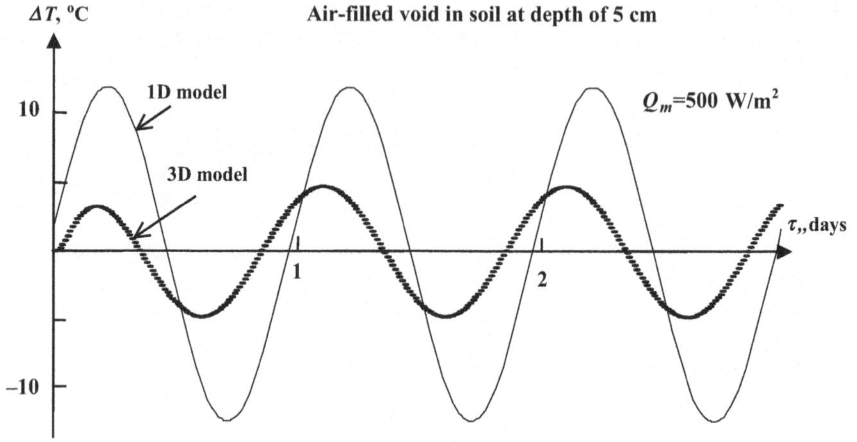

Fig. 4.12 Comparison between 1D and 3D thermal wave models in detecting 5 cm-deep air-filled void in soil

$$T^F(\tau) = \frac{2Q}{e_c\sqrt{\pi}} \begin{bmatrix} \sqrt{\tau} - \sqrt{\tau - \tau_h} + 2\sqrt{\tau}\sum_{n=1}^{\infty}(-\Gamma)^n F(\tau) \\ -2\sqrt{\tau - \tau_h}\sum_{n=1}^{\infty}(-\Gamma)^n F(\tau - \tau_h) \end{bmatrix}; \qquad (4.43)$$

$$F(\tau) = e^{-\tau*/\tau} - \sqrt{\frac{\pi\tau*}{\tau}}\, erfc\left(\sqrt{\frac{\tau*}{\tau}}\right).$$

Step heating of power Q:

$$T^F(\tau) = \frac{2Q}{e_c}\sqrt{\frac{\tau}{\pi}}[1 + 2\sum_{n=1}^{\infty}(-\Gamma)^n \cdot \left\{e^{-\tau*/\tau} - \sqrt{\frac{\pi\tau*}{\tau}} erfc\left(\sqrt{\frac{\tau*}{\tau}}\right)\right\}. \qquad (4.44)$$

Here $\tau* = n^2L^2/\alpha_c$ is the heat transit time with the subscript 'c' being related to the coating, e_c is the coating effusivity, the coefficient of reflection $\Gamma = (e_s - e_c)/(e_s + e_c)$ is defined for the boundary between a coating ('c') and a substrate ('s').

The parameter Γ is an indicator of the thermal match between two materials. $\Gamma = 0$ describes an ideal thermal contact of two identical materials, $\Gamma > 0$ corresponds to a more conductive substrate, while $\Gamma < 0$ means that the substrate is less conductive than the coating. These cases are presented graphically in Fig. 4.13 for step heating. In the initial period, the two-layer structure behaves as a semi-infinite body made of the coating material; the corresponding plot of T *vs.* $\sqrt{\tau}$ is a straight line. Within this period, the boundary between the two media cannot be detected on the surface. At a particular time called the heat transit time, the temperature starts to deviate from the profile of the "no buried boundary" case. The sign of deviation is determined by the sign of Γ.

The above-described technique has been called time-resolved IR radiometry (TRIR) [11].

A sample is heated with laser radiation whose spectrum is outside the spectral band of the IR imager used. Therefore, the front surface temperature can be monitored during the heating stage without interference with the reflected radiation.

Example Consider a thermal model of a turbine blade coating. There are two types of defects that are most common in this structure are delaminations between the coating and the substrate and variations in coating thickness. A three-layer model including a coating, defect and substrate can be used to simulate this structure.

Theoretical relationships between the $\Delta T(\tau)$ signal and defect parameters are given in Fig. 4.14 with l_{ci} being the coating thickness and l_{dj} being the thickness of an air-filled defect (results: courtesy S. Marinetti). Thermal properties are shown in Sect. 3.1.

The shape of optical heat pulses generated by flash tubes is not symmetric. The asymmetrical pulsed heating is approximated by the following function.

$$Q(\tau) = \frac{W\tau}{4\sigma^2} e^{-\tau/2\sigma}, \tag{4.45}$$

where $\sigma = 0.6$ ms and W is the pulse energy (note that the maximum power density takes place at $\tau = 2\sigma$, or at 1.2 ms). The typical coating thickness for a turbine blade is 250 μm. In Fig. 4.15, similar calculated data is presented as the evolution of running temperature contrast $C^{run}(\tau) = \Delta T(\tau)/T_{nd}(\tau)$ in time. Note that coating thinning causes increased heat diffusion into the substrate. The plots in Fig. 4.14 and 4.15 lead to a conclusion that signals produced by delaminations are characterized by higher amplitudes and longer observation times than those caused by variations in coating thickness.

By evaluating the maximum signal C_m and the time of its appearance τ_m, one may retrieve the distribution $\{C_m, \tau_m\}$ shown in Fig. 4.16. This clearly illustrates the ability to discriminate between the two types of defects.

4.2.4.3 Three-Layer Non-adiabatic Plate with Thermally-Capacitive Defect

A 1D model of a sample with a thermally-capacitive defect was discussed in Sect. 3.9.3 as a three-layer plate where each layer is characterized by a set of thermal properties $\{C, \rho, \lambda\}$. The differential temperature signal in this model is given by Eq. (4.24). The solution (3.49) for this problem is fairly cumbersome and requires the use of a computer to obtain practical results (Layer-3 Analytic software). However, such a solution can be used to validate numerical results in the case of infinitely wide defects.

4.2.4.4 A Two-Layer Adiabatic Plate with Thermally-Resistive Defect: Determining the Apparent Effusivity

A simpler solution appears if two adiabatic plates are separated by a thin gap characterized by the thermal resistance $R_d = d/\lambda_d$. The corresponding formulas obtained by Balageas et al. [12, 13] are given in Sections 3.9.2 and 3.9.3.

The solutions for Dirac pulse (flash) heating are quite often used to solve direct and inverse TNDT problems, for example, to determine the so-called *apparent effusivity* [12]. The application of this approach for the evaluation of defect parameters will be discussed in Sect. 5.1. Here we will limit ourselves with the illustration of the apparent effusivity method applied to the inspection of a 100 μm-thick Ni-Cr coating on a 3 mm-thick steel substrate. Thermal conductivities are: $\lambda = 14$ W/(m·K) for the coating and 70 W/(m·K) for the substrate. The densities and heat capacities are the same for both materials: $\rho = 7800$ kg/m^3 and $C = 500$ J/(kg·K).

The basis of the method is that the thermal inertia e of a semi-infinite body subjected to pulsed thermal stimulation can be determined at *any time* by Eq. (3.2): $e = W / \left[T(\tau *)\sqrt{\pi\tau *} \right]$. Evaluating true e value requires the accurate measurement

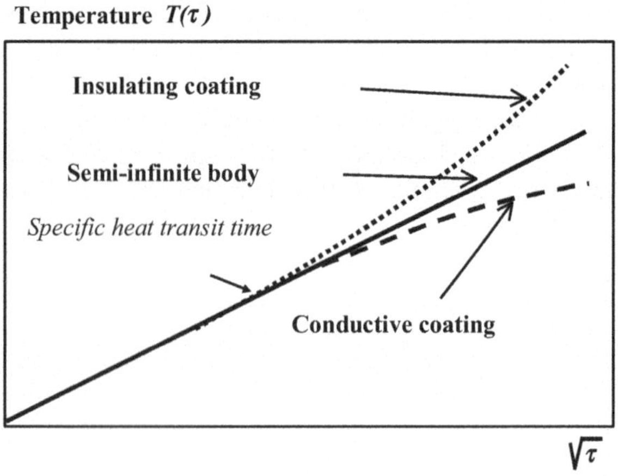

Fig. 4.13 Surface temperature *vs.* time in a coating-substrate structure (heating of semi-infinite body with uniform heat flux)

of absorbed energy, which is technically difficult to achieve. It was proposed to apply this method at shorter observation times ($Fo < 0.1$), when a plate behaves as a semi-infinite body. Since the front surface stationary temperature of a plate is $T_\infty = W/(C\rho L)$, the thermal inertia can be determined as follows:

$$e = C\rho L / \left[T(\tau *) \sqrt{\pi \tau *} \right], \tag{4.46}$$

assuming that C, ρ, L are known.

In the test described above, the heating is provided by a 40 μs heat pulse. We consider Dirac pulse solutions in the following three cases: (1) a steel substrate without a coating, (2) Ni-Cr coating on a steel substrate (ideal thermal contact), and (3) the presence of a 1 μm air gap between the coating and substrate with $R_d^* = \sqrt{C_1 \rho_1 \lambda_1} R_d \sqrt{a_d} / d = 0.43$ being the dimensionless thermal resistance of the gap with respect to the thermal properties of the first layer. The respective plots are shown in Fig. 4.17 as the evolutions of the e_c/e_s ratio versus time. For the steel substrate without the coating, the ratio $e/e_s = 1$ (Curve 1) during the first 100 ms. After that, the front surface temperature starts to be affected by the finite thickness of the plate. If the coating and the substrate are in ideal thermal contact, the thermal inertia of the sample $e/e_s = e_c/e_s = \sqrt{\lambda_c / \lambda_s} = 0.45$ is controlled only by the coating (Curve 2) for the time up to 1 ms. From 10 ms to 100 ms, Curves 1 and 2 approach

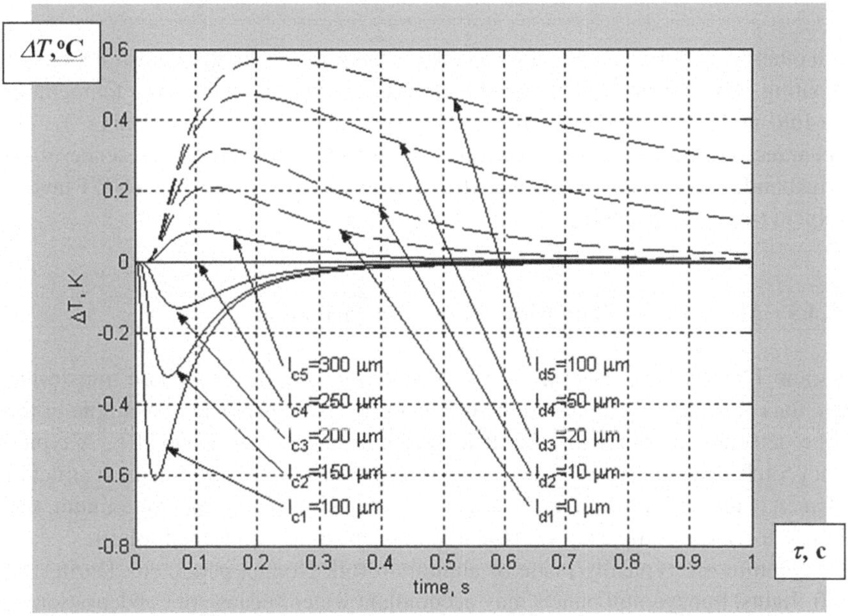

Fig. 4.14 Differential temperature signals over defects *vs.* time (in regard to 250 μm-thick reference coating on AISI 304 steel substrate)

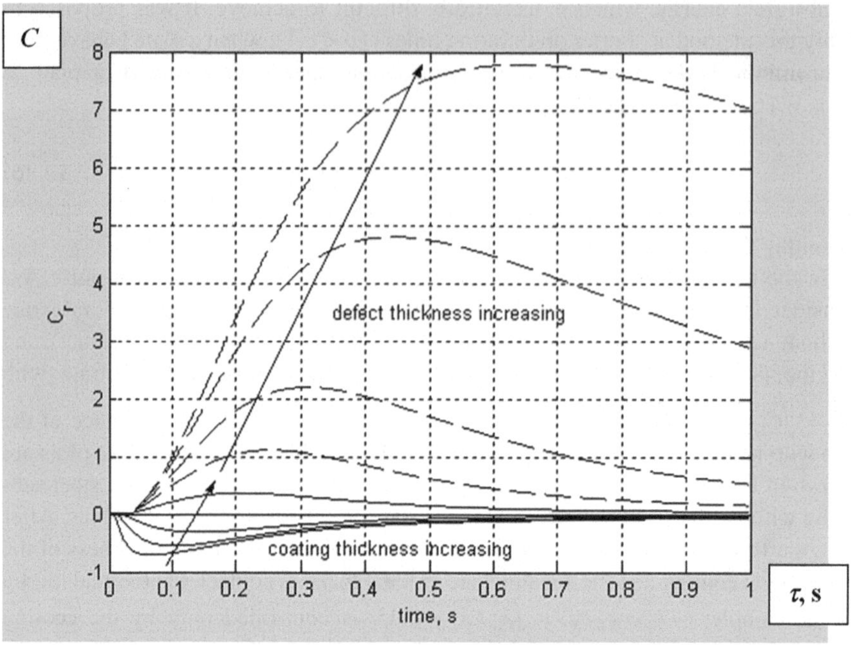

Fig. 4.15 Temperature contrasts over defects *vs.* time (in regard to 250 μm-thick reference coating on AISI 304 steel substrate)

each other and afterwards the sample behaves the same as the steel substrate without a coating. The presence of the air-filled defect increases the surface temperature five-fold at 10 ms, thus twice decreasing the apparent effusivity (Curve 3). The appearance of a minimum point in the $e(\tau)$ evolution reflects the presence of the corresponding $\Delta T(\tau)$ maximum that allows us to solve the respective TNDT inverse problem (see Sect. 5.1) [12].

4.2.4.5 Multi-Layer Plate with Phase Transformation

In some TNDT cases, materials to be inspected may undergo a phase transformation, thus considerably affecting the thermal processes. A practical example of this is the detection of water in aircraft honeycomb panels (see Chap. 10). A typical honeycomb structure involves two skins with a honeycomb reinforcement structure between (Fig. 4.18, left) them. A skin can be made either of metal (aluminum, titanium, steel) or composite (glass/carbon/boron fiber reinforced plastics, etc.). Honeycombs are typically made of aluminum foil, Nomex paper, etc. During aircraft flights, honeycomb panels may accumulate water because of condensation, or exposure to rain, snow, and ice. The presence of water trapped in the honeycomb is a dangerous situation, as it could eventually cause damage to ailerons, fuselage,

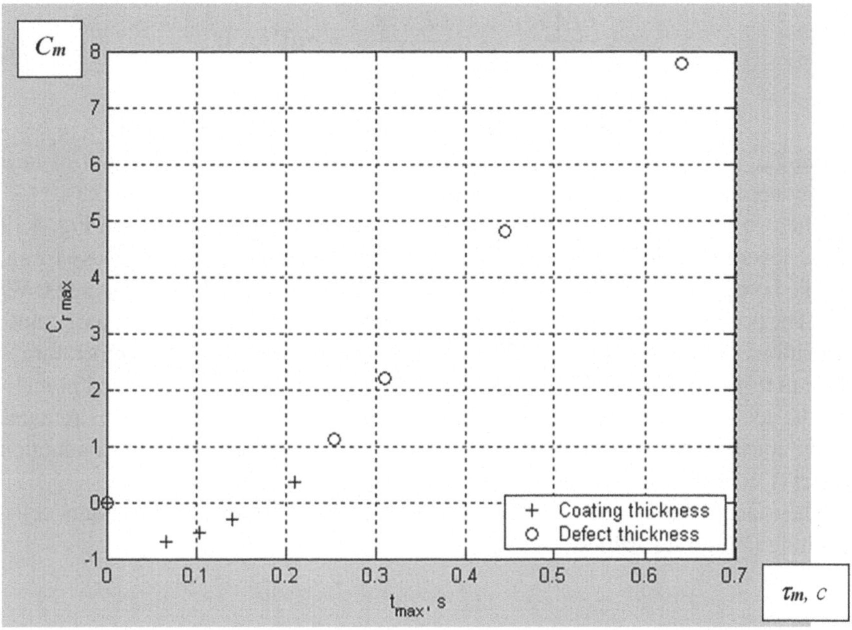

Fig. 4.16 Distribution of maximum contrasts and optimum observation times (coating on AISI 304 steel substrate)

rudders, elevators, etc. There are two mechanisms for this. First, the water may cause corrosion of an aluminum honeycomb and this corrosion may lead to a loss of mechanical strength in the structure. Second, when the water freezes, as it certainly will at cruising altitude, it expands and this expansion may rupture the honeycomb cells or cause separations (disbonds) between the honeycomb and skins After landing, the ice will start melting, unless it's in winter. The model described in Sect. 3.9. may be used to analyze this TNDT problem without considering ice melting. Calculations show that, within the model of 'pure' thermal conduction, optimum observation times are rather short. In particular, in aluminum panels, the 3D heat diffusion quickly eliminates temperature anomalies. However, trapped water may be detected in composite honeycomb panels for several hours (depending on the temperature difference between ground level and at cruising altitude). Detecting water at long observation times can be easily explained by the occurrence of a phase transformation. To melt ice, additional energy must be absorbed; this is called the "latent heat of fusion". Until this quantity of energy is absorbed, the temperature of the ice in honeycombs remains constant, while "dry" panels with no ice continue to warm up. This generates an increasing temperature difference between the two areas.

A TNDT model involving a phase transformation is similar to that in Sect. 3.9.1, but in the middle layer the melting boundary x_b moves along with the ice melting at speed V (Fig. 4.18, right). Boundary condition equations must take into account the continuity condition in the following form:

$$-\lambda_i \frac{\partial T_i(x_b)}{\partial x} + \lambda_{i+1} \frac{\partial T_{i+1}(x_b)}{\partial x} = \rho q V(x_b);$$
$$T_i(x_b) = T_{i+1}(x_b) = T_{phase,} \tag{4.47}$$

where T_{phase} is the phase transformation temperature and q is the amount of heat energy necessary for the phase transformation.

The two models above are illustrated by the temperature evolutions in Fig. 4.19 (three-layer plate simulating an aluminum honeycomb panel, skin thickness 1 mm, middle layer made of 8 mm-thick ice or water, uniform heat flux power $Q = 500$ W/m^2). The problem was solved with the Layer-3 Analytic program. If the phase transformation is taken into account, Curve 3 exhibits a plateau where the temperature is close to 0 °C since a considerable amount of incoming energy is expended in melting the ice. Therefore, the differential temperature signal ΔT (Curve 5) remains nearly constant during a longer period of time than predicted by the pure conduction model (Curve 4).

The model in Fig. 4.19 is 1D, i.e. it is assumed that water occupies many cells, resulting in high ΔT values.

4.2.5 Limits of Applicability for 1D Models

Realistic 1D models should treat the lateral dimensions of the defect $h_x \times h_y$ in Cartesian coordinates or use the defect radius r_d in cylindrical coordinates. Consider the heat diffusion in defect areas by analyzing a 2D cylindrical model. This will be discussed in detail in the next section. Assume that a sample contains a circular defect of radius r_d at depth l. The sample is uniformly heated by a heat pulse of duration τ_h. It is obvious that, at any depth, the 2D differential temperature signal $\Delta T^{(II)}$

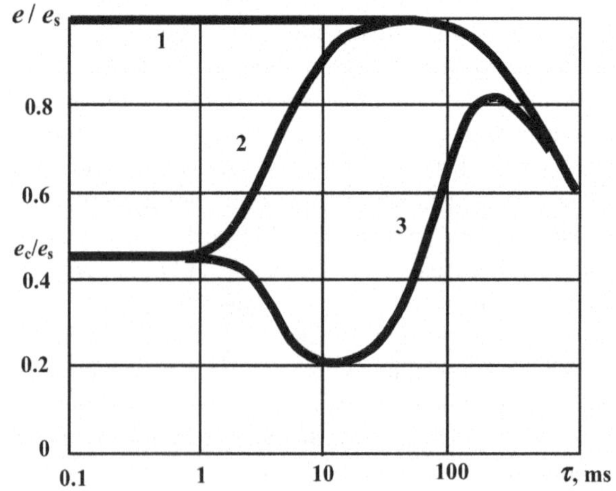

Fig. 4.17 Apparent effusivity *vs.* time (pulsed heating of 100 μm-thick Ni-Cr coating on 3 mm-thick steel substrate, (1) steel coating, (2) ideal thermal contact between Ni-Cr coating and substrate, (3) 1 μm-thick air-filled defect between coating and substrate)

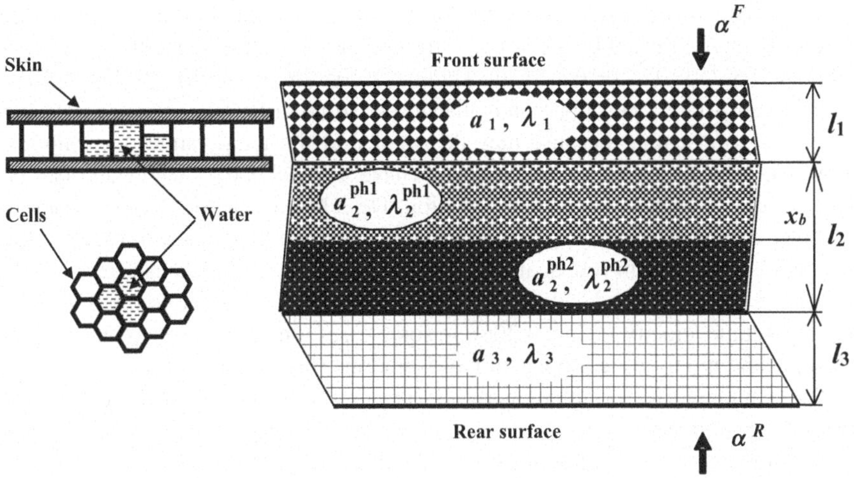

Fig. 4.18 Heating a three-layer plate in the ambient (phase transformation in the middle layer, x_b – moving boundary of phase transform, superscripts "ph1" and "ph2" specify two phase states of the material in the middle layer)

increases with increasing defect radius up to a particular value $\Delta T^{(I)}$ which corresponds to the solution of the corresponding 1D problem for a particular defect geometry in a particular material. Critical r_d/l values are given in Table 4.6 for some defects in different materials. Both defect thickness and depth are 10% and 30% of total sample thickness L. These values correspond to the cases where $\Delta T^{(II)}/\Delta T^{(I)} = 0.9$, i.e. a signal over a real defect achieves the 90% level of the maximum (1D) value.

The data in Table 4.6 shows that the critical values of r_d/l exceed those provided by an accepted "rule of thumb" which is $r_d/l > 1$–2. This leads to the conclusion that 1D models produce unrealistic ΔT values for most defects of practical interest.

4.3 2D TNDT Models

4.3.1 One-Layer Plate with a Channel-like Defect (Cartesian Coordinates)

Consider a sample with an infinitely long linear defect in the y direction. In practice, this means that the defect dimension d_y at a particular depth meets the condition of a transition from a 3D model to the corresponding 1D model with respect to coordinate y (Table 4.6). Hence, such a defect can be simulated by a long groove or channel (Fig. 4.20a) and the problem transitions from 3D to 2D.

If a TNDT model involves the uniform heating of a single defect, due to the x-axis symmetry, the analysis can be performed on just half of the defect; hence, at the boundary $\{x = 0, \quad z = 0 \ldots L\}$, the symmetry condition is $\partial T/\partial x = 0$ (Fig. 4.20b).

Thin subsurface cracks perpendicular to the front surface may not be detected under uniform heating because they do not distort the temperature distributions. However, in some cases, surface cracks behave as blackbody cavities and can be detected due to enhanced thermal radiation. Detection of such defects by using a pure heat conduction mechanism requires the generation of a non-uniform heat flux that creates a heat from propagating *across* the front surface. This can be achieved by heating a sample in a localized area of size S_f which can either be motionless or moving with a velocity V (Fig. 4.20c). In some cases, one can warm up one edge of a sample and monitor the propagation of isotherms (Fig. 4.20d) across it.

The mathematical formulations of all 2D models in Fig. 4.20b–d are identical and not reported here.

A more general 3D Cartesian case will be discussed in Sect. 4.6.

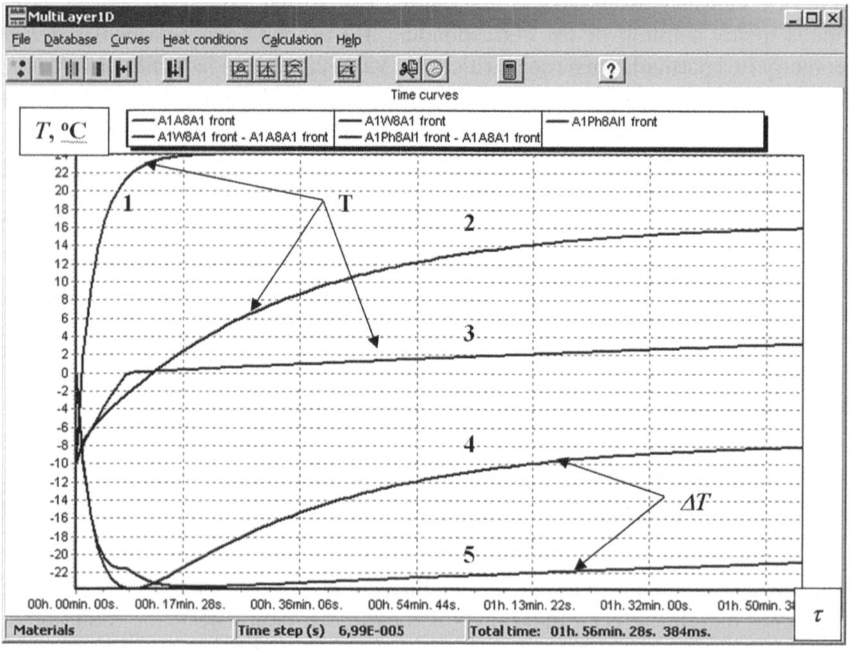

Fig. 4.19 Heating an aluminum honeycomb panel containing hidden water (skin thickness 1 mm, cell height 8 mm, initial temperature −10 °C, heating power density 500 W/m^2, total observation time 1 hr 50 min): (1) non-defect area, (2) area with water (pure conduction model), (3) area with ice/water (phase transformation model), (4) temperature signal (pure conduction model), (5) temperature signal (phase transformation model)

Table 4.6 Critical ratio r_d/l specifying a transition from 2D (cylindrical) to 1D TNDT models (by 90% of ΔT_m)

1D model	2D model
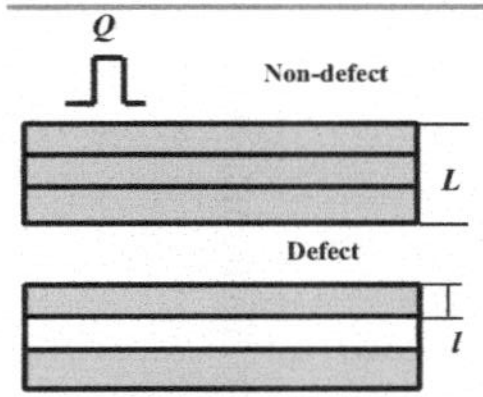	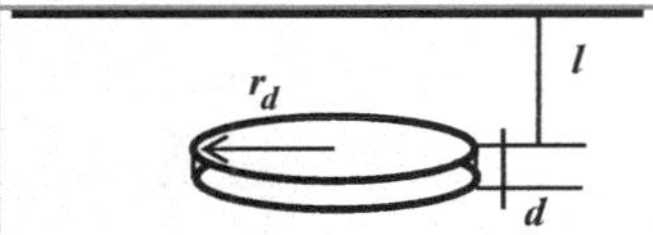

Defect depth l	Defect thickness d	Critical ratio $(r_d/l)_{cr}$
Aluminum: $\lambda = 210W/(mK)$, $a = 8.6 \cdot 10^{-5} m^2/s$, $L = 2\ mm$, $Q = 10^5\ W/m^2$, $\tau_h = 0.01\ s$		
10% L (0.2 mm)	10% L (0.2 mm)	15
	30% L (0.6 mm)	17
30% L (0.6 mm)	10% L (0.2 mm)	7
	30% L (0.6 mm)	7
Carbon Fiber Reinforced Plastic (CFRP) Composite: $\lambda = 0.64\ W/(m\ K)$, $a = 5.2 \cdot 10^{-7}\ m^2/s$, $L = 5\ mm$, $Q = 10^5\ W/m^2$, $\tau_h = 0.01\ s$		
10% L (0.5 mm)	10% L (0.5 mm)	5
	30% L (1.5 mm)	7
30% L (1.5 mm)	10% L (0.5 mm)	3
	30% L (1.5 mm)	5
Soil: $\lambda = 1.1\ W/(mK)$, $a = 7.5 \cdot 10^{-7}\ m^2/s$, $L = 20\ cm$, $Q = 10^3\ W/m^2$, $\tau_h = 12\ hr$		
10% L (2 cm)	10% L (2 cm)	8
	30% L (6 cm)	9
30% L (6 cm)	10% L (2 cm)	4
	30% L (6 cm)	4

4.3.2 *Three-Layer Anisotropic Plate with a Thermally-Capacitive Defect (Cylindrical Coordinates)*

A good compromise between analyzing finite defects and reducing computation time can be achieved by analyzing a cylindrical defect in a cylindrical layered sample in the cylindrical coordinate system. See Figure 4.21. Due to the circular symmetry, the temperature distribution in this model is independent of angle ϕ. A subsurface defect can be modeled as a disk of radius r_d and thickness $d = l_2$ between two other cyclindrical layers. Defect depth is equal to the thickness of the first layer $l = l_1$. Non-adiabatic boundary conditions exist on both the front (F) and rear (R) surface and the side is perfectly insulated. Let thermal conductivity be anisotropic, i.e. being different in coordinates r and z. In such a model, temperature spatial distribution is circular symmetric while the thermal stimulus can have an arbitrary shape.

The mathematical formulation of the problem is as follows:

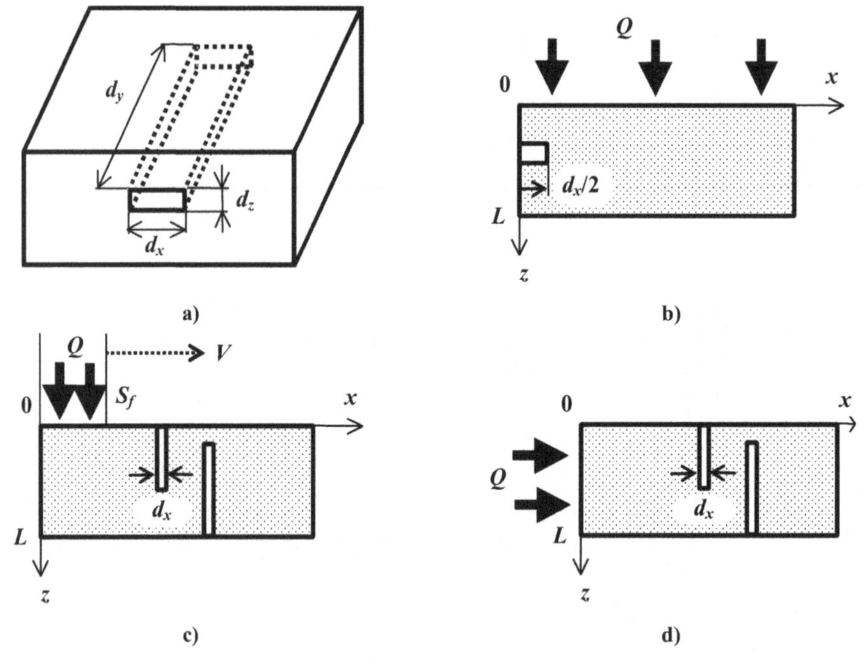

Fig. 4.20 2D TNDT model: (**a**) sample with channel-like defect, (**b**) uniform heating model, (**c**) localized heating model, (**d**) side heating model

$$\frac{1}{a_i}\frac{\partial T_i}{\partial \tau} = \frac{\partial^2 T_i}{\partial r^2} + \frac{1}{r}\frac{\partial T_i}{\partial r} + \frac{\partial^2 T_i}{\partial z^2}; \quad i = 1...3; \quad 0 \le r \le R; \quad 0 \le z \le L; \quad \tau > 0; \tag{4.48}$$

$$T_i(r,z,\tau) = 0; \tag{4.49}$$

$$\frac{\partial T_i(z=0,\tau)}{\partial z} = -\frac{Q(\tau)}{\lambda_1} + \frac{\alpha^F}{\lambda_1} T_i(z=0,\tau); \tag{4.50}$$

$$\frac{\partial T_i(z=L,\tau)}{\partial z} = -\frac{\alpha^R}{\lambda_3} T_i(z=L,\tau); \tag{4.51}$$

$$\frac{\partial T_i(z=R,\tau)}{\partial z} = 0; \tag{4.52}$$

$$\lambda_i \frac{\partial T_i(z,\tau)}{\partial \vec{v}} = \lambda_j \frac{\partial T_j(z,\tau)}{\partial \vec{v}}; \text{at layer defect boundaries} \quad T_i(z,\tau) = T_j(z,\tau). \tag{4.53}$$

Here Eq. (4.48) is the 2D differential equation of heat conduction in cylindrical coordinates,

Equation (4.49) is the initial condition.

Equation (4.50) describes the boundary condition on the front (F) surface (heating/cooling according to Newton's law).

Equation (4.51) describes the boundary condition on the rear (R) surface (cooling).

Equation (4.52) reflects thermally-insulated sample edges.

Equation (4.53) describes the continuity of both temperatures and heat fluxes at the interlayer boundaries.

Fig. 4.21 2D TNDT cylindrical model

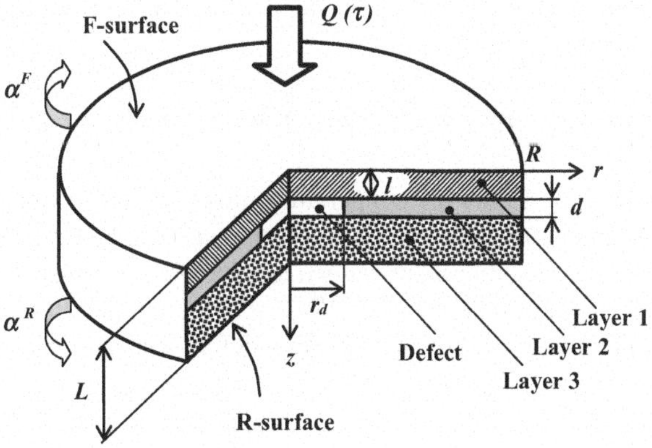

Note that the notation in the model expressed by Eqs. (4.48, 4.49, 4.50, 4.51, 4.52, and 4.53) is identical to the models discussed previously.

Note also that, when modeling anisotropic materials, only the thermal conductivity λ varies with coordinates (direction), while the heat capacity and density are independent of coordinates. Respectively, anisotropic diffusivity is given by $a_i = \lambda_i/(C\rho)$.

The 2D cylindrical model shown in Fig. 4.21 was analyzed with ThermoCalc-2D software (Tomsk Polytechnic University) designed for solving TNDT problems and optimizing test procedures. There are four types of thermal stimulation $Q(\tau)$ which can be modeled in this software: (1) square pulse, (2) single cosine pulse, (3) sinusoidal thermal waves, and (4) arbitrary heating described by a function defined by the user.

4.4 A Simple 3D Model for Detecting a Vertical Surface Crack by Heating a Semi-Infinite Body with a Moving Heat Source

This model represents a thin crack located perpendicular to the surface of a sample. The geometry is shown in Fig. 4.20b and heating is performed by a 3D moving heat source (Fig. 4.22).

In a defect-free area, the problem has an analytical solution shown in Table 2.8 for both point and Gaussian heat sources. As a first simple approximation, a surface crack is simulated by a plane placed at $x = x_0$ perpendicular to the front surface. Assume that such a defect has zero thickness and zero thermal conductivity (Fig. 4.22). The temperature disturbance caused by this defect occurs, at first, along the x axis, and the 3D character of the model is caused by the 3D heat source.

Defect signals can be expressed in a coordinate system that is either based on a moving heat source, or on a stationary sample. If a defect is viewed as a thermal insulator plane, the analytic solution can be obtained by using the *image principle,* which implies that there is an *imaginary* heat source moving opposite to the main source. At any point $x \le x_0$, the temperature in front of the main heat source can be found by:

$$T(\tilde{x}, Fo) = T_b(\tilde{x}, Fo) + T_a\left[\tilde{x} + 2(\tilde{x} - \tilde{x}_0), Fo\right]; \ x \le x_0; \ Fo \le a x_0 / V R^2, \qquad (4.54)$$

where T_b, T_a are the temperatures produced by the main source (subscript "b") and the imaginary heat source (subscript "a") can be found by the formulas in Table 2.8. The dimensionless time $Fo = a x_0/V R^2$ corresponds to the time when the main source arrives at point $x = x_0$. It follows that the temperature over a defect will reach a ΔT value twice as high as that over a defect-free area. However, due to the energy conservation principle, the average temperature in the defect area must be the same as in the non-defect area if the defect is *passive*, i.e. not producing energy. Therefore, the area located behind the defect will be characterized by a temperature drop of similar magnitude.

4.5 3D TNDT Adiabatic Problem

Batsale et al. proposed the solution for a 3D adiabatic TNDT problem of heating an anisotropic plate having a square thermally-resistive defect of size $b \times c$ (Fig. 4.23) [14].

The solution was obtained by applying the Laplace transform to time and the Fourier transform to two lateral spatial coordinates. The problem formulation is classical for 'true' temperatures $T(x, y, z, \tau)$, and we show it in the Laplace domain (p is the Laplace variable which is the analog of the time τ):

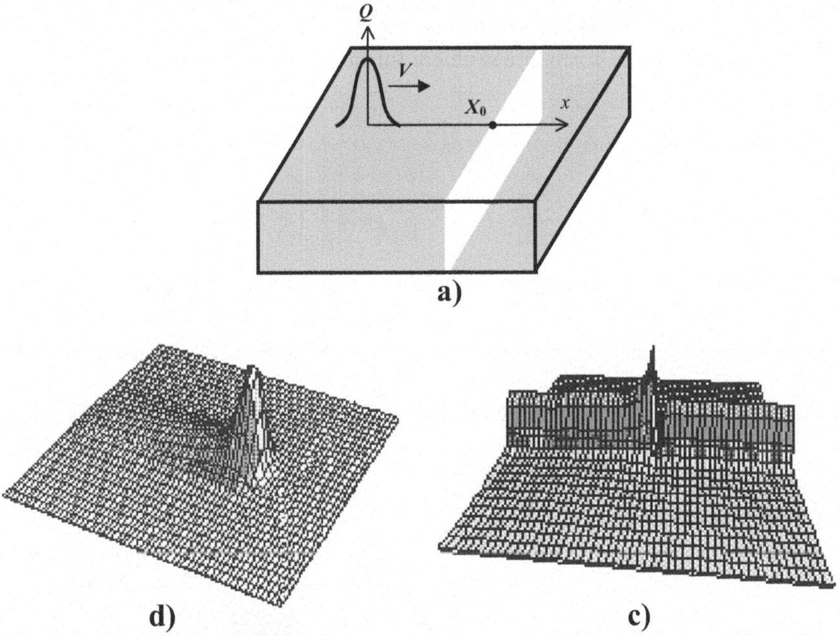

Fig. 4.22 Detecting vertical cracks by using a moving heat source (courtesy P.G. Bison): (**a**) test scheme, (**b**) temperature distribution (no defect, sample-linked coordinates), (**c**) temperature distribution across vertical crack (heat source-linked coordinates)

$$\frac{\partial^2\Theta}{\partial z^2}+\frac{\lambda_x}{\lambda_z}\frac{\partial^2\Theta}{\partial x^2}+\frac{\lambda_y}{\lambda_z}\frac{\partial^2\Theta}{\partial y^2}-\frac{p}{a_z}\Theta=0;$$
$$x=0,\ L_x:\frac{\partial\Theta}{\partial x}=0;$$
$$y=0,\ L_y:\frac{\partial\Theta}{\partial y}=0;$$
$$z=0:-\lambda_z\frac{\partial\Theta}{\partial z}=Q; \tag{4.55}$$
$$z=l:\frac{\partial\Theta^{d.f.}}{\partial z}=\frac{\partial\Theta^{d.r}}{\partial z};\ \Theta^{d.f.}-\Theta^{d.r.}=R_d f(x,y)\left(-\lambda_z\frac{\partial\Theta}{\partial z}\right);$$
$$z=L_z:\frac{\partial\Theta}{\partial z}=0,$$

where Θ is the image of temperature T in the Laplace domain, L_x, L_y, L_z are the plate dimensions, l is the defect depth, $f(x,y)=1$ if $\{x,y\}\in[x_1,y_1]x[x_2,y_2]$, otherwise $f(x,y)=0$; $b=x_2-x_1$ and $c=y_2-y_1$ are the defect dimensions in coordinates x and y; R_d is the thermal resistance of an infinitely-thin defect, and superscripts "*d.f.*" and "*d.r.*" relate to the front and rear defect surface respectively.

In the dimensionless form, the problem formulation is as follows:

$$\Theta*=\Theta/(QL_z/\lambda_z);\ \ b*=\frac{b}{L_z}(\lambda_z/\lambda_x)^{1/2};\ \ c*=\frac{c}{L_z}(\lambda_z/\lambda_y)^{1/2};$$
$$\Psi*=\Psi/Q;\ \ x*=\frac{x}{L_z}(\lambda_z/\lambda_x)^{1/2};\ \ y*=\frac{y}{L_z}(\lambda_z/\lambda_y)^{1/2}; \tag{4.56}$$
$$p*=L_z^2\ p/a_z;\ \ R_d*=R_d/(L_z/\lambda_z);\ \ z*=z/L_z;\ \ l*=l/L_z,$$

where Ψ is the Laplace transform of the z-th component of the heat flux $q_z=-\lambda_z\ (\partial T/\partial z)$.

Below, the subscript '$*$' will be omitted.

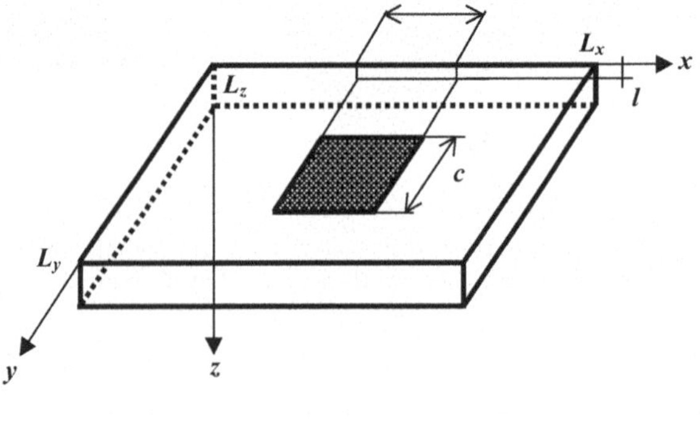

Fig. 4.23 Solving 3D TNDT adiabatic problem

The basic heat conduction equation is:

$$\frac{\partial^2\Theta}{\partial z^2}+\frac{\partial^2\Theta}{\partial x^2}+\frac{\partial^2\Theta}{\partial y^2}-p\Theta=0. \tag{4.57}$$

Let us apply the double cosine-transform to the function Θ in Eq. (4.54):

$$\mathrm{T}(\alpha,\beta,z,p)=\int_0^{L_x}\int_0^{L_y}\Theta(x,y,z,p)\cos(\alpha x)\cos(\beta x)\,dx\,dy, \tag{4.58}$$

where T is the Fourier transform of Θ, and dimensionless spatial pulsations acquire the following discrete values:

$$\alpha_j = j\pi / L_x;\quad \beta_k = k\pi / L_y, \tag{4.59}$$

where j and k are positive round numbers.

In the Fourier domain, Eq. (4.57) can be transformed into:

$$\frac{\partial^2\mathrm{T}}{\partial z^2}-\left(p+\alpha^2+\beta^2\right)=0. \tag{4.60}$$

And the solution to the last equation is:

$$\mathrm{T}=F\mathrm{ch}(uz)+G\mathrm{sh}(uz);\quad u=\sqrt{p+\alpha^2+\beta^2}. \tag{4.61}$$

This solution (4.61) can be derived in an evident form while the return from the Fourier and Laplace domains to the original functions should be done numerically. The advantage of the approach discussed above is that it can be used for characterizing defects while remaining in the space of integral transforms where the corresponding solutions are algebraic.

4.6 General TNDT 3D Model (Three-Layer Anisotropic Non-adiabatic Plate with Arbitrary Thermally-Capacitive Defects)

It is appropriate to simulate defects of arbitrary shapes in 3D Cartesian coordinates. In practice, parallelepiped-like defects can be modeled in a parallelepiped-like body with the following mathematical formulation (Fig. 4.24):

$$\frac{\partial T_i(x,y,z,\tau)}{\partial \tau} = a_i^x \frac{\partial^2 T_i(x,y,z,\tau)}{\partial x^2} + a_i^y \frac{\partial^2 T_i(x,y,z,\tau)}{\partial y^2}$$
$$+a_i^z \frac{\partial^2 T_i(x,y,z,\tau)}{\partial z^2}; i = 1\ldots M+3(\text{three layers} + \text{M defects}); \quad (4.62)$$

$$T_i(\tau = 0) = T_{in}; \quad (4.63)$$

$$-\lambda_1^z \frac{\partial T_1(x,y,z=0,\tau)}{\partial z} = Q(x,y,\tau) - \alpha^F\left[T_1(x,y,z,\tau) - T_a\right]; \quad (4.64)$$

$$\lambda_3^z \cdot \frac{\partial T_3(x,y,z=L_z,\tau)}{\partial z} = -\alpha^R\left[T_3(x,y,z,\tau) - T_a\right]; \quad (4.65)$$

$$\begin{aligned}\frac{\partial T_i(x,y,z,\tau)}{\partial x} &= 0 \text{at} x = 0, \quad y = 0\ldots L_y; \quad x = L_x, \quad y = 0\ldots L_y;\\ \frac{\partial T_i(x,y,z,\tau)}{\partial y} &= 0 \text{at} y = 0, \quad x = 0\ldots L_x; \quad y = L_y, \quad x = 0\ldots L_x;\end{aligned} \quad (4.66)$$

$$T_i(x,y,z,\tau) = T_{i\pm1}(x,y,z,\tau) \text{ and } \lambda_i^{q_j} \cdot \frac{\partial T_i(x,y,z,\tau)}{\partial q_j}$$
$$= \lambda_{i\pm1}^{g_j} \cdot \frac{\partial T_{i\pm1}(x,y,z,\tau)}{\partial q_j} \quad (4.67)$$

at boundaries between layers and defects.

Here T_i is temperature in the i-th area counted from the initial temperature of the sample (i = 1–3 corresponds to the sample layers and $i = 4 - M + 3$ corresponds to Mdefects), T_{in} is the sample's initial temperature, $a_i^{q_j}, \lambda_i^{q_j}$ are the thermal diffusivity and conductivity in the i-th area along coordinate q_j, x, y, z are Cartesian coordinates, q_j is one of three Cartesian coordinates x, y or z ($j = 1 \div 3$), τ is the time, $Q(x, y, \tau)$ is the heat flux density, which generally depends on time and coordinates, α^F and α^R are the heat exchange coefficients on the front and rear surfaces respectively, T_a is the ambient temperature, and L_x, L_y, L_z are the sample lateral dimensions.

By analogy with Eqs. (4.48, 4.49, 4.50, 4.51, 4.52, and 4.53), Eq. (4.62) is a 3D parabolic differential equation of heat conduction, Eq. (4.63) is the initial condition, Eq. (4.64) describes the boundary condition on the front surface including heating and cooling, Eq. (4.65) describes the boundary condition on the rear surface (cooling only), Eq. (4.66) correspond to the adiabatic boundary conditions on the side surfaces (along coordinates x and y), Eq. (4.67) describe continuity of temperatures and heat fluxes at defect boundaries.

Depending on how function $Q(x, y, \tau)$ is specified, this TNDT model may involve several types of heating and cooling, and thus represent various TNDT procedures.

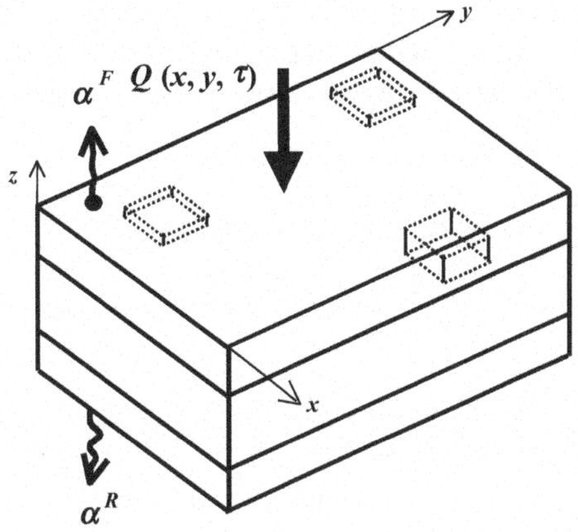

Fig. 4.24 3D TNDT Cartesian model

(4.67)

The 3D Cartesian (ThermoCalc-3D software) model in Fig. 4.24 involved up to 23 sample layers and up to 40 parallelepiped-like defects. In this software, arbitrary heating can be modeled with a piece-wise function, and some other model parameters can be specified as arbitrary functions of time and coordinates.

4.7 Detection Conditions for Subsurface Defects

As mentioned above, a general criterion for comparing TNDT procedures is the signal-to-noise ratio defined by

$$s = \frac{\Delta T(\tau)}{\sqrt{\sum_{i=1}^{N} \Delta \overline{T}_{i\ noise}^{2}(\tau)}}, \tag{4.68}$$

where $\varnothing \overline{T}_{i\ noise}^{2}$ is the dispersion of noisy signals produced by N independent sources (in temperature units). A more detailed review of statistical detection procedures will be presented in Chap. 7. To better understand the optimization of the TNDT process, consider some simple conditions which should be met when choosing parameters of interest for TNDT.

A defect of any type will be reliably detected by an operator or an automatic device if, the signal produced by the defect exceeds noise level:

$$s > 1 \quad \text{or} \quad \Delta T > \sqrt{\sum_{i=1}^{N} \Delta \overline{T}_{i\ noise}^{2}}. \tag{4.69}$$

Noise varies in time; therefore, a maximum noise value s may appear at a time that may not coincide with a maximum value of $\Delta T(\tau)$ or $C = \Delta T(\tau)/T(\tau)$. In this chapter, we will limit ourselves to two extreme cases:

1. TNDT noise is purely additive and fully defined by IR detector noise, or by the temperature resolution of the IR equipment $\Delta T_{res} = \sqrt{\sum_{i=1}^{N} \Delta \overline{T}_{inoise}^{2}}$.
2. TNDT noise is purely multiplicative and defined by absorptivity/emissivity variations.

In the latter case the noise can be characterized by the noise running contrast $C_{noise} = \sqrt{\sum_{i=1}^{N} \Delta \overline{T}_{i\,noise}^{2}}(\tau)/T_{nd}(\tau) = \Delta T_{noise}(\tau)/T_{nd}(\tau)$, which is time-independent due to the linearity of TNDT problems.

In the first case, the optimum observation time coincides with the maximum of $\Delta T(\tau_m)$, and the *first condition* for detecting defects can be expressed in the following obvious form:

$$\Delta T(\tau_m) > \Delta T_{res}. \tag{4.70}$$

In the second case, the time of optimum detection takes place at the time of the maximum contrast $C(\tau_m)$, and the *second condition* for detecting defects is:

$$C(\tau_m) > C_{noise}. \tag{4.71}$$

Note that the highest temperature contrasts appear after short (Dirac) heating.

A principal difference between the two conditions above is that Eq. (4.70) can be met by increasing the heating power Q (energy W), while Eq. (4.71) is independent of Q (W) and depends only on material properties.

A *third condition* of detection is a practical limitation: The *absolute* front-surface sample temperature at the end of heating must not exceed the test sample destruction temperature T_{destr}:

$$T_{abs}(\tau = \tau_h) < T_{destr}. \tag{4.72}$$

Since the sample excess temperature is proportional to W (Q), the latter condition limits absorbed energy. Overheating is mostly likely to occur when applying flash heating. Therefore, each test case requires a judicious choice of heat pulse power and duration.

To summarize, let us assume that subsurface defects can be detected by TNDT if the following three conditions are met:

$$\Delta T(\tau_m) > \Delta T_{res};$$
$$C(\tau_m) > C_{noise};\qquad (4.73)$$
$$T_{abs}(\tau = \tau_h) < T_{destr}.$$

An additional (*fourth*) detection condition is imposed by the IR device. *Temporal* resolution, or the IR camera thermogram (frame) acquisition rate f, must be high enough to store a sufficient number of images prior to the optimum observation time τ_m. This condition can be written as follows:

$$f \geq (5...10)/\tau_m. \qquad (4.74)$$

Equations (4.73 and 4.74) involve parameters of: (1) equipment (ΔT_{res} and f), (2) test material C_{noise}, (3) heating (T_{abs}), and (4) defect (ΔT and C).

4.8 Dependence Between Temperature Signals and Sample Parameters

In the classical TNDT model described above, differential temperature signals ΔT and their derivatives, such as running temperature contrast $C^{run} = \Delta T/T_{nd}$, depend on:

- time $\{\tau, \tau_h\}$,
- heating parameters $\{Q, \tau_h\}$,
- intensity of heat exchange between a sample and the ambient $\{\alpha^F, \alpha^R\}$,
- sample thickness $\{L\}$,
- sample thermal properties $\{\lambda, a\}$,
- defect size $\{h_x, h_y, h_z = d\}$ and depth $\{l\}$;,
- defect thermal properties $\{\lambda_d, a_d\}$.

Therefore, the detectability of defects by TNDT is affected by at least 14 parameters, though some of these, for example $\{\alpha^F, \alpha^R\}$, only weakly influence the inspection efficiency.

4.8.1 Evolution of Temperature Signals in Time

TNDT is often called transient, or dynamic, because defect detectability varies significantly in time. Peculiarities of the temperature evolution of ΔT and C^{run} are of interest because these parameters are used to develop algorithms which allow the enhancement of signal-to-noise ratio.

Some features of temperature evolutions were described in Chap. 2 by analyzing 1D dimensionless classical solutions for heat conduction. In this section, we will discuss the results of the numerical analysis related to the 2D and 3D models described in Sections 4.3 and 4.4. Most illustrations will concern composite materials, such as CFRP, which are widely used in aerospace and thus represent a vast field for applying TNDT. Also, the results discussed below may be useful in understanding general TNDT features.

Front (F) and rear (R) surface temperature evolutions are shown in Fig. 4.25a, b for a 5 mm-thick CFRP plate heated with a heat pulse (material is assumed to be isotropic since ΔT values are mainly affected by the z-direction thermal conductivity component λ_z). If the thermal stimulation duration is short, i.e. the heating period is less than characteristic heat transit times adhered to a particular material ($Fo_h = a\tau_h/L^2 \ll 0.1$), a noticeable difference between defect T_d and non-defect T_{nd} temperatures starts to appear in the cooling stage ($\tau > \tau_h$). In Fig. 4.25b, c, the differential temperature signals $\Delta T = T_d - T_{nd}$ are compared to the corresponding running contrasts $C^{run} = \Delta T/T_{nd}$ (in further consideration, the superscript "*run*" will be omitted). On both sample surfaces, evolutions of ΔT and C parameters reveal maximums appearing at different observation times. On the front surface (Fig. 4.25c), a maximum signal ΔT_m appears earlier than the corresponding maximum of the contrast C_m, if it takes place in the cooling stage. Conversely, if both maximums occur within a heat pulse, C_m surpasses ΔT_m. In many cases, the maximum signal-to-noise ratio coincides with the maximum contrast, hence, $\tau_m(C_m)$ can be regarded as *optimum detection time*. On the rear surface, maximum contrasts typically appear at very short times when excess temperature T and temperature signals ΔT are vanishing. Therefore, in two-sided TNDT (Fig. 4.25d), optimum detection times correspond to the times when the ΔT signals start to exceed noise. Note that, if defect conductivity is lower than that of a host material, differential signals are positive ($\Delta T > 0$) on the front surface and negative ($\Delta T < 0$) on the rear surface. In the case of defects having higher conductivity than the host material, the sign of ΔT will be inverted.

There is an exception of the latter rule called "ΔT *inversion*". This phenomenon occurs on the front surface at longer observation times when the material layer over a defect starts to cool down faster than sound areas, and the sign of ΔT becomes negative. Inversion has been observed over Teflon inserts which are often used to simulate defects in composites. In this case, inversion may be caused by either the presence of air gaps or variations in material thermal properties. The amplitude of inverted signals is typically low, therefore, this phenomenon is rarely observed in experiments. The potential of using inversion to improve signal-to-noise ratio has been scarcely studied and thus not being realized in practice.

Because of lateral heat diffusion, front surface temperature signals are sensitive to defect depth l. With greater l, the differential temperature signals drop down, and the optimum observation times occur later (Fig. 4.25e). Phenomenologically, this can be explained by using the concept of thermal waves: periodic components of a

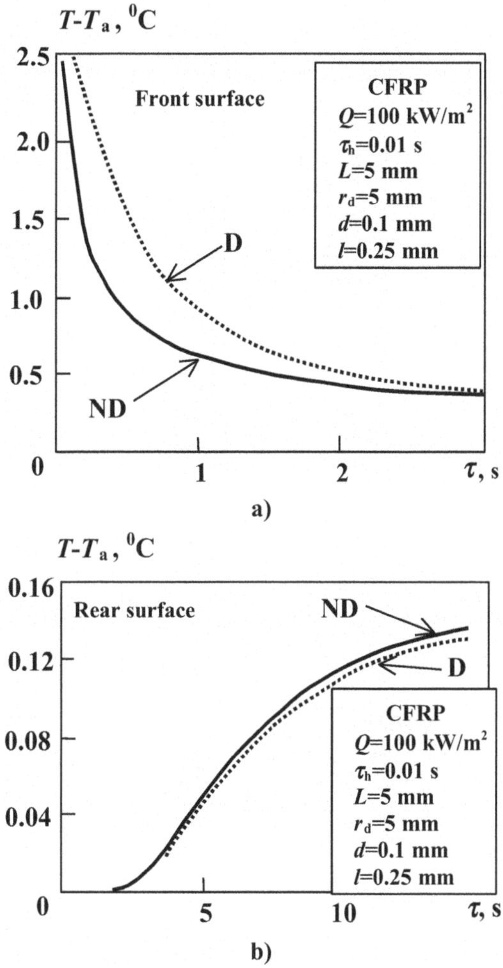

Fig. 4.25 Temperature signals *vs.* time (5 mm-thick CFRP pate heated with 0.01 s heat pulse, ND-non-defect area, D-defect area): (**a**) front-surface temperature, (**b**) rear-surface temperature, (**c**) front-surface temperature signal and contrast, (**d**) rear-surface temperature signal and contrast, (**e**) front-surface temperature signal *vs.* defect depth, (**f**) rear-surface temperature signal *vs.* defect depth

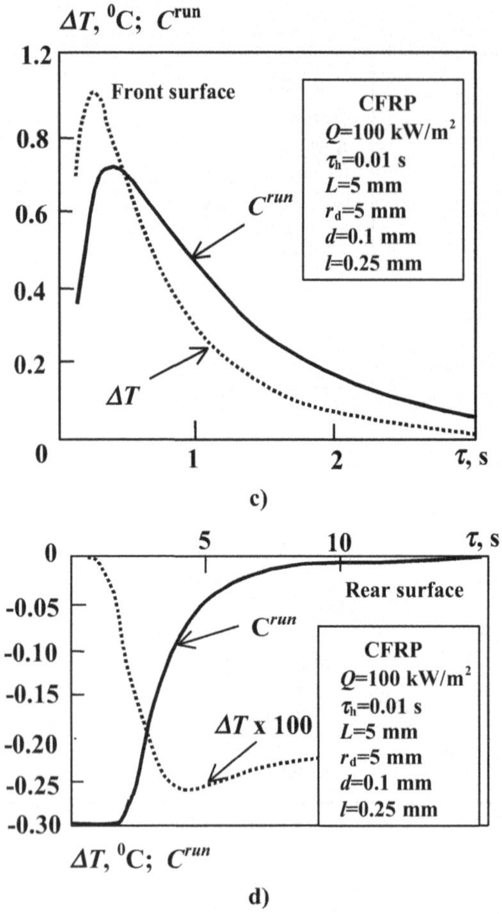

Fig. 4.25 (continued)

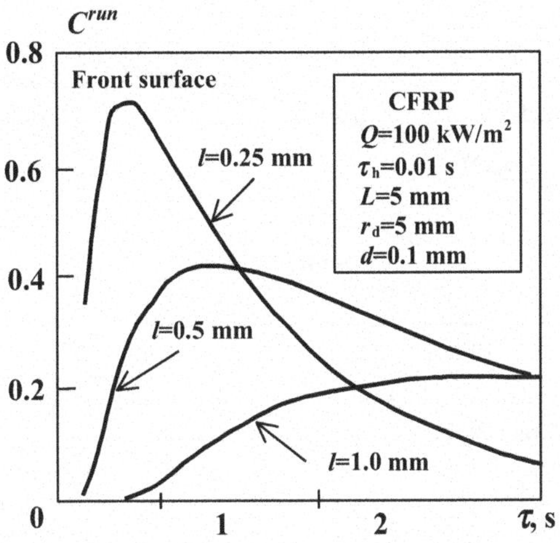

e)

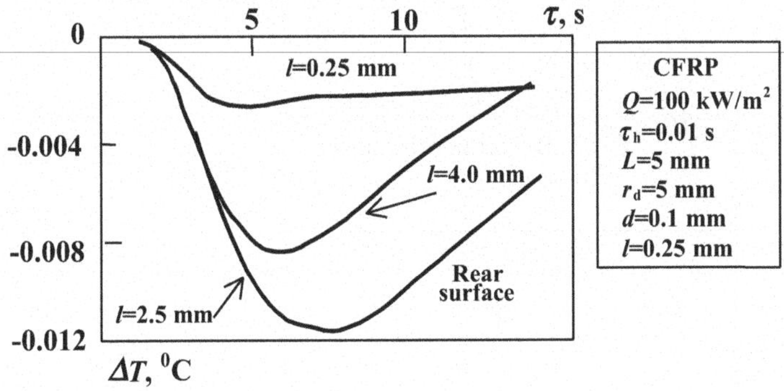

f)

Fig. 4.25 (continued)

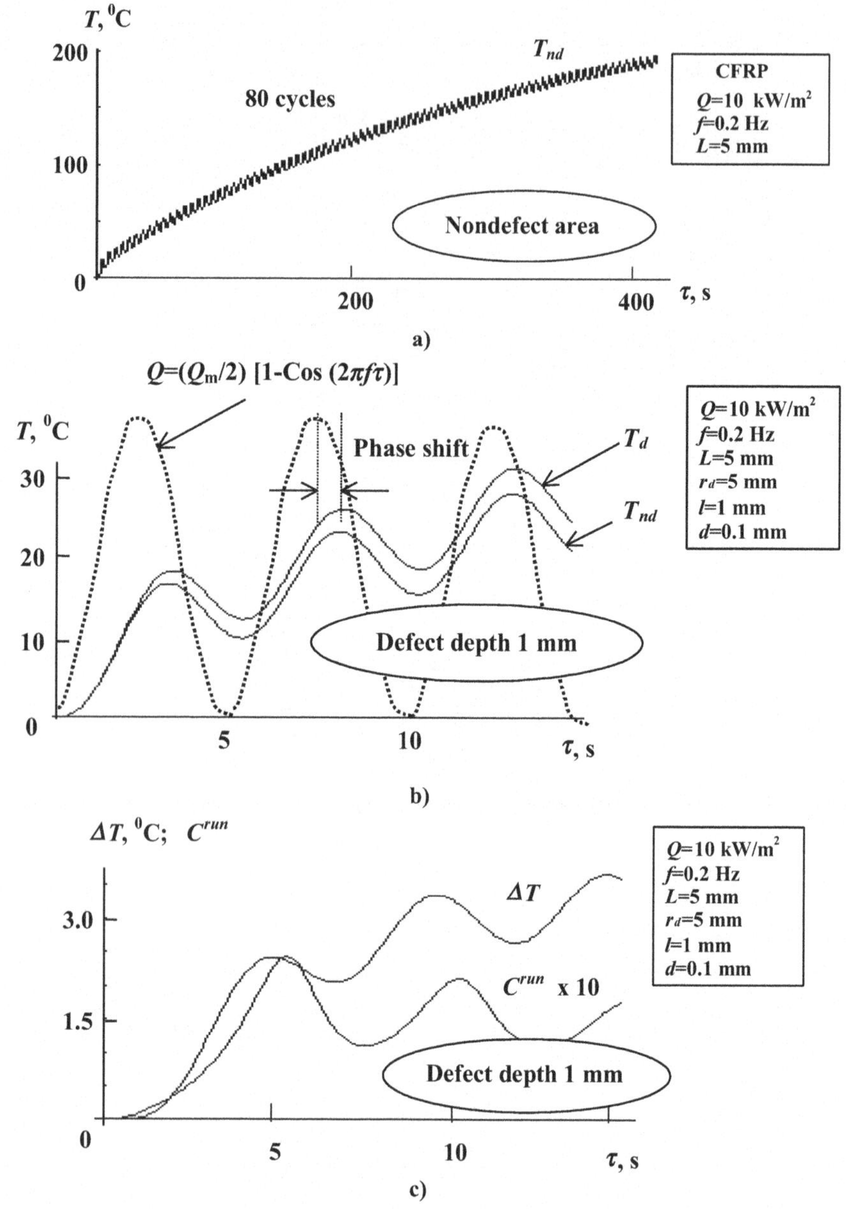

Fig. 4.26 Temperature signals *vs.* time (a 5 mm-thick CFRP sample heated with thermal waves, frequency $f = 0.2$ Hz is close to optimum, defect depth 1 mm): (**a**) front-surface non-defect temperature (80 cycles heating), (**b**) front-surface non-defect and defect temperature (3 cycles heating), (**c**) front-surface temperature signal and contrast (3 cycles heating)

4.stimulating heat pulse penetrate test material up to different depths depending on frequency and the phenomena of reflection, interference and diffusion. Oppositely, in two-sided TNDT, temperature signals are mainly characterized by the sample thickness L and are only weakly-dependent on the defect depth l. It is interesting to note that, in two-sided TNDT, maximum values of both ΔT_m andτ_m correspond to the defects located in the middle of the sample, i.e. at the depth of $L/2$ (Fig. 4.25f).

Thermal wave stimulation is expressed by the following function:

$$Q = \frac{Q_m}{2}\left[1 - \cos\left(2\pi f \tau\right)\right], \tag{4.75}$$

which describes non-zero heat flux. In this case, surface temperature, in both defect and non-defect areas, possesses two components: constant and harmonic (see the example in Fig. 4.26a). Obviously, a quasi-stationary regime will be achieved after a certain number of thermal wave cycles, depending on the type of material and its thickness. The classical *photothermal* method which is typically applied to the inspection of thin materials requires 5–10 cycles to reach a quasi-stationary regime. In TNDT, one often uses a modification of this technique called *lock-in IR thermography*. Unlike the photothermal method, which realizes point-by-point scanning, the IR thermographic implementation of this technique employs both heating and temperature monitoring in relatively big areas. When testing thick samples, such as few mm-thick composites, one might need up to 50–100 cycles to reach the quasi-stationary regime (Fig. 4.26a).

A remarkable feature of surface thermal waves is the presence of the *phase lag* which occurs in response to a stimulating wave (Fig. 4.26b). On the surface of a semi-infinite body, the phase lag of a 'non-defect' thermal wave is equal to −45° in regard to the heating function. A phase lag may be considered as a significant parameter which provides a higher signal-to-noise ratio compared to temperature. Since phase can be expressed in terms of time, analysis of signals in the phase domain is analogous, to some extent, to the time domain analysis. It is known that phase/time measurements are typically more noise-resistant than amplitude measurement, therefore, the corresponding techniques are often used in engineering and NDT.

The concept of absolute temperature signals and dimensionless temperature contrasts is valid also in the case of thermal waves. The examples of ΔT and C evolutions are shown in Fig. 4.26c. Both parameters are subject to periodic variations. The sample temperature grows up in time, at least, during the first five cycles because due to the presence of the constant component in the heating function (see Eq. (4.75)). In its turn, the temperature contrast reaches a maximum in 5 seconds after the start of heating if the inspection is fulfilled with thermal waves of the 0.2 Hz frequency (this frequency is close to optimum when detecting delaminations at the depth of 1 mm in a CFRP composite).

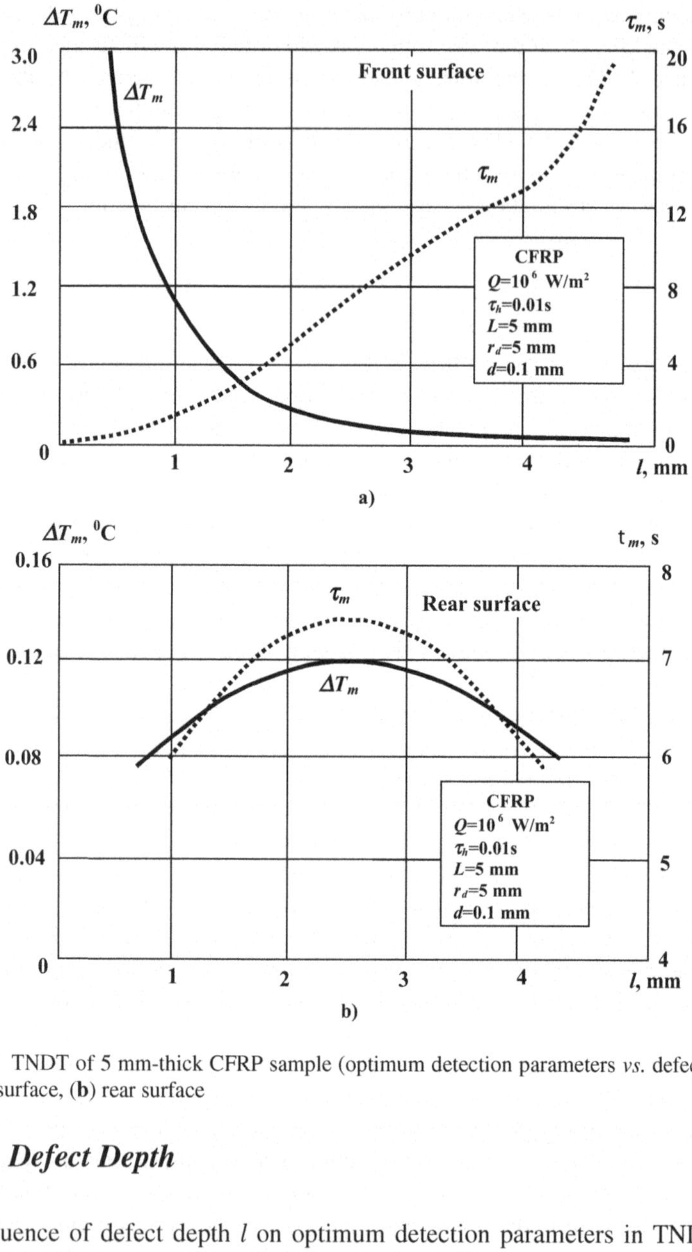

Fig. 4.27 TNDT of 5 mm-thick CFRP sample (optimum detection parameters *vs.* defect depth): (**a**) front surface, (**b**) rear surface

4.8.2 Defect Depth

The influence of defect depth l on optimum detection parameters in TNDT of a 5 mm-thick CFRP composite is illustrated in Fig. 4.27.

On the front surface, the differential temperature signal strongly decays as l increases (Fig. 4.27a). If the temperature resolution of a used IR equipment is 0.1 °C, the maximum depth for defect detection is about 3 mm. This conclusion is valid if the heating power is $Q = 10^6$ W/m^2 and the maximum excess temperature at the end of heating is ~110 °C. A further increase of absorbed energy will enhance ΔT and improve the depth limit but the sample may be damaged because of overheating. This means that material destruction temperature limits possible enhancement of absorbed power when detecting deep defects. It is also true that, in the case of prevailing multiplicative noise, a higher heating temperature will not improve the signal-to-noise ratio because of a proportional noise increase. The optimum observation time τ_m increases with greater l achieving $\tau_m \sim 10$ s at $l \sim 3$ mm. The data in Fig. 4.27a shows that the recommended expression for determining the optimum observation time $\tau_m = l^2/a$, which is independent of defect lateral size, can be used only as a rule-of-thumb.

On the rear surface, both $\Delta T(l)$ and $\tau_m(l)$ plots are symmetrical in respect to the sample middle at l=2.5 mm (Fig. 4.27b). The same conclusion was obtained earlier by analyzing the corresponding analytical solution in the Laplace domain. Hence, in two-sided TNDT, defects located in the middle of samples produce maximum ΔT signals at the longest observation times τ_m.

4.8.3 Defect Thickness

The defect thickness d, that is a defect size in the direction of principle heat diffusion, considerably affects both ΔT and τ_m values. This parameter is important when comparing 1D and 3D (2D) TNDT models. Lateral size of subsurface defects appears on a tested surface as a clearly visible defect contour but the defect thickness has no evident surface indication and should be evaluated by solving the corresponding inverse TNDT problem. As follows from Fig. 4.28, thicker defects are characterized by higher ΔT and longer τ_m values. For defects located in the middle of the sample, i.e. at depth of $l = 2.5$ mm, the corresponding dependencies are similar on both the front and rear surface (Fig. 4.28). Parameters ΔT and τ_m vary significantly at small d; for example, for thin defects $\Delta T \sim d$. For thermally-insulating defects, ΔT and τ_m become nearly independent on d.

We remind that, when developing defect thickness characterization algorithms, the defect thermal resistance $R_d = d/\lambda_d$ serves as a decisive parameter because both defect thickness d and defect thermal conductivity λ_d jointly influence on detection parameters.

4.8.4 Defect Lateral Size and Configuration

It can be assumed that, in a test material, there are two types of heat fluxes propagating both in-depth and in lateral directions. Lateral heat diffusion is most intensive at defect borders thus diminishing ΔT for small defects. Large defects are characterized by 1D heat through defects. This statement is illustrated by the plots of ΔT_m and τ_m *versus* r_d/l presented in Fig. 4.29. It is seen that the lateral heat flux in defect areas

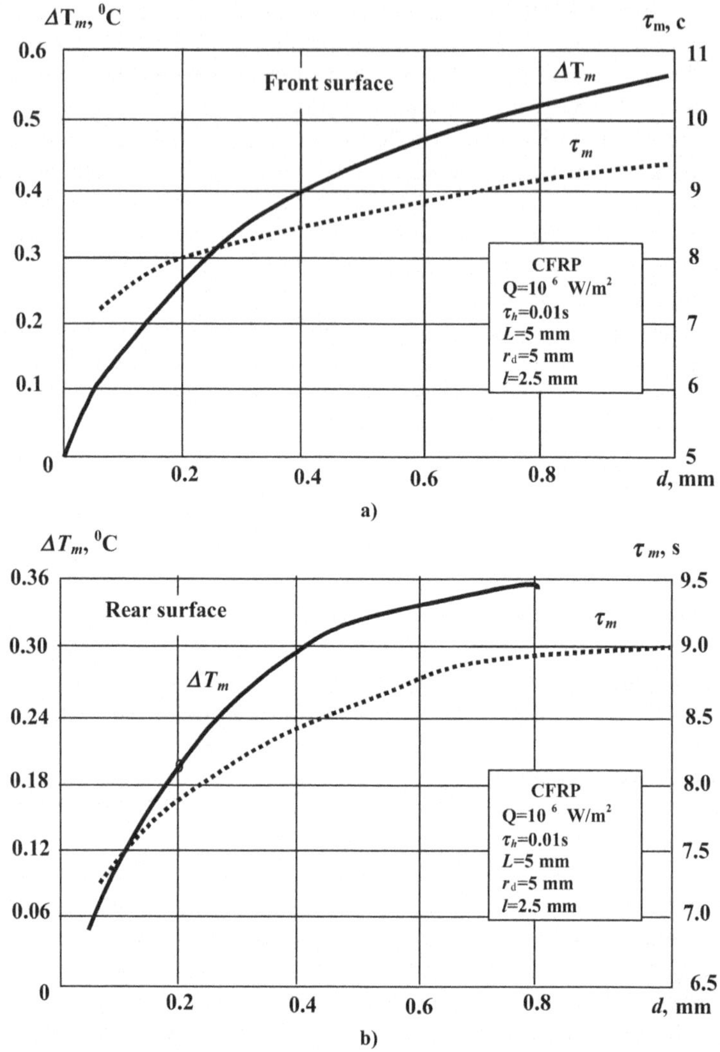

Fig. 4.28 TNDT of 5 mm-thick CFRP sample (optimum detection parameters *vs.* defect thickness): (**a**) front surface, (**b**) rear surface

is negligible in a CFRP composite if $2r_d/l > 5$. In aluminum, this condition becomes more rigid: $2r_d/l > 10$. There is a TNDT rule-of-thumb which requires lateral defect size to be at least twice as defect depth, i.e. $2r_d/l > 2$. The data in Fig. 4.29a confirms this conclusion because ΔT_m starts to drop sharply when $2r_d/l \leq 2$. It is interesting to note that temporal parameters (τ_m) are less dependent on the variation of defect lateral size, therefore, they are often recommended in evaluating l and d.

Another aspect of how defect size influences temperature signals is related to defect configuration, including the influence of neighbor defects. In Fig. 4.30, the

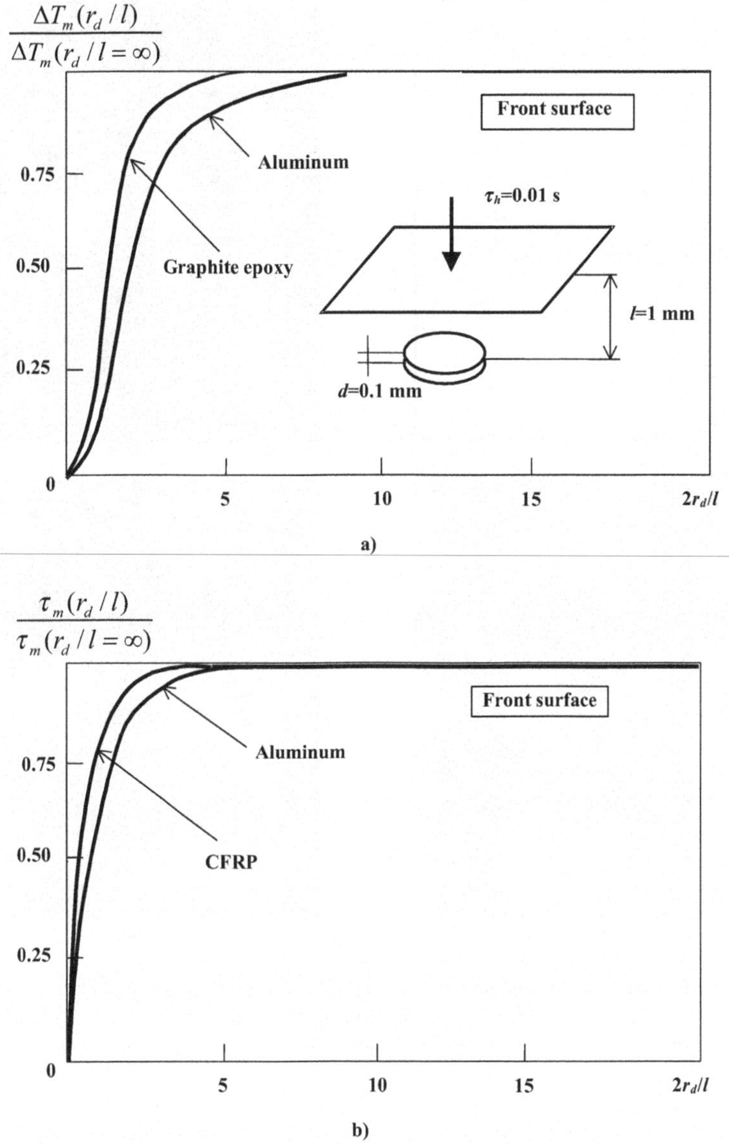

Fig. 4.29 Optimum detection parameters *vs.* defect lateral size: (**a**) maximum temperature signal vs. r_d/l, (**b**) optimum observation time *vs.* r_d/l

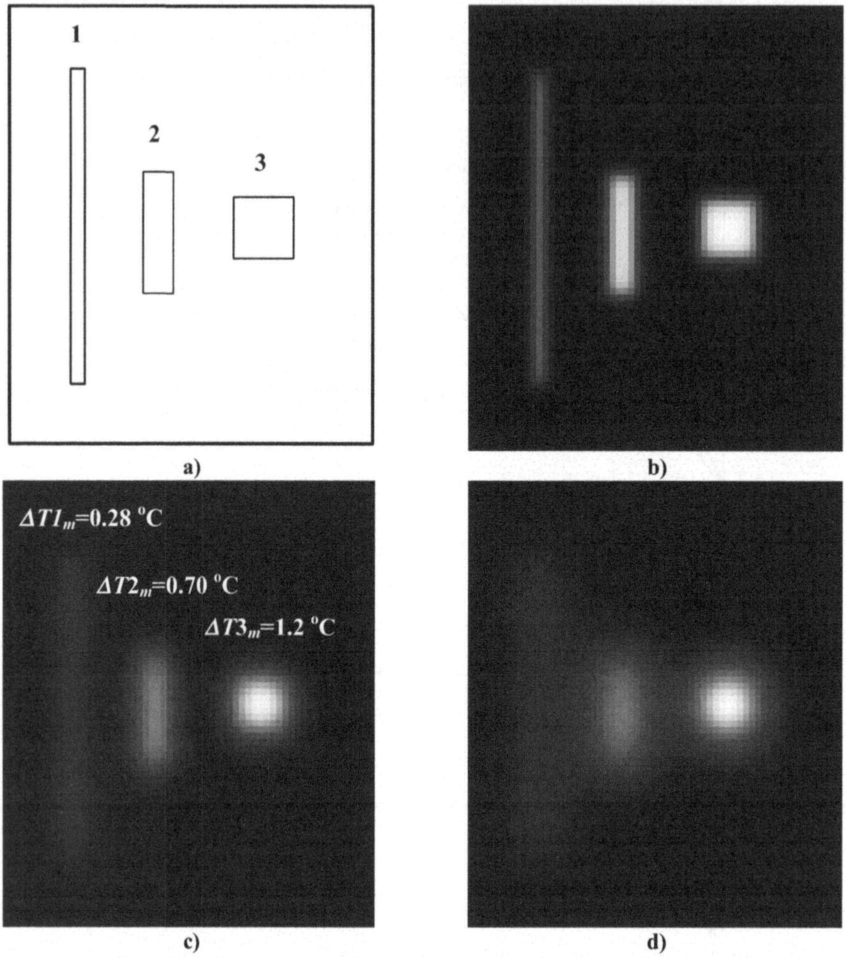

Fig. 4.30 Influence of defect configuration on surface temperature distribution (50% corrosion in 2 mm-thick steel, $Q = 10^6$ W/m^2, $\tau_h = 0.01$ s): (**a**) defect scheme (*Defect 1* – 1 × 25 mm^2, *Defect 2* – 2.5 × 10 mm^2, *Defect 3* – 5 × 5 mm^2), (**b**) temperature distribution at 0.1 s, (**c**) –0.5 s (optimum detection), (**d**) –1 s

calculated IR thermograms show 50% corrosive material loss in a 2 mm-thick steel sample at three observation times. The defect lateral area is 25 mm^2 (Fig. 4.30a). It is seen that defect shape is better identified at shorter observation times (Fig. 4.30b) when lateral heat diffusion is weak and the influence of neighbor defects is minimal. This TNDT peculiarity is used in the *Early Detection* technique. At optimum observation times (Fig. 4.30c), differential temperature signals reach their maximum values ΔT_m but diffusion of heat in lateral directions is also considerable at this time; for example, a five-fold change of defect size leads to an approximately five-fold change in ΔT_m at τ_m, even if the defect lateral surface is constant (compare signals from Defect 1 and 3). At longer times, heat diffusion may essentially ruin surface temperature 'footprints' of sub-surface defects. In anisotropic materials, additional signal distortion will occur because of spatial variation of thermal properties (Fig. 4.30d); for instance, square defects produce elliptical surface patterns where ellipse axes correspond to the principle directions of anisotropic thermal properties.

4.8.5 Heating Protocol

4.8.5.1 Hypothetic Optimal TNDT Procedure

In TNDT, both the excess temperature T and the differential signals ΔT are linearly proportional to the stimulating energy. Maximizing ΔT usually requires more powerful heating. However, the absorbed energy must be limited to a level that does not damage the material; moreover, it is more appropriate to maximize signal-to-noise ratio rather than the differential signal. In many cases, this means maximizing the running temperature contrast $C^{run} = \Delta T/T_{nd}$. In 1975, Karpelson et al. demonstrated that maximum contrasts are provided by point-like instantaneous heat source moving *within* a test sample [15]. The authors investigated extremes of the function obtained as an analytical result of solving a TNDT 3D problem for a solid body with subsurface local variations of thermal properties.

Maximum contrast C^{run} is achieved if an instantaneous point heat source operates at the defect centre. Its value can be estimated by applying the following simple argument. Assume that a body to be tested, consists of many elementary adiabatic volumes ΔV which are one by one heated with energy W. The mean temperature increase in each volume is:

$$T_{nd} = \frac{W}{(c\rho)_{nd}\Delta V}; T_d = \frac{W}{(c\rho)_d \Delta V}; \tag{4.76}$$

The maximum contrast C_m^{run} is:

$$C_m^{run} = \frac{(c\rho)_{nd}}{(c\rho)_d} - 1. \tag{4.77}$$

Maximum concentration and *instantaneous operation* of a hypothetic heat source will ensure adiabatic heat exchange for each volume, i.e. to prevent volumetric diffusion of thermal energy.

A practical implementation of such thermal stimulation method can be hardly imagined but there are some close techniques, namely: (1) surface heating with a localized moving heat source, such as a laser, (2) heating a metallic sample with a short powerful pulse of electrical current, (3) inductive heating of metallic layers in layered structures consisted of both metals and non-metals, (4) powerful volumetric

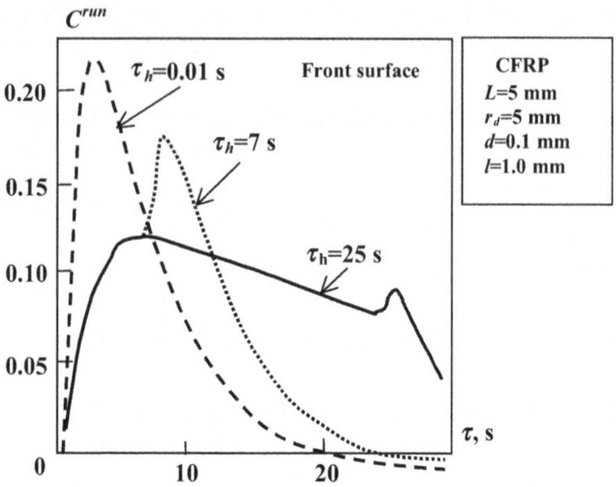

Fig. 4.31 Running temperature contrast *vs.* heating duration

Table 4.7 Maximum temperature contrasts and optimum observation times in the detection of an air-filled defect in CFRP composite ($L = 5\,mm$, $l = 1\,mm$, $d = 0.1\,mm$, $r_d = 5\,mm$)

Heating	Maximum running contrast C_m	Optimum observation time $\tau_m(C_m)$, s
Pulsed heating		
$\tau_h = 0.01$ s	0.230	2.6
$\tau_h = 7$ s	0.185	8.5
$\tau_h = 25$ s	0.127	7.0
Periodical heating		
$f = 0.1$ Hz	0.181	9.7
$f = 0.2$ Hz	0.208	5.3
$f = 0.5$ Hz	0.167	6.2

heating with microwaves. Earlier, in the case of surface heating (see Sect. 4.2.2), we obtained the maximum contrast value as $C_m^{run} = L/l - 1$ that is similar to Eq. (4.77).

In practice, optimization of a heating protocol is accomplished by calculating many variants of heating and temperature recording setups and applying some limitations in order to rank the corresponding test schemes.

4.8.5.2 Practical Optimization of a Heating Protocol

Each test case may benefit from the optimization of heating. However, many test cases will provide satisfactory results even with non-optimized heating. Practical testing is greatly simplified by using a small number of "one size fits all" test conditions.

Some features of one-sided TNDT are presented in Fig. 4.31. The highest contrast is provided by heating with a quasi-Dirac pulse ($\tau_h = 0.01$ s). Under long heating ($\tau_h = 25$ s) the maximum contrast occurs at 7 s. If the heating finishes at 7 s, the contrast will slightly increase due to a sharply decaying surface temperature. In fact, a rear front of a square pulse acts as a Dirac pulse, therefore, a short-term increase

Table 4.8 Influence of heat pulse shape on TNDT parameters (detecting 150 × 40 × 1 mm air-filled defect in 40 mm- thick plaster at depth of 3 mm)

	Optimum detection parameters				
Heat pulse shape	Maximum excess temperature over defect	ΔT_m, °C	$\tau_m(\Delta T_m)$, s	C_m^{run},	$\tau_m\left(C_m^{run}\right)$, s
Pulse 1 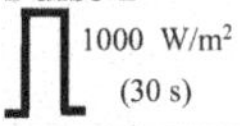	11.6 °C at 30 s	0.934	68	0.274	98
Pulse 2 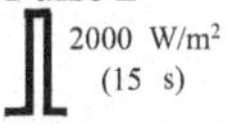	16.3 °C at 15 s	0.952	58	0.275	88
Pulse 3 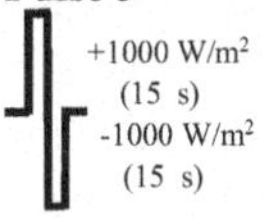	8.2 °C at 15 s	0.256 0.132*	34 22*	→∞	22*
Pulse 4  Total energy 30 kJ/m² (response to **Pulse 1**)	3.89 °C at 124 s	0.526	168	0.185	242
Pulse 5  Total energy 30 kJ/m² (response to **Pulse** *3*)	8.65 °C at 46 s	0.849	94	0.270	120
Pulse 6 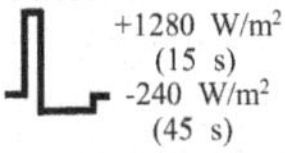	10.44 °C at 15 s	0.500 0.490*	48 51*	→∞	51*

*At time when $T_{nd} = 0$

of running contrast can be always observed after a square pulse is turned off. Note that, at long observation times, both temperature signals and contrasts can become negative due to faster cooling of a material layer over defects (Fig. 4.31).

Heating requirements can be summarized as follows: (1) a heat pulse should be short enough to produce necessary temperature contrasts C; in the case of optical heating, this allows to shift the observation instant out of heating, thus considerably reducing spurious reflected radiation, (2) total heat pulse energy must be high enough to ensure necessary ΔT signals, but (3) not high enough to damage material because of overheating.

Square pulse and harmonic heating procedures are compared in Table 4.7 in the detection of a 1 mm- deep air-filled defect in a 5 mm-thick CFRP composite (one-sided TNDT). It is seen that the highest running contrast is ensured by square pulse-heating.

In practice, there are some other factors influencing the choice of a heating procedure, namely, technical limitations, considerable noise possibly produced by a heater, equipment availability, safety issues, cost, etc.

Winfree suggested to treat TNDT experimental data with a special mathematical filter which is synthesized to present test results as though they were obtained under 'optimum' thermal stimulation [16]. With some simplifying assumptions, it has been concluded that, following the basics of optimal data filtration, a maximum signal-to-noise ratio is achieved if the shape of a heat pulse is identical to a ΔT response in time for a particular defect. In other words, the authors proposed the filtering of experimental data for the expected type of defects.

In practice, the influence of heat pulse shape on TNDT efficiency has been scarcely studied. A theoretical illustration is given in Table 4.8 where calculated optimum detection parameters in the detection of a 150 × 40 × 1 mm delamination under a wall fresco at the depth of 3 mm are presented. Various heating/cooling combinations, including the above-mentioned 'optimum' heating, have been investigated. All ΔT values are normalized by the absorbed energy of $W = 30\ kJ/m^2$, that is equivalent to a 1 kW/m^2 square pulse operating for 30 s.

As introduced earlier, let us optimize test parameters by using two criteria: ΔT and C^{run}. It follows from Table 4.8 that, assuming the temperature resolution of the test equipment to be $\Delta T_{res} = 0.1$ °C, the simulated defect can be detected in all test cases. Since frescos cannot be heated by more than 10 °C above the ambient temperature, the heating with Pulses 1, 3, 5 and 6 is close to the inspection limit, i.e. a further increase in heating power is unacceptable, while the heating with Pulse 2 requires decreasing power by 60% (note that the maximum temperature is to be evaluated over the defect where it is higher than in sound areas). Oppositely, when heating with Pulse 4, the power may be increased by three times. The data in Table 4.8 can be re-calculated by taking into account linearity of TNDT problems: $T \sim Q$ and $\Delta T \sim Q$, then all test cases can be arranged in the following order of diminishing ΔT values: Pulse 3, 4, 5, 1, 2, 6 and 3. Physically, this result is understandable because long heat pulses penetrate deeper into materials and cause minimal overheating, therefore, the total energy of such pulses may be increased.

As mentioned above, by applying 'standard' heating, optimum contrast is achieved with the shortest (Dirac-like) heat pulse (Pulse 2). Hypothetically, combination of both heating and cooling can be considered (Pulses 3 and 6) thus causing the situation where non-defect excess temperature T_{nd} will cross the zero axis and running contrast will tend to infinity due to definition of this parameter (Fig. 4.32). Such heating can be either symmetrical (Pulse 3, Fig. 4.32a) or asymmetrical (Pulse 6, Fig. 4.32b). The advantage of the asymmetrical heating is that, by controlling power and duration of heating, the point $T_{nd} = 0$ may occur at the time of maximum temperature signal ΔT_m (Fig. 4.32b). Experimentally, such scheme of thermal stimulation has not been investigated, moreover, in this case, TNDT efficiency will be limited by a signal-to-noise ratio, as shown in Eq. (4.69), rather than by running contrast.

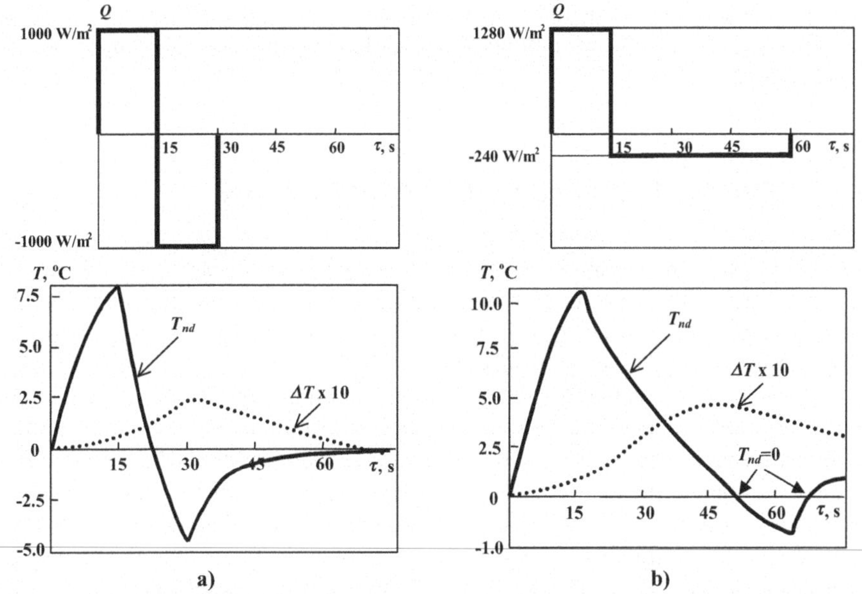

Fig. 4.32 Combined thermal stimulation (model by Table 4.8): (**a**) symmetrical stimulation mode, (**b**) asymmetrical stimulation mode

4.8.6 Material

In the ideal case of no noise, optimal materials for performing TNDT are those producing maximum running contrasts. Low-conductive materials prevent deeper penetration of thermal energy, while high-conductive materials favor strong heat diffusion. Theoretical analysis shows that temperature contrasts are higher in metals in regard to non-metals but higher noise typically inherent to metals reduces signal-to-noise ratio. Time for observing defect is to be within technical performance of used test equipment. For example, corrosion in thin aluminum panels produces considerable temperature contrasts which survive for only short times (10–100 ms). With such observation times, the use of standard IR imagers providing image acquisition frequency up to 30 Hz (serial reading-out IR detector circuitry) may lead to distortion of corrosion images because the temperature in sample different points is recorded at slightly different times. Therefore, for detecting corrosion in thin high-conductive materials, the so-called snap-shot FPA IR imagers with parallel reading-out IR detector circuitry are recommended. Moreover, metals should be typically black-painted to solve a number of problems, namely: (1) enhancing absorbed energy, (2) reducing random fluctuations of radiation across surface, and (3) reducing reflected radiation.

Influence of a type of material on optimum detection parameters is illustrated in Fig. 4.33. It is seen that higher contrasts appear in higher-conductive (Fig. 4.33a) materials at shorter times (Fig. 4.33b). This conclusion is valid for different defect depths, even if volumic heat diffusion is stronger for deeper defects. In fact, for each particular defect, there is an 'optimum' material where the corresponding contrast C is maximal.

4.8.7 Heating Power and Surface Heat Exchange

Since both excess temperatures and differential signals are linearly proportional to absorbed energy W, or power Q, temperature contrasts (both running and normalized) are independent of W (Q). In many cases, more intensive heating provides defect indications of a better quality due to higher signal-to-noise ratio $s = \Delta T/\Delta T_{noise}$, where ΔT_{noise} is the signal conditioned by several types of noise, including the noise of an IR imager determining temperature resolution ΔT_{res}. The ΔT_{res} noise is additive, while surface noise can be considered as multiplicative. The most important components of the multiplicative noise is caused by fluctuations of emissivity which appears as a coefficient in the output signal equation of an IR imager. In the case of strictly multiplicative noise, noise contrast is stable in time thus being a good characteristic of a particular material. However, in general case, noise may vary during heating/cooling because of reflected radiation and some non-linear phenomena.

Intensity of surface heat exchange is expressed by coefficients α^F and α^R on the front and rear surfaces respectively. In the inspection of thin and/or high conductive materials which meet the condition $Bi = \alpha L/\lambda < 0.1$ (adiabatic boundary conditions), temperature signals are weakly dependent on surface heat exchange in a wide range of α. If thick and/or low-conductive materials are tested, intensive cooling of the front surface will bring down temperature contrasts on this surface. In two-sided TNDT, temperature signals are nearly independent of α.

A special case is where intensive surface heat exchange is used for producing a dynamic thermal regime, for example, when cooling hot steel with water jet.

4.8.8 Thermal Properties Anisotropy

Anisotropy phenomena may be essential in the inspection of fiber reinforced composites where single plies (fiber layers) are laid up under different angles, with thermal conductivity along fibers being considerably higher than in perpendicular directions. This is particularly true in the case of unidirectional composites. In Figure 4.34a shows a sample made of three CFRP layers of the 0.2 mm thickness each and containing square-shape delaminations. Thermal properties of each layer are different by coordinates x and y. Heating is performed with a 10 ms heat pulse. At 0.2 s, the square shape of subsurface defects is well reproduced on the front surface with a certain distortion being observed for the shallow defect in the first layer (Fig. 4.34b). At longer times (1 s), defect indications become elliptical because of

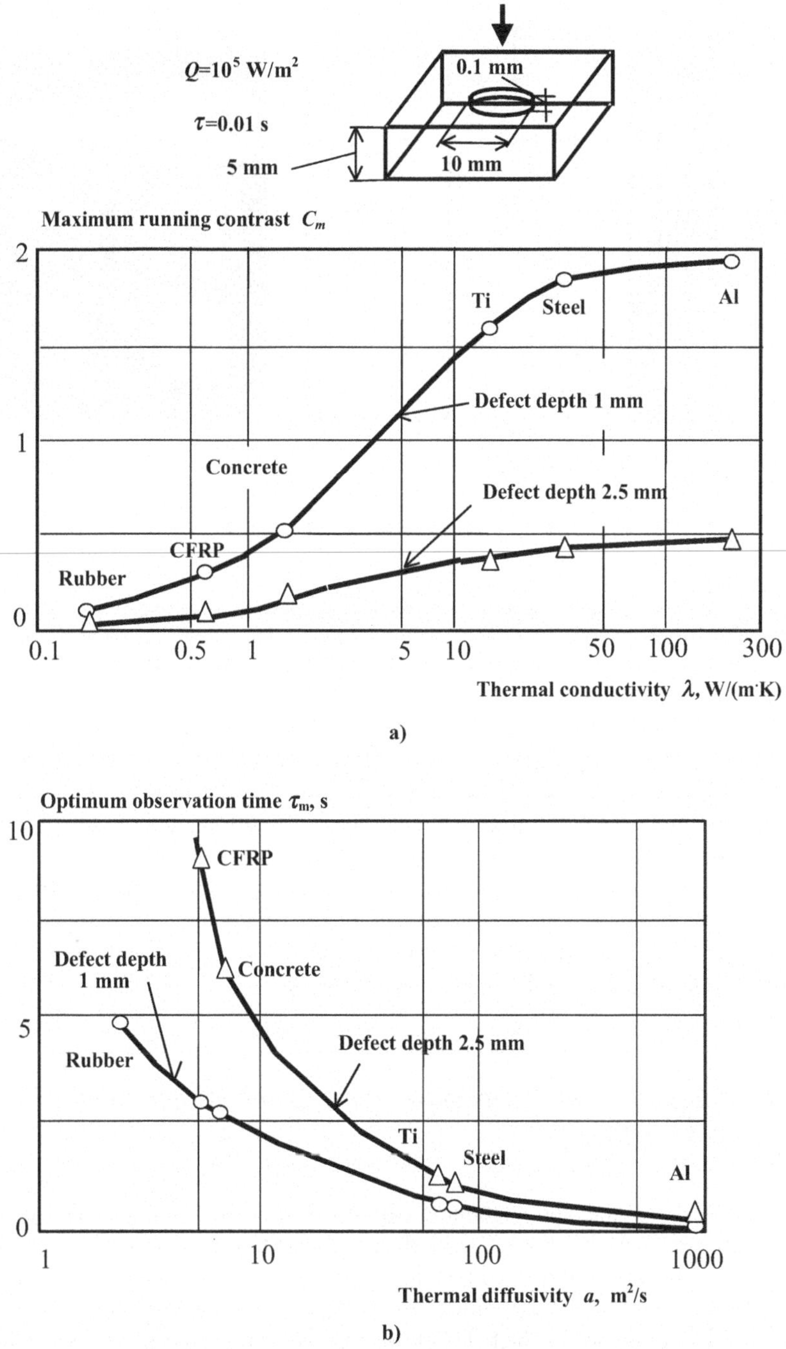

Fig. 4.33 Optimum observation parameters *vs.* material thermal properties: (**a**) maximum running contrast *vs.* thermal conductivity, (**b**) optimum observation time *vs.* thermal diffusivity

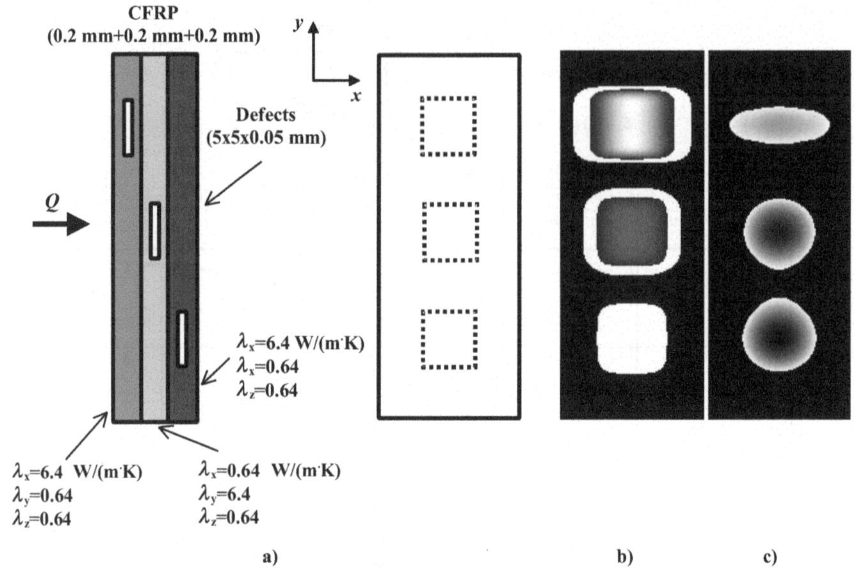

Fig. 4.34 Influence of thermal property anisotropy on temperature distribution over defects in CFRP after flash heating (ThermoCalc-3D software): (**a**) sample scheme, (**b**) IR thermogram at 0.2 s (isothermal presentation), (**c**) IR thermogram at 0.2 s (isothermal presentation)

uneven heat diffusion; the pattern of the first layer defect reveals the significant stretching by x because the x -component of thermal conductivity is 10 times higher than by the y coordinate (Fig. 4.34c). The shapes of other defect indications are also distorted because of the anisotropy of upper layers. Therefore, in thin samples made of CFRP composite, which is a typical anisotropic material, surface temperature signatures may be distorted and even shifted in regard to location of subsurface defects. In thick samples, due to variation of a fiber lay-up angle, overall composite anisotropy is diminishing due to the averaging effect, even if lateral and in-depth thermal properties are very different within each ply.

4.8.9 Temperature Distribution over a Crack Perpendicular to a Surface

A sample made of steel and containing two air-filled defects (Fig. 4.35a) is heated with a line heater during 0.5 s. One defect is located at the depth of 1 mm, thus simulating a subsurface crack, and another defect practically reaches the front surface (being at depth of 10 μm) simulating a surface crack with a close mouth (or filled with paint, mud etc.). Such defects are typical, for instance, in turbine blades [17]. In 0.1 s after turning the heater on, surface temperature distribution is conditioned only by line heating (on the left section of the sample) revealing no defects (Fig. 4.35b).

The temperature profile is regular at this time with temperature dropping from the left to the right. Starting from a particular time, which depends on distance between the heated zone and the cracks, the regular temperature behavior starts to distort because the crack prevents heat flowing from the left to the right and in-depth (Fig. 4.35c). However, the shape and the amplitude of the ΔT signal do not allow reliable detection of both cracks if an operator simply views IR thermograms. It was suggested in [18, 19] to underline the presence of defects by differentiating a spatial profile as shown in Fig. 4.35d–g. The image in Fig. 4.35d was obtained by treating the image in Fig. 4.35c with a *statistical local variance filter*. In this case, the following 3x3 filter mask was used:

$$\sigma^2(i,j) = \frac{1}{N^2}\sum_{i=1}^{N=3}\sum_{j=1}^{N=3}\left[T(i,j) - m\right]^2; \quad m = \frac{1}{N^2}\sum_{i=1}^{N=3}\sum_{j=1}^{N=3} T(i,j),, \tag{4.78}$$

where $T(i,j)$ is the temperature at a point $[i,j]$. The corresponding temperature profile for both defects is shown in Fig. 4.35e. The classical shapes of the temperature derivative are shown in Fig. 4.35f, g as the result of treating the thermogram in Fig. 4.35c with the Laplace filter (3 × 3 mask):

$$3\times 3 \quad \begin{vmatrix} -1-1-1 \\ -1+8-1 \\ -1-1-1 \end{vmatrix} \tag{4.79}$$

The distinct images of both defects in Fig. 4.35d–g are due to the steep slope of the temperature profiles across the cracks. However, in the case of the subsurface defect, the temperature signal is approximately three times lower than for the surface crack. In practice, the quality of such images worsens to compare with theoretical predictions because of enhancement of high-frequency noise as a result of differentiation that, in its turn, reduces signal-to-noise ratio.

The quality of source thermograms can be improved by: (1) localizing surface heating, (2) moving a surface heat source with a constant speed and recording temperature at a particular distance from a heated zone, (3) heating an edge of a sample thus transiting from surface to volumic stimulation.

4.8.10 Forced Sample Cooling

As mentioned above, processes of heating and cooling are mathematically identical and simply characterized by opposite heat flux signs in the corresponding equations. However, technologically, heating is more appropriate, in particular, in small areas and with high power. As an example of intensive cooling, consider the forced

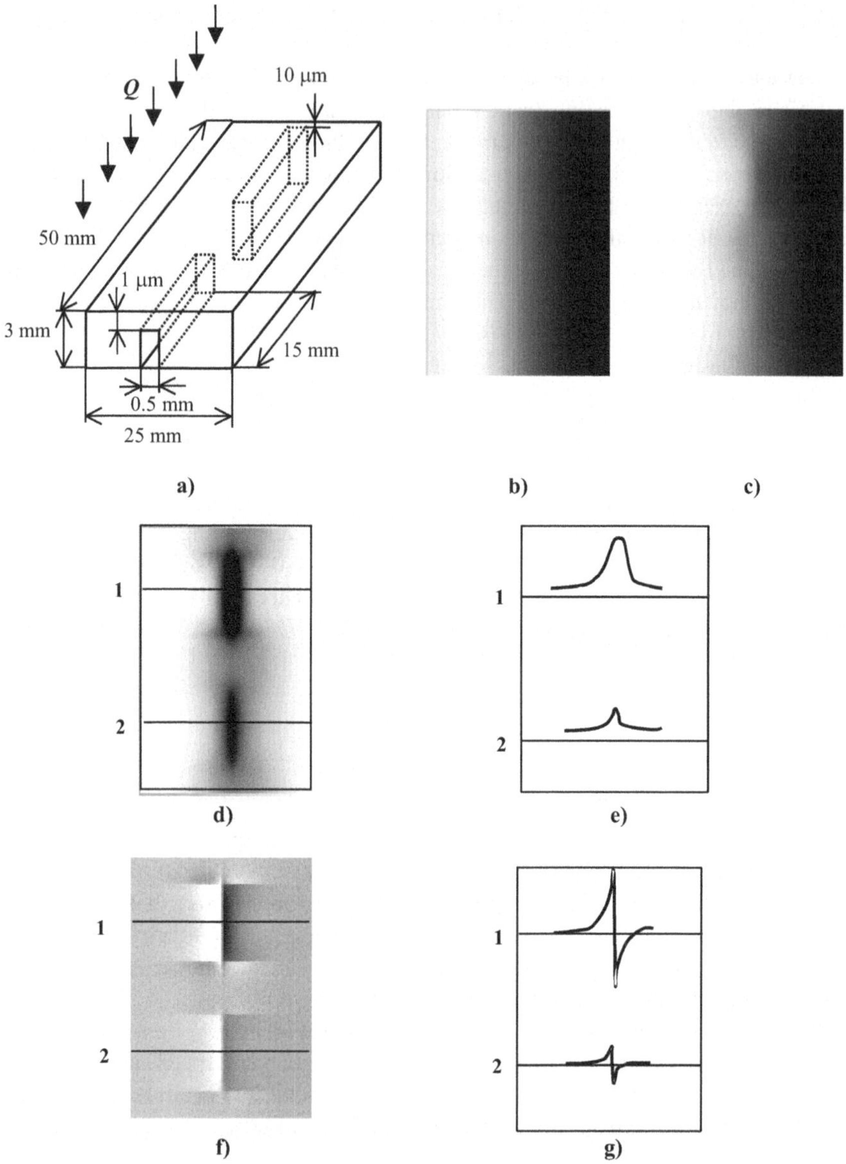

Fig. 4.35 Detecting perpendicular-to-surface cracks in AISI 1010 steel (strip heating for 0.5 s, $\lambda = 63, 9$ W/(m·K), $a = 18.8 \cdot 10^{-6}$ m²/s, ThermoCalc-3D): (**a**) sample scheme, (**b**) temperature distribution at 0.1 s, (**c**) temperature distribution at 2 s, (**d**) same as (c) after processing with statistical filter (inverse presentation), (**e**) same as c) after processing with Laplace filter, (**f**) temperature profile for d) (**g**) temperature profile for e)

cooling of a 3 mm-thick AISI 1010 steel sample, which is pre-heated up to 850 °C, with a cold water jet (such test cases take place in metallurgy). Assume that the sample contains three air-filled defects located at depth of 30 μm (*Defect 1* and *2* which differ by lengthwise dimensions) and 60 μm (*Defect 3*), see Fig. 4.36. The water is delivered onto the sample surface for 0.01 s (equivalent heat exchange coefficient is $\alpha_{forced} = 5000$ W/(m^2·K), afterwards, the sample continues cooling by 'normal' convection with the coefficient $\alpha = 10$ W/(m^2·K). Due to relatively high thermal diffusivity of steel, thermal processes in the sample occur for tens milliseconds (see the example of temperature distribution in Fig. 4.36a). The evolution of the temperature signal in time is similar to that which would occur under short pulse heating, but the ΔT sign is negative because, in this case, defect areas are colder than the

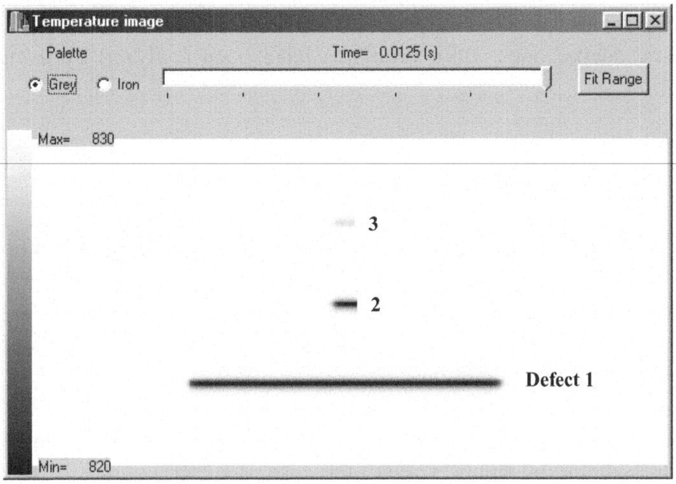

a)

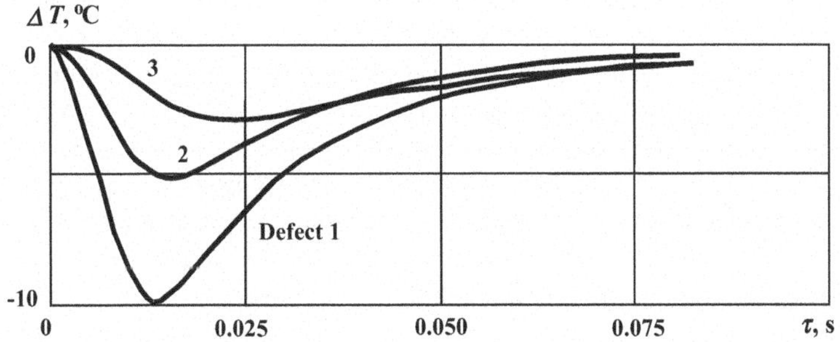

b)

Fig. 4.36 Detecting air-filled defects in a 3 mm-thick AISI 1010 steel sample (forced cooling for 0.01 s, heat exchange coefficient α_{forced} = 5000 W/(m^{2}K), ThermoCalc-6 L): (**a**) thermogram at 0.0125 s (Defect 1: $l = 0.3$ mm, $d = 90$ μm, $h_x \times h_y = 40 \times 0.9$ mm, Defect 2: $l = 0.3$ mm, $d = 90$ μm, $h_x \times h_y = 4.5 \times 0.9$ mm, Defect 3: $l = 0.6$ mm, $d = 90$ μm, $h_x \times h_y = 4.5 \times 0.9$ mm, (**b**) differential temperature signal *vs.* time surrounding (Fig. 4.36b). Obviously, the deeper defect produces the lowest signal. Maximum signals and optimum observation times are:

Defect 1: $\Delta T_m = 10.0$ °C at $\tau_m = 0.0125$ s,
Defect 2: $\Delta T_m = 5.1$ °C at $\tau_m = 0.015$ s,
Defect 3: $\Delta T_m = 2.8$ °C at $\tau_m = 0.0225$ s.

4.8.11 Identifying Neighbor Defects

If two defects are close each to other, the two following problems should be analyzed. First, two defect should be spatially separated by viewing their surface 'footprints' (the same problem appears in optics). Second, shallow defects may overshadow deeper ones. Both problems have been analyzed by using numerical models.

4.8.11.1 TNDT Spatial Resolution

The task is to provide such testing conditions which would allow better resolving neighbor defect in plane. By analogy with miras used in optics, this can be done by analyzing reference samples containing groove-like defects of different width and depth. For a particular reference sample, spatial resolution can be improved by choosing optimum test conditions.

Let us limit ourselves with the illustration of the approach above by analyzing TNDT of CFRP composite and aluminum samples which contain 6 defects at depth of 0.5 mm (scheme by Fig. 4.37a). In the case of CFRP composite, consider two variants: (1) isotropic material ($\lambda_x = \lambda_y = \lambda_z = 0.64$ W/(m·K)); (2) anisotropic material ($\lambda_y = \lambda_z = 0.64$ W/(m·K); $\lambda_x = 6.4$ W/(m·K)). A ratio between the temperature signal ΔT_{res} above two neighbor defects and the maximum signal ΔT (for a single defect) has been chosen as a measure of spatial resolution (Fig. 4.37b). Assume that two neighbor defects will be reliably detected and resolved if $\Delta T_{res} \geq 0.1\,\Delta T$. Then, in the isotropic composite, spatial resolution will be better than 0.5 mm at 1 s that corresponds to the first two defects on the left (Fig. 4.37b). In the anisotropic composite, due to enhanced heat diffusion by the x coordinate, spatial resolution worsens up to 2.5 mm that is the distance between the last defects on the right (Fig. 4.37c). Respectively, in aluminum, spatial resolution is 0.5 mm at 0.03 s (Fig. 4.37d). It is worth reminding that TNDT spatial resolution drops down for deeper defects and with longer observation times; therefore, the best reproduction of defect shape and correspondingly the best resolution of neighbor defects can be achieved by using the so-called Early Detection technique.

Profiles in Fig. 4.37 also show that few defects located close each to other, for example, a conglomerate of pores in a composite, will enhance a mean temperature in the respective defect area and, if spatial and/or temperature resolution of the used

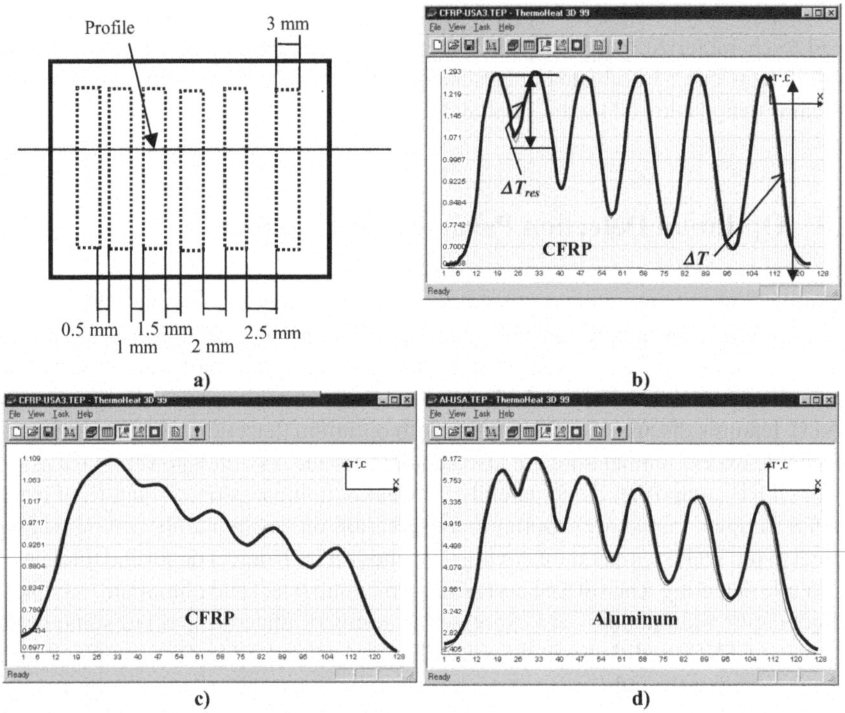

Fig. 4.37 Resolving defects in TNDT (defect lateral size 3 × 25 mm, defect depth 0.5 mm, defect thickness 0.5 mm): (**a**) defect scheme, (**b**) CFRP ($L = 5$ mm, $\lambda_x = \lambda_y = \lambda_z = 0.64$ W/(m·K), $Q = 10^5$ W/m^2, $\tau_h = 1$ s), (**c**) CFRP ($L = 5$ mm, $\lambda_x = 6.4$ W/(m·K), $Q = 10^5$ W/m^2, $\tau_h = 1$ s), (**d**) aluminum ($L = 2$ mm, $Q = 10^6$ W/m^2, $\tau_h = 0.03$ s)

IR imager are low, multiple defects may be identified as one large temperature anomaly (defect).

4.8.11.2 Overlapping Defects

The analysis of this situation performed by Bendada et al. [20] has shown that the summary temperature distribution $\Delta T_{\Sigma}(x, y, \tau)$ over a bunch of small-size defects can be represented as the superposition of signals produced by each single defect: $\Delta T_{\Sigma}(x, y, \tau) = \sum \Delta T_i(x, y, \tau)$, where summation is done in both space and time. In some cases, defect identification is possible by simply viewing IR thermograms. For instance, in the first case shown in Fig. 4.38a, the small deeper defect is seen through the larger shallow one. However, defect characterization is hardly possible in this case because temperature signals caused by deeper defects are typically low and cannot be identified on the noisy background. Theoretically, a summary signal produced by two thin defects behaves slightly different to compare to an equivalent single defect, but experimentally this fact has not been demonstrated. Moreover, the presence of only a single defect may lead to sample rejection, therefore, there is no need to characterize other (deeper) defects. As for the case depicted in Fig. 4.38b, both defects can be evaluated quantitatively except the section of the larger defect which is overshadowed by the small defect.

4.9 Optimum Detection Parameters: Examples

Optimization of experimental procedures is an important goal of the TNDT theoretical analysis. In an optimum test situation, it is expected that defects should produce maximum values of a chosen informative parameter at particular observation times. In Tables 4.9, 4.10, and 4.11, there are three examples which illustrate many TNDT features discussed above. Along with optimum detection parameters, values of sample excess temperature are also supplied for the specified power density.

In CFRP composite, defect lateral size weakly influences on ΔT and C, at least, in thin samples (Table 4.9). Optimum observation times are mainly determined by defect depth while temperature contrast depends on both defect depth and thickness.

While detecting rear-surface corrosion in metallic (steel and aluminum) samples, defect depth and thickness are expressed in terms of material loss but some basic features of TNDT valid also in this case (Tables 4.10 and 4.11), for example, defect lateral size mainly influences on temperature contrasts but not their observation times. The data in Table 4.10 also proves that the three typical TNDT methods, namely, pulsed, step-wise and harmonic heating procedures, provide similar values of detection parameters thus confirming that the general physics behind of these techniques remains the same.

It is worth noting that, in the case of flash heating, the data in Tables 4.10 and 4.11 is well described by Eq. (3.11) which can be also expressed in terms of running contrast:

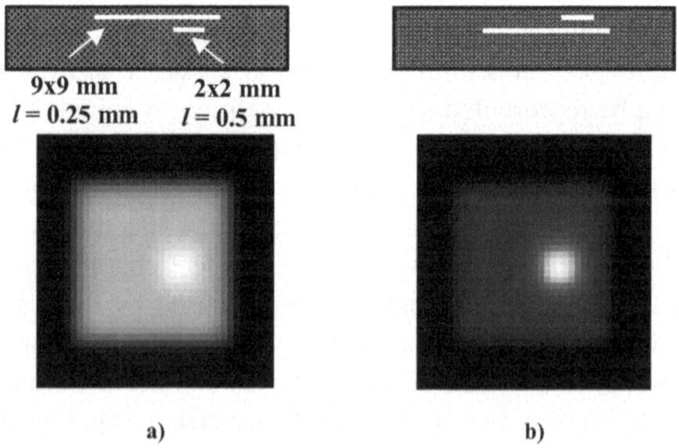

Fig. 4.38 Phenomenon of defect overlapping (CFRP, $L = 5$ mm, $l = 1$ mm, $d = 0.1$ mm, $Q = 10^5$ W/m^2, $\tau_h = 0{,}01$ s, $\tau = 0.5$ s)

Table 4.9 Optimum detection parameters for air-filled defects in composite material (CFRP: $a = 5.2 \cdot 10^{-7}$ m²/s, $\lambda = 0.64$ W/(m·K), air: $a = 5.8 \cdot 10^{-5}$ m²/s, $\lambda = 0.07$ W/(m·K))

Sample thickness L, mm	Defect depth l, mm	Defect radius r_d, mm	Defect thickness d, mm	Maximum running contrast C_m^{run}	Optimum detection time $\tau_m\left(C_m^{run}\right)$, s
Flash heating ($\tau_h = 0.01$ s, $T_{max} \approx 11.3$ °C for $Q = 10^5$ W/m²)					
1	0.25	5	0.05	0.50	0.27
			0.10	0.84	0.36
		10	0.05	0.50	0.27
			0.10	0.84	0.36
	0.50	5	0.05	0.22	0.50
			0.10	0.36	0.57
		10	0.05	0.22	0.50
			0.10	0.36	0.57
	0.75	5	0.05	0.077	0.69
			0.10	0.13	0.83
		10	0.05	0.077	0.69
			0.10	0.13	0.83
2	0.50	5	0.05	0.28	0.77
			0.10	0.49	1.00
		10	0.05	0.28	0.77
			0.10	0.50	1.01
	1.00	5	0.05	0.13	1.73
			0.10	0.22	1.92
		10	0.05	0.13	1.73
			0.10	0.22	1.94
	1.50	5	0.05	0.042	2.45
			0.10	0.076	2.69
		10	0.05	0.042	2.46
			0.10	0.077	2.72
5	1.25	5	0.05	0.12	3.40
			0.10	0.22	3.85
		10	0.05	0.12	3.62
			0.10	0.22	4.00
	2.50	5	0.05	0.047	8.70
			0.10	0.086	9.10
		10	0.05	0.054	9.70
			0.10	0.10	10.3
	3.75	5	0.05	0.014	12.54
			0.10	0.026	12.98
		10	0.05	0.018	14.13
			0.10	0.034	14.75

Table 4.10 Optimum detection parameters in TNDT of corrosion in steel (steel: $a = 7.3 \cdot 10^{-6}$ m²/s, $\lambda = 32$ W/(mK))

Sample thickness L, mm	Material loss, %	Defect radius r_d, mm	Maximum running contrast C_m^{run}	Optimum detection time $\tau_m\left(C_m^{run}\right)$, s
Flash heating ($\tau_h = 0.01$ s, $T_{max} \approx 8.4$ °C for $Q = 10^6$ W/m²)*				
1	90	5	8.79	0.11
		10	8.89	0.15
		20	8.89	0.15
	50	5	0.99	0.14
		10	1.00	0.19
		20	1.00	0.19
	10	5	0.11	0.16
		10	0.11	0.22
		20	0.11	0.23
5	90	5	4.56	0.61
		10	4.78	1.23
		20	8.51	2.05
	50	5	0.58	1.06
		10	0.89	1.57
		20	0.99	2.51
	10	5	0.055	1.49
		10	0.097	1.97
		20	0.11	2.96
10	90	5	2.24	0.86
		10	4.68	2.42
		20	7.52	4.73
	50	5	0.24	3.14
		10	0.58	4.26
		20	0.89	6.25
	10	5	0.017	4.70
		10	0.053	5.51
		20	0.093	7.82

(continued)

Table 4.10 (continued)

Long heating ($\tau_h = 5$ s)				
		5	1.47	5.22
		10	4.12	5.97
	90	20	7.21	7.88
		5	0.19	6.61
		10	0.53	7.60
	50	20	0.86	9.38
		5	0.014	7.88
		10	0.049	9.06
10	10	20	0.090	10.90
Thermal wave heating∗∗				
10	90 (frequency $f = 4$ Hz)	5	8.60	0.26
		10	8.78	0.26
		20	8.78	0.26
	50 ($f = 0.14$ Hz)	5	0.19	7.34
		10	0.50	7.59
		20	0.72	14.69
	10 ($f = 0.1$ Hz)	5	0.013	10.43
		10	0.043	10.93
		20	0.073	11.26

∗Maximum excess temperature at the end of heating is weakly dependent on sample thickness
∗∗Quasi-optimal frequencies are chosen for detecting particular defects

$$\frac{\Delta L}{L} = \frac{C^{run}}{1 + C^{run}}; \quad C^{run} = \frac{\Delta L / L}{1 - \Delta L / L}. \tag{4.80}$$

Equation (4.80) contain no thermal properties thus making this approach very general. For three values of material loss (90%, 50% and 10%) presented in Tables 4.10 and 4.11, the values of the running contrast calculated by this equation are 9, 1 and 0.11 correspondingly thus being close to the table values calculated numerically.

Some comparative results of TNDT of extended (1D) air-filled defects in various materials are presented in Table 4.12 for both flash and long heating. Note the considerable difference in optimum observation times and maximum contrasts. The data in Table 4.12 can be used for predicting TNDT potentials in materials with different thermal conductivities.

Example The data in Tables 4.9, 4.10, and 4.11 allows evaluating realistic potentials of TNDT of different materials if surface noise contrast is known. For instance, it has been reported elsewhere that CFRP composite is characterized by running noise contrast of about 6% (at the level of 2–3 standard deviations of noise), therefore, only defects producing $C^{run} > 0.06$ can be reliably detected. According to Table 4.9, in a 1 mm-thick CFRP composite, the last condition specifies defects of a radius greater than 5 mm, thickness over than 50 μm at depth under 0.75 mm. Respectively, in a 5 mm-thick composite, the detection limits are: by radius – greater

Table 4.11 Optimum detection parameters in TNDT of corrosion in aluminum **(Al:** $a = 8.6 \cdot 10^{-5}$ m²/s, $\lambda = 210$ W/(m·K))

Sample thickness L, mm	Material loss, %	Defect radius r_d, mm	Maximum running contrast C_m^{run}	Optimum detection time $\tau_m\left(C_m^{run}\right)$, s
Flash heating ($\tau_h = 0.01$ s)				
1 ($T_{max} \approx 5.57$ °C for $Q = 10^6$ W/m²)	90	5	8.27	0.02
		10	8.88	0.03
		20	8.91	0.04
	50	5	0.91	0.02
		10	0.99	0.04
		20	1.00	0.05
	10	5	0.10	0.03
		10	0.11	0.04
		20	0.11	0.06
2 ($T_{max} \approx 4.54$ °C for $Q = 10^6$ W/m²)	90	5	7.48	0.03
		10	8.78	0.05
		20	8.91	0.07
	50	5	0.83	0.04
		10	0.98	0.06
		20	1.00	0.09
	10	5	0.088	0.04
		10	0.108	0.07
		20	0.110	0.10

than 5 mm, by defect thickness - over 100 μm and by depth – under 2.5 mm. Obviously, a detection limit by any detection parameter can be improved by worsening other parameters, for example, defects by a radius under 5 mm can be detected either at a lower depth or they should be thicker than 50 μm.

4.10 Advanced TNDT Models

All TNDT models discussed above can be considered classical because they predict principal relationships between involved parameters which are well studied and used in developing characterization algorithms (see Chap. 6). Further improvement of theoretical models is related to: (1) some peculiar problems where specific detection parameters appear; for example, water motion in soil and surface evaporation in the detection of buried landmines may significantly modify temperature distributions, (2) some subtle phenomena which are not decisive for defect detection but important for better defect identification; for example, anisotropy of composites may distort surface temperature signals, (3) possible modification of thermal properties of a tested material caused by presence of defects, (4) combination of

Table 4.12 Optimum detection parameters for extended (1D) air-filled defects in different materials subject to both flash and long heating (defect depth $l = 10\%L$; defect thickness $d = 10\%L$, material properties: Al – $\lambda = 210\ W/(mK)$, $a = 8.6 \cdot 10^{-5}\ m^2/s$, steel – $\lambda = 32\ W/(mK)$, $a = 7.3 \cdot 10^{-6}\ m^2/s$; CFRP – $\lambda = 0.64\ W/(mK)$, $a = 5.2 \cdot 10^{-7}\ m^2/s$, concrete: $\lambda = 1.5\ W/(mK)$, $a = 7.5 \cdot 10^{-7}\ m^2/s$, air defect – $\lambda = 0.07\ W/(mK)$, $a = 5.8 \cdot 10^{-5}\ m^2/s$; heat exchange coefficients on both surfaces $\alpha = 10\ W/(m^2K)$; Layer-3 Analytic software for analytical solving 1D TNDT problems)

		Optimum detection parameters							
		Al		Steel		Concrete		CFRP	
Mate-rial thick-ness, mm	Heating conditions	C_m (τ_m, s)	$\Delta T_m, {}^oC$ (τ_m, s)	C_m (τ_m, s)	$\Delta T_m, {}^oC$ (τ_m, s)	C_m (τ_m, s)	$\Delta T_m, {}^oC$ (τ_m, s)	C_m (τ_m, s)	$\Delta T_m, {}^oC$ (τ_m, s)
1	$\tau_h = 0.01\ s$, $Q = 10^5\ W/m^2$	8.7 (0.015)	3.60 (0.012)	7.8 (0.06)	1.9 (0.04)	2.80 (0.20)	2.80 (0.08)	1.65 (0.13)	3.70 (0.05)
	$\tau_h = 100\ s$, $Q = 10^4\ W/m^2$	7.73 (0.05)	25.1 (98)	5.9 (0.15)	18.5 (100)	1.91 (0.4)	24.8 (80)	1.10 (0.36)	24.5 (50)
5	$\tau_h = 0.01\ s$, $Q = 10^5\ W/m^2$	8.7 (0.3)	0.72 (0.2)	7.8 (1.5)	0.38 (0.6)	2.72 (5)	0.56 (1.2)	1.61 (3.1)	0.73 (1.1)
	$\tau_h = 100\ s$, $Q = 10^4\ W/m^2$	7.62 (1.2)	54.1 (75)	5.96 (6)	54,0 (100)	1.85 (10)	52.2 (55)	1.05 (8)	49.7 (50)

(continued)

Table 4.12 (continued)

Mate-rial thick-ness, mm	Heating conditions	Optimum detection parameters							
		Al		Steel		Concrete		CFRP	
		C_m (τ_m, s)	$\Delta T_m, {}^oC$ (τ_m, s)	C_m (τ_m, s)	$\Delta T_m, {}^oC$ (τ_m, s)	C_m (τ_m, s)	$\Delta T_m, {}^oC$ (τ_m, s)	C_m (τ_m, s)	$\Delta T_m, {}^oC$ (τ_m, s)
50	$\tau_h = 0.01\ s,$ $Q = 10^5\ W/m^2$	8.8 (18)	0.072 (10)	7.14* (80)	0.035* (25)	2.1 (200)	0.049 (100)	1.28 (130)	0.06 (90)
	$\tau_h = 100\ s,$ $Q = 10^4\ W/m^2$	7.99 (111)	65.7 (101)	7.35 (200)	36.0 (120)	2.21 (320)	48.4 (160)	1.30 (250)	57.1 (150)

*Approximate values

thermal and optical phenomena, such as relationship between a viewangle and reflection index, etc.

To take into account the above-mentioned factors, some *advanced* TNDT models have been proposed where the following phenomena have been studied:

- arbitrary dependence of heating power $Q(x, y, z, \tau)$ on time and coordinates,
- thermal property anisotropy,
- phase transformation and mass transfer in materials under test (water evaporation, ice melting),
- roughness of tested materials,
- convective and radiant components of heat exchange varying in time,
- dependence of emissivity on material status and viewangle.

4.10.1 Detecting Teflon Inserts in CFRP Composite

A common practice of simulating defects in composite reference samples is placing Teflon inserts between material plies. Modeling such defects with classical TNDT models is not always possible because the hypothesis of ideal contact between inserts and the host material leads to underestimation of temperature signals in regard to experimental data. Therefore, one should concede the presence of air gaps between inserts and the host material, as well as possible modification of thermal properties in defect areas.

In Fig. 4.39a, there are the schemes of two reference samples made of CFRP composite where Teflon inserts are placed at the depths of 0.5 and 1.3 mm. Experimental ΔT temperature evolutions are presented in Fig. 4.39b in the case of flash heating ($Q = 1.5{\cdot}10^{-6}$ W/m^2, $\tau_h = 10$ ms). In the sample **A**, the maximum positive signal $\Delta T_m = 1.2$ °C occurs at $\tau_m = 0.8$ s, while in the sample **B** the signal amplitude is lost in the noise, even if visually there is a little temperature drop in this defect area. Experimental values of ΔT_m and τ_m for both samples are given in Table 4.13 to compare with the data calculated by a classical model which presumes ideal contact between Teflon and composite. It is important noting that the presence of air gaps around inserts cannot explain the negative sign of ΔT in the sample **B**. Such discrepancy between theoretical and experimental results can be avoided if, in addition to air gaps, to hypothesize modification of thermal properties of the composite that might occur in the process of sample preparation, see Fig. 4.40a. Satisfactory coincidence of results (see the curves in Fig. 4.40b and the data in Table 4.13) can be reached if to assume that both composite density and thermal conductivity are increased by 25%; note that in this case thermal diffusivity remains constant by definition.

It is worth noting that the approach above allows solving inverse TNDT problems by trial and does not guarantee solution uniqueness.

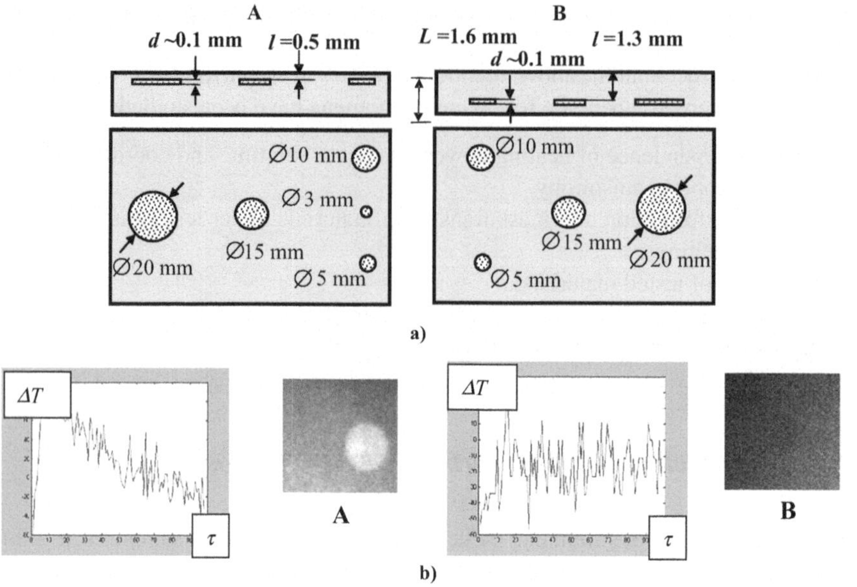

Fig. 4.39 Teflon inserts in CFRP reference sample (**a**) and experimental temperature signals (**b**) under flash heating ($Q = 1.5{\cdot}10^{-6}$ W/m^2, $\tau_h = 10$ ms)

Table 4.13 Comparison between experimental and theoretical data in TNDT of CFRP composite (Fig. 4.39)

Parameters		Experiment	Classical model (ideal contact between Teflon and composite)	Model by Fig. 4.40 (program ThermoCalc-2D)
Sample A	ΔT_m	+1.2 °C	+0.81 °C	+1.3 °C
	τ_m	0.8 c	0.63 c	0.78 c
Sample B	ΔT_m	~−0.2 °C	+0.21 °C	−0.2 °C
	τ_m	~1…2 c	2.23 c	1.1 c

4.10.2 Detecting Buried Landmines

Another example of advanced TNDT models is the detection of landmines buried in soil (see Fig. 4.41). The nomenclature of antitank and antipersonnel landmines is very wide but in theoretical models they are often simulated with a disk made of a homogeneous material and placed in soil at depths from 1 to 15 cm. Soil is heated with solar radiation of whose the diurnal cycle is described with Eq. (3.33): $Q(\tau) = Q_0\ [1 + \cos\ (\omega\ \tau)]$, where the cyclic frequency is related to the day-night variation of solar illuminance ($\omega = 2\pi/24\ hrs$). Most experts believe that detecting buried landmines by using IR thermography is possible due to local variations of soil thermal properties appearing when placing a mine. This is illustrated by the diurnal cycle of ΔT which is predicted by a simple TNDT model and confirmed experimentally. However, IR thermographic contrasts over landmines are

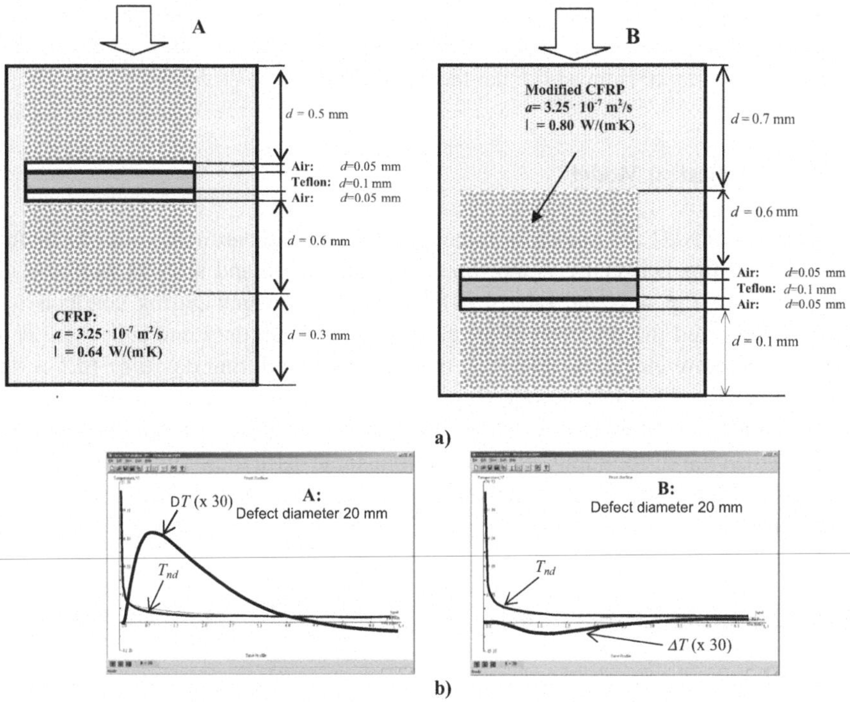

Fig. 4.40 Advanced TNDT modeling (Teflon inserts in CFRP): (**a**) thermal property modification scheme, (**b**) calculated temperature signals

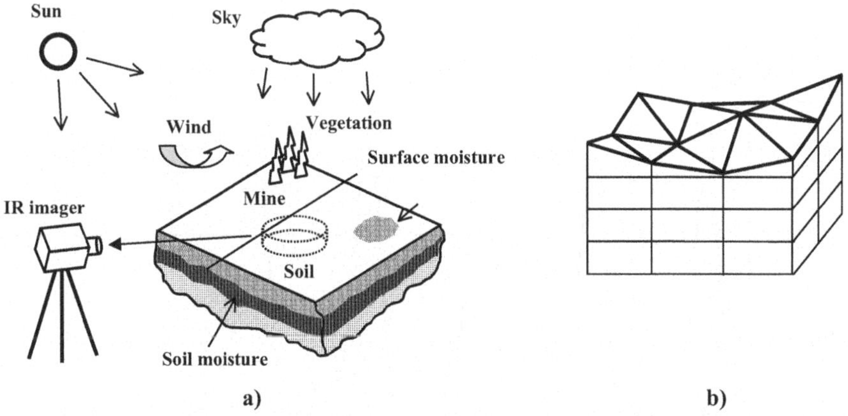

Fig. 4.41 Modeling IR thermographic detection of landmines in soil: (**a**) test scheme and boundary conditions on soil surface, (**b**) surface numerical mesh (modeling soil surface roughness)

conditioned not only by temperature variations but also by changed emissivity of soil due it s aeration and moisture variation. Presence of vegetation which shades true soil temperature is an additional factor influencing the detection.

4.10.2.1 Classical Model

The classical TNDT 2D model formulated in cylindrical coordinates allows obtaining relationships between ΔT and time, landmine depth and size, as well as soil roughness (see Fig. 4.42 and Table 4.14). Due to lower integral thermal conductivity of landmines and the presence of air gaps between landmines and soil, there is a local temperature elevation appearing over a landmine during day time ($\Delta T > 0$), reaching nearly ΔT_m~5 °C at noon (Fig. 4.40). For shallow landmines, there might be inversion of ΔT at night time when defect areas are characterized by higher temperature than the surrounding. As follows from Table 4.14, the simple TNDT model predicts that landmine depth l is a dominating parameter which influences temperature signals ΔT.

4.10.2.2 Advanced TNDT Model

Consider the detection of antipersonnel no-case landmines made of trinitrotoluene by the 20 cm-diameter and located at two depths of 1 and 5 cm. Several types and profiles of heating are presented in Fig. 4.43. The time is counted from 6 a.m. that corresponds to 0 hr in Fig. 4.43; then, a maximum of solar irradiation occurs at 12 a.m. This TNDT problem has been analyzed by using the ThermoCalc-3D

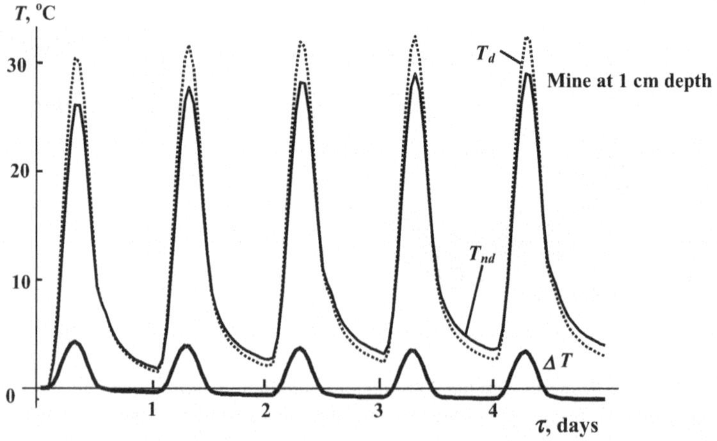

Fig. 4.42 Temperature evolution in TNDT of disk-like no-case landmine placed in soil at 1 cm depth (Q_m = 500 W/m^2; r_d = 10 cm, d = 10 cm, l = 1 cm), (see thermal properties of soil and trinitrotoluene in Table 2.2)

Table 4.14 Optimum landmine detection parameters (classical model, see material thermal properties in Table 2.2)

Model parameters				Optimum detection parameters*	
Material	Landmine diameter, cm	Landmine thickness, cm	Landmine depth, cm	ΔT_m, ^0C**	τ_m, ч***
Soil	5	1	1	+1.0	0***
			5	+0.2	+1.0
			10	+0.03	+2.0
		10	1	+1.5	+0.5
			5	+0.3	+1.25
			10	+0.05	+2.5
	20	1	1	+2.6	0
			5	+1.1	+1.25
			10	+0.4	+2.75
		10	1	+4.4	+0.75
			5	+1.8	+2.0
			10	+0.6	+3.5
Sand, dry	5	1	1	+0.33	+2.8
			5	+0.034	+5.0
			10	+0.0023	+7.2
		10	1	+0.60	+3.6
			5	+0.061	+5.6
			10	+0.0046	+8.2
	20	1	1	−0.54 +0.58	−1 +5.3
			5	−0.15 +0.13	+1 +7.8
			10	−0.036 +0.023	+3.5 +14
		10	1	−1.36 +0.84	+0.3 +7
			5	−0.41 +0.21	+2.3 +12.5
			10	−0.10 +0.053	+5.0 +21.3
Sand, moist	20	10	1	+8.68 −0.92	+1 +11.5
			5	+2.69	+3.3
			10	+0.75	+6.3

*In some cases, there are temperature signals of both signs in the table (particularly, for dry sand which has thermal properties close to trinitrotoluene) appearing in a diurnal cycle

**Sign + or – means that the area over the 'defect' is respectively warmer or colder than the surrounding

***Sign + or – means that the maximum signal appears respectively later or earlier than the maximum of solar irradiation

***With accuracy by one spatial step

software where all model parameters can be presented as step-wise functions varying in time. Modeling results are given in Table 4.15.

Profile 1 simulates heating with a cosine heat pulse analyzed in the previous paragraph (total energy absorbed for 12 hours is 3 kWhr/m^2). The comparison between the data in Table 4.14 and 4.15 shows that optimum detection parameters are close in both cases (some minor discrepancies are explained by the difference between smooth and step-wise heating functions and, presumably, by different defect shapes: square – in ThermoCalc-3D and disk – in ThermoCalc-2D).

4.10.2.3 Wind Speed

In the framework of the accepted model, influence of wind can be reduced to variation of the heat exchange coefficient $\alpha(\tau)$. An example of the time evolution $\alpha(\tau)$ II is shown in Fig. 4.43b; it is assumed that $\alpha(\tau)$ increases from 10 to 30 W/(m^2·K) starting from 12 a.m. that is, according to Eq. (2.7), equivalent to increase of wind speed from 2 to 11 m/s. Obviously, more intensive surface heat exchange reduces temperature patterns produced by landmines; this phenomenon is particularly strong for a landmine placed at depth of 5 cm because α increase has occurred at the time of landmine maximum visibility.

4.10.2.4 Combined Heating by Solar Radiation and the Ambient

Until now, we have been considering the heating with a heat flux Q, with the initial temperature T_0 being equal to the ambient temperature T_a; the boundary condition is given by Eq. (4.48): $\frac{\partial T_i(z=0,\tau)}{\partial z} = -\frac{Q(\tau)}{\lambda_1} + \frac{\alpha^F}{\lambda_1} T_i(z=0,\tau)$. It is easy to demonstrate that variation of ambient temperature is equivalent to delivery of additional energy which will warm up a sample if $T_a > T_0$ and cool it down if $T_a < T_0$. Air temperature is subject to a diurnal cycle, and solar radiation strongly depends on sky cloudiness. In this case, it is convenient to consider an arbitrary heating by both the Sun and the ambient:

$$\frac{\partial T_i(z=0,\tau)}{\partial z} = -\frac{Q(\tau)}{\lambda_1} - \frac{\alpha^F}{\lambda_1} T_a(\tau) + \frac{\alpha^F}{\lambda_1} T_i(z=0,\tau). \tag{4.81}$$

The profile III simulates such combined type of heating; the increase of the ambient temperature T_a occurs in respect to the initial time 6 a.m. However, even at the time of the maximum ambient temperature (8 °C increase from 1 to 2 p.m.), the additional heat flux is only $\alpha^F T_{amax} = 80$ W/m^2, i.e. much lower than the main radiant flux. Therefore, the corresponding ΔT increase is insignificant (see Table 4.15).

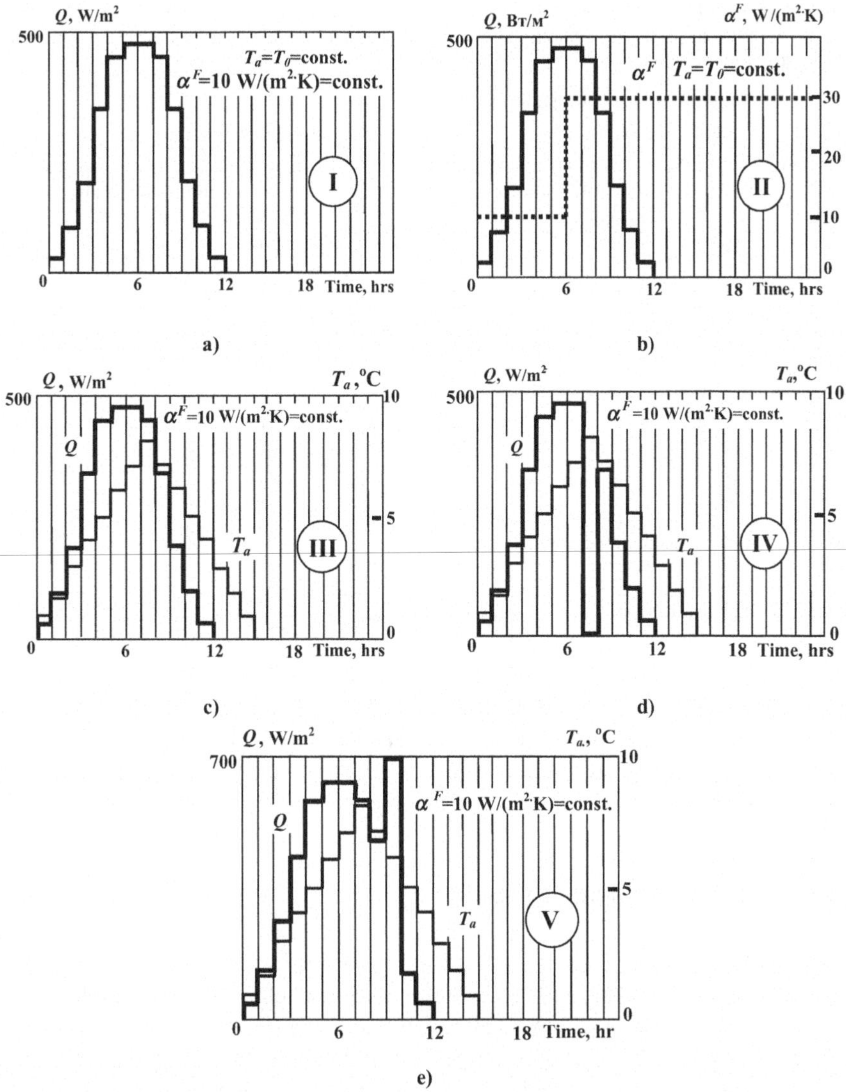

Fig. 4.43 Heating profiles in the detection of landmines in soil: (**a**) radiant heating (solar radiation, clear day, maximum solar irradiance at 12 a.m., (**b**) modeling wind speed increase starting from 12 a.m., (**c**) combined heating by radiant flux and changing ambient temperature, (**d**) combined heating by radiant flux and changing ambient temperature (cloudy Sky: zero solar radiation from 1 to 2 p.m.), (**e**) combined heating by radiant flux and changing ambient temperature (turning on additional heater: radiant flux increased by 500 W/m^2 from 3 to 4 p.m.)

Table 4.15 Numerical modeling of landmine detection (landmine diameter 20 cm, thickness 10 cm, see thermal properties of soil and trinitrotoluene in Table 2.2, heating profiles in Fig. 4.43, ThermoCalc-3D software)

	Optimum detection parameters			
	Landmine depth 5 cm		Landmine depth 5 cm	
Test case	ΔT_m, °C	τ_m, hr* (shift in respect to maximum solar irradiation at 12 a.m.)	ΔT_m, °C	τ_m, hr* (shift in respect to maximum solar irradiation at 12 a.m.)
I-Fig. 4.43a	4.26	1	1.61	2
II-Fig. 4.43b	3.92	0	1.20	0
III-Fig. 4.43c	4.55	1	1.58	2
IV-Fig. 4.43d	4.2	0	1.21	1
V-Fig. 4.43e	4.8	4	1.64	5

*With accuracy by one temporal step (15 min)

4.10.2.5 Cloudy Sky

The example of a diurnal heating function which simulates a short-term drop of solar radiation (cloudy Sky) from 1 to 2 p.m. is shown in Fig. 4.43d (profile IV). It follows from Table 4.15 that even an essential decrease of solar power weakly influences temperature signals because these signals are mainly determined by the energy absorbed during the first half of the day.

4.10.2.6 Using an Additional Heater

The model of arbitrary heating can be used for evaluating efficiency of additional heating in order to enhance temperature signals over landmines; the example of this is shown in Fig. 4.43e (profile V, heat flux density increased by 500 W/m^2 from 3 to 4 p.m.). As expected, a certain improvement in landmine detection occurs with a time delay in regard to the moment when the heater is turned on. A stimulated ΔT value depends on heater power and, in the case of profile V, increases by 3–5%. Obviously, such small signal increase is probably explained by choosing a non-optimal time for turning on the additional heater; it would be more efficient to start additional heating when ΔT~0 °C.

4.10.2.7 Soil Roughness

A regular shape of landmines is helpful in identifying their temperature signatures. However, this shape can be distorted if an IR imager views soil surface under an angle different from normal (Fig. 4.43a). Moreover, in real cases, soil roughness can change both landmine depth and amount of absorbed energy with the latter one alsovaried while the Sun moves across the Sky. For example, soil depressions comparable to landmines by size may cause surface temperature patterns which can be identified by an operator as false indications. Soil roughness was numerically modeled by Sendur and Bartlein [21] (see Fig. 4.41b). A simple example of temperature pattern distortion over a landmine located under small-size soil elevation is shown in Fig. 4.44.

4.10.2.8 Soil Moisture

Influence of varying soil moisture on detectability of objects buried in soil is significant but not well studied because it depends on a variety of factors which may affect temperature signals in opposite way, namely: (1) for example, precipitations which have typically temperature below soil temperature lead to decreasing soil temperature and 'smashing' surface temperature signatures, (2) moisture changes soil thermal properties, (3) water evaporation changes energy balance on soil surface, and (4) soil emissivity (absorptivity) depends on moisture.

Approximately it can be assumed that immediately after precipitations surface temperature becomes close to temperature of precipitation water that strongly smoothes temperature signatures over landmines.

In many cases, moistening soil enhances its thermal conductivity, while thermal diffusivity might reveal a certain extremum due to a particular ratio between varying λ, C and ρ. For example, the moistening of sand may cause inversion of a ΔT sign and increase its amplitude due to enhanced soil conductivity (see Table 4.15).

Surface water evaporation reduces heat flux which penetrates soil and causes the building-up of differential temperature signals. Heating of moist materials in the ambient was analyzed by Luikov [2] to demonstrate that evaporation occurring at constant rate is equivalent to reducing ambient temperature by a ΔT_a value which depends on pressure, moisture and temperature. In the framework of the discussed model, this means introducing the $Q_{evap} = \alpha^F \Delta T_a / \lambda$ evaporation member into Eq. (4.81) with its sign being opposite to that of the main heat flux Q. Values of ΔT_a can be found in psychometric tables (see Appendix 2), T_a is the indication of the 'dry' thermometer and ΔT_a is the difference between indications of the 'dry' and 'moist' thermometers. Under normal conditions, the following analytical formula can be used for evaluating ΔT_a with accuracy about 20%:

$$\Delta T_a = 0.06544\ T_a^{0.4420}\ (100 - W)^{0.8036}, \tag{4.82}$$

where W is relative ambient humidity in %. The graphical presentation of Eq. (4.82) is shown in Fig. 4.45. For example, if $T_a = 20\,^{\circ}\mathrm{C}$, $W = 51\%$, then $\Delta T_a = 6^{\circ}\mathrm{C}$. In the case of moist sand (Table 4.14), the equivalent reduction of heating flux is $Q_{evap} = \alpha\,\Delta T_a = 60\ \mathrm{W/m^2}$ if the main heat flux density is $Q = 500\ \mathrm{W/m^2}$. Respectively,

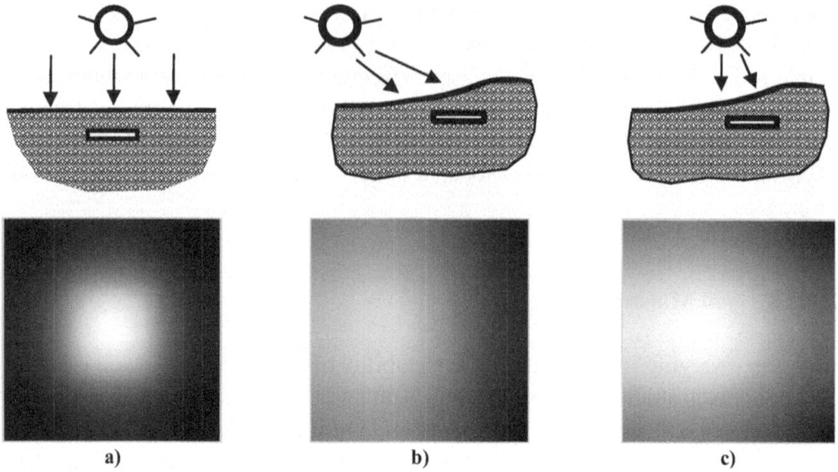

Fig. 4.44 Model of uneven heating due to soil elevation (no-case trinitrotoluene landmine): (**a**) classical model, (**b**) low Sun, (**c**) Sun at the peak of the day

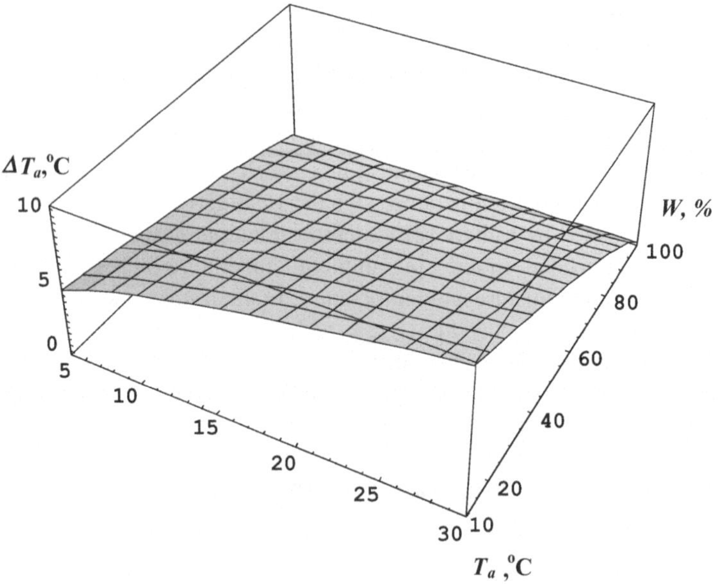

Fig. 4.45 Heating moist materials: equivalent ambient temperature reduction *vs.* ambient temperature and humidity under normal conditions

the temperature signal will drop by 12% (however, moistening sand will cause general signal enhancement due to increasing thermal conductivity).

To summarize, it should be stated that surface water evaporation induces stronger ΔT reduction with weaker stimulating heat flux, lower air humidity and higher ambient temperature.

Finally, it can be stated that influence of moisture on soil emissivity seems to be a phenomenon of the second order because it occurs across large areas (here we do not consider more subtle phenomena may affect some particular test cases, for instance, water may be preferentially in the soil layer above a landmine.

References

1. Carslow, H.S., Jaeger, T.S.: Conduction of Heat in Solids, p. 580. Oxford University Press, Oxford, UK (1959)
2. Luikov, A.V.: In: Hartnett, J.P. (ed.) Analytical Heat Conduction Theory, p. 685. Academic, New York (1969)
3. Maillet, D., Andre, S., Batsale, J.-C., et al.: Thermal Quadrupoles: Solving the Heat Equation through Integral Transforms. John Wiley & Sons Publisher, Chichester (2000)
4. Degiovanni, A., Lamine, A.-S., Houlbert, A.-S., Maillet, D.: Identification of subsurface defects using a sensibility analysis. In: Proc. of Fourth Europ. Conf. on Composite Mater., Stuttgart, Germany, 25–28 Sept, p. 691–695 1990
5. Winfree, W.P., Zalameda, J.N.: Thermographic determination of delaminations depth in composites. Proc. SPIE "Thermosense-XXV". **5073**, 363–373 (2003)
6. Vavilov, V.P.: Thermal NDT of Composite Structures and Radio Electronic Components, p. 162. Radio i Svyaz Publish, Moscow, Russia (1984)., (in Russian)
7. Marinetti, S., Vavilov, V.: Sensitivity analysis of classical heat conduction solutions applied to materials characterization. Int. J. Heat Transfer Eng. **26**(9), 50–60, Taylor & Francis (2005)
8. Almond, D., Patel, P.: Photothermal Science and Techniques. Chapman & Hall, London (1996)
9. Bennett, C.A., Patty, R.R.: Thermal wave interferometry: a potential application of the photoacoustic effect. Appl. Opt. **21**, 49–54 (1982)
10. Mulaveesala, R., Tuli, S.: Theory of frequency modulated thermal wave imaging for nondestructive subsurface defect detection. Applied Physics Letters, **89**(19), art. no. 191913 (2006)
11. Aamodt, L.C., Maclachlan Spicer, J.W., Murphy, J.C.: Analysis of characteristic thermal transit times for time-resolved infrared radiometry studies of multilayered coatings. J. Appl. Phys. **68**(12), 6087–6098 (1990)
12. Balageas, D.L., Krapez, J.-C., Cielo, P.: Pulsed photo-thermal modeling of layered materials. J. Appl. Phys. **59**(2), 348–357 (1986)
13. Delpesh, P.H., Krapez, J.-C., Balageas, D.L.: Thermal defectometry using the temperature decay rate method. In: Proc. Quant. Infr. Thermography QIRT-94, Eurotherm Seminar #42, Sorrento, Italy, 23–26 Aug, p. 220–225 1994
14. Batsale, J.C., Bendada, A., Maillet, D., Degiovanni, A.: Distribution of a thermal contact resistance: inversion using experimental Laplace and Fourier transformations and an asymptotic expansion. In: Proc. 1st Intern. Conf. on Inverse Problems in Engineering, Palm Coast, Florida, USA, 13–18 June, p. 122–129 1993
15. Karpelson, A.E., Popov, Y.U., Upadyshev, A.B.: Determining an optimal procedure of active thermal NDT. In: Defectoscopy (Rus. J. NDT), #8, p. 86–95 (in Russian) 1978
16. Winfree, W.P.: Enhanced thermographic detection of delaminations with computational pulse shaping. In: Rev. of Progress in Quant. NDE, Plenum Press, New York, vol. 17A, p. 441–448 1998
17. Vavilov, V.P., Kasatki, M.A.: Thermal NDT of heat-resistant nickel-based alloys. In: Defectoscopy (Rus. J. NDT), # 4, p. 51–54 (in Russian) 1990

18. Shepard, S.: Temporal noise reduction, compression and analysis of thermographic image data sequences. U.S. Patent No. 6516084 2003
19. Ringermacher, H.I., Howard, D.R., Knight, B.: Thermal imaging NDT at General Electric. 18th WCNDT, 16–20 April Durban, South Africa: 4 (available on CD) 2012
20. Bendada, A., Maillet, D., Degiovanni, A.: Nondestructive transient thermal evaluation of laminated composites: discrimination between delaminations, thickness variations and multidelaminations. In: Proc. Eurotherm Seminar 27 Quant. InfraRed Thermography QIRT'92, 7–9 July, Châtenay-Malabry, France, p. 218–223 1992
21. Sendur, I.K., Baertlein, B.A.: Numerical simulation of thermal signatures of buried mines over a diurnal cycle. In: Proc. SPIE Detection and remediation technologies for mines and minelike targets, Florida, USA, p. 144–150 2000

Chapter 5
Defect Characterization

Abstract This chapter is devoted to the solution of inverse TNDT problems, and performing defect characterization. Most of the common data inversion techniques for the evaluation of defect depth and thermal resistance, such as apparent effusivity, early detection, Laplace and Fourier domain treatment, non-linear fitting and others, are discussed. Methods for determining the lateral size and shape of a defect by its diffused surface temperature pattern are analyzed. There is a special emphasis on the detection of rear-surface corrosion.

In this chapter, we will consider identification of defect size $\{h_x, h_y, d\}$ and depth $\{l\}$ by solving the corresponding inverse problem:

$$\{h_x, h_y, d, l\} = \{T(x, y, \tau)\}^{-1}, \tag{5.1}$$

that is equivalent to the more general problem by Eq. (2.31) where the dynamic surface temperature $T(x, y, \tau)$ serves as the input data.

It is obvious that 1D algorithms allow the identification of $\{d, l\}$ values, while the evaluation of defect lateral size $\{h_x, h_y\}$ requires solving multidimensional heat conduction problems.

There are some approaches to the development of defect characterization algorithms. It was demonstrated in Chaps. 3 and 4 that immediate inversion of direct TNDT problems, as a rule, is impossible. Mathematically, the most general approach is based on the analysis of *functional residuals* produced by comparing experimental and theoretical data. This approach requires searching global extremums of a respective functional formulated in a multidimensional space of TNDT features (see also Sect. 3.7.5); in TNDT publications, this method is often called *non-linear fitting*. In practice, some simple solutions to inverse TNDT problems can be found by inverting numerical solutions of the corresponding direct problems and fitting them with some inversion formulas. To do this, both analytical and numerical direct solutions can be analyzed. The structure of respective inversion formulas can be prompted by direct solutions. Numerical coefficients in inversion formulas are obtained by applying a least-square technique; the corresponding options are available in some commercial software packages, such as Mathematica, Matlab etc.

V. Vavilov, D. Burleigh, *Infrared Thermography and Thermal Nondestructive Testing*, https://doi.org/10.1007/978-3-030-48002-8_5

5.1 Defect Characterization by Analyzing the Temperature Response on a Plate Heated with a Pulse

5.1.1 Apparent Effusivity Method (Dirac Pulse Heating)

We remind that the apparent effusivity technique discussed in Sect. 3.2 is based on the formula (3.2) describing the surface temperature of a semi-infinite bode subjected to Dirac heating: $e = W / \left[T(\tau *)\sqrt{\pi\tau *} \right]$. The same formula is valid in the case of a plate at short observation times ($Fo < 0.1$).

As shown in Sect. 3.2, the thermal effusivity determined by Eq. (3.2) is independent on time in the case of a uniform body, therefore, observable deviations of e from a reference value should be treated as indications of thermal property variations [1–4].

Surface temperature evolution is often drawn in the $Ln(T) - Ln(\tau)$ coordinates (see Sect. 3.1). In the ideal case of a uniform adiabatic body, this should produce a straight line, such as the curve 1 in Fig. 5.1a. In practice, some additional parameters, such as finite heat pulse duration, sample thickness and heat losses, distort the shape of the curve (see curve 1 at the end of the process in Fig. 5.1a). A two-layer body is characterized by the curve 2, while the presence of a subsurface defect leads to the curve 3 (Fig. 5.1a).

In order to exclude some disturbing factors, let us introduce the ratio between a current point and a point (area) chosen as a reference and characterized by the effusivity e_{nd}. The e/e_{nd} ratio represents a straight line parallel to the time axis everywhere except time periods where subsurface defects appear in the temperature evolution. An experimental example is given in Fig. 5.1b. In accordance to the theory described in Chap. 4, deeper defects cause distortions of e/e_{nd} which appear at later times being characterized by smaller amplitudes. A defect can be characterized by two dimensionless parameters: $Bi_d = l/(\lambda R_d)$ and $Fo_d = a\tau/l^2$ (notation corresponds to that accepted in Chap. 3). Each defect of thermal conductivity lower than that of a host material can be characterized with a particular point $\{Fo_{d\,\min}, (e/e_{nd})_{\min}\}$ in the corresponding $e(\tau)/e_{nd}$ curve. This point is related to defect parameters by the following relationships:

$$\begin{aligned} \left(e/e_{nd}\right)_{\min} &= Fo_{d\min}^{-0.528}; \\ Fo_{d\min} &= 1 + 0.62 Bi_d^{-1}. \end{aligned} \tag{5.2}$$

Respectively, the formulas for evaluating defect depth and thermal resistance have been proposed in the following form:

$$\begin{aligned} l &= \sqrt{a\tau_{\min}} \left(e/e_{nd}\right)_{\min}^{0.95}; \\ R_d &= \left(1.161/e_{nd}\right)\sqrt{\tau_{\min}} \left(e/e_{nd}\right)_{\min}^{0.95} \left[\left(e/e_{nd}\right)_{\min}^{-1.89} - 1\right]. \end{aligned} \tag{5.3}$$

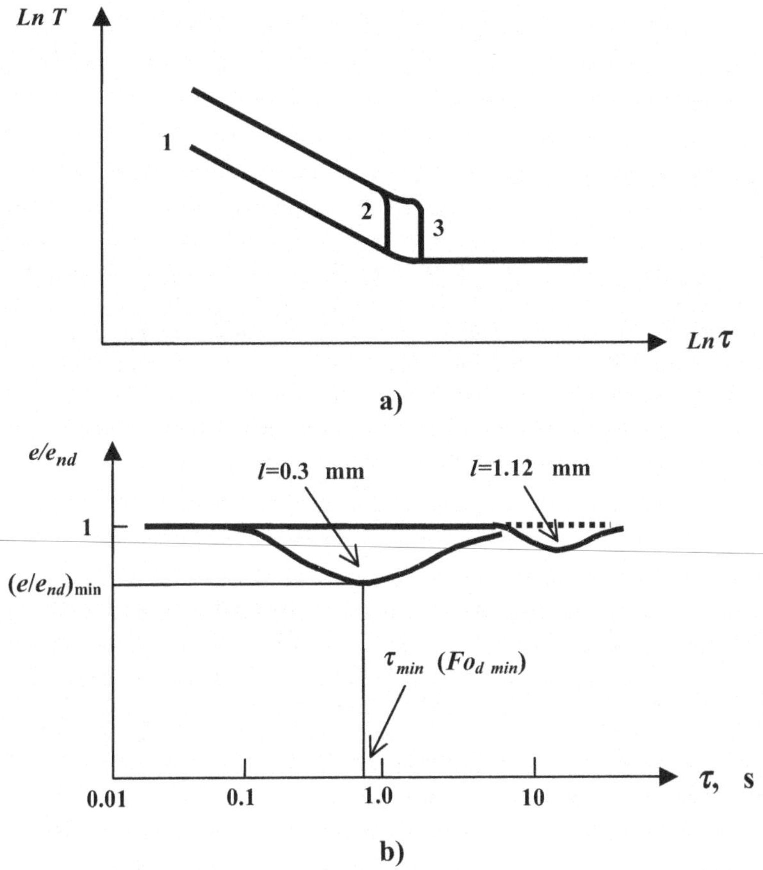

Fig. 5.1 Apparent thermal effusivity method: (**a**) surface temperature evolution after flash heating (1) one-layer sample, (2) two-layer sample with ideal thermal contact between layers, (3) two-layer sample with defect between layers), (**b**) apparent thermal effusivity *vs.* time (CFRP sample with two 20 × 20 mm Teflon inserts)

It is clear that minimum $(e/e_{nd})_{\min}$ values in this model correspond to the corresponding maximum ΔT_m values in the "classical" TNDT theory described in Chap. 4, therefore, the inverse solution (5.3) can be obtained also for $\tau_{1/2}$ or other specific heat transit times.

The method above was used to evaluate defects in CFRP composite. The characterization errors were 20% by defect depth and 40% by defect thermal resistance [2, 3].

Eqs. (5.3) were corrected by Krapez and Balageas [4]:

$$l = \sqrt{a_z \tau_{\min}}\, C_{\min}^{inv\;0.97};$$
$$R_d = 0.0286 \frac{l}{\lambda_z} \left(C_m^{run}\right)^{0.952 + 0.0362\left[\log_{10} C_m^{run}\right]^2}, \tag{5.4}$$

where $C^{inv} = T_{nd}/T_d$ is the *inverted relative temperature contrast.*

Eqs. (5.4) are more general than (5.3) but they are valid only for extended (1D) defects located far from the sample rear surface.

5.1.2 Using Early Detection Times

The term *Early Detection* was proposed by Balageas et al. to specify a time when an evolving $\Delta T(\tau)$ signal starts to exceed noise (see also Sect. 6.2). At this time, the signal-to-noise ratio is lower than at optimum observation time but defect shape is better reproduced due to weaker lateral heat diffusion. Besides, defect thermal resistance (defect thickness) scarcely influences surface temperature in defect areas at earlier times, therefore, the early detection technique can be used for determining defect depth l independently on defect thickness [5, 6]:

$$l = \sqrt{a_z \tau\left(C^{run}\right) Ln\left(2/C^{run}\right)}, \tag{5.5}$$

where $\tau(C^{run})$ is the time when the contrast C^{run} is observed. It was reported that l can be evaluated by Eq. (5.5) with an error not worse than 4% for defects of a higher thermal resistance R_d. The earlier is the determined difference between defect and non-defect areas, the higher is the characterization accuracy. In fact, the authors suggested applying Eq. (5.5) to signals which only start to exceed a noise level, for example, when C^{run}~1 ÷ 3%? Although, in some test cases, noise can be significantly higher than the above-mentioned values.

Another characterization formula was proposed for defects with low R_d:

$$C^{run} = 200\left(1 - \frac{2Fo*}{R* + 2Fo*}\right)\exp\left(-1/Fo*\right), \tag{5.6}$$

where $Fo* = a_z \tau / l^2$; $R* = R_d \lambda_z / l$. Since Eq. (5.6) contains $R*$ in the implicit form, the solution is to be found iteratively in combination with Eq. (5.4) for R_d. The highest accuracy (~2%) has been achieved for low contrasts.

5.1.3 Retrieving Corrosion Profile

A simple algorithm allowing to retrieve a thickness profile of a plate was proposed by Lugin and Netzelmann [7]. The algorithm illustrated in Fig. 5.2 is based on comparison between theoretical and experimental data. Let us use the expression for Dirac heating of a semi-infinite body (Table 2.4):

$$T_{si}(z,\tau) = \frac{W}{\sqrt{\pi C\rho\lambda\tau}} e^{-\frac{z^2}{4a\tau}}; T_{si}(z=0,\tau) = \frac{W}{\sqrt{\pi C\rho\lambda\tau}}. \tag{5.7}$$

The in-depth temperature distribution for the particular time $\tau*$ is shown in Fig. 5.2 with the bold dotted line. In the case of a plate with the thickness L, a pulsed thermal wave is reflected from the plate rear surface thus increasing the temperature at any point $z \leq L$ (solid line in Fig. 5.2). The difference between two temperatures shown with the thin dotted line represents the mirror reflection of the curve "tail" at $z \leq L$ (the time $\tau*$ is small, the lateral heat diffusion is negligible, and only the first reflection of the thermal wave from the plate rear surface is considered). Then, the increase in surface temperature conditioned by the presence of the reflective surface is equal to the temperature of the semi-infinite body at the depth of $2L$:

$$\Delta T(z,\tau) = \frac{W}{\sqrt{\pi C\rho\lambda\tau *}} e^{-\frac{(2L)^2}{4a\tau *}} = \frac{W}{\sqrt{\pi C\rho\lambda\tau *}} e^{-\frac{L^2}{a\tau *}}. \tag{5.8}$$

By introducing the running temperature contrast of a plate temperature in respect to the temperature of the semi-infinite body T_{pl} as $C_{si} = [T_{pl}(\tau*) - T_{si}(\tau*)]/T_{si}(\tau*)$, one may calculate the plate thickness by the following expression:

$$L = \sqrt{-a\tau * Ln\left[C_{si}(\tau *)\right]}. \tag{5.9}$$

When experimentally implementing the algorithm above, one replaces the contrast C_{si} with the experimental running contrast C^{run} calculated in respect to a chosen

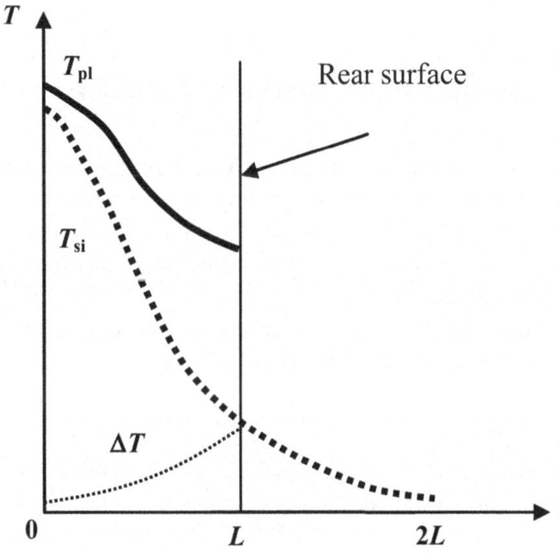

Fig. 5.2 Temperature responses of a semi-infinite body and plate. (Adapted from [7])

reference point. At the points where the contrast exceeds the noise level, one may calculate the conventional sample thickness called *echo defect depth* (EDD). Furthermore, an iterative procedure can be applied to retrieve sample thickness profile. Consecutive estimates of EDD can be introduced in a 1D, 2D or 3D theoretical model. The algorithm has been verified by on a finite element model including 5500 nodes. The reconstruction of a 2D corrosion profile required 3 minutes of personal computer time.

Another approach to the evaluation of corrosion will be described in Sect. 5.8.

5.1.4 *Characterizing Thin High-Conductivity Materials Using Flash Heating*

Delpeche et al. proposed the following approximation of the general expression of front-surface temperature in thin high-conductive materials (3.51) subjected to flash heating [8]:

$$T^F(\tau) = \beta_1 \; e^{-\beta_2} + \beta_3 \; e^{-\beta_4}. \tag{5.10}$$

The variation analysis has shown that coefficients β_1 and β_2 which are determined by the first root of Eqs. (3.52–3.57) are weakly-dependent on parameters of subsurface defects (for example, β_2 is mainly related to the heat exchange intensity α). In their turn, coefficients β_3 and β_4, which are determined by the second root of the transcendent equation, allow the evaluation of defect depth l_1 and thermal resistance R_d. Some peculiarities of the iterative evaluation of coefficients β are described in Sect. 5.6.

5.2 "Individual" Inversion Functions

Vavilov et al. suggested to solve TNDT inverse problems by approximating numerical solutions of the corresponding direct problems with polynomial functions [9]. The term *individual* means that a solution is to be derived for particular materials and test procedures, thus ensuring an enhanced accuracy of defect characterization.

The example of the corresponding inversion formula intended for the evaluation of delaminations in CFRP is as follows:

$$\begin{aligned} l[mm] &= -0.005 + 0.177\tau_m[s] + 0.00424\tau_m^2; \\ d[mm] &= 0.0162\left\{\frac{C_m^{norm}}{1-\exp(-r_d[mm]/6)}\right\}^{0.487}\left\{\frac{\tau_m}{1-\exp(-r_d/4)}\right\}^{1.363}, \end{aligned} \tag{5.11}$$

where the *normalized* C^{norm} (see Table 1.1) appears as a result of signal normalization that is a typical procedure in pulsed TNDT allowing suppression of uneven heating phenomena (see Sect. 6.11).We remind that maximums of C^{norm} coincide with the corresponding maximums of ΔT.

For a 5 mm-thick CFRP sample, the errors in characterizing defect depth l and thickness d are ~5% and ~25%respectively. Note that the last formula in (5.8) includes defect lateral size which is to be visually evaluated by directly analyzing defect indications. However, for larger defects (h>10 mm), one can neglect the exponential member in Eq. (5.11), thus coming to the corresponding 1D case.

5.3 General Inversion Formulas

The ThermoFit Pro software (Tomsk Polytechnic University, Russia) includes the inversion formulas which are valid for a wide range of materials heated with square pulses.

Within a heat pulse ($\tau \le \tau_h$) these formulas are
for running temperature contrast:

$$l\left[m\right]=0.4066\left(a\tau_m\right)^{0.4885}\left(1-\lambda_d/\lambda\right)^{0.1903}\left(L/\lambda\right)^{0.07884}Fo_h^{0.1208}C_{run}^{-0.2869},$$

$$R_d\left[m^2KW^{-1}\right]=1.2390\left(a\tau_m\right)^{0.4043}\left(1-\lambda_d/\lambda\right)^{-0.04380}\left(L/\lambda\right)^{0.1193}Fo_h^{-0.1681}C_{run}^{0.2255},$$

for normalized temperature contrast:

$$l\left[m\right]=0.1694\left(a\tau_m\right)^{0.4937}\left(1-\lambda_d/\lambda\right)^{0.3591}\left(L/\lambda\right)^{0.2699}Fo_h^{0.4137}C_{norm}^{-0.7348},$$

$$R_d\left[m^2KW^{-1}\right]=1.9106\left(a\tau_m\right)^{0.4208}\left(1-\lambda_d/\lambda\right)^{-0.1341}\left(L/\lambda\right)^{0.2222}Fo_h^{-0.002878}C_{norm}^{-0.01405}. \qquad (5.12)$$

If a maximum of the running contrast occurs after the heat pulse ($\tau > \tau_h$):

$$l\left[m\right]=0.7076\left[a\left(\tau_m-\tau_h\right)\right]^{0.4564}\left(1-\lambda_d/\lambda\right)^{0.2849}\left(L/\lambda\right)^{0.05359}Fo_h^{0.05227}C_{run}^{-0.1511},$$

$$R_d\left[m^2KW^{-1}\right]=4.3682\left[a\left(\tau_m-\tau_h\right)\right]^{0.3432}\left(1-\lambda_d/\lambda\right)^{1.2647}\left(L/\lambda\right)^{0.2862}Fo_h^{-0.05300}C_{run}^{0.4256},$$

and for normalized contrast:

$$R_d\left[m^2KW^{-1}\right]=4.3682\left[a\left(\tau_m-\tau_h\right)\right]^{0.3432}\left(1-\lambda_d/\lambda\right)^{1.2647}\left(L/\lambda\right)^{0.2862}Fo_h^{-0.05300}C_{run}^{0.4256},$$

$$R_d\left[m^2KW^{-1}\right]=4.9209\left[a\left(\tau_m-\tau_h\right)\right]^{0.03140}\left(1-\lambda_d/\lambda\right)^{1.3705}\left(L/\lambda\right)^{0.2961}Fo_h^{-0.04242}C_{norm}^{0.4071}.$$

Eqs. (5.12) are fairly universal because they are valid for a wide range of materials and pulsed test procedures.

If $\tau \le \tau_h$, characterization errors by l and R_d are typically less than few percent. If extremums of the informative parameters appear at a cooling stage ($\tau > \tau_h$), the accuracy in determining l is about 10% and 15% for both normalized and running contrast, while for R_d, the accuracy drops down to 40% and 60% respectively.

5.4 Simplified Inversion Formulas

5.4.1 *Determining Depth and Thermal Resistance of Defects Located Between Two High-Conductivity Plates*

Consider a thermally-resistive defect located between two high-conductive materials by thickness of l_1 and l_2. A defect characterization approach proposed in [10] is based on the assumption that the temperature across a host material is constant at any time due to material high thermal conductivity, while a major temperature gradient occurs across a defect. The front-surface temperature following the Dirac-pulse stimulation is:

$$T\left(z=0,\tau\right)=\frac{W}{C_1\rho_1l_1+C_2\rho_2l_2}\left[1+\frac{C_2\rho_2l_2}{C_1\rho_1l_1}\exp\left(-\frac{C_1\rho_1l_1+C_2\rho_2l_2}{RC_1\rho_1l_1C_2\rho_2l_2}\tau\right)\right], \quad (5.13)$$

where $\dfrac{W}{C_1\rho_1l_1+C_2\rho_2l_2}=T_\infty$ is the temperature at the end of the thermal process.

In the case of a square pulse, the solution was derived by means of the Duhamel integral:

$$\frac{T\left(z=0,\tau\right)}{T_\infty}=\{1+\frac{R_d}{\tau_h}\frac{\left(C_2\rho_2l_2\right)^2}{C_1\rho_1l_1+C_2\rho_2l_2}\left[\exp\left(\frac{C_1\rho_1l_1+C_2\rho_2l_2}{RC_1\rho_1l_1C_2\rho_2l_2}\tau_h\right)-1\right]x$$

$$x\exp\left(-\frac{C_1\rho_1l_1+C_2\rho_2l_2}{R_dC_1\rho_1l_1C_2\rho_2l_2}\tau\right)\}. \quad (5.14)$$

In a homogeneous material:

$$\frac{T(z=0,\tau)}{T_\infty}=\frac{R_d}{\tau_h}\frac{C\rho(L-l)^2}{L}\left[\exp\left(\frac{L}{R_dC\rho l(L-l)}\tau_h\right)-1\right]\exp\left(-\frac{L}{R_dC\rho l(L-l)}\tau\right). \quad (5.15)$$

A procedure for determining R_d and l is as follows. In logarithmic presentation, the temperature evolution in time is described with the straight line $Ln\ [T(z=0,\tau)/T_\infty] = Ln\ [T(z=0,\tau=0)/T_\infty] + p\tau$ of which the slope is:

$$p=-\frac{C_1\rho_1l_1+C_2\rho_2l_2}{RC_1\rho_1l_1C_2\rho_2l_2}, \quad (5.16)$$

and the initial value is:

$$\frac{T(z=0,\tau=0)}{T_\infty}=\frac{R}{\tau_h}\frac{(C_2\rho_2l_2)^2}{C_1\rho_1l_1+C_2\rho_2l_2}\left[\exp\left(\frac{C_1\rho_1l_1+C_2\rho_2l_2}{RC_1\rho_1l_1C_2\rho_2l_2}\tau_h\right)-1\right]. \quad (5.17)$$

If

$$k=\exp\left(\frac{C_1\rho_1l_1+C_2\rho_2l_2}{RC_1\rho_1l_1C_2\rho_2l_2}\tau_h\right)-1, \quad (5.18)$$

then

$$\frac{T(z=0,\tau=0)}{T_\infty}\frac{p}{k}=\frac{C_2\rho_2l_2}{C_1\rho_1l_1}\frac{1}{\tau_h}. \quad (5.19)$$

Two unknown parameters $l_1 = l$and R_d are linked by the following expressions:

$$l_1=\frac{L}{1+\frac{C_1\rho_1}{C_2\rho_2}\cdot\frac{p\tau_h\left[T(z=0,\tau=0)/T_\infty\right]}{k}};$$
$$R_d=\frac{C_1\rho_1l_1+C_2\rho_2l_2}{pC_1\rho_1l_1C_2\rho_2l_2}. \quad (5.20)$$

If defect depth is *a priori* known, defect thermal resistance can be found from the second equation, otherwise the system of two equations is to be solved with the two parameters p and $T(z=0,\tau=0)/T_\infty$ being known from the experiment. Typically, three iterations can be enough if to start from $l_1 = L/2$ [10].

The solution for a homogeneous plate is:

$$l = \frac{L}{1 + \dfrac{p\tau_h\left[T(z=0,\tau=0)/T_\infty\right]}{\exp\left(\dfrac{\tau_h}{RC\rho l(1-l/L)}\right) - 1}};$$

$$R_d = \frac{1}{pC\rho l(1-l/L)}. \tag{5.21}$$

The method above has been used for evaluating a bonding layer between metals and composites [10]. The characterization errors were 10–14% by l and 10–95% by R_d.

5.4.2 Determining Defect Depth by Using the Optimum Observation Time Technique

If $\tau_m(\Delta T_m) >> \tau_h$, defect depth can be approximately determined by inverting Eq. (3.11):

$$l = \sqrt{a\tau_m}. \tag{5.22}$$

Eq. (5.22) is recommended for evaluating low-conductive (gas-filled defects) of whose depth is much smaller than the sample thickness.

If duration of heating is comparable but still shorter than τ_m, more realistic estimates can be found by the formula:

$$l = \sqrt{a(\tau_m - \tau_h)}. \tag{5.23}$$

Note the certain resemblance between Eqs. (5.22 and 5.23) and the general inversion formulas (5.8).

Shepard et al. have found the empiric expression which relates the time $\tau_m(\Delta T_m)$ called *peak contrast time* to the depth of a disk-like bottom-hole defect of the diameter D [11]:

$$l = \frac{a}{\pi D\sqrt{2}}\tau_m. \tag{5.24}$$

The authors have introduced the concept of a heat trap which presumes that τ_m is the time of filling with thermal energy a cylinder of the volume $V = \pi D l$; then Eq. (5.24) can be presented as follows:

$$V = \frac{a}{\sqrt{2}} \tau_m.. \tag{5.25}$$

Eqs. (5.22) and (5.25) represent two different relationships between optimum observation times and defect depths: quadratic for 1D defects and linear for finite-size defects.

The concept of using a reference point in TNDT is often criticized because uneven heating and other factors may significantly affect evaluation results. An alternative approach to processing thermogram pixel used in the TNDT equipment from Thermal Wave Imaging involves presentation of temperature evolutions in the logarithmic coordinates: $Ln(T) \;-\; Ln(\tau)$. It is assumed that a test sample can be considered as a semi-infinite body, then, in the logarithmic presentation, the temperature evolutions at non-defect points are characterized by straight lines with the slope of −0.5. Then, any deviation from such 'reference' behavior should be regarded as a defect indication.

5.4.3 Determining Defect Thermal Resistance by Using the Zero-Order Temporal Moment

Bocher et al. proposed this algorithm by considering flash heating of a plate [10]. An integral $M = \int_0^\infty T(\tau)d\tau$ is called a *zero-order temporal moment* of which value tends to infinity in the case of an adiabatic thermal process, i.e. $\{T(\tau \to \infty) \to T_\infty\}$. If a test body contains a defect of the thermal resistance R_d at the depth l, the 'non-defect' temperature $T(\tau)$ is to be replaced with the 'defect' temperature $T_d(\tau)$ which deviates from the reference one within a particular time period called *defect visibility time*. It was shown analytically that variation of a zero-order temporal moment is

$$\Delta M = \int_0^\infty [T_d(\tau) - T_{nd}(\tau)]d\tau = W R_d (1 - l/L)^2, \tag{5.26}$$

where W is the absorbed energy. Eq. (5.26) contains two unknown parameters R_d and l. If $l < < L$(thick sample or shallow defect):

$$\Delta M = W R_d. \tag{5.27}$$

The last expression implies subsequent summing IR thermograms in a captured sequence that also leads to reduction of uncorrelated noise by $\sqrt{N}$ times (N is the number of images in the sequence). However, it should be noted that the necessity to measure the absorbed energy makes usefulness of Eq. (5.27) questionable.

5.5 Defect Characterization in the Laplace Domain (Thermally-Resistive Defects)

As mentioned above, the analytical solutions describing the heating of a multilayer plate are typically obtained by applying the technique of the Laplace transform. Such solutions are bulky and their inversion in the domain of originals in the evident form is impossible because of the presence of infinite series or the error function. Degiovanni et al. proposed determining parameters of thermally-resistive defects in the Laplace domain, i.e. not returning to the domain of original functions (see Sect. 4.1.2). Since the corresponding expressions in the Laplace domain, such as Eq. (4.17), are algebraic, they can be solved in respect to sought values of l and d (R_d).

Primarily, the idea was to identify defect parameters by analyzing differential temperature signals thus recording both defect and non-defect temperatures. The appearance of the differential solutions by Eq. (4.18) prompts that a practical defect characterization procedure should involve applying the Laplace transform to experimental normalized signals $\Delta T/T_\infty$ (normalization is to be done by the sample stationary temperature T_∞ while assuming adiabatic boundary conditions). Since there are two unknown defect parameters, one has to deal either with solutions for two sample surfaces or for one surface but, at least, at two times τ_1 and τ_2 which correspond to the Laplace variables p_1 and p_2. It is known that a system of two equations with two unknowns has a unique solution if these equations are linearly independent. The authors of the approach above have found that equations $\Delta\theta^F(p_1)$ and $\Delta\theta^F(p_2)$ are not fully independent, however, they can be used for characterizing defects if some additional conditions are met.

Below we will describe a defect characterization procedure which is based on the analysis of basic temperature functions $\theta^F(p)$ and $\theta^R(p)$ rather than the corresponding differential signal.

5.5.1 *Analyzing Rear-Surface Differential Temperature Signal*

It was found that a rear-surface solution $\Delta\Theta^R(p)$ is insensitive to the parameter $x* = l/L$ in the middle of a plate and reveals a weak dependence on this parameter at other depths; the same conclusion follows from the relationship ΔT^R (l) presented in Chap. 4. Analytically, the variable $x*$ can be excluded in a particular case of $p_2* = 4\ p_1*$ by using features of hyperbolic functions; this allows determining defect thermal resistance by:

$$\frac{R_d}{L/\lambda}=\frac{M_1^2\left[1+\alpha_2\mathrm{sh}(\alpha_2)M_2\right]\mathrm{sh}(\alpha_1)\mathrm{th}(\alpha_1)}{\left[1+M_1\alpha_1\mathrm{sh}(\alpha_1)\right]\left[-M_1+M_2\mathrm{ch}(\alpha_1)-1.5M_1M_2\alpha_1\mathrm{sh}(\alpha_2)\right)}, \tag{5.28}$$

where $M_{1,2}$ is the Laplace transform of an experimental normalized temperature signal $\Delta T^R/T_\infty$, an the subscripts "1" and "2" are related to two chosen times; $\alpha = \sqrt{p*}$; $p* = L^2 p / a$.

The consideration above confirms the conclusion that two-sided TNDT allows determining only defect thermal resistance (defect thickness) but not defect depth. Due to this reason, the technique of dynamic thermal tomography based on the relationship between defect depth and observation time cannot be realized in a two-sided test.

5.5.2 *Analyzing Front-Surface Differential Temperature Signals*

The analysis of Eqs. (4.17) has proven that one-sided TNDT is fairly sensitive to defect depth but weakly-dependent on defect thickness. By excluding R_d for p_1* and $p_2* = 4\ p_1*$, one can obtain the following transcendent equation for determining l:

$$\begin{gathered}\alpha_1 \operatorname{sh}(\alpha_2) sh(u_1)\left[4\operatorname{ch}(u_1)\operatorname{ch}(\alpha_1 - u_1) - \operatorname{ch}(\alpha_1)\right] + \\ +2\operatorname{sh}(\alpha_1 - u_1)\left[\frac{1}{M_1}\operatorname{ch}^2(\alpha_1) - \frac{1}{M_2}\operatorname{ch}^2(\alpha_1 - u_1)\right] = 0,\end{gathered} \tag{5.29}$$

where $u_1 = \alpha_1 l/L$. For small values of R_d, the last expression can be written in the evident form:

$$l/L = 1 - \frac{1}{\alpha_1}\ln\left\{\left(\frac{M_2}{M_1}\right)^{1/2}\operatorname{ch}(\alpha_1) + \left[\frac{M_2}{M_1}\operatorname{ch}^2(\alpha_1) - 1\right]^{1/2}\right\}. \tag{5.30}$$

5.5.3 *Coating on a Substrate: Two-Sided TNDT*

Let thermal properties be l_1, C_1, ρ_1, λ_1 for a coating and l_2, C_2, ρ_2, λ_2 for a substrate. In 1D approximation, the temperature difference $\Delta T*$ between coated (d) and uncoated (nd) areas is as follows:

$$\Delta T* = T_d / \frac{W}{C_1\rho_1\lambda_1 + C_2\rho_2\lambda_2} - T_{nd} / \frac{W}{C_2\rho_2\lambda_2}. \tag{5.31}$$

The identification problem can be solved by using the expression for the rear surface temperature. The Laplace transform $\Delta\Theta^R$ of the function ΔT^R (here the star which specified the dimensionless signal is omitted) is [12, 13]:

$$\Delta\theta^R = \frac{l_2^2}{a_2}\Delta\theta^R*; \tag{5.32}$$

$$\Delta\theta^R* = \frac{1}{s}\left[\frac{1+k_1k_2}{k_2\,sh(k_1s)ch(s)+sh(s)ch(k_1s)} - \frac{1}{sh(s)}\right],$$

where $s = \sqrt{p*}$; $p* = l_2^2 p / a_2$; $k_1 = \frac{l_1}{l_2}\sqrt{\frac{a_2}{a_1}}$; $k_2 = \sqrt{\frac{C_1\rho_1\lambda_1}{C_2\rho_2\lambda_2}}$..

In fact, we deal with a two-parameter inverse problem $\Delta T^R(\tau, k_1, k_2)$ where the function $\Delta T^R(\tau)$ is known from the experiment. The unknown parameters k_1 and k_2 can be determined if the sensitivity coefficients $x_1(\tau, k_1, k_2) = \partial\Delta T/\partial k_1$ and $x_2(\tau, k_1, k_2) = \partial\Delta T/\partial k_2$ are linearly-independent.

The analysis of the sensitivity coefficients reported in [12] showed their strong correlation, therefore k_1 and k_2 cannot be found experimentally. However, some other coefficients were proposed, such as k_2 and $k_3 = k_1k_2$, which are independent and thus convenient for solving the corresponding inverse TNDT problem.

5.6 Defect Characterization by Non-linear Fitting (Minimizing a Residual Functional)

5.6.1 Using Classical Heat Conduction Solutions

As noted above, a residual functional is defined as a difference between two numerical sequences $\{T_i(\tau_i) - T_i(q_j, \tau_i)\}$ $(i = 1 \div N)$ of which one is experimental and another follows from the corresponding theoretical model which includes parameters of identification: q_j $(j = 1\ldots M)$. These parameters can be evaluated by iteratively finding a global minimum in the M-dimensional space of parameters q_j. For some parameters, such as defect thickness, a global minimum of the corresponding functional can be fairly 'flat', therefore, the corresponding characterization accuracy might be low. Also, it is obvious that the iterative minimization requires introducing a proper initial approximation as a calculation start point, and, in any case, the computation may be time-consuming even on powerful computers, in particular, when processing images of a large pixel format.

By using classical heat conduction solutions as *basic functions* $T_i(q_j, \tau_i)$, it is possible to evaluate material thermal properties and the heat exchange coefficient along with defect parameters. Grinzato and Marinetti applied the algorithm of non-linear fitting for analyzing the heating of a plate of known thermal conductivity and thickness [13]. The solution for heating a non-adiabatic plate and a semi-infinite body with a square pulse was used as an input theoretical model. For iteratively minimizing least squares, the Levenberg-Marquardt method was applied. This method is standard when looking for extremums in a multidimensional space being widely used in TNDT.

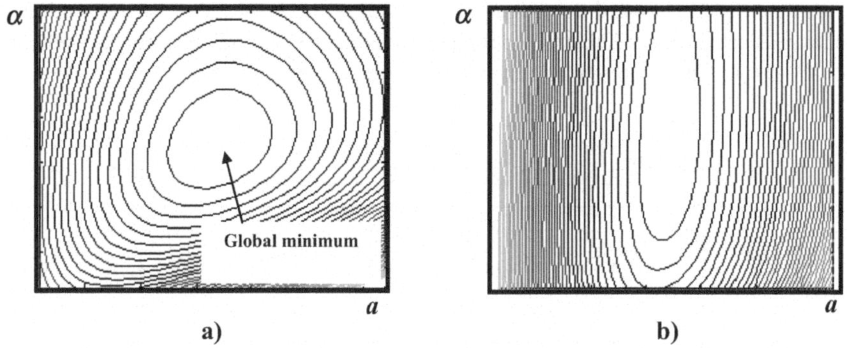

Fig. 5.3 Illustration to non-linear fitting by using the classical solution for heating a non-adiabatic plate (determining thermal diffusivity a and surface heat exchange coefficient h in one- and two-sided procedures)

Thermal diffusivity a was a parameter to be evaluated in the two-sided inspection of a CFRP sample; the Parker technique was chosen to alternatively verify diffusivity values. It has been concluded that non-linear fitting provides a higher characterization accuracy due to taking into account surface heat exchange.

It was also shown that thermal diffusivity can be also evaluated in a one-sided inspection procedure but the computation was longer, and the accuracy was worse in this case.

The approach described above is illustrated in Fig. 5.3 showing a space of two identification parameters, namely, the thermal diffusivity a and the heat exchange coefficient α, which were evaluated by non-linear fitting. As expected, the analysis of the rear-surface temperature ensured the most reliable estimates: the iterative process which started from an arbitrary point $\{a, \ \alpha\}$ always leaded to a global minimum which is well seen in Fig. 5.3a. On the front surface, the minimum looked as a 'valley', therefore, a correct a value can be missed while moving along a values in the vertical direction (Fig. 5.3b).

The semi-infinite body (front-surface) solution allows evaluating the thermal inertia e and the ratio between the absorbed power and the heat exchange coefficient Q_a/α.

It has been found that the algorithm of non-linear fitting is more stable toward high-frequency noise than the polynomial fitting followed by smoothing the data with standard filters.

Winfree and Zalameda data treatment in the Laplace domain in combination with the technique of *orthogonal decomposition*, or *Principal Component Analysis* (PCA) [14]. This approach will be described in Chap. 6. Unlike the analysis of differential temperature signals in the Laplace domain (see p. 5.5), in this case, one has to analyze front- and rear surface temperatures given by Eqs. (4.17). Both defect depth and thermal resistance can be identified by using this approach. Since Eqs. (4.17) include the defect depth l_1 in combination with the thermal diffusivity a, it is impossible to determine l_1 without independent measurement of a. It was proposed to evaluate the relative defect depth that requires the knowledge of a sound area serving as a reference. Assuming defect depth equal to r in such way that $l_1 = rL$ and $l_2 = (1 - r)L$, Eqs. (4.17) may be written as follows:

$$\theta^F(p) = \frac{W\sqrt{a}}{\lambda\sqrt{p}} \frac{2\mathrm{ch}\left(\sqrt{\frac{p}{a*}}\right) + \sqrt{\frac{p}{a*}} R_d^* \left[\mathrm{sh}\left(\sqrt{\frac{p}{a*}}\right) - \mathrm{sh}\left(\sqrt{\frac{p}{a*}} - 2\sqrt{\frac{p}{a*}} r\right)\right]}{2\mathrm{sh}\left(\sqrt{\frac{p}{a*}}\right) + \sqrt{\frac{p}{a*}} R_d^* \left[\mathrm{ch}\left(\sqrt{\frac{p}{a*}}\right) - \mathrm{ch}\left(\sqrt{\frac{p}{a*}} - 2\sqrt{\frac{p}{a*}} r\right)\right]};$$

$$\theta^R(p) = \frac{W\sqrt{a}}{\lambda\sqrt{p}} \frac{2}{2\mathrm{sh}\left(\sqrt{\frac{p}{a*}}\right) + \sqrt{\frac{p}{a*}} R_d^* \left[\mathrm{ch}\left(\sqrt{\frac{p}{a*}}\right) - \mathrm{ch}\left(\sqrt{\frac{p}{a*}} - 2\sqrt{\frac{p}{a*}} r\right)\right]}, \quad (5.33)$$

where $a* = a/L^2$ and $R_d^* = R_d\lambda / L$. The use of a standard non-linear fitting procedure by minimizing least squares results in considerable computational efforts because each evaluated pair of r and R_d^* values requires transiting from the Laplace to the original domain. A principle of orthogonal decomposition involves producing eigenvectors by 10,000 temperature signals calculated for different r and R_d^* [14]. A minimal value of l_1 was taken equal to the thickness of a single composite ply, while the minimal defect depth d was 20% of this value; experimental results were obtained for composites consisted of 10 and 20 plies. Non-linear fitting was applied to interpolated data by using eigenvectors. It was stressed that reasonable characterization results were obtained if there were three mechanical shutters mounted in front of two flash tubes in order to ensure an appropriate heating protocol.

Delpeche applied a standard non-linear fitting technique in TNDT of thin high-conductive samples heated with a square pulse [8]. Defect identification was made by solving Eq. (5.10) where the coefficients β_3 and β_4 were related to defect depth l_1 and thermal resistance R_d. The Levenberg-Marquardt method, which is principally applicable to evaluating all four coefficients β, has proven to be very time-consuming when processing full-format images, therefore, in practice, it was applied only to some chosen points. Faster computations took place in the case of the adiabatic model where $\beta_2 = 0$. Within such approach, a 34 μm-thick air gap between two 1 mm-thick steel plates was evaluated with the accuracy of 5% [8].

A more advanced model of a two-layer plate with a thermally-resistive defect was proposed by Krapez and Cielo [15]. The plate was heated with the heat flux which penetrated the material according to Bouguer's law. The corresponding analytical solution $T_i(q_j, \tau_i)$ was derived by using the Green's function in a rather bulky form; however, the data inversion took only 20 s of computer time. Afterwards, the same approach was also applied to a two-dimensional model [16].

5.6.2 Using Particular TNDT Models

The technique of non-linear fitting was used in thermal tomography [17] by introducing the basic function by Eq. (4.42) describing pulsed heating of a coating on a substrate. The scheme for solving this problem is shown in Fig. 5.4. By minimizing the corresponding functional, the following test parameters were determined:

- heating parameter W/e_c,
- coating thickness parameter α_c/l^2,
- reflection coefficient Γ
- time offset $\Delta\tau_{off}$ as an additional parameter defining possible shift of experimental and theoretical data in time,
- amplitude offset амплитудный сдвиг ΔT_{refl} as a specific parameter defining undesirable amplitude variations, for instance, because of background reflected radiation.

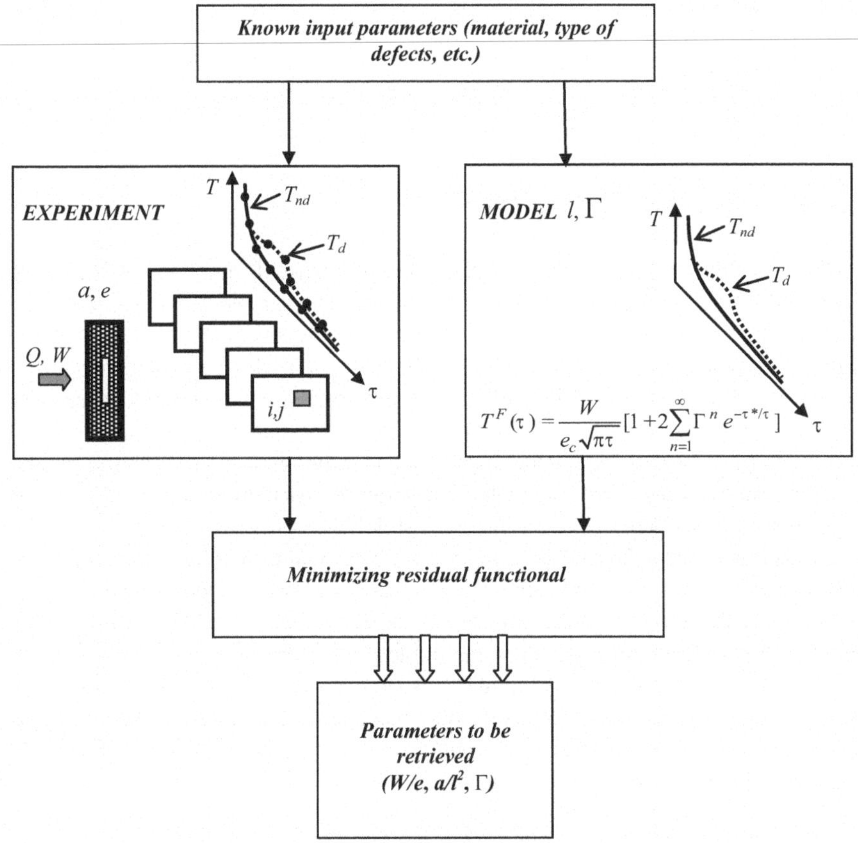

Fig. 5.4 Thermal characterization of coating on substrate by minimizing residual functional

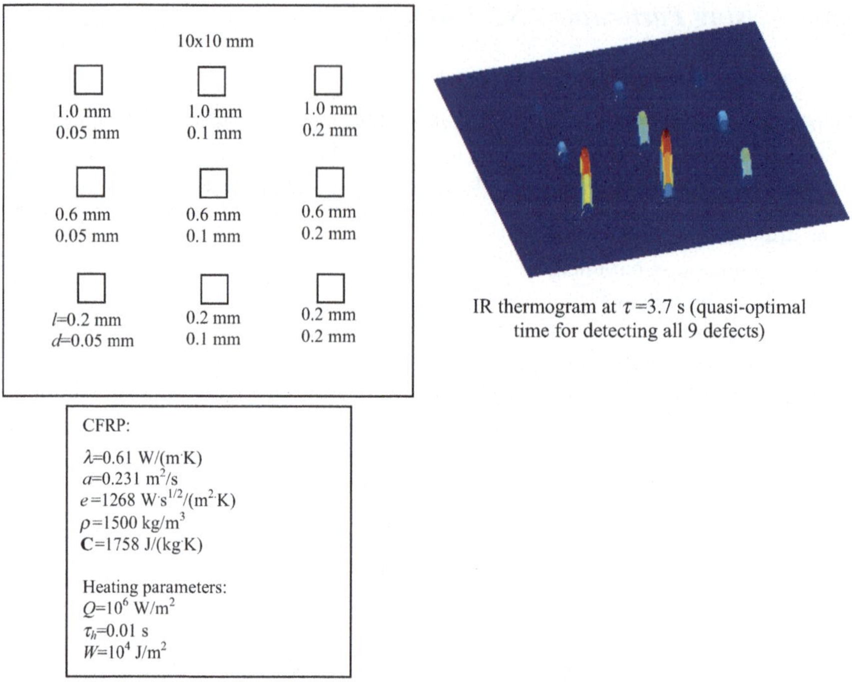

Fig. 5.5 Modeling air-filled defects in CFRP

The algorithm in Fig. 5.4 was validated on a synthetic IR image sequence which described TNDT of a 2 mm-thick CFRP sample. The sample contained 9 defects with the thicknesses of 0.05, 0.1 and 0.2 mm at the depths of 0.2, 0.6 and 1.0 mm (Fig. 5.5). The defect lateral size was 10 × 10 mm. All model parameters are shown in Fig. 5.5. The behavior of the calculated temperature signals in defect areas was classical being characterized by maximal differential signals ΔT_m and their observation times τ_m. These test parameters were used in solving the corresponding inverse problem. An example of the calculated temperature distribution at $\tau_m = 3.7$ s that is the optimum observation time for the deepest defect is presented in Fig. 5.5.

It was found that a decisive parameter for the optimal retrieving of defect parameters is a time period used in fitting. The best results were obtained by analyzing the evolution of the $\Delta T(\tau)$ signal from zero to its maximum value. The results of the evaluation of defect depths are shown in Table 5.1.

A practical use of the algorithm above is limited with a long computation time. At each pixel, the number of iterations may reach several hundreds of seconds thus requiring about 15 s of a personal computer time. Hence, the analysis of a 320 × 240 pixel image may require up to 13 days.

Table 5.1 Efficiency of determining defect depth l in in the model by Fig. 5.5

Defect	Estimate of l, mm	Error, %
$l = 0.2$ mm $d = 0.1$ mm	0.182	9.8
$l = 0.6$ mm $d = 0.05$ mm	0.51	17.6
$l = 0.6$ mm $d = 0.1$ mm	0.52	14.7
$l = 0.6$ mm $d = 0.2$ mm	0.54	11.6
$l = 1.0$ mm $d = 0.1$ mm	0.91	10.3

5.6.3 Using Multi-dimensional Numerical Solutions

In 1988, Kush et al. applied a technique of functional minimization to 2D temperature evolutions. It is interesting that solving an inverse problem required up to 56 iterations and 168 minutes of the processor time on an EC-1055 computer which used to be one of the most powerful in the USSR by that time [18]. The reported characterization accuracy was 20%. In order to reduce computation time up to 3–5 minutes, the authors analyzed basic relationships between defect parameters and expected temperature signals.

Krapez et al. used a 2D cylindrical model similar to that described in Sect. 4.3 [16]. The input experimental parameters were:

- maximum running contrast C_m^{run},
- half-rise time $\tau_{1/2+}$ which was determined on the rising slope of the temperature signal $\Delta T(\tau)$,
- half-drop time $\tau_{1/2-}$ which was determined on the decaying slope of the temperature signal $\Delta T(\tau)$,
- apparent defect lateral size evaluated by maximum surface gradient of the $\Delta T(r)$ signal (see Sect. 4.8.4).

The output (calculated) parameters were:

- normalized defect depth,
- normalized defect thermal resistance,
- defect lateral size,
- heat exchange coefficient.

A least-square method was applied in combination with the Levenberg-Marquardt algorithm which previously proved its validity in solving 1D inverse problems [15]. The accuracy of defect depth and lateral size characterization was 10%.

The approaches suggested for solving multi-parametric inverse problems require using powerful computers, and they may find a wider use in the future due to avalanche-like growth of computer power.

5.7 Determining Lateral Size of a Defect

5.7.1 *Using Spatial Temperature Profiles*

Unlike defect depth and thickness, lateral dimensions of subsurface defects h_x, h_y can be visually evaluated by their surface temperature 'footprints' (note that TNDT commonly deals with laterally-extended defects). By placing a well-visible marker on a test surface, for example, a piece of aluminum foil, it is possible to automatically determine h_x, h_y. In many practical cases, the errors in determining defect size and shape by surface temperature patterns are under 10%. However, the need to automate this process requires establishing a certain temperature threshold ΔT_{thr} and this may significantly affect h_x, h_y estimates.

Another h_x, h_y identification procedure called *Full Width Half Maximum* (FWHM) implies measuring distance between points which correspond to a half of the maximum temperature signal $\Delta T_m/2$ [19, 20].

Vavilov and Shirayev demonstrated that extremums of $T(x, y)$ derivatives along any surface direction (coordinate) precisely correspond to the projections of defect borders onto any sample surface [21] (see also Sect. 4.8.4).

The three approaches above are illustrated in Fig. 5.6 for a 0.5 mm-deep defect with the radius of r_d = 1 mm in both aluminum and CFRP. The estimates of the defect radius are presented in Table 5.2 to demonstrate that the analysis of the extremums of the first derivative $(\partial T/\partial r)_{extr}$ ensures accurate evaluation of r_d in both materials. However, this technique is difficult to apply in the case of noisy temperature profiles. A reasonable accuracy is also achieved by using the $\Delta T_m/2$ criterion. In aluminum, more accurate estimates appear at shorter observation times because of more intensive heat diffusion occurring at longer times. In CFRP, both overestimates and underestimates of r_d values have been observed; the latter phenomenon was called *defect shrinkage* (Almond et al. [20]). Note that the proposed method of

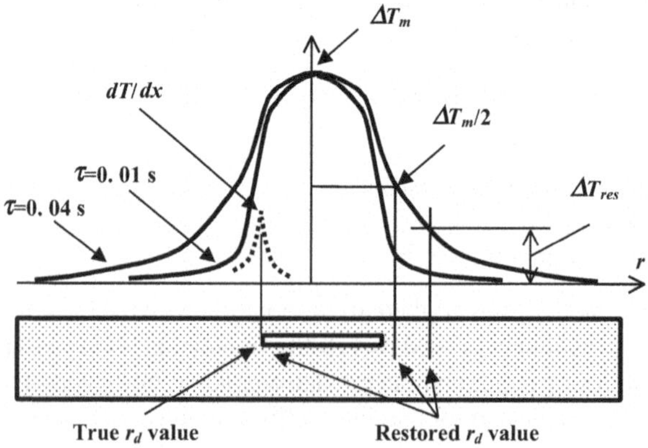

Fig. 5.6 Normalized temperature profiles over air-filled defect in aluminum ($L = 2\,mm$, $l = 0.5\,mm$, $r_d = 1\,mm$, $d = 0.2\,mm$, $Q = 10^6\,W/m^2$, $\tau_h = 0.01\,s$)

Table 5.2 Retrieving lateral size of disk defects

Observation time, s	True defect radius, mm	Reconstructed defect radius, mm by $(\partial T/\partial r)_{extr}$	by $\Delta T_m/2$	by $\Delta T_{thr} = 0.01^0C$
Aluminum: $L = 2\,mm,\ l = 0.5\,mm,\ d = 0.2\,mm,\ Q = 10^6\,W/m^2,\ \tau_h = 0.01\,s$				
0.005		1.0[a]	1.0	2.5
0.01	1.0	1.0[a]	1.0	3.0
0.05		1.0[a]	2.4	1.8
CFRP: $L = 5\,mm,\ l = 1.5\,mm,\ d = 0.5\,mm,\ Q = 10^5\,W/m^2,\ \tau_h = 0.01\,s$				
1.0		5.0[a]	5.4	4.0
2.5	5.0	5.0[a]	4.6	7.0
5.0		5.0[a]	4.3	7.5

[a]With the accuracy of a single spatial step in the numerical mesh

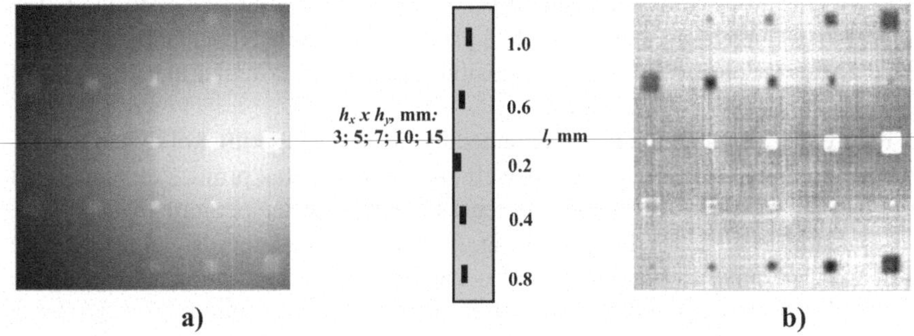

Fig. 5.7 One-sided TNDT of 2 mm-thick CFRP reference sample (defects simulated by 25 Teflon inserts): (**a**) optimal raw IR thermogram, (**b**) image of 2nd principal component

extrapolating $r_d(\tau)$ estimates in time up to $\tau = 0$ has not lead to a higher characterization accuracy in the case of CFRP (see Table 5.2). Finally, the attempt to estimate r_d by the area occupied by the signal $\Delta T(r) > \Delta T_{thr} = 0.01$ °C has resulted in the essential spread of estimates depending on observation time.

Examples

Figure 5.7 shows results of TNDT of a 2 mm-thick CFRP reference sample containing 25 Teflon inserts with lateral size from 3 × 3 to 15 × 15 mm at depths from 0.2 to 1 mm (experimental data: courtesy W. Świderski). In the raw image, a trained operator was able to detect 18 defects (Fig. 5.7a). Lateral size of each defect was evaluated directly on the image by identifying defect border pixels and comparing the corresponding distances with a marker of a known size. By using such simple procedure, the mean error of determining defect lateral size was surprisingly as low as 4.6%. By processing the raw sequence with the technique of principal component analysis, the number of detected defects was increased up to 22 (Fig. 5.7b) with the corresponding mean error of 4.2%.

The high efficiency of evaluating defect lateral size in the example above can be explained by low defect depths and relatively low thermal conductivity of CFRP preventing lateral heat diffusion. The next experiment was done on a 10 mm-thick sample made of AISI 1010 steel where 9 bottom holes simulated rear-surface corrosion. In this case, only 4 defects were reliably detected in the phasegram (Fig. 5.8). Respectively, the mean error in determining defect lateral size was 7.3% (20 and 40 mm defects with material loss 25% and 40%).

The same approach was assumed in the inspection of an icon manufactured according to the Middle Age technology. A 20 mm-thick wooden plate (courtesy E. Grinzato) represented a multi-layer structure which included lacquer, gypsum glue, canvas and wood. TNDT of this sample will be described in Chap. 10. Here we analyze a 27 × 4 mm Teflon insert placed at the depth of about 1 mm. An optimum IR thermogram is shown in Fig. 5.9a. The image scale is 0.84 mm per pixel. The horizontal temperature profile across the defect is shown in Fig. 5.9b and its derivative – in Fig. 5.9c. Since the source image is fairly noisy, that is typical when inspecting art objects, direct characterization of defect lateral size is difficult. The defect extension by $\Delta T_m/2$ criterion is 45 pixels, or 37.8 mm. Note that the use of the $(\partial T/\partial r)_{extr}$ criterion proved to be impossible because of essential enhancement of high-frequency noise (Fig. 5.9c).

Figure 5.9d shows a timegram obtained by using the algorithm of dynamic thermal tomography (see Sect. 6.3). It is well known that spatial profiles of heat transit times are steeper at defect borders in regard to 'normal' temperature distributions. This phenomenon is clearly seen in Fig. 5.9d where the defect lateral size can be evaluated by simply counting pixels occupying the defect "footprint", namely, 45 × 8 pixels. Hence, the defect size determined by the timegram is 37.8 × 6.7 mm thus matching well the estimate by $\Delta T_m/2$.

It is worth noting that the retrieved estimate of defect size is different from the designed one that often appears when simulating defects with some inserts placed between sample layers. If reference defects are produced as bottom holes, grooves etc., the characterization accuracy by using the above-mentioned methods is much higher.

5.7.2 Point-Source Function and Defect Characterization in the Fourier Domain

A general approach to determining size and shape of hidden defects by their surface temperature patterns was developed by Favro et al. in pulsed TNDT [22] and by Walther et al. in thermal wave procedures [23, 24]. This approach was borrowed from astronomy and optics. It assumes that, in the ideal case, defect characterization results are independent of time and defect parameters.

Let the shape of a defect in the plane parallel to the tested surface be described with the function $F(x, y, l)$. The surface temperature response from a point-like defect located at the depth of l in a material heated with a Dirac pulse is called a *point-source function* (PSF) $G(x, y, l, \tau)$ which is an analog of the Green's function widely used in solving differential equations. This function can be defined either in

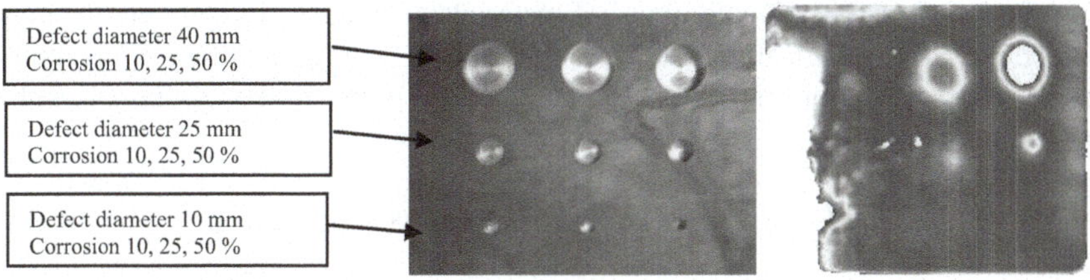

Fig. 5.8 Evaluating corrosion in 10 mm-thick AISI 1010 steel

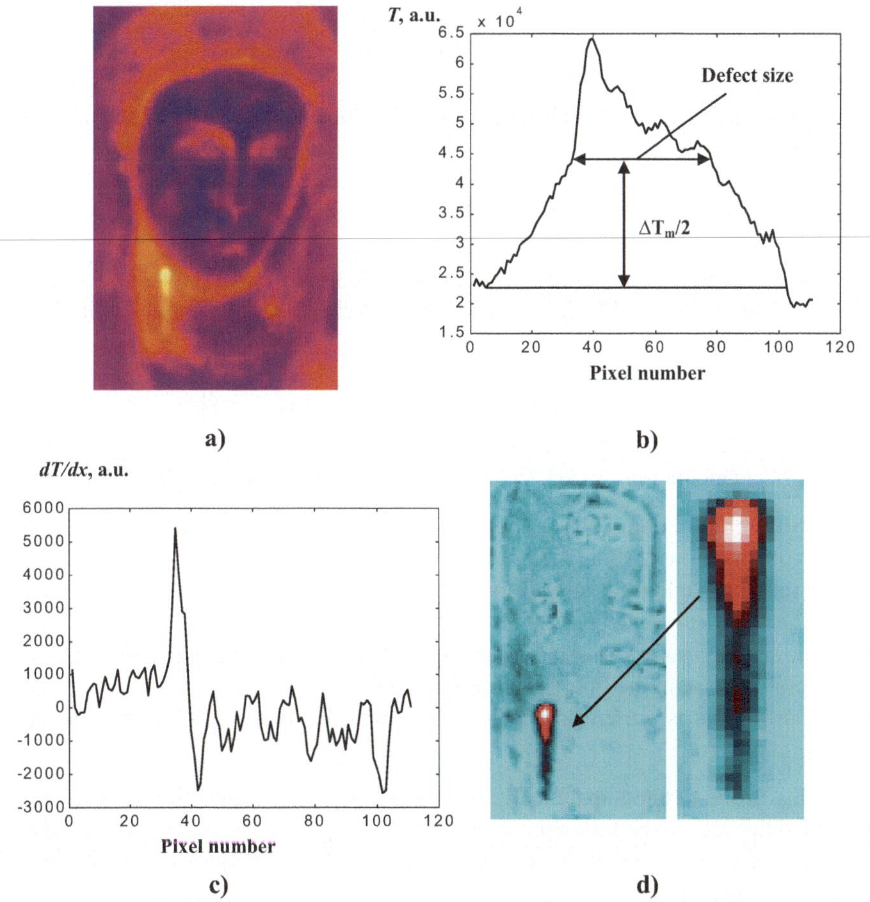

Fig. 5.9 Determining lateral size of Teflon inserts in icon: (**a**) IR thermogram (optimum observation time), (**b**) temperature profile over defect, (**c**) first temperature derivative: profile b), (**d**) timegram

a general way or for particular types of materials and defects in particular heating procedures. Mathematically, in the case of arbitrary heating, a surface temperature signal over an arbitrary defect can be obtained by performing *convolution* of the following functions:

$$\Delta T(x,y,l,\tau) = W(x,y,\tau) ** F(x,y,l) ** G(x,y,l,\tau), \tag{5.34}$$

where $W(x, y, \tau)$ is the heating function.

If to define a PSF for a particular type of heating, the last equation acquires the form:

$$\Delta T(x,y,l,\tau) = F(x,y,l) ** G(x,y,l,\tau). \tag{5.35}$$

Convolution in the domain of originals can be substituted by multiplication in the Fourier domain:

$$\Theta(\upsilon,\varsigma,\tau,l) = \Phi(\upsilon,\varsigma,l) \cdot \Gamma(\upsilon,\varsigma,\tau,l). \tag{5.36}$$

Here, the original functions are replaced with their Fourier transforms. Defect shape can be determined by using the inverse Fourier transform applied to the ratio $\Theta(\upsilon, \varsigma, \tau, l) / \Gamma(\upsilon, \varsigma, \tau, l)$:

$$F(x,y,l) = Fourier^{-1}\left[\Theta(\upsilon,\varsigma,\tau,l) / \Gamma(\upsilon,\varsigma,\tau,l)\right]. \tag{5.37}$$

Practical calculations by Eq. (5.37) may be difficult because of the following:

- a PSF should be defined in an 'optimum' way for particular defects; for example, it is unclear how G should take into account defect thickness,
- the ratio of two Fourier transforms in Eq. (5.37) is a function of time but, by definition, F must be independent of time;
- dividing by Γ values which are close to zero may lead to the appearance of considerable noise in the Fourier spectrum of F,
- the noise that is present in source images tends to increase after applying any operation above.

PSFs can be defined analytically or numerically. In pulsed TNDT, Favro et al. obtained the PSF in the following form by neglecting multiple reflections of pulsed thermal waves between surfaces of a sample and a defect (Born's approximation known in optics) [22]:

$$G(x,y,\tau,l) = -\frac{C}{2\pi}\sqrt{\frac{1}{\pi a\tau}}\frac{\partial}{\partial l}\frac{\exp\left\{-\frac{\left[\sqrt{x^2+y^2+l^2}+l\right]^2}{4a\tau}\right\}}{\sqrt{x^2+y^2+l^2}}, \tag{5.38}$$

where A is a constant (the derivative by l is not fulfilled for simplicity). The authors of the algorithm suppose that Eq. (5.38) fairly well describes propagation and retardation of pulsed thermal waves, however, the problem of choosing an optimum time for performing operation by Eq. (5.37) was not solved (the authors suggested to do this empirically).

In the case of point thermal wave heating (photothermal TNDT technique), Walther et al. proposed the following formulation of the PSF in the cylindrical coordinates [23, 24]:

$$G(r,\tau,l)=\frac{\pi\sigma^2}{2\sqrt{r^2+l^2}}\exp\left[-\sigma\left(l+\sqrt{r^2+l^2}\right)\right]\cdot\left[B+\frac{Al}{\sqrt{r^2+l^2}}\left(1+\frac{1}{\sigma\sqrt{r^2+l^2}}\right)\right], \quad (5.39)$$

where $A=\Delta\lambda/\lambda$ and $B=\Delta(C\rho)/(C\rho)$ have been called parameters of 'thermal strength' of a point-like discontinuity, $\sigma^2 = 2i/\mu$ is the wave number, μ is the thermal diffusion length, r, l are the distances between a defect point and a heating point in the horizontal and vertical direction respectively. Equation (5.39) is similar to Eq. (5.38) with no time variable due to the quasi-stationary character of the thermal wave process.

Defect characterization by Eq. (5.37) can be fulfilled numerically. This technique allows sharpening defect fuzzy 'footprints' but, because it enhances high-frequency noise, it is recommended to filter noise already in the Fourier domain.

5.7.3 Laplacian and Restoring Defect Fuzzy Borders

Determination of the contours of subsurface defects by analyzing spatial derivatives (Sect. 5.7.1) or applying the PSF concept (Sect. 5.7.2) can be explained from the point of view of image processing by considering the Laplace filter. Such filter enables the sharpening of images which "diffuse" in time (see the monograph by Rosenfeld and Kak [25]).

The Laplacian is a linear differentiating operator

$$\nabla^2 T=\frac{\partial^2 T}{\partial x^2}+\frac{\partial^2 T}{\partial y^2}, \quad (5.40)$$

which is invariant in regard to rotation, therefore, the use of the Laplacian is invariant to the direction of filtration of $\Delta T(x,y)$.

Let us demonstrate the relationship between the Laplacian and the diffusion process described by

$$\frac{\partial T}{\partial \tau}=a\nabla^2 T. \quad (5.41)$$

Suppose that temperature distribution at zero time $T(x,y,\tau=0)$ accurately reflects a shape of subsurface defects (it was shown in Chap. 4 that the reproduction of defect shape improves at shorter observation times). Respectively, the function $T(x, y, \tau > 0)$ describes a blurred defect image. Expanding $T(x, y, \tau > 0)$ into a Tailor's series around the time point $\tau*$ leads to:

$$T(x,y,0) = T(x,y,\tau) - \tau * \frac{\partial T}{\partial \tau} + \frac{\tau *^2}{2} \frac{\partial^2 T}{\partial \tau^2} - \ldots \quad (5.42)$$

By neglecting the members of the second and higher order, one can finally obtain:

$$T(x,y,\tau) = T(x,y,\tau = 0) - a\tau * \nabla^2 T.$$

In the numerical form, this specifies the following filtration process:

$$T(i,j) - \nabla^2 T(i,j) = 5T(i,j) - \left[T(i+1,j) + T(i-1,j) + T(i,j+1) + T(i,j-1)\right]. \quad (5.43)$$

5.7.4 *Using the Solution for a 3D TNDT Adiabatic Problem*

The technique of quadrupoles proposed by Batsale et al. for solving 1D heat conduction problems was extended onto 3D problems by considering defect lateral dimensions $b \times c$ along with defect depth l and thermal resistance R_d [26]. The principles of solving direct TNDT problems by using the Laplace and Fourier transforms were discussed in Sect. 4.5. From the point of view of defect characterization, the simple algebraic expressions appear in the case of defects with low R_d. For determining defect size, it is necessary to analyze results of both one- and two-sided inspection. Two-sided procedures allow the evaluation of defect thermal resistance while one-sided tests are convenient for determining defect depth; in both cases, the 1D approach described in Sect. 4.1.2 was used. Then, the concept of a *mean spatial Laplace contrast* which is to be determined experimentally was suggested. As shown in [26], the above-mentioned contrast is proportional to $b \times c$. The proposed algorithm for determining defect lateral size is quite cumbersome and thus represents a rather theoretical interest; however, it may become practical with more powerful computers entering the market.

5.8 Evaluating Hidden Corrosion: General Inversion Formulas

In Sect. 3.7.3, the coefficient taking into account 3D heat diffusion at a particular time, was introduced. Vavilov and Marinetti extended this approach onto inspection of corrosion in 3–5 mm-thick metals.

An optimum observation time is to be found as a function of dimensionless heating time Fo_h, expected material loss $\Delta L/L$ and characteristic defect size D/L:

$$Fo_m = 0.540303\, Fo_h^{0.284477} (\Delta L / L)^{-0.205735} (D / L)^{0.463687}. \tag{5.44}$$

This formula is valid for square- and round-shape defects in metals heated with a square pulse of any duration ensuring the accuracy of determining Fo_m about 15%. It is important that heating cannot be longer than optimum observation time, i.e. $Fo_h \le Fo_m$. It follows from Eq. (5.44) that there is the particular combination of $\Delta L/L$ and D/L leading to $Fo_m = Fo_h$:

$$Fo_m = Fo_h = 0.422999 (\Delta L / L)^{0.287531} (D / L)^{0.64804}. \tag{5.45}$$

t is worth noting that maximum temperature contrasts appear after having applied a short heat pulse but, in practice, longer heating may be needed in order to ensure the warming of a whole test sample and producing higher differential signals over corroded areas.

Material loss is evaluated by the following formulas:

$$\frac{\Delta L}{L} = \left[1 - \frac{T_{nd}^{ap}(\tau_m)}{T_d^{ap}(\tau_m)}\right] / k(D / L, \tau); \tag{5.46}$$

$$k(D / L, \tau) = 0.114792\, Fo_m^{-0.94743} (D_{\max}^{ap} / L)^{0.166544} (D_{\min}^{ap} / L)^{0.898941} (T_{nd}^{ap} / T_d^{ap})_{\min}^{-0.034607} \tag{5.47}$$

in the case of flash heating, and

$$k(D / L, \tau) = 0.161392\, Fo_h^{-0.319918}\, Fo_m^{-0.371239} (D_{\max}^{ap} / L)^{0.262664} (D_{\min}^{ap} / L)^{0.842875} (T_{nd}^{ap} / T_d^{ap})_{\min}^{-0.316916} \tag{5.48}$$

in the case of longer heating,

where the superscript "*ap*" specifies apparent, i.e. experimentally evaluated values, $D_{\max}^{ap}$, $D_{\min}^{ap}$ correspond to minimum and maximum apparent lateral dimensions of corroded areas, $Fo_h = Fo_m$, if optimum observation time appears during heating, otherwise $Fo_h < Fo_m$. When using Eqs. (5.46, 5.47, and 5.48), the estimation accuracy has proved to be 2–25%.

For square defects ($D_x = D_y$):

$$k(D / L, \tau) = 0.114792\, Fo_m^{-0.94743} (D^{ap} / L)^{1.065485} (T_{nd}^{ap} / T_d^{ap})^{-0.034607} \tag{5.49}$$

in the case of flash heating, and

$$k(D / L, \tau) = 0.161392\, Fo_h^{-0.319918}\, Fo_m^{-0.371239} (D^{ap} / L)^{1.10554} (T_{nd}^{ap} / T_d^{ap})_{\min}^{-0.316916} \tag{5.50}$$

in the case of longer heating.

The inverse formulas above are to be used in the following way. First, an optimum observation time is to be chosen by Eq. (5.44), then, a sequence of IR thermograms is to be experimentally collected. Next, an operator, or an automatic device, has to identify areas of corrosion and determine minimums of the T_{nd}^{ap} / T_d^{ap} functions. Finally, lateral dimensions of corroded areas are to be evaluated to allow the calculation of material loss by substituting experimental values into Eqs. (5.46, 5.47, and 5.48).

The analytical expressions above allow to evaluate the limits of corrosion detection which are conditioned by surface noise running contrast C_n. By substituting Eq. (5.44) into Eqs. (5.47 and 5.48) and assuming $r = \left\{ \frac{T_{nd}}{T_d} \right\}_{\min} = \frac{1}{1+C_n}$, one can obtain the following expression for evaluating $\{\Delta L/L\}_{\min}$:

$$\{\Delta L / L\}_{\min} = 3.75623\, Fo_h^{0.225557} (D/L)^{-0.52403}\, r^{0.0289617} (1-r)^{0.836876} \tag{5.51}$$

in the case of flash heating, and

$$\{\Delta L / L\}_{\min} = 4.40248\, Fo_h^{0.395323} (D/L)^{-0.867169}\, r^{0.294428} (1-r)^{0.929043} \tag{5.52}$$

in the case of longer heating.

In the limiting case of $Fo_m = Fo_h$:

$$\{\Delta L / L\}_{\min} = 3.62736 (D/L)^{-0.689343}\, r^{0.332188} (1-r)^{1.04819}. \tag{5.53}$$

The inversion formulas (5.46, 5.47, and 5.48) have been verified in the inspection of a 10 mm-thick steel sample which contained two areas of 50% material loss by diameter of 20 and 40 mm (Fig. 5.10). The sample was heated for 20 s, but the maximum contrasts C_m = 0.66 (D1) and C_m = 0.36 (D2) were observed at 11 and 6 s respectively (see thermograms in Fig. 5.10). The thermal diffusivity of the steel $a = 13\ 10^{-6}$ m^2/s was determined by using Parker's technique. Apparent defect dimensions were 43 mm ($D/L = 4.3$) and 21 mm ($D/L = 2.1$). By substituting experimental values into Eq. (5.48) and assuming $Fo_m = Fo_h$, the 53% and 55% estimates of material loss $\Delta L/L$ have been retrieved (compare with the true value of 50%). Note that 1D estimation of corrosion by Eq. (3.11) produces less accurate values of 40% and 26% of material loss respectively because of the essential influence of lateral heat diffusion.

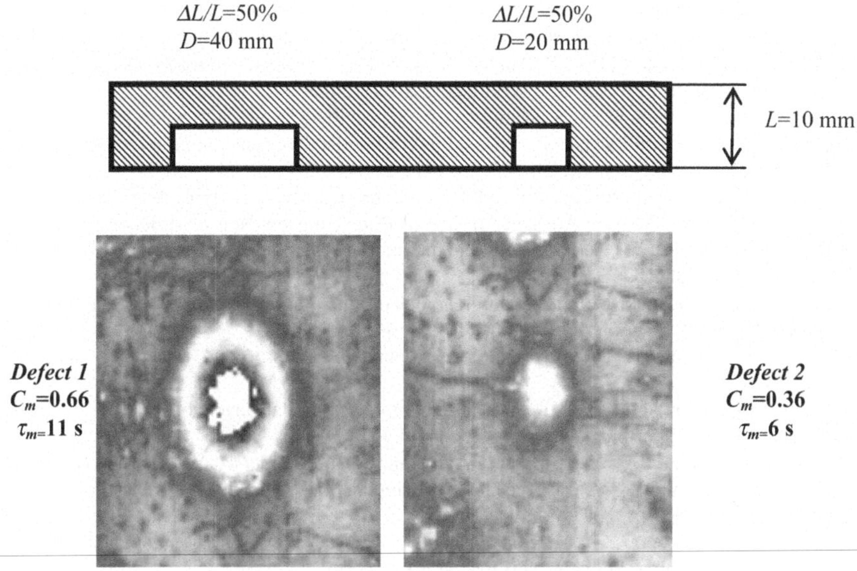

Fig. 5.10 Detecting 50% material loss in AISI 1010 steel sample (20 s heating)

References

1. Balageas, D.L., Krapez, J.-C., Cielo, P.: Pulsed photo-thermal modeling of layered materials. J. Appl. Phys. **59**(2), 348–357 (1986)
2. Delpech, P., Balageas, D.L.: Mesure par thermographie infrarouge stimulee de resistances thermique d'interface dans des structures bonnes condutrices de la chaleur. Journee d'Etude du 9 Janvier 1991, Soc. Fr. des Thermiciens, Paris, France. 12 p
3. Balageas, D.L., Deom, A.A., Boscher, D.M.: Characterisation and NDT of carbon epoxy composites by a pulsed photothermal method. Mater. Evaluation, April 1987, p. 461–465
4. Boscher, D.M., Balageas, D.L., Deom, A.A., Gardette, G.: Nondestructive evaluation of carbon epoxy laminates using transient infrared thermography. Proceedings of 16th NDE Symposium, San Antonio, TX, USA, 21–23 April, 1987. p. 322–328
5. Krapez, J.-C., Balageas, D.L.: Early detection of thermal contrast in pulsed stimulated infrared thermography. Proceedings of Quantitative InfraRed Thermography QIRT-94, Eurotherm Seminar #42, Sorrento, Italy, August 23–26, 1994. p. 260–266
6. Krapez, J.-C., Balageas, D.L., Deom, A., Lepoutre, F.: Early detection by stimulated infrared thermography. comparison with ultrasonics and holo/shearography. In: Maldague, X. (ed.) Advances in Signal Processing for Non Destructive Evaluation of Materials NATO ASI Series, Series E: Applied Sciences, vol. 262, pp. 303–321. Kluwer Academic Publish
7. Lugin, S., Netzelmann, U.: A defect shape reconstruction algorithm for pulsed thermography. NDT&E Intern. **40**, 220–228 (2007)
8. Delpesh, P., Krapez, J.-C., Balageas, D.L.: Thermal defectometry using the temperature decay rate method. Proceedings of Quantitative InfraRed Thermography QIRT-94, Eurotherm Seminar #42, Sorrento, Italy, August 23–26, 1994. p. 220–225
9. Vavilov, V., Grinzato, E., Bison, P.G., Marinetti, S.: Thermal characterization and tomography of carbon fibre reinforced plastics using individual identification technique. Mater. Eval. **54**(6), 604–611 (1996)

10. Boscher, D.M., Deom, A.A., Gardette, G., Balageas, D.L.: Thermal NDT of carbon epoxy laminates. Proceedings of Eurotherm-4 Seminar, Nancy, France, June 28–July 1, 1988. p. 46–50
11. Shepard, S.M., Lhota, J., Hou, Y., Ahmed, T.: Blind characterization of materials using single-sided thermography. Proc. SPIE Thermosense-XXVI. **5405**, 442–446 (2004)
12. Degiovanni, A., Lamine, A.-S., Houlbert, A.-S., Maillet, D.: Identification of subsurface defects using a sensibility analysis. Proceedings of Fourth European conference on Composite Materials, Stuttgart, Germany, Sept 25–28, 1990. p. 691–695
13. Grinzato, E., Marinetti, S.: Materials NDE by non linear filtering applying heat transfer models. In: Maldague, X. (ed.) Advances in Signal Processing for Non Destructive Evaluation of Materials NATO ASI Series, Series E: Applied Sciences, vol. 262, pp. 117–132. Kluwer Academic Publish (1994)
14. Winfree, W.P., Zalameda, J.N.: Thermographic determination of delaminations depth in composites. Proc. SPIE Thermosense-XXV. **5073**, 363–373 (2003)
15. Krapez, J.-C., Cielo, P.: Thermographic NDE: data inversion procedure (Part I: 1D analysis). Res. NDE. **3**(2), 69–100 (1991)
16. Krapez, J.-C., Maldague, X., Cielo, P.: Thermographic NDE: data inversion procedure (Part II: 2D analysis and experimental results). Res. NDE. (2), 101–124 (1991)
17. Vavilov, V.P., Nesteruk, D.A., Shiryaev, V.V., Ivanov, V.A., Świderski, W.: Thermal (infrared) tomography: terminology, basic procedures and applications in nondestructive testing of composite materials. Rus. J. NDT. **3**, 3–15 (2010)
18. Kush, D.V., Rapoport, D.A., Budadin, O.N.: Inverse problem of automated thermal NDT. Defectoscopiya (Soviet J. NDT). **5**, 64–68 (1988)
19. Avdelidis, N.P., Almond, D.P., Dobbinson, A.: *Aircraft composites* assessment by means of transient *thermal NDT*. NDT&E Intern. **36**(6), 433–439 (2004)
20. Almond, D.P., Saintey, S., Lau, S.K.: Edge effects and defect sizing by transient thermography. Proceedings of Quantitative InfraRed Thermography QIRT'94, Eurotherm Seminar #42, Sorrento, Italy, August 23–26, 1994. p. 247–252
21. Vavilov, V.P., Shiryaev, V.V.: The method for determining defect size in thermal NDT. Defectoscopiya (Rus. J. NDT). **11**, 63–65 (1979). (in Russian)
22. Favro, L.D., Crowther, D.J., Kuo, P.K., Thomas, R.L.: Inversion of pulse-echo thermal wave images. Proc. SPIE Thermosense-XV. **1933**, 138–141 (1993)
23. Walther, H.G., Seidel, U.: Some remarks on definition, resolution and contrast in photothermal imaging. Proceedings of Eurotherm Seminar #27 Quantitative InfraRed Thermography-QIRT 92, Chatenay-Malabry, France, July 7–9, 1992. p. 251–258
24. Friedrich, K., Seidel, U., Walther, H.G., et al.: Proposal for photothermal characterisation of boundaries between layer and substrate. Res. NDE. **5**, 31–39 (1993)
25. Rosenfeld, A., Kak, A.: Digital Picture Processing, vol. 1. Academic, Orlando (1982). 434 P
26. Batsale, J.C., Bendada, A., Maillet, D., Degiovanni, A.: Distribution of a thermal contact resistance: inversion using experimental Laplace and Fourier transformations and an asymptotic expansion. Proceedings of the 1st International Conference on Inverse Problems in Engineering, Palm Coast, FL, USA, June 13–18, 1993. p. 122–129

Chapter 6
Data Processing in TNDT

Abstract Numerous mathematical techniques to enhance the signal-to-noise ratio in IR image sequences are discussed. It is shown that the most efficient method is to use temporal phenomena accompanied by the time evolution of pixel-based temperature functions. Standard image processing algorithms are summarized. Also, advanced processing techniques including pulse phase thermography, thermal wave lock-in technique, principal component analysis, polynomial fitting, correlation analysis, thermal tomography, neural networks and more exotic methods, are discussed. Some novel TNDT techniques, including vibrothermography, eddy current, and electromagnetic IR thermography are described. Particular attention is paid to the analysis of thermo-mechanical phenomena in solids.

Data processing in TNDT is related to the analysis of temperature evolutions $T(i,j,\tau)$ in both defect and non-defect areas (Fig. 6.1). A standard approach involves choosing a proper *time gate* (observation interval) where a differential temperature signal $\Delta T(\tau)$ should be captured. A particular processing algorithm should be used within a chosen time gate in order to enhance defect visibility and perform defect characterization (see some examples in Fig. 6.1).

6.1 Optimum Observation Time

There are few implementations of pulsed TNDT which can be defined as a method of *optimum observation time*. In fact, this method is obvious because it requires capturing sample surface temperature at the time τ_m when differential signals ΔT over defects, or other related parameters, such as running contrast or signal-to-noise ratio, reach their maximum values. An optimum IR thermogram is to be chosen in a recorded sequence or captured directly at the time τ_m which is to be determined *a priori*. Such approach is realized in *synchronous*, *lock-in* (*box-car*, *time-resolved*) and other TNDT techniques.

In the frame of all these techniques, it is important how optimum observation times are "positioned" in regard to the end of heating ($\tau = \tau_h$). The typical recommendation in TNDT requires $\tau_m > \tau_h$, otherwise reflected radiation can be a serious

V. Vavilov, D. Burleigh, *Infrared Thermography and Thermal Nondestructive Testing*, https://doi.org/10.1007/978-3-030-48002-8_6

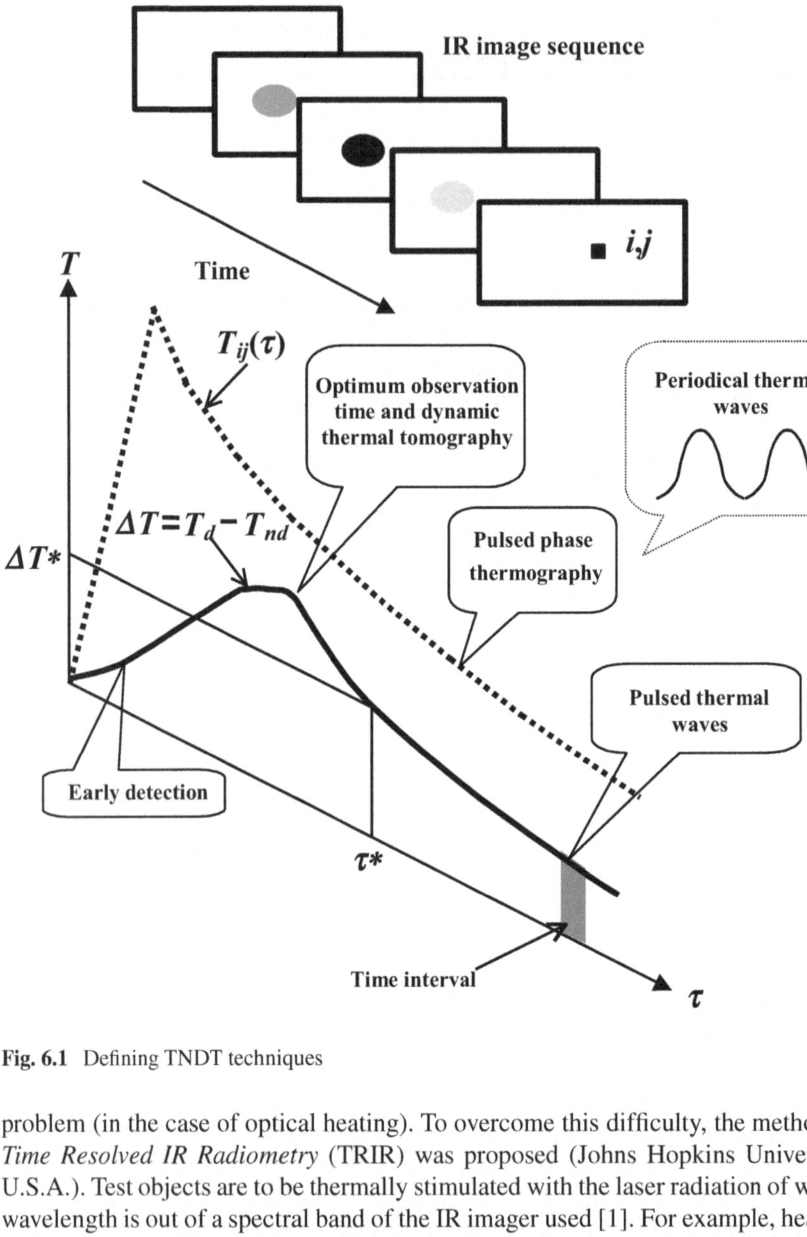

Fig. 6.1 Defining TNDT techniques

problem (in the case of optical heating). To overcome this difficulty, the method of *Time Resolved IR Radiometry* (TRIR) was proposed (Johns Hopkins University, U.S.A.). Test objects are to be thermally stimulated with the laser radiation of which wavelength is out of a spectral band of the IR imager used [1]. For example, heating may be performed continuously in the visual or near-IR range while sample thermal radiation should be collected in the 7–14 μm wavelength band. From the point of view of ensuring a maximum running contrast, the TRIR method is inferior to flash heating but in some practical tasks it can be acceptable if not optimal. Recently this technique has gained a new interest due to appearance of commercial powerful LED panels, of which the most interest is represented by the so-called VCSEL (Vertical Cavity Surface Emitting Laser) devices.

6.2 Early Detection Technique

The *Early Detection* technique proposed by Krapez and Balageas stipulates that temperature signals ΔT should be recorded at $\tau \ll \tau_m$ (see Sect. 5.1.2) [2, 3]. Obviously, this causes *SNR* values lower than those provided by the method of optimum observation time, therefore, the early detection technique is applicable to the detection of defects which produce fairly high signals, i.e. large-size and/or shallow. The main advantage of early detection is a good reproduction of defect projection (defect shape) on tested surfaces due to weak lateral diffusion of heat. For example, being applied to the inspection of aluminum and composite aviation panels, this method ensures the quality of IR thermograms which is comparable to ultrasonic C-scan images.

6.3 Dynamic Thermal Tomography

6.3.1 Basic Principles

The term "tomography" is derived from two Greek words that mean respectively "slice", or "section", and "to write", therefore, tomographying is sectioning. There are about 40 types of tomographic imaging techniques introduced in the last few decades and differing in underlying physical principles.

In publications of the last decades, the combination of the terms "infrared", "thermographic", "thermal" and "tomography" specifies a number of different techniques intended for reconstructing internal structure of opaque solid objects by analogy with computed X ray tomography.

Emission IR thermographic tomography of semi-transparent gases and plasma, as well as *IR tomography* of charge carriers in semiconductors, are the techniques which are most close to classical computed tomography, i.e. they realize a principle of rectilinear propagation of information carriers, such as electromagnetic radiation and/or charged particles.

The first approach is implemented in optical tomographic systems where CCD-devices operating in the visible or near-IR spectral range are replaced with IR imagers, thus allowing visualization of spatial temperature cross-sections in gas (plasma) by viewing its volume from 2 to 12 directions [4, 5].

The second method was elaborated for evaluating life time and diffusion length of charge carriers in silicon samples by size of up to 1 m. It is based on probing a tested object with crossed laser beams [6]. A pumping laser (1.17–1.18 μm spectral range) injects electrons and holes, and a probing laser (3.39 μm) investigates spatial/temporal evolution of excess charge carriers. In this case, the accuracy of determining spatial coordinates is about 1 cm.

In biological tissues, the temperature can be directly measured by using a technique of microwave radiometry first proposed by Barrett et al. in the detection of breast cancer [7]. In Russia, the parallel research was fulfilled by Troyitsky et al. [8].

Microwave thermography can be remote and applicative. The first technique is analogous to IR thermography ensuring non-contact temperature measurement; however, the most common is the applicative technique which is realized with antennas placed on the surface of test objects. Depending on a calibration regime, frequency range and transducer design, this technique allows measuring both absolute and differential in-depth temperatures across biological objects by taking into account that microwave penetration depth is strongly dependent on frequency. In Russia, some basic research was fulfilled by using a "radiothermograph" RT-20 developed at Institute of Radio Electronics, Russian Academy of Science. In 1997, a computerized microwave radiometer RTM-01 was elaborated at All-Russia Institute of Radio Techniques. Such units can be employed as thermal tomographic devices in probing materials which are transparent in the microwave spectral band.

Another technique intended for determining in-depth temperature is *passive acoustic thermotomography* proposed by Guliayev et al. and later investigated by Pasechnik et al. [9]. Temperature profiles at depths of few centimeters in biological tissues can be retrieved through solving the corresponding inverse problem by using either the Tikhonov's regularization principle or 'standard source' technique.

In this book, the term "dynamic thermal tomography" (DTT) is referred to the evaluation of material thermal properties by analyzing transient surface temperature distributions, i.e. by applying active TNDT. This approach differs principally from the classical computed tomography by the fact that sample images obtained under some geometrical angles are substituted with IR thermograms captured at different observation times.

Primarily, the term "thermal tomography" appeared in the 1980s in the thermal wave studies. The contemporary methodology and instrumentation in this area is traced to the work of Rosenzwaig et al. [10]. An estimate of thermal wave penetration depth, or defect depth detection limit, is given by the thermal diffusion length $\mu = \sqrt{0 / \pi f}$. A classical photothermal method implies utilizes high-frequency thermal waves which penetrate solids to depth of few mm. The last studies combine the use of low-frequency periodical waves, as well as pulsed thermal stimulation which produces packets of thermal waves of multiple frequencies and thus requires no frequency re-adjustment when probing materials up to different depths [11]. Mandelis et al. extended the photothermal approach onto tomographic presentation of subsurface micro-defects and identification of thermal diffusivity distribution by using a couple of processing techniques, to mention spatial Laplace spectral decomposition, ray-optic reconstruction and Thermal-Wave Slice Diffraction Tomography [12–17]. In 1984, Busse and Renk proposed a simple "geometrical" idea of two-sided thermal tomography but it received no further development [18].

Since the 1990s, thermal tomography has received multiple developments. Kaiplavi et al. applied a technique of truncated-correlation photothermal coherence tomography to the detection of early mineral loss in bones by introducing a thermal wave occupation index [17]. A similar concept was used in medical diagnostics for identification of living tissues [19]. A review on active thermal tomography techniques compiled by Milovanović and Pečur contains interesting results obtained on thick concrete structures under natural heating [20].

The idea of *Dynamic Thermal Tomography* was suggested by Vavilov and Shiryaev in 1985 on the wave of a general explosive interest to tomographic presentation of solids by means of electromagnetic radiation or particles propagating through materials [21–24]. The first experimental thermal tomogram was obtained by Vavilov in 1989 in collaboration with the research team from Wayne State University (USA) by analyzing an experimental image sequence captured in pulsed TNDT of a plastic sample [22]. In Russia, the parallel research was conducted by Kush et al. who were seeking tomographic solution as a result of iteratively solving TNDT inverse problems [25]. It is known that the invention of computed X ray tomography and its use in medical and technical diagnostics proved to be revolutionary in the same way as the invention of X rays. In NDT, X ray tomography allows observing low-contrast details behind high-contrast ones by examining solid bodies under various geometrical angles. Unlike particles and quantums of electromagnetic radiation, thermal energy propagates in solids through diffusion, therefore, a purely geometrical approach to the analysis of material internal structure, that is characteristic for computed X ray tomography, was substituted in dynamic thermal tomography by the analysis of temperature temporal evolutions. Dynamic thermal tomography is based on the fact that, in one-sided TNDT, deeper material layers are characterized by longer time delays of the thermal response. According to the theory described in Chaps. 4 and 5, it follows that dynamic thermal tomography can be realized only in one-sided TNDT; in two-sided TNDT, the $\Delta T(l)$ dependence is ambiguous, as demonstrated in Sect. 4.8. In fact, dynamic thermal tomography can be considered as a special type of data presentation of active TNDT results by using the $l(\tau*)$ relationship, where $\tau*$ defines a heat transit time chosen as a specific testing parameter.

Thermal tomography allows:

- performing layer-by-layer analysis of solids (see Sect. 4.8),
- relax the influence of surface clutter on IR thermogram quality,
- improving defect detectability,
- evaluating defect depth with a reasonable accuracy (up to 15%).

Dynamic thermal tomography involves a standard TNDT procedure presented in Fig. 6.2a. An operator is supposed to choose a reference (non-defect) point (i_{ref}, j_{ref}) on the heated surface of a test body, and all other points (pixels) will be further examined in respect to this point for their possible belonging to defect areas. By other words, each pixel can be characterized by a differential temperature signal $\Delta T(i,j,\tau*) = T(i,j,\tau*) - T(i_{ref},j_{ref},\tau*)$ defined for a chosen observation time $\tau*$.

As an example, consider the case of $\tau* = \tau_m$. Each pixel (i, j) can be characterized by two parameters $\Delta T_m(i, j)$ and $\tau_m(i, j)$ which produce images called *maxigram* and *timegram* respectively. Depending on a defect type and a chosen reference area, $\Delta T_m(i, j)$ signals can be positive, negative or zero. A basic concept of such approach is that a source image sequence is replaced with only two images. A maxigram shows defects in their 'best' presentation and a timegram can be regarded as an encoded image of material in-depth structure (Fig. 6.2b).

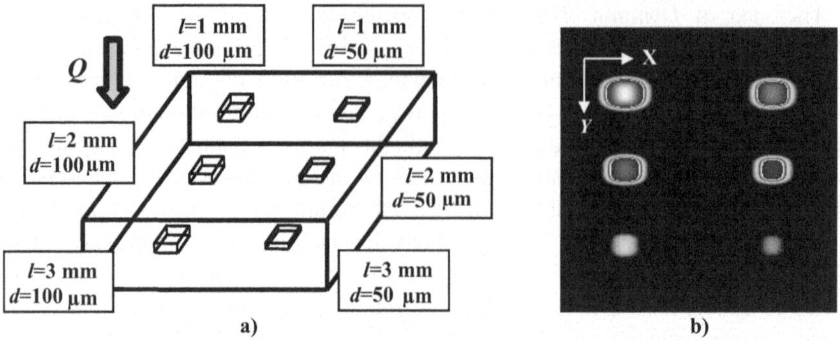

Fig. 6.2 One-sided pulsed TNDT of 4.9 mm-thick CFRP sample with air-filled defects ($Q = 10^5$ W/m^2; $\tau_h = 10$ ms): (**a**) Defect location, (**b**) IR thermogram at 4.6 s

Optimum results in determining layer coordinates can be achieved by calculating a calibration function $\tau_m(l)$. Then each interval of optimum observation times $\Delta\tau_m$ will correspond to the depth interval Δl (Fig. 6.2b), thus allowing to produce *thermal tomograms*. Simple estimates of layer coordinates can be done by Eq. (5.19). It is worth noting that, strictly speaking, a particular calibration function $\tau_m(l)$ is attached to a particular type of defects, and the use of Eq. (5.19) is possible when a coefficient of thermal wave reflection at defect borders (gas-filled defects in solids) is close to unity.

Example Consider TNDT of a 4.9 mm-thick carbon fiber reinforced plastic (CFRP) sample containing 3 pairs of 10 × 10 mm air-filled defects at depths 1, 2 and 3 mm (Fig. 6.2a). The thickness of the defects in each pair is 50 and 100 μm. An example of the IR image calculated with ThermoCalc-6L is shown in Fig. 6.2b (note that defect indications are stretched in the *X* direction because of composite anisotropy).

Thermal tomography utilizes the following peculiarities of transient TNDT illustrated with Fig. 6.3 and Fig. 6.4 (see details in Sect. 4.8): (1) any subsurface defect is characterized by the optimum observation time τ_m and maximum differential signal ΔT_m (Fig. 6.3), (2) τ_m increases with greater defect depth l, (3) defect thickness d strongly affects ΔT_m and weakly – τ_m, and (4) ΔT_m is linearly proportional to the absorbed energy.

The dependence $\tau_m(l)$ shown in Fig. 6.4a serves as a calibration characteristic when evaluating defect depth. The maxigram in Fig. 6.4b exhibits subsurface defects at their best observation times no matter at what times occur ΔT_m values. In its turn, the timegram in Fig. 6.4c reflects distribution of optimum observation times for defects at different depths. By choosing a particular $\tau_{m1} - \tau_{m2}$ interval, one can produce a thermal tomogram (Fig. 6.4d) which shows the sample "thermal structure" within the $l_1 - l_2$ layer, and the maxigram can be used for filtering noise by thresholding ΔT_m values.

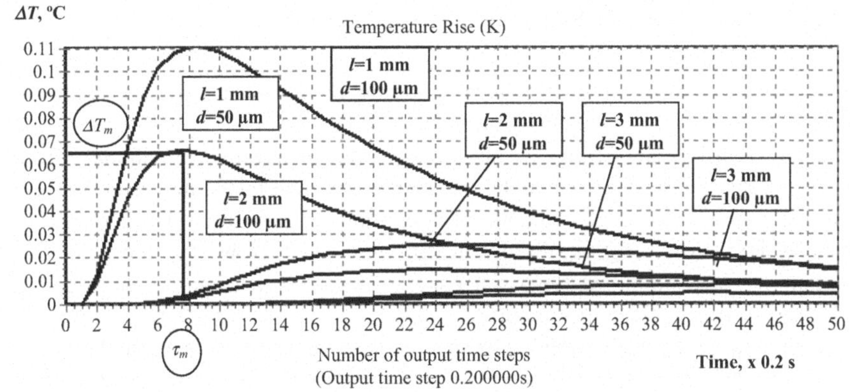

Fig. 6.3 Differential temperature signal vs. time for 10 × 10 mm defects in CFRP sample heated with heat pulse ($Q = 10^5$ W/m², $\tau_h = 10$ ms, l = defect depth, d = defect thickness)

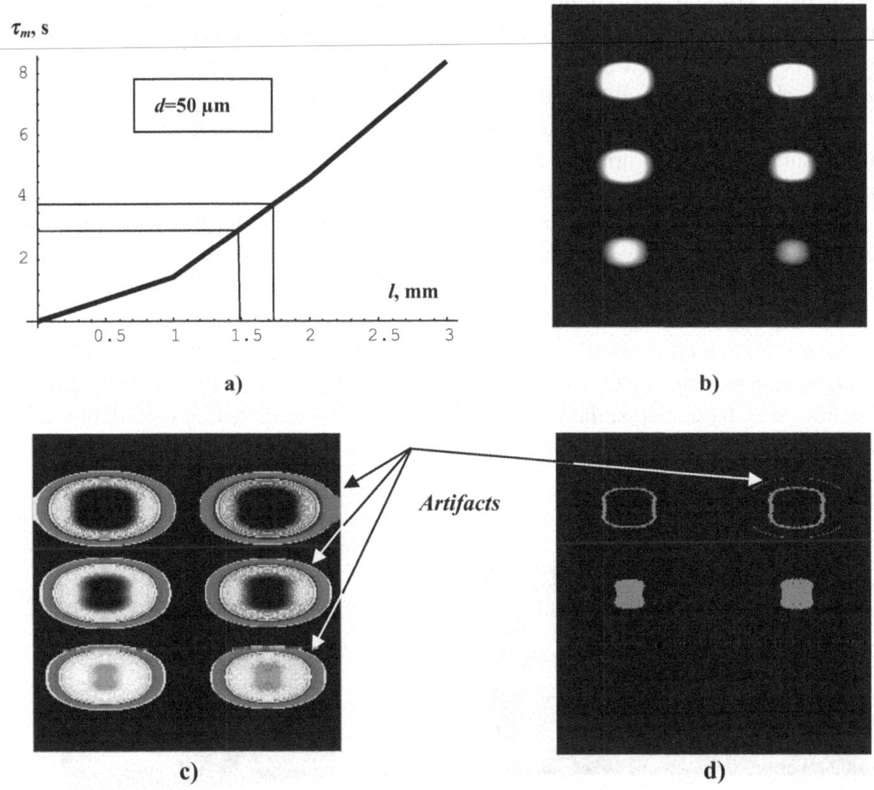

Fig. 6.4 TNDT of CFRP: (**a**) Calibration curve $\tau_m(l)$, (**b**) maxigram ΔT_m, (**c**) timegram τ_m, (**d**) thermal tomogram, defects at depth of 2 mm

6.3.2 Maxigram, Timegram and Tomogram

Maxigrams *optimally* exhibit structural inhomogeneities of materials in regard to a chosen reference and are typically characterized by *maximum* signal-to-noise ratio values in regard to raw IR images. By other words, maxigrams contain more information about all defects in a sample than each individual image in a sequence. It is worth stressing an artificial character of maxigrams to compare to IR thermograms because they show pixel amplitudes at different times. However, maxigrams are characterized by smoothed (Gaussian) shape of spatial distributions over defects $\Delta T_m\,(i, j)$ similar to IR thermograms.

A timegram reflects distribution of specific heat transit times in a tested object in regard to a reference. In fact, each pixel in a timegram contains the number of the image where the maximum value of ΔT occurs. To compare to the corresponding maxigrams, timegrams are characterized by steep spatial profiles of $\tau_{m\ i,j}$ and thus better reflect true size and shape of hidden defects. Profiles of $\tau_m\,(i, j)$ are nearly "flat" over the central part of defects with some spikes (higher delay times) at defect edges, and this phenomenon causes the appearance of artifacts. This feature of timegrams is illustrated in Fig. 6.5 with the first experimental timegram of a plastic sample obtained by Vavilov et al. in 1989 [22].

Visually, maxigrams look like IR thermograms; in fact, they are thermograms. A timegram is a result of non-linear data transformation, therefore, timegram appearance depends on a chosen reference point and may be unpredictable.

Thermal tomograms can be obtained by "slicing" timegrams. Non-zero pixels in a tomogram contain τ_m values in a chosen interval $\tau_{m1} - \tau_{m2}$, while "zero" pixels are characterized by τ_m values out of a chosen interval. A thermal tomogram can be "binarized" in order to produce a map of defects. Theoretically, a maximum number of layers for which a sample can be separated should be equal to the number of images in a sequence. But, in practice, because of noise and a limited temperature resolution of IR cameras, the thickness of resolve layers increases with depth, and the total number of resolved layers does not exceed 4–6.

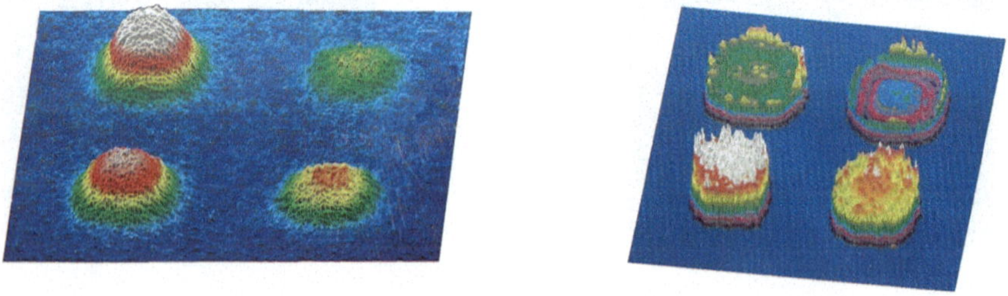

a) b)

Fig. 6.5 IR thermogram (**a**) and timegram (**b**) of defects in a plastic sample (experiments at Wayne State University, USA, 1989) [22]

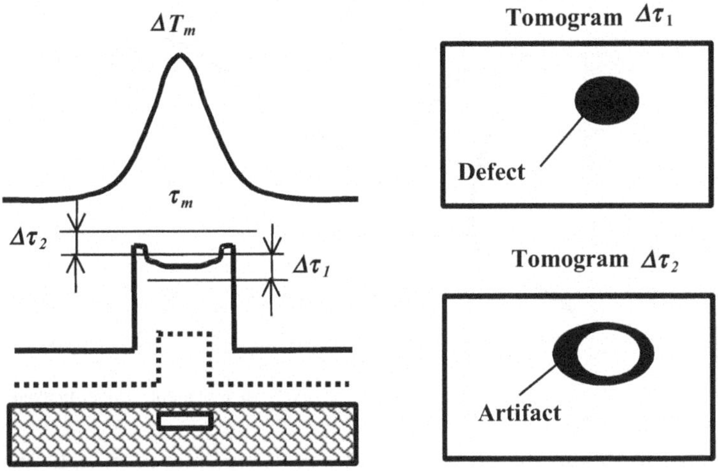

Fig. 6.6 Artifacts in dynamic thermal tomography

6.3.3 *Artifacts*

A specific type of τ_m (i, j) distributions over defects makes artifacts appeared in thermal tomograms. They look like traces of the defects located out of the chosen layer. Appearance of artifacts is illustrated with Fig. 6.6. A temperature distribution over a hidden defect ΔT is bell-shaped, while the corresponding distribution of τ_m has some spikes on the projection of defect edges (note that the ideal profile is shown in Fig. 6.6 with the dashed line). Hence, in a certain interval of τ_m one may observe appears of defects located in other layers (Fig. 6.6).

Fortunately, "high" amplitudes of τ_m spikes correspond to low ΔT signals, therefore, artifacts can be effectively suppressed by *thresholding* ΔT thus making the corresponding timegrams and maxigrams user-friendly. According to the decision making theory, a higher amplitude threshold will decrease the probability of false alarm but simultaneously decrease the probability of correct detection. It is recommended establishing a threshold equal to $\Delta T_m = (2 - 3)\sigma_n$,where σ_n is the standard deviation of noise which is typically evaluated in a defect-free area, for instance, by analyzing the relevant histogram.

6.3.4 *Uneven Heating*

While implementing the algorithm of DTT, uneven heating may seriously distort appearance of both maxigrams and tomograms because of necessity to choose a reference point. For example, choosing the T_{nd1} value in Fig. 6.7 as a reference will lead to reasonable results, while choosing the T_{nd2} value will make the thermogram unacceptable because of a lack of an evident extremum in τ_m. The typical recom-

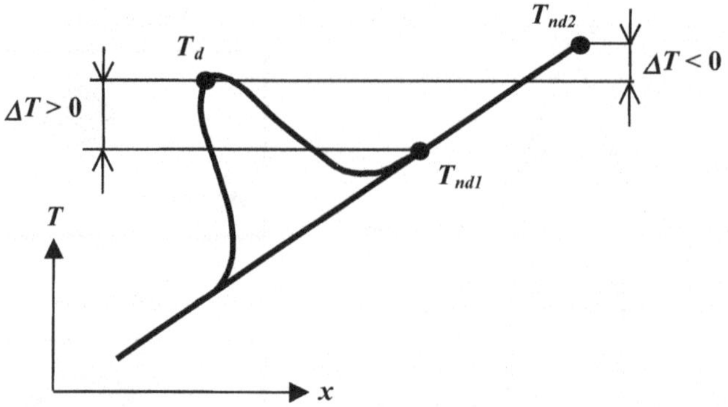

Fig. 6.7 Influence of choosing a reference point on maxigram appearance under uneven heating

mendation is to choose a reference point close to a region of interest. In practice, a sequence should be often normalized prior applying the algorithm of thermal tomography (see Sect. 6.11).

One more example of influence of uneven heating is shown in Fig. 6.8. Tomographic presentation of the left top defect is possible if a reference point is chosen near the defect (Fig. 6.8a), but other defects can be missed in this case (Fig. 6.8b). The problem can be solved by proper data treatment, for example, by applying: (1) normalization, i.e. division a source sequence by a properly chosen reference image (Fig. 6.8c), (2) Fourier transform with subsequent analysis in the phase domain, and (3) 3D filtration (see below).

6.3.5 Influence of Noise on the Appearance of the Maxigram and Timegram

Spatial/temporal fluctuations of temperature lead to noisy maxigrams and timegrams because of the following factors (Fig. 6.9):

- non-defect points are characterized by a certain dispersion of both ΔT_m and τ_m because they are not identical to a chosen reference point,
- chaotic distribution of ΔT_m extremums in pixel temporal evolutions leads to noisy pixel values in timegrams, and these values are not associated with defect depth.

6.3.6 Thermal Tomography of Impact Damage in Composites

Impact damage is a typical type of in-use defects occurring in aviation composites. Low-energy damage is characterized by impact energy from 5 to 40 J. In this energy range, impacted surface may reveal no visible indications of defects, particularly in

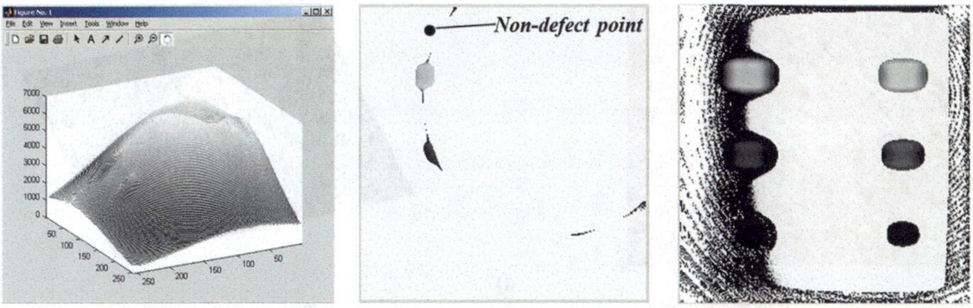

a) b) c)

Fig. 6.8 Dynamic thermal tomography under uneven heating (TNDT of the sample from Fig. 6.2a, single flash tube): (**a**) Source IR thermogram at 4.6 s, (**b**) timegram of source sequence, (**c**) timegram after normalizing source sequence

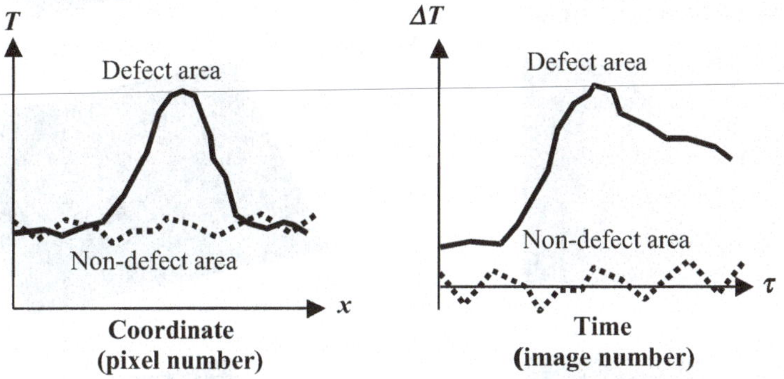

Fig. 6.9 Influence of noise on appearance of maxigrams and timegrams

CFRP, while the main defect body may appear within the bulk material and often close to the sample rear surface. In practice, only one-sided inspection on impacted surface is practically feasible thus generally leading to weak defect signals over deeper defects. The DTT technique may be helpful in the comparative analysis of impact damage in GFRP and CFRP.

Vavilov and Kuimova used the DTT technique in the inspection of two 6 mm-thick glass fiber reinforced plastic (GFRP) and CFRP samples subjected to a standard 34 J impact and tested on both sides by applying a one-sided TNDT procedure [26]. Best optimal images and the corresponding thermal tomograms are shown in Fig. 6.10.

There was no visible damage mark on the impacted surface of the CFRP sample, but the IR thermogram of Fig. 6.10a (left) shows a faint defect indication which can hardly be used for further analysis. The thermal tomogram obtained by the DTT algorithm, when inspecting the impacted sample surface (Fig. 6.10a, right), clearly shows that the main body of the delamination is located within the 0.82–1.45 mm layer. The results obtained on the surface opposite to the impacted one show the known "butterfly-like" defect indication (Fig. 6.10b). Test results obtained on the GFRP sample reveal a defect structure that is fairly different from that in a very

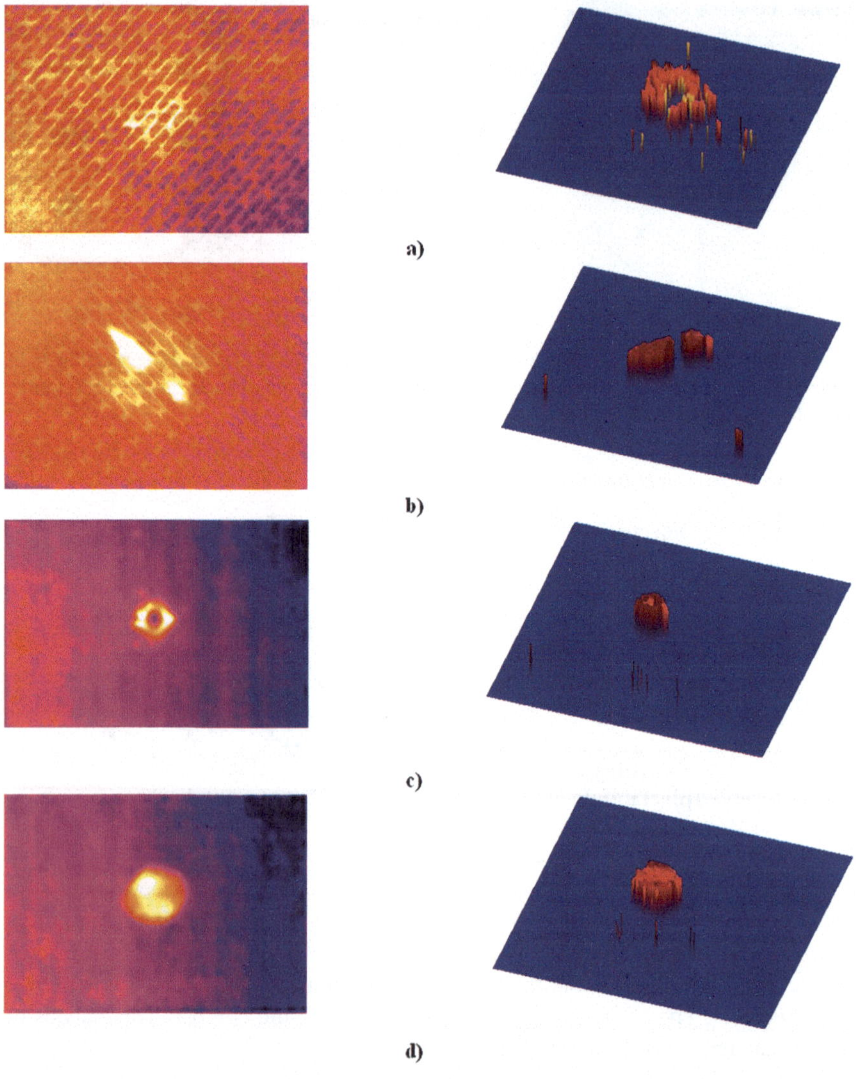

Fig. 6.10 Detecting 34 J impact damage in 6 mm-thick CFRP/GFRP samples. (**a**) CFRP, impacted surface: optimal IR thermogram (left) and tomogram of 0.82–1.45 mm layer (right), (**b**) CFRP, surface opposite to impact: optimal IR thermogram ((left) and tomogram of 0–0.65 mm layer (right), (**c**) GFRP, impacted surface: optimal IR thermogram (left) and tomogram of 0–0.3 mm layer (right), (**d**) GFRP, surface opposite to impact: optimal IR thermogram (left) and tomogram of 0–0.7 mm layer (right). (Adapted from [26])

anisotropic CFRP composite (Fig. 6.10c). The damage area looks round-shaped on both surfaces but it seems bigger on the rear surface (compare the images in Fig. 6.10c, d). Note that the dark central spot in Fig. 6.10c corresponds to a visible surface indentation caused by impact.

6.4 Pulsed Phase Thermography

6.4.1 Basic Principle

The technique of pulsed phase thermography (PPT) proposed by Maldague, Marinetti and Couturier is based on applying the Fourier transform to 1D temperature evolutions $T_{ij}(\tau)$ [27, 28]. This method can be regarded as a generalization of the method of modulated (periodical or pulsed) thermal waves which penetrate materials experiencing specific spatial/temporal distortions in defect areas. It was shown in Sect. 4.2.3 that, under periodical heating, a temperature signal at any surface point is also described by a periodical function of the same frequency as the exciting wave thus thus being characterized by a particular phase and magnitude. A simple estimate of defect detection limit is given with the thermal diffusion length $\mu = \sqrt{\alpha / \pi f}$. Obviously, deeper defects are to be detected with slower waves. In the photothermal method, μ values are in the range from micrometers to millimeters. It is possible to use slower waves but experiments become longer because several wave periods are to be recorded to produce meaningful results. However, in some special cases, such slow waves as produced by day-night temperature variations have been used in the detection of specific defects. Diurnal waves penetrate low-conductive materials up to depths of 5–15 cm thus allowing detection of hidden objects in soil and building constructions. It is also worth mentioning super-long thermal waves which are excited by seasonal temperature variations and penetrate soil up to 1 m. Phase analysis of such slow waves is hardly possible.

Pulsed phase thermography realizes the idea of considering *total* differences in temperature evolutions of defect and non-defect areas. It is known that a pulse of thermal energy (square- or Dirac- shaped) absorbed by a solid is characterized by a spectrum of frequencies which penetrate a solid experiencing dissipation by energy (amplitude) and retardation in time (phase). Single spectrum components compete in a complicated way to produce bell-shaped temperature signals over defects. As mentioned in Chap. 4, this process can be considered as propagation of heat packets traveling between sample surface and internal defects. With increasing material depth, the amount of high-frequency components in the dynamic temperature spectrum drops down. Hence, material layers play the role of specific frequency filters. Qualitatively, this means that a near-surface material layer can be probed with relatively high-frequency thermal waves (photothermal technique), while deeper defects are to be detected at lower frequencies; the latter approach is typically used in IR thermographic NDT. In all cases, a phase shift between an exciting and response wave serves as an informative parameter.

Phenomenologically, the preference of using PPT in a one-sided test procedure can be illustrated with the following consideration. It is known that: (1) deeper defects are characterized by low temperature signals, (2) in the Fourier spectrum of pulses low-frequency components carry more energy than high-frequency. Taking into account that the detection of deeper defects requires lower-frequency thermal waves, the statements above mean that stimulation with square (Dirac) heat pulses is optimal for detecting defects through the sample. Of course, in practice, there are some additional limitations related to available power of heaters, optical radiation spectrum, possible sample overheating (surface burning), presence of noise, etc.

Unlike standard image processing applications, where the 2D Fourier transform is often used, in TNDT, this mathematical procedure is applied to temporal excursions of temperature thus being essentially 1D. By other words, the Fourier transform is used in TNDT for analyzing temperature transients in IR image sequences. A certain difficulty in identifying images of phase (*phasegrams*) and magnitude (*ampligrams*) is that Fourier processing results depend on both an image acquisition interval and the number of such intervals. This feature prevents obtaining general calibration relationships which can be used, for example, in performing phase-based thermal tomography.

Let us use the Fourier transform definition accepted in physics:

$$\begin{aligned} \text{Fourier}_s &= \frac{1}{\sqrt{N}} \sum_{n=1}^{N} T_n e^{2\pi j\,(n-1)\,(f-1)/N} = R(f) + jI(f), \\ M(f) &= \sqrt{R(f)^2 + I(f)^2}, \\ \Phi(f) &= ArcTan\left[\frac{I(f)}{R(f)}\right], \end{aligned} \tag{6.1}$$

where T_n is the discrete source function containing N read-outs (N is the number of images in a sequence), f is the frequency, j is the imaginary unit, n is the read-out number, $R(f)$ and $I(f)$ are the real and imaginary component respectively, $M(f)$ and $\Phi(f)$ are the magnitude and the phase respectively.

Suppose that the analyzed sequence contains 20 temperature read-outs starting from zero time ($\tau = 0$), i.e.:

- zero time $\tau = 0$ corresponds to ($\leftrightarrow$) *Point 1*,
- time $\tau = 0 + \Delta\tau \leftrightarrow$ *Point 2*,
- time $\tau = 0 + 2\Delta\tau \leftrightarrow$ *Point 3*,
- ....
- time $\tau = 0 + 19\Delta\tau \leftrightarrow$ *Point 20*,

where $\Delta\tau$ is the acquisition interval (sampling rate) which is regular through the sequence.

The sequence T is presented in Fig. 6.11a. Applying the Fourier transform to this sequence yields in the magnitude M and the argument (phase) Φ shown in Fig. 6.11b, c.

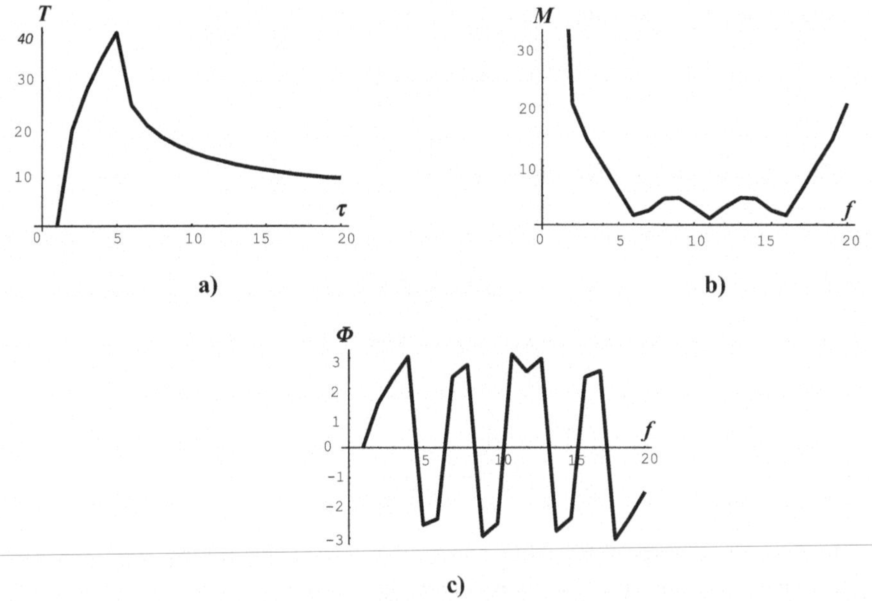

Fig. 6.11 Typical surface temperature evolution (**a**) and its Fourier transform parameters: magnitude (**b**) and phase (**c**)

Like the source sequence, the graphs of Fourier module and phase also include 20 read-outs with *Point 1* corresponding to the zero frequency. Due to the aliasing phenomenon, the "useful" information is carried by frequencies from 1 to 10 (*Points 11–20* are mirror reflections of the first 10 points).

If the total number of points is *N*, the frequency increment corresponding to the interval $\Delta\tau$ is:

$$\Delta f = \frac{1}{\Delta\tau}\,\frac{1}{N-1}. \tag{6.2}$$

Thus, the *n*-th point corresponds to the frequency:

$$f_n = \Delta f\left(n-1\right) = \frac{1}{\Delta\tau}\,\frac{n-1}{N-1}. \tag{6.3}$$

The lowest meaningful frequency, except zero, is defined for *Point 2*:

$$f_{\min} = \frac{1}{\Delta\tau}\,\frac{1}{N-1}. \tag{6.4}$$

The highest frequency in the spectrum corresponds to *Point N*:

$$f_{max} = \frac{1}{\Delta\tau}. \tag{6.5}$$

Then the sequence of frequencies will be:

$$\begin{aligned} &f_1 = 0;\\ &f_2 = \frac{1}{\Delta\tau}\,\frac{1}{N-1};\\ &f_3 = \frac{1}{\Delta\tau}\,\frac{2}{N-1};\\ &\dots\dots\dots\dots\dots\dots\\ &f_{10} = \frac{1}{\Delta\tau}\,\frac{9}{N-1}. \end{aligned} \tag{6.6}$$

It is worth stressing that, unlike some other TNDT algorithms, such as early detection or dynamic thermal tomography, PPT requires recording temperature read-outs during long time to allow the analysis of low frequencies that is necessary for detecting deeper defects. On the other hand, the acquisition interval should be short enough to provide high-frequency components in the Fourier spectrum necessary for detecting shallow defects (in this case, the central part of the spectrum in Fig. 6.11b drops to nearly zero). Therefore, PPT often deals with sequences consisting of several hundreds and even thousands of images.

By now, the feasibility of PPT has been demonstrated mainly on a qualitative level, however, the sensitivity of this technique to material structural inhomogeneities is so high that it is often regarded as a primary technique in validating other TNDT approaches. Moreover, the use of PPT requires no reference point, therefore, this technique can be easily automated and implemented in TNDT devices operated by unexperienced operators.

6.4.2 *Quantitative Approach to Pulsed Phase Thermography*

Consider TNDT of impact damage in a 2 mm-thick CFRP sample as an experimental illustration of the quantitative approach to PPT. The sample was heated with three flash tubes with the total energy of 9 kJ and pulse duration 10 ms [29]. The dynamic thermal process was recorded for 5 s with the 66.7 ms interval (74 images in the sequence). This allowed analyzing the following Fourier frequencies: 0 Hz, 0.202 Hz, 0.404 Hz etc. The source image captured at the best observation time is shown in Fig. 6.12a to display the well-known ‘butterfly’ shape of temperature patterns over impact damage. The ultrasonic C-scan image revealed five defects located at different depths and oriented along carbon fibers. The thermal tomographic anal-

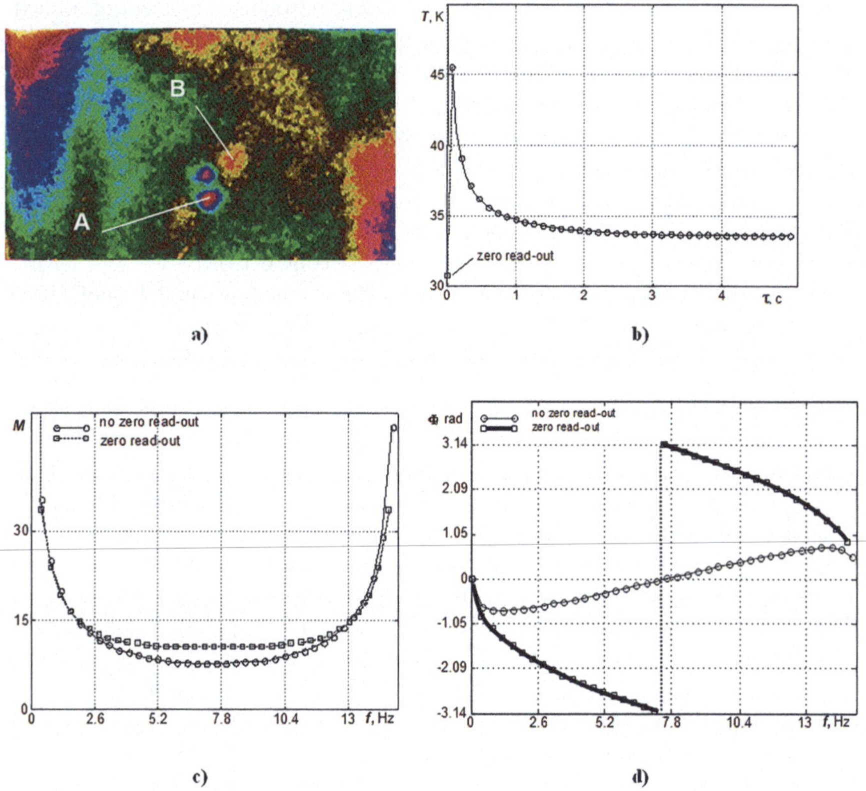

Fig. 6.12 Fourier analysis in pulsed TNDT of a 2 mm-thick CFRP sample [29]: (**a**) Source IR image (1.3 s after heating), (**b**) non-defect temperature evolution, (**c**) thermal wave module (with and without zero read-out), (**d**) thermal wave phase (with and without zero read-out)

ysis of this sample showed four defective areas of which two (*Area A* in Fig. 6.12a, depth from 0.3 to 0.7 mm, and *Area B* in Fig. 6.12a, depth from 1 to 1.5 mm) were visible in the source images.

An example of the temperature evolution $T(i_{ref}, j_{ref}, \tau)$ in a non-defect area is given in Fig. 6.12b (the zero read-out was taken prior heating when the sample temperature was equal to the ambient one). The spectrums of the Fourier magnitude M and the phase Φ calculated for this evolution are shown in Fig. 6.12c, d. Below we will mainly deal with phase characteristics, although further development of PPT may also require using amplitude spectrums and then returning to the original time domain.

In the example above, the Fourier transform was applied to the decaying part of the temperature signal. However, the appearance of the Fourier parameters depends on whether zero read-outs (prior heating) are taken into account or not (compare graphs in Fig. 6.12b–d). In particular, the presence of a zero read-out before the temperature rise causes a distinct extremum in the phase *vs.* frequency relationship

(Fig. 6.12d). This has allowed some researchers to introduce a concept of a maximum phase. However, the presence of such extremums may affect the physical sense of using independent thermal wave frequencies for in-depth probing of solids.

The plots in Fig. 6.13 demonstrate that the presence of the zero read-out in the source sequence weakly influences on the phase shift $\Delta\Phi$ between defect and non-defect areas. This is explained by the fact that the low-frequency spectrum components are those which mainly contribute to the build-up of temperature signals, while the absence of a zero-read-out simply decreases the share of high-frequency components appearing due to sharp signal increase (Fig. 6.13).

By applying the Fourier transform to the results of computer modeling of TNDT in the inspection of the CFRP above, the calibration relationships of the Fourier module and phase on defect depth l have been calculated (Fig. 6.14). It is seen that:

- the modules of Fourier transforms decay quickly with defect depth and reach a "plateau" for deep defects (Fig. 6.14a); it seems that the plateau amplitude is related to the sample thickness specifying the transit to a semi-infinite body geometry; note also the appearance of a weak minimum which probably has the same origin as that in Fig. 4.14 (see Sect. 4.2.3);
- in the general case, phases of Fourier transforms reveal two extremums (Fig. 6.14b); approximately, it can be assumed that, at any frequency, the phase decreases within a certain near-surface layer reaching a minimum, and this minimum transits to deeper defects for slower waves; afterwards, the phase increases to reach the plateau similarly to the behavior of the Fourier module;
- the presence of the plateau in both relationships proves that the thermal probing of materials is limited by certain depths.

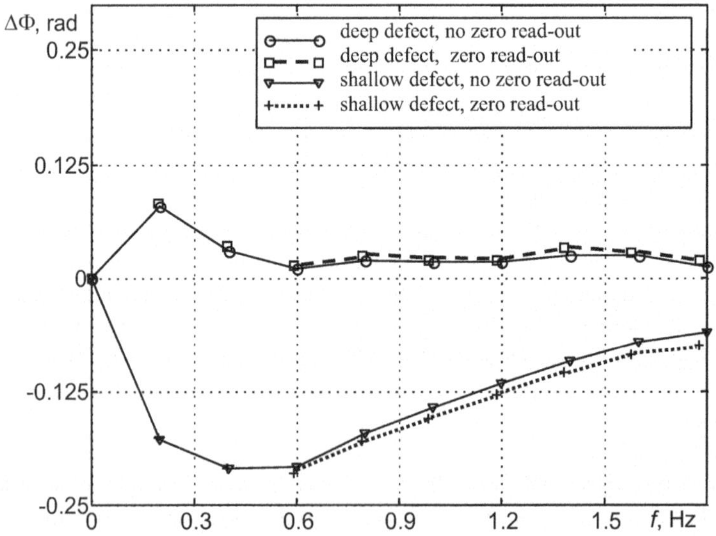

Fig. 6.13 Thermal wave phase shift for shallow and deep defect (with and without zero-read-out)

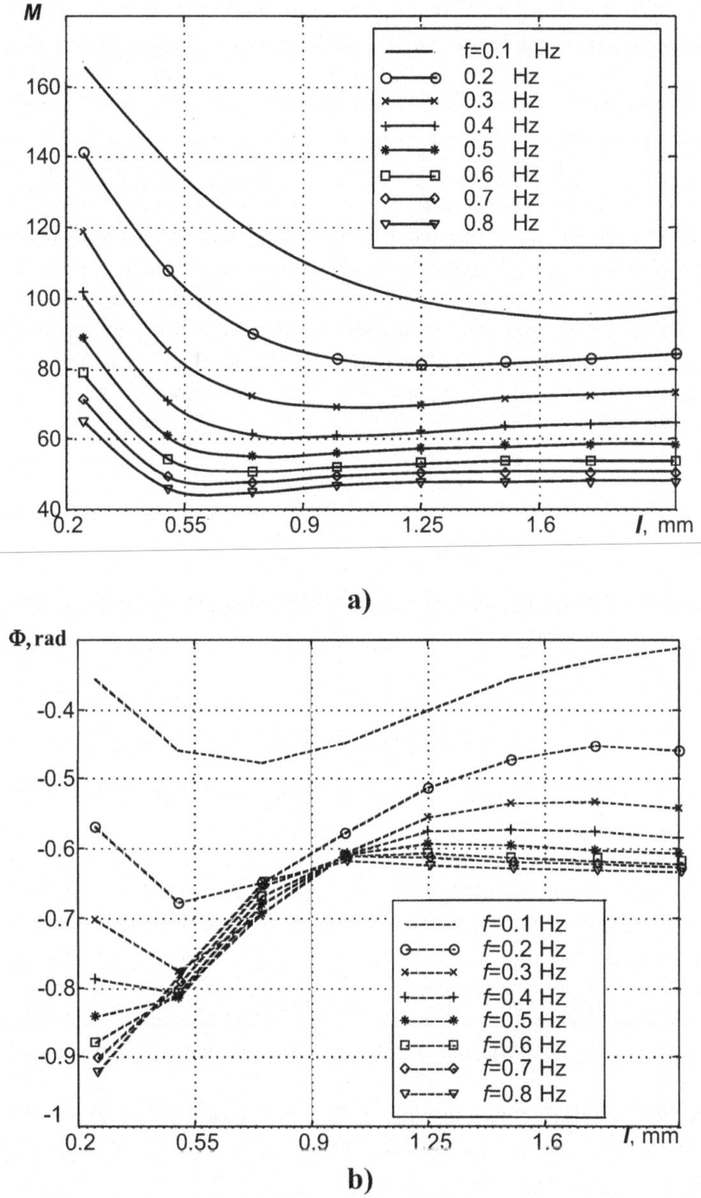

Fig. 6.14 Thermal wave module (**a**) and phase (**b**) at different frequencies *vs.* defect depth (TNDT of 2 mm-thick CFRP sample)

The fact that the phase *vs.* defect depth relation reveals extremums could make defect characterization difficult. Let us evaluate defect depth detection limits by analyzing the Fourier module and/or phase. Assume that maximum detection depth corresponds to the beginning of the plateau in the respective relationships. Such assumption seems to be reasonable in the case of the Fourier module (Fig. 6.14a) but it is dubious in the case of phase because of the specific character of the phase *vs.* depth relationship. Suppose that both the growing and decaying parts of this relationship can be used for profiling defects by depth (Fig. 6.14b), although the decaying part seems to be less sensitive, in particular, at lower frequencies. Considering the notes above, it can be concluded that phase components of the Fourier spectrum penetrate deeper into solid materials than the Fourier module. By applying the least-square method, the following estimates for maximum penetration depth of phase $L_{\max}^{Phase}$ and module $L_{\max}^{Ampl}$ have been obtained [29]:

$$L_{\max}^{Ampl} \approx 0,82\sqrt{\alpha}\, f^{-0,44};$$
$$L_{\max}^{Phase} \approx 1,27\sqrt{\alpha}\, f^{-0,34}. \tag{6.7}$$

Equations (6.7) have been proposed for CFRP. However, they are valid for other materials. In the general case, the ratio between the respective penetration depths can be expressed with the following formula:

$$L_{\max}^{Phase} / L_{\max}^{Ampl} \approx 1,55 f^{0,1}, \tag{6.8}$$

which shows that the phase penetrates by 55% deeper than the module, and the ratio weakly depends on frequency.

The curves presented in Fig. 6.14b allow the elaboration of *defect depth vs. phase* inversion functions [29]. The phasegram presented in Fig. 6.15a was obtained in TNDT of CFRP at the frequency 0.2 Hz that is close to optimal for this test case. In regard to the source image (Fig. 6.12a), the phasegram reproduces better the defect shape. The preference of the phase image was proven by comparing the signal-to-noise ratio in *Area B*: $S = 7.2$ in the phasegram and $S = 2$ in the source image. The phase thermal tomograms shown in Fig. 6.15b–d will be discussed below.

Another approach to characterizing defects by phase was developed by Ibarra-Castanedo and Maldague who suggested the concept of a "blind frequency" [30]. This concept is illustrated with the graphs in Fig. 6.16 where the relationship between the Fourier phase shift $\Delta\Phi$ and defect depth l and frequency f is presented. It is seen that, for each l, one can determine a limiting ("blind") frequency f_b, and no defect can be detected at frequencies higher than f_b. It has been shown that $1/\sqrt{f_b}$ values are linearly proportional to l. Therefore, by finding a f_b value along with accompanying coefficients for a particular material, one can evaluate defect depth.

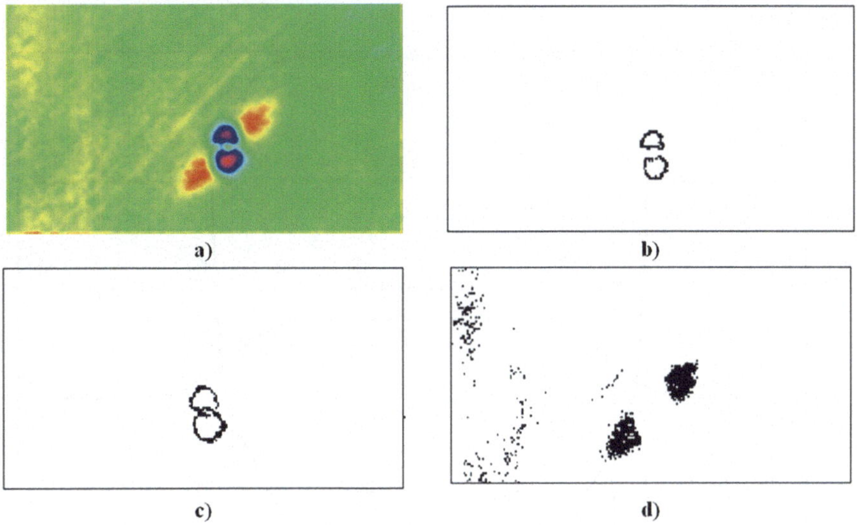

Fig. 6.15 Phasegram (**a**) and phase thermal tomograms (**b–d**) of impact damage in CFRP sample

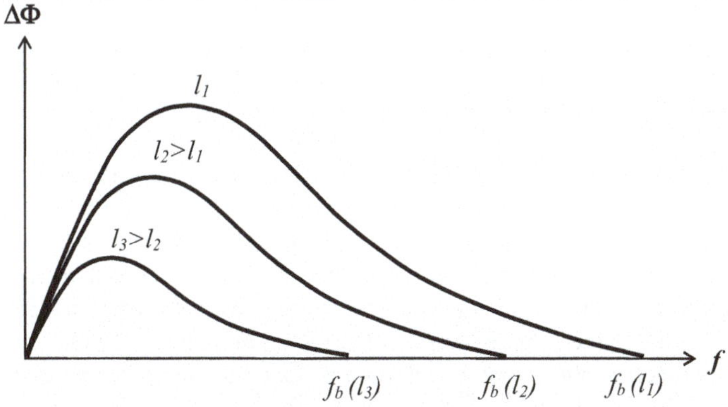

Fig. 6.16 Thermal wave phase shift $\Delta\Phi$ *vs*. frequency f for a defect at different depths l

6.5 Reference-Free Thermal Tomography

6.5.1 Pulsed Phase Tomography

Phase tomograms of particular material layers can be synthesized if a relationship between phase shift and defect parameters (depth and thickness) is known. For example, the tomograms in Fig. 6.15b–d were synthesized by processing the phasegram in Fig. 6.15a, i.e. by selecting pixels within a particular phase range thus

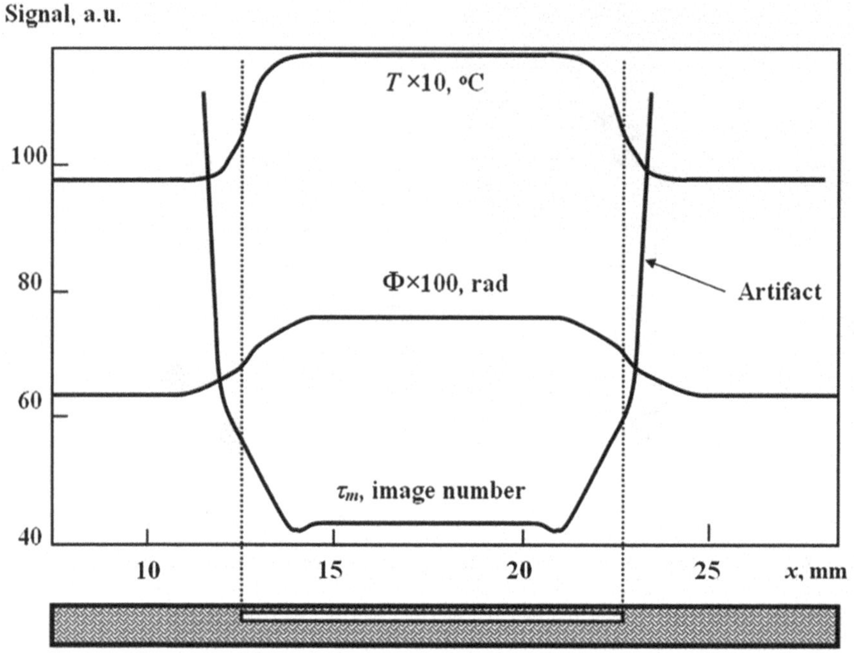

Fig. 6.17 Comparing spatial profiles of temperature T, characteristic time τ_m and phase Φ ($10 \times 10 \times 0.05$ mm defect at 0.5 mm depth in CFRP; T and φ values are normalized to make results comparable). (Adapted from [31])

specifying particular layer coordinates [29]. The nearly linear character of the respective calibration function allows transforming phase into depth by simple proportional re-calculation. Thus, the tomogram in Fig. 6.15b shows a near-surface section of the sample in the phase range $\Phi = 1.35$–1.45 rad that corresponds to defect depths l from 0.82 to 0.94 mm. The tomograms in Fig. 6.15c, d are respectively associated to the sample layers $l = 0.94$–1.06 mm and $l > 1$ mm.

The approaches above are based on the fact that temperature distributions $T(x, y)$ over defects can be replaced with the corresponding distributions of heat transit time τ^* (τ_m) or phase Φ. The comparison of these three parameters are shown in Fig. 6.17 where the spatial profiles of temperature T (source image), characteristic time τ_m (timegram) and phase Φ (phasegram) over a $10 \times 10 \times 0.05$ mm defect at the depth of 0.5 mm in CFRP (defect depth 0.5 mm) are presented. It appears that the profiles of temperature and phase look similarly with the signal plateau over the defect projection. This means that such defect can be considered as 1D, and heat diffusion takes place only at the defect borders. For example, at the defect borders, the temperature signals decrease by 70% in regard to their maximal value over the defect center. The corresponding profile of τ_m is also characterized by the plateau but it is more complicated due to the fact that timegrams represent a non-linear

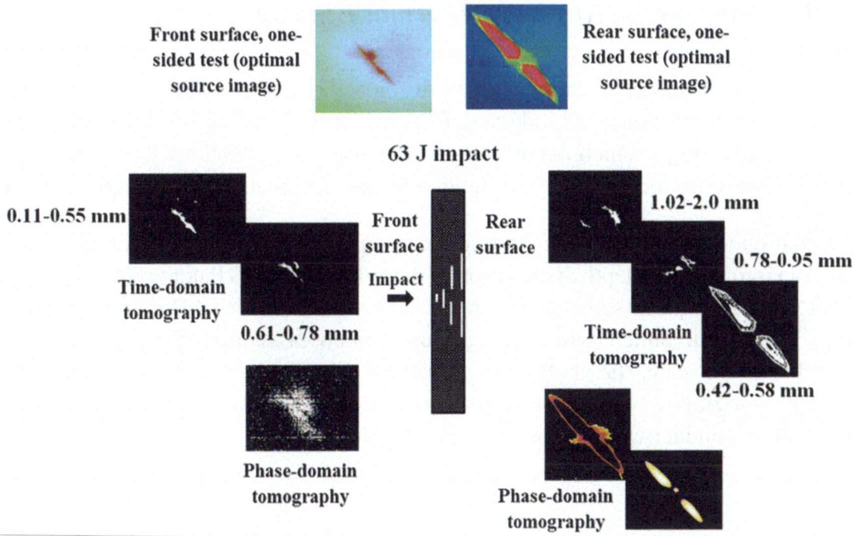

Fig. 6.18 Time- and phase-domain thermal tomography of impact damage in 5 mm-thick CFRP sample (63 J impact energy) [31]

result of processing maxigrams. The values of τ_m, first, slightly drop in the areas where lateral heat diffusion starts, then, increase as ΔT values diminish up to zero in defect-free areas. Experimentally, in non-defect areas, τ_m acquires random values from 1 to N because of a noisy character of ΔT. The noise can be either experimental or computational depending on whether experimental or synthetic images are processed. This peculiarity of producing timegrams causes round-shaped artifacts around defects when choosing particular $\tau_{mi} - \tau_{mj}$ intervals, as discussed above. We also remind that the fact that ΔT signals tend to zero far from defects is used for thresholding τ_m profiles.

Figure 6.18 shows tomographic images obtained by processing experimental results in both time- and phase domains [31]. A 5 mm-thick CFRP sample was impacted with the energy of 63 J and then inspected on both front and rear surfaces by applying the one-sided TNDT procedure. The front-surface damage was detected as a surface crack but the tomograms revealed some faint delaminations under the point of impact. Note that the phase-domain tomogram showed a larger damaged area compared to the time-domain data but the time-domain tomography allowed separating the damaged area for two layers: the superficial one where a thin surface crack was seen, as well as the deeper defect overshadowed by the shallower one. On the rear surface, three time-domain tomograms showed sections of the whole defect located in some layers of the composite characterized by different fiber layup angles. In this test case, the phase tomograms have proven to be less informative in regard to the time-domain images.

6.5.2 Determining Apparent Thermal Effusity

The concept of apparent thermal effusivity $e = \sqrt{\lambda C \rho}$ was considered in Sect. 5.1.1. It was shown that, over hidden defects, the corresponding $e(\tau)$ profiles reveal specific signal drops which occur at particular times τ_{min}. Obviously, variations in relative effusivity behave in the same way as differential $\Delta T(\tau)$ signals, and $\tau_{min} = \tau_{max}$.

Sun proposed a technique of thermal tomography by converting surface temperature data into a spatial depth distribution of thermal inertia (US Patents No. 7365330, 8465200) [32–34]. The same approach has been used to perform 3D thermal tomography in cancer treatment with layer resolution of about 25 μm [35].

In the last decade, the analysis of effusivity approach was used as a means of thermal tomography of asteroids to provide evidence that asteroid material density and thermal conductivity increase with depth [36].

6.5.3 Analyzing $T\cdot\tau^n$ Function

The main idea of this approach is to create some specific (inflection) points in a front-surface temperature response and use their observation times for characterizing materials by depth of material discontinuities. For example, such points appear while analyzing an artificial function $T(\tau)\cdot\tau^n$, where T is the front-surface temperature, and n should be less than 0.5 [26]. Originally this approach was proposed for determining material diffusivity in a one-sided test (see Sect. 3.7.6). It was found that, in the heating of an adiabatic plate, $T(\tau)\cdot\tau^n$ functions are characterized by minimums which occur at particular times τ_{min} dependent on the presence of subsurface defects. This phenomenon can be used for performing reference-free thermal tomography but, as shown in [26], efficiency of this technique is low.

6.6 Wavelet Analysis

Wavelet analysis, or the technique of wavelet transforms, was first developed in the 1980s for detecting pulsed signals on a noisy background, for example, radar signals, etc. This technique is a sort of alternative to the Fourier transform when analyzing temporal (spatial) data series with hidden "irregularities". Unlike the Fourier transform, which localizes frequencies but does not resolve an analyzed process in time, the wavelet analysis, which uses a movable self-adjusting spatial/temporal window, enables the detection of both low- and high-frequency signal characteristics within different time scales. The wavelet analysis has proven to be so efficient in data compression and noise reduction that sometimes one talks about the "wavelet revolution" occurred at the end of 1980s. Since this method successively precises process parameters by transiting from a large scale to small, it is called sometimes a "*mathematical lens*", or "*mathematical microscope*". Note that, at the end of the 1970s, for analyzing seismological data, Huberman et al. (Russia) developed a similar algorithm which was called "God Damn Details". A condensed review of wavelets was published in Russia by Levkovich-Masliuk [37]. In TNDT, this technique was suggested by Galmiche et al. as an alternative to the Fourier transform [38, 39].

The wavelet transform of a real continuous function $f(\tau)$ in regard to a real wavelet function g, called *basic*, or "*moth*er", is defined in the following form:

$$U(a,b) = \frac{1}{w(a)} \int_{-\infty}^{+\infty} g\left(\frac{\tau - b}{a}\right) f(\tau) d\tau, \tag{6.9}$$

where a is the *dilation* parameter, b is the *location* parameter and $w(a)$ is the *weight* function which allows the visualization of transform results. The wavelet transform can be also regarded as cross-correlation between a signal and a set of wavelets. It is commonly assumed that $w(a) = a$ or $w(a) = \sqrt{a}$. The term "mathematical lens" originates from the fact that the parameters a^{-1} and b correspond to signal magnification and location respectively. By other words, the wavelet transform allows the detection of periodical structures related to a finite range of a.

The discrete presentation of Eq. (6.9) for a temporal sequence $x(i)$ is:

$$U(a,b) = \frac{1}{w(a)} \sum_{i=1}^{N} g\left(\frac{i - b}{a}\right) x(i). \tag{6.10}$$

The user has a certain freedom in choosing a type of a wavelet, however the following conditions are to be met:

- a wavelet must possess finite energy, i.e. $\int |g(\tau)^2| \, d\tau < \infty$,
- if $G(\omega)$ is the Fourier transform of $g(\tau)$, then $\int |G(\omega)^2| (1/\omega) d\omega < \infty$, i.e. a wavelet must possess no zero frequency (no constant signal level).

Complex wavelets must meet the following condition:

- the Fourier transform must be a real function becoming negligible at $\omega \leq 0$.

In many applications, a "Mexican hat" wavelet is often used:

$$g(y) = (1 - y)^2 e^{-y^2/2}. \tag{6.11}$$

The discrete presentation of the wavelet transform by using the 'Mexican hat' is simple:

$$U(a,b) = \frac{1}{w(a)} \sum_{i=1}^{N} x(i) \left[1 - \left(\frac{i - b}{a}\right)^2\right] \exp\left[-\frac{1}{2}\left(\frac{i - b}{a}\right)^2\right]. \tag{6.12}$$

In TNDT, the *Morlet* basic function has been also used. In this case, the above-mentioned transform parameters are called the *translation factor Tr* and the *scale factor S*:

$$W(S,Tr) = \int_{-\infty}^{+\infty} T(\tau) h_{STr}(\tau) d\tau, \tag{6.13}$$

where W specifies the wavelet transform, and h_{STr} is the "*daughter*" function which relates to the mother function by the expression:

$$h_{STr}(\tau) = \frac{1}{\sqrt{S}} h\left(\frac{\tau - Tr}{S}\right). \tag{6.14}$$

Since a scale factor is related to frequency and a translation factor adheres to time, in the wavelet method, there is no loss of time information which is necessary for characterizing defect depth.

The wavelet transform can be also considered as the correlation between a signal and a scaled mother function. Thus, if $F_T(\omega)$ is the Fourier transform of the signal $T(\tau)$, and $F_h(\omega)$ is the Fourier transform of the function $h(\tau)$, the wavelet transform can be considered as the inverse Fourier transform of the product of two Fourier transforms:

$$W(S,Tr) = \frac{\sqrt{S}}{2\pi} \int_{-\infty}^{\infty} F_T(\omega) F_h(\omega) e^{-j\omega T} d\omega. \tag{6.15}$$

The product of time and frequency intervals has to meet the Heisenberg uncertainty principle:

$$\Delta\tau \ \Delta\omega \geq 1/2. \tag{6.16}$$

The Morlet mother function is defined as follows:

$$h(t) = \exp(-j\omega_0 \tau) \exp\left(-\frac{\tau^2}{2}\right), \tag{6.17}$$

where ω_0 is the *size* of the Morlet function. This function is the product of the Fourier transform basis, which is responsible for frequency analysis, and the Gaussian window which describes time parameters. Figure 6.19 shows the Fourier transform of the Morlet function. It includes two Gaussian functions shifted by ω_0 and $-\omega_0$. In this way, a scale factor allows analyzing various frequencies determined by a translation factor. By taking into account the Heisenberg uncertainty principle, each wavelet transform is defined around S and Tr within a $\Delta\tau\Delta\omega$ rectangle. Such limitation in positioning wavelets affects the choice of ω_0. Uncertainties in ΔTr and l values increase with greater ω_0 and smaller $\Delta\omega$.

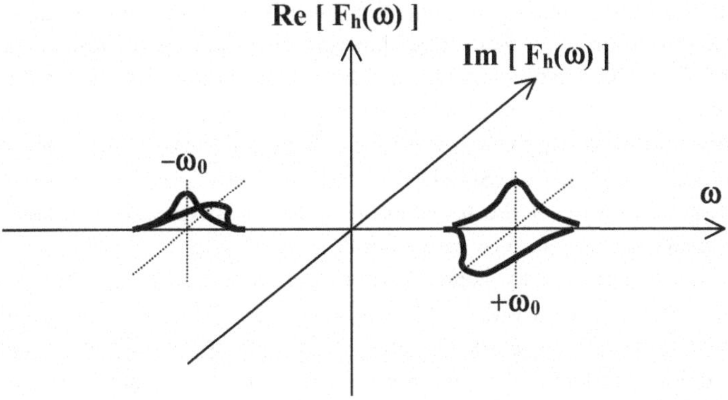

Fig. 6.19 Fourier transform for the Morlet function

The wavelet transform performs decorrelation of a 1D signal (a function of time) into a 2D signal (a function of time and the scale factor) that requires considerable computation efforts. To reduce computations, it has been suggested to use two following expressions known in TNDT:

$$\text{Observation time } vs. \text{defect depth}: \tau = \frac{l^2}{\alpha} \text{ and}$$
$$\text{Thermal diffusion length } vs. \text{frequency } \mu = \sqrt{\frac{2a}{\omega}}. \quad (6.18)$$

If $l = \mu$, the combination of these expressions leads to:

$$\tau = \frac{2}{\omega}. \quad (6.19)$$

The translation factor *Tr* corresponds to the observation time τ, while the scale factor *S* corresponds to ω_0/ω. The connection between these two parameters is given by the expression:

$$S = \frac{\omega_0}{2} Tr. \quad (6.20)$$

The last expression allows keeping the dimension of analyzed signals by calculating a scale factor *S* for each *Tr* value which is limited by observation time. Therefore:

$$W(S, Tr) = \int_{-\infty}^{\infty} T(\tau) \frac{1}{\sqrt{0.5\omega_0 Tr}} h\left(\frac{\tau / Tr - 1}{0.5\omega_0}\right) d\tau. \quad (6.21)$$

Similarly to Fourier transforms, wavelet transforms contain both a real and imaginary parts, thus allowing to determine phase characteristics in the transform domain (due to this fact, the wavelet analysis has been classified as the implementation of PPT). Wavelet transforms are characterized by the same features as Fourier transforms. Wavelet transform phase is used for detecting defects, while defect segmentation is done by applying the Sobel operator to the corresponding 'phasegrams'. In its turn, the calibration of the translation factor (as difference of pixel-based *Tr* values) allows the evaluation of defect depth [39].

In fact, a translation factor is an analog of time and ensures the best "visibility" of particular defects. To avoid double calibration by defect size and depth, it has been proposed to use the early detection technique. As shown in Chap. 4, this ensures weak dependence of temperature signals on defect lateral size and thickness, while defect depth always appears a strongly-influencing parameter.

In TNDT, the wavelet analysis is poorly explored and its advantages are disputable because evolutions of differential signals $\Delta T(\tau)$ in time represent slowly-changing functions rather than pulse-like signals, and this makes the determination of *a* and *b* parameters difficult.

6.7 Defect Thermal Characterization

In Chap. 5 we discussed the algorithms of defect characterization intended for evaluating parameters of materials and defects by applying some mathematical models. It is obvious that accuracy of identification depends on how accurately are defined those model parameters which are assumed to be *a priori* known.

In general, we can state that: (1) most of identification algorithms are developed for one-sided TNDT, (2) lateral dimensions of subsurface defects can be determined directly (or by using some simple procedures) by surface temperature 'footprints' of subsurface defects with accuracy better than 5–10%, (3) defect depth can be evaluated with accuracy of 20–30%, and (4) identification of defect thickness, or thermal resistance, is characterized by accuracy of 30–60%. In practice, there are two groups of methods intended for solving TNDT inverse problems: (1) using inversion formulas, and (2) applying an iteration least-square process to match experimental and theoretical data (functional minimization technique, or non-linear fitting).

Defect characterization may result in a special type of images where each pixel is expressed in units of a parameter to be identified; for instance, when determining defect depth and thickness, the terms "*depthgrams*" and "*thicknessgrams*" have been proposed.

An example of defect characterization by using maxigrams and timegrams and applying Eq. (5.9) is presented in Fig. 6.20. A 2 mm-thick CFRP panel with stringers on the rear surface contained two delaminations and a local thinning (Fig. 6.20a). The best source image is shown in Fig. 6.18b. Both the depthgram and the thicknessgram presented in Fig. 6.20c, d were produced by using the ThermoFit Pro software (Tomsk Polytechnic University). In a sound area, depthgram pixel values

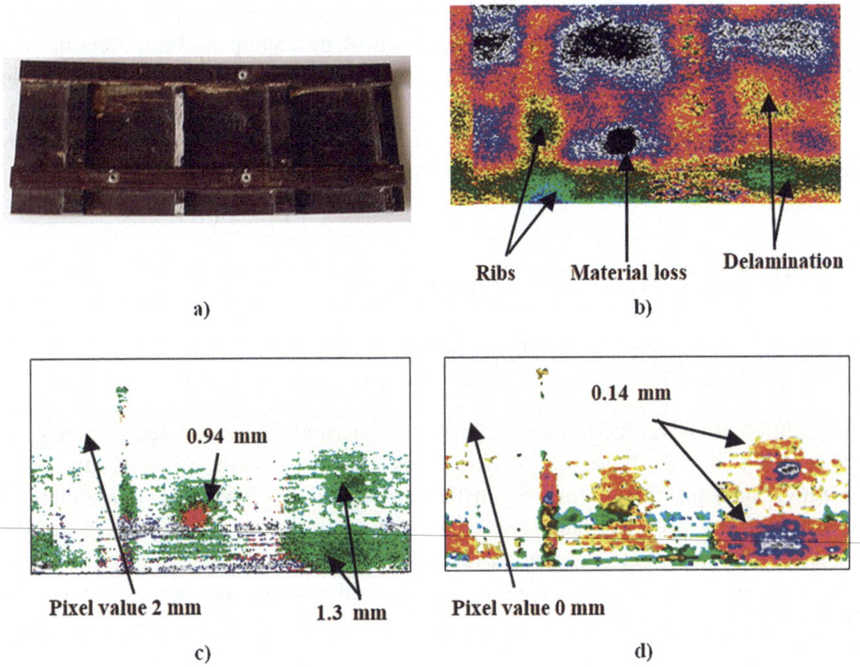

Fig. 6.20 Characterizing defects in 2 mm-thick CFRP sample: (**a**) Sample, (**b**) optimal IR thermogram, (**c**) depthgram, (**d**) thicknessgram

are 2 mm thus corresponding to the sample thickness, while the corresponding thicknessgram values are zero thus specifying defect-free areas. In the area of thinning, the material remained thickness was 1 mm that coincided well to the retrieved value of 0.94 mm (note that the corresponding thicknessgram value had no sense because the model was not able to identify rear-surface material loss). Finally, the depth of both delaminations was estimated as ~1.3 mm that was confirmed after the destruction of the sample.

When applied to noisy data, defect characterization formulas may produce unrealistic values of defect parameter, therefore, in practice, first, a region of interest (ROI) is to be chosen and a respective threshold is to be established, as it was done in the example above. However, it is worth noting that the thresholding of noise prevents the detection of small defects.

6.8 Quantitative Evaluation of Hidden Corrosion

Quantitative evaluation of hidden corrosion (material loss) is a good example of the practical application of defect characterization. In Sect. 3.5, we reported the simple inversion formula (3.11): $\Delta L/L = 1 - T_{nd}/T_d$, which allows the determination of relative material loss $\Delta L/L$ by experimental temperature contrast T_{nd}/T_d. This formula has proven its validity in the inspection of thin samples, both metallic and non-metallic, where the size of corroded areas is essentially greater than sample thickness [40].

To take into account 3D heat diffusion (finite size of corroded areas), some correction coefficients have been introduced into Eq. (3.11), see Sect. 3.7.3.

6.9 Thermal Wave IR Thermography

6.9.1 Lock-in Optical Stimulation

As mentioned above, the longer are periodical thermal waves, the deeper they penetrate into materials; however, the detection of particular defects may require optimization of a probing frequency. In the classical (high-frequency) photothermal radiometry, one uses thermal waves of which penetration depth is under few millimeters [41]. Using long thermal waves in a point-by-point procedure may require unacceptably long inspection time: for example, producing an image of the 320 × 240 format would require 9 days of inspection if each point is to be analyzed for 10 s that is a frequent case.

With the appearance of computerized IR thermography, it has become possible to inspect large areas by synchronizing pixel responses with a periodical heating function while heating large areas. The combination of the thermal wave technique and IR thermography was proposed by Carlomagno and Berardi in 1976 [42] and further improved by other researchers [43–48]. This method was called optical *lock-in* thermography. Obviously, the best results are expected if a test data is treated in the phase domain. Phasegrams are typically obtained by applying the Fourier transform. The corresponding simple technology was patented by Busse et al. [43]. It assumes that four equidistant thermograms S_{1-4} are to be recorded within a single thermal wave period, afterwards, phasegrams and ampligrams can be synthesized according to the following expressions:

$$\Phi = arctg\frac{S_3 - S_1}{S_4 - S_2}; \tag{6.22}$$

$$M = \sqrt{(S_3 - S_1)^2 + (S_4 - S_2)^2}. \tag{6.23}$$

By using the ratio of pixel amplitudes in Eq. (6.23), one can reduce the influence of uneven heating/absorption on TNDT results that is characteristic for phase data treatment. A commercial equipment involving this technique was produced by AGEMA Infrared Systems (now FLIR Systems). A useful feature of thermal wave IR thermography is its capability to inspect large areas, such as airplane panels, which, because of their curved geometry, are susceptible to uneven heating when applying pulsed TNDT [47]. Heating was performed with six 1 kW lamps of which radiation was modulated with the 0.015 Hz frequency. Delaminations between the skin and stringers which lead to a loss of structural stiffness were detected within 4 minutes while inspecting areas of few square meters. This method enabled producing images of the horizontal stabilizer of a Do-328 airplane from the distance of 3 m. A similar TNDT system was employed for evaluating thermal resistance between cast iron (1 mm) and aluminum (3 mm) [48]. The modulation frequency varied from 0.01 to few Hz, and IR thermograms were collected during 20–200 s with the frequency of 15 Hz. The results obtained were promising enough to conclude that this method can be used for evaluating thermal resistance between a gasket and a cylinder barrel in combustion engines.

When comparing thermal wave and pulsed techniques, it is worth noting that the treatment of signals in the time domain leads to results which are similar to those obtained in the frequency domain.

6.9.2 Lock-in Ultrasonic Stimulation (Periodic and Pulsed)

Hypothetically, an optimal stimulation in TNDT should enhance material temperature only in defect sites. In the 1990s, a technique of *vibrothermography* was proposed to analyze surface temperature distributions in objects subjected to periodical mechanical loading. Both sonic and ultrasonic frequencies can be used to excite local temperature signals by generating frictional heat.

Presently, ultrasonic stimulation in the frequency range from 20 to 40 kHz is typically used. This technique combines the principle of "optimal" heating and the feasibility of the lock-in data analysis. It is being developed by some world research teams, merely to mention Busse et al. (Germany) who use the term "*ultrasonic lock-in thermography*" (ULT) [49–52], Thomas et al. (USA) who prefer the term "*sonic IR imaging*", or "*thermosonics*" [53], and Burke and Miller (USA) who have introduced the terms "*VibroIR*" and "*SonicIR*" [54].

The difference between surface optical and volumic ultrasonic stimulation is illustrated in Fig. 6.21. Under optical heating, sample temperature increases not only in defect areas but throughout a sample, therefore, in sound areas, there might appear noisy signals caused by material inhomogeneities. Oppositely, ultrasonic heating produces temperature signals only in defect sites, while in sound areas the temperature remains close to ambient ("*dark field*" principle). As a result, the probability of correct detection increases, and thus the requirements to operators' experience might be relaxed. It is important stating that temperature patterns can be very different under optical and ultrasonic stimulation (see Fig. 6.21); in many cases, only certain subsurface defects generate heat due to edge friction, therefore, the topography of surface temperature indications is associated to developing cracks rather than to "stable" defects.

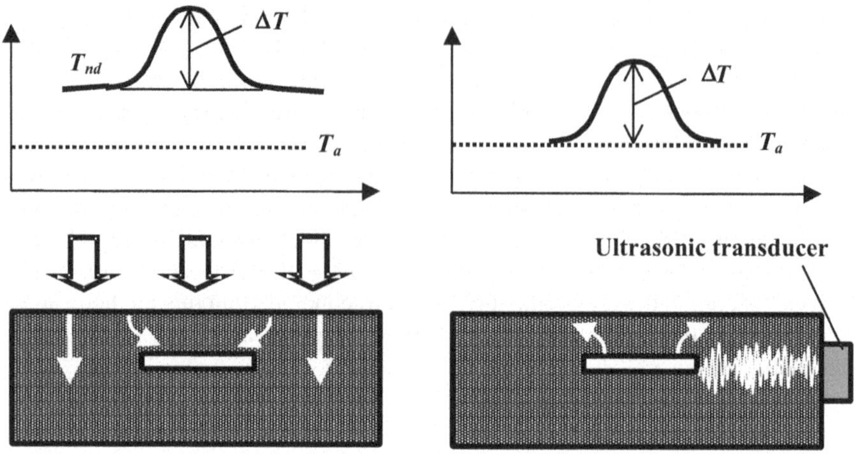

Fig. 6.21 Thermal wave IR thermography (optical and ultrasonic stimulation)

An experimental unit which allowed ultrasonic stimulation with 20 kHz elastic waves modulated with frequency up to 1 Hz was described in [51, 52]. The stimulation power was few hundred Watts, and temperature was recorded with a Jade II IR imager (320 × 240 image size, 3–5 μm spectral range, 200 Hz frame frequency, up to 3 min test time). Composite materials (carbon- and glass fiber reinforced plastics, and ceramics) were chosen for inspection. Some known advantages of ultrasonic stimulation, such as selectivity to different types of defects and high efficiency in using stimulating energy, were demonstrated. In particular, this technique proved to be very efficient in detecting delaminations and cracks in aileron stringers made of CFRP by using the 0.4 Hz modulation frequency. Another convenient test object appeared to be graphite fiber reinforced silicon carbide which is widely used in re-entry space vehicles, high-speed train brakes and other application areas where materials have to withstand high temperatures. The authors believe that ultrasonic stimulation is convenient for detecting cracks while optical heating suits better the evaluation of areas with abnormal thermal properties [50]. When inspecting riveted lap joints, which still are typical in aviation, it has been found that standard thermograms contain distinct indications of rivets and are not affected by the presence of cracks. In its turn, the ultrasonic stimulation lead to local temperature signals around cracks and loosened rivets. Ultrasonic IR thermography was used to verify results of eddy current NDT which revealed the crack growing between the rivets. This crack was thermographically well detected at the 0.06 Hz frequency (600 W ultrasonic stimulation power), and its apparent extension was larger on both the phasegram and ampligram to compare to the eddy current indications.

Further improvement of competitiveness of ultrasonic IR thermography requires expanding inspected areas (now under 0.2–1 sq.m.), e.g. by enhancing injected power, but this may lead to sample damage at a point of stimulation. When choosing an ultrasonic "heater", some measures are to be taken against standing waves which are seen in thermograms as regular strips (defects cannot be detected at standing wave knots).

One of main drawbacks of periodical ultrasonic stimulation is the necessity to keep a good contact between a sample and a transducer for few minutes in order to ensure detecting deeper defects. In order to overcome this problem, by analogy with optical heating, it was proposed to use short ultrasonic pulses with a high-frequency filling and capture temperature evolutions at a cooling stage. This emerging technique was called *phase angle thermography with ultrasound burst excitation* [49, 51]. Experimental results have been obtained in the detection of cracks growing between rivets in aviation aluminum panels by applying 2 kW 100 ms-long ultrasonic excitation. The cooling stage was analyzed for 3 s at the frequency of 15 Hz. Cracks were detected at 270 ms due to internal friction but the identification of IR images was difficult due to noise and intensive material heating at the point of stimulation. Optimal results appeared in phasegrams which were obtained as a result of applying the Fourier transform. The profiles of corresponding signals and their spectrums are given in Fig. 6.22.

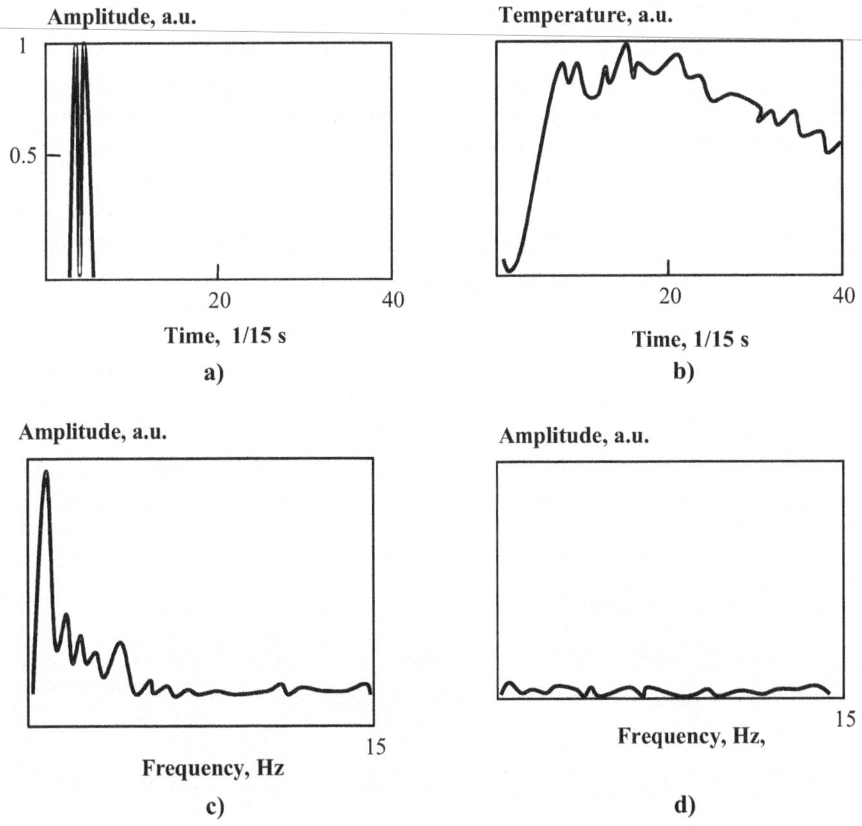

Fig. 6.22 Pulsed ultrasonic stimulation. (**a**) Ultrasonic pulse, (**b**) surface temperature signal, (**c**) spectrum of temperature signal in defect area of aluminum stringer (crack in rivet joint), (**d**) same as (**c**) in non-defect area. (Adapted from [49])

Like in the case of optical heating, pulsed ultrasonic excitation allows stimulating materials at frequencies which can hardly be obtained in the periodical mode. It is possible to produce a large number of phasegrams and ampligrams corresponding to the Fourier spectrum of a stimulating ultrasonic pulse. The main disadvantage of pulsed ultrasonic IR thermography is a high stimulation power that makes a nondestructive character of testing questionable and requires additional protection of the staff.

Some comparative results on the efficiency of optical and ultrasonic stimulation in the inspection of aerospace structures were reported in [52]. The parameters of used ultrasonic pulses were as follows: power 600 W, pulse duration 100 ms, carrier frequency 20 kHz. Results of ultrasonic IR thermography have proven to be most effective in the inspection of composites, turbine blade ceramic coatings and riveted joints.

The research at Wayne State University (U.S.A.) has been targeted toward the detection of cracks by applying pulsed ultrasonic stimulation [53]. A Branson ultrasonic gun generated pulses with the power of 1 kW and duration from 30 to 200 ms (20 kHz carrier frequency). Heating pulses were injected through a 1.3 cm-diameter steel waveguide attached to a sample with immersion paste. Subsurface cracks appeared as local temperature elevations on the background of the 'cold' sample (Fig. 6.23a). The phenomenon of heat generation in cracks has been particularly well seen by animating captured sequences (an example of a 1 mm-long growing crack is shown in Fig. 6.23b).

At Lawrence Livermore National Laboratory (USA), ultrasonic vibrothermography was used for testing soldered joints in parts of a complicated shape. Defective joints used to reveal specific temperature signatures which appeared after test objects were excited for 100 ms with 300 kW pulses (30 kHz carrier frequency) [54]. The advantage of this technique was a weak dependence of results on a position of the injection point, and the main drawback was possibility of inducing new defects by applying ultrasound of such high power. The same team compared potentials of acoustic microscopy, radiography, ultrasonic NDT and vibrothermography when detecting defects in a space shuttle panel made of a 36-ply CFRP laminate and covered with silicon carbide. Vibrothermographic results proved to be promising.

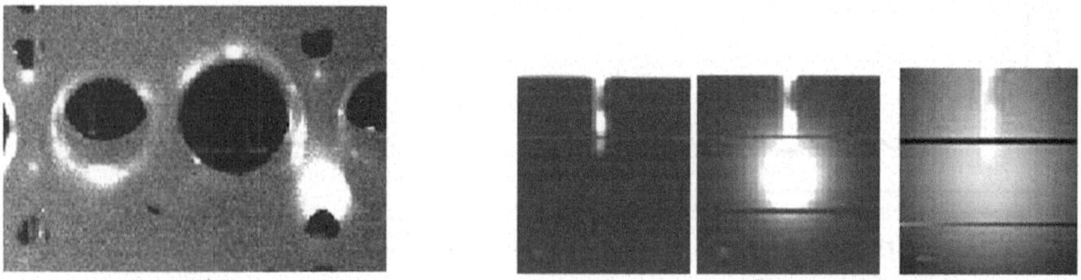

a) b)

Fig. 6.23 Pulsed ultrasonic IR thermography. (**a**) IR thermogram of plentiful cracks in an aluminum panel of the combustion engine cylinder block, (**b**) crack propagation in a 3 mm-thick aluminum panel (notch-like defect, crack length less than 1 mm). (Courtesy R. Thomas)

In order to reduce injected ultrasonic power, some basic research on studying the interaction between acoustic waves and structural defects has been conducted to demonstrate the presence of the so-called "local defect resonance" (LDR) which appears at characteristic frequencies of applied acoustic waves [55, 56]. Resonance stimulation of damaged materials provides efficient pumping of acoustic energy into defect areas thus causing intensive local vibrations [57].

As a conclusion, it can be stated that the combination of IR thermography and ultrasonic stimulation is prospective but its features are not well explored; for example, in some experiments, where dynamic temperature signals were expected, they did not appear. Transient temperature responses depend upon excitation energy, material properties and how a test sample is secured. A frequency of ultrasonic stimulation should be close to inherent frequencies of a sample in order to improve the efficiency of heat generation.

6.10 Fitting Temperature Evolution Functions

The fitting of pixel-based temperature functions $T_{ij}(\tau)$ is becoming increasingly popular due to the following factors:

- high-frequency noise can be effectively subdued,
- some mathematical operations, such as division, differentiating, Fourier transform, etc., may be more efficient when applied to fitted data,
- raw IR image sequences of arbitrary length can be replaced with few images of fitting coefficients (typically, less than six); a source sequence can be restored by these coefficients but without high-frequency noise,
- defect depths can be associated to some fitting coefficients in the same way as Fourier spectrum components.

The idea of matching a particular mathematical model to experimental temperatures was realized by Vernotte in 1937 in the experiments on determining thermal properties of materials [58]. Presently, procedures of non-linear filtration allow the simultaneous evaluation of several TNDT parameters, for example, defect size and depth, absorbed energy, surface heat exchange, etc. (see Sect. 5.6).

Typical fitting procedures use polynomial functions of a degree less than six [59], as well as the combinations of some exponential functions [60].

6.10.1 Polynomial Fitting

A type of the core function to be used in polynomial fitting can be chosen from the corresponding heat conduction solution (Table 6.1).

It is expected that the formulas in Table 6.1 are placed by order of their validity for fitting non-defect temperature evolutions. However, in practice, fitting efficiency

Table 6.1 Polynomial fitting in TNDT

Classical solution as fitting core	Polynomial presentation of temperature evolution
Logarithmic fitting, adiabatic case (LFA) Adiabatic pulsed heating of semi-infinite body $\frac{T}{W}=\frac{1}{e\sqrt{\pi\tau}}$, T – surface temperature, W – absorbed energy, $e=K/\sqrt{a}$ – thermal inertia, K – thermal conductivity, a – thermal diffusivity, τ – time	$T(\tau)\rightarrow Ln\ T(\tau)=A_0+A_1\ x+A_2\ x^2+\cdots$ $(x=Ln\ \tau)$ $T=e^{A_0+A_1\ x\ldots}$ (6.24)
Quadratic fitting, adiabatic case (QFA) Square pulse adiabatic heating of semi-infinite body $\frac{T}{Q/\lambda}=2\sqrt{\frac{\alpha\tau}{\pi}}-2\sqrt{\frac{\alpha(\tau-\tau_h)}{\pi}}$, τ_h – heat pulse duration	$T(\tau)=A_0+A_1\ t+A_2t^2\cdots\left(t=\sqrt{\frac{\tau}{\tau_h}}-\sqrt{\frac{\tau}{\tau_h}-1}\right)$ (6.25)
Quadratic fitting, non-adiabatic case (QFNA) Square pulse non-adiabatic heating of semi-infinite body $\frac{T}{Q/\alpha}=e^{H^2a(\tau-\tau_h)}erfc\left(H\sqrt{a(\tau-\tau_h)}\right)-$ $-e^{H^2a\tau}erfc\left(H\sqrt{a\tau}\right)$, α – heat exchange coefficient, $H=\alpha/\lambda$	$T(\tau)=A_0+A_1\frac{1}{\sqrt{\tau}}+A_2\frac{1}{\sqrt{\tau-\tau_h}}+A_3\frac{1}{\left(\sqrt{\tau}\right)^3}+$ $+A_4\frac{1}{\left(\sqrt{\tau-\tau_h}\right)^3}+A_5\left(\sqrt{\tau}\right)^3+A_6\left(\sqrt{\tau-\tau_h}\right)^3+$ $+A_7\left(\sqrt{\tau}\right)^5+A_8\left(\sqrt{\tau-\tau_h}\right)^5\cdots$ (6.26)

also depends on other factors, such as heat pulse shape, 3D heat diffusion, unstable heat exchange with the ambient, reflected radiation during and after heating. A simple graphical illustration can be associated to logarithmic fitting. As shown in Sect. 5.1 (see also Fig. 5.1), surface temperature excursion in adiabatic "non-defect" area subjected to Dirac heating is described with a straight line in the $Ln\ T\ -\ Ln\ \tau$ coordinates, and deviations from this line are treated as indications of subsurface defects. In order to achieve the same accuracy of fitting, the number of fitting coefficients should be typically higher in a defect area.

When applying a least-square method, the approximation accuracy can be evaluated by the χ^2 criterion:

$$\chi^2=\sum_{k=1}^{N}\left[\frac{T_k-T_f\left(\tau=\tau_k\right)}{\sigma_k}\right]^2. \tag{6.27}$$

Here T_k is the experimental temperature value at the k-th point; T_f is the fitting function value at the time τ_k, σ_k is the temperature standard deviation in the k-th point.

It is often assumed that signal standard deviation is constant through an IR image sequence. Then, the accuracy of fitting will be:

$$\chi^2 = \sum_{k=1}^{N}\left[T_k - T_f\left(\tau = \tau_k\right)\right]^2. \quad (6.28)$$

It is not for granted that a fitting function which provides minimal χ^2 is also optimal from the point of view of defect detection. The LFA and QFA algorithms from Table 6.1 have been compared by χ^2 for different values of Bio and Fo. The dependence of χ^2 on the number of polynomial coefficients is presented in Fig. 6.24. The main conclusions are: (1) with few exceptions, more polynomial coefficients ensure better approximation accuracy, (2) LFA provides the minimal value of χ^2 for *Case 1* (Thin plate, Short pulse), while the maximal value occurs in *Case 4* (Thick plate, Long pulse). The QFA algorithm has seemed to be mostly appropriate in *Case 2* (Thin plate, Long pulse) and inefficient in *Case 3* (Thick plate, Short pulse). As expected, while dealing with a plate, the QFA algorithm has been most efficient because it better describes heat conduction in a plate.

The results in Table 6.2 are presented for both defect and sound areas in TNDT of a 1 mm-thick CFRP sample with an air-filled defect at the depth of 0.5 mm. It is seen that there is no specific rule in producing both amplitudes and signs of fitting coefficients, therefore, it seems that polynomial coefficients can be hardly used for characterization purposes.

There is a certain analogy between polynomial coefficients and spectral components in the Fourier analysis: low-order coefficients carry more "spectral energy", while higher-order coefficients are responsible for small deviations of the temperature evolution from a "normal" behavior.

Data Normalization

The important characteristic of the LFA algorithm is automatic data normalization when producing images of polynomial coefficients by order higher than A_0. The example of applying logarithmic fitting to a synthetic IR image sequence simulating TNDT of an icon-on-wood is presented in Fig. 6.25. It is assumed that the red area on the sample surface absorbs twice energy in regard to the black area (Fig. 6.25a). Six defects are located at two depths (1 and 2 mm). Identification of TNDT results in such samples is difficult due to a 'spotty' character of standard thermograms. In particular, the image of the A_0 coefficient (A_0-image) keeps all peculiarities of the source thermograms (Fig. 6.25b), while the A_1-image smoothes signals in both areas, thus normalizing the source data (Fig. 6.22c).

Unlike LFA, the QFA algorithm operates with "true" temperatures in the changed time scale, therefore, it provides no data normalization (Fig. 6.26). For example, the images of A_0 and A_2 exhibit uneven heating (Fig. 6.26a, b), while the A_6-image in Fig. 6.26c is characterized by the distorted defect pattern, thus illustrating such important feature of fitting coefficients as the change of coefficient signs for defects located at different depths.

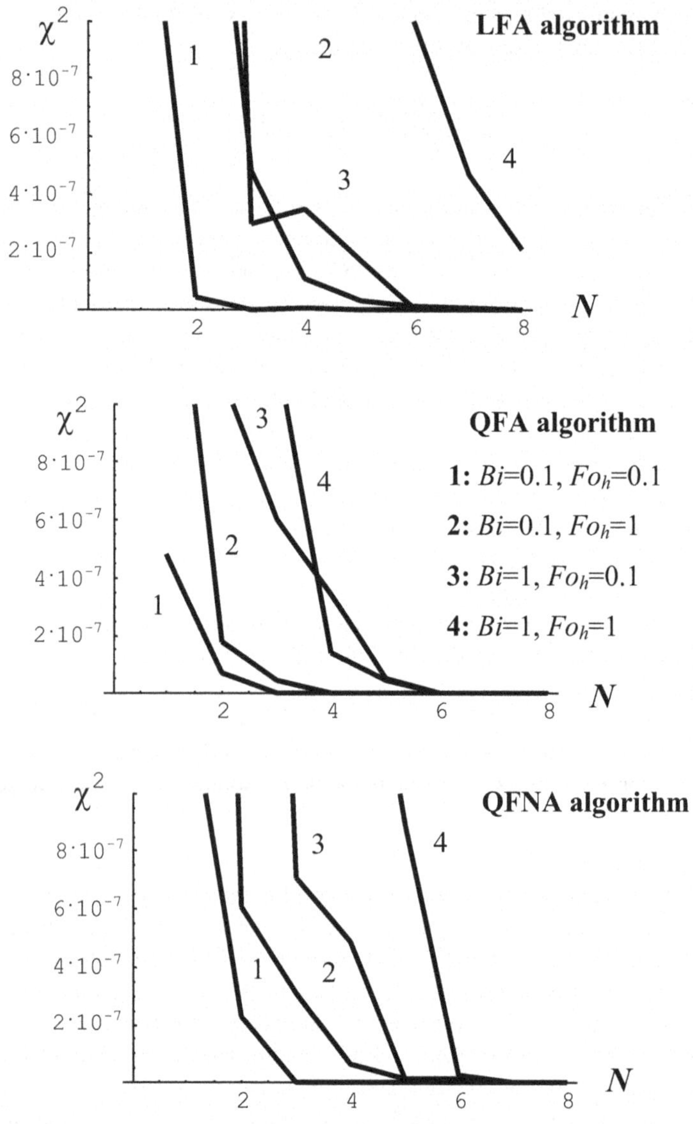

Fig. 6.24 Criterion χ^2 vs. number of polynomial coefficients N used in fitting by the algorithms from Table 6.1 (1: Thin plate, short pulse; 2: Thin plate, long pulse; 3: Thick plate, short pulse; 4: Thick plate, long pulse)

Table 6.2 TNDT of 1 mm-thick CFRP sample (0.1 mm-thick air-filled defect at 0.5 mm depth, heating for 1 s, acquisition step 0.1 s, total inspection time 10 s, cooling stage is only analyzed)

Algorithm	Parameter	A_0	A_1	A_2	A_3	A_4	A_5	A_6
LFA	D[a]	−0.056	−0.145	0.0359	0.0156	−0.0099	−0.002	0.00082
$\chi^2 = 2.2 \cdot 10^{-7}$	ND[b]	−0.231	−0.045	0.0266	−0.015	−0.0024	0.0012	−0.0001
QFA	D	−0.241	13.9	−73.2	197	−271	182	−46.5
$\chi^2 = 1.3 \cdot 10^{-7}$	ND	0.148	6.82	−31.7	82	−126	111	−42.8
QFNA	D	4.98	−0.0094	0.575	−48.7	−39.5	4.48	0.225
$\chi^2 = 1.3 \cdot 10^{-9}$	ND	−0.49	0.00676	−0.348	42.0	33.7	−2.43	-3.12

[a]D – defect area
[b]ND – non-defect area

Fig. 6.25 Polynomial logarithmic fitting (LFA) of artificial image sequence, 20 mm-thick icon-on-wood, 6 defects, heating time 5 s. (**a**) Optimum source IR thermogram (upper row defects at depth 1 mm, bottom row – 2 mm), (**b**) A_0 image (third order polynomial), (**c**) image A_1 (third order polynomial)

Fig. 6.26 Applying the QFA algorithm to the data from Fig. 6.25: (**a**) A_0 image (third order polynomial), (**b**) A_2 image (third order polynomial), (**c**) A_6 image (third order polynomial)

Defect Detectability

After polynomial fitting, further processing can be applied either to a restored sequence or to coefficient images. As mentioned above, polynomial coefficients can be compared with Fourier spectral components, therefore, one may expect that particular defects can better be detected in particular coefficient images; in any case, the fitting performance depends on both a particular coefficient and a total number of coefficients.

The technique of polynomial fitting was applied to the results of TNDT of an icon-on-wood (Fig. 6.27). The 3D plot presented in Fig. 6.28 shows the relationship between the signal-to-noise ratio *S*, polynomial degree and coefficient number in the detection of the deepest *Defect 1*; the corresponding thermogram is shown in

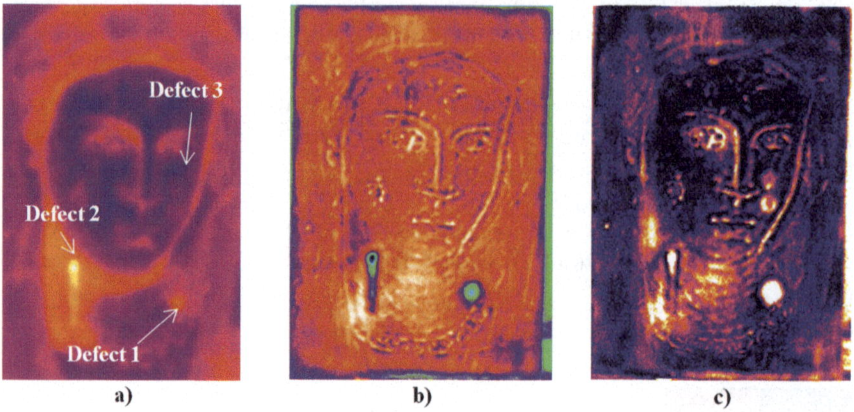

Fig. 6.27 Detecting defects in an icon-on-wood reference sample: (**a**) Optimum source IR thermogram at 2.5 s after 5 s heating ($S = 1.3$ for *Defect 1*), (**b**) A_2 image (QFA, second degree polynomial function, $S = 5.8$ for *Defect 1*), (**c**) χ^2 image (QFA, $S = 9.9$ for *Defect 1*)

Fig. 6.27a. The icon was heated for 5 s with a heat flux power of 3 kW/m^2, thus enhancing sample temperature by 7–10 °C. A number of artificial and natural defects were located in the first 2.2 mm-thick icon layer made of gypsum. According to Fig. 6.28, an optimal visibility of defects is supplied by the A_2-image of the second-degree polynomial (Fig. 6.27b).

It was mentioned above that low-order fitting better suits non-defect temperature evolutions, while defect areas require higher-order approximation. Hence, according to Eq. (6.28), an image expressed in χ^2 value can be also used for identifying defects; it should be higher in the case of "powerful" defects because of a stronger temperature deviation from a classical behavior. The image of $\chi^2(i, j)$ in Fig. 6.27c is characterized by the maximum value of $S = 9.9$ when detecting the deepest *Defect 1*. Note that, in this case, the χ^2 criterion is defined as the difference between an experimental sequence and a first-degree polynomial.

Like phasegrams, images of polynomial coefficients may produce signals of different signs for defects located at different depths even if they are of the same physical nature. This makes using the criterion of signal-to-noise ratio difficult; an example is presented in Table 6.3 when detecting all three defects in the icon-on-wood (Fig. 6.27a). It is seen that: (1) the difference between values of S in the images of χ^2 and A_2 that is well seen in Fig. 6.27 becomes smaller when considering all three defects, (2) defect indications may appear as both "warm" and "cold" areas in regard to the background, therefore, even if they are well detected, the corresponding S values, averaged by all defects, may be low (see Fig. 6.29a), (3) the highest S value is provided by the maxigram in Fig. 6.29b.

"Reference-Free" Tomography

The efficiency of the "classical" thermal tomography discussed in Sect. 6.3 significantly depends on a chosen reference temperature value, i.e. heating uniformity and/or quality of data normalization.

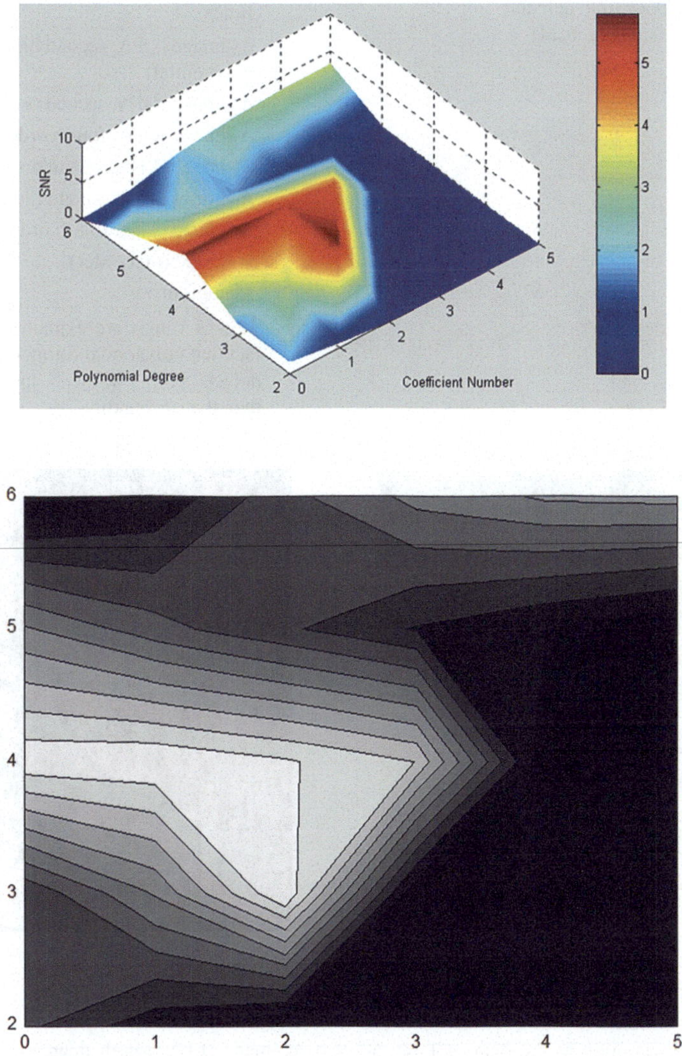

Fig. 6.28 Signal-to-noise ratio *vs.* polynomial coefficient number and polynomial function degree in both 3D and contour presentation (applying the QFA algorithm to the TNDT of the icon-on-wood, see inspection results in Fig. 6.24)

It is obvious that no quantitative estimates are possible without a proper reference. However, in the case of TNDT, it is possible to consider a model of the sample "thermal" behavior which is supposed to be valid in all "non-defect" pixels. Then only deviations from the reference behavior are to be analyzed as candidates to defect indications.

Table 6.3 Detecting defects in icon-on-wood (Fig. 6.24)

Image	S
A_2-image (QFA, second order polynomial)	3.3
A_1-image (QFA, second order)	2.5
A_3-image (QFA, fourth order)	0.8[a]
A_2-image (LFA, second order)	2.6
A_1-image (LFA, second order)	0.2[a]
A_3-image (LFA, fourth order)	0.8[a]
χ^2-image (first order)	3.8
Maxigram	4.4

[a]Low S values are explained with the fact that polynomial fitting might make defects both "warmer" and "colder" than the surrounding

a) b)

Fig. 6.29 Illustration to the data in Table 6.3: (**a**) A_3 image (LFA, fourth degree polynomial, $S = 0.8$), (**b**) maxigram ($S = 4.4$)

Consider a model of TNDT of the icon-on-wood (Fig. 6.27). In Fig. 6.30a, the "classical" values of T and ΔT calculated for a cosine heat pulse are presented. The maximum signal $\Delta T_m = 0.45$ °C appears at $\tau* = \tau_m = 6.8$ s. The evolutions of both defect and non-defect temperatures were fitted with two types of polynomials: by the second and sixth order, and the differential signal was defined as $\Delta T_{ij\,f} = T^{(6)}{}_{ij\,f} - T^{(2)}{}_{ij\,f}$, where the superscript specifies a degree of the polynomial. It is important that the approach above can be applied to each pixel without introducing a reference; in fact, it is assumed that the second-order polynomial describes

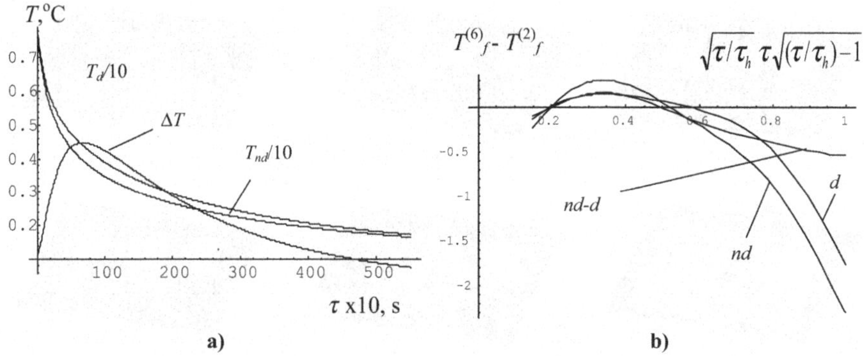

Fig. 6.30 Differential temperature signals *vs.* time (TNDT of an on wood, Q = 3000 W/m^2, $\tau_h = 5$ s, $L = 20$ mm, defect depth 2.2. mm, defect thickness 0.2 mm): (**a**) T and $\Delta T(\tau)$, (**b**) $T^{(6)}{}_f - T^{(2)}{}_f$ (QFA algorithm)

non-defect evolution, while the sixth-order polynomial includes both the "classical" signal behavior and deviations in defect areas. The corresponding curves are shown in Fig. 6.30b including the $\Delta[\Delta T_{d\,f} - \Delta T_{nd\,f}]$ function of which maximum equal to 0.16 °C takes place at $\sqrt{\tau/\tau_h} - \sqrt{(\tau/\tau_h)-1} = 0.34$, i.e. at τ = 13.5 s, if to count from the start of heating, or τ = 8.5 s, if to count from the end of heating. Since differential temporal evolutions involve some local extremums (see Fig. 6.30b) caused by hidden defects; this approach can be implemented for fulfilling "reference-free" tomography.

Consider the principle of "reference-free" tomography in application to the artificial sequence from Fig. 6.25 where IR thermograms are characterized by strongly uneven heating. The algorithm by the formula $\Delta T_{ij\,f} = T^{(6)}{}_{ij\,f} - T^{(2)}{}_{ij\,f}$ being applied to each pixel revealed evolutions similar to those in Fig. 6.27b. The thermal tomogram in Fig. 6.31a corresponds to the layer with two defects at different depths. In this case, the reference level is zero because the corresponding functions are already differential, similarly to Fig. 6.25. Then, the chosen layer was separated for two more layer to exhibit defects located at different depths (Fig. 6.31b, c). Note that the results in Fig. 6.31a, b have been derived with no amplitude threshold. The tomograms in Fig. 6.31d–g were obtained by using the algorithm of standard thermal tomography with a particular amplitude threshold applied. The tomogram of Fig. 6.31d is identical to that in Fig. 6.31a. It is seen that the "classical" tomogram contains artifacts which appear both around defects (crown-shaped areas) and between areas with different absorbed energy. The indications of defects located under the transition zone are fairly distorted in the "classical" tomogram. Respectively, the tomogram in Fig. 6.31e is similar to that in Fig. 6.31b containing a lot of false defect indications due to no thresholding. Introducing the threshold significantly improves the image quality (see Fig. 6.31f).

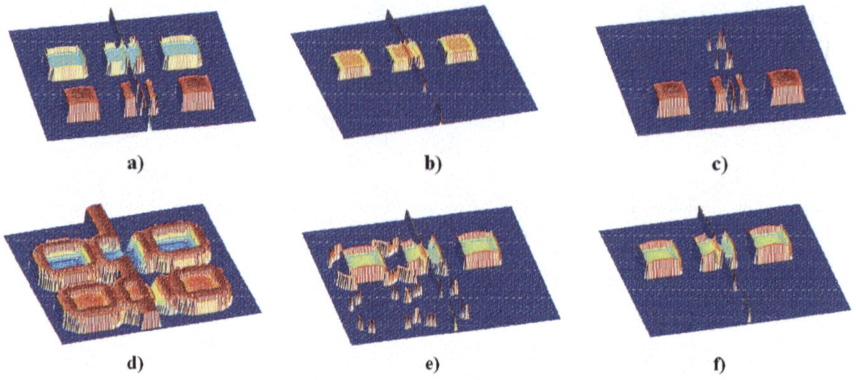

Fig. 6.31 Dynamic thermal tomography (artificial sequence from Fig. 6.25, TNDT of icon-on-wood): (a) Tomography by using algorithm $T^{(6)}{}_f - T^{(2)}{}_f$ (chosen layer includes all defects), (b) same as (a) (upper layer chosen), (c) same as (a) (bottom layer chosen), (d) 'classical' tomogram (chosen layer includes all defects, no amplitude threshold), (e) same as (d) (upper layer chosen, no amplitude threshold), (e) same as (d) (upper layer chosen, 2% amplitude threshold)

6.10.2 Exponential Fitting (Thin High-Conductivity Materials)

A temperature response on the surface of thin high-conductive materials subjected to square- or Dirac-pulse heating can be fitted with the core function by Eq. (5.7):

$$T^F(\tau) = \beta_1 e^{-\beta_2} + \beta_3 e^{-\beta_4}, \tag{6.29}$$

where coefficients β_3 and β_4 are associated with defect depth l and thermal resistance R_d respectively. Within this approach it is possible to produce coefficient images where pixel values reflect the respective defect parameters [60].

6.10.3 Temperature Derivatives and Thermographic Signal Reconstruction (TSR)

In one-sided pulsed TNDT, temperatures over defects change slowlier than in sound areas due to less intensive heat diffusion through defects. Hence, not only a basic temperature function $T(\tau)$ but its derivatives $(\partial T(\tau)/\partial\tau)$ and $(\partial^2 T(\tau)/\partial\tau^2)$ might bear useful information about hidden defects. Mathematically, peculiarities of derivatives are well known but Shepard et al. found them useful in TNDT when analyzing heat diffusion in defect areas. The authors proposed the technique of the so-called *Thermographic Signal Reconstruction* (TSR) which has been implemented in the commercial hardware from Thermal Wave Imaging (the term "*Synthetic Data Processing*" (SDP) is also used [59, 61, 62]).

According to this concept, in the absence of defects, heat flux is essentially 1D being governed by the equation:

$$\frac{\partial^2 T}{\partial z^2} = \frac{1}{a}\frac{\partial T}{\partial \tau}, \tag{6.30}$$

of which solution for a uniformly-heated semi-infinite bode has the canonic form (see Table 2.4):

$$T(z=0) = \frac{Q}{e\sqrt{\pi\tau}}. \tag{6.31}$$

Over low-conductive defects, the heat flux falls into a "trap" and starts diffusing mainly in lateral directions toward lower "non-defect" temperatures. In this case, the heat diffusion can be considered 2D according to the equation:

$$\nabla^2_{x,y} T = \frac{1}{a}\frac{\partial T}{\partial \tau}. \tag{6.32}$$

Thus, an overall process of 3D heat diffusion can be conventionally separated for "normal" and "lateral" phases. The transition from one phase into another is closely related to the concept of optimum observation time (peak slope time); and, within the TSR technique, observation times are analyzed by the behavior of temperature derivatives.

Since taking derivatives of noisy functions enhances high-frequency noise, an essential feature of the reported technique is the prior fitting of experimental functions $T(\tau)$ by using the LFA algorithm (see Sect. 3.1). Logarithmic fitting: (1) relaxes the influence of finite heat pulse duration, (2) reduces the behavior of "non-defect" pixels to the straight line with a slope of –0.5, and (3) operates as a high-frequency filter while leaving defect signals undistorted. The second step of the procedure is restoration of the source sequence by using some polynomial coefficients by Eq. (6.24). This operation step has the following advantages: (1) "synthetic" functions are free of a high-frequency noise, (2) these functions can be processed by applying such mathematical operations as differentiation, integration, etc., (3) specific heat transit times can be analyzed algebraically, and (4) the number of polynomial coefficients may be low, therefore, data processing is faster and source data may occupy less computer memory.

Another implementation of the TSR technique is the analysis of dynamic histograms of logarithmic temperature derivatives on the sample surface [61]. By transiting from basic functions $T(i,j,\tau)$ to histograms of $d[Ln(T)]/d[Ln(\tau)]$, it is possible to realize the above-mentioned advantages of "synthetic" data treatment thus making results invariant on random displacements of the IR imager and the reference. An example of surface temperature evolution is shown in Fig. 6.32a in the

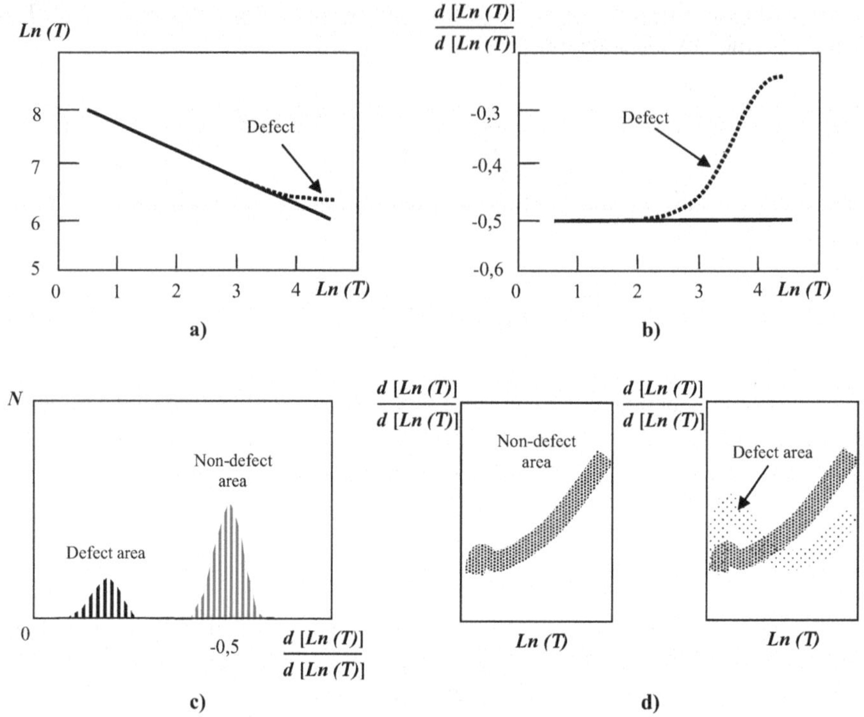

Fig. 6.32 Thermographic signal reconstruction. (**a**) $Ln(T) - Ln(\tau)$, (**b**) logarithmic temperature derivative vs. time, $[Ln(T)]/d[Ln(\tau)] - Ln(\tau)$, (**c**) defect and non-defect histograms, (**d**) logarithmic temperature derivative 'footprints'. (Adapted from [59])

logarithmic coordinates; the first derivative of this function is given in Fig. 6.32b. In the presence of air-filled defects, the temperature deviates from the straight line that is stressed by the derivative. The histograms of defect and non-defect areas are well separated with the "non-defect" histogram being sharply peaked around the −0.5 value (Fig. 6.32c). If to choose some characteristic histogram modes and display their dependence on $Ln(\tau)$ through the sequence, one may produce some specific images shown in Fig. 6.32d for a defect and non-defect area. These images are called "*thermographic fingerprints*". When making a decision on quality of particular sample sections, the left image in Fig. 6.32d may serve as a reference. It is important that, as mentioned above, thermographic fingerprints are invariant on displacement and rotation of tested areas. Moreover, a whole sequence can be reduced to a single image, or even a single curve if to consider single histogram peaks. As each subsequent image appears, it can be compared to a master image of the histogram sequence projection.

A deep theoretical and experimental analysis of using TSR (first and second logarithmic derivatives, polynomial coefficient images, etc.) in the case of step-heating in comparison to pulsed heating was fulfilled by Balageas and Roche [63, 64].

The results obtained with pulsed heating were generally better because of some unpleasant features of step-heating, namely, low signal-to-noise ratio in the beginning of the process and high surface temperature at the end of the experiment. It was also demonstrated that the optimal degree of the logarithmic polynomials was 7, while 3 "best" coefficient images were used for implementing a new technique of data presentation called *RGB projection* [65]. The analysis of RGB composite images has also shown advantages of pulsed heating even for deeper defects (up to 4 mm in CFRP).

The TSR (SDP) processing technique has been successfully applied to metallic samples with and without coatings allowing obtaining high-quality images in the inspection of turbine blades, aircraft panels, etc.

6.11 Data Normalization

In TNDT, normalization specifies dynamic division of IR thermograms that is being done in order to suppress influence of uneven heating. First, this term was suggested by Degiovanni et al. and then used in the ThermoFit Pro software (Tomsk Polytechnic University). There are two types of normalization: (1) normalization of all images in a sequence by a chosen image with the latter one being called *normalizing*, and (2) 3D normalization that implies the division of two sequences where one is experimental and another - synthetic.

6.11.1 Normalization Based on One Image

Consider an IR mage sequence $T(i,j,\tau)$ captured in a pulsed TNDT procedure (τ_h- heating duration). In many cases, optimal detection conditions take place at the cooling stage, i.e. at $\tau_m > \tau_h$. The normalization formula by a single image is as follows:

$$C_{norm}(i,j,\tau) = \frac{T(i,j,\tau) - T_{init}(i,j)}{T_{norm}(i,j,\tau_{norm}) - T_{init}(i,j)}, \tag{6.33}$$

where $C_{norm}(i,j,\tau)$ is the normalized contrast, $T_{init}(i,j)$ is the sample initial temperature (typically equal to the ambient temperature T_a), and $T_{norm}(i,j,\tau_{norm})$ is the normalizing temperature. If a normalizing image is chosen at such time when defect signals have not yet appeared, all normalized images will be characterized by a uniform background with defect indications appearing at times close to τ_m. For example, in the case of short-pulse heating, the normalizing image is often taken at the end of heating. A principle of normalization by a single image is presented in Fig. 6.33.

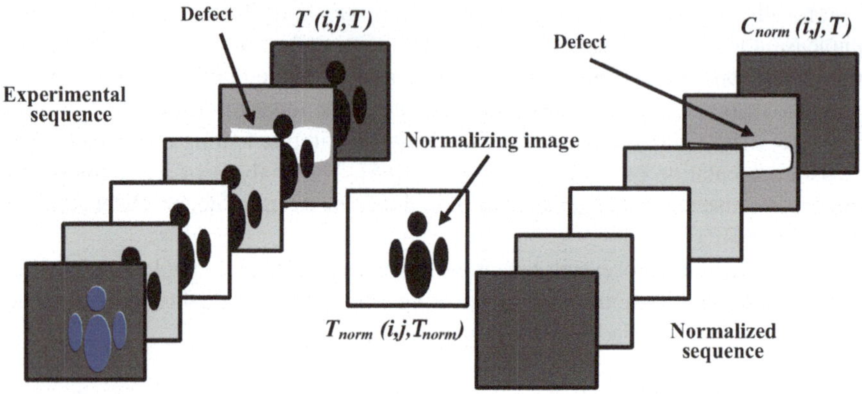

Fig. 6.33 Normalizing dynamic IR image sequence by a single image

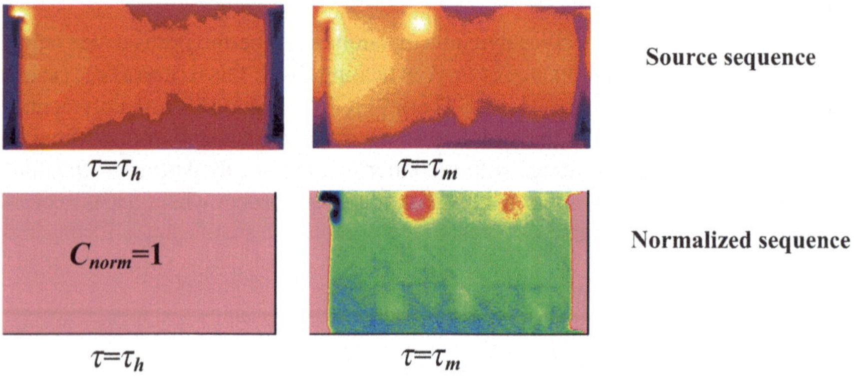

Fig. 6.34 Normalization in TNDT of CFRP

Efficiency of normalization is illustrated with the results of pulsed TNDT of CFRP presented in Fig. 6.34. At the end of heating, the source image clearly shows uneven heating of a sample with two lamps, more powerful – on the left (see the left top image in Fig. 6.34). Note that the uneven heating pattern survives until the optimum observation time $\tau = \tau_m$. Normalization was done by the end of heating ($\tau = \tau_h$). At $\tau = \tau_h$, the normalized contrast is unity (see the left bottom image), while at $\tau = \tau_m$ the normalized image shows defects clearer than the optimal raw image.

6.11.2 3D Normalization

When normalizing by a single image, acceptable results appear if sample surface is relatively uniform, and the unevenness phenomena occur in IR thermograms due to variations in heating energy. If the sample surface is "spotty", i.e. contains areas with different absorptivity, the significant lateral heat diffusion may diminish

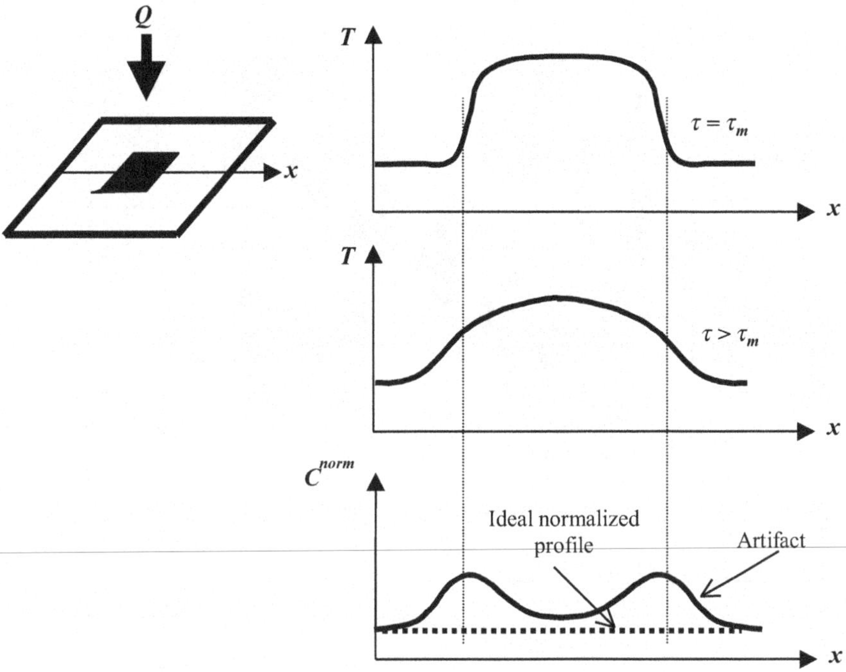

Fig. 6.35 Influence of 3D heat diffusion on efficiency of standard normalization

efficiency of 1D normalization. This statement is illustrated with the temperature profiles in Fig. 6.35. Let a test object, for example, a wall fresco, has an area with high absorptivity (dark color); then, at the end of heating, this area acquires higher temperature with clear boundaries. With time, the boundary temperature profiles become smoother due to lateral heat diffusion. Therefore, the application of 1D normalization will cause the appearance of boundary artifacts which may be recognized by operators as defect indications. In this case, the profiles of normalized contrast C^{norm} might essentially differ from the expected uniform profile shown in Fig. 6.35 with a dashed line.

Normalization can be performed sequence-by-sequence [66]. A normalizing sequence is to be a non-defect replica of the corresponding source sequence obtained in assumption of no defects. A replica sequence can be obtained by calculating expected temperature patterns on the sample surface and using one of experimental source images as a heating mask $Q(i,j)$. The main requirement to a mask image is that it should reveal no visible defect indications, thus reflecting only heating/absorbance phenomena; for example, the $T(i,j,\tau_h)$ thermogram can be used as the mask. To summarize, 3D normalization represents the division of two sequences of which one is experimental, thus containing noise and defect signals, and another one is synthetic but calculated with an experimental mask, thus containing only patterns of noise (Fig. 6.36).

3D normalization has proven its validity in TNDT of wall frescos where optical heating leads to essentially uneven temperature patterns (see Chap. 11).

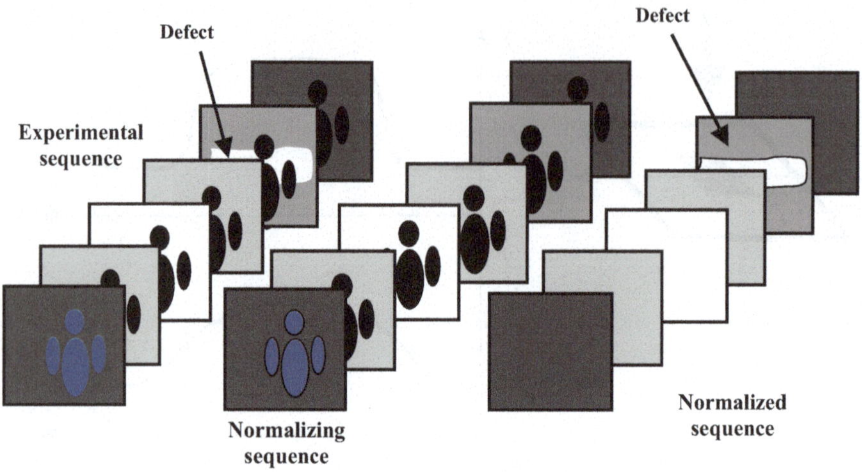

Fig. 6.36 Normalizing dynamic sequence by another sequence (3D normalization)

6.12 Moving Heat Source

6.12.1 Continuous Heating

A method of a continuously-moving heat source has been developed in several modifications. Its basic feature is that a test sample is heated in a localized area and scanned area-by-area. In such case, the heating is essentially 3D thus allowing to detect crack-like defects located perpendicularly to the scanned surface. In TNDT, both spot and line heating principles are employed with the temperature being respectively recorded either in a point (Fig. 1.3a) or along a line (Fig. 1.3b). Temperature profiles along a heat source path can be analyzed in two coordinate systems: either related to the sample (Fig. 6.37a) or to the heat source (Fig. 6.37b). Typically, a thermogram obtained in the coordinate system related to a moving heat source is analyzed. In this case, all surface points are under identical conditions in respect to heating, and temperature fluctuations are determined either by absorptivity/emissivity variations or by presence of defects. In particular, thermal energy is stored in front of surface and near-surface cracks thus enhancing material temperature; respectively, there is energy deficiency behind the cracks that causes the mirror-like temperature decay (Fig. 6.37b).

By analogy with uniform heating, any TNDT procedure involving a moving heat source can be characterized by an optimum observation time (delay time) τ_m which depends upon material thermal diffusivity and defect depth. When using this technique, a main critical parameter is the distance between a heated area and a temperature recording point b defined as:

$$b = V\tau_m, \tag{6.34}$$

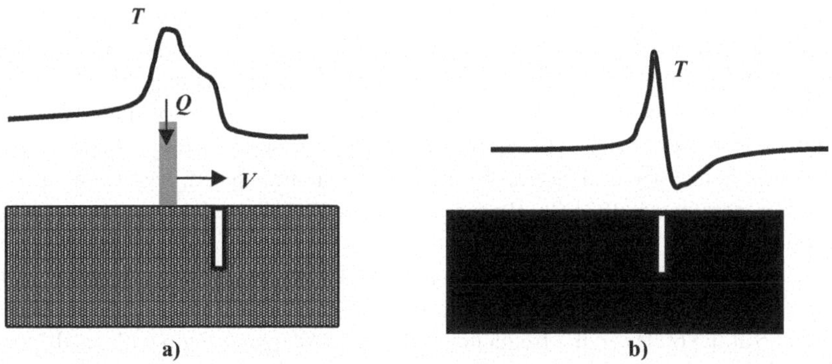

Fig. 6.37 Moving heat source technique (temperature profiles): (**a**) Object coordinate system, (**b**) heat source coordinate system

where V is the heat source speed. Estimates of τ_m can be obtained by using the corresponding solution for heating a test object with a uniform (extended) heat flux, although the most accurate estimates can be derived by solving a moving heat source problem (see Sect. 4.4). It is worth mentioning a paradoxical fact that, in high-conductive materials with shallow defects, an optimum observation point can be located *in front* of the heat source; in this case, delay times τ_m which are typically counted from the beginning of heating are "negative".

Heat source speed may vary significantly in the range from millimeters to meters per second depending on material thermal properties and required depth of material probing. Respectively, the inspection hardware may implement either mechanical or opto-mechanical scanning. For example, a "flying spot" technique implies both remote heating and temperature recording by using swinging mirrors [67]. Heating can be done with a laser beam, however, because of high speed of scanning, the average absorbed energy might be small. Building materials having thickness of few centimeters have been inspected under low scanning speed ("crawling spot" technique) [68]. In TNDT of composite materials, scanning speed may reach few centimeters a second.

A moving heat source technique is selective in regard to defect parameters, particularly, defect depth, therefore, reliable detection of defects at different depths may require repeating the test. However, in practice, it is recommended to choose b optimal for the deepest defects.

A TNDT procedure shown in Fig. 1.3a is characterized by the highest sensitivity toward hidden defects but it is low productive. A compromising procedure employs line heating combined with line-scanning IR radiometry. In Russia, such procedure was successfully used by Storozhenko et al. in the inspection of wound shell glass fiber composites [69]. It was reported that delaminations between fiber bundles were detected at depths up to 10–15 mm. Advantages of line heating are: (1) high efficiency of stimulation due to small distance between a heater and a sample, (2) more uniform heating, and (3) low cost of line-scanning systems in regard to IR imagers.

The interest to moving line heating revived, mainly, due to research of Cramer, Winfree and Woolard (U.S.A.) who suggested a concept of a "thermal photocopier" [70, 71]. By involving advanced computer data treatment, it is possible to produce synthetic IR images at several time delays after a single travel of the heater. The analytical model of line scan thermography (LST) was suggested by Khodayar et al. to investigate test performance as a function of scanning speed, heat source power and other parameters [72]. In Russia, a similar TNDT method was patented by Shirayev et al. in 1984 [73].

An experimental unit developed at NASA (USA) included an InSb-based IR imager which enabled the temperature resolution of 0.025 °C (image format of 256 × 256 at the 60 Hz frame frequency) [70]. A 1 kW quartz lamp with an elliptical reflector was used as a line heater at scanning speeds up to 30 cm/s. Synthetic IR thermograms consisted of 256 pixels in the vertical direction and up to 1200 pixels in the direction of scanning.

Lesniak and Boyce proposed a technique of *forced diffusion thermography* (FDT) based on the stimulation by a dynamic heat pattern [74]. The thermal flux emitted by an extended heater, for example, a 500 W lamp, passes through a grid mask which allows spatial modulation of the heat flux. A test object is monitored with an IR imager (Fig. 6.38) to produce a full-field image. The mask spatial/temporal frequencies can be optimized to detect specific flaws. The heat flux determined by a moving line pattern is mathematically described with the following expression [74]:

$$Q(x,\tau)=\frac{Q}{2}\left[1+\cos\left(2\eta\pi x+\omega\tau\right)\right], \tag{6.35}$$

where Q is the heat flux density, η is the line density in the mask, and ω is the angular frequency. At a particular value of x, the flux experiences continuous oscillations with the frequency ω and the phase shift $2\eta\pi x$.

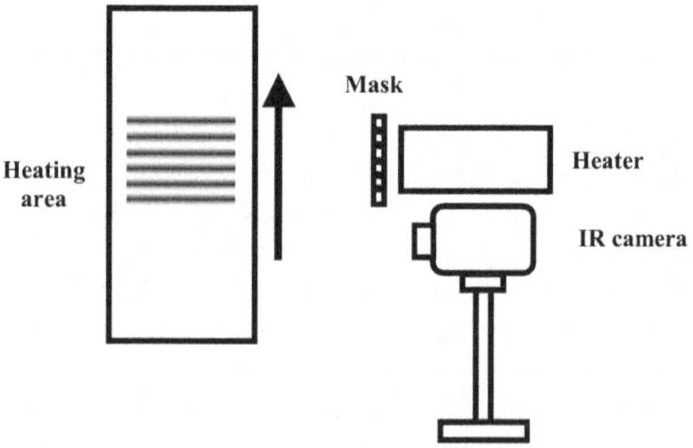

Fig. 6.38 IR thermography by forced heat diffusion

The flux spatial gradient is also subjected to harmonic oscillations:

$$\frac{\partial}{\partial x} Q(x,\tau) = \eta \pi I \cos(2\eta \pi x + \omega \tau). \tag{6.36}$$

Moving strip-like temperature patterns are typically analyzed with *differential IR imagers* which record temperature increments above a certain level. Each point in the image to be analyzed is characterized by a particular phase. By expanding the harmonic function in Eq. (6.36), it can be demonstrated that a moving line pattern can be considered as a superposition of the in-phase and out-of-phase temporal components, for instance, as follows:

$$Q(x,\tau) = \frac{I}{2}\left[1 + \cos(2\eta\pi x)\cos(\omega\tau) + \sin(2\eta\pi x)\sin(\omega\tau)\right]. \tag{6.37}$$

If a crack appears within a moving field of heating, all temperature components start to experience deviations from a regular behavior thus allowing defect detection (Fig. 6.39). By comparing temperature profiles in Figs. 6.37 and 6.39, it can be stated that the difference between a moving heat source technique and FDT is of the same order as between pulsed and thermal wave inspection techniques.

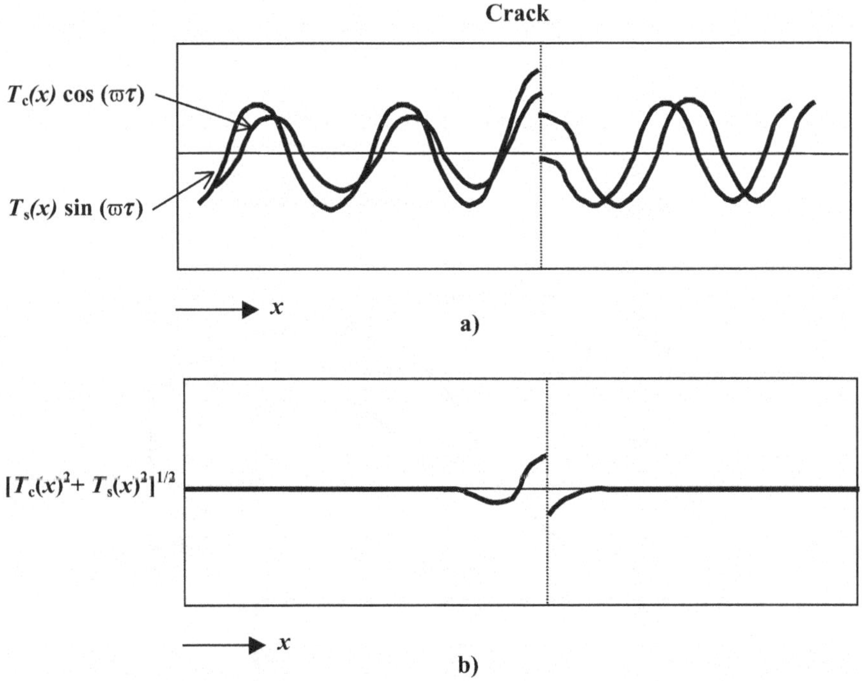

Fig. 6.39 Forced heat diffusion in application to TNDT of a sample with a crack. (**a**) Temporal component, (**b**) amplitude component. (Adapted from [74])

6.12.2 Photothermal Technique (Thermal Waves and Pulsed Heating)

In the classical photothermal technique where periodical thermal waves are used, the scanning is discrete and up to 10 wave periods are to be excited at each surface point in order to reach a quasi-stationary regime. This technique is typically applicable to thin high-conductive materials with thermal wave frequency up to few kHz. The penetration depth of such waves is in a micrometer range. Analyzed areas are in the same range, while a heated area should be much larger to provide 1D flowing of heat that is mathematically described by simple analytical solutions. In such way, only small-size areas can be inspected. The photothermal method combines advantages of 3D probing and phase analysis and also allows material depth profiling. The main application areas are: (1) NDT of coatings and films (lacquers, galvanic coatings etc.), and (2) evaluation of steel hardening depth.

An example of a photothermal microscopic system is presented in Fig. 6.40 [75]. The heating is performed in a 1 mm^2 spot by means of an argon laser of which radiation is modulated with an acoustooptical device, and the temperature is measured within an area of few tens sq. μm. This unit has been employed for evaluating steel carbonization and inspecting quality of bonding of a 1.2 μm-thick Ni-Cd coating on a 0.3 mm-thick copper substrate [75].

The arrangement by Fig. 6.40 can be realized also in a pulsed regime; then, the frequency analysis is to be replaced with the phase analysis.

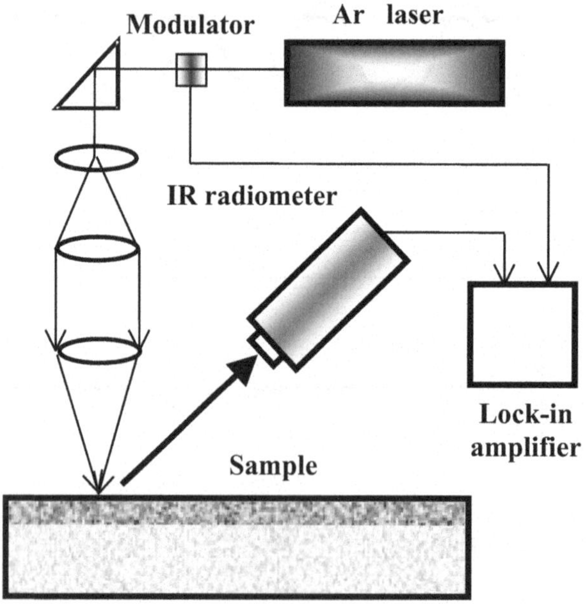

Fig. 6.40 Photothermal microscope

Recently, a combination of the classical photothermal technique and IR thermography was developed [76]. The experimental scheme was similar to that in Fig. 6.40 but the IR radiometer was substituted with the IR imager which enabled the spatial resolution of 30 μm with the image format being 320 × 240. The unit was used for the identification of thermal conductivity and optical transparency of a 40 μm-thick epoxy resin on a 1.5 mm-thick aluminum substrate.

6.13 Combining TNDT and Other NDT Techniques (Data Fusion)

Combining TNDT and other inspection techniques makes sense if: (1) TNDT serves as a screening technique which forestalls the use of other, presumably, more reliable NDT methods, (2) inspection hardware is characterized by mutual components, such as a heater, and (3) data fusion is not a simple summation but leads to a new quality of inspection.

The example of the first-type applications is the assessment of water ingress in aviation sandwich panels, as well as the detection of impact damage and delaminations in composites. TNDT allows localizing suspicious areas, and afterwards, a more accurate technique, for example, ultrasonics, should be applied to better identify defects. In the same way, TNDT can be used to evaluate quality of riveted joints in aircraft frames, with the eddy current method to is used as a basic technique.

The second-type combination of NDT techniques takes place, for example, while using IR thermography and laser interferometry, or shearography. The principle of shearography is based on the analysis of mechanical displacements of the test sample surface, and the resulting images appear by mixing two slightly shifted speckle-images. The first shearographic image is to be obtained in the absence of loading. Then, the sample must be stimulated by changing air pressure, or applying the heating or vibration, and the second shearographic image is to be recorded. The superposition of two shearographic images results in the moiré picture associated to the $\partial z/\partial x$ distribution, where ∂z is the difference between the displacements of two points, and ∂x is the controlled displacement of the IR camera. Sound areas are more rigid, therefore, a specific interferometric picture, where pixel magnitudes are proportional to the displacements, appears in defect areas. In general the efficiency of shearography worsens for thick and rigid materials. However, by stimulating samples with a pressure gradient, it has been possible to detect fairly deep defects. The above-mentioned features of IR thermography and shearography allow developing such NDT hardware where the sample is optically stimulated, and sample response is analyzed by two channels: IR and interferometric. In Table 6.4, the comparison between two techniques is given in application to aero space structures to prove usefulness of combination of these techniques. Some prototypes of the commercial equipment implementing both IR thermography and shearography were developed by Laser Technology, USA.

Table 6.4 Comparison between thermal NDT and shearography in application to aero space

Material (defect)	TNDT	Shearography
Carbon and glass fiber reinforced plastics (thin)	++	+
Carbon and glass fiber reinforced plastics (thick)	+−	+−
Uncured laminates	++	+
Laminate/metal (high conductive) cells	++	+
Laminate/Nomex cells, foam or balsa core	−	+
Thin rubber on dense substrates	+	++
Thin rubber on lightweight substrates	−	++
Rubber laminates (tires)	+−	++
Metal facesheet/metal honeycombs	++	+
Shiny metal samples	−	+
Matte surfaces	+	+−
Trapped water in honeycomb	+	−
Honeycomb core damage	−	+
Core splices in metallic honeycomb panels	+	+−
Foreign material in laminates/composites	+	−
Thickness variation in honeycomb skin and core	+	+−
Kissing defects	+−	+
Porosity (depends on density)	+−	+

Adapted from [77]
+ Acceptable method
++ Best method
− Unacceptable method
+− Reasonable method

The combination of the techniques of the third type is related to a new approach to NDT in areas featured by variety and complexity of tested objects. It is assumed that data fusion means no simple summation of inspection results obtained with different NDT methods but produces a new inspection quality. Fusing data obtained with different sensors is well known in military. In NDT, the first research can be traced to 1995 [78], although many publications arrived in the later period [79, 80]. A short review of image fusion methods was recently done by Wronkowicz who also proposed the technique of fusion of RGB and IR images called Speeded Up Robust Features (SURF) [81]. This technique allows matching pairs of images by resizing and rotating.

In the aerospace industry, data fusion has been applied by combining three complementary techniques: ultrasonic, X-ray and shearographic [79]. Since these NDT methods are principally different, it has been believed that identification of test results is to be done by trained operators who are able to discriminate between defects and artifacts. While using these methods sequentially, financial and time losses increase considerably. This can be avoided by fusing data and interpreting resulting images properly. In [79], an automatic data fusion algorithm was described. The inspection procedure was based on comparing NDT results with a sample model produced according to Computer-Aided Design (CAD) and including the following steps:

- preliminary treatment of individual images in order to exclude their possible distortions,
- choosing informative signatures on the basis of a proper model in order to derive a set of basic parameters for further comparison between individual images and the computer model,
- detecting informative features in each individual image according to the set obtained on the computer model,
- primary comparison of features for preliminary matching of individual image features and those obtained on the computer model,
- final matching of features,
- decomposition of matched images for individual areas each of which is characterized by a particular structure,
- fusion of multiple image features in the wavelet domain (see Sect. 6.6).

Let us consider peculiarities of each data fusion step in short.

Image distortion may occur due to particular test geometry, as well as because of some factors of an electronic nature. Besides, in aero space, vast areas are to be inspected, therefore, special software for producing the so-called *mosaic* images may be needed. In the above-mentioned study [79], a honeycomb aircraft panel was covered with a single ultrasonic image; the X-ray image included 3 sub-regions, while for producing the shearographic image up to 32 sub-regions were inspected. The ultrasonic image was clearly related to panel thickness variations, and skin-cell delaminations were weakly detected, while the very honeycomb structure was scarcely seen. In its turn, the X-ray image being of a high spatial resolution demonstrated well a cell honeycombs structure but was practically insensitive to delaminations. Shearography proved to be efficient in detecting delaminations accompanied with cell damage.

Rectilinear segments which are invariant to perspective distortions were chosen as parameters of a 3D wireframe panel model. The chosen particular segments depended on each individual image where they were well distinguished. The detection of parameters was preceded by the following operations: filtration, deletion of insignificant details and binarization. Rectilinear segments were determined in binary images by applying the iterative Hough algorithm. Then, by using least-square fitting, a set of rectilinear segments derived from the individual images were compared with the set obtained on the wireframe model.

After each individual image obtained with a particular NDT technique was matched to the corresponding computer model, some separate areas differed by material, thickness, shape, etc. were picked up. For example, in a honeycomb panel, three areas were distinguished: a CFRP structure, a thick CFRP structure and transition areas.

The purpose of data fusion as an emerging NDT technique is performing analysis of inspection results in an integrated form to allow their simple and reliable identification. A data fusion technique is supposed to:

- keep information provided by each NDT method,
- prevent appearance of false defect indications,
- suppress noise,
- represent multi-parametric data in such way that they could easily be identified by a trained operator.

Final data fusion can be done pixel-by-pixel but this does not always provide optimal results. Specifically, data fusion can be fulfilled by using the discrete wavelet transform which enables presentation of image features with wavelet coefficients in a compact form. By merging these coefficients, it is possible to synthesize a final image.

Thermal images can be successfully complemented to optical, X-ray, ultrasonic and others. It is believed that particularly efficient may be fusion of IR and shearographic images obtained in a single inspection procedure. As mentioned above, object stimulation can be done by optical heating, while temperature and mechanical displacements can be monitored in parallel. The advantage of data fusion is that, while TNDT is more sensitive to shallow defects, shearography allows better examination of deeper flaws.

Another example of using data fusion in the inspection of impact damage in CFRP was reported in [80]. Results of IR thermographic and eddy current NDT reduced to a unique image size were compared pixel-by-pixel. Some data fusion algorithms were analyzed. The simplest algorithm involved choosing a pixel with maximum magnitude from two available images. More reliable results were obtained by using logical fusion (operator AND), as well as by averaging signal amplitudes in a mask chosen on both images.

A new approach to the evaluation of thermal wave phase images by using scatter plots for data fusion was proposed by Spiessberger et al. [82]. This approach enabled extraction of thermal features and defect identification regardless of their location in the original images. Additional information on lateral heat diffusion phenomena allowed mapping of defect edges and evaluation of reflection coefficients.

A simple example of data fusion is shown in Fig. 6.41. Optical stimulation of impact damage in CFRP reveals larger delaminations characterized by essential thermal resistance (Fig. 6.41a). In its turn, ultrasonic stimulation is more convenient for detecting minor delaminations, or "kissing" defects (Fig. 6.41b). By fusing both images, one may better visualize defect structure (Fig. 6.41c).

a) b) c)

Fig. 6.41 Data fusion in the inspection of impact damage in CFRP: (**a**) Optical stimulation, (**b**) ultrasonic stimulation, (**c**) fusion result

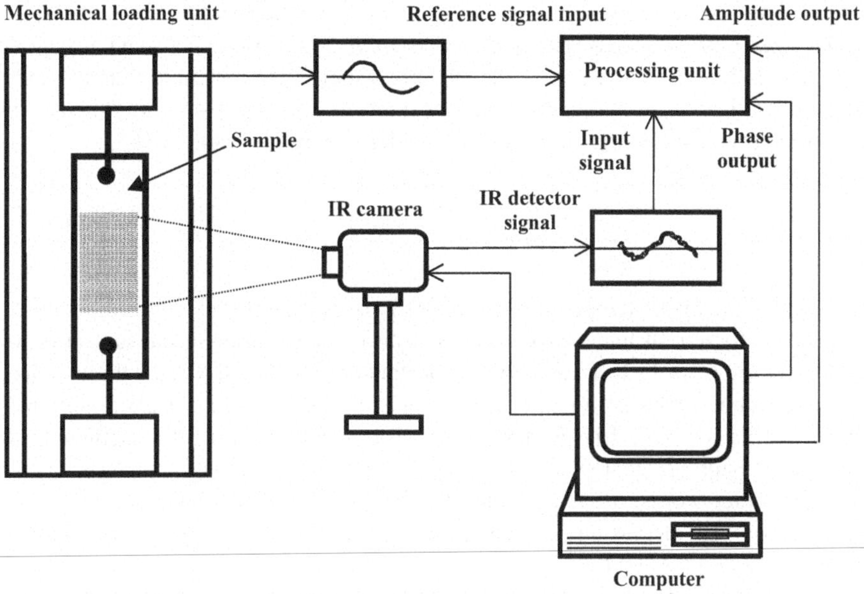

Fig. 6.42 Thermographic stress analysis (TSA). (Adapted from [74])

6.14 Thermomechanical Effects in Solids

Mechanical stretching and compressing, as well as damage of solids, are accompanied with temperature phenomena conditioned by transformation of mechanical energy into thermal. Heat is generated either due to mechanical hysteresis or plastic deformation which takes place when cracks appear and grow up. The first mechanism can be either integral or local following periodical loading of objects. Static (monotonous) loading does not cause essential temperature phenomena if it is applied within material elasticity limits. The second mechanism shows up when, because of concentration of mechanical stresses, areas of plastic deformation appear around material structural discontinuities (crack tips).

6.14.1 Vibrothermography and Thermoelasicity

If a periodical load operates within material elasticity limits and the stress rate is high, heat losses due to thermal conduction are negligible and the sample returns to initial shape and temperature after unloading. Such process is reversible. For example, in steel, temperature signals caused by thermoelastic deformations are about 0.001 °C under the load of 1 MPa. The corresponding NDT technique was called *vibrothermography*; similar physical principles have been also used in ultrasonic IR thermography.

In the 1990s, an implementation of this technique called *Stress Pattern Analysis by Thermal Emission* (SPATE), or *Thermographic Stress Analysis* (TSA), was proposed. It is based on the use of the thermoelasticity phenomenon which relates dynamic temperature variations and mechanical loading (see the scheme in Fig. 6.38) according to the classical formula derived by Thomson in 1853:

$$\frac{\delta T}{T_o} = \frac{\beta \delta \sigma}{\rho C_p} = K \delta \sigma, \tag{6.38}$$

where δT is the cyclic temperature variation, T_o is the sample absolute temperature, β is the coefficient of linear expansion, K^{-1}, $\delta\sigma$ is the variation of the sum of principle stresses, C_p is the heat capacity (at constant pressure), and K is the thermoelasticity constant, which is, for instance, equal to $0.86 \cdot 10^{-11}$ 1/Pa for aluminum and $28 \cdot 10^{-11}$ 1/Pa for steel. Elastic materials with positive values of the coefficient of linear expansion are characterized by negative values of the thermoelasticity constant; by other words, the temperature on the surface of such adiabatic materials drops down with increasing mechanical stress, therefore, Eq. (6.42) may contain a negative sign on the right.

An important advantage of the thermoelasticity analysis is that it allows the evaluation of stresses rather than deformations. This method cannot be applied to static processes, and the frequency range of measured stresses is from fractions of Hz to few kHz. Thermoelastic effects cause temperature signals in the range from 0.001 to 0.1 °C; these signals are determined by a sum of principle stresses σ. For example, in aluminum, mechanical stresses $\sigma = 2$ MPa correspond to the relative deformation of 5 μm/m.

Mechanical stresses are typically observed with differential IR imagers. Temperature oscillations which appear due to the thermoelastic phenomenon are determined by analyzing an alternating component of the temperature distribution. Sensitivity of differential IR imagers can be further enhanced by averaging signals in time or/and using a lock-in principle of detection. In the latter case, a reference signal is generated by a stress gauge resulting in two categories of images which are in-phase and out-of-phase in regard to the reference signal. A distribution of mechanical stresses is obviously associated to the corresponding in-phase image.

Since temperature variation is proportional to the sum of deformations, being invariant of their direction, the thermoelasticity analysis cannot provide information on separate deformation components, as well as deformation direction. It is seen from Eq. (6.38) that $\delta\ T$ is independent of a mean stress level. However, the recent research has shown that a thermoelasticity constant depends on stresses due to the relationship between the elasticity module and temperature [83].

A more precise form of the thermoelasticity equation, which connects temperature variation rate and stress in a homogeneous Hooke's material, was proposed by Wong et al. in 1988 (adiabatic thermal process) [84]:

$$\rho\ C_\varepsilon\ \frac{\partial T/\partial\tau}{T}=-\left[\beta+\left(\frac{\nu}{E^2}\ \frac{\partial E}{\partial T}-\frac{1}{E}\ \frac{\partial \nu}{\partial T}\right)\ I_1\right]\left(\partial I_1/\partial\tau\right)+ \\ +\left[\frac{\left(1+\nu\right)^2}{E^2}\ \frac{\partial E}{\partial T}-\frac{1}{E}\ \frac{\partial \nu}{\partial T}\right]\sum_{i=1}^{3}\sigma_{ii}\ \left(\partial\sigma_{ii}/\partial\tau\right), \tag{6.39}$$

where T is the thermodynamic temperature, K; σ_{ii} are the principle stresses, I_1 is the sum of the principle stresses, C_ε is the heat capacity, J/(kg·K), E is the Young's modulus, Pa, and ν is the Poisson coefficient. Unlike the classical Eq. (6.38), in the last expression, the modules of thermoelasticity are functions of time, therefore, the effective values of K are given with [84]:

$$K=\left(\beta-\frac{1}{E^2}\ \frac{\partial E}{\partial T}\ \sigma_m\right)\left(\rho C_\varepsilon\right)^{-1}, \tag{6.40}$$

where σ_m is the mean stress. An important conclusion which follows from Eq. (6.39) is that a temperature response of a material, that is subjected to harmonic loading with frequency of ω, contains two components with the frequencies ω and 2ω. The first harmonic component is a function of both cyclic load amplitude and mean load, while the second component is determined by the squared load amplitude. Hence, by analyzing both components, one can evaluate both cyclic load amplitude and mean load. Experimental results obtained in the framework of the revised thermoelasticity concept by using IR thermography were reported in [84].

The appearance of temperature gradients depends on a frequency of mechanical loading that leads to the concept of *local mechanical resonance* [85]. For example, at certain loading frequencies, delaminations in composites vibrate out of phase in regard to the host material, therefore, the analysis of the corresponding resonant frequencies may supply a useful information on defect size and depth.

6.14.2 Materials Destruction Caused by Energy Input

IR thermography has proven to be useful in analyzing processes of appearance and growth of cracks under both cyclic and static loading. An increase in a load enhances stress concentration on material micro-defects, then, after reaching a critical stress level, appeared micro-cracks start to unite in backbone cracks which cause ultimate materials fracture. On the tips of growing cracks, as well as in other areas of appeared plastic deformation, heat is generated being an indicator of both crack presence and direction of their propagation. The underlying theory and some experimental results are described in [85].

According to the Griffith crack theory, fracture is a process of "energetical" relaxation of a structure where appearance of cracks leads to changes in the total structure energy. The following parameters are introduced within this approach: the surface energy of fracture γ and the characteristic crack size l equal to the half of total crack length L. The basic Griffith formula determines the contribution of elastic energy ΔW to the system energy in respect to crack size:

$$\Delta W = -\pi\sigma^2 l^2 / E, \tag{6.41}$$

where E is the Young modulus and σ is the border stress.

A critical stress causing crack growth is determined by:

$$\sigma_{cr} = \sqrt{(2E\gamma)/(\pi l)}. \tag{6.42}$$

Assume that the fracture area is $S = LH$, where H is the crack depth and $\gamma = \Delta W/S$. Then, the relationship between crack length and generated energy is given by:

$$L = \frac{1}{\sigma_{cr}}\sqrt{\frac{4E\Delta W}{H}}. \tag{6.43}$$

Metals

Industrial installations are frequently subjected to periodical (sign-alternating) loading or deformation. Some parts of these installations can be of configuration which may cause such concentration of mechanical stresses in certain areas which may exceed an elasticity limit; as a result, the durability (lifetime) of these parts will worsen. Progressing degradation of strength because of periodical loading is called *fatigue*.

The energy of plastic deformation plays an important role in fatigue processes. As a rule, fatigue cracks begin to grow up in areas of surface defects. Therefore, the analysis of surface stresses and surface thermal phenomena is of a special interest. The IR thermographic aspect of this research has been developed by Kurilenko, Moyseychik et al. in Russia [86–88] and Luong in France [89].

Being subjected to cyclic loading, metals can collapse under stresses which are lower than in the case of static loads because of the fatigue phenomenon. Failure stresses decrease with more cycles according to the so-called "*S-N*" curve (stress amplitude *versus* the number of cycles to failure). By other words, material may keep their strength parameters only during a particular number of loading cycles, and afterwards they failure.

Information on strength parameters is often obtained by swirling a sample that causes maximum stresses in a material surface layer. Fatigue failure is a local process in the sense that a fatigue crack appears at a certain point and, while growing, may lead to sample failure even if the rest of the material keeps high strength.

Plastic deformation energy per one loading cycle ΔW is the area under the hysteresis curve, and the total energy for N cycles is $W_f = \sum_{i=1}^{N} \Delta W_i$. The phenomenon of fatigue damage is characterized by the energy of plastic deformation: each material absorbs energy until a certain limit, afterwards fatigue failure begins. In experiments with a controlled load, fatigue failure is often defined in terms of crack length. Oppositely, if deformation is controlled, fatigue failure is defined through a peak load.

It has been noted that the energy ΔW is weakly dependent on the number of cycles. However, even if cyclic loading is fully reversible, the deformation depends on sample pre-history, therefore, the ΔW value changes during sample lifetime. In softening materials, the magnitude of plastic deformation and the deformation energy ΔW per cycle grow up, while hardening materials are characterized by diminishing values of the above-mentioned parameters.

In crystals, fatigue cracks propagate along shear bands where stresses concentrate and temperature enhances. Initiation of a fatigue crack is accompanied by a temperature rise in the zone of shear bands. IR thermography enables evaluating quantitatively processes of appearance and growth of fatigue cracks, as well determining stress limits and developing recommendations on preventing sample collapse. It also allows localizing failure areas and follows their building up. IR thermographic analysis has been successfully used in investigation of fatigue failure of a number of materials, as well as in the detection of plastic deformation areas on the top of growing cracks in steel under monotonous loading [89]. In the same work, a general thermomechanical equation including the members responsible for thermal conduction, heat sources and sinks, as well as thermoelastic phenomena and energy dissipation by viscose/plastic phenomena, was proposed [89].

A practical study on the detection of fatigue cracks in steel bridges in field conditions have been fulfilled by Sakagami et al. [90] Steel bridges experience frequent and heavy wheel loadings conditioned by traffic. A mass of heavily-loaded trucks may exceed an allowed load thus endangering performance of bridge constructions. It was experimentally demonstrated that the TSA analysis enables the detection of cracks due to thermoleastic temperature elevations with the amplitude up to 1 °C appearing around crack tips Fig. 6.43). The temperature in non-defect points can serve as a reference allowing the use of the technique called self-reference lock-in thermography.

Example In collaboration with Sharkeev and Belyavskaya, the authors have fulfilled a comparative analysis of the failure of standard samples made of polycrystallic and nano-structured titanium. The samples were subjected to tensile tes on an Instron machine. The dynamic temperature distributions were recorded by means of a NEC Avio TH9100 IR imager at the acquisition frequency of 10 Hz. The imager ensured the temperature resolution of 60 mK and was placed at the distance of 0.4 m from the samples. The field of view was 17 × 12.8 with the pixel size of 0.5 mm. It has been found that, in accordance with the Thomson formula, deformation the curves correlated well with the temperature change in areas of damage. The analysis of IR thermograms has shown that the destruction of nano-titanium is characterized by a uniform distribution of multiple localized areas of maximum temperature corresponding to the distribution of failure points across the sample cross-section (Fig. 6.44a). Oppositely, in polycrystallic titanium appeared a single powerful zone of destruction (Fig. 6.44b).

Fig. 6.43 Thermal NDT of bridge welding [90]. (**a**) Test area photo; (**b**) temperature distribution in the area of fatigue crack

Fig. 6.44 Images of temperature derivatives at time of failure: (**a**) Nano-titanium, (**b**) polycrystallic titanium

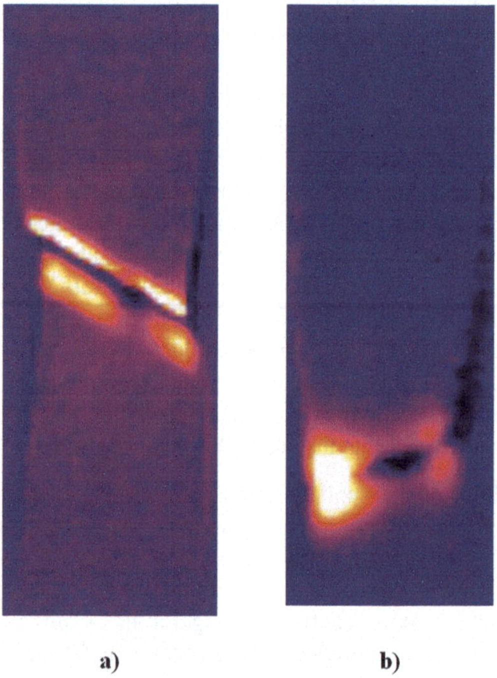

Building Materials

In the last decade, the world witnessed catastrophic collapses occurred in some building constructions, such as envelopes, tunnels, bridges etc. In many cases, it happened because of some external factors, such as tectonic processes, human activity, violation of exploitation norms, etc. In other cases, a fracture reason has not been identified but, from the point of failure mechanics, it was clearly related to deterioration of material structural integrity by involving the appearance, first, of microdefects, then macrodefects, which were growing up thus causing final failure. In practice, there is a great interest to finding precursors of such catastrophes. IR thermography, due to its non-contact operation principle and high inspection productivity, has attracted attention of institutions which are appealed to inspect potentially-dangerous systems in industry. However, potentials of IR thermography still remain disputable; experts have not yet agreed about a magnitude of temperature signals which appear on the surface and in the volume of building materials subjected to cyclic loading. The estimates made by the Thomson formula lead to negligible signals even in the case of total material destruction. But some laboratory experiments fulfilled by Luong have demonstrated that, under specific load conditions, temperature signals can reach few degrees Celsius [91]. In "live" constructions, this conclusion has not been confirmed, and some other fragmentary results (see Chap. 11) prompt that, under normal conditions, the magnitude of temperature precursors of catastrophes does not exceed some fractions of the Celsius degree. Therefore, detection of such signals is possible only in well-controlled laboratory experiments and is hardly possible in outdoor conditions, where the Sun, rain and external heat source may seriously corrupt inspection results.

Example In collaboration with Ilyushenov, the authors have conducted IR thermographic analysis of human bone damage under a static load. Ten 7 mm-thick samples of a thigh-bone were subjected to mechanical compression with a constant rate. Loading was applied perpendicularly to the sample surface by using a cylindrical 2.4 mm-diameter steel punch till sample crashed. The surface temperature distribution was captured with a Thermovision-570 IR camera with the frequency of six images per second. An example of temperature patterns on the surface of one of samples is presented in Fig. 6.45 (dashed line denotes the sample, dark areas to the left and to the right correspond to the press bites).

There are microcracks beginning to appear in the bone material with increasing stress. First, no visible temperature variations appear (see the first IR thermogram in Fig. 6.45). Then, in structurally-weak zones, micro-cracks are embodying into macro-cracks; this process is accompanied with energy generation and temperature elevation already at 1/6 s after applying a load; at this time, maximum excess temperatures occur in the fracture centres (13 and 7.7 °C signals respectively in the second image of Fig. 6.45). The available acquisition frequency has not allowed effective visualization of the process in time. Phenomenologically, already in the next IR thermogram, i.e. in 2/3 s after visible damage indications appeared, the temperature in the fracture centres began decaying that corresponded to a kind of 'energetical discharge' of the material (see two last images in Fig. 6.45). It is interesting to note the presence of a micro-cracks 'bridge' which connect two powerful fracture centres and is characterized by the weaker temperature rise of about 3 °C (Fig. 6.45).

Let us show that observed temperature signals cannot be determined by the mechanical hysteresis even if applying optimal cyclic loading. The bone thermal properties have been assumed identical to glass: $\rho = 2442$ kg/m^3, $C = 837$ J/(kg·K), $\lambda = 0.88$ W/(m·K) and $a = 0.43 \cdot 10^{-6}$ m^2/s. The elasticity module is $E_1 = 18$ GPa

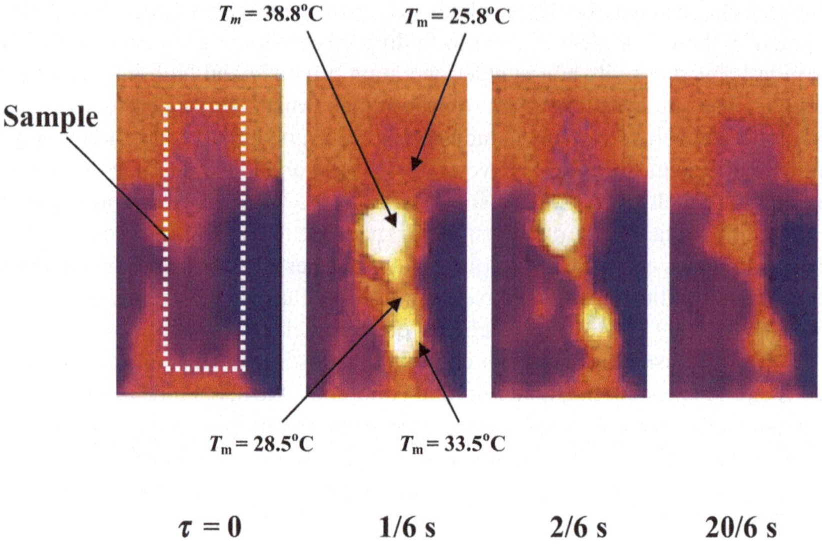

Fig. 6.45 Fracture of human bone (maximum temperatures in defect and non-defect areas are indicated)

(large human cannon-bone). Hence, a value of thermoelasticity constant can be evaluated as $K \sim 4.4{\cdot}10^{-12}$ 1/Pa. According to the published data, the fracture stress limit (under compression) is about 128 MPa [92]. Then, by Eq. (6.38), a temperature signal equivalent to bone fracture is $\Delta T \sim 0.17$ °C if the sample temperature is 300 K. In principle, such signals can be thermographically detected but they are significantly lower than those observed in the experiment.

Estimate L and γ by using the Griffith theory and assuming that, in one of fracture centres (see Fig. 6.39), the temperature rise was $\Delta T \approx 13$ °C at $\tau \approx 1/6$ s ($\sigma_{cr} = 128$ MPa). Suppose that the appearance of a crack is equivalent to the appearance of a pulsed point-like heat source of energy ΔW that causes the following temperature elevation:

$$\Delta T = \frac{\Delta W / \rho C}{4\left(\pi a \tau\right)^{3/2}} e^{-\frac{r^2}{4a\tau}}, \tag{6.44}$$

By using Eq. (6.44), it can be shown that, with the thermal properties accepted above, the temperature rise $\Delta T = 13$ °C in the point $r = 0$ at $\tau \approx 1/6$ s is stipulated by energy discharge $\Delta W = 11.4{\cdot}10^{-3}$ J. If to assume a crack depth equal to the sample thickness, i.e. $H = 7$ mm, Eq. (6.43) gives the estimate of crack size $L \sim 2.7$ mm. Respectively, the surface fracture energy is $\gamma \approx 0.6{\cdot}10^3$ J/m^2 or $6{\cdot}10^5$ erg/cm^2 that, by the order of magnitude, fits well the published data for materials like polymethylmethacrylate and polystyrene [93].

6.15 Electromagnetic IR Thermography

A technique of *ElectroMagnetic InfraRed* (EMIR) thermography was proposed by Balageas et al. in the 1990s [94]. A 'photothermal target' made of a dielectric film of the thickness from 10 to 25 μm and covered with very thin electrically-conductive coating absorbs microwave radiation (in the millimeter/centimeter range) and converts it into heat due to a photothermal effect. Thus, the temperature distribution, which is induced in the film and monitored by an IR imager, reflects the distribution of incident microwave radiation. The film temperature increase is estimated as:

$$\Delta T = \frac{W}{C\rho L + 2h}, \tag{6.45}$$

where W is the absorbed power density, C, ρ, L are the film thermal properties and thickness, h is the heat exchange on the film surface.

EMIR has been originally elaborated for evaluating energy distributions in powerful microwave beams (a similar method was developed in the former U.S.S.R. for analyzing laser beams). The implementation of a one-sided NDT procedure is shown in Fig. 1.2f. A temperature distribution induced on the surface of a film represents the superposition of the incident and reflected electromagnetic waves. Due to its physical principle, this method is applicable to a limited number of materials. Experimental results have been obtained in the detection of metallic inserts in CFRP and landmine surrogates buried in soil. In the latter case, a 150 W microwave source operating at the frequency of 2.45 GHz was used [95].

6.16 Eddy Current IR Thermography

The only difference of eddy current (induction) IR thermography from standard active TNDT is in the means of heating. According to the scheme of Fig. 1.2c, an inductor excites eddy currents in an electrically-conductive material. Thermal energy which is generated due to material electrical resistance propagates in the sample thus enabling defect detection like in the case of traditional TNDT. To realize a principle of induction lock-in thermography (ILT), a carrier frequency of eddy currents, which is few hundred kHz, is modulated with the frequency of 0.01–1 Hz. Inductor power can reach few kW. Recorded IR image sequences can be processed, for example, by using the Fourier transform to result in phasegrams with a signal-to-noise ratio higher than in the case of optical heating. Except metals, ILT is also applicable to CFRP composites and C/C-SiC ceramics.

Eddy current IR thermography has proven to be successful in the detection of cracks in compressor blades, toothed gear wheel and steel samples used in fatigue investigation [96–98]. It can be applied to both ferro-magnetic and non-magnetic materials allowing also the estimation depth of cracks [97]. Netzelmann et al. described the first use of induction IR thermography from a moving train in order to detect surface cracks in the rails at a train speed up to 15 km/h [98].

A range of identified material thicknesses (defect depths) depends on the modulation frequency and material properties, such as magnetic permeability and electrical conductivity. In metals, the penetration depth of eddy currents $l_{e.c.}$ is low due to the skin-effect and can be determined by [96]:

$$l_{e.c.} = \sqrt{\frac{2}{\pi f \mu k}} = \sqrt{\frac{2}{\pi f \mu_r \mu_o k}}, \quad (6.46)$$

where f is the cyclic frequency, μ_r is the relative permeability, $\mu_o = 4\pi \cdot 10^{-7}$ V·s/((A·m) is the magnetic constant (permeability of vacuum), k is the electrical conductivity, Sm/m. For example, assuming for steel $k = 10^6$ Sm/m and $\mu_r = 500$, at the frequency of 100 kHz we obtain $l_{e.c.} = 0.07$ mm that is significantly smaller than the corresponding thermal diffusion length. In CFRP, $l_{e.c.}$ may reach 50 mm ($k = 10^3$ Sm/m, $\mu_r = 1$, $f = 100$ kHz) that ensures volumic rather than surface heating. In ceramics, the presence of carbon fibers leads to strong anisotropy of k, therefore, IR thermograms exhibit well fiber orientation.

Another implementation of the ILT technique involves the heating by a moving inductor (Fig. 6.46). This approach was realized in TNDT of aviation honeycomb structures made of aluminum and boron. The penetration depth for eddy currents with the 100 kHz frequency was 5 mm for aluminum honeycombs and up to 300 mm for boron [99]. A typical vertical stabilizer of an F-15 airfighter includes a boron-epoxy skin and aluminum cells. Due to such combination of materials, IR thermograms clearly exhibit disbonds having` size of a single cell.

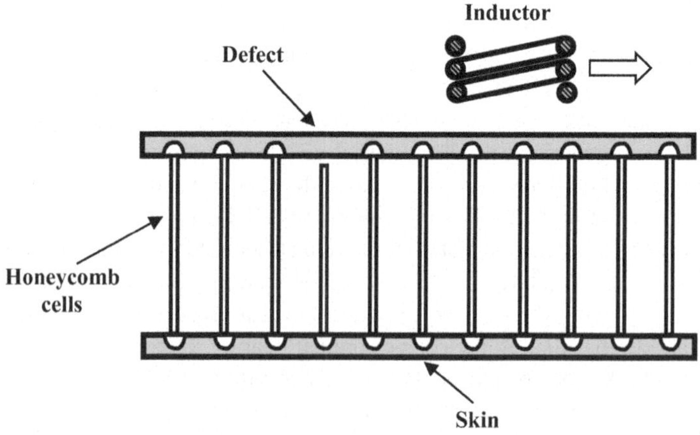

Fig. 6.46 Induction (eddy current) IR thermography (NDT of honeycomb structure)

6.17 Artificial Intelligence (Neural Networks) in TNDT

A neural network can be considered as a set of simple processors with multiple interconnections. The processors possess an internal memory acquired as a result of training and used for performing some simple operations. Typically, neural networks contain some layers; this approach makes the use of neural networks convenient for solving multi-parametric problems, including *pattern recognition*.

The scheme depicted in Fig. 6.47 shows how a single neuron percepts information. Such neuron has two inputs and one output, and the corresponding scheme is often called *perceptron*. The input signals $x_1, x_2, x_3, \ldots, x_n$ are converted into a scalar value y. The neuron multiplies input signals by weight coefficients $w_1 \ldots w_n$ and combines them into a linear function $S = w_1x_1 + w_2x_2 + \cdots + w_nx_n$. The output signal is regarded zero until its value is under the threshold y_{thr}. In fact, a threshold value activates the neuron: $y = f_{sig}(S - y_{thr})$, where f_{sig} is the sigmoid function, for example, $y = 1/[1 + \exp(-kx)]$. The neuron shown in Fig. 6.39 may be a structural element of different neural networks.

A process of network training involves determining weight coefficients. Two cases are possible: (1) each input data vector corresponds to a required or expected vector of output data, or (2) classification of an output data vector is performed by a neural network itself.

The structures like shown in Fig. 6.47 allow solving relatively simple problems. More advanced neural networks contain additional layers. As an example, Fig. 6.48 exhibits a three-layer network classified as a multi-layer perceptron. There are two hidden layers in this network which are not directly connected to the output. The

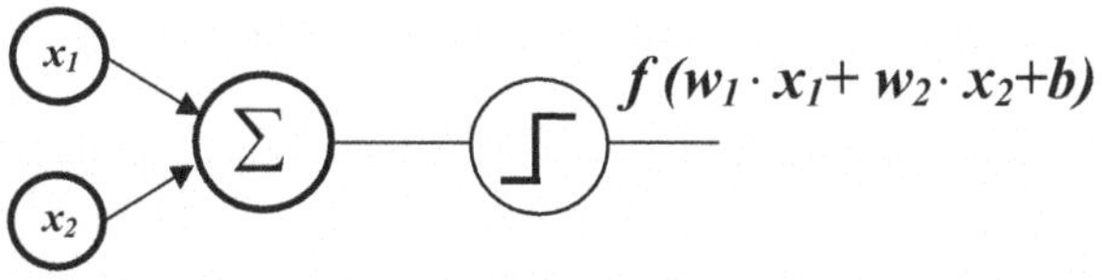

Fig. 6.47 Simple perceptron with two inputs and one output

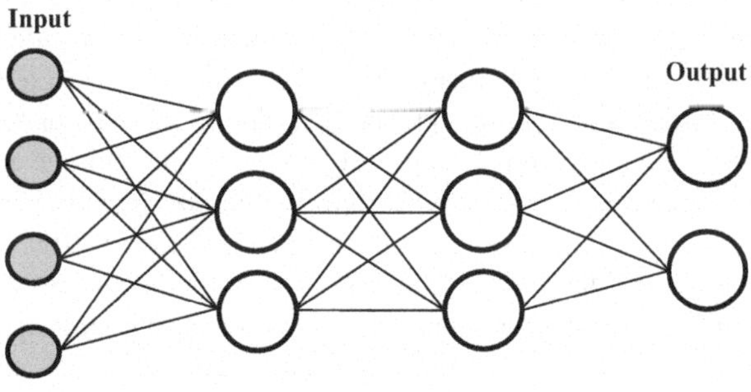

Fig. 6.48 Three-layer neural network

presence of hidden layers complicates neural network training because of difficulties in evaluating identification inaccuracies, in particular, when the activation function is non-linear. This problem is typically solved by using the so-called *back-propagation algorithm*. First, at the stage of propagation, or relaxation, some random values are assigned to weight coefficients, and the appeared input-output combination is being evaluated by the neural network. An error is first calculated as a squared signal difference for each output neuron and then it propagates backward through the network to be used for correcting weight coefficients. The process is repeated with a new input-output combination until the least-square error will become less than a prescribed level or after the prescribed number of iterations is completed.

The advantage of using neural networks for classifying defects is their ability of learning, i.e. sensing small differences between classes which are to be identified. This is achieved by training a network on a set of data which can be obtained experimentally or theoretically.

Intuitively, it is supposed that TNDT that is typically accompanied by plentiful noise represents a vast field for applying neural networks. They could be an effective tool for defect characterization; for example, the input data can be supplied by pixel values from corresponding maxigrams and timegrams, and the output can be expressed in defect depths.

Prabhu et al. studied feasibility of neural networks in NDT of aircraft panels [100, 101]. The work [100] describes the detection of disbonds in lap joints of aluminum sheets. Panels were heated with quartz lamps, next, the captured 256 × 256 IR thermograms were first averaged, then, temperature gradients in time (temporal derivatives) were computed. Gradient images may be of independent interest but a more efficient noise reduction has been achieved by applying a neural network with 20 input neurons and a single output neuron. A switch-off status of the output neuron (value 0.75) corresponded to sound areas and a switch-on status (value 0.25) specified defect areas. Binarization was done by the threshold equal to 0.5. It is interesting that the neural network was trained on experimental data (36 reference functions) obtained in the inspection of a commercial airplane. Since the training set was small, the training process was fast and the results obtained with the neural network matched well the results of ultrasonic inspection.

Two types of neural networks were used for detecting corrosion in aluminum panels. The first network called a "*flaw detector*" [101] was identical to that described in [100] and allowed detection of 25% material loss in a 3.1 mm-thick aluminum sheet which was lap-joined to another aluminum sheet of the same thickness (such defects are dangerous because of possible water storage and premature corrosion). The second network called a "*flaw estimator*" included 10 output neurons intended for classifying testing results by material loss: 0–10%, 10–20% etc. The most impressive results were obtained by combining the two networks of which the first one supplied a mask of defects to the second one. With this approach, material loss from 5% to 10% in a 1 mm-thick aluminum sample was successfully detected.

The neural network described in [102] used to process TNDT results by applying the Fourier transform. It was trained on calculated data, and experimental defect depths were evaluated by analyzing 32 images of both phase and magnitude. Up to 80 architectures of the neural network have been tried. A network with two layers produced significant errors in evaluating small defect depths because of small phase variations. By using the network with three layers, the characterization accuracy by defect depth was better than 10%.

6.18 Principal Component Analysis

Principal Component Analysis (PCA) introduced into TNDT by Rajic [103] is a useful statistical procedure which is being increasingly used in pattern recognition and data compression. Mathematically, it is often regarded as implementation of the so-called *singular values decomposition* (SVD technique which allows extracting spatial/temporal information from a matrix of source data. The idea of the technique is in calculating a *covariation* matrix of input data, as well as *eigenvectors* and *eigenvalues* of this matrix. Eigenvectors are placed by order of their magnitude thus supplying components of the analyzed statistical set. By neglecting low-order components, it is possible to reduce a problem metrics, for example, to exclude influence of convective heat exchange in multi-parametric thermal models. The most significant eigenvectors produce a feature vector which is a matrix column of eigenvectors. Finally, multiplying a transposed feature vector by the corresponding vector of input data, one obtains a new set of data expressed in terms of those eigenvectors which are recognized significant.

Let a source data be represented as a matrix A of whose size is $M \times N$ $(M > N)$. Decomposition is based on the following data presentation:

$$A = USV^T \tag{6.47}$$

where S is the diagonal matrix $N \times N$, and singular values A are diagonal, U is the matrix with the size of $M \times N$, V^T is the transposed matrix $N \times N$. The process by Eq. (6.47) is illustrated in Fig. 6.49. The columns of the U matrix represent a set of empirical orthogonal functions (EOF) which describe spatial variation of the source data. On the other hand, the so-called principal components (PC) which represent signal variations in time are the line values of the V^T matrix.

The PCA method is intended for extracting both spatial (EOF) and temporal (PC) peculiarities from IR thermographic sequences of source data (source matrix). A 3D matrix is first converted into the corresponding 2D matrix, where time readouts are placed in columns and spatial information is displaced as shown in Fig. 6.50a. Then, this 2D matrix is decomposed and a resulting matrix can be represented again as a sequence of principal component images (Fig. 6.50b).

A PCA procedure looks like follows. First, the covariation matrix of source data is calculated along with its *eigenvectors* and *eigenvalues*. Eigenvectors are placed

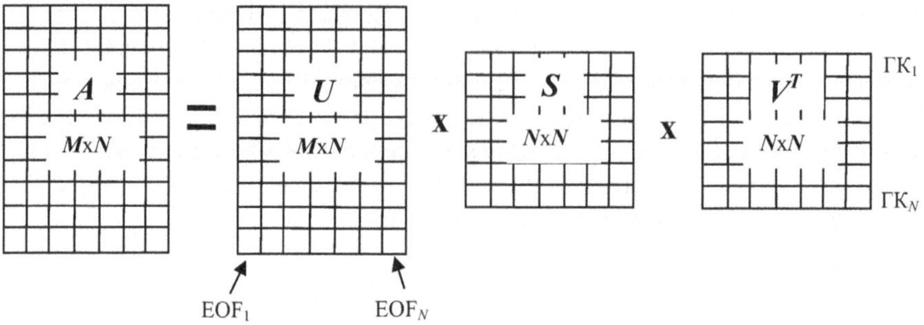

Fig. 6.49 Decomposition of matrix A singular values by matrices U, S and V^T

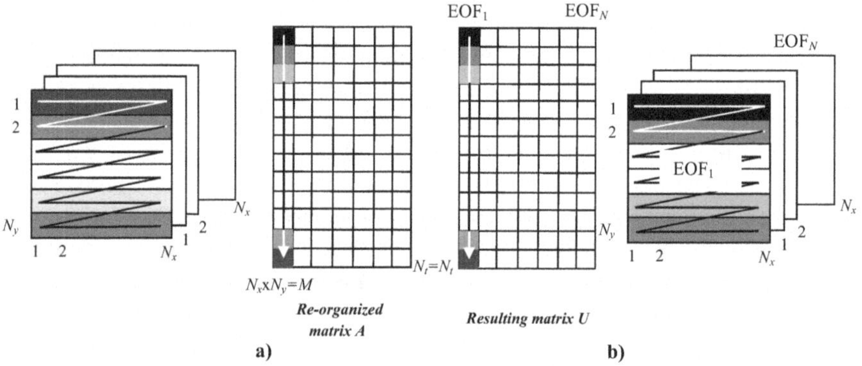

Fig. 6.50 Converting IR thermographic data from 3D matrix into 2D matrix A (**a**) and subsequent conversion of A into the 3D matrix containing empirical orthogonal functions – EOF (**b**)

by order of their significance; therefore, by neglecting some minor eigenvectors, it is possible to reduce the problem dimension. By choosing the most significant eigenvectors, the so-called *feature vector* can be derived. This vector represents a matrix consisted of a column containing eigenvectors. Finally, by multiplying the transposed feature vector by the vector of source data, it is possible to obtain a new set of data expressed in terms of those eigenvectors which are considered significant.

In criminalistics, the PCA method is used for identifying human faces. The comparison of current face features is done by analyzing discrepancies between current and reference eigenvectors, thus essentially accelerating identification.

PCA is being increasingly used in TNDT where its efficiency is close to that of pulse phase thermography and may be recommended as a first-try technique when processing experimental data. For example, it was successfully used for evaluating porosity of CFRP composites used in aircraft [104]. The preliminary data treatment was performed by applying the 2D wavelet transform to the time domain data. The porosity was evaluated indirectly by determining thermal diffusivity in each sample point. At ONERA, France, this method was used for identifying results of the industrial inspection of metallic samples by using the "flying spot" technique.

In the automated mode, the reliability of detection of open cracks was enhanced due to PCA of input data [105]. Recently, a novel approach based on choosing random combinations of meaningful PC images was proposed by Gavrilov and Maev [106]. Normally, PCA is fairly efficient at the stage of defect detection while its applicability to defect characterization is questionable.

6.19 TNDT of Objects Having Complicated Shape

Most TNDT procedures have been developed in the assumption that objects to be tested are of a plane geometry. When inspecting samples of complicated shape, some specific problems may appear: (1) no access to some sample zones, (2) inadequate heating in shadowed areas, and (3) appearance of false indications caused by multiple reflections. It is known that emission and absorption of thermal radiation is maximal in the normal direction to the object surface (in diffuse emitters, this fact is expressed by Lambert's law). Identification of IR thermograms becomes difficult when heating and monitoring objects at steep angles of view. For example, the detection of *Defect B* shown in Fig. 6.51 is less reliable than in the case of the identical *Defect A*. This requires applying some correcting measures, such as point-like heating, videothermal stereometry, proper data calibration, etc. Maldague developed a technique of shape restoration called *Shape-from-Heating*. It requires no additional hardware and employs principles of machine vision [107]. Object shape is determined by the first images in a sequence where defect indications have not yet appeared. In fact, such technique is close to data normalization discussed in Sect. 6.11. Data processing is done by three stages: segmentation, shape identification and correction.

Recently, a deeper analysis of thermal excitation conditions taking into account distribution of heating energy, as well as distance between a heater and a test sample and orientation of a heater was performed by Mayre and Hendorfer [108].

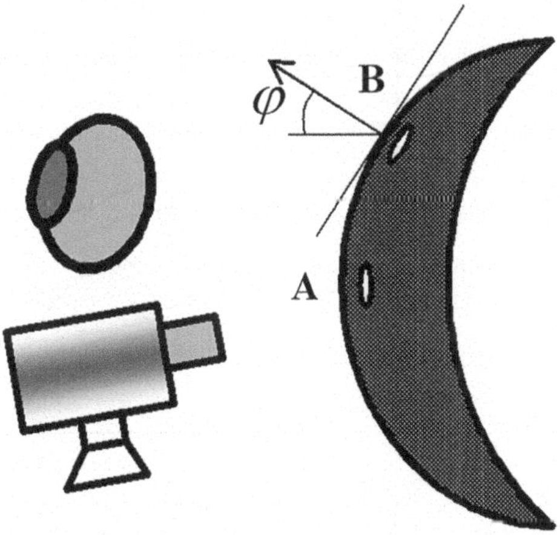

Fig. 6.51 TNDT of objects of a complicated shape

6.20 Correlation Technique

A correlation technique is intended for detecting similarity between temperature evolutions at a reference point and each point of sample surface. A reference point is to be chosen by the thermographer although its location is not crucial as, for instance, in the case of maxigrams/timegrams. The resulting correlation image shows pixel-based values of the correlation coefficient $r(i,j)$ according to the formula:

$$r(i,j)=\frac{\sum_{k=1}^{N}T(i,j,k)T\left(i_{ref},j_{ref},k\right)-\frac{1}{N}\sum_{k=1}^{N}T(i,j,k)\cdot\sum_{k=1}^{N}T\left(i_{ref},j_{ref},k\right)}{\sqrt{\left\{\sum_{k=1}^{N}T^{2}(i,j,k)-\frac{1}{N}\left[\sum_{k=1}^{N}T(i,j,k)\right]^{2}\right\}\cdot\left\{\sum_{k=1}^{N}T^{2}\left(i_{ref},j_{ref},k\right)-\frac{1}{N}\left[\sum_{k=1}^{N}T\left(i_{ref},j_{ref},k\right)\right]^{2}\right\}}},\qquad(6.48)$$

where $T(i,j,k)$ is the pixel value in the k-th image and $T(i_{ref},j_{ref},k)$ is the pixel value in a reference point and N is the number of images in the sequences analyzed.

For example, in the ThermoFit Pro software (Tomsk Polytechnic University, Russia), there are four types of correlation used for enhancing defect visibility: (1) between a current and reference pixel profiles, (2) between each profile and the profile calculated by applying the corresponding heat conduction solution, (3) self-correlation, and (4) autocorrelation. The principle of correlation is illustrated with Fig. 6.52. Assume that a sample is unevenly heated, and defect indications T_{d1} and T_{d2} appear at different non-defect temperature levels, e.g. T_{nd1} and T_{nd2}. It is obvious that the correlation coefficient $r(i, j)$ between T_{nd1} and T_{nd2} profiles will be close to unity while the most significant difference will take place between T_{d1} and T_{nd1} and/or T_{nd2}, or T_{d2} and T_{nd1} and/or T_{nd2}. As a result, if a reference point is chosen in a defect free area, this area will be characterized by $r(i, j) \sim 1$ while defect areas will exhibit lower values of $r(i, j)$. It is worth noting that variations in $r(i,j)$ values

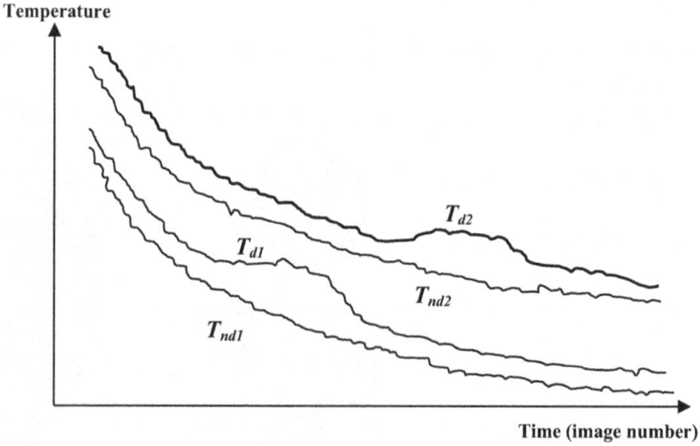

Fig. 6.52 Temperature evolutions in defect T_d and defect-free T_{nd} areas (correlation technique)

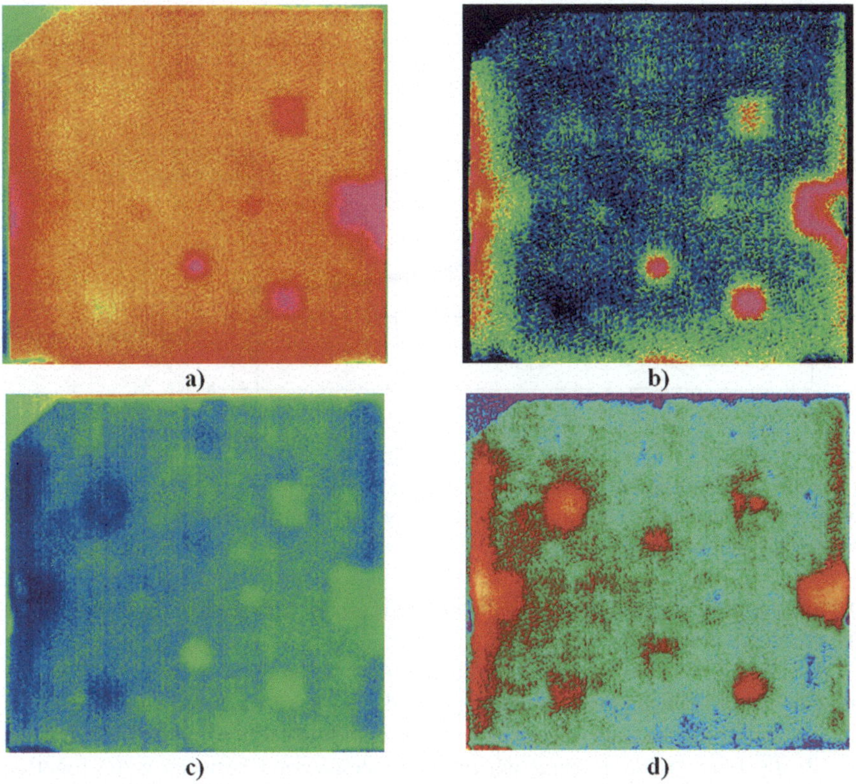

Fig. 6.53 Correlation technique in TNDT of a CFRP reference sample with multiple defects of different nature embedded between composite plies: (a) Raw image, (b) correlation in respect to a reference point, (c) autocorrelation image, (d) self-correlation image

because of defect presence can be enhanced by choosing shorter time intervals where temperature deviations caused by defects are observed best of all. As an additional option, a non-defect profile can be calculated theoretically, for example, by using the classical solution which describes heating a plate with a square pulse. Very often, correlation images provide essential enhancement of defect visibility as illustrated in Fig. 6.53, see also [109]. A similar approach called cross-correlation maximization technique was described in application to TNDT of brazing joints [110].

The self-correlation technique involves the pixel-by-pixel calculation of the correlation between the each pixel profiles and the corresponding "approximation profile" at the same pixel. This algorithm is illustrated by Fig. 6.54. The thermographer has to place a few points (five points in Fig. 6.54) on raw temperature curve and choose how many polynomial coefficients are to be used for approximating the original curve (black solid line) with a smoothed (approximation) curve shown with the dashed line. Then the correlation between these two curves is calculated. The chosen points for approximation are automatically and uniformly distributed along a pixel-based profile. In this way, some defect signals (thin solid line) may appear

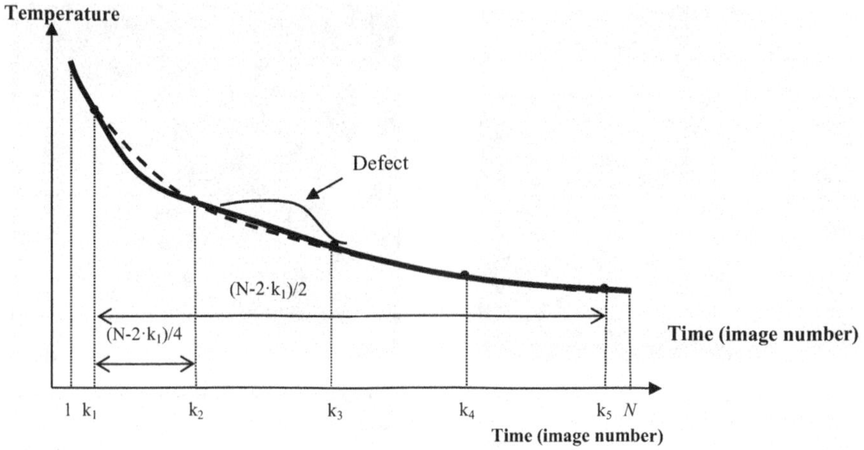

Fig. 6.54 Self-correlation algorithm (N is the number of images in a sequence, k_1 is the shift of the first point from the first image; $k_3 = (N–2·k_1)/2$ is chosen in the middle of the sequence; $(N–2·k_1)/4$ is the distance between points)

between approximation points (as shown in Fig. 6.54) thus producing deviations in the corresponding correlation values.

A technique of autocorrelation requires no reference point because $r(i,j)$ values are calculated between points of each particular temperature profile separated by a chosen time interval. A classical autocorrelation function is defined by the $r = \int_{-\infty}^{\infty} f(t) f(t-\tau) dt$ integral which shows connection between a $f(t)$ function and the same function shifted in time by τ.

An autocorrelation function is calculated as the correlation between two sequences $\boldsymbol{x}$ and $\boldsymbol{y}$, where $\boldsymbol{x} = x_1, x_2, x_3, \ldots, x_{n-t}$ and $\boldsymbol{y} = y_1 + t, y_2 + t, y_3 + t, \ldots, y_n$; here t is the time interval (shift) and n is the number of function readouts (profile points). First, one calculates, pixel by pixel, mean values $T_1(x,y)$ starting from the first significant point (image) and up to the image $T_{n-a}(x,y)$, then mean values of $T_a(x,y)$ are determined starting from the shifted image and up to the last image $T_n(x,y)$. Finally, the autocorrelation function r_a is determined as follows:

$$r_a = \frac{j = \sum_{j=1}^{n-a} \left(T_{j+a} - \bar{T}_a\right) \cdot \left(T_j - \bar{T}_1\right)}{\sum_{j=a}^{n} \left(T_j - \bar{T}_{j+a}\right)^2 \cdot \sum_{j=1}^{n-a} \left(T_j - \bar{T}_1\right)^2}, \tag{6.49}$$

where $\bar{T}_1 = \frac{\sum_{j=1}^{n-a} T_1}{n-1}$, $\bar{T}_a = \frac{\sum_{j=a}^{n} T_j}{n-a}$.

The self-correlation image in Fig. 6.53d provides the best visibility of embedded defects in the CFRP sample, however, peculiarities of this technique are not explored.

6.21 Standard IR Image Processing

The processing algorithms described above are based on the analysis of temporal signal evolutions $T\ (i,j,\tau)$ in each pixel; hence, they are 1D and applied to image sequences. The main purpose of data treatment is the enhancement of signal-to-noise ratio defined for object areas which are regarded as either sound or defective. In some cases, this purpose can be achieved by processing single images thus dealing with pixel morphology rather than with their time histories. The image processing theory is well-developed, and below we will describe in short only general algorithms widely used in TNDT. Following the general theory, we define "IR image processing" as the *manipulation* and *analysis* of IR images by means of a digital computer. Philosophically, it should be stated that any treatment "distorts" raw images, therefore, only such processing is advisable which introduces signal distortions that are "useful" from the user's point of view.

A digital image is composed of a finite number of elements *called picture elements*, *image elements*, *pels*, or *pixels*. Pixels can be clearly seen by zooming an image. Image size may vary from 128 × 128 to 1024 × 1024; until now, most popular IR image formats are 320 × 240, 640 × 480 and 640 × 512 (Fig. 6.55). It is clear that larger formats ensure better image quality at a constant viewangle. Pixel amplitude can be represented either *arithmetically*, for example, in terms of true temperature, or in the *Byte* form; in the latter case, the number of bits per pixel can be 8; 12; 14 or higher. Such images are also called *intensity* images, and their elements have integer values in a particular range, e.g. from 0 to 65,535 in the case of the 16-bit format. The higher is the number of bits per pixel, the larger is a dynamic range of an IR imager. Both image format and pixel digital "depth" define a size of a file occupied by each individual image, or image sequence in total. For example, if 1000 images by size of 150 kB each are recorded during a test, the total size of the data to be processed is 150 MB, and this may cause some computational difficulties.

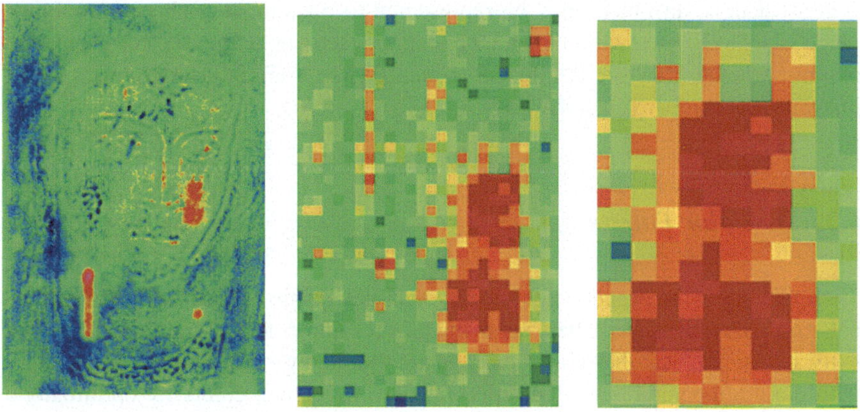

Fig. 6.55 Zooming a digital image reveals its pixel structure

Image processing involves the following stages:

- data digitization and compression,
- enhancement, restoration and reconstruction,
- matching, description and recognition.

6.21.1 Image Enhancement

The definition of image quality depends on a purpose for which such image is produced (for instance, a utilitarian look at fuzzy pictures of impressionists may prompt the necessity of their processing with high-pass filters in order to make images sharper. Therefore, image enhancement procedures include such processing algorithms which make images more acceptable to end-users. The statistical goal of NDT is a search for certain patterns (defect indications) which are observed on a noisy background. Therefore, image quality is often evaluated by special test objects which contain areas of step- and/or grid-wise signal variations (such reference objects are typical in optics). An example of a grid-wise test object used in TNDT was presented in Sect. 4.8.11.

6.21.2 Histogram Processing

In contemporary IR cameras, a full signal can be digitized by up to 14 bits, however, each particular image shows pixel amplitudes in a rather narrow range in order to reveal more lower amplitude details. To achieve optimal data presentation, a histogram of pixel values should be stretched between a minimum and maximum values in the area of interest. Often, images contain areas of abnormally high (spikes) or low (background) signals which bear no useful information. In order to view low-contrast details, an image may be represented to operator in a chosen amplitude range which is located within the total histogram (see Fig. 6.56). The corresponding options, whether they are automatic or manual, are often built in commercial IR imagers or accompanying software.

6.21.3 Color Palette

Thermal images are pseudocolor because the relationship between temperature and color is defined by a user. There are many palettes implemented into IR imagers and software. In practice, the typical palettes are "Rainbow" and "Iron". Sample areas with close temperatures can be represented with contrast colors by optimizing a

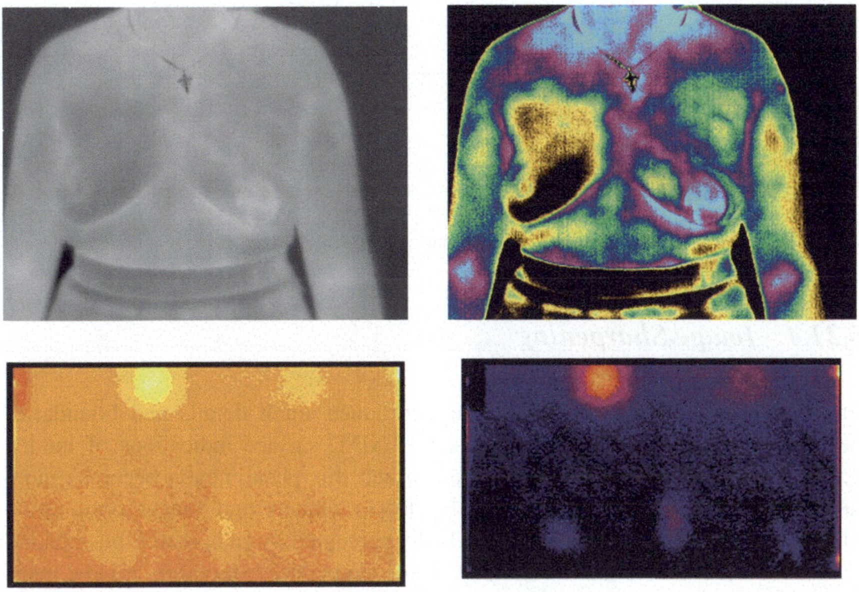

Fig. 6.56 Image enhancement by stretching histogram and optimizing color palette (medical IR thermographic diagnostics and TNDT of CFRP)

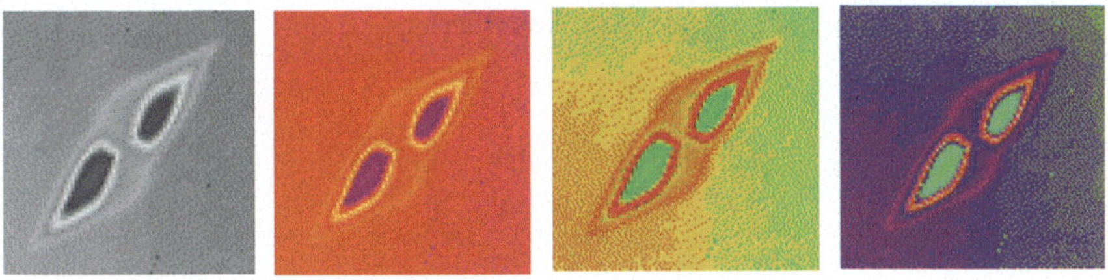

Fig. 6.57 Using color palettes in the TNDT of impact damage in CFRP

palette (see Figs. 6.56 and 6.57), but importance of color thermogram presentation should not be exaggerated, in particular, if one deals with defect detection. Very contrast color images, sometimes called "cock tail", are badly percepted by an operator if they contain a variety of small spatial details. Therefore, at the initial stage of IR thermography, many operators used to prefer black and white (B&W) images. A special study performed by Vavilov (see Chap. 7) has shown that reliability of defect detection weakly depends on how an IR image is represented to an operator, including images in different color palettes and their B&W replicas. To some extent, the "Iron" palette can be considered as a compromise between color and B&W thermogram presentations. Defect maps are presented in a binary format (see below), and some binary-like color palettes may make TNDT results very illustrative.

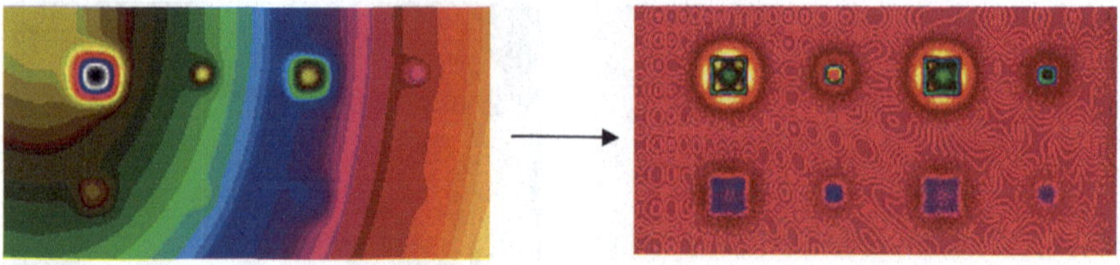

Fig. 6.58 Image sharpening by applying high-pass filter (artificial image, specific image pattern is due to peculiarities of numerical calculations)

6.21.4 Image Sharpening

Image *sharpness* defines the ability to reproduce small details and boundaries between areas of different amplitudes. In TNDT, surface indications of hidden defects are smoothed due to heat diffusion and this phenomenon becomes more intensive for high-conductive materials, deeper defects and longer observation times. Blurred temperature patterns are badly percepted by an operator thus making reliable defect characterization difficult. Since temperature smoothing can be interpreted as integration by spatial coordinates, it is assumed that images can be sharpened by applying differentiation. In fact, images are often processed with high-pass filters which selectively pass signals with high gradients and suppress low-frequency trends caused by uneven heating, reflections, material inhomogeneities, etc. (see in Fig. 6.58). An example of a simple high-pass filter is:

$$3\times 3: \quad \begin{vmatrix} -1 & -1 & -1 \\ -1 & +9 & -1 \\ -1 & -1 & -1 \end{vmatrix}, \tag{6.50}$$

where 3 × 3 is the mask size. Filtration by Eq. (6.50) is 2D because it is fulfilled by two spatial coordinates. Similar but 1D filters can be applied to the temporal coordinate when processing image sequences. It is worth noting that spatial filtration does not imply automatic filtration in time and *vice versa*.

A drawback of using simple high-pass filters is a possible change of a derivative sign. This inconvenience is overcome by subtracting from a source image the Laplacian $\nabla^2 T = \frac{\partial^2 T}{\partial x^2} + \frac{\partial^2 T}{\partial y^2}$, which is a linear differentiating operator invariant of direction. In a digital form, this filter is defined as:

$$T(i,j) - \nabla^2 T(i,j) = 5T(i,j) - \left[T(i+1,j) + T(i-1,j) + T(i,j+1) + T(i,j-1)\right]. \tag{6.51}$$

Unfortunately, high-pass filters tend to enhance high-frequency noise. In TNDT, it is recommended to apply such filters to smooth noisy temperature evolutions. In the case of the 1D temporal coordinate, data fitting ensures smooth presentation of noisy data, while 2D spatial fitting is cumbersome and has not been used in TNDT.

Another type of filters underlining low contrast-signals is a *statistical filter* which calculates signal dispersion within an area limited by a chosen mask.

6.21.5 Image Smoothing

Smoothing is an operation reverse to sharpening. It is typically used if an image contains high-frequency noise unpleasant for visual perception. Such noise indications are to be smaller than sought defects not to "smash" defect indications. Smoothing can be performed by applying some types of filters among which the most common are: (1) low-pass filter, (2) median filter, and (3) temporal averaging.

An example of a low-pass filter:

$$3\times3:\ \frac{1}{9}\cdot\begin{vmatrix}1&1&1\\1&1&1\\1&1&1\end{vmatrix};\quad 5\times5:\ \frac{1}{25}\cdot\begin{vmatrix}1&1&1&1&1\\1&1&1&1&1\\1&1&1&1&1\\1&1&1&1&1\\1&1&1&1&1\end{vmatrix}. \tag{6.52}$$

The larger is a mask size, the greater is an area where temperature is being smoothed; masks larger than 12 × 12 are rarely used because they may seriously distort a source image.

Low- and high-pass filters change an image histogram, while a median filter leaves it untouched. This filter is quite efficient in eliminating single-pixel ("salt and pepper") noise which used to be typical in analog-to-digital converters (ADC) of a low quality. Let us illustrate median filtering in 1D implementation: if pixel readouts are placed in time as 35 255 74, then a median filter will re-arrange them in the following order: 35 74 255.

Temporal averaging improves temperature resolution because this process diminishes uncorrelated ("white") noise but worsens temporal resolution.

6.21.6 Boundary Enhancement

Boundary enhancement is a typical problem of image segmentation which is being solved by using gradient filters. A 2D spatial Sobel filter determines gradients in a gray-level source image. Applying such filter results in a dark image which contains clearly-seen indications of maximum signal gradients.

For a 3 × 3 mask

$$\begin{vmatrix}P_1&P_2&P_3\\P_4&P_5&P_6\\P_7&P_8&P_9\end{vmatrix},$$

a practical formula for computing gradients is:

$$G=\left|\left(P_1+2P_2+P_3\right)-\left(P_7+2P_8+P_9\right)\right|+\left|\left(P_3+2P_6+P_9\right)-\left(P_1+2P_4+P_7\right)\right|. \tag{6.53}$$

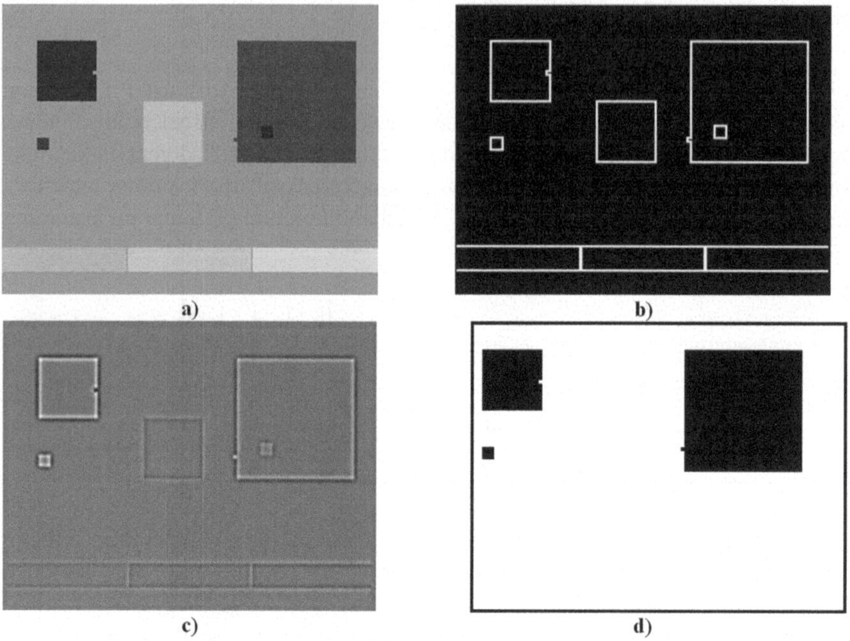

Fig. 6.59 Image segmentation. (a) Test image, (b) after processing with Sobel filter, (c) after processing Laplacian of the Gaussian function, (d) threshold binarization

The example of applying a Sobel filter to the source image from Fig. 6.59a is shown in Fig. 6.59b.

The Laplacian is also an efficient gradient filter but it is quite sensitive to noise. Sometimes, the combination of the Laplacian and the Gaussian (LoG filter) can be used. The Gaussian is:

$$h(r) = e^{-\frac{r^2}{2\sigma^2}}, \tag{6.54}$$

where $r^2 = x^2 + y^2$, or $r^2 = i^2 + j^2$.

The filter mask which is to be convoluted with a source image appears as a result of applying the second derivative (Laplacian) to the Gaussian:

$$\nabla^2 h(r) = -\left[\frac{r^2 - \sigma^2}{\sigma^4}\right] e^{-\frac{r^2}{2\sigma^2}}. \tag{6.55}$$

Convolving an image with $\nabla^2 h(r)$ is equivalent to, first, smoothing with the Gaussian and, second, computing second derivatives thus leading to a new image where area boundaries are represented with double lines. An exact boundary locates between these lines, and it can be found by using a filter which senses zero crossings.

A LoG filter is often zero-centered:

$$LoG(x,y) = -\frac{1}{\pi\sigma^4}\left[1-\frac{x^2+y^2}{2\sigma^2}\right]e^{-\frac{x^2+y^2}{2\sigma}}. \tag{6.56}$$

A mask of the LoG filter for $\sigma = 1.4$ is given below; the example of its application is illustrated with Fig. 6.59c.

0	0	3	2	2	2	3	0	0
0	2	3	5	5	5	3	2	0
3	3	5	3	0	3	5	3	3
2	5	3	−12	−23	−12	3	5	2
2	5	0	−23	−40	−23	0	5	2
2	5	3	−12	−23	−12	3	5	2
3	3	5	3	0	3	5	3	3
0	2	3	5	5	5	3	2	0
0	0	3	2	2	2	3	0	0

A simple method to enhance borders is binarization. Either a global or local (floating) threshold can be chosen from the histogram; then, all pixels which are located to the right from the threshold are represented, for example, by white color (unity value), respectively, all pixels to the left are shown in black (zero value). An example of binarization is shown in Fig. 6.59d.

6.21.7 Morphological Filtration

Mathematical morphology is a tool for extracting image components, such as boundaries, skeletons and the convex hull. In its classical form, morphological filtration is a logical operation which being applied to binary images changes configuration of regions of interest. Pixels are modified within the so-called *structuring element* which can be of different size and shape; the simplest element is a 3 × 3 mask. Morphological treatment is based on the combination of two fundamental operations called "*erosion*" and "*dilation*". Dilation "grows" or "thickens" objects, if a structuring element covers at least one unity pixel, because there is unity in the center of a structuring element. Respectively, erosion "shrinks" or "thins" objects because a structuring element adds zero pixels to objects if this element covers at least one zero pixel of the background. Erosion converts a unity pixel into zero, if this pixel is surrounded by few zeros, therefore, this operation effectively removes single pixel noise. Since erosion shrinks edges of unity areas (Fig. 6.60), it is often followed by dilation which "eats" up boundary pixels but does not restore eliminated single pixels.

Morphological filtration can be also applied to gray level images. In this case, dilation means the allocation in the centre of a structuring element of a maximum value from that image section which is covered with the element. Correspondingly, erosion allocates a minimum value.

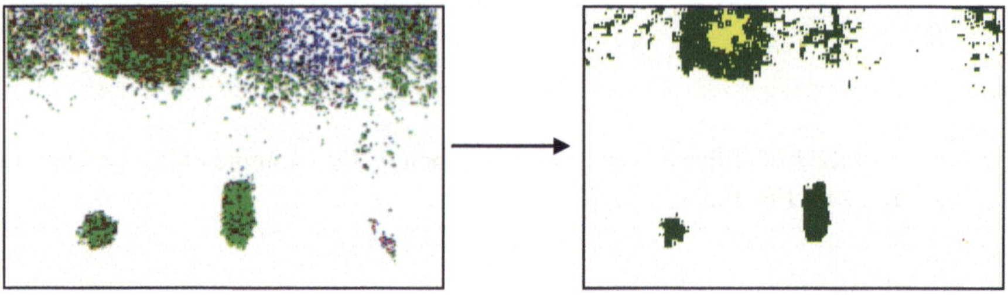

Fig. 6.60 Morphological filtration (erosion)

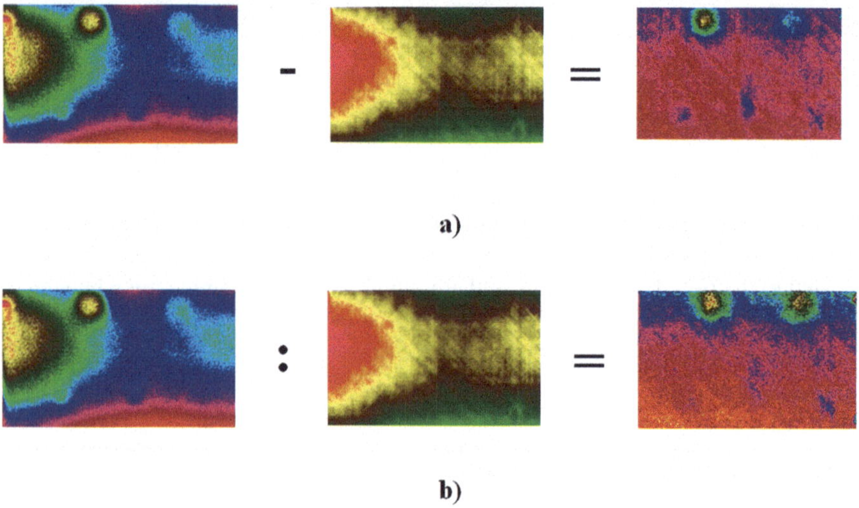

Fig. 6.61 Image subtraction (**a**) and division (**b**) while detecting Teflon inserts in CFRP sample

6.21.8 Image Restoration

The purpose of restoration is to improve a given image in a particular predefined sense. Image restoration means matching it under a certain mathematical model, e.g. in TNDT this is often done by fitting temporal evolutions, see Sect. 6.10.

6.21.9 Image Subtraction and Division

In some cases, even simple arithmetic operations, such as subtraction and division can significantly reduce the noise which adheres to many TNDT procedures (Fig. 6.61). For instance, subtraction of two thermograms is efficient in the removal of additive noise (reflected radiation) which is present in both images to be processed. In its turn, division reduces multiplicative noise caused by emissivity/absorptivity fluctuations. In this case, one has to deal with excess temperatures and reconcile with increasing high-frequency noise.

References

1. Aamodt, L.C., Maclachlan Spicer, J.W., Murphy, J.C.: Analysis of characteristic thermal transit times for time-resolved infrared radiometry studies of multilayered coatings. J. Appl. Phys. **68**(12), 6087–6098 (1990)
2. Krapez, J.-C., Balageas, D.L.: Early detection of thermal contrast in pulsed stimulated infrared thermography. In: Proc. Eurotherm Seminar No. 42, "Quant. IR Thermography QIRT-94", Sorrento, Italy, August 23–26, 1994, pp. 260–266
3. Krapez, J.-C., Balageas, D.L., Deom, A., Lepoutre, F.: Early detection by stimulated infrared thermography. comparison with ultrasonics and holo/shearography. In: Maldague, X. (ed.) Advances in Signal Processing for Non Destructive Evaluation of Materials NATO ASI Series, Series E: Applied Sciences, vol. 262, pp. 303–321. Kluwer Academic Publishers, Dordrecht (1994)
4. Pikalov, V.V., Melnikova, T.S.: Plasma Tomography. Nauka Publishers, Novosibirsk (1995)., 210 p. (in Russian)
5. Vorontsov, S.S., Pikalov, V.V., Preobrazhensky, N.G., et al.: The algorithm for projections self-matching in IR thermographic tomography. In: Automation of Aerodynamic Experiments, pp. 63–74. Institute of Theoretical and Applied Mechanics, Russian Academy of Sciences, Novosibirsk (1985). (in Russian)
6. Akhmetov, V.D., Fateev, N.V.: Infrared tomography of the lifetime and diffusion length of charge carriers in semi-conductive silicon ingots. Phys. Tech. Semicond. **35**(1), 40–47 (2001). (in Russian)
7. Barrett, A., Myers, P.S., Sadowsky, N.L.: Detection of breast cancer by microwave radiometer. Radio Sci. **12**(68), 167–171 (1977)
8. Troyitski, V.S.: To the theory of contact radiothermometric measurement of inner body temperature. Izv. Vuzov Radiophys. Ser. **24**(9), 1054–1058 (1981). (in Russian)
9. Pasechnik, V.I., Anosov, A.A., Barabanenkov Yu, N., et al.: Determining in-depth temperature of biological objects by using passive acoustical thermo-tomography. In: Proc. Nizhny Novgorod Acoustical Scientific Session, pp. 375–378. NNGU (2002). (in Russian)
10. Rosenzwaig, A., Gersho, A.: Thermal-wave imaging. Science. **218**, Washington, DC, American Association for the Advancement of Science, 223–228 (1982)
11. Mulaveesala, R., Tuli, S.: Theory of frequency modulated thermal wave imaging for nondestructive subsurface defect detection. Appl. Phys. Lett. **19**(19), 191913 (2006)
12. Mandelis, A., Mieszkowski, M.: A thermal wave sub-surface defect imaging and tomography apparatus. U.S. Patent No. 4950897 (1990)
13. Mandelis, A.: Theory of photothermal wave diffraction tomography via spatial Laplace spectral decomposition. J. Opt. Soc. Am. **A6**, 298 (1991)
14. Mandelis, A.: Theory of photothermal wave diffraction tomography via spatial Laplace spectral decomposition. J. Phys. A: Math. Gen. **24**, 2485 (1991)
15. Munidasa, M., Mandelis, A., Ferguson, C.: Resolution of photothermal tomographic imaging of sub-surface defects in metals with ray-optic reconstruction. Appl. Phys. **54**, 244–250 (1992)
16. Pade, O., Mandelis, A.: Computational thermal wave tomography of aluminum with ray-optic reconstruction. Rev. Sci. Instrum. **64**, 3548 (1993)
17. Kaiplavi, S., Mandelis, A., Wang, X., Feng, T.: Photothermal tomography for the functional and structural evaluation, and early mineral loss monitoring in bones. Biomed. Opt. Express. **5**(8), 2488–2502 (2014)
18. Busse, G., Renk, F.: Stereoscopic depth analysis by thermal wave transmission for NDE. Appl. Phys. Lett. **42**(4), 366–368 (1983)
19. Nowakowski, A., Kaczmarek, M.: Active dynamic thermography – problems of implementation in medical diagnostics. Quant. Infrared Thermogr. J. **8**(1), 89–106 (2011)
20. Milovanović, B., Pečur, I.B.: Review of active IR thermography for detection and characterization of defects in reinforced concrete. J. Imaging. **2**(11), 27 p (2016)
21. Vavilov, V.P., Shiryaev, V.V.: Thermal tomography. USSR Patent No. 1266308 (1984) (in Russian)

22. Vavilov, V.P., Jin, H., Thomas, R., Favro, L.: Experimental thermal tomography of solids by using pulsed one-sided heating. Defectoscopiya (Russ. J. NDT). **12**, 122–128 (1990). (in Russian)
23. Vavilov, V., Maldague, X.: Dynamic thermal tomography: new promise in the IR thermography of solids. Proc. SPIE "Thermosense-XIV". **1682**, 194–206 (1992)
24. Vavilov, V., Bison, P.G., Bressan, C., et al.: Some new ideas in dynamic thermal tomography. In: Proc. Eurotherm Seminar No. 27, "Quant. IR Thermography QIRT-92", Chatenay-Malabry, France, July 7–9, 1992, pp. 259–255
25. Kush, D.V., Rapoport, D.A., Budadin, O.N.: Inverse problem of automated thermal NDT. Defectoscopiya (Soviet J. NDT). **5**, 64–68 (1988)
26. Vavilov, V.P., Kuimova, M.V.: Dynamic thermal tomography of composites a comparison of reference and reference-free approaches. J Nondestr Eval. **38**(1), 36 (2019)
27. Maldague, X., Marinetti, S.: Pulse phase infrared thermography. J. Appl. Phys. **79**, 2694–2698 (1996)
28. Couturier, J.-P., Maldague, X.: Pulsed phase thermography of aluminum specimens. Proc. SPIE "Thermosense-XIX". **3056**, 170–175 (1997)
29. Vavilov, V.P., Marinetti, S.: Pulsed phase thermography and tomography by using the Fourier transformation. Defectoscopiya (Russ. J. NDT). **2**, 58–72 (1999). (in Russian)
30. Ibarra-Castanedo, C., Maldague, X.: Defect depth retrieval from pulsed phase thermographic data on Plexiglas and aluminum samples. Proc. SPIE "Thermosense-XXVI". **5405**, 348–365 (2004)
31. Vavilov, V.P., Shirayev, V.V., Kuimova, M.V.: Time- and phase-domain thermal tomography of composites. Photonics. **5**(4), 31 (2018)
32. Sun, J.G.: Method for thermal tomography of thermal effusivity from pulsed thermal imaging. US Patent No. 7365330 (2006)
33. Sun, J.G.: Method for implementing depth deconvolution algorithm for enhanced thermal tomography 3D imaging. U.S. Patent No. 8465200 (2013)
34. Sun, J.G.: Quantitative three-dimensional imaging of heterogeneous materials by thermal tomography. J. Heat Transfer. **138**, 112004–1–10 (2016)
35. Three-Dimensional Thermal Tomography (3DTT) Advances Cancer Treatment: US Dep. Energy, Argonne National Lab. U.S.A (Sept. 2012), 2 p
36. Harris, A.W., Drube, L.: Thermal tomography of asteroid surface structure. Astrophys. J. **832**(2), 15 p (2016)
37. Levkovich-Masliuk, L.: Wavelet-analysis digest. Computerra. **8**, 31–37 (1998)
38. Galmiche, F., Vallerand, S., Maldague, X.: Wavelet transform applied to pulsed phase thermography. In: Grinzato, E., Bison, P., Mazzoldi, A. (eds.) Proceedings of the V-th Workshop on Advances in Infrared Technology and Applications, pp. 117–122. CNR, Venice (1999)
39. Galmiche, F., Maldague, X.: Depth defect retrieval using the wavelet pulse phased thermography. In: Proc. Eurotherm Seminar No. 64, "Quant. IR Thermography", Reims, France, July 18–21, 2000, pp. 194–199
40. Grinzato, E., Vavilov, V.: Corrosion evaluation by thermal image processing and 3D modeling. Rev. Gen. Therm. **37**(8), 669–679 (1998)
41. Kanstad, S.O., Nordal, P.E.: Experimental aspects of photothermal radiometry. Can. J. Phys. **64**, 1155–1164 (1986)
42. Carlomagno, G.M., Berardi, P.G.: Unsteady thermophototopography in nondestructive testing. In: Proc. 3rd Biannual Exchange, U.S.A., St. Louis, 1976, pp. 33–39
43. Busse, G., Wu, D., Karpen, W.: Thermal wave imaging with phase sensitive modulated thermography. J. Appl. Phys. **71**, 3962–3965 (1992)
44. Wu, D., Salerno, A., Malter, U., et al.: Inspection of composites by using lockin thermography. In: Proc. Eurotherm Seminar No. 27, "Quant. IR Thermography QIRT-96", Châtenay-Malabry, France, July 7–9, 1992, pp. 251–254
45. Beaudoin, J.L., Marienne, E., Danjoux, R., Egee, M.: Numerical system for infrared scanners and application in the subsurface control of materials by photothermal radiometry. Proc. SPIE "Infrared Technology and Applications". **590**, 287 (1985)

46. Kuo, P.K., Feng, Z.J., Ahmed, T., et al.: Parallel thermal wave imaging using a vector lock-in video technique. In: Hess, P., Pelzl, J. (eds.) Photoacustic and Photothermal Phenomena, pp. 415–418. Springer-Verlag, Heidelberg (1987)
47. Zweschper, T., Dillenz, A., Riegert, G., Busse, G.: Thermography with excitation by elastic waves: comparison of techniques (pulse, burst, lockin). In: Proc. Int. QIRT Conf. "Quant. IR Thermography", Dubrovnik, Croatia, Sept. 24–27, 2002, pp. 47–52
48. Laloue, P., Nigon, F., Offermann, S., et al.: Photothermal infrared thermography applied to the characterization of thermal resistances at metal-metal interfaces. In: Proc. Eurotherm Seminar No. 64, "Quant. IR Thermography", Reims, France, July 18–21, 2000, pp. 206–211
49. Dillenz, A., Zweschper, T., Busse, G.: Elastic wave burst thermography for NDE of subsurface features. Insight. **42**(12), 815–817 (2000)
50. Zweschper, T., Dillenz, A., Busse, G.: Ultrasound lock-in thermography – a defect selective method for the inspection of aerospace components. Insight. **43**(3), 173–179 (2001)
51. Zweschper, T., Dillenz, A., Busse, G.: Ultrasound lockin thermography – an NDT method for the inspection of aerospace structures. In: Proc. Eurotherm Seminar No. 64, "Quant. IR Thermography", Reims, France, July 18–21, 2000, pp. 212–217
52. Dillenz, A., Zweschper, T., Busse, G.: Phase angle thermography with ultrasound burst excitation. In: Proc. Eurotherm Seminar No. 64, "Quant. IR Thermography", Reims, France, July 18–21, 2000, pp. 247–252
53. Favro, L.D., Han, X., Ouyang, Z., et al.: IR imaging of cracks excited by an ultrasonic pulse. Proc. SPIE "Thermosense-XXII". **4020**, 182–185 (2000)
54. Burke, M.W., Miller, W.O.: Status of VibroIR at Lawrence Livermore National Laboratory. Proc. SPIE "Thermosense-XXVI". **5405**, 313–321 (2004)
55. Solodov, I., Bai, J., Bekgulyan, S., Busse, G.: A local defect resonance to enhance acoustic wave-defect interaction in ultrasonic nondestructive evaluation. Appl. Phys. Lett. **99**(21), 211911 (2011)
56. Solodov, I., Bai, J., Busse, G.: Resonant ultrasound spectroscopy of defects. Case study of flat-bottomed holes. J. Appl. Phys. **113**(22), 223512 (2013). https://doi.org/10.1063/1.4810926
57. Solodov, I., Rahammer, M., Gulnizkiy, N.: Kreutzbruck. Noncontact sonic NDE and defect imaging via local defect resonance. J Nondestr Eval. **35**(3), 48–48 (2016)
58. Vernotte, P.: Mesure de la conductibilite thermique des isolants. methode de toushau. Chaleur Ind. **208**, 331–337 (1937). (in French)
59. Shepard, S.: Advances in pulsed thermography. Proc. SPIE "Thermosense-XXIII". **4360**, 511–515 (2001)
60. Delpesh, P., Krapez, J.-C., Balageas, D.L.: Thermal defectometry using the temperature decay rate method. In: Proc. Eurotherm Seminar No. 42, "Quant. IR Thermography QIRT-94", Sorrento, Italy, August 23–26, 1994, pp. 220–225
61. Shepard, S.M., Lhota, J.R., Rubadeux, B.A., Wang, D., Ahmed, T.: Reconstruction and enhancement of active thermographic image sequences. Opt. Eng. **42**, 1337–1342 (2003)
62. Shepard, S., Lhota, J., Hou, Y., Ahmed, T.: Gold standard comparison of thermographic sequence data. Insight. **46**(4), 210–213 (2004)
63. Balageas, D.L., Roche, J.-M.: Common tools for quantitative time-resolved pulse and step-heating thermography – part I: theoretical basis. Quant. Infrared Thermogr. J. **11**, 43–56 (2014)
64. Balageas, D.L., Roche, J.-M.: Common tools for quantitative time-resolved pulse and step-heating thermography – part II: experimental investigation. Quant. Infrared Thermogr. J. **12**(1), 1–23 (2015)
65. Roche, J.-M., Leroy, F.-H., Balageas, D.L.: Images of TSR coefficients: a simple way for a rapid and efficient detection of discontinuities. Mater. Eval. **42**, 73–82 (2014)
66. Pawar, S.S., Vavilov, V.P.: Applying the heat conduction-based 3D normalization and thermal tomography to pulsed infrared thermography for defect characterization in composite materials. Int. J. Heat Mass Transf. **94**, 56–65 (2016)
67. Gruss, C., Lepoutre, F., Balageas, D.: Nondestructive evaluation using a flying-spot camera. In: Proc. 8th Int. Thermo Conf., Budapest, Hungary, June 2–4, 1993, 6 p

68. Bison, P.G., Bragiotti, A., Bressan, C., et al.: Crawling spot thermal NDT for plaster inspection and comparison with dynamic thermography with extended heating. Proc. SPIE "Thermosense-XVII", Orlando, USA. **2423**, 145–152 (1995)
69. Storozhenko, V.A., Vavilov, V.P., Volchek, A.D.: Nondestructive testing of industrial products by using active thermal method. Tekhnika Publisher, Kiev (1988)., 128 p
70. Cramer, K.E., Winfree, W.P.: Thermographic detection and quantitative characterization of corrosion by application of thermal line source. Proc. SPIE "Thermosense-XX". **3361**, 291–297 (1998)
71. Woolard, D., Cramer, K.: The thermal photocopier: a new concept for thermal NDT. Proc. SPIE "Thermosense-XXVI". **5405**, 366–373 (2004)
72. Khodayar, F., Lopez, F., Ibarra-Castaneda, C., Maldague, X.: Parameter optimization of Robotized line scan thermography for CFRP composite inspection. J Nondestr Eval. **37**(1), 5–9 (2018)
73. Shirayev, V.V., Vavilov, V.P., Ivanov, A.I.: The method of active thermal nondestructive testing. U.S.S.R. Patent No. 1075131 (1984)
74. Lesniak, J.R., Boyce, B.R.: Differential thermography applied to structural integrity assessment. Proc. SPIE "Thermosense-XVII". **2473**, 179–187 (1995)
75. Ritter, R., Schmitz, B.: Photothermal inspections of adhesion strengths and detection of delaminations. In: Proc. Eurotherm Seminar No. 27, "Quant. IR Thermography QIRT-96", Châtenay-Malabry, France, July 7–9, 1992, pp. 251–254
76. Horny, N., Henry, J.-F., Offermann, S., et al.: Photothermal infrared thermography applied to the identification of thin layer thermophysical properties. In: Proc. Eurotherm Seminar No. 64, "Quant. IR Thermography", Reims, France, July 18–21, 2000, pp. 36–41
77. Burleigh, D.: A portable, combined thermography/shearography NDT system for inspecting large composite structures. Proc. SPIE "Thermosense-XXIV". **4710**, 578–587 (2002)
78. Gros, X.E., Strachan, P., Lowden, D.W., Edwards, I.: NDT data fusion. In: Proc. 6th European Conf. NDT, vol. 1, 1994, pp. 355–364
79. Shark, L.K., Matuszewski, B.J., Smith, J.P., Varley, M.R.: Automatic feature-based fusion of ultrasonic, radiographic and shearographic images for aerospace NDT. Insight. **43**(9), 607–615 (2001)
80. Gros, X.E., Bousigue, J., Takahashi, K.: NDT data fusion at pixel level. NDT E Int. **32**, 283–292 (1999)
81. Wronkowicz, A.: Automatic fusion of visible and infrared images from different perspective for diagnostics of power lines. Quant. Infrared Thermogr. J. **13**(2), 155–169 (2016)
82. Spiessberger, C., Gleiter, A., Busse, G.: Data fusion of lockin-thermography phase images for innovative nondestructive evaluation. Quant. Infrared Thermogr. J. **6**(2), 149–161 (2009)
83. Gyekenyesi, A.L.: Testing static and dynamic stresses in metallic alloys using thermoelastic stress analysis. Mater. Eval. **60**(3), 445–451 (2002)
84. Wong, A.K., Sparrow, J.G., Dunn, S.A.: On the revised theory of the thermoelastic effect. J. Phys. Chem. **49**, 395–400 (1988)
85. Tenek, L.H., Henneke, E.G. II: Flaw dynamics and vibrothermographic-thermoelastic NDE of advanced composite materials. In: Proc. SPIE "Thermosense-XIII", 1991, pp. 252–259
86. Kurilenko, G.A.: Testing and prediction of individual fatigue resistance in mechanical engineering by using the kinetics of passive thermal distributions. D.Sc. thesis, Novosibirsk State Technical University (2000), 402 p. (in Russian)
87. Moyseychik, E.A., Vavilov, V.P., Kuimova, M.V.: Infrared thermographic testing of steel structures by using the phenomenon of heat release caused by deformation. J Nondestr Eval. **37**(2), 1–10 (2018)
88. Moyseychik, A.E., Vavilov, V.P.: Analyzing patterns of heat generated by the tensile loading of steel rods containing discontinuity-like defects. Int. J. Damage Mech. **27**(6), 950–960 (2018)
89. Luong, M.P.: Infrared thermography of fatigue in metals. Proc. SPIE "Thermosense-XIV". **1682**, 222–232 (1992)

90. Sakagami, T., Yzumi, T., Mori, N., Kubo, S.: Development of self-reference lock-in thermography and its applications to remote nondestructive inspection of fatigue cracks in steel bridges. In: Proc. 10th International Conference on Quantitative InfraRed Thermographlecture – QIRT'2010, Quebec, Canada, Sept. 2–7, 2010, 7 p
91. Luong, M.P.: Infrared thermographic evaluation of fatigue behavior of concrete. In: Transact. 14th Int. Conf. Structural Mech. in Reactor Technol. (SMiRT 13), Lyon, France, Aug. 17–22, 1997, pp. 155–162
92. Obraztsov, I.F. (ed.): Strength Problems in Biomechanics. Vysshaya Shkola Publishers, Moscow (1988)., 310 p. (in Russian)
93. Yankovski, V.E., Goryainov, O.P., Klevno, V.A., et al.: Micro-failure of bone tissue. Orthop. Traumatol. Prosthet. **10**, 9–12 (1991). (in Russian)
94. Balageas, D., Levesque, P.: EMIR: a photothermal tool for electromagnetic phenomena characterization. Rev. Gen. Therm. **37**, 725–739 (1998)
95. Balageas, D., Levesque, P.: Mines detection using the EMIR method. In: Proc. Quant. IR Thermography QIRT'02, Dubrovnik, Croatia, Sept. 24–27, 2002, pp. 71–78
96. Riegert, G., Zweschper, T., Busse, G.: Lockin thermography with eddy current excitation. Quant. Infrared Thermogr. J. **1**(1), 21–32 (2004)
97. Oswald-Tranta, B.: Induction thermography fro surface crack detection and depth determination. Appl. Sci. **8**(2), 257 (2018)
98. Netzelmann, U., Walle, G., Ehlen, A., Lugin, S., Finckbohner, M., Bessert, S.: NDT of railway components using induction thermography. AIP Conf. Proc. **1706**, 150001 (2016)
99. McCullough, R.W.: Transient thermographic technique for NDI of aerospace composite structures. Proc. SPIE "Thermosense-XXVI". **5405**, 390–401 (2004)
100. Prabhu, D.R., Howell, P.A., Syed, H.I., Winfree, W.P.: Application of artificial neural networks to thermal detection of disbonds. In: Thompson, D.O., Chimenti, D.E. (eds.) Review of Progress in Quantitative Nondestructive Evaluation, vol. 11, pp. 1331–1338. Plenum Press, New York (1992)
101. Prabhu, D.R., Winfree, W.P.: Neural network based processing of thermal NDE data for corrosion detection. In: Thompson, D.O., Chimenti, D.E. (eds.) Review of Progress in Quantitative Nondestructive Evaluation, vol. 12, pp. 1260–1265. Plenum Press, New York (1993)
102. Maldague, X., et al.: A study of defect depth using neural networks in pulsed phase thermography: modelling, noise, experiments. Rev. Gen. Therm. **37**(2), 708–716 (1998)
103. Rajic, N.: Principal component thermography for flaw contrast enhancement and flaw depth characterization in composite structures. Compos. Struct. **58**, 521–528 (2002)
104. Ciliberto, A., Cavaccini, G., Salvetti, O., et al.: Porosity detection in composite aeronautical structures. Infrared Phys. Technol. **43**, 139–143 (2002)
105. Hermosilla-Lara, S., Joubert, P.-I., Placko, D., et al.: Enhancement of open-cracks detection using a principal component analysis/wavelet technique in photothermal nondestructive testing. In: Proc. Int. Conf. Quant. Infrared Thermography QIRT'02, Dubrovnik, Croatia, Sept. 24–27, 2002, pp. 41–46
106. Gavrilov, D., Maev, R.G.: Extraction of independent structural images for principal component thermography. Appl. Sci. **8**, 459–474 (2018)
107. Maldague, X.: Nondestructive Evaluation of Materials by Infrared Thermography, 440 p. Springer, London (1993)
108. Mayre, G., Hendorfer, G.: Characterization of defects in curved carbon fiber reinforced plastics using pulsed thermography. Quant. Infrared Thermogr. J. **7**(1), 3–16 (2010)
109. Klein, M.T., Ibarra-Castaneda, C., Bendada, A., Maldague, X.P.: Thermographic signal processing through correlation operators in pulsed thermography. Proc. SPIE "Thermosense-XXX". **6939**, 693915-1 (2008)
110. Sriyubol, T., Nontakaew, U., Chantrasmi, T.: Analyzing sizing defects in brazing joints using pulse phase thermography with a cross-correlation maximization technique. Quant. Infrared Thermogr. J. **13**(2), 144–154 (2016)

Chapter 7
Basics of Thermal Radiation

Abstract This chapter introduces the basic physics of thermal (IR) radiation as it relates to the analysis of IR thermography and solving the challenging emissivity problem. The possible use of dual-band IR thermography is discussed. The influence of atmosphere absorption and spurious reflected radiation on the results of TNDT is discussed.

7.1 Short Review of IR Technology and IR Thermography

In the scientific literature, the credit to the discovery of infrared (IR) energy in 1800 is paid to Sir William Herschel who was a royal astronomer for King George III. Hershel worked on the protection of his eyes against sunlight and observed the highest temperature increase, by using a thermometer, actually beyond red light. The existence of invisible thermal rays had been hypothesized by Titus Lukretius Carus (c. 99-c. 55 BCE), a Roman poet and the author of the philosophic epic "De Rerum Natura" ("On the Nature of the Universe"). In 1696, Della Porta, an Italian observer, noted that when a candle was lit and placed in front of a large silver bowl in a church, he could sense the heat on his face. Hoffmann appears to have been the first who collected the invisible heat of a stove into a focus by the reflection of one or more concave mirrors. Afterwards, in 1770, Pictet, a French scientist, published the description of his famous experiment on focusing heat and cold. Well before Hershel's research, Lomonosov, a founder of the Russian science, had been working on the problem of night vision. In 1758, Lomonosov invented a night vision telescope (*tubo nyctoptico*), and in 1762 he developed a mirror with high reflectivity. In 1829, Nobili proposed the first thermocouple based on the thermoelectrical effect which was discovered by Seebeck in 1821. Few years later, Melloni succeeded to detect the presence of a person by thermal radiation from the distance of 3 m. A prototype of IR imagers was an evaporograph proposed by John Hershel, William's only son, who focused with a lens solar radiation onto a suspension of carbon particles in alcohol. It is remarkable that in 1840 J. Hershel called a thermal image "thermogram", the term still in use today. In 1857, Svanberg invented a new type of IR detectors called the bolometer. Improvement of this detector allowed Longley detecting animals at distances up to 400 m. In 1892, Dewar proposed the construction of a vessel for keeping liquefied gases which became an important part of IR

V. Vavilov, D. Burleigh, *Infrared Thermography and Thermal Nondestructive Testing*, https://doi.org/10.1007/978-3-030-48002-8_7

detectors for the next years. A year later, Golitzyn introduced the term "radiation temperature". In 1895, Lebedev who became famous for the discovery of light pressure invented a vacuum thermoelement which appeared to be for a long time the most sensitive IR detector. Another Russian physicist Stoletov developed a gas-filled photo-cell which used the principle of the photoemissive effect, or the external photoeffect (1898). The second birth of thermal sciences dates back to 1900 thanks to fundamental studies of Planck, Einstein, Kirchhoff, Golitzyn, Wien and others, from one hand, and due to quick progress in IR detector technology, on the other hand. The patent literature of the beginning of the XXth century contains numerous ideas on using IR sensing devices for detecting naval ships, airplanes and personnel, as well as for developing tracking and communication systems. The first operating systems were developed in 1914–1918. The equipment developed by Hoffman (1919) allowed the detection of a person from 200 m and an airplane - from 1600 m. During the First World War, a popular thallium sulfide photodetector was perfected. In 1917–1920, Case developed photoresistors for the near IR range. In Germany, this type of IR detectors was used by military in the communication system accepted in 1935. In 1927, Czerny refined the idea of Hershel's evaporograph by using sublimation of naphthalene and camphor instead of oil evaporation. The most sensitive IR devices of that time used different kinds of bolometers. Before the Second World War, systematic studies of semiconductor IR detectors began; for example, the industrial production of lead sulfide detectors in Germany reached 4000 units a month. The first electrooptic convertor was developed by Holst in 1934. In those years, Vengrinov and independently Goley proposed the construction of photo-acoustic (pneumatic) IR detector. In the USA, such convertors, improved by Zvorykin and Morton, were used for night driving, as well as night sights ("sniper-scopes" and "snooperscopes") ensuring aimed fire at distances up to 90 m. The first equipment for night vision operated in the near IR range, i.e. they required scene illumination by 100–200 W IR projectors (lamps) protected by IR filters opaque for visible rays. Passive IR systems developed in those times detected high-temperature objects, such as airplane engine exhausts, at distances up to 32 km. In Russia, Luchinin developed the first high-speed semiconductor bolometer (sensitive area 0.1×0.12 mm, time constant 0.003 s, threshold heat flux 10^{-7} W/Hz$^{1/2}$). The first Russian experiments on the IR detection of warships were fulfilled by Pokrovsky et al. in 1927. By 1934, the Russian coastal IR systems were developed where the focusing optics reached a diameter of 1.5 m.

The experimental work on the development of IR visualizing systems by using different physical principles, including evaporography, was conducting before the Second World War in some countries, including the former USSR. In the USSR, during the WWII IR detecting devices were used for protecting allies convoys, and the Soviet army began using night vision equipment developed by Nikolayev. In 1950, an evaporograph intended for military and industrial purposes was developed in the USA. The IR "vision" found a wide military use in the beginning of the 1950s during the war waged by the USA in Korea. The first IR thermographs used IR detectors produced on the basis of lead salts and operated in the near IR range. They required the additional illumination of a scene that reduced their efficiency. After the WWII, in the USSR, Faerman et al. built an industrial evaporograph. Khrustalev, Kuzmin et al. perfected semiconductive IR detectors. In parallel with the photonic detection of IR radiation, other physical principles were explored to develop various types of IR converters. In 1946, the luminescent devices ("*metascopes*") appeared but got no further development due to the competition from opto-mechanical systems. Scanning IR imagers were developing during WWII in Germany, Great Britain and France. It is often believed that a real prototype of an airborne opto-mechanical IR imager was developed by Barnes, USA, in 1954. This unit initiated the development of Forward Looking Infrared (FLIR) systems mounted on aircraft. FLIR systems employed only line scanning because the frame scan was performed due to airplane movement. The first portable IR imagers using linear cooled detectors on the basis of lead selenide (PbSe) were tested in the USA in the 1960s and then commercially manufactured as AN/PAS systems. For example, an *imaging* system AN/PAS 110 used a linear PbSe IR detector (2–5 μm wavelength band) and an oscillatable mirror as a scanner, thus providing the frame frequency of 30 Hz and the temperature resolution of 0.1 °C. This model was of 10 kg weight and included a belt with batteries. Primary applications were in technical diagnostics and coast-guard. In the USSR, in 1961 Miroshnikov demonstrated the thermal trace of a person lying on a wooden floor which was keeping for 30 minutes after the person left. Even the person contour was detectable due to the unique (0.03 °C) for those years temperature sensitivity of the IR thermometer. The first scanning IR radiometers which were able to measure temperature appeared in the 1960s as a bypass of military systems. Notice that now the term "imaging" system often means no temperature measurement option, while the very word "radiometer" typically specifies measuring systems. The energy crisis of the 1970s ensured a state support to two Swedish companies AGA (now FLIR Systems) and Bofors which released on the market the first portable IR radiometers in the design which became widely-recognized for many next years. A Thermovision 650 unit from AGA reminded a telescope due to a large-diameter optics used. The next commercial model Thermovision 665 weighed 35 kg and required the cooling of the IR detector with liquid nitrogen. This model served as a prototype for one of the first mass-scale Soviet IR imager TV-03 which was producing without significant modifications until the "perestroika". The Swedish model Thermovision 680 with changeable lenses became the first commercial model of a wide use. Then, the introduction of a battery-based power supply into the next Thermovision 750 model made IR imagers really compact and portable. However, the necessity to measure true temperatures with such units still required placing a reference body within the camera viewangle. In 1978, the model Thermovision 780 began to implement a built-in temperature reference and record IR thermograms on a videotape.

The basis of the Soviet (Russian) IR thermography was founded by Miroshnikov, Timofeev, Arkhangelsky and others in the 1960s. Miroshnikov and Sobakin performed the fundamental research on medical applications of IR thermography by using an IR imager with the temperature resolution of 0.3 °C and the frame time of 15 minutes. This investigation followed the pioneer research by Krasnogorsky (1942) and Lawson (1956) on IR thermographic medical diagnostics.

From 1970 to 1990, the progress in civilian applications of IR thermography chiefly owed to the activity of AGA (later AGEMA Infrared Systems, now FLIR Systems). In 1986, this company refused using liquid nitrogen and introduced a thermoelectric cooler, as well as a built-in processor for calculating true temperatures in a real time, into a Thermovision 870 model. In 1988, a Thermovision 400 model was marketed. Weighing 7 kg, this IR imager was remaining unsurpassed in outdoor applications for many years.

The Soviet research in the 1980–1990s was developing in few directions. At State Optical Institute by S.I.Vavilov, Miroshnikov et al. worked up the original theory of optoelectronic devices and in cooperation with the Azov Optomechanical Plant produced several models of IR cameras for both military and industrial applications ("Filin", "Rubin", "Raduga", "Stator", "Vulcan" and their modifications) [1]. In parallel, Zhukov from Istok Research Institute was exploring the concept of an IR imager TV-03 which originated from some early models of AGA but required the creation of high-tech components, namely, InSb and HgCdTe IR detectors, optical materials, such as high-quality silicon and germanium, and precise optomechanics. A more simplified approach to designing IR imagers, e.g. by reducing frame frequency, was undertaken at the Moscow Institute for Radio Technics, Electronics and Automation to result in a limited series of ATP units. The team from the Spektr enterprise attempted to develop a native IR imager IF-10TV by using alternative Soviet components. A specialized IR camera intended for the analysis of energy distribution in powerful laser beams were developed at the All-Union Institute for Optophysical Measurements and Institute of Applied Mechanics. A number of airborne IR systems for military purposes were developed at State Institute of Applied Optics, Orion enterprise and some other organizations but their performance remained unknown to publics. The unpleasant feature of the contemporary status of the IR thermography in Russia is a lag in the development of uncooled IR detector arrays (initial research in this area has been fulfilled by Kravchenko).

In the late 1980s, the US military released the Focal Plane Array (FPA) technology into the commercial marketplace. In 1995, AGEMA Infrared Systems began to produce IR imagers of a new generation (Thermovision-500) implementing FPA IR detectors. A US company Inframetrics invented a miniature Stirling cooler which was able to cool down IR detectors up to −200 °C. IR imagers of the 500th (later – 600th) series and, finally, cameras using Quantum Well IR Photodetectors (QWIP) commemorated the appearance of "IR vision" as it has been anticipated by the analogy with standard TV.

In the mid 1990s, three known producers of IR cameras FLIR (USA), Inframetrics (USA) and AGEMA Infrared Systems (Sweden) merged and later, after having joined Indigo Systems and CEDIP (France), the company FLIR Systems has become one of the major producers of civilian (and military) IR hardware in the world. This company manufactures IR cameras for a variety of applications: (1) scientific research (SC 8000 model with the capability of capturing IR thermograms in a real time with the temperature resolution better than 0.025 °C), (2) outdoor surveys and technical diagnostics of industrial installations (ThermaCam P60 model with an uncooled IR detector, digital data recording and parallel video-channel; from 2002, an economic ThermaCam E2 model of 0.7 kg weight and its plentiful modifications, including a recent InfraCAM, has been produced), (3) night vision (e.g. a "police" version ThermaCAM Scout), and (4) airborne surveying (ThermaCAM 1000 model with high spatial resolution and availability of accessories allowing placing a remotely-controlled IR camera in gimbal). There are several other world-rank companies supplying high quality IR imagers in noticeable quantities, merely to mention Raytheon (USA), Santa Barbara Focal Plane (USA), Mikron, later Fluke (USA), NEC Avio (Japan), AEG Infrarot-Module and Jenoptik (Germany). In 2002, two Chinese companies (WuHan Guide Electronic Industrial and Guangzhou Sat Infrared), which mainly use French ULIS FPA detectors, manifested on the world market.

In general, the world market has undergone serious changes for the last decade conditioned by the finish of the Cold War and introduction of a new generation of IR detectors (FPA). Many modern IR cameras can be regarded as dual-use equipment equally convenient for both military and civilian applications.

In 1999, the world market for IR imaging systems, both commercial and dual-use, was estimated by 660 million US dollars. Fullop, a leading U.S. analytic in this area, predicted that the market size will reach 2.6 billion U.S. dollars by 2006, mostly due to boosting sales of night-driving units. This application was pioneered by Raytheon which first mounted an IR imager on Cadillac DeVille 2000. Even if there are no reliable figures published on the world sales of IR cameras, it is expected that, due to above-mentioned factors, a market size will be permanently increasing.

During the last years, the Russian research in the field of IR thermography technologically repeated the ideas of the earlier years but put a strong accent on the implementation of advantages supplied by digital data treatment. For example, a well-known Russian model IRTIS-2000 continues using slow opto-mechanical scanning, that makes it helpless in capturing fast thermal events, but provides an acceptable IR thermogram quality (by the formats 256 × 256 and 640 × 480) while being used in technical diagnostics. The competitiveness of this model, which has been keeping due to its reasonable price, is currently becoming questionable because of the appearance of economic FPA cameras using 160 × 120 arrays (also 320 × 240). Recently, some Russian models employing both Western (TH-4604 model from the Spektr enterprise) and domestic (TKVr-IFP from Institute for Physics of Semiconductors) FPAs appeared on the market but the problem of accurate temperature measurements has been not yet solved in these cameras. During 2005–2015, in Russia, tank IR thermographic sights, as well as riffle sights, were developed by using both Russia-made and foreign IR detectors. In 2016, commercial production of IR microbolometric arrays in Russia was announced to reach the turnover of 10 thousand pieces a year.

In addition to IR imagers, in the recent years have become increasingly popular the so-called IR thermographic modules (both imaging and radiometric) which have no built-in monitors but have a small weight and can be easily connected to computers.

To finish this short historic review, let us provide one of accepted IR imagers classification by their generations. The *first* generation used single IR detectors and opto-mechanical scanning devices along 2 spatial coordinates. In the equipment of the *second* generation, IR arrays of a small format (160 × 120, 320 × 240) were used allowing the recognition of a human face in IR color thermograms at distances up to 1 km. The *third* generation can be considered as night vision equipment of a high quality (640 × 480). Finally, the term '*fourth* generation' often specifies multi-spectral military IR cameras.

A short chronology of IR technology and IR thermography is presented in Table 7.1.

7.2 Basics of Thermal Radiation Theory

7.2.1 Electromagnetic Spectrum and Units

Maxwell suggested the existence of electromagnetic waves in a free space and in material media. Waves in a free space are classified in accordance with their frequency f and wavelength λ, these being related to the free-space propagation velocity $c = 3 \cdot 10^8$ m/s by the expression $c = f\lambda$. The radiant energy of wavelengths between 0.4 and 0.8 μm is appreciated by the eye as light of various colors over the visible spectrum between violet (the shorter wavelength) and red (the longer wavelength). Waves shorter than visible are the ultra-violet, which may excite visible fluorescence in appropriate materials. At the long-wave end of the visible spectrum, there is IR radiation, felt as heat. The range of wavelengths of a few millimeters and upward is used in radio communication. The electromagnetic wave spectrum is shown in Fig. 7.1.

IR thermography, or *IR imaging,* is the technique for non-contact recording and visualizing thermal radiation from objects, first of all, intended for analyzing surface temperature distributions. Electromagnetic (thermal, or IR) radiation occurs in

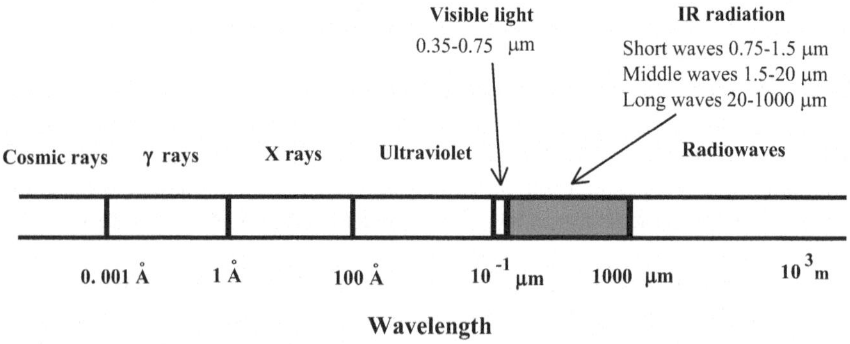

Fig. 7.1 Electromagnetic spectrum (not in scale)

Table 7.1 IR thermography chronology [2–4][a]

1758	Lomonosov invented a telescope for night vision
1762	Lomonosov made mirrors with high reflectivity
1800	William Herschel discovered IR radiation and called it "calorific rays"
1821	Nobili invented the first thermocouple
1833	Melloni made the first thermopile by connecting many thermocouples
1840	John Herschel obtained a visual "footprint' of IR radiation by focusing solar radiation onto a suspension of carbon particles in alcohol
1880	Svanberg invented the bolometer (perfected by Longley and Abbot)
End of the XIXth century	Lebedev and Stoletov started research on the external photoeffect
1900	Max Planck (Nobel Prize winner in 1918) refined the theory of thermal radiation
1905	Photofilm sensitive up to 800 nm developed.
1927	In the USSR, Pokrovsky, Kozyrev and Gourov detected warships
1929	Czerny elaborated an evaporograph as the first IR thermograph
1932	First evidence of CO_2 presence in the Venus atmosphere obtained by Adams and Danham by photographing reflected solar radiation
1934	Electrooptical converter for wavelengths up to 1.5 μm developed by Holst et al.
1942	The first military use of an IR viewing system is demonstrated at Fort Belvoir, USA
1946	The US military developed the first IR line scanner. Frame time was 1 hour.
1954	The first IR imaging scanner is developed in the USA Frame time was 45 min.
1958	The first IR sight for military purposes was developed in the USA
1950–1960	The first commercially available evaporograph is developed in the USA
1960	In Russia, IR thermographs produced images for 5 min; Miroshnikov, Timofeev, Arkhangelsky and Sobakin founded the Soviet IR thermography.
1961	Lindberg and Halmberg invented the opto-mechanical scanner using rotating prisms.
1964	Texas Instruments produces the first forward-looking IR (FLIR) system, USA
1965	AGA, Sweden, introduced the first commercial IR imager AGA-650 intended for predictive maintenance.
1967	AGA produced the first color electronic thermogram, Sweden.
1972	An imaging radiometer Dynarad 210 implementing a dual-band principle was introduced, USA
1973	The first battery-operated portable system AGA 750 was introduced in Sweden.
1974	The first commercially available handheld IR imager Probeye, USA
1977	The first TV compatible IRT imager Inframetrics 510, USA
1978	The first dual-band real-time IR imager AGA 780 for R&D applications, Sweden.
1970–1980	Russian IR thermographs "Fakel:", "Rubin", "TV-03", "Raduga", "IF-10TV", "ATP", "Laserovisor" were developed..
1980	LW pyroelectric vidicon was commercialized by ISI VideoTherm, USA
1984	The first battery-operated portable IR camera Inframetrics 600 with on screen temperature measurement, USA
1987	A single-piece, thermoelectrically-cooled IR radiometric camera AGEMA 470 with onboard digital storage was introduced by AGEMA Infrared System, Sweden.

(continued)

Table 7.1 (continued)

1987	The first FPA IR imager Mitsubishi IR-5120 with a Stirling cooler, Japan.
1990	In the USSR, cooled linear IR detector arrays were developed for military use.
1993	A small-size FPA IR camera InfraCAM using camcorder batteries was developed by Inframetrics, USA
1995	A fully-featured camcorder-type FPA camera ThermaCAM with measurement and digital storage (Inframetrics, USA).
1997	First low-cost, linear array thermoelectrically-cooled snap-shot camera (Infrared Solutions IR Snapshot, USA)
1997	An uncooled microbolmeter radiometric camera AGEMA 570 entered the market, Sweden.
1999	The first commercial QWIP detector IR imager ThermaCAM SC 3000 was introduced by FLIR Systems, USA
2000	An IR camera ThermaCAM PM 695 (then ThermaCAM P60) incorporating both visual and IR imaging was marketed by FLIR Systems, USA
2003–2008	Low-cost FPA imaging radiometers (image format 320 × 240 and 160 × 120), as well as IR thermographic modules, were manufactured by FLIR Systems, NEC, Raytek (Fluke), Irisys, Land Guide and others. Acquisition frequency in research systems increased up to several hundred and thousand Hz. Large format (640 × 480) imagers appeared on the market.
2011	IR imager SC 8000 from FLIR Systems is supplied with an IR detector with the format of 1240 × 1024.
2012–2016	FLIR Systems and Opgal commercialized smartphone-based IR thermographic modules.

[a]The idea of this table and some information have been prompted by Lyon and Orlove [4]

solids, liquids and gases due to oscillations of atoms in a lattice or rotary/oscillatory movement of molecules. IR radiation occupies a wide band of the electromagnetic spectrum from 0.75 μm (400 THz), or 750 nm, to 1000 μm (300 GHz), or 1 mm, between visible light and radiowaves (see Fig. 7.1). There are several definitions of IR bands in the technical literature. For example, Wikipedia introduces near-IR (0.75–2.5 μm), mid-IR (2.5–10 μm) and far-IR (10–1000 μm) wavelength bands. In IR thermography, the terms "short wavelength" (SW) and "long wavelength" (LW) typically specify the bands from 3 to 5.5 μm and from 7 to 14 μm respectively which are associated with the spectral sensitivity of common InSb and CdTeHg IR detectors, as well as, with the corresponding atmosphere transparency windows.

The time rate of *radiant* energy is measured in *watt* and expressed in *radiant flux* or *radiant power*. If the energy is emitted from the surface, it is called *emissive power*. If the energy is received by the surface, then it is called *irradiance*. If the energy is leaving the surface, whatever the physical cause of which, it is called *radiosity*. Since the definitions used by different authors can be confusing, it is necessary to follow carefully the *units* that give a good prompt to the physical content of a term. The description of the most important radiometric terms and units is presented in Table 7.2.

Table 7.2 Radiometric terms and units

Term	Definition	Unit
Radiant power, radiant flux	$\Phi = \int_0^\infty \Phi_\lambda d\lambda$	W
Radiant energy	$W = \int_0^\tau \Phi(\tau) d\tau$	J
Radiant intensity	$I = d\Phi/d\Omega$	W/sr
Radiant exitance	$R = d\Phi/dF$	W/m^2
Irradiance, dose-rate	$E = d\Phi/dF$	W/m^2
Radiance	$L = \dfrac{I}{dF \cos\Theta}$	W/(m^2·sr)

[a]The subscript λ relates to spectral (monochromatic) characteristics. For example, R_λ is radiant exitance at a particular wavelength (Planck function) and its dimension is [W/(m^2·μm)]

7.2.2 *Thermal Radiation Laws*

According to the *Planck law*, the spectral (monochromatic) radiant exitance, i.e. the spectral power irradiated by a physical body per unit surface, is fully defined by its temperature T and spectral emissivity ε_λ:

$$R_\lambda(T,\varepsilon_\lambda) = \frac{\varepsilon_\lambda C_1}{\lambda^5 \left(e^{C_2/\lambda T} - 1\right)} [W/\left(m^2 \cdot \mu m\right], \qquad (7.1)$$

$$C_1 = 3.7418 \cdot 10^8 \ W \cdot \mu m^4 / m^2; !_2 = 1.4388 \cdot 10^4 \ \mu m \cdot K$$

Planck's law was first derived for a blackbody (BB) which emits maximum radiant energy at a particular temperature; this takes place if $\varepsilon_\lambda = \varepsilon_\lambda^{BB} = 1$. Therefore, the classical formulation of the Planck law is:

$$R_\lambda^{BB}(T) = \frac{C_1}{\lambda^5 \left(e^{C_2/\lambda T} - 1\right)}. \qquad (7.2)$$

For $\lambda T < 5000$ (μm·K), the *Wien approximation* is valid:

$$R_\lambda^{BB}(T) = \frac{C_1}{\lambda^5 e^{C_2/\lambda T}}. \qquad (7.3)$$

For $\lambda T > 10^5$ (μm·K), one can use the *Rayleigh-Jeans formula*:

$$R_\lambda^{BB}(T) = 2C_1 T / \left(C_2 \lambda^4\right). \qquad (7.4)$$

In TNDT, λT varies from 800 to 5000 μm·K that corresponds to wavelengths from 3 to 14 μm and temperatures from 0 to +100 °C, hence, one can use the Wien formula (7.3).

In practical calculations, the dimensionless presentation of the Planck formula is convenient (Fig. 7.2b):

$$y = 142.32\,x^{-5}\left[\exp(4.9651/x)-1\right]^{-1};\, x = \lambda/\lambda_m;\, y = R_\lambda^{BB}/R_{\lambda m} \tag{7.5}$$

A curve family for the Planck function R_λ^{BB} by Eqs. (7.2) and 7.5) is presented in Fig. 7.2a for some blackbody temperatures. It follows that: (1) for each temperature, a maximum of radiant power appears at a particular wavelength which becomes longer for lower temperature, (2) radiant power drastically decays with lower temperature. The features of thermal radiation are expressed with the *Wien* and *Stefan-Boltzmann laws*.

The maxima of the Planck function λ_m is given by the *Wien displacement law*:

$$\lambda_m\left[\mu m\right] = \frac{2898}{T\left[K\right]} \approx \frac{3000}{T}. \tag{7.6}$$

For example, assuming the temperature of a person in clothing equal to 300 K, we obtain the wavelength of the maximum emitted power equal to 10 μm.

The fraction of the total emission in a spectral band λ_1-λ_2 is obtained by integration of the Planck function between these two limits:

$$R(T,\varepsilon_\lambda) = \int_{\lambda_1}^{\lambda_2} R_\lambda(T,\varepsilon_\lambda)\,d\lambda\left[\mathrm{W/m^2}\right]. \tag{7.7}$$

Spectral behavior of material emissivity may be intricate, particularly, in gases which are characterized by a line emission spectrum. Emissivity of many solid and liquid materials called "gray" is constant through the spectrum, therefore:

$$R(T,\varepsilon) = \varepsilon\int_{\lambda_1}^{\lambda_2}\frac{C_1}{\lambda^5\left(e^{C_2/\lambda T}-1\right)}d\lambda = \varepsilon R^{BB}(T,\tau). \tag{7.8}$$

Integral values of R^{BB} in various spectral bands are tabulated [5]. As mentioned above, in IR thermography, one typically deals with the SW (3–5.5 μm) and LW (7–14 μm) bands. Note that, according to Wien's law, major radiation power density emitted by bodies with temperature from 0 to 100 °C is concentrated in these wavelength bands.

The total radiant exitance (wavelengths from 0 to ∞) is given with the well-known *Stefan-Boltzmann law:*

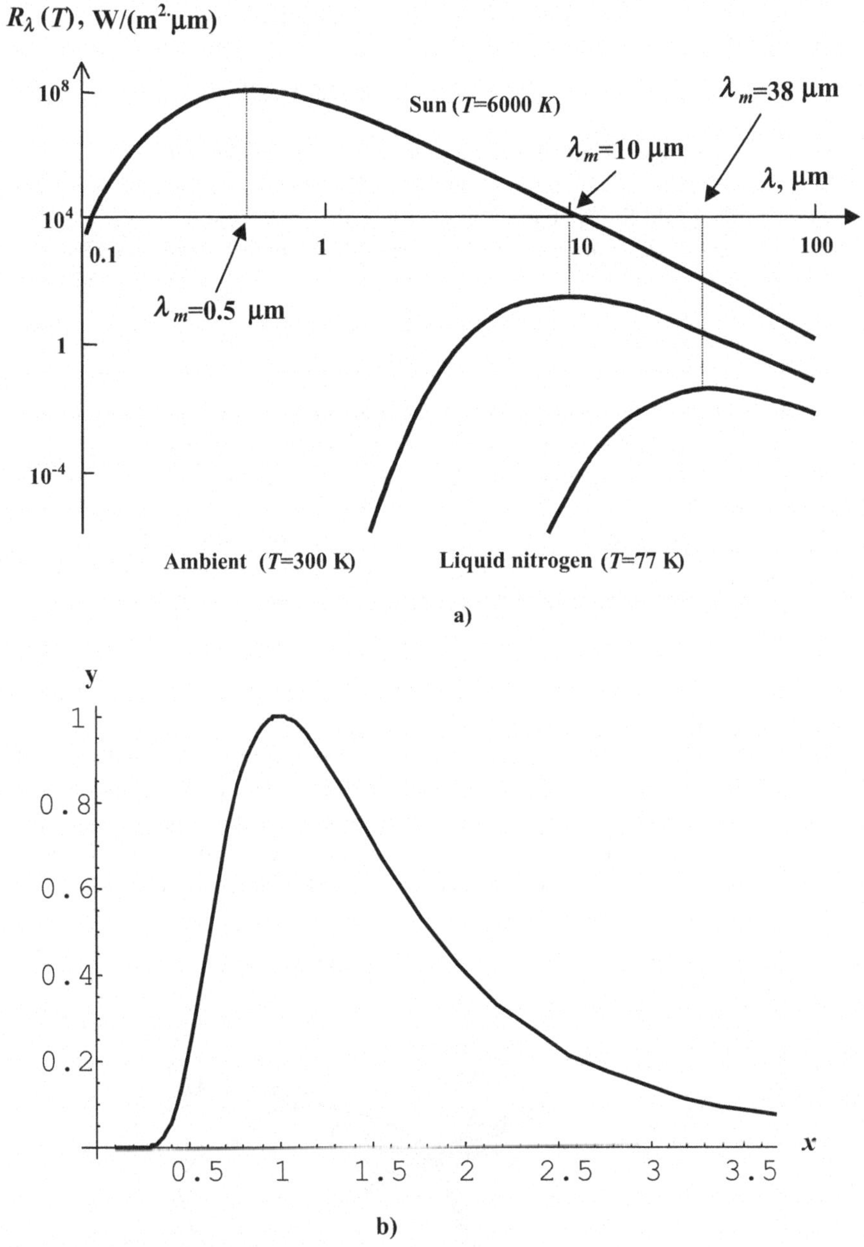

Fig. 7.2 Planck's law: (**a**) in absolute units, (**b**) dimensionless presentation

$$R^{BB}(T,\lambda = 0\ldots\infty) = \sigma T^4 \text{ for a BB,} \tag{7.9}$$

$$R(T,\varepsilon,\lambda = 0\ldots\infty) = \varepsilon\sigma T^4 \text{for a graybody.} \tag{7.10}$$

Here $\sigma = 5.67 \cdot 10^{-8}$ W/(m^{2}K^4) is the Stefan-Boltzmann constant.

Thermal radiation, of which power density is described by Eqs. (7.7, 7.8, 7.9, and 7.10), is emitted into the spatial angle of π steradians. According to the *Lambert law*, an IR imager will capture only a fraction of this radiation within the Ω angle:

$$J_\Omega = R\Delta S \frac{\Omega}{\pi} \cos\varphi, \tag{7.11}$$

where ΔS is the area viewed by an IR imager within its instantaneous field of view, ϕ is the angle between a viewing direction (angle Ω) and the normal direction to the scanned surface. Both ΔS and Ω values are determined by the IR detector area and lenses parameters.

If a body is illuminated with optical radiation, the following phenomena take place (Fig. 7.3): (1) absorbance with the coefficient α_λ, (2) reflectance with the coefficient ρ_λ, and (3) transmittance with the coefficient τ_λ. It is obvious that

$$\alpha_\lambda + \rho_\lambda + \tau_\lambda = 1, \tag{7.12}$$

that is the expression of the energy conservation law.

The *Kirchhoff law* expresses an important fundamental feature of media providing a link between emissivity and absorbance. In each surface point of any 'thermal emitter', at a particular temperature and wavelength, a spectral emissivity is equal to spectral absorbance for oppositely-directed non-polarized radiation. In practice, this means that coefficients of absorbance and emissivity are numerically equal:

$$\alpha_\lambda = \varepsilon_\lambda. \tag{7.13}$$

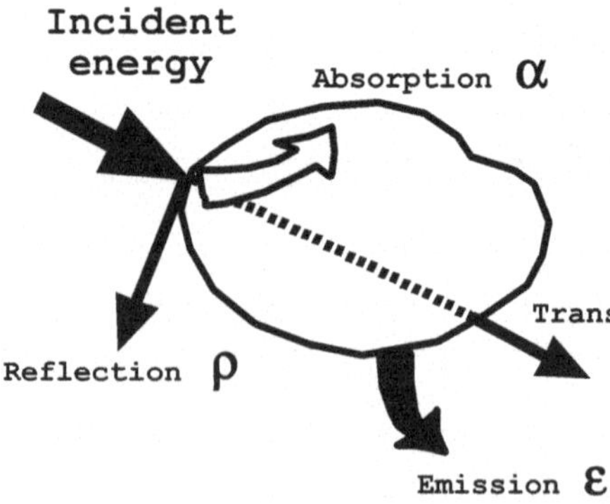

Fig. 7.3 Interaction between a physical body and incident electromagnetic radiation

The Kirchhoff law also means that a body is capable to emit and absorb equal amounts of energy at the same wavelength. This law is sometimes verbalized as "good absorbers are good emitters" [6]. In IR thermography this statement allows the better interpretation of the phenomena of absorbance/emission and the correct identification of the readings of IR thermometers and imaging radiometers which are typically calibrated by blackbody references.

When measuring temperature by thermal radiation, a change in the spectral radiance as a function of temperature is of interest. The temperature derivative of the Planck function is:

$$\frac{\partial}{\partial T} R_\lambda^{BB}(T) = \frac{C_2\, e^{C_2/\lambda T}}{\lambda T^2 \left(e^{C_2/\lambda T} - 1\right)} R_\lambda^{BB}(T). \tag{7.14}$$

These values are tabulated in [5]. For a graybody, in the total spectrum:

$$\frac{\partial}{\partial T} R(T,\varepsilon) = 4\varepsilon\sigma T^3, \tag{7.15}$$

or in finite differences:

$$\Delta R(\Delta T = T_1 \ldots T_2, \varepsilon) = 4\varepsilon\sigma \left(\frac{T_1 + T_2}{2}\right)^3 \Delta T. \tag{7.16}$$

In radiation thermometry, it is convenient to use the following monochromatic presentation of the Planck function:

$$R_\lambda^{AYT}(T) = K(\lambda) T^n \text{ or } R_\lambda(T) = \varepsilon_\lambda K(\lambda) T^n, \tag{7.17}$$

where $K(\lambda)$ is the function of wavelength, $n = 5/\beta$ at $\beta \le 2.5$, $n = 1 + 2.5/\beta$ at $\beta \ge 2.5$, $\beta = \lambda/\lambda_m$. It follows from Eq. (7.15) that

$$n = \frac{\partial R_\lambda(T)/R_\lambda(T)}{\partial T/T}. \tag{7.18}$$

Hence, the number exponent n characterizes a change of $R_\lambda(T)$ in percent if temperature changes by 1%. Eq. (7.17) contains an apparent contradiction because it seems that $R_\lambda(T)$ increases unlimitedly with shorter wavelengths and greater n. However, in fact, $R_\lambda(T)$ decreases because of a quicker diminution of $K(\lambda)$. Therefore, Eq. (7.17) means that, when passing to a short wavelength section of the spectrum, only a *relative* increment of radiant exitance increases.

In any spectral band, the following formula similar to Eq. (7.17) is valid:

$$\int_{\lambda_1}^{\lambda_2} R_\lambda^{BB}(T)\,d\lambda = K\,T^n, \tag{7.19}$$

where K is a numerical coefficient, and the number exponent n depends upon wavelength. For example, in the most important wavelength bands 3–5.5 and 7–14 μm $n = 10.11$ and $n = 4.83$ respectively for temperatures from −20 to +80 °C. Equation (7.19) allows obtaining some important conclusions which are related to a thermography scheme and calibration of IR imagers (see Sect. 7.2.4).

7.2.3 IR Thermography Scheme

Since a test sample is always surrounded by other objects and the atmosphere, its total radiation represents a sum of *emitted*, *reflected* and *transmitted* fluxes. In many cases, IR thermography deals with opaque bodies ($\tau_\lambda = 0$) and Eq. (7.12) acquires the simpler form:

$$\alpha_\lambda + \rho_\lambda = 1, \text{or}\ \varepsilon_\lambda + \rho_\lambda = 1 \tag{7.20}$$

The scheme for thermographying opaque bodies is shown in Fig. 7.4. The thermal flux captured by an IR camera includes the radiation emitted by the test sample Φ_{em}, as well as the fraction of the radiation emitted by surrounding objects and reflected from the sample surface Φ_{refl}:

$$\Phi = \Phi_{em} + \Phi_{refl} = \Gamma_1 \int_{\lambda_1}^{\lambda_2} \varepsilon_\lambda\, \tau_\lambda^{atm}\, R_\lambda^{BB}(T) + \Gamma_2 \int_{\lambda_1}^{\lambda_2} \rho_\lambda\, \varepsilon_{\lambda 0}\, \tau_\lambda^{atm}\, R_\lambda^{BB}(T_0), \tag{7.21}$$

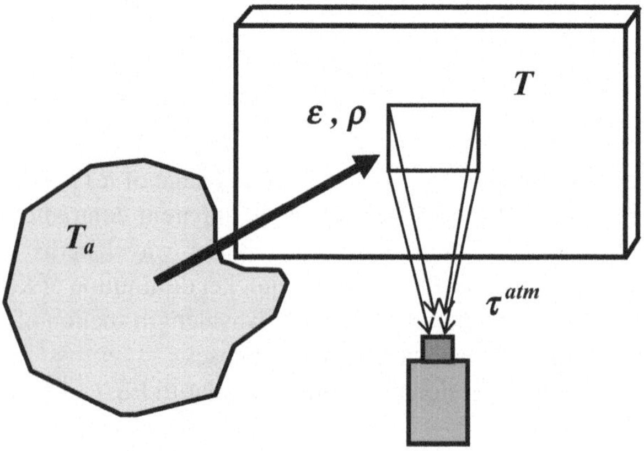

Fig. 7.4 IR thermography scheme

where Γ_1, Γ_2 are the geometrical factors which describe attenuation of radiation due to a particular experimental geometry and optics parameters, τ_λ^{atm} is the transmittance of the atmosphere for a particular distance between the test object and the IR camera, and the subscript "*a*" relates to the ambient including surrounding objects.

A careful analysis of thermal fluxes sensed by an IR detector is important in determining IR camera detection limits. The total flux is consisted of the radiation which comes from the sample, as well as the radiation of IR camera components, including the IR detector itself. A simple illustration of the mutual influence of neighbor objects is the case of two infinite parallel "black" plates with the temperatures T_1 and T_2. The resulting thermal flux in the space between two planes (no-absorbance case) is:

$$\Delta R = \sigma \left(T_1^4 - T_2^4\right). \tag{7.22}$$

For example, consider a person with the surface temperature +27 °C in the ambient with the temperature +20 °C. The person emits 460 W/m^2 but, because of the counter ambient radiation, the resulting flux is only 40 W/m^2. Note that, if a person surface is about 2 m^2, the total heat loss is 80 W.

The interaction between bodies of various geometrical shapes and different emissivities is described by the coefficient Γ and the effective emissivity ε_{eff}. Below we will limit ourselves with the simplified Eq. (7.21).

7.2.4 Emissivity Problem

Without loosing generality, one can assume that the ambient emits as a BB, then Eq. (7.20) can be written as:

$$\Phi = \Gamma_1 \int_{\lambda_1}^{\lambda_2} \varepsilon_\lambda \tau_\lambda^{atm} R_\lambda^{BB}(T) + \Gamma_2 \int_{\lambda_1}^{\lambda_2} (1-\varepsilon_\lambda)\tau_\lambda^{atm} R_\lambda^{BB}(T_0). \tag{7.23}$$

Equation (7.23) clearly shows that, when monitoring real objects, IR imager readings are affected not only by the test object temperature, but also by its emissivity and atmosphere transmittance. The parameters T, ε and τ^{atm} affect on IR thermogram appearance in a complicated way, thus making difficult the data interpretation in terms of true temperatures.

Tables 7.3, 7.4, 7.5, and 7.6 contain the emissivity values of some common materials. Note the characteristic difference between metals and non-metals. Metals, in particular, polished, behave as mirrors which reflect well and emit badly thermal radiation. In many cases, non-metals can be considered as graybodies of whose thermal radiation is close to BBs by magnitude and spectrum. Another interesting fact is that some materials, such as human skin, snow and light paints, are good reflectors in the visible range but good absorbers (emitters) in the IR spectrum.

Table 7.3 Integral emissivity of building materials at 20 °C

Material	Emissivity
Concrete	0.84…0.95
Graphite, oxidized	0.98
Gypsum	0.8…0.9
Clay, burnt	0.91
Wood, chipped	0.8…0.9
white raw	0.7…0.8
polished	0.5…0.7
Brick, red	0.86…0.03
Brickwork, plastered	0.94
Brick, fireclay	0.95
Paint, oil (different colors)	0.94
Lacquer, black mat	0.96…0.98
Lacquer, white	0.8…0.96
Marble, polished	0.93
Asphalt felt	0.93
Soot	0.95…0.97
Soot with liquid glass	0.96
Soot on solid surface	0.96
Glass mat	0.96
Plaster	0.91
Paper black, mat	0.94
Ebonite	0.90
Plastic	0.92
Timber (pine)	0.86
Rubber	0.94
Asbestos sheeting	0.92
Textolite	0.93
Teflon	0.95
White ceramics	0.97
Vitreous ceramics	0.91
Alabaster	0.89
Polyurethane, foamed (rough)	0.97
Polyurethane, foamed (smooth)	0.98

The influence of emissivity on IR thermography results is two-fold. First, uncertainties in an emissivity values makes true temperature measurements inaccurate. In TNDT, exact knowledge of ε is not required but surface fluctuations of ε produce indications which could be interpreted by an operator or an automatic device as defects. A typical solution is to establish a decision making threshold; however, false signals may be so high that TNDT may loose the competition with other NDT techniques if no special data treatment algorithms are applied.

In general, one should avoid thermographying unpainted metallic objects, in particular, if hot objects there are nearby, and their thermal radiation may reflect from the viewed surface to produce deceptive flares. There are many different types of surface noise indications (clutter) appeared in this case being conditioned by oxides,

Table 7.4 Material emissivity recommended by FLIR Systems (AGEMA Infrared Systems) for some spectral bands

Material	Wavelength, μm	Emissivity
Aluminum	SW[a]	0.83–0.94
Asbestos	SW	0.96
Brick	SW	0.87–0.86
Brick, red	SW	0.90
Concrete, dry	5	0.95
Soil, frozen	LW[a]	0.93
Glass	5	0.97
Granite	5	0.96
Gravel	LW	0.28
Iron	SW	0.9–0.96
Limestone	5	0.96
Lime	SW	0.87
Plastic, black	SW	0.95
Plastic, white	SW	0.84
Cardboard	5	0.81
Plaster	SW	0.86–0.90
Veneer	SW	0.83–0.98
Polypropylene	SW	0.97
Rubber	5	0.97
Wallpaper	SW	0.90

[a]SW = 2–2.5 μm
LW = 6–20.0 μm

scratches, mud, paint, fingerprints, etc. Detection of hidden defects on the noisy background is a difficult task except the cases where defect areas significantly exceed false indications either by magnitude or by size. True temperature of metals can be effectively measured by depositing high-emission coatings, such as paints, soot, etc., on the surface of test objects, Sometimes, attaching a piece of an opaque adhesive tape can be a good solution.

Consider the relationship between the temperature and emissivity increments (ΔT and $\Delta \varepsilon$) which is predicted by the thermal radiation laws. For graybodies, it follows from the Stephan-Boltzmann law that:

$$\frac{\Delta T}{T} = \frac{1}{4}\frac{\Delta \varepsilon}{\varepsilon}. \quad (7.24)$$

For example, if a material is characterized by the mean emissivity $\varepsilon = 0.9$ with the practical variations $\Delta \varepsilon = \pm 0.02$, the equivalent temperature variations will be $\Delta T = \pm 1.7^{\circ}C$ at $T = 300\ K$; such noise fluctuations considerably exceed the temperature resolution of typical IR imagers (up to 0.03 °C). However, in practice, equivalent ΔT values will be lower due to the compensating radiation from the ambient, see Eq. (7.33) below.

Equation (7.17), which includes ε_λ as a multiplier, is convenient for analyzing the relationship between ΔT and spectral emissivity fluctuations $\Delta \varepsilon_\lambda$. The following expression can be easily obtained:

Table 7.5 Emissivity of radio electronics components [1]

Component	Emissivity
Resistors:	
carbonic film	0.85
metallic film	0.85–0.90
glass tubular	0.90
wound wire	0.87
Capacitors:	
alternating	0.85–0.95
electrolytic	0.28–0.36
ceramic disk-wise	0.90–0.94
cylindrical	0.90
film	0.90–0.93
mica	0.90–0.95
glass	0.91–0.92
Transistors	0.90
Diodes	0.89–0.90
Pulsed transformers	0.91–0.92
Smoothing inductors	0.89–0.93
Board:	
mica and epoxy resin	0.86
Teflon	0.80
epoxy resin	0.80
Plates	
gold-plated copper	0.30
copper	0.35
Wires:	
tin-plated	0.28
copper insulated with formaldehyde resin	0.87–0.88
Steatite leads	0.87

Table 7.6 Landscape emissivity in some spectral bands (by Kriksounov)

	Emissivity		
Landscape element	1.8–2.7 μm	3–5 μm	8–13 μm
Foliage, green	0.84	0.90	0.92
Leaves, dry	0.82	0.94	0.96
Leaves, green	0.67	0.90	0.92
Branches coniferous	0.86	0.96	0.97
Grass, dry	0.62	0.82	0.88
Sand, various	0.54–0.62	0.74–0.82	0.93–0.98
Wood bark	0.75–0.78	0.87–0.90	0.94–0.97

$$\frac{\Delta R_\lambda(T)}{R_\lambda(T)} = n\varepsilon_\lambda \frac{\Delta T}{T}, \tag{7.25}$$

which is similar to Eq. (7.24). For shorter wavelengths, the magnitude of n increases and respectively increases the sensitivity of IR systems to temperature variations.

Another important conclusion follows from Eqs. (7.17) and (7.25) when analyzing the signal-to-noise ratio S defined as the ratio between signal variations $\Delta R_\lambda(T)$, which are conditioned by fluctuations in T and ε respectively:

$$S = \frac{(\partial R_\lambda(T)/\partial T)\,\Delta T}{(\partial R_\lambda(\varepsilon)/\partial T)\,\Delta\varepsilon} = n\frac{\Delta T/T}{\Delta\varepsilon/\varepsilon}. \tag{7.26}$$

In TNDT, one can assume that $n = 5/\beta = 5\,\lambda_m/\lambda$. The sample absolute temperature can be represented as the sum of the absolute initial temperature T_{init} and the excess heating temperature T_h, specified as T in the previous chapters. Then, Eq. (7.26) can be written as:

$$S = 5\,\frac{\lambda_m}{\lambda}\,\frac{1}{1+T_{init}/T_h}\,\frac{\Delta T/T_h}{\Delta\varepsilon_\lambda/\varepsilon_\lambda}. \tag{7.27}$$

According to the Wien law, $\lambda_m \approx 3000/(T_{init} + T_h)$. Hence, finally:

$$S = 5\,\frac{3000}{\lambda}\,\frac{T_h}{(T_{init}+T_h)^2}\,\frac{\Delta T/T_h}{\Delta\varepsilon_\lambda/\varepsilon_\lambda}, \tag{7.28}$$

where $\Delta T/T_h = C^{run}$ is the running temperature contrast introduced earlier. It follows from Eq. (7.28) that, for a particular defect and a certain noise level, the magnitude of S increases with shorter wavelengths and higher excess temperature. For example, if T_h varies from +10 to +100 °C ($T_{init} = 300\ K$), S magnitude increases six-fold. Therefore, the optimization of an optimal heating temperature, i.e. heating power, requires analyzing spectral behavior of $\Delta\varepsilon_\lambda/\varepsilon_\lambda$.

If IR thermography is performed in the presence of the emitting ambient (surrounding objects) with the temperature T_a, one can obtain the following relationship between ΔT and $\Delta\varepsilon$ variations by neglecting the both the geometrical factors and the spectral character of ε and τ^{atm} in Eq. (7.22):

$$\Phi \sim \varepsilon\tau^{atm}\int_{\lambda_1}^{\lambda_2} R_\lambda^{BB}(T)\,d\lambda + (1-\varepsilon)\tau^{atm}\int_{\lambda_1}^{\lambda_2} R_\lambda^{BB}(T_a)\,d\lambda. \tag{7.29}$$

By taking into consideration Eq. (7.19), Eq. (7.29) will acquire the following form in a particular spectral interval:

$$\Phi \approx \varepsilon \tau^{atm} T^n + (1-\varepsilon)\tau^{atm} T_a^n. \qquad (7.30)$$

It is obvious that an IR system calibrated by a reference BB will show the apparent (radiation) temperature of a graybody T_{ap} according to the equation:

$$T_{ap}^n = \varepsilon \tau^{atm} T^n + (1-\varepsilon)\tau^{atm} T_a^n. \qquad (7.31)$$

Equation (7.31) is used in contemporary IR imaging radiometers for the correction of temperature readings, if the material emissivity ε and the ambient temperature T_a, as well as the distance to a test object, are inputted by the user. Note that two latter parameters determine the atmosphere transmission τ^{atm} (see Sect. 7.2.5).

In the "still" atmosphere, at distances less than 30–50 m, one can assume that τ^{atm} =1. Then, the basic thermography equation can be simply written as:

$$T_{ap}^n = \varepsilon T^n + (1-\varepsilon) T_a^n. \qquad (7.32)$$

Assuming Eq. (7.32), the formula for the evaluation of the relationship between variations in temperature and emissivity will be:

$$\left|\frac{\Delta T}{T}\right| = \frac{1}{n}\left[1-\left(\frac{T_a}{T}\right)^n\right]\left|\frac{\Delta \varepsilon}{\varepsilon}\right|. \qquad (7.33)$$

We remind that for the spectral bands 3–5.5 and 7–14 μm and the temperature range from −20 to +80 °C n= 10.11 и n= 4.83 respectively.

It is worth noting that, at $T = T_a$, the full compensation of the emissivity will take place. In practice, this situation appears when surveying building envelopes of which external temperature is close to the outdoor ambient temperature. Then, according to Eq. (7.33), temperature readings are weakly dependent on materials emissivity.

Example Let a building envelope be decorated with plaster is at the temperature T = 263 K (−10 °C). The temperature of the ambient and the cloudy Sky is T_a = 258 K (−15 °C). What is an error in determining the plaster wall temperature by using a 7–14 μm IR radiometer if the emissivity for brick (ε = 0.94) is inputted as a correction factor? It follows from Table 7.3 that the difference between the emissivities of brick and plaster is $\Delta\varepsilon$ = 0.94 – 0.91 = 0.03. Then, the use of Eq. (7.33) yields $\Delta T \approx$ 0.15 °C for both spectral bands 3–5.5 and 7–14 μm, while Eq. (7.24) which does not take into account the compensating radiation of the ambient produces $\Delta T \approx$ 2.1 °C. Physically, the obtained difference in ΔT values means that the ambient is a source of compensating radiation which diminishes the influence of emissivity fluctuations on IR thermographic measurements.

When monitoring thermographically an object with the temperature T_o on the background with the temperature T_b, the thermal contrast for a particular spectral band can be introduced [2]:

$$TC = \frac{\int_{\Delta\lambda} R_\lambda(T_o)d\lambda - \int_{\Delta\lambda} R_\lambda(T_b)d\lambda}{\int_{\Delta\lambda} R_\lambda(T_o)d\lambda + \int_{\Delta\lambda} R_\lambda(T_b)d\lambda}. \quad (7.34)$$

As follows from Eq. (7.25), in the SW band (3–5.5 μm), the thermal contrast is twice higher than in the LW band (7–14 μm).

In optical radiometry, a plenty of methods to correct emissivity influence has been developed but most of them are unacceptable in IR thermography and TNDT where a high efficiency of processing all pixels, instead of single points, is required. Besides, many ingenious correction techniques are oriented toward objects with stationary temperature and have never been implemented in the IR thermography practice. The classification proposed in [1] includes two groups of the correction techniques shortly described in Table 7.7. The methods of the first group require some knowledge about test objects or, at least, involve, preliminary object treatment; the second group is based on proper processing of radiant flux.

The most common method of the first group is to paint samples black that ensures a low noise level but is technologically unpractical. As "black" coatings, one can use lacquers and paints of different types and colors (non-obligatory black), sprays and suspensions of graphite powder in water, engraving enamel, black rubber soot, as well as water, oil and some other materials which are characterized by high emissivity and can be easily deposited and removed. However, while applying such coatings, some additional noise may appear because of small inclusions and air bubbles, as well as due to variations in coating thickness. Thin coatings can be transparent for both heating and emitted thermal radiation thus worsening test efficiency. Moreover, when heating a material with short powerful pulses, which are generated by flash tubes and lasers, coatings may evaporate because of a very high temperature at the end of heating. Another significant disadvantage of "black painting" is a drop in inspection productivity, therefore, this method is often used in laboratory experiments and very rarely in practice. It is worth noting that TNDT was first successfully used in the USA in the inspection of aviation panels on military aircraft typically painted green. Inspecting unpainted surfaces of civilian airplanes represents a serious difficulty but the modern tendency to paint passenger liners in color may relax this problem. When testing steel cases of chemical reactors and aluminum cages, true temperatures are better measured in the areas with paint, dust, rust, etc. In such cases, it is recommended to stick on a test surface a piece of an adhesive tape, of which emissivity is about 0.95, and then determine emissivity values; afterwards, the temperature readings on the natural surface can be corrected.

Table 7.7 Emissivity correction in IR thermography and TNDT

Method	Description
Preliminary sample treatment	
"Black painting"	Lacquers and paints of different types and colors (non-obligatory black), sprays and suspensions of graphite powder in water, engraving enamel, black rubber soot, as well as water, oil and some other materials can be used as 'black' coatings. This method is low-productive and unacceptable for objects of complicated shape, such as electronic components.
Radiation "enforcement"	Radiation of real bodies is to be approached to that of a blackbody by producing surface cavities or placing some special elements, such as spheres, plates etc., close to a monitored surface. True temperature is to be determined mathematically. This method is low-productive and used for the correction of emissivity only in single areas.
Varying temperature of IR detector	The method implies the third law of thermodynamics which prohibits the energy exchange between media with identical temperatures. The practical implementation of this method is cumbersome and hardly applied to follow transient thermal events.
Comparison with reference	Absolute emissivity values cannot be determined with this method. Defects are detected by analyzing temperature deviations in regard to reference values. This method is expedient in the inspection of single-type test objects, e.g. electronic components, electrical installations, etc.
Capturing temperature signals at different times	This is the modification of the 'comparison with reference' technique. It is used in transient TNDT where dynamic temperature signals survive in defect areas during within certain time intervals.
External illumination	This method employs Eq. (7.29). Emissivity is determined by the formula $\varepsilon = 1 - \rho$. The spectral composition of illuminating radiation should match that of emitted radiation. This method can be hardly realized in the case of objects with changing temperatures.
Heating objects in a "thermal camera"	Emissivity across the sample surface is determined in a 'thermal camera' with an elevated temperature in the absence of defects. Then, the obtained emissivity map is used for correcting current temperature readings. This method is common in TNDT of electronic components. Years ago, it was realized in the EQUAL software from AGEMA Infrared Systems (now FLIR Systems).
Thermal radiation treatment	
Operating in the SW band	Relative temperature sensitivity of IR imagers increases with shorter wavelengths, see Eq. (7.25). The use of this technique is limited with worsening temperature resolution and increasing spurious reflected radiation.
Bi-color pyrometry	The principle of bi-color pyrometry is based on the division of temperature signals in two different wavelength bands, see Eq. (7.37) below. This method is typically used in high-temperature thermometry. In low-temperature IR thermography it can be scarcely applicable due to the weak dependence of the signal ratio upon temperature.
Polarization methods	This group of methods is used only in high-temperature pyrometry.
Mathematical treatment of spectral components (multi-channel optical pyrometry)	In the cases, where the Wien approximation is valid, one produces a new signal distribution by wavelength and temperature, thus leading to the necessity of solving N equations with N unknowns. Implementation of this technique is cumbersomes, its applicability to TNDT is unclear.

7.2.5 *The Relationship Between True and Apparent Temperatures*

Equation (7.32) allows determining apparent temperature values which would appear if an IR imager calibrated by a BB monitors a "gray" object of the emissivity ε in the absence of the atmosphere:

$$T_{ap} = \left[\varepsilon T^n + (1-\varepsilon)T_a^n\right]^{1/n}. \tag{7.35}$$

In contemporary IR cameras, the temperature correction is done by introducing the material emissivity value, the ambient temperature and the distance to an object to be tested (in some models, atmosphere humidity can be also taken into account). The two latter parameters are used for calculating the atmosphere transmittance but in many industrial surveys it can be neglected. Then, the sample emissivity and the ambient temperature remain the most important parameters influencing IR thermography results. In particular spectral bands, the relationship between all involved parameters is given with the formula which follows from Eq. (7.35):

$$T_{:>@@} = \left\{\frac{1}{\varepsilon}\left[T_{ap}^n - (1-\varepsilon)T_a^n\right]\right\}^{1/n}. \tag{7.36}$$

Equation (7.36) allows the evaluation of temperature measurement errors if values of ε and T_a are inaccurately defined.

Example Let the object emissivity be $\varepsilon = 0.7$ and the ambient temperature be $t_a = 27$ °C ($T_a = 300$ K). The IR thermographic survey is performed with a SW (3–5.5 μm) camera, i.e. $n = 10.11$. If the user defines the emissivity from 0.5 to 0.9, the temperature read-outs t_{ap} will change depending on the true temperature t_{true}, as shown in Fig. 7.5a, even if the correct value of T_a is introduced. Respectively, if the ambient temperature is introduced with an error, the measurement will be also inaccurate, even if the emissivity is correct (Fig. 7.5b, true ambient temperature $t_a = 47$ °C).

As follows from the plots in Fig. 7.5a, the emissivity underestimate leads to the temperature overestimate, and *vice versa*. In the same way appears the influence of the ambient temperature, and the respective errors grow up with lower object temperature (Fig. 7.5b).

Note that exam questions related to evaluating efficiency of IR thermographic measurements in the case of inaccurate ε and T_a values are typical in the Level I, II and III certification tests conducted by American Society for Nondestructive Testing (ASNT).

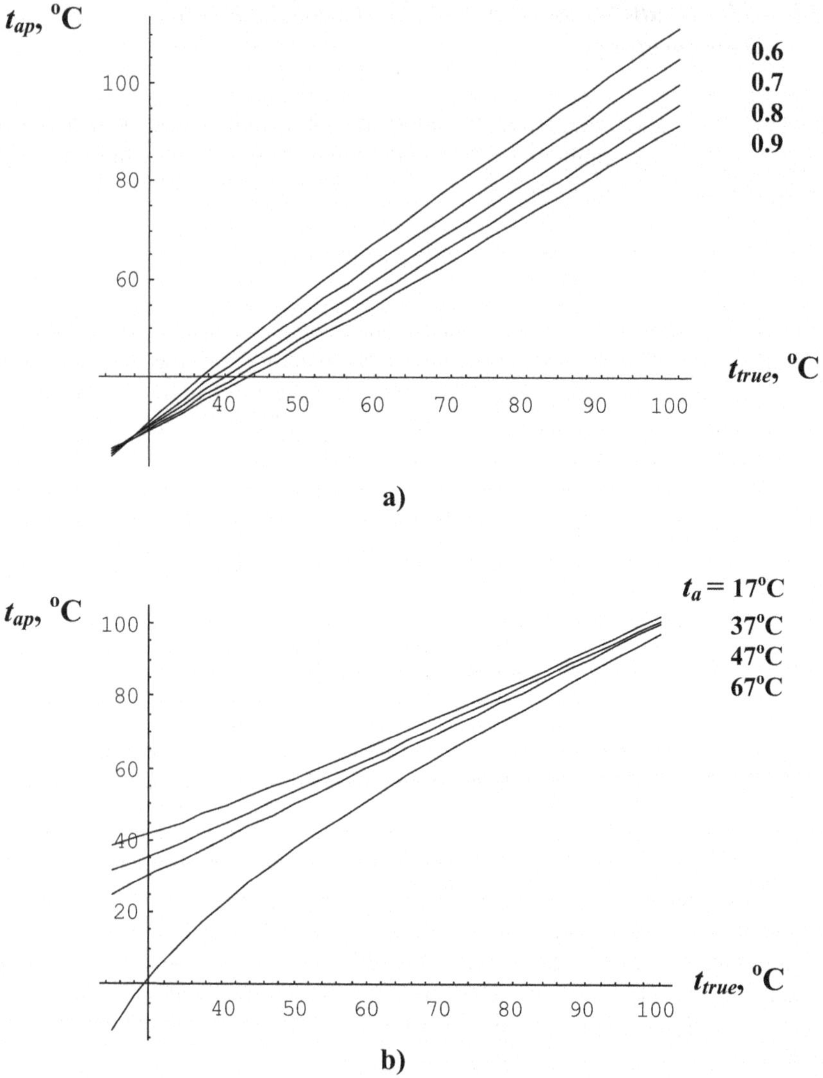

Fig. 7.5 Temperature measurement errors by using IR thermography in the SW band (3–5.5 μm) under varying object emissivity and ambient temperature: (**a**) true values: $\varepsilon = 0.7$ and $t_a = +27$ °C; user's values: $t_a = +27$ °C and ε varies from 0.5 to 0.9, (**b**) true values: $\varepsilon = 0.7$ and $t_a = +47$ °C; user's values: $\varepsilon = 0.7$ and t_a varies from +17 to +67 °C

7.2.6 Dual-Band IR Thermography

By summarizing features of the SW (3–5.5 μm) and LW (7–14 μm) bands, one can state that:

- the SW band is relatively more sensitive to temperature variations,
- the LW band is typically characterized by the highest temperature resolution that is important when thermographying objects with temperature close to ambient,
- in outdoor surveys, the reflected solar radiation represents a more serious problem in the SW band,
- the above-mentioned wavelength bands correspond to the atmosphere transparency windows; the LW band is "wider" and passes more thermal energy from objects with temperature close to ambient,
- in aerial surveys (IR reconnaissance), up to 24 spectral bands are used thus allowing the analysis of various physical effects which show up in different ways at different wavelengths; for instance, the ageing of crops should be followed in the near-IR band,
- in TNDT, optimization of an IR band is not evident; a spectral behavior of noise in these bands is a decisive factor,
- a ratio of the signals in two spectral bands can be used for reducing the influence of emissivity on temperature measurement.

The latter feature of dual-band IR thermography is borrowed from high-temperature bi-color pyrometry. The basic idea is to produce the ratio between two signals:

$$\frac{U_1}{U_2} = \frac{\varepsilon_1 \int_{\Delta\lambda 1} R_\lambda(T)d\lambda}{\varepsilon_2 \int_{\Delta\lambda 2} R_\lambda(T)d\lambda}. \tag{7.37}$$

Being applied to objects with high temperature (>800 °C), where the most energy is emitted at short wavelengths, bi-color pyrometry has proven its validity, however, the implementation of this technique in low-temperature measurement has not proven to be successful.

One of few examples of dual-band IR thermographic systems was a Thermovision 900 camera from FLIR Systems (AGEMA Infrared Systems) which included two scanners operating in the SW and LW bands. The recently-appeared hybrid IR detectors include two different radiation-sensitive elements integrally mounted one under another, thus allowing to combine IR images recorded in two spectral bands.

DelGrande et al. described the use of a Thermovision 900 system for producing maps of ε and T [7]. The method called by authors "*dual-band infrared imaging*" implied two spectral bands with effective wavelengths 5 and 10 μm:

$$U_5 = \varepsilon\ T^{10}; U_{10} = \varepsilon\ T^5, \tag{7.38}$$

that is also prompted by Eq. (7.17). The following formulas have been proposed for producing images E and Θ which respectively reflect distributions of ε and T:

$$\theta = \left(\frac{T}{T_{av}}\right)^5 = \frac{U_5 / U_5^{av}}{U_{10} / U_{10}^{av}};$$
$$E = \frac{\left(U_{10} / U_{10}^{av}\right)^2}{U_5 / U_5^{av}}, \tag{7.39}$$

where U_5^{av} and U_{10}^{av} are the mean pixel values in the corresponding images.

A deeper analysis can be done by using Eqs. (7.17) and (7.37). It follows from Eq. (7.37) that the closer are objects to "gray" the more efficient is the suppression of emissivity iuncertainties. In the extreme case, if $\varepsilon_{\lambda 1} = \varepsilon_{\lambda 2}$, U_1/U_2 is independent of ε. In its turn, for relaxing the influence of temperature, it is necessary to process the function $U_1 / U_2^{n_1/n_2}$. For two above-mentioned cases, the expressions for the signal-to-noise ratio defined by Eq. (7.26) acquire the form:

$$S(U_1 / U_2) = (n_1 - n_2)\,\frac{T_h}{T_{init} + T_h}\,\frac{\Delta T / T_h}{\Delta\varepsilon_{12} / \varepsilon_{12}}; \tag{7.40}$$

$$S\left(U_1 / U_2^{n_1/n_2}\right) = 0, \tag{7.41}$$

where $\varepsilon_{12} = \varepsilon_1/\varepsilon_2$. Equations (7.27), (7.40) and (7.41) allow the comparison of single- and dual-band IR imaging techniques. The main difference between Eqs. (7.40) and (7.41) is that the zero result of processing by Eq. (7.1) means the full suppression of the temperature influence, thus allowing the creation of emissivity maps. However, the ratio of signals by Eq. (7.40) will still keep the dependence on ε, therefore, the most efficient suppression of emissivity will take place only for graybodies.

Let us consider some peculiarities of dual-band IR thermography on the example of a Thermovision 900 system which used to be calibrated by using the following signal model in the i-th spectral band:

$$U_i = \varepsilon_i\,\frac{R_i}{\left[\exp(B_i / T)\right] - F_i}, \tag{7.42}$$

where R_i, B_i, F_i are the empirical constants to be determined during calibration.

By taking into account Eq. (7.42), Eqs. (7.27), (7.40) and (7.41) become as follows:

$$S(U_i) = \frac{T_h}{(T_{init} + T_h)^2}\,\frac{B_i e^{B_i/T}}{e^{B_i/T} - F_i}\,\frac{\Delta T / T_h}{\Delta\varepsilon_i / \varepsilon_i}; \tag{7.43}$$

$$S(U_1/U_2) = \\ = \frac{T_h}{(T_{init}+T_h)^2} \frac{B_1 e^{B_1/T}\left(e^{B_2/T}-F_2\right)-B_2 e^{B_2/T}\left(e^{B_1/T}-F_1\right)}{\left(e^{B_1/T}-F_1\right)\left(e^{B_2/T}-F_2\right)} \frac{\Delta T/T_h}{\Delta\varepsilon_{12}/\varepsilon_{12}}; \tag{7.44}$$

$$S\left(U_1/U_2^k\right) = \\ = \frac{T_h}{(T_{init}+T_h)^2} \frac{B_1 e^{B_1/T} - kB_2 e^{B_2/T} \dfrac{e^{B_1/T}-F_1}{e^{B_2/T}-F_2}}{e^{B_1/T}-F_1} \frac{\Delta T/T_h}{\Delta\varepsilon_{12,k}/\varepsilon_{12,k}}. \tag{7.45}$$

where k is the analog of n_1/n_2 in Eq. (7.41).

Obviously, by choosing k properly, it is possible to make $S\left(U_1/U_2^k\right)$ independent of temperature. Such k value can be found from the equation:

$$B_1 e^{B_1/T} - kB_2 e^{B_2/T} \frac{e^{B_1/T}-F_1}{e^{B_2/T}-F_2} = 0. \tag{7.46}$$

Calculations show that for two common spectral bands 3–5.5 and 7–14 μm, a k value is close to 2; see Eq. (7.19). This proves that the gravity centers of these two spectral bands are at about 5 and 10 μm respectively. Thus, dual-band IR thermography allows producing images of object emissivity which are weakly dependent on object temperature, but the determination of true temperatures depends on the behavior of emissivity in particular spectral bands.

7.2.7 *Propagation of IR Radiation Through the Atmosphere*

IR radiation changes its power and spectral composition by propagating through the atmosphere due to absorption and scattering on molecules of gas, aerosols, rain, snow and suspensions, such as smoke, smog, fog, etc. The main absorbing components are water vapor, carbon dioxide and ozone. The atmosphere humidity is characterized by the water amount deposited along a radiation path. For instance, at the air temperature +20 °C and humidity 60%, the thickness of the water layer is 13 mm/km. At any wavelength and particular atmosphere status, the attenuation of IR radiation is described by the *Bouguer* law [2, 8]:

$$\tau_A(\lambda) = \exp\left[-\gamma(\lambda)R\right], \tag{7.47}$$

where γ is the medium absorption coefficient which depends on a wavelength, and R is the distance, or path. Sometimes, a transmitting medium can be characterized by optical density D:

$$D = \log \frac{1}{\tau\left(\lambda\right)} = 0.43\gamma\left(\lambda\right)R. \quad (7.48)$$

The phenomenon of scattering on atmosphere particles changes the spatial distribution of transmitted radiation energy; the main role in this process is played by particles of which size is comparable to wavelengths. Note that the atmosphere scattering explains the blue color of the Sky in daytime, when the Sun is at the zenith, and the red color at the sunset.

In the humid atmosphere, suspended particles of size about 0.5 μm agglomerate water molecules, thus creating fog. When droplet size reaches about 0.25 mm, the droplets become too heavy and it rains. Water vapor absorption is noticeable at 2.6 μm and over 20 μm.

Due to solar heating, turbulences are created by air convection in the atmosphere influencing the atmosphere refraction index according to the *Gladstone law*:

$$n_{air} = 1 + k_n \rho_{air}, \quad (7.49)$$

where k_n is a constant and ρ_{air} is the air density. Normally $n_{air} = 1$. When determining atmosphere transmittance, it is often assumed that turbulences are homogeneous and isotropic.

For a given wavelength and distance, the practical formula for calculating global atmosphere transmittance is:

$$\tau_A\left(\lambda\right) = \tau_{H_2O}\left(\lambda\right)\tau_{CO_2}\left(\lambda\right)\tau_{diff}\left(\lambda\right)\tau_{rain}\left(\lambda\right), \quad (7.50)$$

where $\tau_{H_2O}\left(\lambda\right)$, $\tau_{CO_2}\left(\lambda\right)$, $\tau_{diff}\left(\lambda\right)$, $\tau_{rain}\left(\lambda\right)$ are the transmittance coefficients determined respectively by water vapor, carbon dioxide, particle diffusion and deposited water (rain); their values for various scenarios of the atmosphere status are tabulated [2, 8]. Equations (7.47), (7.48) and (7.50) are also valid in particular spectral bands if one can neglect the dependence of transmittance on wavelength.

An example of the atmosphere transmittance profile is presented in Fig. 7.6 for a 1.8 km distance. The presence of two transparency windows (3–5 and 7–14 μm) is well seen. Atmosphere absorption can be typically neglected at distances less than 20–30 m but it can be very important, for example, in satellite surveys. Atmosphere phenomena may be of a certain interest in the inspection of electrical transmission line, smokestacks and other objects located far from a thermographer. The enhanced absorption of thermal radiation may occur if there is a significant amount of water vapor and/or dust in some industrial environments.

The equations like (7.47) and (7.50) are employed in contemporary IR imagers for automatic re-calibration of temperature readings by assuming a typical atmosphere scenario and using a distance value inputted by the user. In fact, a variety of atmosphere scenarios has been developed, and discrepancies between models vary up to about 10%. Some commercial and freeware versions of the corresponding computer programs are available from various vendors.

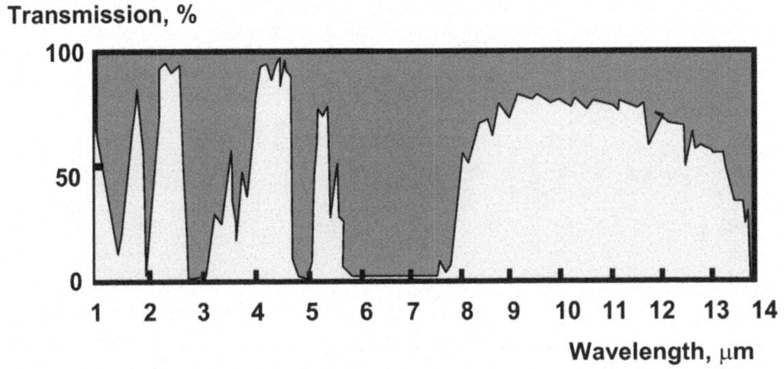

Fig. 7.6 Atmosphere transmission (1.8 km path)

References

1. Vavilov, V.P.: Thermal NDT of Composite Structures and Radio Electronic Components, p. 162 (in Russian). Radio i Svyaz Publish, Moscow (1984)
2. Hudson, R.D.: Infrared System Engineering. Wiley-Interscience, New York (1969)
3. Nondestructive Testing Handbook. Vol. 3 "Infrared and Thermal Testing", Bellingham, A.S.N.T (2001)
4. Lyon, Jr. B.R., Orlove, G.L.: A brief history of 25 years (or more) of infrared imaging radiometry. Proc. SPIE "Thermosense-XXV" Vol. 5078, pp. 17–19 (2003)
5. Bramson, M.A.: Infrared Radiation: a Handbook for Applications, p. 623. Plenum Press, New York (1968)
6. Daniels, A.: Field Guide to Infrared Systems, p. 120. SPIE Press, Washington (2006)
7. DelGrande, N., Clark, G.A., Durbin, P.F. et al.: Buried object remote detection technology for law enforcement. Proc. SPIE "Surveillance Technologies". Vol. 1479. pp. 335–352 (1991)
8. Accetta, J.S., Shumaker Exec, D.L. (eds.): The Infrared and Electro-Optical Systems Handbook. V. 3 "Electro-Optical Components", p. 668. ERIM and SPIE Optical Engineering Press, Bellingham/Washington (1993)

Chapter 8
Equipment for Active TNDT

Abstract This chapter contains descriptions of TNDT hardware for both commercial and laboratory testing. In addition, TNDT heating methods are presented.

8.1 Designing Hardware for Active TNDT

The technology of TNDT schematically depicted in Fig. 8.1 involves some interrelated procedures, thus requiring a system approach to designing test equipment.

A start point is the description of an object to be tested and the simulation of such inspection procedures which can be technically realized. Here the term "technically realized" means a procedure which can be implemented by taking into account both inspection manipulations with test objects and realistic technical performance of inspection systems.

Modeling inspection procedures starts from the analysis of a test object. Material thermal properties, as well as defect depth and size, define both magnitudes and optimum observation times of temperature signals in defect areas. Detection limits by defect parameters are determined by noise, and the most important source of noise is a test object itself. Therefore, optimization of TNDT parameters, such as power and duration of heating, as well as acquisition frequency, is mainly related to test object characteristics. Thus, simulation of TNDT procedures allows optimizing experimental parameters by taking into consideration practical limitations. The result of an optimized experiment is the temperature evolution $T(i, j, \tau)$ defined at each sample point (see Chaps. 2, 3, and 4). Experimental data is analyzed with specialized software which will enable the detection of defects with prescribed statistical detection parameters, such as probabilities of false alarm and correct detection. Since a defect is detected, one may evaluate its parameters, i.e. solve the corresponding TNDT inverse problem. Inspection should be completed by producing a defect map which is a binary image where, for instance, the *unity* value is assigned to pixels belonging to defect areas, while sound areas are marked with *zeros*.

With the above-mentioned approach in mind, the principal components of TNDT systems are: (1) a heating source, (2) a computerized IR thermographic system, and (3) specialized software for modeling and data processing. Both modeling and

V. Vavilov, D. Burleigh, *Infrared Thermography and Thermal Nondestructive Testing*, https://doi.org/10.1007/978-3-030-48002-8_8

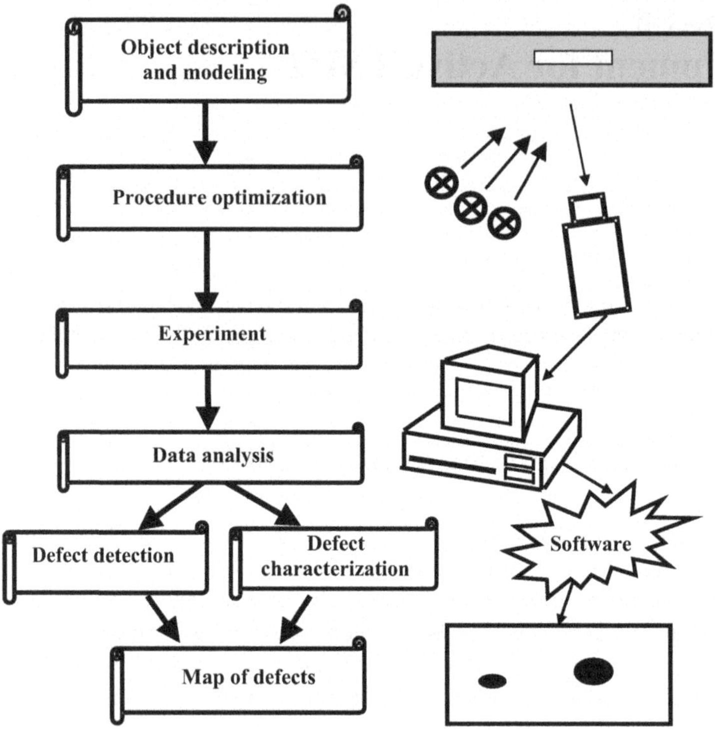

Fig. 8.1 Active TNDT technology

processing approaches, as well as defect characterization algorithms, were described in the previous chapters. In this chapter, we will consider commercial TNDT systems and, in more details, heat sources which are typical in thermal testing. The Chap. 9 will be devoted exclusively to IR imaging systems.

8.2 Commercial TNDT Systems

Active TNDT of raw materials, as well as of intermediate and final products, can be considered as a routine procedure only in a few countries, first of all, in the USA, France, Canada, Great Britain, Italy, Germany and, to some extent, in Russia. The inspection systems described in many related publications have been developed mainly for a laboratory use and rarely manufactured in noticeable quantities. The short description of TNDT equipment is presented in Table 8.1.

Some currently available commercial TNDT systems are presented in Fig. 8.2. All systems implement the same basic principle of combining a heater and an IR imager, as well as a particular data processing algorithm (or a set of algorithms).

Table 8.1 Commercial systems for active TNDT

System (manufacturer)	Description
Sapphire[a] (CEDIP, France)	Computerized IR thermographic system, based on using the Fourier analysis.
TIP[a] (Bales Scientific, U.S.A.)	Computerized, pulsed TNDT system implementing a specialized processor.
Lock-in thermal wave TNDT system (AGEMA Infrared Systems, Germany)	System uses a Thermovision 900 IR imager, low-power harmonic heating, lock-in detection principle and phase analysis.
Aladin[a] (Siemens, Germany)	Thermal wave IR thermographic microscope.
CompuTherm[a] (EDO, U.S.A.)	TNDT in microelectronics.
Barnes RM-2A[a] (EDO, U.S.A.)	TNDT in microelectronics.
Altaïr LI[a] (CEDIP, France)	System is intended for the visualization of mechanical stresses in the frequency range from 0.1 to 1000 Hz. Stress range (by aluminum) ±2000 MPa, temperature resolution up to 0.02 °C. Lock-in and phase IR thermography detection principles.
FDM 2000A[a] (CEDIP, France)	Portable device for determining thermal diffusivity by Parker's method.
MECIR[a] (University of Reims and Avion Marcel Dassault, France)	Pulsed TNDT of aerospace components on the basis of AGEMA Infrared Systems (FLIR Systems).
ThermoSoniX[a] (Indigo Systems, U.S.A.)	System for lock-in ultrasonic IR thermography, spectral range 3–5 μm, temperature resolution <0.02 °C, frame frequency 60 Hz, 256 × 256 format, 20 kHz carrier frequency, ultrasonic stimulation power up to 400 W.
Thermal Bond Inspection Systems (NASA, U.S.A.)[a]	Pulsed TNDT of aerospace components.
Portable TNDT system from University of Helsinki (Finland)[a]	System includes a manually-operated line scanner and a control unit. Laser heating power 15 W, inductive heating power up to 200 W. Inspection of CFRP composites and coatings on metals.
INTROTHERM[a] (Laser technology center, Russia)	Photothermal principle, spatial resolution 50 μm, scanned area 200 × 200 mm, instantaneous area of view 6 × 25 mm, testing productivity up to 3 cm^2/s.
INFRATEST[a] (Moscow State University Scientific Park, Russia)	Laser photothermal inspection. Optical waveguide and diode laser of 1 to 10 W power, wavelength 0.98 μm.
TNDT system for rolled metals[a] (VEMO Institute, Russia)	System developed for metallurgical plants. TNDT of rolled metals is performed by heating a moving sheet of metal with quartz lamps or gas-jet and recording temperature with a commercial IR imager.
TNDT system from VIAM (Russia)	Detecting turbine blade channels.

(continued)

Table 8.1 (continued)

System (manufacturer)	Description
TNDT system from Salut (Russia)	Detecting turbine blade channels.
ThermoScope, EchoTherm,VoyageIR, T³S models), (Thermal Wave Imaging, U.S.A., *www.thermalwave.com*)	Portable systems for inspecting composites, metals and ceramics (flash heating). Circular and/or tubular flash tubes and quartz lamps. Integrated system including pulsed heating, data collection and analysis. Synthetic data processing algorithm.
(ThermoSpectorWELD THERMOSENSORIK, Germany, *www.thermosensorik.de*)	QWIP (FPA) systems for pulsed TNDT, in particular, of turbine blades. The Fourier transform as a basic processing algorithm to produce phasegrams.
COMPOSITE-CHECK (moviTHERM, U.S.A., *www.movitherm.com*)	Advanced IR thermographic solution for Remote Monitoring, Automation and Non Destructive Testing.
C-CheckIR (Automation Technology, Germany, www.c-checkir.com)	A mobile unit for TNDT in aerospace, automotive, at wind energy plants, etc. Equipped with a vacuum fixture tool for easy mounting, even on vertical surfaces.
EDEVIS OTvis (EDEVIS, Germany, *www.edevis.com*),	A unit for lock-in IR thermographic NDT, mainly, intended for characterization of composites.
(InfraTec, Germany, *www.infratec.de*	Pulsed TNDT systems (in collaboration with CEDIP, now FLIR Systems). Passive TNDT by using Varioscan IR imagers.
NDTherm NT/LF (Opgal, Israel, *www.opgal.com/thermography-ndt-testing*)	A portable NDT unit implementing classical TNDT.

[a]Obsolete system

In all cases, the operators of such systems must be properly trained, especially when applying the equipment and data processing options to a new class of test objects. Some robotic and automatic inspection units have also been developed. Automatic defect recognition has been explored in some test cases but in most cases TNDT remains a manual procedure.

8.3 Thermal Stimulation Systems for Active TNDT

The nature of thermal stimulation often decides about practical implementation of TNDT procedures. Since cooling is typically low-intensive, heating is more common (cooling a hot metal with water jet and spraying with liquid nitrogen are the exceptions). The following types of heating are used: (1) optical, (2) inductive, (3) microwave, (4) electric current, (5) mechanical, and (6) convective (with hot gases or liquids).

According to the TNDT theory, described in the previous chapters, the optimum detection of subsurface defects is provided by short powerful heating. This basic condition can be met by all above-mentioned types of stimulation, perhaps, except heating with gases or liquids.

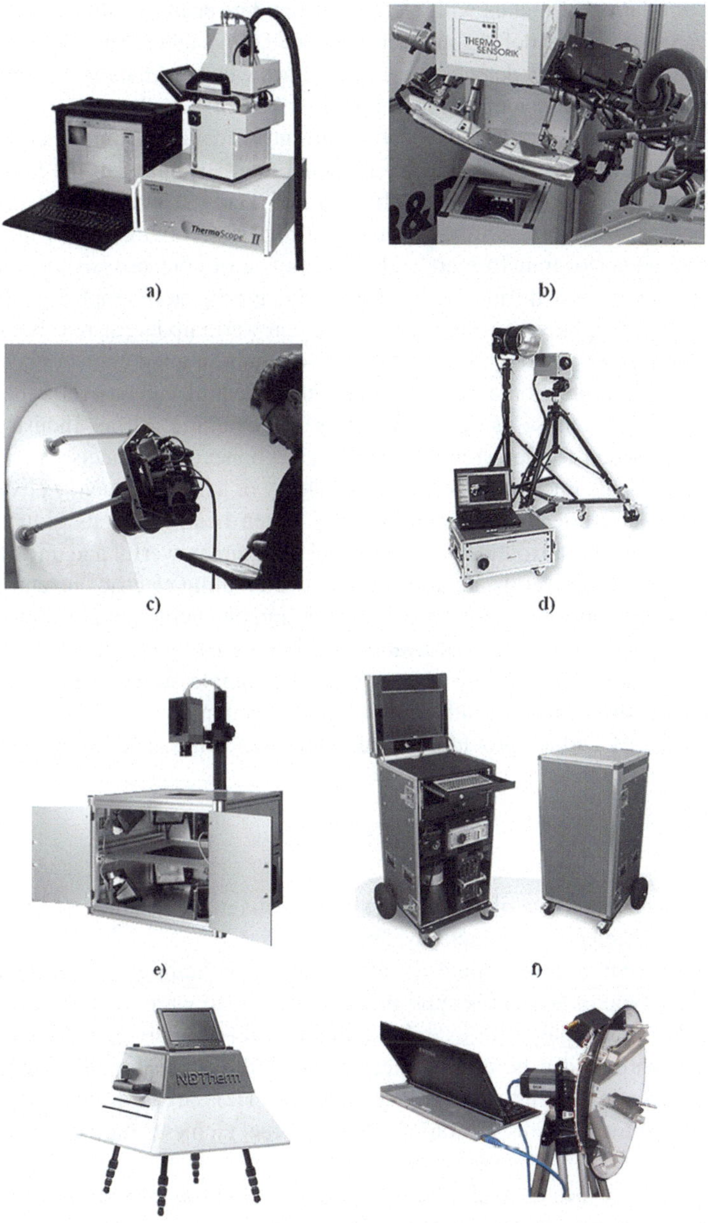

Fig. 8.2 Commercial systems for active TNDT: (**a**) ThermoScope (Thermal Wave Imaging, U.S.A.,), see also EchoTherm, VoyageIR and T[3]S models), (**b**) ThermoSpectorWELD (Thermosensorik, Germany, www.thermosensorik.de), (**c**) C-CheckIR (Automation Technology, Germany, www.c-checkir.com), (**d**) – EDEVIS OTvis (EDEVIS, Germany, www.edevis.com), (**e**) active TNDT device (InfraTec, Germany, www.infratec.de), (**f**) COMPOSITE-CHECK (moviTHERM, U.S.A., www.movitherm.com). (**g**) NDTherm NT/LF (Opgal, Israel, /www.opgal.com/thermography-ndt-testing). (**h**) ThermoDish (Tomsk Polytechnic University, Russia)

The stimulation of metals with electric current has been mainly used in laboratory research, however, defects in electrical installations and electronic (microelectronic) units are typically detected in work regimes where each class of components is characterized by a specific temperature distribution. *Contact electrical heaters* either supply powerful pulses of electric current to metallic samples or warm up samples by direct mechanical contact. The important advantage of the first type of heating is short excitation time (up to fractions of seconds) and a lack of emissivity noise; the drawback of such the scheme is a high current strength in the second winding of powerful transformers and the presence of contact resistance between the sample and the transformer outlet. The second type heaters supplied with a thick copper/aluminum plate ensuring even heating can warm up large areas but they are inertial, produce intensive emissivity noise and require a good contact between the heater and the sample. Some drawbacks of contact heating can be overcome in flexible heaters, e.g. in those using carbon fibers. For example, Airbus Industry uses a 'hot blanket' heater for detecting water in aircraft honeycomb panels.

In inductive and microwave sources, the energy of a high-frequency electromagnetic field is converted into thermal energy. When heating metals inductively, a sample should be be surrounded by a wire coil to ensure powerful and uniform heating, however, one-side coils are also used. The alternating electric current which is passing through the coil creates the alternating magnetic field. Eddy currents induced in metals cause material heating. With a higher electromagnetic field frequency, the current density increases toward the sample surface (the skin-effect). Energetically, the quality of inductive heating is evaluated by a minimum current frequency f_{min}, Hz, which provides a required test efficiency. The following formula has been recommended [1]:

$$f_{min} = 8/(L\mu\rho), \tag{8.2.1}$$

where L is a mean size of a sample, m, μ is the magnetic conductivity, Hn/m, and ρ is the specific electric conductivity, Sm/m.

The inductive heating, which is ubiquitous in metallurgy, is rarely used in TNDT. This type of heating is appropriate if one has to warm up a metal behind a non-metal (for example, when examining rubber-steel structures); in this case, interface delaminations can be easily detected. Steel alloys can be effectively heated with the current of the industrial frequency (50/60 Hz). Green successfully applied inductive heating in the inspection of cylindrical fuel elements used in nuclear engineering [2].

Dielectrics are heated with alternating electric field due to the re-polarization phenomenon which occurs by surmounting friction forces. Heating efficiency improves with increasing dielectric loss tangent (*tg* δ). The specific power W, W/cm^2, which is supplied by a source of high-frequency energy to a sample placed between two plates of a capacitor, is determined by the formula [1]:

$$W = 0,55 \cdot 10^{-12} \varepsilon tg\delta fE^2, \tag{8.2.2}$$

Table 8.2 Industrial microwave sources for heating dielectrics [1]

Material	Frequency, kHz	Specific power, W/cm^3	Field strength, W/cm	Heating duration
Timber (drying)	300	0.04	400	up to 30 hrs
Cane fiber mass (drying)	20,000	1.5	500	5–30 min
Paper (drying)	30,000	5.0	800	up to 3 hrs
Vinyl (welding)	75,000	600	10,000	0.5–20 s
Plastics (welding)	200,000	1500	30,000	0.03 s

where f is the frequency, Hz, E is the electric field strength, V/cm.

For heating dielectrics, a medium- (f = 3–3000 kHz), short- (f = 3–30 MHz) and meter-wavelength (f = 30–300 MHz) ranges can be used. Some parameters of industrial microwave heaters are presented in Table 8.2.

From the TNDT point of view, the additional advantage of microwave stimulation, similarly to electric current heating, is its volumic character that may enhance temperature contrasts in defect sites. Microwave heating is particularly appropriate in the detection of moistened areas in porous non-metals, such as wood, foam thermal insulation, etc..

Gas arc heaters called *plasmotrons* utilize the plasma which appears between the cathode made of a high-temperature resistant material and the anode; the sample itself or the orifice serve as a capacitor. A gas of whose atoms or molecules are ionized at high temperatures is supplied to the orifice typically made of copper and cooled by water. In such devices, which are used for cutting metals, the power density may reach $2{\cdot}10^{10}$ W/m^2. Plasmotrons can be used for fast scanning heating but flame instability, optical noise and high cost do not make them perspective in TNDT.

Cathode ray guns which are used in welding provide power density up to 10^{13} W/m^2 but require vacuum better than 10^{-5} Pa (under normal pressure, free length of electrons is only a few millimeters). They also are not very practical in TNDT.

In the inspection of layered, composite and honeycomb structures, mainly used in aero space, the optical heating is still most popular (see Fig. 1.7, 1.8). This type of heating can be realized with: (1) lasers, (2) flash tubes, (3) halogen lamps of continuous operation, and (4) standard electrical bulbs. *Lasers* are monochromatic and allow changing size of heated areas with power density reaching 10^{21} W/m^2 (in very short pulses. The monochromatic character of laser radiation can be important because it may prevent the appearance of reflected noise in the spectral range of an IR imager. However, in practical testing, laser heating is rarely used because its high power density is achieved only in small-size areas. By using the 'flying-spot' technique, the laser power is distributed throughout larger areas thus decreasing the average absorbed energy. Laser efficiency is typically low, therefore, the use of lasers in TNDT is limited in the framework of using the Parker technique for determining thermal diffusivity of thin high-conductive materials.

In the units manufactured by Thermal Wave Imaging (U.S.A.) and mainly used in aero space, *xenon flash tubes* are employed as heaters. Such tubes are used for pumping lasers and in pavilion photography. Commercial photographic systems typically include two tubes with energy up to 3 kJ and a power supply unit; flash duration is about 5 ms. Such units are manufactured by Balcar, Calumet, Hensei et al. Xenon lamps are characterized by a uniform emission spectrum from 0.2 to 1.4 μm with some spikes at certain wavelengths. In specialized TNDT systems, several tubular Xenon lamps with total energy up to 18–20 kJ can be used.

When applying flash (0.1–1 ms) powerful heating, material overheating at the end of a heat pulse may be a serious problem. The temperature of a near-surface layer or coating may reach several hundred and even thousand degrees Celsius, thus possibly causing material evaporated (appearance of plasma). To overcome this difficulty, the heating duration should be extended up to 10–100 ms. Optical heating of metals can be ineffective because of their high reflectance, therefore, such metals as aluminum, copper, stainless steel etc. can be inspected thermographically only if they are painted.

Quite technological are IR heaters which are classified for: (1) high-temperature (IR electrical bulbs, quartz halogen lamps etc., up to 1500 °C); (2) middle-temperature (quartz, glass, ceramic, metallic etc., 450–1000 °C); (3) low-temperature (glass, ceramic, metallic, under 450 °C). *Quartz halogen lamps* with power from 0.5 to 2.5 kW are produced in various designs and dimensions. They are often used in TNDT for heating materials during 5–30 s. Standard electrical bulbs provide similar results but with longer heating times. Normally, a set of lamps is used to heat up a large sample area. Also, using a single electrical bulb of a higher power (up to 1 kW) is possible in projector-type heaters. Single lamps with focusing reflectors provide power density up to 50 kW/m^2 and match well spot IR radiometers. Note that in some applications one may use solar radiation of which power density can reach 1 kW/m^2 in middle latitudes.

Optical radiation can be easily modulated, for example, when implementing the thermal wave technique. A shape of a single heat pulse is not very important when optimizing TNDT results. In some cases (inspection of metals and shiny composites, the residual radiation of lamp balloons after switching a heater off may represent a serious problem. In modeling, the pulse shapes presented in Table 1.2 are typically considered.

In order to realize a high efficiency of IR thermography, a heat source should provide a necessary power density across a large test area. For a particular heater, a heated area can be defined empirically by heating a "black" target made of a low-conductive material.

IR line scanning systems are well combined with *line resistive heaters* made of alloys with high electrical resistance (nichrome, constantan etc.). Such heaters can be made as long ceramic tubes with wire winding providing power density up to100 kW/m^2 with controlled radiation uniformity along the heater length. Very similar are industrial tubular heaters where windings are mounted in a metallic case. The drawbacks of linear heaters are, first, the necessity of placing them closely to test objects, second, they allow relatively small heated areas to compare to extended heaters, and, third, they are characterized by a high thermal inertia. Recently, NASA successfully used linear heating for the detection of corrosion in aluminum and steel samples. IR imagers can be combined with *plane heaters* made of high-resistive materials which are deposited on thermally-resistant glass.

One of the simplest heating techniques is the blowing of test samples with hot air produced by a *fan* (*air gun*). The sample surface can be thermographically monitored at the heating stage but the absorbed energy is typically low (500–1000 W/m^2). This type of heating is applicable to non-metals where temperature gradients evolve for a long time, and the use of more powerful heaters may damage the sample surface due to overheating. Weak uniform heating can be performed by moving samples into a premise with a higher or lower temperature. In 1965, such method was described by Beller in the inspection of rocket missile cases [3].

Thermal stimulation can be also fulfilled by cooling objects down, e.g. by spraying them with liquid nitrogen or placing them in a refrigerator. Water can be detected in aircraft honeycomb panels after landing due to a natural temperature gradient between a cruising altitude and the ground level.

The heating by mechanical hysteresis is used in experiments with low-frequency periodical loading. This kind of stimulation is low-power and applied in studying material fatigue damage. The outdoor implementations of mechanical loading, such inspection of large-size components under wind loading, have not been very successful because of a low signal-to-noise ratio. However, in the last decade, IR thermography has been successfully used in the combination with ultrasonic excitation. This technique allows the selective stimulation of some specific defects, such as 'kissing cracks', while the host material remains 'cool'.

To summarize the principles of choosing an optimum heat source in TNDT, one should consider the following factors: (1) a type of the contact between a sample and a heater, (2) required duration of heating, (3) possibility of modulating thermal flux, (4) emitted power (energy), (5) power (energy) be effectively absorbed by a test object, (6) spectral range of heating and residual radiation, particularly, in the case of optical heating, (7) noise level, (8) efficiency, and (9) safety issues.

References

1. Bulyubash, B.V., Gurvich, V.Z.: Electricity and Heat, p. 172 (in Russian). "Nauka" Publish, Moscow (1978)
2. Green, D.R.: Emissivity-independent infrared thermal testing method. Mater. Eval. **23**(2), 79–85 (1965)
3. Beller, W.S.: Navy sees promise in infrared thermography for solid case checking. Missiles Rockets. **22**, 1234–1241 (1965)

Chapter 9
Infrared Systems

Abstract This chapter summarizes the basics of IR detectors, optics, and modern IR imagers, and the methods of storing IR images and image sequences. Software for data analysis and compiling reports is discussed. Also, line scanners, pyrometers and Terahertz (not IR) imaging systems are briefly described.

9.1 IR Visualization Systems

Analyzing spatial temperature distributions by one or two coordinates requires either mechanical (opto-mechanical) scanning of sample surface or using staring (array-like) detectors of IR radiation.

In the second half of the XXth century, opto-mechanical scanning was the principle feature which separated visualization systems in the visible (near-IR) and middle-IR wavelength bands. In opto-mechanical systems, the view angle is deflected by using rotating or oscillatable optical elements (mirrors or prisms), thus requiring a high precision of their manufacturing and assembling. In the 1970–1990s, there were two main principles of opto-mechanical scanning employed in commercial IR imagers. The first principle was implemented in the units manufactured by AGEMA Infrared Systems (Sweden) involving perpendicularly-rotating silicon prisms which deflect IR radiation by a necessary angle thanks to a high refraction index of silicon. The second principle utilized oscillatable mirrors; the corresponding device was patented and commercialized by Inframetrics (USA).

The principle of electronic scanning was used in photodetector arrays developed at the end of the XXth century. These IR detectors are often called Focal Plane Arrays (FPA) since they are placed in focal planes of the optics. They represent 2D arrays of single IR detectors thus allowing to produce thermal images consisted of pixels. The main advantages of FPAs are a lack of moving components and the possibility of staring at all scene points at once (strictly speaking, the latter feature is implemented only in the so-called *snap-shot* arrays).

V. Vavilov, D. Burleigh, *Infrared Thermography and Thermal Nondestructive Testing*, https://doi.org/10.1007/978-3-030-48002-8_9

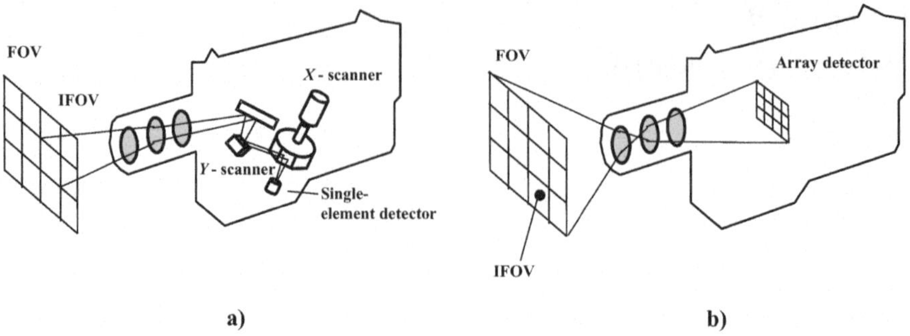

Fig. 9.1 Schematic presentation of IR imagers with opto-mechanical (**a**) and electronic (**b**) scanning. (Adapted from Ref. [1])

Some simplified schemes of IR imagers using opto-mechanical and electronic scanning are presented in Fig. 9.1. Opto-mechanical units contain a lot of movable details while, FPA imagers in fact represent TV tubes operating in the IR band of optical radiation.

Modern IR arrays use both photonic and thermal detectors. By price, the most affordable are imaging and radiometric systems which implement uncooled microbolometric (ferroelectric) arrays. In special cases, where high frame frequency and/or high temperature resolution and/or spectral selectivity are needed, IR imagers may use cooled arrays. Typically, such arrays are made on platinum silicide (PtSi), indium antimonide (InSb), HgTeCd alloys and some semiconductors, such as gallium arsenide (GaAs), which utilize a principle of the so-called *quantum-wells*.

IR imagers are considered as opto-electronic systems which include:

- IR detector,
- lens,
- scanner (except FPA systems),
- cooling device (uncooled microbolometric systems are also available),
- built-in temperature reference,
- electronic unit,
- display,
- software.

Commercial IR imagers also include:

- image recording/storing unit,
- means of transportation,
- power supply and batteries,
- auxiliary devices (contact thermometer, GPS navigator et al.).

Depending on applications, IR cameras are differed by technical parameters, but they implement similar physical operation principles discussed below.

9.2 IR Detectors

The detectors of electromagnetic radiation which are intended for converting radiation energy into other types of energy are classified for *biological* and *physical.* A typical biological photodetector is a human eye which is sensitive to the radiation with wavelengths from 0.38 to 0.78 μm of which brightness exceeds 11 orders of magnitude. A human eye, well-adapted to darkness, may see objects with temperature over 450 °C. Hence, under some special conditions (full darkness), it is possible to see directly overheated electrical joints, hot metals, etc.

Physical detectors generate electrical signals which are proportional to radiation power absorbed by their receiving elements. IR detectors include:

- sensing element (s),
- entrance window which separates the ambient from the volume around the sensing element,
- aperture which limits a viewangle (it can be cooled or thermally-stabilized),
- Dewar flask for storing cooling agents or other types of cooling systems, for example, based on Stirling engine cryocoolers and thermoelectrical coolers; some IR detectors operate at room temperature.

Detectors of optical (IR) radiation are typically classified for two big groups: *thermal* and *photonic.*

An operation principle of thermal detectors is based on one of four effects: (1) *bolometric* effect (change of electrical resistance of a sensing elements heated with absorbed radiation), (2) *thermovoltaic* effect (output voltage is generated by a heated junction of two different metals), (3) *thermopneumatic* effect (detector temperature variation changes gas volume), (4) *pyroelectric* effect (temperature variation changes a dipole moment of a crystallic sensing element).

In photonic detectors, the quantums (photons) of absorbed radiation quantums increase the number of free carriers of electric charge, thus changing an electrical status of a sensing element. If the *photoconductivity* phenomenon is used, the absorbed IR radiation changes electrical conductivity of a sensing element. In the case of the *photovoltaic* effect, the absorbed radiation produces electron-hole pairs near the p-n junction, thus generating electric current. In some IR detectors, *photoelectromagnetic* effect can be also used.

9.2.1 Thermal Detectors

In bolometers, the electrical resistance of semiconductors or metals is modulated by the absorbed thermal radiation. Historically, bolometers were one of the first types of detectors but their use in imaging IR systems was limited because of high thermal inertia. At the end of the XXth century, low-inertia microbolometric and ferroelectric arrays emerged. The microbolometer technology has been developed by

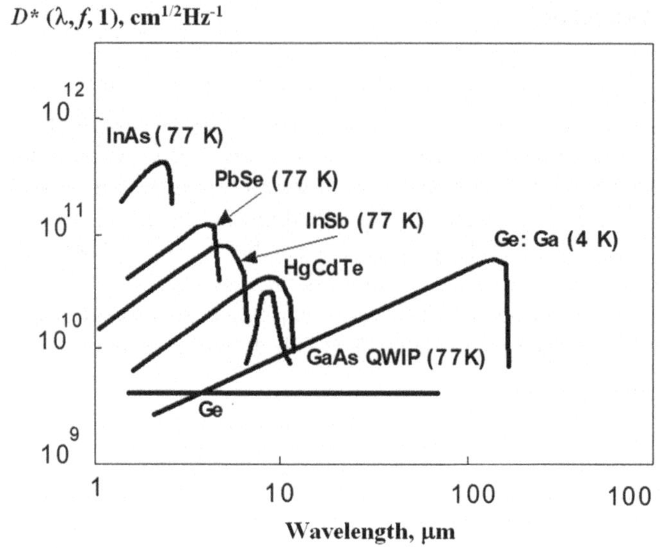

Fig. 9.2 Spectral response ($D*$) curves of common IR detectors

Honeywell and others; ferroelectrics have been developed by Raytheon and GEC Marconi. Both types of IR detectors have been implemented in uncooled FPA IR imagers that have become so popular nowadays.

Pyroelectric IR detectors are manufactured of materials called ferroelectrics, e.g. barium titanate. Such materials are characterized by permanent electric polarization which is a strong function of temperature. For a long time, economic IR imagers have been produced on the basis of pyroelectric vidicons (pyrovidicons), but presently this class of IR equipment is nearly ousted by FPA IR cameras.

IR thermometers (pyrometers) frequently use thermoelements of which operation is governed by the Seebeck effect. Metallic thermoelements employ antimony-bismuth junctions, silver, tellurium, iron, constantan, chromel and their alloys. In semiconductor thermoelements, antimony, silicon, tellurium and selenium are used.

Thermal detectors are characterized by "flat" behavior of the spectral detectivity $D*$ (see definition of $D*$ in Sect. 9.2.6), i.e. they are non-selective and require no cooling, however, their detecting performance values is lower than that of photonic detectors (Fig. 9.2).

9.2.2 *Photoemissive Detectors*

In photoemissive devices, the flow of electrons producing electric current in a primary circuit appears due to the *external photoeffect* which takes place on a photocathode subjected to optical irradiation. A maximum wavelength which can be detected by such a device is determined by photoelectric work function that is specific for each photocathode material; for oxygen-silver-cesium photocathodes, it is 1.3 μm. Photoemissive detectors include vacuum and gas-filled photoelements and photomultiplier tubes, or image intensifier tubes. In the latter devices, electrons are accelerated in the electric field and multiplied on dynodes due to secondary emission. In multi-stage devices, a multiplication factor can reach 10^5 to 10^7. A typical time constant is about 10 ns. Photomultiplier tubes are typically used for the detection of low-intensity light in a microwatt range; high-power radiation may damage last dynodes.

In IR imagers, photoemissive detectors are not used because their spectral sensitivity is in the visual band.

9.2.3 Photonic Detectors

Photonic, or quantum, detectors realize the *internal photoeffect.* According to the zone theory of solids, charge carriers do not leave the detector material but pass into the conduction band either from valence or impurity levels which are located in the forbidden band. The spectral response of some typical IR photonic detectors is presented in Fig. 9.3.

Lead sulfide (PbS) was the first practical IR detector developed in Germany before the WWII. In the 1950s, the elaboration of new photoconductive materials, such as lead selenide (PbSe), lead telluride PbTe and indium antimonide (InSb), allowed coping with the spectral range up to 3–5 μm. The further use of impurity levels of copper, zinc and gold broadened the spectral range up to 30 μm. At the end of 1950s, HgCdTe alloy IR detectors were proposed in the Great Britain; along with

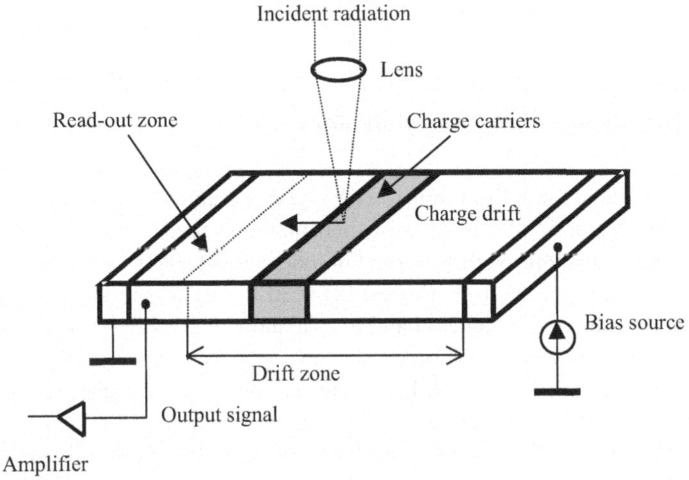

Fig. 9.3 The SPRITE detector operation principle

InSb, they used to remain as very typical IR receivers due to their stability and ability to control detection parameters. By the beginning of the 1990s, the technology of *photolithography* appeared to allow the development of line and 2D array detectors first based on PbS, PbSe and InSb. By the 1970s, most linear arrays used by the military were made of HgCdTe. It is interesting noting that the first linear arrays included up to 200 sensing elements and each had an independent wire lead. With improving parameters of IR detectors, a SPRITE (Signal Processing In The Element) technology was elaborated by Elliott et al. [2]. This kind of detectors cooled up to cryogenic temperatures is still in intensive use until today (Fig. 9.3). SPRITE detectors are made of a HgCdTe photoconductor strip deposited on a sapphire substrate. The strip is stretched in the direction of scanning. The operation principle involves applying a bias current in such way that the charge carrier drift rate is equal to the scanning rate. Since an image point is moving along the strip, the induced charges are moving synchronously and stored by the end of the strip. Hence, the main advantage of SPRITE detectors is their inherent ability to integrate signals, thus decreasing noise.

Long-wave photonic detectors typically should be cooled down to cryogenic temperatures in order to suppress the thermal emission of the sensing element on itself and bring down the depletion of energy levels which occurs because of thermal excitation of impurity atoms. Therefore, the first photonic detectors were mounted in Dewar flasks filled with liquid nitrogen (cooling down to 77 K). Depending on flask volume and design, one had to add liquid nitrogen each 2–8 hours (some devices for continuous nitrogen supply were also suggested). In order to avoid problems related to the use of cooling liquids, Stirling machines were developed on the basis of closed circulation of liquid helium. Such cooling devices were used in the 1980–1990s by Inframetrics (now FLIR Systems). In the case of not very "deep" cooling (over 190 K), quite practical are thermoelectric coolers, but they are power-hungry.

A time constant of photonic IR receivers varies from 10^{-11} seconds in super-fast silicon devices to few milliseconds in PbS detectors.

9.2.4 *Quantum-Well Photodetectors*

In the last decade of the XXth century, a new type of commercial IR detectors which are based on the principle of the so-called quantum wells was developed. Such detectors called Quantum Well Infrared Photodetectors (QWIP) [2, 3] use the effect of photo-excitation of electrons (holes) between the basic and the first excitation levels in the conduction (valence) band of a quantum well (Fig. 9.4). The structure of a quantum well is such that photoexcited charge carriers manage to get out of the well and produce electric current. For instance, one of the first commercial QWIP array detectors used a quantum well of GaAs of the height 45 Å and an AlGaAs barrier of the height 500 Å (see Fig. 9.4). The basic levels of electrons in the quantum well were created by Si impurity. The QWIP detectors using GaAs allow the detection of IR radiation with wavelengths longer than 6 μm.

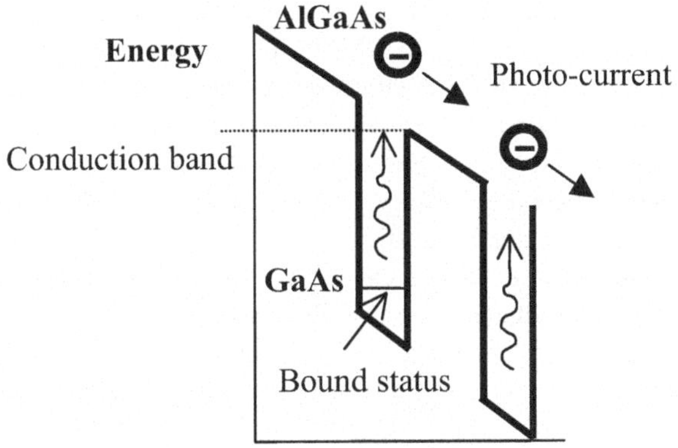

Fig. 9.4 Operation diagram for IR QWIP GaAs-AlGaAs detector in external electric field (absorbed photons move electrons from basis status into conduction band)

QWIP array detectors are becoming increasingly popular in advanced IR imagers, (see Sect. 9.2.5).

9.2.5 Arrays

Photodetector arrays (FPA) which are installed in a focal plane of a lens include a large number of sensing elements placed by lines and columns and thus representing a matrix. Sensing elements do not entirely fill the matrix, therefore a *fill factor* is one of important characteristics of array detectors. The better is a fill factor, the higher is the image quality, in particular, when monitored temperature is close to ambient, and/or temperature gradients are low. In the best arrays, a fill factor can reach 90%.

There are two basic technologies of manufacturing arrays: *monolithic* and *hybrid*. Monolithic matrices are cheaper and simpler in production because their sensing elements are mounted together with signal outputs in the same material, but their typical fill factor is under 55%. In monolithic arrays, it is difficult to realize some advanced built-in algorithms of data processing, e.g. a variable integration time. In hybrid arrays, sensing elements and electrical outputs are placed in two different layers separated by the indium contacts through which *multiplexing* of element signals becomes possible. Such array detectors ensure the highest fill factor and the excellent image quality with temperature resolution up to 20 mK.

A multiplexer is a device which organizes and formats electrical signals generated by sensing elements and brings them into a special processor. From the point of view of the connection to a processor, there are two types of devices: (1) charge-coupled-device (CCD), and (2) complementary metal-oxide semiconductors (CMOS). In CCD detectors, a signal from each element is produced by electrons which are collected up to the last column, afterwards, a resulting signal is read-out. In such architecture, the so-called *charge-coupled transfer loss phenomenon* may appear to cause partial signal loss. Another unpleasant phenomenon is *blooming*, i.e. image blurring due to flowing of excessive electrons from one element to another when monitoring objects with high temperature. Cooled CCDs are more power-hungry to compare to CMOS, and they are widely used in imaging IR cameras where their above-mentioned drawbacks are not crucial. In radiometric systems, some special measures against signal loss and image blurring are needed. CMOS detectors must be effectively cooled down to −200 °C that is typically achieved by using miniature Stirling machines; in this case, the highest measuring capabilities of IR cameras are ensured.

Most IR detectors can be considered as photon counters, therefore, the next important characteristic of IR arrays is the *quantum efficiency* which defines the ability of a detector to collect quantums of electromagnetic radiation and convert them into electrical signals. It is worth noting that the quantum efficiency of very common IR arrays made of PtSi is less than 1%. In general, IR detectors with higher quantum efficiency provide a higher temperature resolution and a better image quality.

Another important characteristic of IR arrays is the *integration time* which is typically 16 or 20 ms that corresponds to a full frame (60 or 50 Hz). Detectors with variable integration time collect IR photons for shorter times, thus allowing observing temperatures up to +450 °C. IR imagers having no variable integration time require using spectral filters for capturing temperatures in the range from −20 to +2000 °C.

The optics of FPA IR imagers is also characterized by some peculiarities. The most advanced is the *reimaging optics* which creates two focal points: at one point, a detector array is placed, and at another point, one installs a viewangle diaphragm which prevents *off-axis stray radiation*. Standard (non-reimaging) optics creates only one focal point and is used mainly in imaging IR cameras where off-axis stray radiation does not affect the measuring performance.

An important drawback of IR arrays is non-uniform sensitivity of array elements. Non-uniformity correction is done in some ways: (1) by measuring temperature of a lens cap (the simplest method), (2) periodically placing in the optical path an object with uniform known temperature by which the response of independent sensing elements is corrected, and (3) performing continuous correction of temperature read-outs by means of a built-in processor (this method is the most metrological).

One more problem in designing high-quality radiometric IR cameras, which is particularly important in the case of uncooled FPA imagers, is the necessity to compensate variations of ambient temperature. Typically, this problem is solved by mounting within an IR camera some contact temperature sensors which perform necessary temperature correction.

9.2.5.1 Uncooled FPAs

The common FPA manufacturing technologies are: (1) vanadium oxide technology, first developed by Honeywell (U.S.A.), (2) ferroelectric BST technology (Texas Instrument, U.S.A.), and (3) amorphous silicon technology (Texas Instrument, U.S.A.). The technologies of microbolometric and ferroelectric FPAs have won fast recognition because such detectors do not require cooling and ensure a reasonable

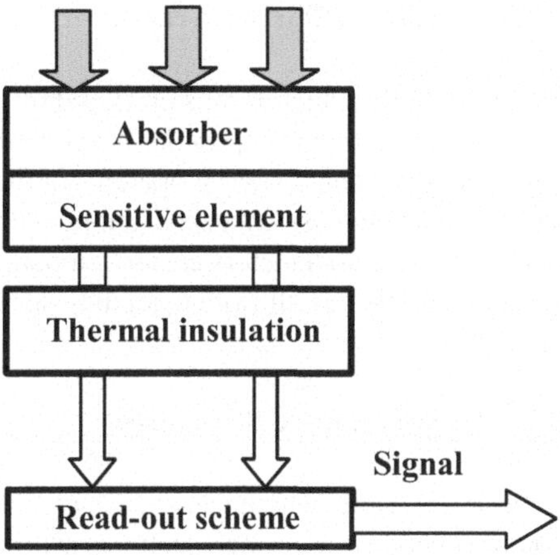

Fig. 9.5 The structure of microbolometric arrays

image quality even operating at the ambient temperature. Main drawbacks of uncooled array detectors, to compare to photonic arrays, are their inferior sensitivity and lower metrological stability.

A scheme of an uncooled array is shown in Fig. 9.5. Each sensing element realizes a principle of resistive, pyroelectric or ferroelectric detector. In resistive elements, the absorbed IR radiation changes electrical resistance that is sensed by electronics. In pyroelectric detectors, at temperatures under the Curie point, temperature variations modulate surface electric charge, thus generating electric current. The pyroelectric effect can be enhanced by imposing electrical field that leads to the ferroelectric phenomenon. Modern arrays possess a layer which absorbs IR radiation; its thickness is about 1 μm. Independent array elements are thermally-insulated each from other to enable maximum spatial and temperature resolution. Similarly to photonic detectors, "thermal" arrays use two architectures: hybrid and monolithic. In hybrid matrices, the sensing elements and the electronic read-out circuitry are manufactured separately and then assembled together. In monolithic devices, the detector is mounted over the read-out circuitry during production that ensures better detection parameters.

A typical commercial microbolometric array consists of 320 × 240 or 640 × 480 elements and provides temperature resolution about 70 mK in ferroelectric detectors and up to 20 mK in silicon microbolometers (in laboratory units, temperature resolution up to 10 mK has been achieved). In 2013, Defense Advanced Research Projects Agency (DARPA), U.S.A., announced the introduction of a 1280 × 720 long-wave FPA operating at 5 μm. Contemporary FLIR X8400sc cameras provide the resolution of 1280 × 1024 elements.

9.2.5.2 Fotonic Arrays

Common photonic arrays use the Shottky barrier, superlattice, intrinsic conductivity and Z-planar technologies.

9.2.5.3 Shottky Barrier Arrays

This type of IR detectors was proposed by Shepard and Yang in 1973 [4]. Presently, many manufacturers propose arrays consisted of up to 512 × 512 elements, mostly based on PtSi. This material operates in the 3–5 μm wavelength range being stable and resistant against enhanced temperatures. In spite of low quantum efficiency (~1%), PtSi photodetectors ensure temperature resolution up to 50 mK. Introducing gallium silicide has allowed extending spectral sensitivity up to 8–16 μm.

9.2.5.4 Superlattice Structures

In these detectors, which can be also QWIP-type, alternate layers of different semiconductors allow tailoring spectral sensitivity. The most perspective technology utilizes GaAs-GaAlAs structure developed in 1987. A cut-off wavelength in superlattices can vary from 6 to 11 μm with detectivity from 10^{10} to 10^{11} $cm{\cdot}Hz^{1/2}{\cdot}W^{-1}$ (cooling from 50 to 70 K). Having quantum efficiency from 5% to 10% at 9.5 μm, these IR detectors enable a very high temperature resolution (15 mK).

QWIP FPA detectors of a large format have allowed developing high-sensitivity IR imagers operating in the long-wave IR range.

9.2.5.5 Arrays on Semiconductors with Intrinsic Conductivity

In these arrays, common photoconductive and photoelectric (photovoltaic) detectors are used. As it is with superlattices, most perspective is the hybrid technology which presumes implantation of sensing element into a silicon substrate of a readout chip. Typical are arrays made of HgCdTe (format up to 256 × 256) and InSb (format up to 640 × 512). The quantum efficiency of InSb detectors can be 80–90% but it is not always reached in practice due to overflowing of quantum wells with electrons. Another disadvantage of these IR receivers is their relatively low stability in time, particularly, when they are subjected to cooling cycles.

9.2.5.6 Z-Planar Arrays

This technology was proposed in order to add simple detection of IR radiation with some functions of preliminary signal processing. Silicon chips are mounted on a thin ceramic plate (thickness about 100 μm), while the IR detector is attached to the side surface of the plate. The plates are assembled together to produce a matrix. In Z-planar detectors, it is possible to perform signal convolution, differentiation and built-in analog-to-digital conversion.

Modern array IR detectors are characterized by the following parameters:

Table 9.1 IR array detectors from Santa Barbara Focalplane, USA

Model	Format	Material	Spectral band, μm	Pixel size, μm	Maximum frame frequency, Hz[a]
SBF184 (snap-shot)	1024 × 1024	InSb	1.0…5.3	19.5	142
SBF184Q (snap-shot)	1024 × 1024	LWIR QWIP	7.8–9.4	19.5	142
SBF180/194 (snap-shot)	320 × 256	InSb	1.0–5.3	30	457
SBF135	256 × 256	InSb	3.0–5.3	30	142
SBF161 (snap-shot)	128 × 128	LWIR HgCdTe	2–10	40	2012

[a]Higher frequency can be achieved in the windowing mode

- HgCdTe: format up to 480 × 640 (array element size from 20 μm to 1 mm); operating temperature T_{op} = 80 K,
- PtSi: format up to 1024 × 1024, T_{op} = 80 K,
- InSb: format up to 1024 × 1024, T_{op} = 80 K,
- uncooled arrays: format up to 640 × 480.

Commercial QWIP arrays of the formats 320 × 256 and 640 × 480 operate at temperature 70 K and provide temperature resolution up to 20 mK in the 8–12 μm wavelength range. Cooled QWIP arrays are manufactured by QWIP Technologies, Santa Barbara Focalplane and BAE Systems (USA), Acreo (Sweden), XenICs (Belgium) and ULIS, a daughter company of SOFRADIR (France). Table 9.1 contains parameters of FPA detectors from Santa Barbara Focalplane, USA The first uncooled microbolometric arrays used in IR imagers were produced by Boeing, by the Honeywell technology. Then, Boeing sold the technology for manufacturing both cooled and uncooled array to DRS (USA). Uncooled microbolometric arrays on the basis of amorphous silicon are produced by ULIS, Acreo and XenICs. The NEC company (Japan) has mastered production of uncooled VOx arrays with the temperature resolution 70 mK in the 640 × 480 format. In Germany, a major producer of IR detectors of both military and dual use is AIM which is specialized on cooled arrays, in particular, made of PtSi. A known German developer of cooled QWIP arrays is Fraunhofer Institute of Applied Solid State Physics. In Israel, InSb matrices are supplied by SCD Semi-Conductor Devices. Chinese home-made IR arrays recently entered the market.

The Russian research in the area of cooled and, in particular, uncooled IR arrays, has remained fragmentary and has not met market requirements. In this context, it is worth mentioning cooled array and respectively IR cameras from "Electron-Optronic" and "NPP Silar", as well Institute of Semiconductor Physics. Uncooled arrays are being also developed at the "Orion" enterprise.

Some commercial array IR detectors and their short description are presented in Fig. 9.6.

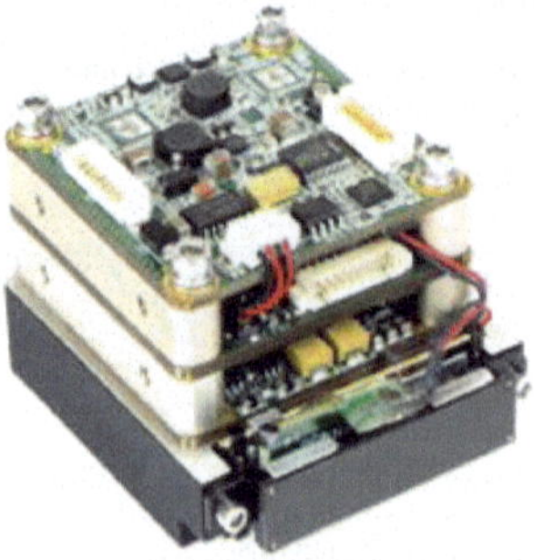

Opgal, Israel

EyeR 640 17 Thermal core
NETD 50 mK
Array 384×288
Spectral range 7.5-14 μm
Frame rate 30 Hz
www.opgal.com

ULIS, France

Pico384P 8-14 mm
NETD 70 mK
Array 384×288
Spectral range 8-14 μm
Frame rate 60 Hz
www.ulis-ir.com

ULIS Pico1024, France

NETD 50 mK
Array 1024×768
Spectral range 8-14 μm
Frame rate 120 Hz
www.ulis-ir.com

AIM Infrarot Module, Germany

NETD 50 mK
Array 640×512
Spectral range 0.8-5.2 μm
Frame rate 100 Hz
Integral Rotary cooler Snapshot operating mode
www.aim-ir.com

Fig. 9.6 IR array detectors and cores

9.2.6 Parameters of IR Detectors

IR detectors are sophisticated opto-electronic devices which are characterized by several parameters of which basic are:

- detectivity $D*$, cm $Hz^{1/2}$ W^{-1},
- spectral sensitivity, μm
- integral (voltage) sensitivity which determines magnitude of electrical signals at the detector output, V/W,
- area of a sensitive element A_d, cm^2,
- time constant, s.

9.2.6.1 Detectivity

The *detectivity* is an important spectral parameter of IR detectors. It was introduced by Jones as:

$$D^* = \frac{(A_d \Delta f)^{1/2}}{NEP}, \tag{9.1}$$

where *NEP* is the *Noise Equivalent Power*, Δf is the frequency band, Hz. In Russia, according to the national standard 21934-83, NEP is the called *sensitivity limit* being expressed in Watt. Often, this parameter is expressed in W/Hz$^{1/2}$, thus taking into account the white noise collected within a particular frequency band. The most correct definition of D^* requires indicating a wavelength, modulation frequency and a unit frequency band: $D^*(\lambda,\ f_m,\ 1) = D^*_\lambda(f_m,1)$.

The *efficiency* of an IR detector by a particular emitter is expressed in the following form:

$$k(T) = \frac{\int_0^\infty s(\lambda) R_\lambda(T)\, d\lambda}{\int_0^\infty R_\lambda(T)\, d\lambda}, \tag{9.2}$$

where $s(\lambda) = D^*_\lambda / D^*_m$ is the relative spectral sensitivity of IR detectors, D^*_m is the maximum detectivity (in the spectral sensitivity band).

When designing IR imaging devices, an analog of detector efficiency called the *comparison coefficient* is introduced:

$$g^{-1} = \int_0^\infty \frac{\partial R_\lambda(T)}{\partial T} D^*_\lambda\, d\lambda \tag{9.3}$$

This parameter allows comparing efficiency of different IR detectors used in radiometric IR devices.

In the former USSR, the nomenclature of commercial (single-element) IR detectors was fairly wide (see the handbook by Kriksounov [5]). After "perestroika" the production of Russian IR detectors has significantly dropped down; moreover, the unpleasant tendency of the last decade has been a permanently increasing lag in regard to the world level, particularly, in manufacturing IR arrays and imaging modules.

9.2.6.2 Spectral Sensitivity

The *spectral sensitivity* of an IR imager is determined by spectral characteristics of a used IR detector (see Fig. 9.2). Common imaging devices, such as TV tubes, operating in the visible and near IR wavelength band, can be considered as temperature evaluating devices only when monitoring objects with relatively high temperatures.

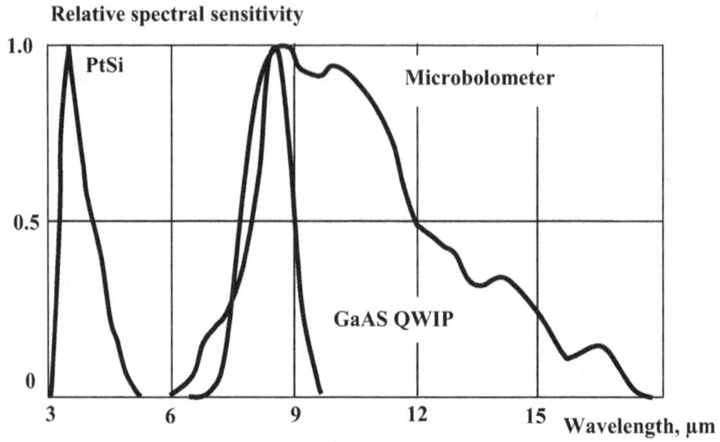

Fig. 9.7 Spectral sensitivity of IR imagers with three common IR detectors

However, the term "IR imager" is typically applied to the devices with the spectral sensitivity in the 3–14 μm band (see Fig. 9.2). Historically, middle-wave (SW) IR imagers were based on InSb detectors operating in the 3–5 μm band, while long-wave (LW) cameras often implemented HgCdTe alloy detectors (7–14 μm).

The relative spectral sensitivities of three common types of IR imagers operating in different spectral bands are compared in Fig. 9.7.

9.2.6.3 Cooling Systems

In order to prevent irradiation by an IR detector by itself and diminish IR detector noise, a sensing element should be cooled down. For a long time, liquid nitrogen has served as the most important cooling agent capable to decrease temperature up to −200°C. Depending on Dewar flask volume, one flask filling was enough for up to 6 hours of work, therefore, an operator had to carry a Dewar vessel or, at least, an everyday thermos. With transition to thermoelectric coolers, the necessity of using liquid nitrogen dropped off. Since thermoelectric devices enable temperature decrease only from −70 to −90°C, in some models of IR imagers, miniature compressors, also known as Stirling machines, are used to cool detectors down to the liquid nitrogen temperature.

The first commercial IR imager of the Thermovision-500 series which operated at the room temperature was marketed by AGEMA Infrared Systems (now FLIR Systems). In this imager, a microbolometric FPA IR detector was used. The necessity to take into account varying ambient temperature required the introduction of some novel technological solutions which represented "know how" of manufacturers. The use of high-sensitive photonic FPA detectors still requires their cooling in order to achieve the best temperature resolution.

9.3 Optics of IR Imagers

In all high-quality IR imaging systems, an optical objective is a sophisticated and expensive component including a set of lenses and mirrors which are made of fragile and precisely-processed materials, such as germanium, silicon and special IR glasses.

9.3.1 Focal Length, Magnification, Aperture Ratio and Optical Efficiency of Optical Objectives

An optical scheme of any complexity can be reduced to a single lens with the *focus length* $F = F'$. Strictly speaking, for a thin lens in air, the focal length is the distance from the center of the lens to the principal foci, or focal points. Such lens produces the image Y' of the object Y as shown in Fig. 9.8. The displacements of the Y and Y' planes of the focal points F and F' are respectively X and X'. The basic equation of a thin lens is:

$$XX' = F^2. \tag{9.4}$$

The focal length is a measure of how strongly it converges (focuses) or diverges (diffuses) light. A system with a shorter focal length has greater optical power than one with a long focal length.

The lens *magnification* is the ratio between the object size in an image and the object true size:

$$Y' / Y = S' / S, \tag{9.5}$$

where S is the distance between the object and the lens, S' is the distance between the image and the lens.

The *aperture ratio* is the ratio between the diameter of the entrance pupil and the optics focal length D/F. This parameter is expressed as the fraction where the

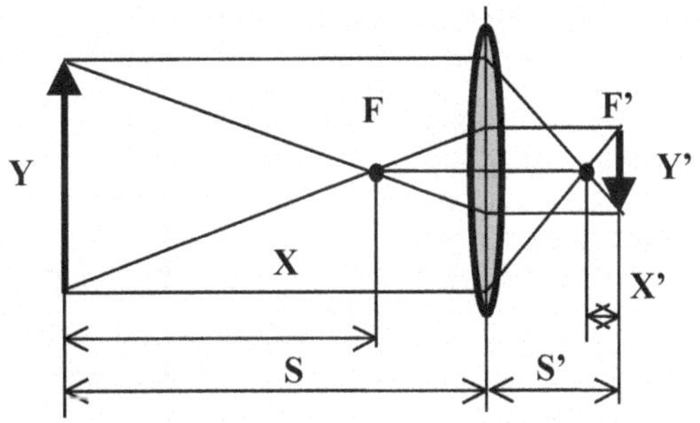

Fig. 9.8 Imaging with a single lens

numerator is unity, e.g. 1:1.5. The maximum possible value of aperture ratio is 1:0.5, in this case, the full conical angle of the rays coming together into the rear focus point is 180°. The reverse quantity *F*/*D* is called the *F-number.*

The *optical efficiency* characterizes the ability of an objective to collect optical radiation:

$$J = \frac{\pi}{4}\left(\frac{D}{F}\right)^2. \tag{9.6}$$

Hence, to compare optical efficiency of two objectives, one should compare their squared aperture ratios.

9.3.2 Instantaneous Field of View and Field of View

If *Y* is the size of an IR detector sensing element, the magnitude $Y/S' = Y/S$ is called the *instantaneous field of view* (IFOV) which determines spatial resolution of an IR system (Fig. 9.9). A scanning imager using a single IR detector at each moment collects radiation within its IFOV which continuously changes its position in the space, thus performing scanning and producing a full *field of view* (FOV). In FPA IR systems, a FOV is related to matrix size while an IFOV is defined by size of a single sensing element.

A FOV of an optical system is schematically shown in Fig. 9.9. When observing remote objects, long-focus narrow-angle objectives are necessary (airborne IR reconnaissance, inspection of electrical transmission lines, etc), while a wide-angle optics is convenient when test objects are near to the operator and larger areas are to be monitored (R & D, building, power production applications). Manufacturers supply IR imagers with changeable optics which can be ordered for particular applications. In many practical cases, a reasonable spatial resolution is ensured by standard objectives with a FOV about 18–20°. Typical changeable objectives provide the following FOVs: 2,5°; 7°; 12°; 20° and 40°.

Table 9.2 shows the size of areas observed with different objectives at the 10 m distance. To determine this parameter for any other distance, one has to perform simple proportional calculation, e.g. the 20° FOV at the 100 m distance defines the size of the observed area as 3.5 · (100/10) = 35 m.

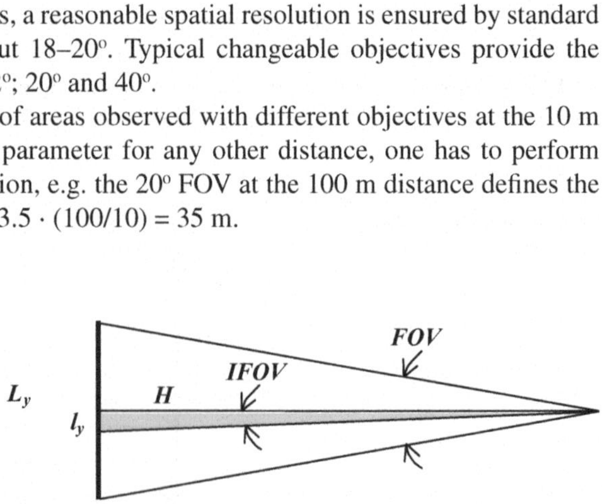

Fig. 9.9 Field of view (FOV) and instantaneous field of view (IFOV)

Table 9.2 Test area size for objectives with different FOVs

Objective	Area size at 10 m, m
7°	1.2
12°	2.1
20°	3.5
40°	8.2

Geometrically, the two parameters above can be defined by two scanned coordinates, e.g.:

$$FOV = 2arctg\frac{L_y}{2H}, \tag{9.7}$$

$$IFOV = 2\ arctg\frac{l_y}{2\ H} \approx \frac{l_y}{H},$$

where L_y is the size of a test area by the coordinate Y defined by FOV, l_y is the size of an area viewed within IFOV (also by Y); usually, $L_x \neq L_y$ and $l_x = l_y$.

9.3.3 Slit Response Function and Modulation Transfer Function

Let a slit of variable width d be placed at the distance H from an IR camera (Fig. 9.10). There is a background object with the temperature higher than ambient behind the slit. By measuring the output signal U as a function of variable slit width, the *slit response function* (SRF) $U(d/H)$ is obtained. An angular slit size which corresponds to the 50% signal drop defines spatial resolution of the system at the 50% modulation. However, while comparing different optical systems, it is important to provide identical test conditions rather than choosing a modulation depth. If a slit angular width is less than the IFOV, the output signal profile corresponds to the pulsed response of the system toward a line; this function is called *line dispersion function* (LDF). The module of the LDF Fourier transform is called *modulation transfer function* (MTF). The MTF is a measurement of the optical system ability to transfer contrasts from the sample to the intermediate image plane at a specific resolution. It can be obtained by determining the contrast of a mira; such the contrast decreases with a higher spatial frequency. A disadvantage of MTF as the quality criterion of IR imagers is the fact that this function determines modulation depth irrespectively noise (see Sect. 9.4.2).

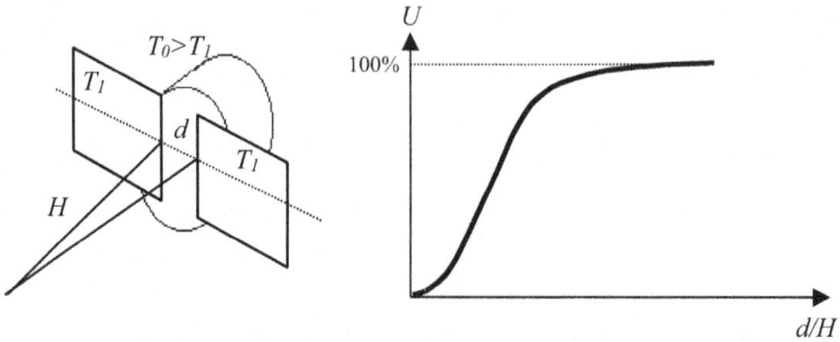

Fig. 9.10 Slit response function

9.4 Spatial and Temperature Resolution of IR Imagers

9.4.1 Image Format and Frame Rate

In FPA IR imagers, image format corresponds to the number of elements in the array. For example, if the typical detector array 320 × 240 corresponds to the 24 × 18^0 FOV, the magnitude of the IFOV will be identical by both coordinates: $IFOV = \frac{24^0}{320} = \frac{18^0}{240} = 4.5' = 1.3$ mrad. Respectively, the spatial resolution in the horizontal direction will be characterized by 320 elements per each of 240 lines.

In digital image processing, the number of *pixels* (picture elements) is introduced. It can be more, equal or less than the corresponding number of scanned IFOVs. For instance, a line signal defined by 320 IFOVs can be digitized for 640 computer pixels; however, such operation will be artificial, and the real spatial resolution will be still equal to 320 resolvable image elements, as before.

The *frame rate* f_{fr} is determined by the FOV scanning total time τ_{sc}. For instance, if each element output signal is electronically scanned for 1/30 s, the corresponding frame rate will be $f_{fr} = 1/\tau_{sc} = 30$ Hz. The same definition of f_{fr} is valid for opto-mechanical IR imagers. In earlier models, two FOVs were scanned and then imposed to create *interlacing* fields. For example, if a FOV consisted of 70 lines is scanned for 1/50 s, the full frame will include 140 lines analyzed for 1/25 s, i.e. $f_{fr} = 25$ Hz.

IR image sequences are stored in either analog or digital form. Analog recording involves using standard video recorders which are connected to IR imagers having PAL or NTSC outputs. Two different formats (thermographic and TV) are matched by means of built-in buffer processors.

Digital image storage can be realized with a rate determined by the IR camera and the analog-to-digital-conversion (ADC) rate. Real time recording is possible if an ADC time is less than an IFOV analysis time. FPA systems are characterized by a high frame rate, for example, in TNDT are becoming common units with f_{fr}~1200 Hz; moreover, the use of the *windowing* principle allows digital recording in the 14-bit format with a frame rate up to few tens kHz, e.g. in the 32 × 32 window.

Modern IR thermographic computerized systems allow storing very long image sequences which may include several thousands of IR thermograms. Recording can be done with a maximum rate, i.e. capturing all images coming from an IR camera or involve a controlled acquisition interval. For instance, a sequence of 300 images recorded with the 30 Hz rate corresponds to the total *observation time* 300/30 = 10 s with the acquisition interval being 1/30 s. By increasing the interval between neighbor images up to 1 s will result in the observation time of 300 s.

9.4.2 Temperature Resolution

9.4.2.1 Noise Equivalent Temperature Difference

The *noise equivalent temperature difference* (NETD) is an important characteristic of IR imagers which is determined as a temperature signal ΔT_{NETD} equivalent to the IR detector noise. Let the temperature T_{ref} of a target considered as a reference be constant. Any IR imager is an electronic device of which output signal can be characterized in terms of both electrical and temperature units. Noise is typically expressed in terms of the temperature standard deviation $\sqrt{\overline{\varnothing T}^2}$. A value of NETD is determined as the temperature variation ΔT_{NETD} which ensures the signal-to-noise ratio S equal to unity: $S = \Delta T_{NETD} / \sqrt{\Delta \overline{T}^2} = 1$ (sometimes, ΔT_{NETD} is accepted to be equal to 2 or 3 standard deviations).

NETD as a parameter of IR detectors is typically reported as follows: $\Delta T_{NETD} = 0.07°\text{C}$ at the reference temperature $T_{ref} = 30°\text{C}$.

In a particular IR imager, a value of ΔT_{NETD} can be evaluated by the formula [6, 7]:

$$\Delta T_{NETD} = \frac{4}{\pi} \left[\frac{FN^2 \sqrt{\Delta f}}{D* \sqrt{A_d} \dfrac{\partial R}{\partial T}} \right] \tag{9.8}$$

where $FN = F/D$ is the F-number of the optical system, F is the focal length, D is the effective diameter of the optical system (entrance pupil), $D*$ is the detectivity, A_d is the effective area of the sensing element, $\frac{\partial R'}{\partial T} = \frac{1}{\pi} \frac{\partial R}{\partial T}$ is the Planck function derivative by temperature related to a unity space angle, W/(m^2·K·sr).

A drawback of the ΔT_{NETD} parameter as a comparison criterion of IR imagers is that the enhancement of temperature resolution by using sensing elements of a larger area A_d is accompanied by the respective deterioration of spatial resolution $IFOV = S'/Y' = \sqrt{A_d}/Y'$.

Eq. (9.8) is valid in the case of the so-called *real* IR detector which is not limited by background noise. In ideal IR detectors, noise is limited only by fluctuations of the photon flux. In this case, the formula for determining ΔT_{NETD} can be found in [6–8].

9.4.2.2 Minimum Resolvable Temperature Difference

The *minimum resolvable temperature difference* (MRTD) ΔT_{MRTD} is assessed by an operator analyzing a bar target. A curve of MRTD *versus* spatial frequency characterizes both temperature and spatial resolution of an IR imager; it depends on operator skills, his motivation and influence of the ambient. Usually, ΔT_{MRTD} values are to be averaged by some operators.

The magnitude of $\Delta T_{MRTD}$is proportional to ΔT_{NETD} and reversely proportional to the MTF, thus determining the ability of an IR imager to resolve details in thermal images. This parameter is considered to be more subjective than ΔT_{NETD}.

9.4.2.3 Signal Dynamic Range

The *dynamic range* (*DR*) of electronic devices is a ratio between the smallest (U_{min}) and largest (U_{min}) possible signal values which are transferred (analyzed) without distortion. Usually, a *DR* is expressed in decibels, dB:

$$DR = 10 \log \frac{U_{\max}}{U_{\min}} \tag{9.9}$$

For example, if an IR detector senses thermal flux of the power from 10^{-6} to 10^{-1} W, its $DR = 10 \rightleftarrows \log \frac{10^{-1}}{10^{-6}} = 10 \cdot 5 = 50$ dB, or 5 orders of magnitude.

In IR imagers, 8-, 12-, 14- and 16-bit signal presentation is used that corresponds to separation of a *DR* by 2^8, 2^{12}, 2^{14} and 2^{16} levels with each level determining a minimum resolvable temperature signal.

Example Let an IR imager measures temperature in the range from −20 to +1200 °C. For the sake of simplicity, assume that the electrical signal is proportional to the measured temperature (strictly speaking, this is incorrect because of a non-linear character of Planck's law). Hence, the total analyzed temperature range is 1220 °C. In the case of 8-bit signal presentation, the signal range will be separated for $2^8 = 256$ levels, of which each level corresponds to the temperature change of 1220/256 = 4.77 °C. Such 'depth' of digitization is insufficient, if the IR detector enables NETD, for example, equal to 0.1 °C. Therefore, the 8-bit presentation does not cover the total *DR* and the manual range adjustment will be necessary while operating such IR camera. Modern IR cameras use 12- or 14-bit digitization level that enables analyzing wide dynamic ranges without loosing temperature resolution and adjusting temperature ranges. For example, the use of the 14-bit (($2^{14} = 16{,}384$)) format provides the digitization step equal to 1220/16,384 = 0.074 °C. However, ensuring the best temperature sensitivity still requires the use of temperature measurement ranges in IR radiometric cameras.

9.5 Modern IR Imagers

9.5.1 Scanner

Until 1995, most radiometric IR imagers used single-element IR detectors, including SPRITE, which, at each time moment, observed only a tiny section of the scene defined by the IFOV. The scene was scanned by means of rotating/oscillating prisms/mirrors. The apparent simplicity of such scanning principle, however, was compensated by the necessity to provide precise performance of opto-mechanical components.

In parallel, electronic scanning has been used in pyroelectric IR imagers that are imaging rather than radiometric devices. With the development of FPA technologies, the problem of reducing the pixel format of IR images to the standard TV format is being rapidly solved. By now, both imaging and radiometric IR cameras of large formats (up to 1280 × 1024 pixels and more) are available.

9.5.2 Electronics

Processing of electrical signals in IR imagers is described in [2, 5–8]. In the built-in electronics of IR cameras, some preliminary processing operations can be performed in real time to present resulting signals in the form which is convenient either for direct visualization (NTSC and/or PAL video formats) or further digital processing (as 8-, 12- or 16-bit matrices).

On-line signal processing by means of built-in microprocessors is typically intended for maintaining required metrological parameters and sometimes for implementing special algorithms, for example, averaging pixel-based values in time, calculating signal derivatives, etc. The purpose of averaging is to reduce the "white" noise of the IR detector. For instance, averaging by 4 frames in an IR imager with the temperature resolution of 0.1 °C will decrease the effective frame frequency by 4 times, i.e. imaging of quickly moving objects and/or transient thermal processes might be distorted, but improve the temperature resolution by $\sqrt{4} = 2$ times, i.e. becoming 0.05°C. Advanced image processing, particularly, when using high-speed IR systems, is fulfilled with the help of specialized computer programs operating off-line.

Images of IR thermographic formats which can be unique in particular IR imagers are typically converted in the TV format in order to facilitate connecting IR cameras to standard TV equipment. Most contemporary IR cameras are supplied with a standard video output thus allowing image reproducing/recording on standard TV monitors/video recorders. IR imagers are supplied with either B&W or color monitors. Also, there are the so-called thermal imaging modules which are to be connected to video cameras; in this way, one can observe either visible or IR image on the screen. Some IR imagers, e.g. FLIR X8400sc, detachable LCD monitors are implemented to enable IR monitoring in places of difficult access.

9.5.3 Black and White and Color Thermogram Presentation

Color presentation of IR images is done in *pseudo-colors*, i.e. each image can be represented with different colors by operator's preference. There are many color palettes proposed in IR thermography. Some palettes, such as "Rainbow", allow the presentation of close temperatures with contrast colors. They are useful when underlining weak temperature nuances. However, an image with many weak temperature details may look deceptive, therefore, professionals use palettes like "Rainbow" mainly for demonstration purposes, while in practical surveying they may prefer the "Gray or similar palettes. In the last decade, the palette called "Iron" has become popular because it combines advantages of both B&W and color image presentation.

It is worth noting that digitization depth in modern IR imagers is from 12 to 16-bit, thus covering broad temperature ranges, e.g. from −20 to +2000 °C, without the necessity to manually adjust temperature ranges during IR camera operation. However, it should be remembered that a dynamic range of typical monitors is more narrow than a potential digitization depth. Therefore, subtle temperature signals may be effectively visualized but a full signal may go out of the used temperature range. In this case, some target areas with a higher temperature will look "white" while lower temperature areas will be "black". However, it is worth noting that, due to deep digitization, no loss of temperature information occurs if such full signals are digitally recorded.

9.5.4 Metrology of IR Imagers

High metrological characteristics of IR imagers are achieved by using built-in temperature references and sensors mounted at some critical (heated) points of a camera. A special microprocessor system keeps temperature readings stable in time and independent (weakly-dependent) on changes of the ambient temperature.

In Russia, all foreign radiometric IR systems supplied to the national market must have certificates issued by Federal Agency for Technical Regulation. A manufacturer calibration certificate supplied with an IR imager is typically valid for 2 years, afterwards, each unit is to be periodically verified. Verification is done by using certified blackbody references. The same certificates should have Russia-made IR imagers.

Temperature measurements with an IR imager are fulfilled by the operator either directly on the monitor screen by placing the corresponding marker or by means of a computer as the post-processing procedure. Some simple measurement options, such as producing spatial/temporal profiles and evaluating area statistics, can be also realized in IR imagers or by using a computer.

It is worth noting that in precise physical experiments, where a high temperature resolution is required and the observation time is long, the problem of maintaining stability of temperature readings is not fully solved; particularly unstable are uncooled array imagers. In some models of IR imagers, the periodical self-calibration which is performed inside the camera, may cause saw-like signal changes reaching amplitude of some fractions of the Celsius degree, thus complicating the processing of image sequences in time.

9.5.5 Recording and Storing IR Thermograms

IR thermograms can be recorded in many ways:

- by analog video recording with the loss of true temperature readings (in some earlier models, digital recording on a magnetic tape was possible),
- digital recording on a floppy-disk (obsolete method today), PCMCIA or flash-card depending on a camera driver; this allows saving true temperature values, and further data processing can be done on a computer (flash-card recording is most common in portable IR cameras used in IR surveying),
- real-time recording images on a high-speed hard disk which is to be acquired separately from the IR imager (this type of data storage was used in the Thermovision 900 model from FLIR Systems),
- digital recording on a computer hard disk by means of a special interface program; in the last years, high-frequency IR images are transferred onto a computer in real-time through the USB or Fire Wire interface (such type of image recording is used in IR imagers intended for research).

Hard copies can be done by:

- off-screen photographing (obsolete at present time),
- video printing,
- computer printing.

Computer programs supplied with IR imagers allow issuing reports containing results of IR surveys. These programs are general to meet inspection requirements in power production, building and industry. Typically, they contain the date of the survey, information about meteorological conditions, IR images with defect indications and their short description or conclusions.

9.5.6 Power Supplies

IR imagers can operate from an alternate current network system or from batteries. Both specialized and standard video camera batteries with operation time reaching some hours are used.

9.5.7 *Software for Data Analysis and Report Compilation*

TNDT software allows inputting data into a computer, as well as specialized data treatment and report compilation. For example, AGEMA Infrared Systems (now FLIR Systems) used to accompany each new IR imager model with a respective software package that allowed a user purchasing a computerized IR thermographic system rather than a simple IR imager: (1) DIPS program with earlier 700th model, (2) TIC-8000 system with the CATS software for the 700th and 800th models, (3) BRUT system for analyzing images from the 800th model in a real time, IrWin package for serving the 400th, 500th and 600th models, as well as the last versions of the ThermaCam Explorer, ThermaCam Reporter, Quick Viewer, FLIR Tools, FLIR Insite App and ThermaCam Researcher programs. Specifically for equalizing sample emissivity the company used to supply the EQUAL computer program which was designed for inspecting electronic equipment by pre-heating a test component to a certain temperature.

Computer programs allow analyzing IR thermograms, in particular, changing IR camera settings, such as emissivity, temperature range, ambient temperature, etc., drawing signal spatial/temporal profiles, fulfilling statistical analysis of IR images, etc. When compiling reports, it is possible to add visible images obtained with a digital photo or video camera (in many ultimate IR imager models, a built-in video camera is incorporated to observe test objects practically under the same viewangle as in the IR channel). Such programs allow using standard Windows options that makes them flexible in compiling and editing documents. Some novel models of IR imagers allow working with several images simultaneously. For example, the IRT Stitch 2.0 software from Grayess allows performing image composition (mosaic) and image fusion (collage). Resulting mosaic images are not distorted and enable inspecting large target areas. IR cameras use specific image formats which represent know-how of manufacturers. A stack of IR mages can be stored either as a sequence of image files or as one big file which can be further separated for single images.

Special software has been developed for working with line-scanning IR radiometers, such as Thermoprofile from FLIR Systems, allowing producing color IR thermograms of revolving objects.

Advanced IR systems are supplied as sets of hardware/software including specialized computer programs which are compatible only with particular models (image formats). Imaging IR cameras enable data recording as standard digital ∗.jpeg or/and ∗.avi images which can be opened with many Microsoft Office programs. However, identifying true temperature values in recorded images requires using specialized software. For example, FLIR Systems has used image recording in an ∗.img or the so-called radiometric ∗.jpeg formats, while NEC Avio, Japan, has used a ∗.svt format. In the USA, a public non-radiometric ∗.fts format, of which description can be found in Internet, is popular.

FLIR Systems supplies free programs ThermaCAM Explorer and ThermaCAM QuickView intended for reading image files, as well as Report Viewer aimed for processing reports. The specialized programs have been ThermaCAM Reporter 7 (for report compilation), ThermaCAM DataBase (for data storage and archiving), ThermaCAM Researcher (for recording fast thermal processing and advanced data treatment) and ThermaCAM RTools (for advanced processing of radiometric images). In the last version of the Image Reporter software, the user can input mathematical formulas for processing temperature data.

The above-mentioned principle of supplying both free software for image reading and high-cost specialized software for advanced data treatment is used by many manufacturers.

Recently, the *Fusion* technology, first proposed by Fluke, has become popular. This technology allows matching visible and IR images in order to better localize thermal findings.

A special class of TNDT software is intended for processing thermal testing results by using some sophisticated algorithms which can be related to peculiarities of heat conduction in test objects. For example, CEDIP, France (now FLIR Systems, U.S.A.) used to supply an Altair LI system intended for the non-contact analysis of mechanical stresses in solids. Tomsk Polytechnic University, Russia, has developed a ThermoFit Pro package which is intended for processing results of active TNDT. The program allows performing preliminary data filtering, polynomial fitting, Fourier transform, principal component analysis (PCA), dynamic thermal tomography, analyzing image statistics and evaluating defect parameters. Other available computer programs ensure the analysis of transient processes, evaluation of heat dissipation, compensation of test object displacements during a test, etc.

9.5.8 Accessories

Many producers of IR cameras propose a variety of accessories of which the most important and expensive are changeable optics and spectral filters (high temperature, flame, atmosphere, solar, carbon dioxide, plastic and long wave filters). Some cameras can be complemented with a removable shield preventing camera overheating in hot atmosphere, batteries, charging devices, software, devices for combining IR and visible images, etc. Changeable optics broadens applications of IR thermography. For instance, building surveys can be done in many cases with a standard $24 \times 18^{\circ}$ lens, while inspecting smokestacks and upper sections of buildings may require using 4, 7 or 12° optics.

9.5.9 Classification of IR Imagers

By technical performance IR imaging systems can be classified for:

- line-scanning, or Forward Looking Infrared Systems (FLIR), and staring systems with scanning by two directions,

- opto-mechanical (first generation) and FPA IR imagers (second generation),
- systems with a cooled (with liquid nitrogen, Stirling machine or thermoelectric cooler) or uncooled IR detector,
- imaging and radiometric systems,
- stationary, portable and gimbal-mounted systems,
- short wave (SW) and long wave (LW).

Kaplan proposed a "commercial' principle of the classification of IR imagers by three categories taking into account technical parameters, application areas and cost.

The first category includes IR imagers of standard applications which use uncooled FPA detectors. A good representative of this category is InfRec R300SR from Nippon Avionics with the following main parameters:

- temperature resolution (NETD): 0.03 °C,
- spectral range: 7.5–13 μm,
- temperature measurement range −40…2000 °C,
- image format: 320 × 240 (640 × 480 in the super-resolution mode by means of InfRec Analyzer NS9500 software),
- frame rate: 8.5; 60 Hz.

The second category includes IR imagers implementing cooled detectors, e.g. on the base of InSb, which operate in the middle IR range. Such is a FLIR GF 335 model from FLIR Systems:

- temperature resolution (NETD): 15 mK,
- spectral range: 3–5 μm,
- image format: 320 × 240 (InSb detector),
- frame rate: 60 Hz.

Finally, advanced IR imagers of flexible configuration mainly used in research are rated to the third category. For instance, a snapshot FLIR A6750 camera from FLIR Systems is characterized with the following parameters:

- temperature resolution (NETD): 18 mK,
- spectral range: 3–5 μm,
- image format: 640 × 512 (InSb FPA detector cooled with a FLIR Closed Cycle Rotary),
- frame rate: 125 Hz (max frame rate 4175 Hz).

Modern IR imaging systems of the second generation realize a module principle and use FPA detectors, both cooled and uncooled. Supplying an IR detector module with a lens converts it into what is called a "thermal module" (a core) which in turn becomes a non-radiometric IR imager after being added with a monitor. However, introducing a measuring temperature function requires undertaking considerable technical efforts and represents a know-how of manufacturers; development of the corresponding technologies is particularly challenging in the case of FPA detectors. The last models of IR thermographic cores, for instance, TAU, Quark and Lepton cores from FLIR Systems, are radiometric.

9.5.10 Commercial IR Imagers

The general information about world producers of IR imagers can be found in Web: *www.directindustry.com/industrial-manufacturer/infrared-camera.* Basic technical parameters of many models are presented in Table 9.3. It is worth noting that some Russian cameras presented in the table are out of date, and some other models have been produced in limited quantities. Below we will shortly consider some typical commercial models which represent IR cameras of different classes (Fig. 9.11). Leading world manufacturers apply a flexible price strategy to make IR imagers strongly dependent on system basic components and accessories, mainly, lenses, filters and software. Approximately it can be assumed that ex work price for typical *imaging* (night vision) systems may be from US$1,0 K to US$3,0 K. *Radiometric* systems of a camcorder design may be in the range from US$10,0 K to US$30,0 K, and the price of specialized models, e.g. mounted on gyro-platforms of flying vehicles, as well as research cameras, may exceed US$150,0 K. Recently, some economic models appeared on the market, i.e. FLIR E53, FLIR ADK, Fluke Ti49FT, Testo 875, etc., pricing from US$1,0 K to US$5,0 K. It is important that, due to intensive competition on the world market, IR cameras of the same class but from different manufacturers are characterized by close prices which also depend on accessory availability and a level of equipment service in a particular country. In the last decade, some Chinese manufacturers of economic IR imaging devices entered the world market. They use mostly IR detectors from ULIS, France, or home-made detectors produced under foreign license.

A variety of IR imagers intended for various applications is produced by FLIR Systems. A low-cost IR camera E4 (~US$1,0 K) is recommended for the inspection of electrical installations. The next price niche is occupied by the models E40, 50, 60 (up to US$4,0 K) which can include changeable optics and communicate with tablet computers thus expending diagnostic possibilities. A medium-price model is T425 (US$14,0 K), while universal diagnostic IR cameras of high quality are the models T620, T640 and T1020 (from US$20,0 K to US$40,0 K including accessories) implementing 640 × 480 and 1024 × 768 array IR detectors. The IR module ThermoVision A655 with the array of 640 × 480 can be used for continuous temperature monitoring in computerized diagnostic systems, including TNDT units where such module is to be added with a built-in heater. The advanced research system is SC3000 (320 × 240 QWIP array, Sweden), as well as SC4000/6000/7000/8000 (US-made IR arrays). The model SC5000 can be added with a microscopic attachment to allow the analysis of small-size objects, while the cutting-edge SC7000 research system can include different types of IR detectors to ensure temperature resolution of 18 mK and the measurement accuracy of ±1% or ± 1°C. Some specialized models of IR imagers ("IR security cameras") are available for both stationary and portable indoor/outdoor surveillance systems (protection of government buildings and airports, anti-terroristic and search/rescue operations, etc.): ThermoVision Ranger III, ThermoVision IS, ThermoVision Security HD, ThermoVision Scout and others. The IR camera ThermoVision WideEye has a

Table 9.3 Technical performance of IR imagers (information by autumn 2019)[a]

Model, manufacturer, country	Temperature measurement range, °C	Temperature sensitivity, °C	Spectral range, μm	Image format	Frame frequency, Hz	Work temperature, °C	Mass, kg	Price[b]
IRTIS 2000, IRTIS, Russsia	−60 …300 (up to 1700)	0.05	3–5 8–12	640 × 480	1/0.8 1/1.6 1/3.2	−20 – +50	1.4 (camera)	18,000$
TKVr-IFP, Institute of Semiconductor Physics, Russia	+20 … + 42 °C (medicine)	0.025 (0.007 in averaging mode)	2.6–3.05	128 × 128	Up to 100	–	3.5	–
TH-4604 MB, Spektr-KSK, Russia	−50…500	0.15	7–14	160 × 120	25	–	2.5	–
TH-4604MP, Spektr-KSK, Russia	Built-in IR thermometer	0.12	7–14	320 × 240	25	–	3.5	–
Boson (core). FLIR Systems. U.S.A.	–	0.04 °C	7.5–13.5	320 × 256 640 × 512	<9. 30. 60	−40 – +80	7.5 g	–
Lepton (core). FLIR Systems. U.S.A.	–	0.05 °C	8–14	80 × 60	9	−10 – +65	0.55 g	–
FLIR C2. FLIR Systems. U.S.A.	−10…150	0.1 °C	7.5–14	80 × 60	9	−10 – +50	0.13	700 $
FLIR T640. FLIR Systems. U.S.A.	−40…500 (up to 2000)	0.04 °C (±2% ± 2°C)	7.5–13	640 × 480 3 lenses	60	−15 – +50	1.7	27,000 $
FLIR T460. FLIR Systems. U.S.A.	−20…1500	0.03 °C (±2%. ± 2°C)	7.5–13	640 × 480 3 lenses	60	−15 – +50	1.7	14,000 $
FLIR T1020. FLIR Systems. U.S.A.	−40 … + 2000	0.02	7.5–14	1024 × 768	30	–	1.9	41,000$
FLIR A325 (module). FLIR Systems. U.S.A.	−20 …350 (up to 1200)	0.07	7.5–13	320 × 240	60 (9)	–	0.7	7000 $
FLIR A655 (module). FLIR Systems. U.S.A.	−20 …120 (up 1200)	0.07	7.5–13	640 × 480	50	–	–	–

FLIR GF 320 (for detecting methane and volatile organic compounds). FLIR Systems. U.S.A.	−40…350	0.08	3.2–3.4	InSb	30	−20 – +50	–	65,000 $
FLIR SC 7000 (research). FLIR Systems. U.S.A.	−20 … + 3000	0.018 (±1%. ± 1°C)	3–5 8.0–12	MCT InSb QWIP 640 × 512	380 (up to 30 kHz)	−20 – 55	4.95	–
FLIR SC 8400 (research). FLIR Systems. U.S.A.	5 … + 300 (up to 3000)	0.020 (±1%. ± 1°C)	1.5–5.1	1280 × 1024	106	−20 – 50	5	=
FLIR One (for Android and OiS smartphones). FLIR Systems U.S.A.	−20…120	0.15	7.5–13	160 × 120	8.7	0–35	32 g	250 $
Testo 875 (Testo AG. Germany)	0 … + 280 (−20 … + 100)	0.1	8–14	160 × 120 32 × 23° 9 × 7°	9	−15 – +40	0.9	–
Testo 880–3 (Testo. Germany)	−20 … + 100 0 … + 350	<0.1	8–14	160 × 120	33 (9)	−15 – +40	0.9	–
Ti 55FT (Fluke. U.S.A.)	−20 … + 600	0.05	8–14	320 × 240	60	−10 – +50	1.95	–
Ti 90 (Fluke. U.S.A.)	−20 … + 250	015	9–15	80 × 60	9	−10 – +50	0.73	–
T6-P. Dali, China	−20…350 (up to 600)	0.06	8–14	384 × 288	50	−15 – +50	0.91	–
TH-3108. NEC Avio. Japan	0 ÷70 (medicine)	0.1		HgCdTe 261 × 207	1/0.8	+5 – +40	3 (camera) 3.8 (unit)	59,000 $

(continued)

Table 9.3 (continued)

Model, manufacturer, country	Temperature measurement range, °C	Temperature sensitivity, °C	Spectral range, μm	Image format	Frame frequency, Hz	Work temperature, °C	Mass, kg	Price[b]
MiKroScan-9100WV. or Mikron. U.S.A.	−40 … + 2000	0.06	8–14	320 × 240	60	−15 – +50	1.6	–
MikroScan-7515. Mikron. U.S.A.	−40 … + 500	0.08	8–14	320 × 240	30; 60	−15 – +50	–	35,000 $
TS-7302. NEC Avio. Japan	−40 … + 2000	0.08	8–14	320 × 240	60	−15 – +50	1.4	–
PalmIR Pro. Raytheon. США	Staring	0.1 °C	7–14	320 × 240	30	−20 – +49	1.8	18,000 $
Series 300. Raytheon. U.S.A. (last version ControlIR 300D)	Staring	0.08	7–14	320 × 240	30	−20 – +49	0.68	13,000 $
ProtectIR 400. Raytheon. U.S.A.	Staring	–	7–14	–	50; 60	−40 – +55	4.5	–
IRI 1011. IRISYS. Great Britain	−10 … + 300	0.3	8–14		8	−5 – +50	0.6	–
SAT-GM18. Guangzhou Sat Infrared technology. China	Staring	0.25	8–14	–	–	–	2.5	20,000 $
SAT-HY6800. Guangzhou Sat Infrared Technology. China	−10 … + 2000	0.08	8 … 14	320 × 240	50	–	2.5	48,000 $
Ir913. Wuhan Guide Electronic Industrial Co. China	−20 … + 1500	0.06	8–14	320 × 240	50	–	–	20,000 $
DI-9800. Digital Imaging Infrared. U.S.A.	Staring	0.05	Optional	320 × 240	–	0 – +70	1.8	28,000 $
μCAM. Germany	Staring	0.1	3.2–5	PtSi 256 × 256	–	–	1.7	–

VARIOSCAN-3022. 3021. Jenoptik Laser Optik Systeme GmbH. Germany	−10 … + 1200	0.12	8–12	HgCdTe 360 × 240	1/0.9	−10 - +40	5	–
VarioTHERM. Jenoptik Laser Optik Systeme GmbH. Германия	−25 … + 1200	0.1	3.4–5	PtSi 256 × 256	50	−10 … +50	3.9	–
VarioCAM Inspect 170/25 и 130/25. Jenoptik Laser Optik Systeme GmbH. Germany	−40 … + 2000	0.1	8–13	320 × 240	50	−15 – +50	2.2	62,000 $
VisIR Ti 100. Ti 200 Indigo Systems and Thermoteknix Systems. U.S.A.	−50 … + 500	0.085	7.5–13	160 × 120	30	–	1.6	20,000… 30,000 $
MicroIR LTC550. BAE Systems. U.S.A.	Staring	0.1	8–14	320 × 240	60	−20 – +50	2.5	15,000… 25,000 $
IR Snapshot. ISI. U.S.A.	−30 … + 600 °C	0.1	8–12	120 × 120	1/1.5	0 – +50	<2	–
Miricle 110K (370K). Thermotechnics Systems. Great Britain	–	0.06 (0.05)	8–12	384 × 288 (640 × 480)	50/60 (240)	−20 – +50	0.166	–

[a]The most representative models are described. Some Russian models have been produced in limited quantities.
The information in this table is not provided for advertisement purposes and it should not be taken as an endorsement of the authors.
[b]Some prices are given on the base of private information and cannot be taken as official; prices depend on accompanying software and accessories.

Fig. 9.11 Examples of modern IR imagers: (**a**) MilSight HISS S100 thermal weapon sight (U.S.A.). (**b**) Opgal Mini Tavor IR imager for night driving (Israel), (**c**) FLIR Traffic One traffic IR sensor (U.S.A.), (**d**) PalmIR Pro IR staring camera (U.S.A.), (**e**) TH460MP IR staring camera (Russia), (**f**) Testo 869 radiometric imager (Germany), (**g**) FLIR T425 radiometric imager (U.S.A.), (**h**) FLIR 640 IR radiometric imager (U.S.A.), (**i**) Boson IR radiometric core (U.S.A.), (**k**) FLIR SC 6000 radiometric system (U.S.A.), (**l**) IRTIS 2000 radiometric system (Russia), (**m**) KTIPM IR imager (Russia)

super-wide viewangle (180°) and enables identification of human beings at the distance up to 150 m. With a proper lens, missed persons can be detected even over a distance of several kilometers. Another system of continuous operation is the model ThermoVision Sentinel which allows auto-focusing of thermal images; this camera is supplied with the InstAlert option which highlights warm objects, such as people and vehicles, in red for immediate operator attention. The module system Triple QWIP Module is sensitive in the 8–9 μm spectral band and therefore cannot be blinded by enemy's laser weapon. This system allows three FOVs with the respective three IFOVs (0.054, 0.3 and 1.37 mrad). The modules ThermoVision Integration Series can be integrated with the standard observation system Pelco EH2508. In Russia, Pergam-Engineering developed a Titan system which allows crispy thermal images under various metrological conditions (full darkness, sunny days, fog, snowfall, etc.). Its unique feature is the ability to operate at ambient temperatures up to −65 °C.

For users' convenience, IR imagers from FLIR Systems are specified by special letters depending on an application area. The models E, C and T which effectively combine both IR and visible channels are oriented toward industrial experts being typically used in the inspection of electromechanical equipment, building envelopes, smokestacks and other potentially-dangerous industrial objects. A price-to-quality ratio for these IR imagers is well justified when inspecting electrical installations; for example, the introduction of three measurement points on the IR imager monitor facilitates the checking of three-phase equipment, and the Thermal Fusion option enables superposition of the IR and visible images to improve defect localization. Moreover, communication with these units can be performed via Wi-Fi thus accelerating preparation of reports and transferring test results to a host computer and some mobile devices, such as iPad, iPhone and iPod Touch. The model line specified as "T" is convenient for the professional servicing of industrial installations, i.e. fulfilling predictive maintenance and repair. The LS and H types of IR imagers are related to night vision equipment. To continuously control technological processes, one may use IR modules of the series "A", which can be easily built in automated industrial systems as non-contact thermometers. The specification "SC" relates to the units which are mostly recommended for research; such IR cameras are characterized by the highest temperature resolution and frame frequency. Finally, the IR imager models specified as "GF" (GasFindIR) allow the detection of gas leaks from piping in a real time with a leak being visualized as a "smoke indication" on the IR camera monitor. In the last decades, IR thermography has been increasingly used in civil engineering as a tool of energy audit in building. Although, there are no specific "building" IR cameras, one may summarize some features which are indispensable for the successful use of such cameras in energy audit. To allow determination of some quantitative parameters of building thermal protection, an IR imager should: (1) ensure a relatively high temperature resolution and good measurement accuracy, (2) measure temperatures in the range from −20 to +120 °C (in northern areas, the measurement level should be as low as −40 °C), (3) operate under low ambient temperatures, (4) include changeable optics for surveying tall buildings, such as skyscrapers and smokestacks, (5) have an IR detector array not worse than 160 × 120 and image frequency higher than 8 Hz, (6) include a hygrometer and thermal fluxmeter to measure both a dew point and thermal flux in building envelopes, (7) allow producing mosaic images by using the corresponding software.

9.5.10.1 Night Vision Units

IR thermographic night vision cameras are widely spread in military applications. IR cameras are mounted on helicopters, airplanes, vessels, tanks, drones, quadrocopters and used as thermal weapon sights to detect, recognize and identify targets from extreme ranges. Such sights can be attached and re-attached without affecting the boresight of the day scope and allow performing sniper and reconnaissance missions with targets out to 2 km. An important characteristic of military IR imaging equipment is its ability to operate within a broad range of ambient temperatures. Figure 9.11a shows the MilSight HISS S100 long range thermal weapon sight from FLIR Systems which ensures high-sensitive mid-wave thermal imaging and weighs 1.6 kg (*www.flir.com/surveillance*).

9.5.10.2 Opgal Mini Tavor IR Imager for Night Driving

In the last years, the Opgal company, Israel, has become active on the world market of IR equipment. For night driving in complete darkness, through smoke and dust, the Opgal Tavor DV60 system operating in combination with a CCD camera was developed. This system uses a 640 × 480 IR detector and a wide-angle (60°) optics (Fig. 9.11b). For the same purpose, FLIR Systems developed a FLIR Path Finder unit. It is worth noting that IR imagers of such class and in fact any other IR imagers can be considered as night vision units.

In 2016, FLIR Systems announced the manufacturing of the FLIR Traffic One sensor including a long-wave Lepton micro-camera using an FPA of 80 × 60 pixels. This intellectual sensor is installed outdoors for traffic applications with on-line data transmission to the technical control room where the traffic manager will receive a visual image of the scene when an alarm is generated (Fig. 9.11c). FLIR TrafficOne is independent on sun glares and shadow, and its main applications are: intersection control, pedestrian safety and mobility, highway monitoring, automatic incident detection, data collection and flow monitoring, fire detection in tunnels, trackside monitoring, etc.

9.5.10.3 PalmIR Pro IR Imaging Camera

A small-size PalmIR Pro unit manufactured by Raytheon, U.S.A., is a typical representative of low-cost IR imaging cameras intended for both night vision and technical diagnoctics (Fig. 9.11d). This camera includes a long-wave 320 × 240 IR array detector and a 2.5-inch monitor; IR images are recorded on a Compact Flash card. Unit mass is 1.8 kg.

9.5.10.4 TH460MP IR Imaging Camera

The Russian company Spektr-KSK elaborated this IR imager by using a US-made 320 × 240 IR array detector and a high-quality Russian-made optics. By technical performance, this unit corresponds to the world level but a lack of a temperature measurement option prevents its using in quantitative IR diagnostics and TNDT.

9.5.10.5 Testo 869 and 875 Radiometric Imagers

The international company Testo developed the economic IR imager Testo 869 (Fig. 9.11f) which ensures the temperature resolution of 120 mK with the spatial resolution 160 × 120 and frame frequency 9 Hz (temperature measurement range is from −20 to +280 °C, viewangle 34 × 26°). Such low acquisition rate is conditioned not by technical limits but by the company policy to supply this IR camera on the international market without a license from the corresponding U.S. body (State Department or Department of Commerce). Such license is indispensable in the case of high-speed IR imagers with large detector arrays. IR cameras with low spatial resolution are used in technical diagnostics of electrical installations, while their use in building applications is limited. In addition, economic IR imagers cannot typically operate under low ambient temperature that is the case in building.

The next Testo 875 model of this series is characterized by the following parameters: temperature resolution 80 mK, spatial resolution 160 × 120 and 32 × 23° standard optics.

9.5.10.6 FLIR T425 Radiometric Imager

The FLIR T425 imager (Fig. 9.11g) is supplied with the sensoric display and the option called Picture-in-Picture (PiP). It is recommended for industrial applications by professionals specialized on the inspection of electromechanical installations, With the IR array of 320 × 240 pixels this imager ensures high temperature resolution (50 mK), can measure temperature up to 1200 °C and possesses a number of useful functions, such as quick report compilation, periodic recording and video recording. This is the first IR camera from FLIR Systems which has a multifunctional sensoric display and allows making schemes and defect marking directly on the screen by using a stilus.

9.5.10.7 FLIR T640 IR Radiometric Imager

The FLIR 640 model is considered to be the first professional IR imager on the base of a 640 × 480 element IR array combining an IR camera, GPS binding of IR images, an integrated 3.2 megapixel digital video camera with a lighting lamp, and a laser pointer (Fig. 9.11h). A supply set may also contain accessory lenses: 5 × 19°,

15° × 11° and 45° × 34°. With its broad temperature measurement range (from −40 to 2000 °C, high temperature resolution (40 mK) and affordable price (under US$30,0 K), this unit is one of the best universal cameras with preferable applications in technical diagnostics. FLIR SC660 version includes the FLIR Thermal Fusion option for visual/IR image fusion and the PiP option allowing adjusted image sizing and location.

9.5.10.8 Boson, Lepton and Muon IR Radiometric Cores

The Boson longwave IR camera core is characterized by a remarkable technical performance: vanadium oxide (VOx)-based 320 × 256 and 640 × 512 IR array, temperature resolution 40 mK, and embedded industry-standard interfaces (Fig. 9.11i). This unit may operate under ambient temperatures from −40 to +80°C, and one of eight optional lenses provides a viewangle from 4 ° to 92 °. Its weight and size revolutionarily reached 7.5 grams and 4.9 cm^3.

However, miniaturization reached its reasonable limit in the Lepton core. This core is a base for the smartphone FLIR One camera for Android and iOs; it is also used in the hybrid FLIR TG165 device which combines features of an IR pyrometer and imager, as well as in the FLIR A × 8 device intended for the control of industrial processes. The main Lepton characteristics are: wavelength range 8–14 μm, temperature resolution <50 mK, spatial resolution 80 × 60, mass 0.68 (!) grams and size 8.5 × 8.5 × 5.9 mm.

FLIR Muon thermal camera subassembly was developed for thermal product OEMs being optimized for size, weight and power (640 × 512 VOx microbolometer, 7.5–15 μm spectral range, 60 mK temperature resolution, 60 Hz frame frequency).

Thermal cores can be mounted on helicopters, drones, quadrocopters or other types of Unmanned Aerial Vehicles (UAV) that requires from a pilot to have the corresponding certificate.

9.5.10.9 FLIR SC 6000–8000 Radiometric Systems

This IR thermographic system intended for research purposes (Fig. 9.11k) possesses temperature resolution<25 mK (typical value 18 mK), image format 640 × 512 pixels and frame frequency from 1 to 125 Hz in the full-image mode (in the windowing regime, frame frequency may reach 36 kHz). The IR camera contains no built-in shutter normally used for self-calibration, therefore, the output signal is not corrupted with sharp signal drops which appear in shutter-containing imagers during long observation time. The next model of this series is FLIR SC7000 added with the CNUC and Hypercal technologies which allow avoiding the correction of array non-uniformity and working in a wide dynamic range. This system is the development of a high-speed IR Phoenix imager which was manufacturing by Indigo (U.S.A.), the company recently purchased by FLIR Systems. In the last

FLIR SC8400 model of this series, the 1280 × 1024 array is used to ensure thermal sensitivity of 20 mK.

9.5.10.10 IRTIS Radiometric System

For the last two decades, the Russian small-business company called IRTIS successfully developed ideas of earlier Soviet IR imagers which used the concept of two-coordinate opto-mechanical scanning. The most known IRTIS 2000 (Fig. 9.11) model is announced as a high-precision scanning IR camera recommended for both visualization and quantitative analysis of relatively slow temperature distributions. The enhanced metrology of this camera is conditioned by the fact that each pixel signal value is related to the single IR detector thus ensuring uniform temperature sensitivity across the whole viewangle. IRTIS 2000 is based on either InSb (3–5 μm) or CdTeHg (8–12 μm) IR detector; the spatial resolution is 640 × 480 and the temperature resolution is 50 and 20 mK respectively. It is worth noting that such high temperature sensitivity is determined by the long frame time (0.8, 1.6 and 3.2 s) and cooling of the IR detector down to the liquid nitrogen temperature (once filled Dewar flask ensures camera operation for 6 hours).

For scientific research purposes is of interest the IRTIS 2200 imager which has few world analogs. This model allows obtaining three simultaneous images: two images in the 3–5 μm and 8–12 μm wavelength bands and one visible image. All three images can be outputted on the monitor screen, recorded on the hard disk and further processed by the computer.

9.5.10.11 Russian FPA IR Cameras

In Russia, FPA cooled detectors were developed at Institute of Semiconductor Physics, Russian Academy of Science. The small-size KTIPM IR imager (Fig. 9.11m) uses a 384 × 288 FPA operating in the 3.4–4.8 μm spectral band (IR detector work temperature is 77 K, temperature sensitivity 40 mK). The recent development is a cooled 640 × 512 IR array on the base of InSb. In the last years, Russian microbolometric IR arrays were also developed. The corresponding IR imagers with weight about 1.5 kg provide temperature resolution of 50 mK and 50 Hz frame frequency.

9.5.11 Choosing an IR Imager

Since IR imagers are universal devices for a contactless analysis of objects by their thermal radiation, users may try to use them for solving a variety of practical measurement and diagnostic tasks, However, while acquiring such sophisticated equipment, one should clearly define its potential application areas in order to optimize a quality/price ratio, For example, the use of an IR imager as a means of the long-distance high-sensitivity night vision device is hardly possible without a narrow-angle long-focus optics, but the temperature measurement option is not obligatory in this case, A thermal imaging monocular Scout II from FLIR Systems can serve as a simple means of IR thermographic observation, however, the best solution would be an advanced FLIR T1020 system. In the military area, various IR imagers from Raytheon are fairly popular. In electrical engineering, particularly, in indoor inspections, an optimal unit can be FLIR E53, In some application areas, availability of spectral filters may be important to reduce solar reflections or improve accuracy of temperature measurements of some specific substances, such as hot glass, plastics, flames, etc. Also, in the industry, IR imagers should be typically portable and battery-energized, while in the medical diagnostics IR thermographic equipment can be stationary and AC-energized. Finally, the scientific research requires using advanced IR thermographic systems, such as ThermaCAM SC 6000 or 8000 or other models which enable capturing image sequences with a high frequency.

In the last decades, on the international market appeared NEC Avio, Japan (TH7800 and TH9100 models), Jenoptik, Germany (VarioCAM Inspect 130/25 and 170/25 models), Fluke, U.S.A. (Ti30 and Ti55 models), and the Chinese producers Guangzhou Sat Infrared Technology and Wuhan Guide Electronic Industrial. The latter company surprised customers with a MobIR model of which design reminds a mobile phone; the drawback of this IR imager was the necessity of its periodical calibration by the imager cap temperature, As mentioned above, due to strong competition between leading manufacturers, technical parameters of similar models, as well as available software options, are close, and many customers make decision by a unit cost which strongly depends on the type of a used array. High-quality IR uncooled arrays were developed by Boeing, afterwards, the respective technology was sold to DRS. In the U.S.A. and some other countries, export of advanced IR arrays requires obtaining export licenses of different levels, e,g. from U.S. Department of Commerce. Therefore, unlicensed arrays have been commercialized in some countries. IR imagers using no-license IR arrays can be supplied without limitation (such imagers typically have a special marking). In Europe, IR array detectors from ULIS, France, are quite popular; export limitations imposed on the sale of these arrays are less fastidious than in the case of U.S. IR detectors. In technical diagnostics, the differences between licensed and no-license IR arrays can be insignificant but in scientific research the use of high-quality IR detectors, which ensure excellent measurement capabilities within a whole field of view with the temperature resolution up to 30 mK, is preferable.

9.6 IR Systems for the Visualization of Gas Leaks

In addition to standard IR imagers, IR systems which utilize the phenomenon of selective absorption of thermal radiation in gas media have been developed. The use of corresponding spectral filters allows the detection of particular gases in a cloud.

IR leak detection devices can be *passive* and *active* (see Chap. 11). The illumination of a scene with a laser operating at an absorption wavelength is expedient because the power of thermal radiation emitted by physical bodies is very low in narrow wavelength intervals, and high-sensitive spectrometric systems are necessary to ensure reliable detection.

Especially for the petrochemical industry FLIR Systems developed a series of ThermaCAM GasFinder IR imagers which, due to built-in spectral filters, allow the reliable detection of leaks of up to 19 gases at reasonable distances.

Active IR thermographic systems implementing laser illumination (CO_2 laser) were developed by Inframetrics (now FLIR Systems). A Gas Vue TG-5 model was intended for the operation on technological lines, heat exchangers, tanks and hermetic vessels. The probing gas SF_6 is to be inputted into an object to be tested. The spatial distribution of the reflected laser radiation is monitored at distances up to 5 m. A minimal detectable leak is 50 g/year. This system was portable and consisted of three units: a laser IR thermographic camera, a power supply with a cooling device and consoles. A MG-30 model was intended for detecting leaks from tanks, piping, vessels, etc. At distances up to 30 m, this unit enables the detection of 77 gases of which 5 were chosen for calibration purposes. A leakage cloud has been seen on the monitor as a dark area corresponding to enhanced absorption of the reflected laser radiation. A minimal detectable leak is 1 kg/year.

9.7 Imaging Systems in the Terahertz Wavelength Band

Terahertz wavelengths are in the range from 10 μm to few mm, therefore, the corresponding visualization systems can be called *Terahertz imagers* which occupy an intermediate position between IR and microwave systems. The interest to these devices has recently boosted due to increased development of anti-terroristic technical means.

The illustration of using terahertz systems will be given in Chap. 11.

9.8 IR Line Scanners

By technical performance and cost, line-scanning IR systems occupy the intermediary position between IR imagers and thermometers. The term "scanning" is applied to these systems because, for decades, they employed the principle of opto-mechanical scanning. Contemporary units implement linear photodetector arrays, thus significantly simplifying scanners design.

Line IR systems can be classified for two large groups: (1) airborne systems mounted on aircraft, satellites and space stations, and (2) industrial systems.

First-type units are characterized by narrow viewangles. Their scanning system is to be linked to the flight altitude *H* and speed *V*, or more correctly, to the ratio

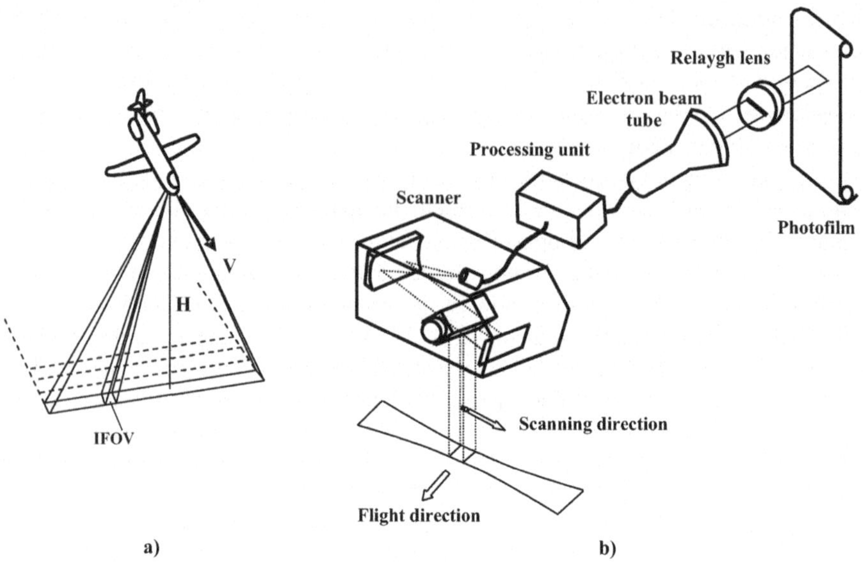

Fig. 9.12 Line scanners mounted on flying vehicles: (**a**) aerial surveying scheme, (**b**) air borne IR opto-mechanical scanner

V/H, in order to provide the compact lay-up of scanned lines (Fig. 9.12a). Too high *V/H* values will lead to missing some areas on the scanned surface, while a low *V/H* ratio leads to line overlapping. Such units are being developed for performing IR reconnaissance, mainly, for military purposes. Noticeable world producers of line scanners are FLIR Systems, Raytheon, Digital Imaging Infrared and others. Opto-mechanical IR scanners of a high technical performance were developed in the former USSR, merely to mention an airborne line-scanning IR imager "Vesuvium-EK" (further modification of the "Voulkan" system) developed for geophysical surveys. With the viewangle of 84°. this IR system ensured spatial resolution 1 mrad (best value 0.2 mrad) and operated in four spectral bands. Thermal images were continuously recorded on the photofilm (Fig. 9.12b) or into the electronic memory and further transmitted to a tracker station. In satellite monitoring, multi-spectral systems are widely used to monitor the electromagnetic radiation of the Earth in several (up to 24) wavelength bands. Satellite and aircraft-based IR reconnaissance systems may implement various scanning schemes different from that in Fig. 9.12. Their description is beyond of the scope of this book (e.g. see [9] for details).

In the systems intended for industrial use (line-scanning IR thermometers), a scanning frequency is constant and the accompanying software allows producing continuous thermal images of moving (revolving) objects. The major consumers of such systems are in metallurgy, cement and glass production, as well as in the pulp-and-paper industry (see Table 9.4 and Chap. 11). Figure 9.13a shows the appearance and some technical parameters of a Thermoprofile-50 unit from FLIR Systems. Some models of IR line scanners have been developed in Russia (Tomsk Polytechnic University, Omsk State Technical University et al.) for monitoring temperature on

Table 9.4 Line IR systems

Producer	Measurement range, °C	Temperature resolution, °C	Spectral range, μm	Line frequency, Hz, and/or image format and viewangle	Mass, kg
Airborne systems (mounted on gyro-platforms)					
Thermovision 1000/Argus 350, FLIR Systems, USA	–	0.08	8–12	5-element SPRITE detector 800 × 445	25 kg – gyro-platform Argus 350
DI – 2000, Digital Imaging Infrared, USA (on the basis of DI-9800 system)	–	0.08	8–12	320 × 240	–
IR scanner, Russia	Imaging	0.1	3–5 8–14	–	–
Vesuvium-EK, Russia	Imaging multi-spectral aircraft mounted	0.1…0.3	0.4–1.0 3–5 8–12	12.5 25 50 100	80
Industrial systems					
Thermoprofile-HT FLIR Systems, USA	+75 … +600	1	3–5	8–20 Hz 90°	–
Pyroline 256, DIAS/CMV, Germany	0 … +80 50 … +550 450 … +1250 600 … +1300	0.5	8–12 1.4–1.8 4.8–5.2 3–5	136 Hz 272 544	3.2
HZK 256, DIAS/CMV, Germany (brake disks diagnostics)	250 … +900	0.5	1.3–1.8	5000 Hz	–
HRP 250, DIAS/CMV, Germany (building)	50 … +180	0.5	3–5	Up to 18 kHz	–
Pyroline 128 L, DIAS/ CMV. Germany	0 … +80 50 … +550	0.2	8–14	Up to 256 Hz	3.2
Thermoprofile-50 FLIR Systems, USA	0 … +1600	–	3–5	48 Hz 256 elements/ line	–
GS100, Raytek, A Fluke Company, USA (glass production)	+20 … +950	–	–	36 or 48 Hz	–

(continued)

Table 9.4 (continued)

Producer	Measurement range, °C	Temperature resolution, °C	Spectral range, μm	Line frequency, Hz, and/or image format and viewangle	Mass, kg
MP50, Raytek, Fluke, USA (glass production)	+20 … +1200	–	1, 1.6, 3.43, 3.9, 5.0, 3–5	36 Hz	–
ICSA-04, Russia	+100 … +450	–	2–20 (pyroelectric)	2 Hz 150 elements/line	8
Introcon-05, Russia	+50 … +600	–	1.5–5.3 (PbSe)	2 Hz 360 elements/line	6
STK-1, Russia	+100 … +500	Accuracy 3%	2.5–5.5	4 Hz	5
TSS-1, Russia	+200 … +650	1	–	150 elements/line Viewangle 120°	
Ivolga, Russia	−20 … + 170	0.1	3–5	150 elements/line Viewangle 40°	1.5

Temperature measurement range: 0 … +1600°C
Line frequency: 48 Hz
FOV: 90°
Elements/line: 256

a)

Temperature measurement range: +50 … +600°C
Line frequency: 8 Hz
FOV: 105°
Elements/line: 360

b)

Fig. 9.13 IR line-scanning systems: (**a**) Thermoprofile-50, FLIR Systems (obsolete unit), (**b**) Introcon (Russia)

surface of revolving kilns, e.g. widely used in cement production. An Introcon-05 system is shown in Fig. 9.13b.

9.9 IR Thermometers (Pyrometers)

IR thermometers (optical pyrometers) are intended for non-contact temperature measurement in local zones chosen on the surface of objects to be tested. In TNDT, this kind of IR equipment is not ubiquitous, except some specialized systems where

IR thermometers are combined with spot heaters. However, they are typical means of passive technical diagnostics in building, electrical and thermal engineering. There are three types of these devices: portable IR thermometers, built-in temperature sensors and research TNDT units, including IR microscopes. Like IR imagers, IR thermometers used in industry are to be certified by an appropriate body and checked every 1–2 years.

The main technical characteristics of IR thermometers are:

- temperature measurement range (standard range is from −30 to +900°C but it can be expanded beyond these limits, for example, up to +3000°C),
- Distance-to-Spot (D:S) value, or the ratio between the distance to the test object and a diameter of the monitored spot; for example. The value 80:1 means that at the distance of 8 m the diameter of the monitored spot is 10 cm,
- principal measurement accuracy (in high-level units it might reach ±1% or ±1°C),
- time constant (depending on a type of the used IR detector, it can be from few μs to hundreds ms),
- operating temperature (many units operate at ambient temperatures from 0 to +50°C, therefore, their usage at sub-zero temperatures may require some "tricks", for example, keeping a unit in clothing or performing measurements for short time; in any case, sudden changes in ambient temperature worsen the metrology of IR thermometers),
- presence of a laser pointer defining a true size and shape of a test area, type of battery, mass and size.

Optical pyrometry as an applied technical field is well explored in both theoretical and practical aspects. The main R & D customers are in metallurgy and engine production. In metallurgy, the problem of measuring temperature of shiny metals, such as aluminum, copper, etc., is still waiting a decent solution. While developing and operating engines, e.g. in aviation, there is a need to determine temperature of turbine blades revolving with a high speed, as well as to assess hot gas temperature in combustion chambers. In order to reduce the influence of emissivity on measurement of high temperatures, one uses multi-channel systems (bi-chromatic, or color pyrometry), as well as illumination of test surface with a laser. Fiber-optic (endoscopic) IR thermometers have been also elaborated to be used in the cases where the thermal radiation from the test zone is transmitted to the IR detector via flexible IR light guides.

9.9.1 Portable IR Thermometers

Many IR thermometers are supplied with a laser pointer. A laser beam shows either a center of a monitored spot (single-spot laser) or its true size (as a circle or circle diameter). In day time inspection, a laser beam is often helpless, therefore, optical sights can be used at longer distances. Portable IR thermometers of a high level. e.g. the 3i series from Fluke (former Raytek) are supplied with digital memory, computer interface and corresponding software intended for data storing and processing.

The cost of most IR thermometers is from US$100.00 to US$1.000. In the West, a considerable segment of the market is occupied by Raytek and Optris which supply a wide nomenclature of IR thermometers. Some models of fairly good quality are being produced in Russia (Russian units are characterized by a narrow FOV because they employ photonic IR detectors instead of radiation thermocouples).

The technical performance of some IR thermometers is described in Table 9.5 and Fig. 9.14.

Table 9.5 Portable IR thermometers

Model	Temperature measurement range, °C	Distance-to-Spot	Additional options	Mass, kg
Russian units				
C-110 Fakel, Russia	−20 – +200	100:1	Single-spot laser	–
C-210 Salut, Russia	−20 – +600	100:1	Single-spot laser	–
C-300 Favorit, Russia	−20 – +200	100:1 150:1	Single-spot laser	–
C-500 Samotsvet, Russia	+400 – +1600	100:1 150:1	Optical sight	–
C-600, Russia	+700 – +1900 +800 – +2200	180:1	Optical sight	–
C-3000, Russia	+1100 – +1900	60:1	Optical sight Automatic emissivity correction	–
Center-358, Russia	−18 – +315	5:1	Fixed emissivity value	0.1
KTA, Russia	+50 – +2500	30:1	Multi-channel	0.5
Dieltest-TV, Russia	+500 – +1800 +700 – +3000 +200 – +1200	–	Memory for 63 values	
Astroterm, Russia	−20 – +200	10:1. 30:1. 40:1. 120:1. 150:1	Laser or optical sight	
Kelvin, Russia	−30 – +400 −20 – +600 50 – +700 −30 - +1300 −20 - +1800 +200 – +1300 +500 – +1500 +700 – +1800 +800 – +2200	90:1 120:1 150:1 180:1 200:1 250:1 400:1	Laser or optical sight Stationary units available	1.1

(continued)

Table 9.5 (continued)

Model	Temperature measurement range, °C	Distance-to-Spot	Additional options	Mass, kg
PIT-500, Russia	+300 – +600 +500 – +700 +650 – +1100 +1000 – +1500 +1300 – +1800 +1500 – +2500	From 1:25 to 1:500	Bi-color Automatic emissivity correction	–
Pentermic-5, Russia	+1250 – +1750 +650 – +3500	50:1	Multi (5)-channel Automatic emissivity correction	–
PP-1, Russia	−20 – +400 +100 – +1200 +400 – +2000	10:1 15:1 30:1	–	–
CT-1, Russia	+150 – +2000	–	–	–
Smotrich-7, Russia	+150 – +900	150:1	–	–
MT2, Raytek, U.S.A.[a]	−18 – +275	8:1	–	–
MT4, Raytek, U.S.A.	−18 – +275	8:1	Single-spot laser	–
MiniTemp FS, Raytek, U.S.A.	−30 – +200	4:1	Single-spot laser Fixed emissivity value	0.34
Raynger ST20, Raytek, U.S.A.	−30 – +400	12:1	Single-spot laser Fixed emissivity value	0.32
Raynger ST30, Raytek, U.S.A.	−30 – +545	12:1	Single-spot laser Fixed emissivity value	0.32
Raynger ST60, Raytek, U.S.A.	−30 – +600	30:1	Circle-spot laser Adjustable emissivity value Thermocouple	0.32
Raynger ST80 (80-IS), Raytek, U.S.A.	−30 – +760	50:1	Circle-spot laser Adjustable emissivity value Thermocouple	0.32
Mx2, Raytek, U.S.A.	−30 – +900	60:1	Circle-spot laser Adjustable emissivity value Optionally: digital output and built-in memory for 100 values	0.48
Mx4+, Raytek, U.S.A.	−30 – +900	60:1	Circle-spot laser Adjustable emissivity value Digital output Software	0.48
PhotoTemp Raytek, U.S.A.	−30 – +900 (−50 – +500)	60:1	Circle-spot laser Adjustable emissivity value Digital output Software Digital photocamera	0.58

(continued)

Table 9.5 (continued)

Model	Temperature measurement range, °C	Distance-to-Spot	Additional options	Mass, kg
3i. Raytek, U.S.A.	−30 ... +1200 (1M model up +1800°C. 2M model up +3000°C)	From 25:1 to 180:1	Circle-spot laser Adjustable emissivity value Digital output Software Spectral filters for glass and plastics	0.8…1
FoodPro and FoodPro Plus, Raytek, U.S.A.	−30 ... +200 (−35 ... 275)	2.5:1	Inspecting food	0.1 (0.15)
ULTIMAx-10, IRCON, U.S.A.	+900 … +3000	–	Viewfinder RS 232	–
ULTIMAx-20, IRCON, U.S.A.	+600 ... +3000	–	Viewfinder RS 232	–
ULTIMAx-40, IRCON, U.S.A.	−50 ... +1000	–	Viewfinder RS 232	–
ULTIMAx-400, IRCON, U.S.A.	−20 ... +300	10:1	Single-spot laser	0.18
Optris LaserSight, Optris, U.S.A.	−35 ... +900	75:1	Laser USB Thermocouple	–
Optris LaserSight. Optris, U.S.A.	−32 ... +530	20:1	Laser Alarm[b]	0.15

[a]In 2005, Raytek merged with Fluke. Nomenclature changed. Models MT-2, 4, ST-20, 30, 60 are produced as Fluke 61, 63, 65, 66, 68 (parameters may be different from those in the table). Raytek is specialized in the production of stationary units. Optris produces portable units
[b]Alarm is available in many other units

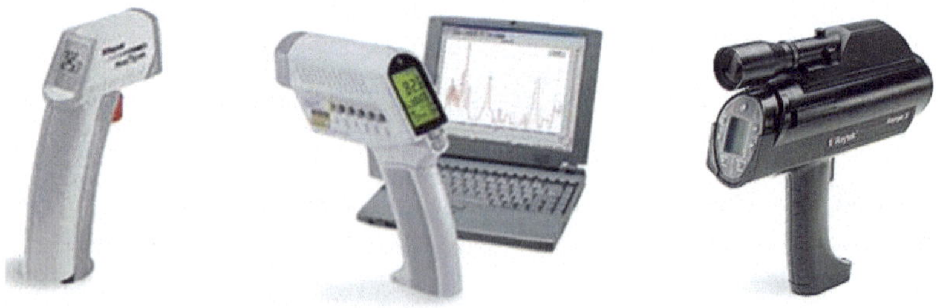

MiniTemp FS
Temperature measurement range: -30 … +200°C
Distance-to-Spot: 4:1
Single-spot laser

MX4+
Temperature measurement range: -30 … +900°C
Distance-to-Spot: 60:1
Circle-spot laser
Digital output and computer processing

3i
Temperature measurement range: -30 … +3000°C
Distance-to-Spot: 180:1
Circle-spot laser
Digital output and computer processing

Fig. 9.14 Portable IR thermometers from Raytek (A Fluke Company), USA

9.9.2 Mounted Non-contact Temperature Sensors

Contact temperature sensors, mainly involving thermocouple, are commonly used in many industries. Their known drawbacks are a relatively short lifetime (without special protective clothing) and high thermal inertia (particularly, when using protective clothing), as well as the impossibility of measuring temperature far away from their installation site. In some applications, these drawbacks can be overcome by using IR sensors which are low-inertial (time constant is fractions of second) and can be easily built in automated control systems. Analog output signal of such sensors, as a rule, matches signals produced by standard thermocouples; a digital link to computers can be done via RS-232 or RS-485 interfaces. Some models, except sensors, may include a processing hardware allowing to adjust emissivity, determine out-of-limits signals and give a signal of alarm. If necessary, such units can be supplied with a protective shroud cooled by air or water. The cost of built-in sensors is close to that of portable IR thermometers. A special type of temperature sensors is bi-color (spectral ratio) pyrometers which enable, to some extent, to solve the emissivity problem in the pyrometry of liquid metals (however, accurate measurements of temperature of liquid aluminum still represent a serious difficulty).

Some common built-in temperature sensors are described in Table 9.6 and Fig. 9.15.

Table 9.6 Built-in temperature sensors

Model	Temperature measurement range, °C	Distance-to-Spot
Raytek, USA		
Compact series		
MID LT	−40 – +600	2:1 and 10:1
MID G5	+150 – +850	2:1 and 10:1
MID MTB	+200 – +1200	2:1 and 10:1
CI	0 – +500	4:1
GP (with laser sight)	−18 – +540	50:1
Thermalert series	−1 – +3000	60:1 and 90:1
Marathon MM. MR. FR. FA series (including color and fiber-optics pyrometers)	−40 - +3000	From 60:1 to 300:1
IMPAC (Infratherm series), France		
ISQ 10-LO	+600 – +3300 °C	Fiber-optics light guide (up +250°C) Measurement spot 0.45 mm
IS 10/IGA 10	+300 – +3300 °C	Measurement spot 0.2 mm
IS 100/IGA 100	+30 – 1800 °C	Measurement spot 0.35 mm
ISQ 5 (color)	+600 – +3000 °C	–
Optris units. USA		
CT	−40 – +900°C	20:1

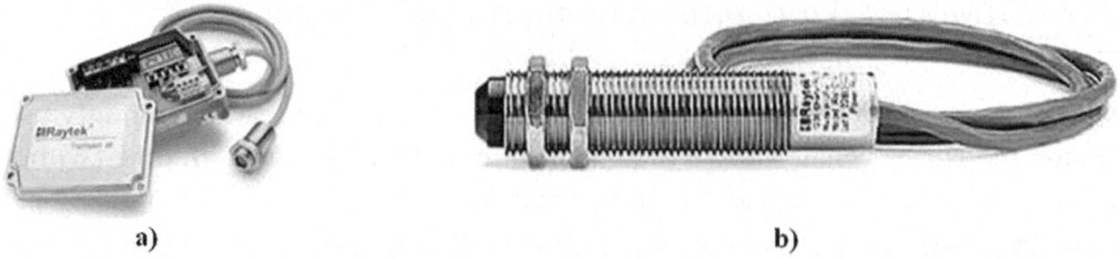

a) b)

Fig. 9.15 Built-in temperature sensors from Raytek (A Fluke Company), USA: (**a**) Compact MID series, (**b**) Compact CI series

9.10 Reference Temperature Sources (Blackbody Models)

In Russia, the primary calibration of IR radiometric imagers and thermometers performed by a manufacturer is valid for 1–2 years depending on a model. Calibration schemes for these devices are well elaborated and documented but they are not always practically realized due to a shortage of low-temperature reference sources (blackbody models). In fact, only few metrological and other specialized organizations in Russia are able to perform calibration and certification of IR equipment. It is worth noting that in practice one often checks only a range and accuracy of temperature measurement, although IR equipment, in particular, IR imagers, is characterized by a set of technical characteristics. A certification procedure may vary depending on application; for example, in Russia there is a special scheme proposed for checking medical IR imagers (see the medical branch standard OSDT 3–4408-82 "Medical thermal imagers. Methods for measuring parameters").

Reference sources for IR imagers and thermometers are similar by their principal schemes and design. However, to evaluate an IR imager, one requires a blackbody with a large emitting area (over 5–10 cm) while in the case of IR thermometers it may be under 2–4 cm. Description of such devices is well beyond of the scope of this book (see [5, 9, 10]), therefore, we will limit ourselves with the technical characteristics of some commercial blackbody models presented in Table 9.7.

9.11 Contact Temperature Sensors

Since high inspection productivity is the important advantage of IR thermographic NDT, contact thermometers play an auxiliary role in TNDT. In this paragraph, a short review of contact sensors borrowed from the handbook [10] is presented.

Four major classes of temperature sensors are described in Table 9.8. Also, there are temperature-sensitive coatings (films) which being deposited on (attached to) a test object surface are able either to visualize the object dynamic temperature distribution (liquid crystals) or witness when a certain surface temperature is reached (melting thermo-indicators).

Table 9.7 Reference sources for checking IR radiometric imagers and thermometers

Model	Temperature measurement range, °C	Accuracy	Temperature resolution, °C	Stability	Aperture diameter, mm	Emissivity
BB1000, Raytek, U.S.A.	−20 – +150	±0.3 °C	0.1	0.1 °C/8 hr	51	0.98
BB2500, Raytek, U.S.A.	+10 – +300	±1 °C or 0.25%	1	0.5 °C/8 hr	76	0.98
BB5000, Raytek, U.S.A.	+200 – +1150	±1 °C or 0.25%	1	0.5 °C/8 hr	51	0.999
BB5000, Raytek, U.S.A.	+300 – +1700	±1 °C or 0.25%	1	1 °C/8 hr	25	0.99
IR-2100, Infrared Systems Development, U.S.A.	−5 – +140	–	–	–	50	–
IR-160, Infrared Systems Development, U.S.A.	From ambient to +350	–	–	–	250	–
BB-165/40/10, Russia	+40… + 100	–	0.01 (display)	0.1 °C/30 min	165	0.98
BB-100/110010, Russia	+100… + 1100	–	–	0.5 °C/1 min	35	0.985
BB-16/900/250010, Russia	+900… + 2500	–	–	5 °C/15 min	16	0.98

Table 9.8 Contact temperature sensors [10]

Advantages	Disadvantages
Thermocouple:	
Self-powered Simple Rugged Inexpensive Wide temperature range Wide variety of physical forms	Non-linear Low voltage Reference required Least stable Least sensitive
Resistance temperature detector:	
Most stable Most accurate More linear than thermocouples	Expensive Slow Current source required Small resistance change Four-wire measurement
Thermistor:	
High output Fast Two-wire ohm measurement	Non-linear Limited measurement range Fragile Current source required Self-heating
Integrated circuit sensor	
Most linear Highest output Inexpensive	T < 250°C Power supply required Slow Self-heating Limited configurations

9.11.1 Thermocouples

The thermoelectric effect discovered by Seebeck in 1821 is responsible for the fact that a circuit of two wires composed of dissimilar metals produces electric current if one of wire joints is heated. The thermoelectric voltage is linearly proportional to temperature:

$$e_{AB} = \alpha \ T \tag{9.10}$$

where α is the thermoelectric coefficient. The properties of some common materials used for producing thermocouples are described in Table 9.9.

A main feature of thermocouples is the appearance of *additional* thermocouples while connecting a volt meter to a measuring device. For instance, the connection of a volt meter to a copper-constantan thermocouple (type T) with copper leads produces no voltage in the copper-copper junction but makes it appeared in the copper-constantan junction (Fig. 9.16a). A common method to overcome this difficulty is the thermal stabilization of an additional junction. e.g. by putting it into an ice bath (Fig. 9.16b). There are different circuits of connecting thermocouples which allow enhancing measurement accuracy.

Table 9.9 Thermocouple properties [10]

Type	Metal (proportion)	Electrical resistance. Ohm/305 mm at 293 K	Thermoelectric coefficient (at 273 K), μV/K
B	(+) 0.7 platinum, 0.3 rhodium (−) 0.94 platinum, 0.06 rhodium	0.22	5.96
E	(+) 0.9 nickel, 0.1 chromium (−) constantan	0.71	58.67 (at 873 K)
J	(+) iron (−) constantan	0.36	50.38
K	(+) 0.9 rhodium, 0.1 rhodium (−) nickel	0.59	39.45
T	(+) copper (−) constantan	0.30	38.75

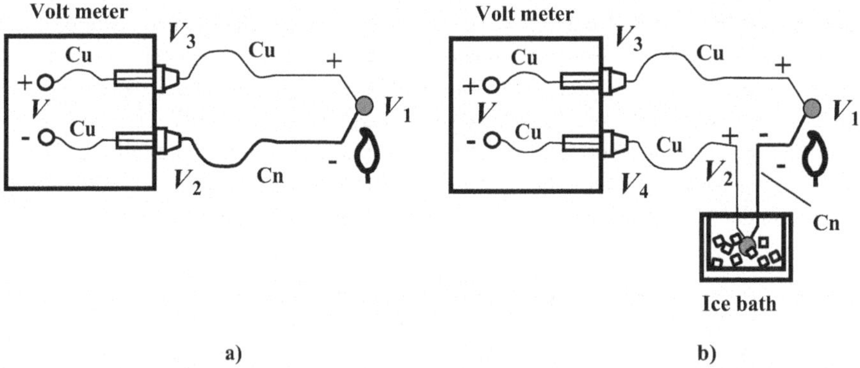

Fig. 9.16 Connecting thermocouples: (**a**) – direct connection ($V_1 = 0$), (**b**) – by using reference junction ($V_3 = V_4 = 0$)

Unlike thermistors and resistance temperature detectors, thermocouples do not measure absolute temperature because of the necessity to produce a reference circuit. Main advantages of thermocouples are wide measurement range and possibility of operating in various, including aggressive, media. Thermocouples are welded to metallic parts or clamped with a screw.

9.11.1.1 TNDT Devices Using Near-Surface Thermoelectric Sensors

In the former USSR, Gavinsky developed specialized TNDT devices called "thermal defectoscopes" and intended for inspecting cylindrical objects made of composites and rubber-metal structures [11]. These units implemented multi-element thermoelectric 1 × 1 cm sensors which were clamped to the inspected surface via Teflon gaskets with a 0.1 mm air gap. The heat transfer in such thin gaps occurs by pure heat conduction; a large time constant makes these near-surface sensors appropriate for detecting large defects which are characterized by observation times of few seconds. The simple design and low cost of such sensors used to stimulate their use in the Russian aerospace industry, but the appearance of economic IR imagers prevented their further development.

9.11.2 Resistance Temperature Detectors

In parallel with the discovery of thermoelectricity by Seebeck, Davy revealed that electrical resistance of metals exhibits noticeable temperature dependence on temperature. In 1871, Siemens suggested to use platinum as the element in a resistance thermometer. A classical design of platinum thermometers was elaborated by Meyers in 1936. In industrial thermometers, a double platinum wire is winded on a ceramic substrate and coated with glass. In the last years, film-based devices gained popularity. Comparing to wire thermometers, film devices are characterized by low inertia, smaller size and cost. They also possess higher electrical resistance that reduces errors related to finite resistance of leads. Normally, a resistance thermometer is connected to a bridge circuit along with a power supply and a volt meter.

Resistance of metals (see Table 9.10) increases with temperature. The most common resistance thermometers are made of platinum, nickel or nickel alloys. These devices are more linear than thermocouples. The Callendar-Van Dusen equation is recommended to approximate the resistance thermometer calibration curve:

$$R_T = R_0\alpha\left[T - \delta\left(\frac{T}{100}-1\right)\left(\frac{T}{100}\right) - \beta\left(\frac{T}{100}-1\right)\left(\frac{T^3}{100}\right)\right] \tag{9.11}$$

where R_T is the resistance at the temperature T, R_0 is the resistance at $T = 0$ °C, α is the temperature coefficient at $T = 0$ °C, $\delta \approx 1.49$, $\beta = 0$ when $T > 0$ °C and $\beta = 0.11$ when $T < 0$ °C. More accurate values of α, β, δ are to be determined by testing resistance temperature detectors at four temperatures and then solving the resultant equations.

Table 9.10 Electrical resistance of common metals [10]

Metal	Resistance, Ohm·mm^2·m^{-1}
Gold	21.61
Silver	14.63
Copper	15.39
Platinum	98.08
Tungsten	49.87
Nickel	59.85

In practice, the following features of resistance thermometers are to be taken into consideration: (1) they are more fragile than thermocouples, and some precautions must be taken to protect them, (2) the electrical current while passing through these devices causes joule heating which changes their temperature and produces self-heating measurement errors (this drawback can be overcome by using low currents and high-resistance metals, as well as immersed detectors in a thermally-conductive medium), (3) attaching a resistance detector to the tested object may cause the effect of 'thermal shunting' when object temperature may alter due to relatively big physical mass of the detector, (4) due to the thermoelectricity phenomenon, e.g. when connecting copper and platinum, a thermal offset voltage may appear thus requiring its compensation.

9.11.3 Thermistors

Thermistors are similar to resistance detectors but they are generally composed of semiconductor materials of which temperature dependence can be very steep. Therefore, thermistors are the most sensitive contact thermometer, while thermocouples seem to be the most universal, and resistance detectors are the most stable.

The electrical resistance of most thermistors decreases with increasing temperature, although there are some semiconductors with a positive temperature coefficient. Thermistors are much less standardized to compare to other contact thermometers; three examples of thermistors curves are presented in Fig. 9.17. It is clearly seen that thermistors are non-linear with individual curves being approximated with the Steinhart-Hart equation [10]:

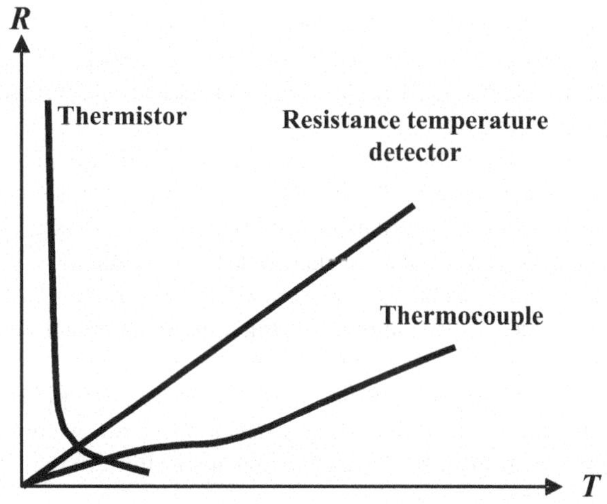

Fig. 9.17 Contact thermometers: resistance *vs.* temperature

$$T = \frac{1}{A + B(LnR) + C(LnR)^3}. \tag{9.12}$$

where T is the temperature, K, R is the thermistor resistance, Ohm. A, B, C are the coefficients determined by three reference temperatures. Common thermistors, of which electrical resistance at the ambient temperature is about 5 kOhm, are less susceptible to some measurement errors which are characteristic when using other types of contact thermometers. A thermistor temperature range is generally limited to a few hundred Kelvins, and their long exposures require periodical re-calibration.

9.11.4 Integrated Circuit Sensors

This type of contact thermometers is relatively new. Commercially available are integrated sensors with both current (1 μA/K) and voltage (10 mV/K) outputs. These devices ensure a very linear output *vs.* temperature but share all disadvantages of thermistors because they also use semiconductors in a limited temperature range. Integrated circuit sensors are often used to compensate thermocouple measurement errors, as a well as to build up multi-channel automated systems for measuring temperature in industry.

9.11.5 Liquid Crystals

Discovered more than a century ago, liquid crystals which are intermediaries between crystallic solids and liquids, have been considered as a sort of curiosity in physics. In the 1970s, the boosted interest to these chemical substances used to lead to a wide range of innovations and industrial applications of contact temperature imaging devices.

As temperature sensors, one uses liquid crystals which reveal a temperature-dependent type of liquid phase. With temperature, molecules of these chemical compounds may orient in a particular order thus creating the so-called *nematic*, *cholesteric* and *smectic* mesophases. By diffusing white light, such temperature detectors produce a color (from red to blue) picture of temperature distributions. To avoid back diffusion of a substrate, inspected materials can be black-coated, or liquid crystal films can be blackened themselves.

The first liquid crystals were susceptible to rapid degradation by contamination and ultraviolet light. Many of their problems were eliminated by introducing a micro-encapsulation process that also allowed reducing color variations with a changing viewangle.

By 1975, a certain experience in using liquid crystal was gained in the inspection of electronic equipment, aircraft structures, heat exchangers and in medical diagnostics.

Common liquid crystal films have time response of about ten milliseconds; sub-millisecond and even microsecond liquid crystals are also available.

Thermochromic liquid crystals are commercially supplied as: (1) sprays, (2) coatings for solid surfaces, (3) microencapsulated liquid crystal slurries to be used as tracer particles in flow field studies (capsule size 5–10 μm). (4) polyester sheets coated with microencapsulated liquid crystals for flat solid surfaces, and (5) unsealed liquid crystal compounds and mixtures for microscopic evaluation of small objects (in the latter case, the spatial resolution may reach few μm).

The chemical composition of liquid crystals determines their color response toward temperature changes. Manufacturers typically provide a two-color temperature code: the starting temperature and the red-to-blue interval. Liquid crystals have been produced in the temperature range from −30 to +100°C with the temperature resolution about 5°C. In the former U.S.S.R., a wide nomenclature of liquid crystals was available (see the book [12]).

By obvious reasons, quantitative temperature measurements with liquid crystals are difficult but, fortunately, many thermal tests are limited by qualitative evaluations. For instance, liquid crystals allow the illustrative inspection of electronic components at the stage of design and operation. The estimation of temperature values is possible when using narrow-band crystals with noticeable color transitions.

When using liquid crystals, the color interpretation is rather subjective, although some TV systems for monitoring liquid crystals, which use the Red-Green-Blue (RGB) model of colors, have been developed.

The following guidelines have been proposed [10]:

- determining maximal and minimal expected temperatures,
- choosing an appropriate thermochromic liquid crystal mixture for a particular temperature range,
- applying a black coating and liquid crystals to the test surface,
- providing adequate optical access to the surface and minimizing spurious reflections,
- providing online lighting and viewing arrangement,
- minimizing IR and ultraviolet radiation onto the test surface,
- calibrating the color *vs.* temperature response (if necessary),
- evaluating temperature distribution by using the appropriate color *vs.* temperature calibration.

9.11.6 Materials with Calibrated Melting Points (Phase Change)

In the industry, certain materials which indicate temperature according to their predictable melting points are widely used. These materials represent mixtures of organic and non-organic substances and are available in several forms, typically as lacquer, pellets or sticks. A temperature-dependent marker is applied to the surface of a work piece. When the work piece attains the predetermined melting point of the indicator mark, the mark instantly liquefies, thus notifying that the work piece reached that temperature. Such thermal indicators are available in the range from +38 to +1370°C. Their sensitivity varies from 14 to 28°C but a step of temperature evaluation may reach 3.4°C. Temperature-indicating materials are removed with water or alcohol but, in some cases, mechanical cleaning may be needed.

The major application areas of materials with calibrated melting points are in metallurgy, welding, power production and other industrial processes where temperature is an important indicator of quality.

References

1. Dumpert, D., West, L.: Focal Plane Array – A New Generation of IR Camera. P/PM Technology, Feb 1995. p. 122–134
2. Elliott, C.T., Day, D., Wilson, B.J.: An integrating detector for serial scan thermal imaging. Infr. Physics. **22**, 31–42 (1982)
3. Gunapala, S.D., Bandara, S.V., Liu, J.K., et al.: Quantum well infrared photodetector research and development at Jet Propulsion Laboratory. Infr. Physics Technol. **42**, 267–282 (2001)
4. Sheppard, F.D., Yang, A.C.: Silicon Shottky retinas for infrared imaging. IEDM (Intern Electron Devices Meeting) Technical Digest, New York, USA, Institute of Electrical and Electronic Engineers, Dec 1973. p. 310–313
5. Kriksounov, L.Z.: Handbook of Infrared Techniques. "Sovetskoye Radio" Publish, Moscow (1978). 400 P. (in Russian)
6. Lloyde, J.M.: Thermal Imaging Systems. Plenum Press, New York (1979). 456 P
7. Gaussorgues, G.: La Thermographie Infrarouge. Technique et documentation, Paris, France (1981). 386 P
8. Hudson, R.D.: Infrared System Engineering. Wiley-Interscience, New York (1969)
9. The Infrared and Electro-Optical Systems Handbook. V. 3 "Electro-Optical Components". Joseph S. Accetta, David L. Shumaker Exec. Editors. ERIM and SPIE Optical Engineering Press, Bellingham, WA, USA, 1993, 668 P
10. Nondestructive Testing Handbook. Vol. 3 "Infrared and Thermal Testing", Bellingham, USA, A.S.N.T, 2001
11. Vavilov, V.P.: Thermal NDT of Composite Structures and Radio Electronic Components. Radio i Svyaz Publish, Moscow (1984). 162 P. (in Russian)
12. Abramovich, B.G., Kartavtsev, V.F.: Color Temperature Indicators. "Energia" Publish, Moscow (1978). 214 P. (in Russian)

Chapter 10
Statistical Data Treatment and Decision Making in TNDT

Abstract In this chapter, elements of statistical data treatment and decision making in TNDT are presented. Noise sources in TNDT are described, along with establishing defect detection limits based on an analysis of surface clutter in TNDT procedures.

10.1 Defect Detection Parameters in TNDT

As mentioned above, any NDT method involves two stages: defect *detection* and defect *characterization*. Detection is considered completed when an operator or an automatic device recognizes the test object either non-defective, i.e. corresponding to some approved standards (norms), or detects some anomalies identified as defects. At this inspection stage, defect parameters are not quantitatively evaluated, and decision making represents a typical statistical procedure. At the characterization stage, one characterizes identified defects quantitatively by using solutions to the appropriate inverse diagnostic problems and decides how severe are discovered defects for further object operation; in some cases, object lifetime (work resource) can be evaluated. In TNDT, the above-mentioned procedures are poorly explored, in particular, if object lifetime is to be predicted. In many cases, decision making is to be done by end-users on the base of approved norms and standards.

In active TNDT, object quality is evaluated by analyzing: (1) amplitude of temperature anomalies (differential temperature signals and their contrast derivatives), (2) peculiarities of temperature evolutions in time (optimum observation time, half-rise time, etc.), (3) morphology of temperature patterns.

When detecting defects by their surface temperature indications, it is supposed that: (1) thermal properties of defects are different from those of the host material, (2) defects disturb a regular flow of the heat flux within the test body. The latter requirement explains the fact that TNDT is mostly applicable for the detection of extended gas-filled defects which are located perpendicularly to the main heat flux delivered by a heat source. Such defects are: (1) delaminations between sample layers and vast discontinuities in a host material, (2) plane cracks, (3) inclusions, (4) local variations in thermal properties, including moistened areas and excessive porosity, (5) zones of local material loss (corrosion), and (6) the so-called 'kissing' defects where two materials are in contact but characterized by enhanced thermal resistance.

V. Vavilov, D. Burleigh, *Infrared Thermography and Thermal Nondestructive Testing*, https://doi.org/10.1007/978-3-030-48002-8_10

10.2 Statistical Decision Making Parameters

Consider the TNDT scheme presented in Fig. 10.1. Input information is supplied by inspection results obtained by an operator or an automated device implementing one of available TNDT techniques. The output information involves four possible results of the diagnostics, or two pairs of conclusions on whether the sample is defective or non-defective. The conclusion can be true-positive – TP (truth: "defect is present", conclusion: "defect is present"), false-negative – FN (truth: "defect is present", conclusion: "defect is absent"), true-negative – TN (truth: "defect is absent", conclusion: "defect is absent") and false-positive – FP (truth: "defect is absent", conclusion: "defect is present"). In the probability theory, the sums of the corresponding conclusions give 100%: TP + FN = 100% and TN + FP = 100% (Fig. 10.1). A true-positive conclusion is characterized by the *probability of correct detection* $P_{c.\,d.}$ while a false-positive conclusion is related to the probability of *false alarm* $P_{f.\,a.}$. Note that $P_{c.\,d.} + P_{f.\,a.} \neq 100\%$. As it will be demonstrated below, in NDT,

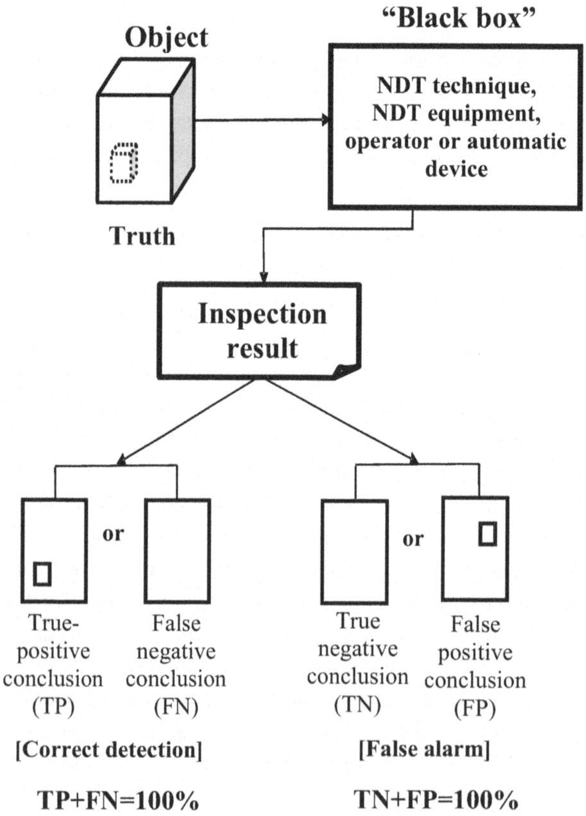

Fig. 10.1 NDT scheme

one often takes decision by using the so-called *Neumann-Pearson strategy* which sets up a value of $P_{f.\,a.}$ thus resulting in a particular value of $P_{c.\,d.}$.

The simplest case of decision making in a 1D space of significant parameters, also called features, is presented in Fig. 10.2. The decision is made at a single pixel level by considering temperature as a significant parameter. Any area on the surface of a test object is characterized by a specific distribution of pixel amplitudes; hence, each area can be classified either as "defect" or "non-defect". In such way, the classification procedure finishes with a *binary map* of defects.

In statistics, the decision making is reduced to checking a null (H_0) and alternative (H_A) hypothesis: $H_0 : \overline{T}_d = \overline{T}_{nd}$ and $H_A : \overline{T}_d \uparrow \overline{T}_{nd}$, where $\overline{T}_d$, $\overline{T}_{nd}$ are the mean temperatures in the defect and non-defect areas respectively. A number of statistical criteria can be used in this process depending on the amplitude distribution of signals and noise. Some of these criteria are described in Table 10.1.

The signal-to-noise ratio, which is defined by

$$S = \frac{\overline{T}_d - \overline{T}_{nd}}{\sigma_{nd}}, \tag{10.1}$$

often serves as a typical statistical criterion of defect detection (we remind that σ_{nd} is the standard deviation of a significant parameter in a chosen non-defect area).

If the distribution of a significant parameter is governed by the *normal* (*Gaussian*) law, one often uses a *Z-criterion* [1]:

$$\hat{z} = \frac{\left|\overline{T}_d - \overline{T}_{nd}\right|}{\sqrt{\frac{\sigma_d^2}{n_d} + \frac{\sigma_{nd}^2}{n_{nd}}}}, \tag{10.2}$$

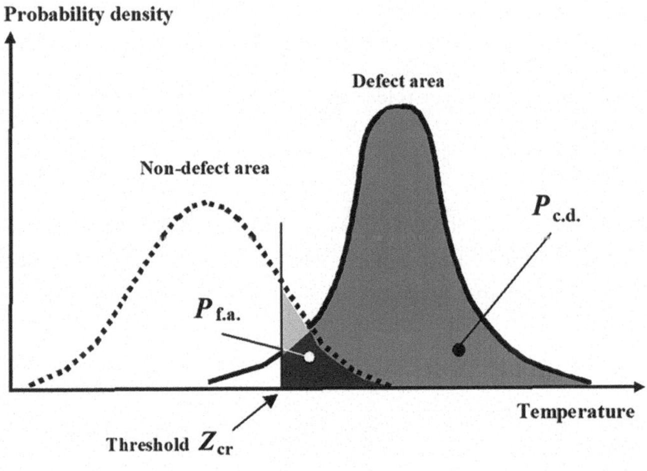

Fig. 10.2 Decision making statistics (1D space of significant parameters)

Table 10.1 Statistical criteria in the comparison of two statistical populations

Criterion	Formula	Note
Signal-to-noise ratio S	$S = \frac{\bar{T}_d - \bar{T}_{nd}}{\sigma_{nd}}$, where σ_{nd} is the standard deviation in a non-defect area.	$S > 1$ specifies a statistical difference between two distributions; this criterion is appropriate in the case of signal normal distribution.
Z-criterion	$\hat{z} = \frac{\lvert \bar{T}_d - \bar{T}_{nd} \rvert}{\sqrt{\frac{\sigma_d^2}{n_d} + \frac{\sigma_{nd}^2}{n_{nd}}}}$, where n_{nd} and n_d are the numbers of pixels in the areas to be compared, σ_d is the standard deviation in a defect area	A value of the Z-criterion is to be compared to the critical value z_{cr} which is obtained from the equation: $\alpha = P_{f.a.} = 1 - F(z_{cr})$, where $F(z_{cr}) = \frac{1}{\sqrt{2\pi}} \cdot \int_{-\infty}^{z_{cr}} e^{-x^2/2} \cdot dx$ is the distribution function, α is the probability of $z \geq z_{cr}$ (probability of false alarm $\alpha = P_{f.a.}$). This criterion is mainly applicable to signal normal distribution.
D-criterion (Kolmogorov-Smirnov test)	$\hat{D} = \max \left\lvert \left(\frac{F_d}{n_d} - \frac{F_{nd}}{n_{nd}} \right) \right\rvert$, where F_d and F_{nd} are the empiric cumulative frequencies.	Maximum value of the D-criterion is to be compared to the critical value D_{cr} which can be determined by the formula (if $n_d + n_{nd} > 35$) [1]: $D_{cr} = K_\alpha \sqrt{\frac{n_d + n_{nd}}{n_d \cdot n_{nd}}}$, where K_α is the constant depending on the significance level α (tabulated in [1]); e.g. $K_\alpha = 0.136$ if $\alpha = 0.05$. The D–criterion can be applied to any distribution.
Probability of false alarm $P_{f.a.}$	$P_{f.a.}$ is numerically equal to the area in the *'non-defect'* distribution which is located to the right from a chosen threshold z_{cr} (Fig. 10.2).	Decision on a statistical difference between two areas is taken with *Risk I* (significance level) α which is equal to $P_{f.a.}$.
Probability of correct detection $P_{c.d.}$	$P_{c.d.}$ is numerically equal to the area in the *'defect'* distribution which is located to the right from a chosen threshold z_{cr} (Fig. 10.2).	*Risk II* to accept the null false hypothesis is $\beta = 1 - P_{c.d.}$.

where n_{nd} and n_d are the numbers of pixels in the analyzed areas (defect and non-defect). A $\hat{z}$ value defined in this way is to be compared to the critical value z_{cr} calculated by the following equation:

$$\alpha = P_{f.a.} = 1 - F(z_{cr}), \tag{10.3}$$

where $F(z_{cr}) = \frac{1}{\sqrt{2\pi}} \int_{-\infty}^{z_{cr}} e^{-x^2/2}$ dx is the distribution function and α is the probability of the standard normal variable z to exceed z_{cr} (α is the probability of false alarm $P_{f.a.}$).

Two populations (inspected areas) are regarded statistically different if $\hat{z} \geq z_{cr}$. It is worth noting that such approach is not always justified in NDT where areas analyzed contain a plenty of pixels. Because of a great number of pixels in images, a *Z*-test always confirms a statistical difference between compared areas of interest. Therefore, it can be mainly applied in the comparison between different inspection techniques or/and processing algorithms. In fact, a best technique (algorithm) should provide a maximum value of $\hat{z}$.

When comparing statistical populations which do not belong to the normal distribution, some other statistical parameters can be used, for example, a *D*-criterion of the Kolmogorov-Smirnov test (Table 10.1).

The statistical distributions shown in Fig. 10.2 serve as a simple graphical illustration to the above-introduced probabilities of false alarm $P_{f.\,a.}$ and correct detection $P_{c.\,d.}$ which can be determined after having established a decision making threshold z_{cr}. A value of $P_{f.\,a.}$ is numerically equal to the area in the "non-defect" distribution to the right from a chosen threshold z_{cr}. This means that the *Risk I* α of decision making is equal to $P_{f.\,a.}$. Respectively, a value of $P_{c.\,d.}$ is numerically equal to the area in the "defect' distribution also to the right from a chosen threshold z_{cr}. The *Risk II* β of decision making is related to the acceptance of the null false hypothesis being equal to $\beta = 1 - P_{c.\,d.}$.

If a chosen significant parameter is governed by the normal distribution, the theory of statistical decisions provides the following expressions:

$$P_{c.d.} = 1 - \Phi\left(Z = \frac{\bar{U}_d - U_{thr}}{\sigma_{nd}}\right);$$
$$P_{f.a.} = \Phi\left(Z = \frac{U_{thr} - \bar{U}_{nd}}{\sigma_{nd}}\right), \qquad (10.4)$$

where $\Phi\left(Z\right) = \frac{1}{2\sqrt{\pi}} \int_{Z}^{\infty} e^{-x^2/2}\, dx$ is the tabulated Laplace integral, $\bar{U}_d$, $\bar{U}_{nd}$ are the mean values of the signal (significant parameter) in the non-defect and defect areas respectively, and U_{thr} is the decision making threshold. Some values of the probability integral are given in Table 10.2. For example, to obtain $P_{f.\,a.} = 5\%$ one has to set the decision making threshold equal to $\bar{U}_{nd} + 1{,}6\ \sigma_{nd}$.

In TNDT, the absolute or excess surface temperature *T* typically serves as a defect criterion only in passive IR thermographic diagnostics where test object temperature is conditioned by object operation, thus being relatively stable in time and space. In active TNDT, the sample temperature varies in time and dramatically depends on a heating power. Therefore, differential temperature signals ΔT or running contrasts C^{run} are typically used to distinguish between defect and non-defect areas (variations in signal phase or observation times are more advanced significant parameters but their statistical analysis is conducted in the same way as temperature). Unpainted metallic surfaces produce false defect indications conditioned by rust, mud, scratches, glares, etc. It is important that statistical signal distributions can be similar in both non-defect and defect areas, hence, defects can hardly be

Table 10.2 Probability integral values for one-sided Z-test

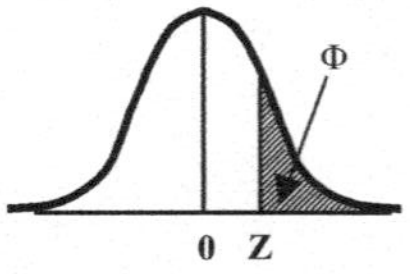

Probability integral values for one-sided Z-test.

Z	0.0	0.1	0.2	0.3	0.4	0.5	0.6	0.7	0.8	0.9	1.0	1.5	2.0	2.5	3.0
Φ	0.500	0.460	0.421	0.382	0.345	0.309	0.274	0.242	0.212	0.184	0.159	0.067	0.023	0.006	0.001

detected. The opposite test situation appears when distributions are absolutely separated, e.g. in the case of large defects, but it is of a weak practical interest in active TNDT. In intermediary cases, the decision making is to be done by a chosen threshold that leads, as mentioned above, to a particular combination of $P_{c.\,d.}$ and $P_{f.\,a.}$ which acquire values from 0 to 1.

The relationship between $P_{c.\,d.}$ and $P_{f.\,a.}$ can be expressed by the *operating characteristic* of the "detector". The parameter of operating characteristic represents a grade of the shift between defect and non-defect distributions expressed in terms of their standard deviations. An example of the operating characteristic is presented in Fig. 10.3.

When comparing performance of NDT techniques in practice, one often uses the dependence of $P_{c.\,d.}$ on defect size called the *probability of detection* (POD). This parameter is averaged by some operators thus taking into account a human factor. An example of such approach is shown in Fig. 10.4 where the probabilities of detection *vs.* defect size are given for ultrasonic, IR thermographic and holographic NDT of delaminations between 0.15 mm-thick honeycomb skin and cells, all made of Inconel titanium alloy. Such structures were used in the thermal protection of the US space shuttle X-33 [2], see also Chap. 11. The statistical data in Fig. 10.4 was obtained on 12 samples of which each contained 18 artificial defects. NDT results were averaged by two operators. It is seen that the ultrasonic technique proved to be

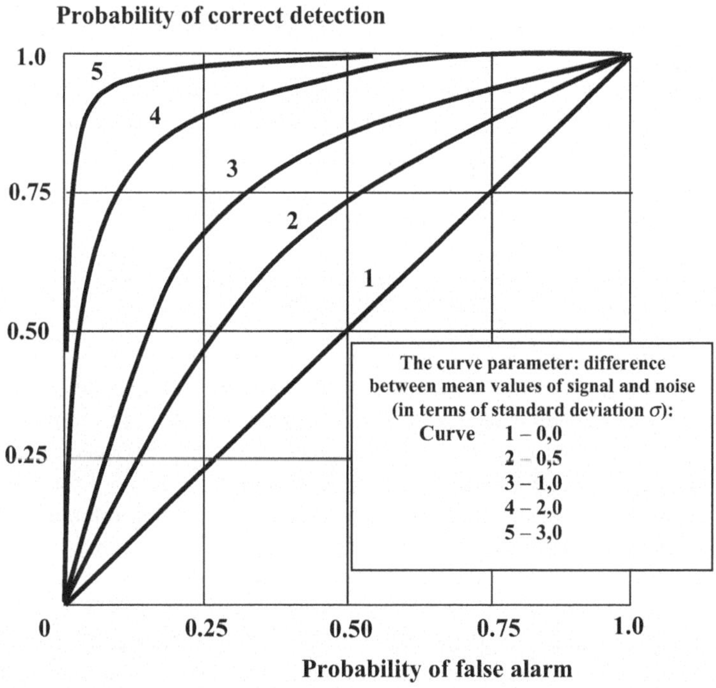

Fig. 10.3 Detector operating characteristic

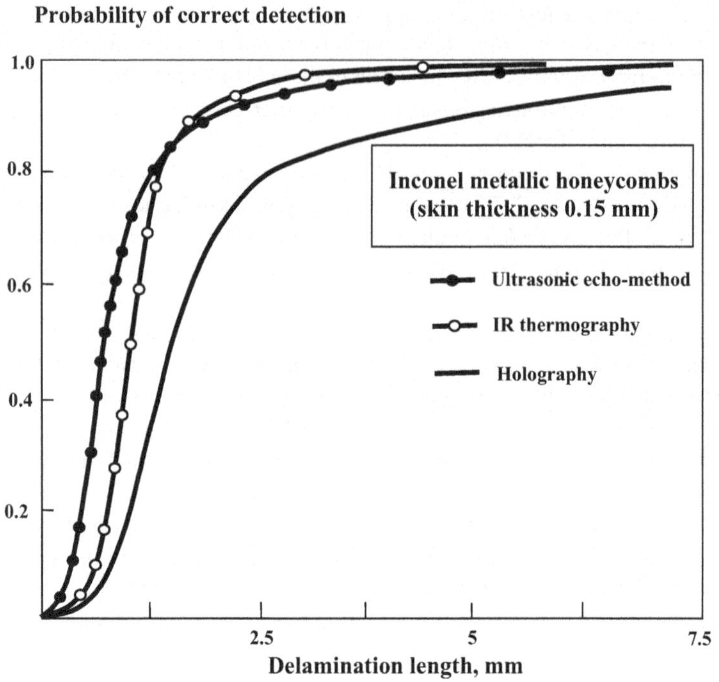

Fig. 10.4 Probability curves in NDT of metallic honeycombs. (Adapted from Ref. [2])

more efficient in detecting small-size defects (<1.5 mm), but the thermal NDT seemed to be preferable in the case of larger defects. The results of holographic evaluation were worse for all types of defects.

The so-called *Tanimoto criterion* known in the pattern recognition theory was proposed as a simple characteristic of sample "quality" [3]:

$$TC = \frac{N_{r.d.} - N_{m.d.}}{N_{r.d.} + N_{f.d.}}, \tag{10.5}$$

where $N_{r.\,d.}$, $N_{m.\,d.}$, $N_{f.\,d.}$ are the numbers of true, missed and false defects detected by an operator in the image analyzed. The interesting feature of the Tanimoto criterion is that it combines missed and false defects. Note that TC becomes equal to 100% only if $P_{c.\,d.} = 100\%$ and $P_{f.\,a.} = 0$. This criterion can be used for comparing various NDT methods and processing algorithms by analyzing reference samples with *a priori* known defects.

The Tanimoto criterion was used when processing results of one-sided TNDT of a graphite/epoxy sample with 9 Teflon inserts by the size of 3 × 3, 6 × 6 and 12 × 12 mm placed at the depth of 0.25, 1.25 and 2.5 mm [3]. The statistical results averaged by 11 operators are presented in Table 10.3. The following conclusions were made: (1) unlike intuitively assumed, only few false defects were detected;

Table 10.3 Using Tanimoto criterion in TNDT of CFRP [3]

Image description	*TC*, %	σ_B, %
Source IR image	8.9	3.4
Source IR image after histogram equalization (color)	52.2	1.6
Source IR image after histogram equalization (B & W)	57.8	1.4
Source IR image after histogram equalization (B & W copy)	44.4	0
After normalization by the end of heating	63.3	3.2
Timegram	65.6	1.1

however, this can be explained by a regular location of defects in the inspected reference sample, (2) various forms of data graphical presentation (color, B & W and their copies) did not influence much the defect detection reliability, and (3) the highest *B* values corresponded to the timegram and its derivatives.

10.3 Map of Defects

A defect map is a binary (hit-miss) image where unity values are prescribed to defect areas and zero – to non-defect areas. Each map produced for a particular sample section is characterized by a certain combination of $P_{f.\,a.}$ and $P_{c.\,d.}$ values which appear after introducing a decision making threshold. As mentioned above, the Neumann-Pearson statistical procedure is often used in NDT. According to this strategy of decision making, one has to set up a $P_{f.\,a.}$ value that is equivalent to setting the corresponding threshold z_{cr}, afterwards $P_{c.\,d.}$ value will be automatically determined. The simplest algorithm for producing a defect map is:

$$T(i,j) = 1, \text{if } T(i,j) \geq T_{thr}, \text{otherwise } T(i,j) = 0, \tag{10.6}$$

where $T(i, j)$ is the accepted significant parameter, and T_{thr} is the decision making threshold. Again, since T_{thr} is established, the corresponding defect map will be characterized by a particular combination of $P_{f.\,a.}$ and $P_{c.\,d.}$.

A simple illustration to the statistical data treatment in medical diagnostics is shown in Fig. 10.5. To evaluate a grade of pathology, an operator should define two areas on the patient's body (rectangular areas in Fig. 10.5) and then make decision about their statistical difference. For example, a healthy patient (Fig. 10.5a) is characterized by the signal-to-noise ratio $S = 0.064$. This means that temperature distributions in the chosen areas are statistically similar. The third-stage breast cancer accompanied with tissue necrosis and sharp temperature decrease (Fig. 10.5b) provides $S = 3.45$ thus proving that there is a significant difference between two areas. The respective defect map shown in Fig. 10.5c clearly exhibits the area of pathology.

Some more examples of defect maps obtained in TNDT of a CFRP sample (see also Table 10.3) are presented in Fig. 10.6. The source image in Fig. 10.6a shows three rectangular defects and a marker script on the sample surface made with a

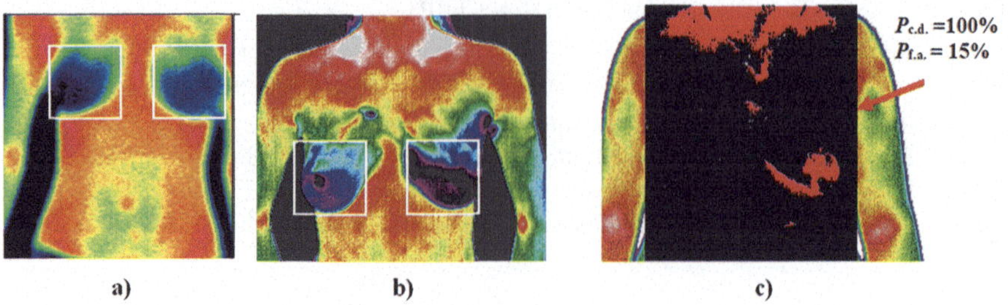

Fig. 10.5 Statistical data treatment in IR thermographic medical diagnostics: (**a**) no-pathology case ($S = 0.064$), (**b**) left breast IIIrd stage cancer ($S = 3.45$), (**c**) 'defect' map by image (**b**)

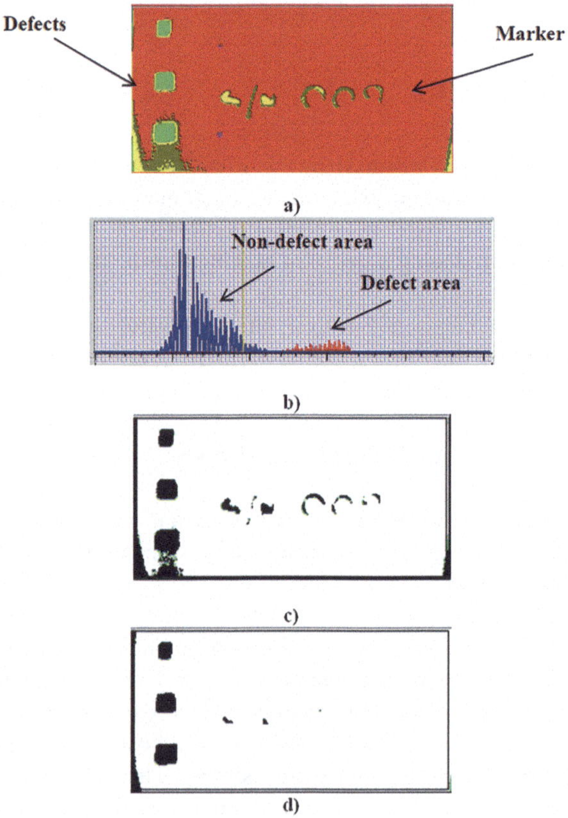

Fig. 10.6 Statistical data treatment in TNDT of CFRP: (**a**) raw IR thermogram, (**b**) distribution of pixel amplitudes, (**c**) defect map ($P_{f.a.} = 10\%$, $P_{c.d.} = 100\%$, $S = 5.5$) (**d**) defect map ($P_{f.a.} = 5\%$, $P_{c.d.} = 100\%$, $S = 5.5$)

white paint. The signal-to-noise ratio calculated by all three defects is $S = 5.5$ (Fig. 10.6b). By setting a threshold between the non-defect and defect distributions, it is possible to determine $P_{f.\,a.}$ and $P_{c.\,d.}$ values. Decreasing the probability of false alarm by moving the threshold to the right (Fig. 10.6b) allows deleting the temperature indication of the marker but simultaneously reduces the visible size of the defect areas (compare the thermograms in Fig. 10.6c and Fig. 10.6d).

10.4 Pattern Recognition in TNDT

As noticed above, in the simplest cases, decision making is performed by a single significant parameter, e.g. by the surface temperature. Let us consider TNDT of corrosion in a steel sample (Fig. 10.7a). Of a practical interest is the case when non-defect and defect histograms are partially overlapping, thus decreasing the reliability of decision making. The inspection efficiency can be enhanced by introducing the second significant parameter, e.g. optimum observation time $\tau_m(\Delta T_m)$. It is obvious that, in a general case, non-defect and defect areas may overlap by one parameter but be well separated by another. For example, in one-sided TNDT, deeper defects are usually characterized by low signals ΔT_m and long observation times τ_m, while the surface clutter will produce random combinations of these parameters. A graphical illustration of decision making in the space of two significant parameters is shown in Fig. 10.7b. In this case, a boundary between two areas in the 2D space is represented with a curve of which parameters are to be defined by the accepted combination of $P_{c.\,d.}$ and $P_{f.\,a.}$. One more significant parameter can be a characteristic size of temperature patterns h (see Fig. 10.7a); this can improve the discrimination between large-size defects and small-size surface clutter, such as scratches.

Generally, any area on the test sample surface, whether it is non-defect or defect, can be characterized by N NDT significant parameters, hence, decision making can be done in the N-dimensional space. Within this approach, the optimum detection of defects represents a standard problem of *pattern recognition* in a multi-dimensional space. Mathematically, this is reduced to evaluating parameters of a hyper-plane which separates non-defect and defect areas.

The important condition which favors the introduction of additional significant parameters is the necessity to have a weak correlation between parameters. For instance, the non-defect and defect areas in Fig. 10.7b are characterized by similar distributions of pixel amplitudes but are very different in terms of τ_m. In TNDT, the analysis of correlation between various significant parameters has been scarcely explored.

Defects can be detected heuristically or automatically. In the first case, decision is made by an operator who implies algorithms which are not yet well-understood. However, it is clear that such algorithms are related to the analysis of a multi-dimensional space of features. In IR thermography, the relationship between the deterministic technique of obtaining IR images and their visual perception has been thoroughly discussed by Lloyd [4]. Automatic decision making, in its turn, is often

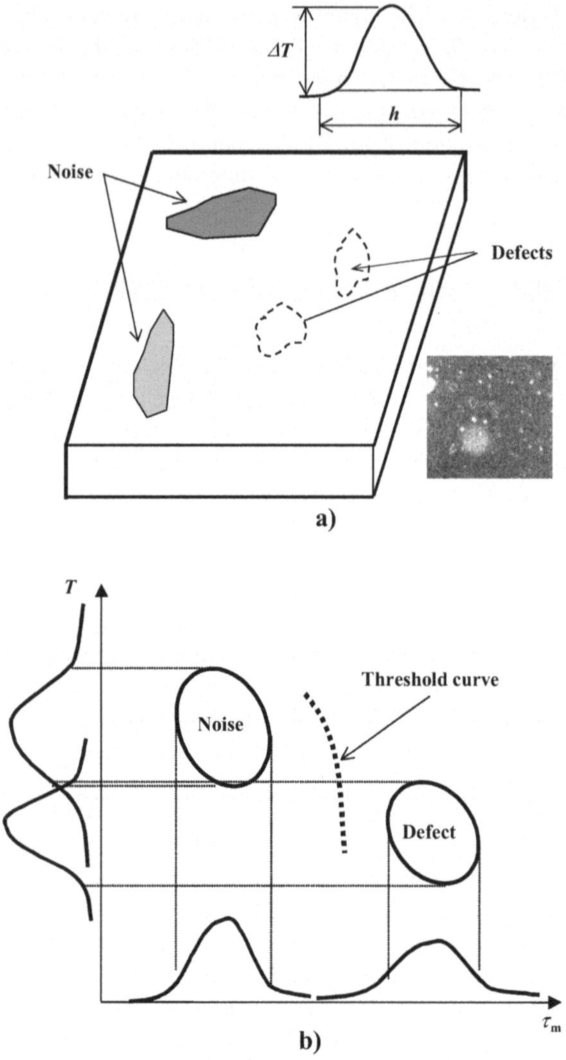

Fig. 10.7 Pattern recognition in TNDT: (**a**) detecting hidden corrosion, (**b**) 2D space of significant parameters

reduced to the use of simple thresholding devices, although some more sophisticated algorithms have been also developed. For instance, in the last decade, neuron networks which are to be trained like human operators have become popular in NDT (see Chap. 6).

10.5 Noise in TNDT

In TNDT, a high level of noise is what one pays for the universal character of this technique. The noise appears in test samples, as well as in the environment and inspection equipment. It may add to a defect temperature signal T (*additive noise* $\tilde{A}$) and/or multiply it (multiplicative noise $\tilde{M}$):

$$u(x,y,\tau) = \tilde{M}\,T(x,y,\tau) + \tilde{A}, \tag{10.7}$$

where u is the test unit output. It is obvious that $u \equiv T$ only if $\tilde{M} \equiv 1$ and $\tilde{A} \equiv 0$.

The best inspection procedure is such one where the detection limit is determined by the noise of the IR detector, i.e. when $\tilde{M} \equiv 1$ and $\tilde{A} \to$ min. A limit value in the detection of temperature signals is expressed in the IR detector temperature resolution ΔT_{res} which may be as good as 0.01°C in the contemporary IR imagers. A further improvement of temperature resolution is possible, for example, by averaging IR images in a recorded sequence. However, in real test cases, convection heat fluxes and outside heat sources produce noise over 0.1 °C; therefore, the latter value can be accepted as an approximate TNDT sensitivity limit.

Each type of noise can be expressed in terms of temperature. Conventionally, the noise can be classified for *structural* ($\sqrt{\varnothing\overline{T}^2_{str}}$), *apparatus* ($\sqrt{\varnothing\overline{T}^2_{app}} = \Delta T_{res}$) and *external* ($\sqrt{\varnothing\overline{T}^2_{ext}}$). Then, a signal-to-noise ratio can be defined as:

$$S = \frac{\Delta T(\tau)}{\sqrt{(\Delta\overline{T}_{res})^2 + (\Delta\overline{T}_{str})^2 + (\Delta\overline{T}_{ext})^2}}. \tag{10.8}$$

When using temporal significant parameters, the equivalent noise can be evaluated by the expression:

$$\Delta T = \frac{\partial T(\tau)}{\partial \tau}\Delta\tau. \tag{10.9}$$

The environmental noise is produced by surrounding objects which emit their thermal radiation directly into IR cameras or reflect the radiation of other heat sources. The main sources of this noise are the Sun, external heaters, electrical lamps, furnaces, heaters, etc. The direct radiation can be suppressed by using shields, blends, filters etc. It is more difficult to reduce the radiation reflected from the surface of test objects, because it is mixed with the radiation emitted by test objects themselves. In active TNDT, stimulating heat sources often produce a major noise. For example, when inspecting metals, the residual heat source radiation reflected by test objects may seriously distort IR thermograms and lead to incorrect temperature read-outs and, hence, wrong quantitative estimates of defect parameters. If object emissivity is low and black-painting is impossible for some reasons, it is necessary to evaluate reflected radiation and then correct transient pixel-based temperature values *prior* applying processing algorithms.

Figure 10.8 shows the temporal profiles of the flash thermal radiation produced by a 2.4 kJ Bowens tube. The profile $\gamma(\tau)$ in Fig. 10.8a recorded with a long-wave IR camera ThermaCam SC 3000 shows that a considerable energy is delivered to a sample during the first second, even if the total flash energy $\phi(\tau)$, which is mainly concentrated in the visible spectral band, is within 10 ms (Fig. 10.8b); note that the curves are given in the normalized form and do not reflect actual energy of radiation. The residual radiation is conditioned by the hot lamp balloon; its reflections from the surface of a test sample may seriously worsen the efficiency of identification of defect parameters. The following approximations have been found useful for the heating functions (in relative units):

$$\gamma(\tau) = 0{,}8933 \cdot e^{-4{,}761\tau} + 0{,}09082 \cdot e^{-0{,}08477\tau}; \tag{10.10}$$

$$\phi(\tau) = \frac{-0{,}00174\tau^2 + 3{,}52 \cdot 10^{-5}\tau + 4{,}39 \cdot 10^{-9}}{\tau^4 + 0{,}00272\tau^3 + 2{,}40 \cdot 10^{-5}\tau^2 + 6{,}02 \cdot 10^{-8}\tau + 7{,}08 \cdot 10^{-11}}, \tag{10.11}$$

where τ is expressed in [s].

In the passive IR diagnostics, external sources of thermal radiation may produce false signals which may be treated by the operators as defect indications. External radiation is reflected not only from metallic but also from many non-metallic materials. Often such problems appear when a smooth ("shiny") non-metallic surface should be thermographically monitored; for instance, in building thermography, operators may watch their own thermal image reflected from the window glass. Some values of methodical uncertainties of temperature measurements in the presence of external noise radiation are given in Table 10.4.

In outdoor IR surveys, noise radiation comes from the Sun, the Sky and the Earth. Solar radiation reaches the Earth ground in the spectral range from 0.3 to 3 μm delivering the power density about 1–1.3 kW/m^2 in a cloudless day. In some TNDT applications, this radiation can be used for thermally stimulating test objects,

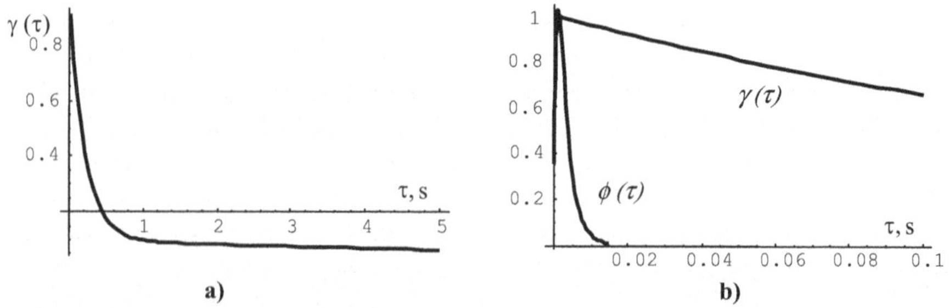

Fig. 10.8 Temporal profiles of Bowens flash tube radiation: (**a**) $\gamma(\tau)$- heating profile recorded by long-wave IR imager ThermaCam SC3000, (**b**). $\phi(\tau)$ – heating profile recorded by "all-waves" pyroelectric detector

Table 10.4 Relative methodical uncertainties of IR thermographic temperature measurement in the presence of external noise radiation [5]

	Measurement uncertainty, %	
Test object-heat source distance	Human person	Electrical bulb
1	6.7	3.1
2	1.9	0.8
3	0.8	0.3
4	0.4	0.2

but in other cases such additional heating may mask temperature signals to be detected, e.g. in the inspection of smokestacks (see Chap. 11). The atmosphere diffuses solar radiation and emits thermal radiation at wavelengths of absorption of carbon dioxide, ozone and other gases. Temperature read-outs supplied by an IR radiometer viewing the cloudless Sky typically shows rather low temperatures (up to −40 °C depending on a viewangle). Clouds emit thermal radiation as a blackbody with the ambient temperature. The ground radiation consists of the emitted and reflected components of the radiation from natural objects which can be regarded gray in respect to their emissivity. Statistical properties of the background radiation are well studied in military applications [4, 6–9].

The "quality" of TNDT results mainly depends on the structural noise produced by test objects. This noise appears due to both surface and volumic variations of material physical properties. Typically, the volumic noise can be neglected, even if some structural defects themselves, such as pores, excessive composite filling, can be regarded as components of such noise. In TNDT, the most unpleasant problems are supplied by surface emissivity/absorptivity fluctuations which are wavelength-dependent and may vary in time and space.

A typical phenomenological approach to ther evaluation of noise is based on the determination of standard deviation of dynamic temperature $\sigma_n(\tau)$ in °C, as well as the noise running contrast $C_n^{run} = \sigma_n(\tau)/T(\tau)$, in a chosen non-defect area. In the case of purely multiplicative noise, the noise running contrast should not vary in time. However, in practice, because of the impact of several factors, including additive noise, the noise running contrast varies in time when monitoring transient thermal events. In some cases, surface noise can be considered time-independent thus being a characteristic of a particular material.

Figure 10.9 shows the temporal evolution of noise running contrast in pulsed TNDT of the ceramic protection coating deposited on a turbine blade. It is seen that the coating noise is not fully multiplicative, probably, because of the considerable additive noise which takes place during the first 25 ms after the flash being conditioned by the residual radiation of the lamp balloons. Hence, maximal values of the signal-to-noise ratio can be expected at later times (after 25 ms), when $C_n \sim 1\%$. It is worth noting that in experiments with other blade coatings the resulting noise reached 5%.

The noise values which are given in Table 10.5 for some materials depend on a type of a material, its surface finish and size of a chosen area where noise was averaged. Since there is a certain freedom in choosing a non-defect areas, the estimates

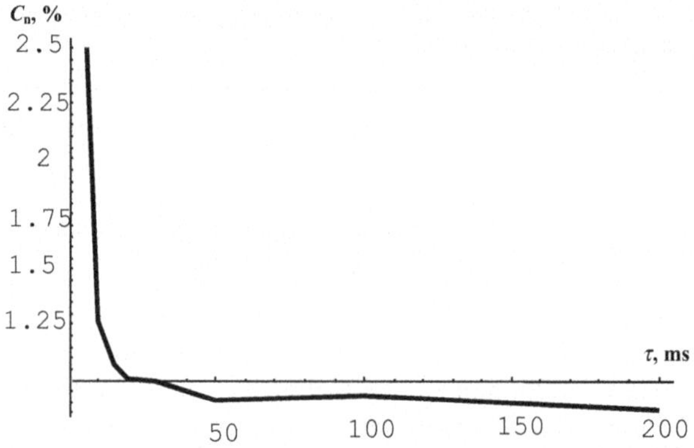

Fig. 10.9 Evolution of noise running contrast in TNDT of 250 μm-thick turbine blade coating (JADE IR imager, heating with a Xenon flash tube)

Table 10.5 Noise running contrasts (C_n) on the surface of some materials (active TNDT, open filament electrical heater, InSb IR detector, scanned spot size 0.8 mm) [10]

Material and surface conditions	C_n, %
Steel	7.2
Steel with rust	17.1
Steel with rust (scanned spot increased up 2.5 mm)	11.1
Copper oxidized	2.5
Copper painted with black lacquer	5.1
Copper painted with oil paint (red color)	4.4
Copper coated with black soot	2.8
Copper coated with black soot (scratched)	14.0
Aluminium	7.2
Textolite	1.1
Glass fiber reinforced plastic	1.5
Carbon fiber reinforced plastic *	6.4
Carbon fiber reinforced plastic (low-frequency filtering, 3 × 3 mask)	5.2
Carbon fiber reinforced plastic (high-frequency filtering, 3 × 3 mask)	13.6

of signal-to-noise ratio vary even when evaluating results of a particular test case. Also, the noise is greater in the case of non-filtered pixel functions $T(i, j, \tau)$. Note that the analogy between noise and emissivity ε is fundamental because namely fluctuations of emissivity and absorptivity are the major contributors to the noise in IR thermographic measurements. The data in Table 10.5 demonstrates that: (1) minimum noise occurs on non-metals and soot coatings, (2) maximum noise appears on corroded metals, (3) clean metals can be characterized by low noise but they reflect well external radiation, and (4) the quality of spatial filtering depends on a type of noise; typically, low-frequency filtering is preferable.

When optimizing a TNDT procedure, it is recommended to analyze a temporal evolution of the signal-to-noise ratio $S(\tau) = \Delta T(\tau)/\sigma_n(\tau=) = C(\tau)/C_n(\tau)$ rather than each independent parameter $\Delta T(\tau)$, $C(\tau)$, $\sigma_n(\tau)$ and $C_n(\tau)$. The time when a maximum of S appears is called the *optimum observation time*. This statement is illustrated with Fig. 10.10 (TNDT of a 6 mm-diameter delamination in a CFRP sample at the depth 1.25 mm). The abscissa coordinate is expressed in the image numbers, therefore, the plot in Fig. 10.10 allows choosing the "best" image that is No.12.

Another example of the evolution of TNDT parameters is presented in Fig. 10.11 (detecting a defect in a wall fresco, see Chap. 11) [11]. Immediately after heating the fresco for 120 s by using four electrical lamps, the appearance of the

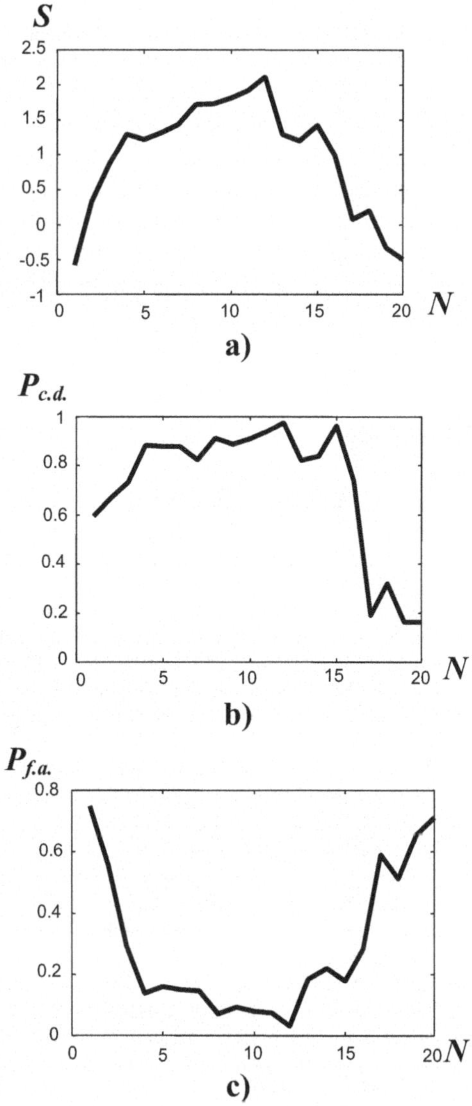

Fig. 10.10 Choosing optimum observation time in TNDT of CFRP (N is the image number): (**a**) signal-to-noise ratio, (**b**) probability of correct detection, (**c**) probability of false alarm

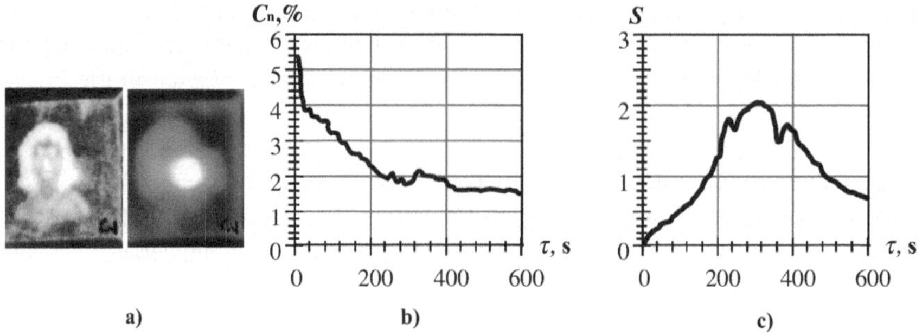

Fig. 10.11 TNDT of wall frescos: (**a**) IR thermogram at the end of heating (on the left) and at the optimum observation time (on the right), (**b**) noise running contrast, (**c**) signal-to-noise ratio

corresponding IR thermogram is essentially affected by the fresco structure; fresco colors lead to the strongly uneven absorption of stimulating optical radiation (left image in Fig. 10.11a), thus producing a very noisy image. At the cooling stage, the noise contrast decays in time due to high fresco emissivity in the IR band and heat diffusion (Fig. 10.11b). The maximal signal-to-noise ratio takes place at 300 s (Fig. 10.11c), even if the maximum differential temperature signal ΔT occurs at 180 s.

References

1. Sachs, L.: Applied Statistics, 220 p. Springer, New York (1982)
2. Taylor, J.O., Dupont, H.M.: Inspection of metallic thermal protection systems for the X-33 launch vehicle using pulsed infrared thermography. Proc. SPIE Thermosense-XX. **3361**, 301–310 (1998)
3. Vavilov, V.P.: Infrared techniques for materials analysis and nondestructive testing. In: Maldague, X. (ed.) Infrared Methodology and Technology, Monograph Series "Intern. Advances in NDT", pp. 230–309. Gordon & Breach Science Publish., USA (1994)
4. Lloyde, J.M.: Thermal Imaging Systems. Plenum Press, New York (1979). 456 P
5. Drozdov, V.A., Sukharev, V.I.: Building Thermography. Stroyizdat, Moscow (1987). 238 P. (in Russian)
6. Hudson, R.D.: Infrared System Engineering. Wiley-Interscience, New York (1969)
7. The Infrared and Electro-Optical Systems Handbook. V. 3 "Electro-Optical Components". Joseph S. Accetta, David L. Shumaker Exec. Editors. ERIM and SPIE Optical Engineering Press, Bellingham, WA, USA, 1993. 668 P
8. Kriksounov, L.Z.: Handbook of Infrared Techniques. "Sovetskoye Radio" Publish, Moscow (1978). 400 P. (in Russian)
9. Gaussorgues, G.: La Thermographie Infrarouge. Technique et documentation, Paris (1981). 386 P
10. Vavilov, V.P.: Thermal NDT of Composite Structures and Radio Electronic Components. Radio i Svyaz Publish, Moscow (1984). 162 P. (in Russian)
11. Grinzato, E., Bison, P.G., Marinetti, S., Vavilov, V.: Non-destructive evaluation of delaminations in fresco plaster using transient infrared thermography. Res. NDE, Springer, New York,. **5**(4), 257–271 (1994)

Chapter 11
Applications of Thermal/Infrared NDT

Abstract This chapter summarizes applications of both passive and active IR thermography and TNDT, including power generation, civil engineering and buildings, smokestacks, electrical installations, aerospace, petrochemical, automotive industry, art objects, medicine and many others. The discussions are supported by numerous illustrations and an extensive bibliography.

11.1 Introduction

As stated in Chap. 1, applications of IR thermography can be classified as either steady-state (or quasi steady-state), or transient (dynamic). In the latter, the temperature distributions change rapidly as a function of time. The applications of steady-state IR are well-developed and widely used, with established procedures and guidelines. The primary application of transient IR is TNDT (Thermal nondestructive testing) of materials. TNDT has been in development in various countries around the world, primarily in the aerospace industry. It is possible to apply the principles of TNDT to some non-aerospace applications, such as the inspection of wall frescos, the detection of landmines, etc. This chapter covers many applications of infrared, both steady-state and transient.

In Russia, a price list for IR inspection was compiled by the Ministry of Energy. According to this document, the cost of an IR inspection varies from $300 to $3000 for applications such as electrical substations, residential buildings, industrial smokestacks, hot water piping, etc. Leading Russian government organizations have acknowledged IR as a routine technique for inspecting electrical installations. Also, it is being increasingly used for evaluating smokestacks, furnaces, boilers and liquid ammonia storage tanks. Additionally, IR is accepted by governmental agencies as an effective tool for the evaluation of building heat losses and construction quality. The Russian Federal Service of Ecological, Technological and Atomic Supervision in Industries (Rostekhnadzor) considers IR thermography to be a supporting technique for the evaluation of potentially dangerous industrial structures such as smokestacks, storage tanks, chemical reaction vessels, buildings etc.

In Sweden, for many years, electrical substations have been subjected to IR inspections, and up to 150 thousand electrical connections are inspected each year. It has been stated that there is no alternative to IR thermography because electrical

V. Vavilov, D. Burleigh, *Infrared Thermography and Thermal Nondestructive Testing*, https://doi.org/10.1007/978-3-030-48002-8_11

connections degrade continuously, so they require periodic inspections. As early as the 1970s, it was reported in Belgium that, due to the systematic inspection of electrical installations, the number of electrical failures declined from 2.35% in 1971 down to 0.24%. In Russia, the use of IR cameras for inspecting high voltage transmission lines saved up to $20 K per region, due to the reduced number of failures and power outages. The U.S. food company H-E-B Bryan, because of a failure in the commutation system and a resulting fire, suffered damage of about $100 K. Afterwards they introduced a program for the IR inspection of electrical installations. On inspecting the repaired equipment, a pre-failure situation was discovered again and duly eliminated. At one US paper manufacturer, an IR survey detected a defective transformer which could have caused losses of $200 K to $300 K worth of product. In the UK power industry, the loss caused from one defective steam valve is estimated to be one thousand pounds sterling per year, and these devices can be thermographically checked in a few minutes. In general, up to 0.8% of piping is assumed to be defective, and the resulting financial loss is about $1.2 K per kilometer of pipe length. In metallurgy, prolonging the lifetime of a ladle refractory allowed British Steel Corporation to save 160 and 180 thousand pounds sterling in just two test cases. The French company Usinor saved about one million francs in a similar situation. The British company Pilkington reported saving about 250 thousand pounds sterling by applying IR thermography to evaluate wear in furnaces used in glass production. It was acknowledged that IR thermographic surveying is 25 times more effective than visual inspection. The Shell (Dutch) and ICI (Great Britain) companies reported that the average payback period for using a single IR camera at a petrochemical plant is about 1 year. IR inspection programs have been implemented at all US nuclear power stations and petrochemical plants.

Information on applications of IR technical diagnostics and its economic impact is presented in Table 11.1.

11.2 Administrative, Industrial and Residential Buildings

11.2.1 Preliminary Notes

One of the most effective ways to save energy in buildings is to reduce the heat losses through the building envelopes as this may comprise up to 30% of total energy losses. IR thermographic monitoring of buildings has proven to be one of the most important diagnostic techniques, both during and after building construction [1–18]. The first systematic research in building IR thermography was performed in Sweden and other Scandinavian countries, where energy saving was necessary due to the harsh climate. Similar research was performed in the USA, Germany and Canada. It is interesting to note that in warm areas, such as the southern United States, the problem is reversed; heat gain into buildings should be minimized to reduce heat input and to minimize the cost of air conditioning. Mathematically this

Table 11.1 Efficiency of IR diagnostics in industry (including information from FLIR Systems, USA)

Equipment	Defects that can be detected	Economic impact
Electrical installations		
Feeder switchboard, distributing devices, capacitors, ground connection, circuit breakers, joints, breaker switches	Loose or corroded connections, non-operating capacitor, serving-out ground connection, bad contacts in circuit breakers, overload, damage connectors in electrical circuits	Overheating, sparking or arcing, latent burning or open flame, possible breaks in damaged connectors. Non-operating capacitors do not provide protection against voltage differentials that leads to early equipment outage. Replacement and repair: $1 K–$100 K. Safety reasons.
Various electrical devices, commutators, switchboards, switches, control panels	Loose or corroded connections, faulty contacts, unbalanced load, overloading and overheating	Sparking or arcing, latent burning or open flame. 25% of failures and fails occur because of loose connections. Repair of main switchboard or control panel: $10 K–$70 K, replacement: $80 K–$100 K. Safety reasons.
Transformers (converters)	Loose or corroded connections, faulty contacts, overheated buses, unbalanced three-phase loading, blocked cooling pipes, level of liquids	Arc, short circuits, fire. Rewinding 5000 kW: up to $70 K, replacement: $80 K–$140 K.
Electric motors and generators	Overheated bearings, unbalanced load, short-circuit or broken coils, overheated brushes, rings and collectors, general overloading and overheating, blocked ventilation channels	Defective bearings can cause catastrophic failure starting with damage to motor shafts and windings. Faulty brushes can cause damage to commutator rings or windings. Rewinding of electrical motor (5000 horsepower): $50 K–$100 K Replacement: $100 K–$200 K Safety reasons.
Reserve electrical generators, batteries, contacts, automatic backup systems.	Wrong battery polarity, dead batteries, broken contacts or commutators	Loss of reserve electricity in the supply systems for key industries, including hospitals, telephonic communication, computers etc.
Boilers and steam systems		
Boilers and furnaces	Damage to insulation and refractory, scale, hot gas leaks, blockage of furnace tubes, leaks in safety valves	Enhanced heat losses and mechanical damage. Section replacement: $1,5 K–$5 K. Boiler replacement: $8 K–$30 K. Possible replacement of boiler. Safety reasons.
Steam systems	Leaks in valves and flanges, poor thermal insulation, leaks from underground pipes, blocked piping.	Steam and energy loss. Steam leakage cost: $10 K–$30 K per year. Equipment damage due to leakage. Repair: $0.1 K–$50 K.

(continued)

Table 11.1 (continued)

Equipment	Defects that can be detected	Economic impact
Steam condensers and valves	Faulty steam condensers, wrong size of steam traps	Open steam condenser, steam and energy loss. Steam leakage cost: up to $30 K per year. Closed steam condenser: transport of water, air and gases to equipment. Repair: $100–$50 K
Mechanical equipment		
Internal combustion engines	Faulty valves and injection systems, blocked radiator tubes and oil cooling systems	Overheating, damage to valves, pistons, studs and cylinder. Damage to power trains. Repair: $1 K–$40 K.
Air heating/cooling and conditioning systems	Air leaks, electric energy loss, obstruction of heat exchanger piping. Leaks of a cooling agent, hot air and steam, air-locks.	Inoperative functioning and energy loss, switching off.
Conveyors, mounting bearings	Overheated bearings or rollers, axis or roller curvature, uneven pressure	Defective bearings cause equipment damage, uneven pressure or curvature cause damage of bearings or rollers and possible failure of conveyor. Repair: up to $500 K.
Pumps, compressors, ventilators, compressor guns	Overheated bearings, excessive temperature of compression and/or oil, broken or defective valves	Damage of bearings and components, possible failure of assembly
Turbines of mechanical driving, small turbine generators	Excessive temperature of lubricants and bearings, blockage of exhaust valves and steam traps, uneven heating of metals, leaks in cylinder gaskets	Water ingress, damage of bearings and components, possible failure of assembly
Calcining kilns	Refractory breakdown, overheating of bearings and reducers	Energy loss, damage of bearings and reducers. Kiln replacement: up to $1 million.
Pulp-and-paper production	Moisture variations, overdrying, hogging of side-wall rollers, fails in steam distribution systems. Breakage of vacuum chambers. heat sinks, exhaust hoods. Deterioration of pulp forming process, wrong steam distribution.	Out-of-range moisture leads to energy loss, low paper quality, damage of fibrous mass because of overdrying.
Furnaces, stacks, refractory, thermal insulation	Torch surge on fire-chamber walls, scale formation on internal surface of piping, fails in furnace operation	Energy loss, low product output, overheating and buckling of tubes may cause explosion with necessity of replacing tubes or fire chambers. Safety reasons.

(continued)

Table 11.1 (continued)

Equipment	Defects that can be detected	Economic impact
Equipment operating in hard conditions, bearings, brakes, hydraulics, rollers, drives, AC/DC motors	Overheating of brakes, bearings, rollers, drives, hydraulics blockage, curvature of drives and rollers	Brake fail, bearing breakage; unadjusted cog-wheels in gearing. Repair: up to $1 million Safety reasons.
Building constructions		
Electrical utilities	See *Electrical installations*	
Envelopes	Heat loss, building defects, air leaks, moist areas, windows installation, architectural solutions	Heat loss, thermal discomfort, structural problems because of moisture, alligatoring and bricks fall-out, masonry ageing. Repair: up to a few hundred dollars.
Reinforced concrete constructions	Detection of reinforcement and evaluation of its wear, inspection of parking heating systems, detection of hot & cold water piping, steam pipes inspection, evaluating heat losses	Undetected duly reinforcement corrosion and concrete detachment causes serious structural problems; idle heating equipment at parking may cause accidents
Heating and air condition systems	See *Air heating/cooling and conditioning systems*	Heat losses, thermal discomfort
Roof	Leaks and moist penetration areas on flat roofing	Heat losses and water leaks cause further expensive damage Repair: $1 K–$10 K.
Power plant and industrial stacks	Cracks in stack shaft, refractory collapse, deterioration of thermal insulation	Untimely fail of stacks. Repair: $5 K–$20 K.
Other applications		
Tanks for storing liquids and gases	Determining level of liquids, poor insulation	Overfilling tanks leads to liquid spills, energy losses, gas leaks into the environment. Potentially fatal.
Drying cylinders	Forming condensate, high temperature of bearings	Damaged bearings cause failure of rollers and cylinder fall-out. Repair: up to $2 million
Blast-furnace, ladle annealing and heat treating	Refractory degradation, inefficient or broken heaters	Energy losses, low productivity, spoilage, overheating and buckling of structures.
Aerospace structures: fuselage, engines (internal combustion and jet), turbine blades, brakes and other components.	Delaminations, voids, moisture ingress. Blockage of turbine blade cooling paths, operation of anti-ice systems, leaks in hydraulic systems, inspection of porous structures and heating devices.	Trapped moisture can cause damage to metallic structures and result in excessive fuel consumption.

(continued)

Table 11.1 (continued)

Equipment	Defects that can be detected	Economic impact
Automotive industry	Evaluation and diagnosis of problems in: Engine components, drive train bearings, brakes,tire design and performance evaluation, windshield defroster systems, radiators and engine cooling systems. air conditioning systems.	IR is used by: 1. Automotive design institutions. 2. Auto repair shops, especially on radiators 3. Automobile racing teams

is the same problem as reducing heat losses with the exception of a minus or plus sign here and there. From 1980 to 2000, several standards and guidelines on using IR thermography in building were issued worldwide (see Chap. 12). In Russia, several research bodies, including the All-Union Research Institute of Building Physics, Moscow Building Institute et al., were involved in the investigations of thermal efficiency of buildings. The Russian experience in this field was summarized by Drozdov, Sukharev and other authors [5, 6, 8–12, 19]. However, in those years, due to the low cost of energy, and low governmental requirements for thermal performance of buildings, the economic motivation for implementing IR thermography in the Russian building industry was low.. Research in this area intensified in 1998 following the introduction of the new Building Norms and Rules (SNiP) II-3-79∗ "Building thermal engineering" (since 1.10.2003: SNiP 23-02-03 "Thermal protection of buildings") which increased the requirements for wall thermal resistance by a factor of almost 3. It was quickly recognized that these new standards could not be met by using standard Russian building materials and designs. This stimulated the development of new designs, materials and constructions methods, as well as tools for checking the quality of thermal protection systems in building envelopes. In recent years, the administrations of many Russian cities, in cooperation with federal agencies responsible for buildings and architecture, developed significant interest to IR thermography as a tool for solving problems in Russian housing and communal services. These included the renovation of older buildings, improvement in the quality of new buildings, reduction of heating costs, etc. It is worth mentioning that, since 1998, in the City of Seversk (Tomsk region in Siberia) the mayor has required the use of IR thermography as a mandatory diagnostic tool for overall inspection of all new buildings. A set of documents encouraging the introduction of IR diagnostics was recently approved in Moscow and and the Moscow region.

According to the guidelines approved by the former Russian Federal Committee for Standardization (now Rostekhregulirovaniye) in 2000 [6], IR diagnostics of buildings involves:

- Determination of total or partial heat losses,
- Detection of hidden defects,
- Evaluation of wall thermal resistance.

Another potential application of IR thermography is predicting the collapse of buildings by detecting "thermal precursors" of catastrophes. In Russia, as well as around the world, unexpected building catastrophes periodically occur, resulting in financial loss and loss of lives. This makes this application an important socio-economic issue. However, there had been very little investigation into observable temperature anomalies or thermal mechanisms that would be precursors to catastrophic building failures. Moreover, according to experts from the Central Research Building Institute, Russia, even the full collapse of concrete buildings do not provide temperature signals detectable with IR cameras. However, in laboratory experiments performed by Luong in France it was demonstrated that in concrete samples subjected to cyclic mechanical loading temperature signals could reach few degrees Celsius [7]. An example of the experimental analysis of a real catastrophe that occurred in 1997 in the City of Tomsk (Siberia, Russia) will be reported in Sect. 11.2.8 [8]. In general, the possibility of predicting the collapse of a building with an operating heating system seems doubtful, as construction damage or defects only create very small temperature signals and these are very difficult to detect, especially on a very noisy thermal background.

11.2.2 Estimating Energy Losses

In Russia, the characteristics used to evaluate the thermal protection of residential buildings are

- average outdoor air temperature, which is specific for each geographic region
- indoor air temperature (20–22 °C)
- indoor relative humidity (50–60%)
- temperature difference between the indoor air temperature and the indoor outer wall temperature (4–6 °C)
- indoor wall temperature, which must exceed the dew point.

The most important characteristic of a building is the energy consumption (in kW·hr/m^2 or other units) per year during the normal heating period.

In Fig. 11.1, a diagram of building heat balance is presented for an annual heating period [9]. The income part depends on: (1) the heating system, (2) electrical devices, and (3) solar radiation. The expenditures involve: (1) heat losses through the building envelope including walls, windows, roof and basement, (2) energy to make up for cold air leaks, (3) energy for warming up the water (if present). A significant portion (up to 50%) of total heat losses is caused by cold air infiltration, both naturally and through the ventilation system.

As a technique for evaluating thermal engineering systems, IR thermography allows the analysis of ventilation systems, evaluating air infiltration and detecting deficiencies in building thermal protection systems which may occur because of errors or violations in standards or codes of the following:

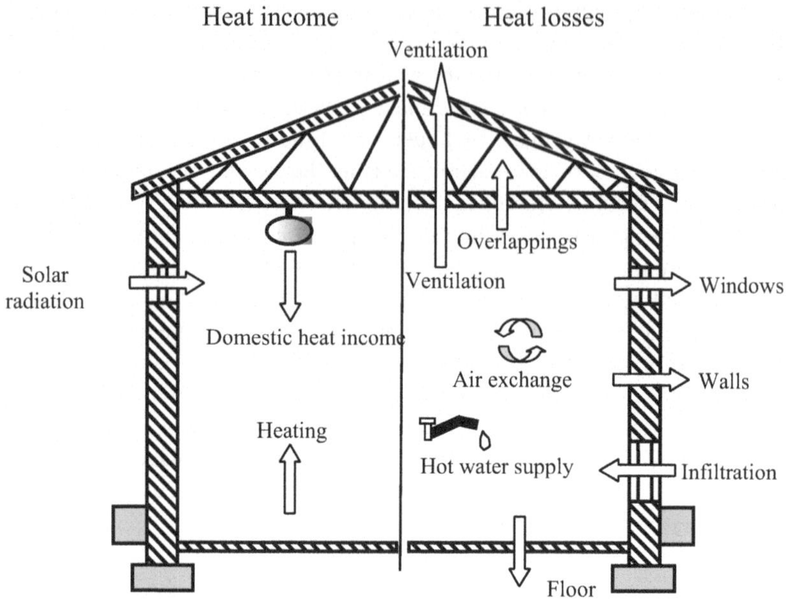

Fig. 11.1 Heat balance of a building. (Adapted from [9])

- Design
- Manufacturing technology,
- Materials handling, including transportation and storage
- Building technology
- Use of materials

The factors above cause premature degradation of thermal protection systems in buildings due to weather (wind, precipitation) and climate (moisture, warm-cold cycles). This degrades the indoor climate, causes an excessive use of energy, and results in financial loss.IR thermography helps to eliminate design errors that cause problems in maintaining the indoor temperature at a desired value. Thus, IR thermography may serve as an important *energy audit* tool. Energy surveys may be required by various government agencies. In Russia, these results may constitute a basic part of a building "energy consumption passport", which are required for (1) industrial consumers (state standard GOST P 51379-99), (2) energy supply systems and their elements (industrial standards approved by the United Electrical Networks of Russia, a major producer of electric energy in the country) coordinated with the Federal Supervision Service in Power Production (Gosenergonadzor), (3) civil buildings, including residential, and (4) other buildings approved by Gosenergonadzor and other relevant bodies. Energy consumption passports contain both design goals and actual thermal performance data of buildings, and IR thermography is considered to be a powerful tool to measure or verify those parameters.

Compared to standard techniques of thermal engineering analysis, IR thermography can evaluate entire sections of a building in a short amount of time. However, IR thermography allows only an approximate, non-quantitative, or comparative evaluation of heat losses through a building envelope. Quantitative results can be obtained by measuring the heat flux in the same way as in determining wall thermal resistance (see Sect. 11.2.4). Heat fluxmeter readings obtained at a few points can be used to calibrate IR heat flux measurements across the whole building.

It is important to evaluate heat losses at the time of an actual inspection. Then, annual losses can be extrapolated by taking into account a standard indoor temperature, average climate conditions and the duration of the normal heating period. In Russia, only side walls of building envelopes are thermographically surveyed, while heat losses through the roof and basement are evaluated by using standard thermophysical recommendations, as shown in Table 11.2. In some countries, roof inspections are being done by using small airplanes, helicopters and, recently, drones and quadrocopters. By some estimates, errors in IR thermographic evaluation of heat losses can be 20–40% of the true value, which can be confirmed *a posteriori* after receiving heating bills.

Typically, the cost of thermal energy losses calculated from IR thermographic data correlates satisfactorily with actual heating bills. The corresponding methodology is as follows.

First, by performing direct measurements or using building plans or other documents, one calculates the area of an envelope S. Also, most IR systems have non-contact, optical methods to measure dimensions of a target by using a reference of a known size.

Then one determines the mean envelope temperature by averaging temperatures in single thermograms and taking into account the contribution of particular building sections. Measurement results depend on the distance to the building being tested and the presence of spurious reflected radiation from the Sun, the cold Sky, other buildings, etc. Heat losses Q can be calculated approximately as:

$$Q = \alpha_{ext} S \left(T_{ext}^{w} - T_{ext}^{air} \right), \tag{11.1}$$

where α_{ext} is the outside heat exchange coefficient, T_{ext}^{w} is the outdoor wall temperature, and T_{ext}^{air} is the outdoor air temperature. The magnitude of the heat exchange coefficient can be evaluated by using a heat fluxmeter (Russian standard GOST 7076-78 or other). Both the convective and radiant components of the heat exchange

Table 11.2 Heat losses through building envelope sections [5]

Building element	Heat loss %
Walls	32
Windows	29
Roof	24
Air exchange	9
Basement	6

between a building and the environment are taken into account. The sensitivity of heat fluxmeters, i.e. the dependence of the output electrical signal on temperature, can be very different; through-the-wall heat flux (5–30 W/m^2) is typically measured on inside walls where heat fluxmeter readings are not influenced by wind and solar radiation. As a result, weak heat fluxes can be evaluated.

When determining total heat losses for a certain period of time, e.g. for a heating period, one should deal with an effective Q_{eff} value which should be calculated using the actual ($\Delta T = T_{\text{int}}^{air} - T_{ext}^{air}$) and the normative ($\Delta T_{norm} = T_{norm}^{\text{int}} - T_{norm}^{ext}$) temperature difference. Here: T_{norm}^{int} is the required indoor air temperature (see Russian Building Norms and Rules SNiP 23-02-03 "Thermal protection of buildings"); T_{norm}^{ext} is the mean winter air temperature specified for a particular geographical region; e.g., in the city of Tomsk (Siberia) this value is −8.4 C. The formula for calculating Q_{eff} is:

$$Q_{eff} = Q \cdot \frac{\Delta T_{norm}}{\Delta T}. \tag{11.2}$$

Equation (11.2) is valid for a temperature difference that would exist if the indoor temperature met an approved standard.

Finally, energy losses are determined by:

$$W = Q\tau, \tag{11.3}$$

where τ is the period duration.

Relative heat losses can be determined if the heat exchange coefficient is unknown and a reference area is chosen:

$$Q / Q_{ref} = \left(T_{ext}^{w} - T_{ext}^{air}\right) / T_{ext\ ref}^{w} - T_{ext}^{air}), \tag{11.4}$$

where the subscript "*ref*" specifies a reference area.

The methodology above involves only heat losses caused by poor thermal insulation of a building. In practice, significant energy losses occur due to air infiltration. The corresponding loss estimates are based on knowing the pressure and temperature differences. However, the accuracy of IR thermographic estimates of building air tightness seems to be low.

A popular technique for estimating air infiltration heat loss involves the use of a "blower door" ("air-door"). This technique is described in ASTM (American Society for Testing and Materials) E1827-11 (2017), "Standard Test Methods for Determining Airtightness of Buildings Using an Orifice Blower Door". These doors are commercially available.

The financial cost of energy losses is determined by calculated values of Q and the cost of energy. IR diagnostics simply reveal the presence and distribution of heat losses; this is valuable information that can be used to stop the losses. IR thermography can also evaluate the effectivness of building repair and renovation, as well as modifications to optimize heat and ventilation in buildings,.

When evaluating building envelopes, it is important to perform an IR evaluation of windows, as maximum heat losses occur in or around windows. Measuring the true temperature of window glass requires using some special techniques because IR camera readings are affected by the viewing angle and reflected radiation of the sun and the sky, as well as hot or cold neighboring objects. Some modern windows have up to three layers of glass filled with argon, and some have special coatings. These may require a more complicated analysis to determine the radiant component of heat loss.

11.2.3 Detecting Hidden Defects

Hidden defects decrease the thermal comfort in buildings and, in rare cases, may cause degradation of the structure. In building engineering, similar to the NDT of materials, the term "defect" means a deviation of a specified parameter from its design value. The following defects can be detected in buildings by means of IR thermography:

- Air and water leaks,
- Cold or hot "thermal bridges (shorts)"; this is a deterioration in a wall: either the thermal resistance (a lack of thermal insulation, excessive moisture, poor quality masonry, incorrect architectural and design solutions, etc.),
- Defective panels (deviation for the design insulation, either the type or thickness, water ingress, collapse or compression of insulation materials, etc.),
- Detachments between a support wall and plaster or decorative coatings, etc.

Modern IR cameras allow the easy and reliable location of defective areas in building. And the quantitative evaluation of defects can be performed under laboratory conditions using mathematical analysis and computer software.

The detection of hidden defects is generally accomplished by looking for thermal anomalies or by comparing inspected areas to reference (non-defect) areas. A reference area can be chosen either by knowledge of the structure or by analyzing test results, such as by determining wall thermal resistance. There are many international standard procedures for using IR to evaluate buildings. In most cases, IR cameras are used to measure the surface temperature, while the heat flux (heat exchange coefficient) is measured by a heat fluxmeter.

The following is a list of some standards; there are other standards from Sweden, England and other countries.

- Russian standard GOST 26629-85. "IR thermographic inspection. Thermal insulation of building envelopes (in Russian), 1985".
- ASTM (American Society of Testing and Materials) C 1060-90 "Thermographic Inspection of Insulation Installations in Envelope Cavities of Frame Buildings".
- ASTM C1046-95 (2007) "Standard Practice for In-Situ Measurement of Heat Flux and Temperature on Building Envelope Components".

- ASTM E 1186 "Air Leakage Site Detection in Building Envelopes and Air Barrier Systems.
- ASTM E1827 "Standard test methods for determining air tightness of buildings using an orifice blower door".
- ISO 6781 "Thermal Insulation - Qualitative Detection of Thermal Irregularities in Building Envelopes, Infrared Method".
- Canadian Standard NMS Section 02 27 13 "Thermographic Assessment – Building Envelope".
- Canadian GSB 149-GP-2MP – "Manual for Thermographic Analysis of Building Enclosures".

Hidden defects in buildings can be detected both indoor and outdoor. In general, outdoor inspection is more appropriate for evaluating heat losses and for detecting defects that influence the temperature distributions on the outer walls. Indoor inspection provides more details when analyzing the thermal performance of specific buildings. There is a method for performing indoor inspections while depressurizing a building using a "blower door". The lowered pressure increases air leaks into the building, which increases the thermal "signature" of the leaks. But temperature indications over hot/cold bridges remain at the same level. As a result, this allows the discrimination between these two types of defects. In summer time, when the indoor-outdoor temperature difference is generally low, the temperature difference may be increased by the use of interior heaters. In this case, measurements may be performed a few days after the heat is turned on.

In the outdoor inspection of brick buildings, the primary attention should be paid to checking vertical seams, looking for areas of moist masonry and evaluating architectural elements such as windows, as these are potential thermal bridges caused by the use of metallic parts. While performing indoor surveys, poor insulation between walls and in floor support structures, as well water leaks around windows, doors and balconies, can be reliably detected. IR thermography is particularly effective in the inspection of buildings made of prefabricated panels. Many Russian residential buildings built in the 1960's, called "Khrushchev's houses", used this construction method. In these buildings, the common defect type is faulty (unninsulated) inter-panel seams, and the economic benefit can be easily evaluated, as the repair cost is proportional to seam length (about $10 per meter in Russia). Inspection experience shows that seams that look defective visually may still contain insulation, and may only require cosmetic repair. Conversely, seams that visually appear to be OK may be empty of insulation, and residents will complain about cold areas in the building. When temperature differences between inside and outside the building are large, IR thermography can visualize the internal structure of building panels and thus evaluate the presence and quality of internal insulation. Obviously, IR thermography serves as a good tool for inspecting the quality of insulation that is added as a standard practice during building renovation.

In recent years, the necessity of IR inspection of industrial buildings has become increasingly evident, especially in the inspecting of power plant buildings. Russian thermal power plants built in the 1960–1970's are deteriorating. In one example, the steel reinforcement bars ("rebar") in reinforced concrete panels widely used in these buildings is corroding; this corrosion causes fracture of the concrete and the ultimate structural failure of the panels. The extent of the corrosion is related to panel moisture which, can be assessed by IR thermography.

Results of IR inspection of buildings are represented as black and white and/or color images which can be either single IR thermograms or the *panoramic* (*mosaic*) images. It is generally valuable to produce a map of defects in a building and compile a list of defects that are to be repaired.

11.2.4 Evaluating Wall Thermal Resistance

The theory for the evaluation of wall thermal resistance was presented in Sect. 3.7.9. The stationary thermal flux flowing through a multi-layer plate which separates two media with the temperatures T_{in}^{a} and T_{out}^{a} respectively, is given by the expression (Fig. 11.2):

$$Q = \frac{T_{in}^{a} - T_{out}^{a}}{R}, R = \frac{1}{\alpha_{in}} + \frac{l_1}{\lambda_1} + \frac{l_2}{\lambda_2} + \frac{l_3}{\lambda_3} + \frac{1}{\alpha_{out}}, \quad (11.5)$$

$$Q = \alpha_{in}\left(T_{in}^{a} - T_{in}^{w}\right), \quad (11.6)$$

$$Q = \alpha_{out}\left(T_{out}^{w} - T_{out}^{a}\right), \quad (11.7)$$

where the notation corresponds to that in Sect. 3.12.

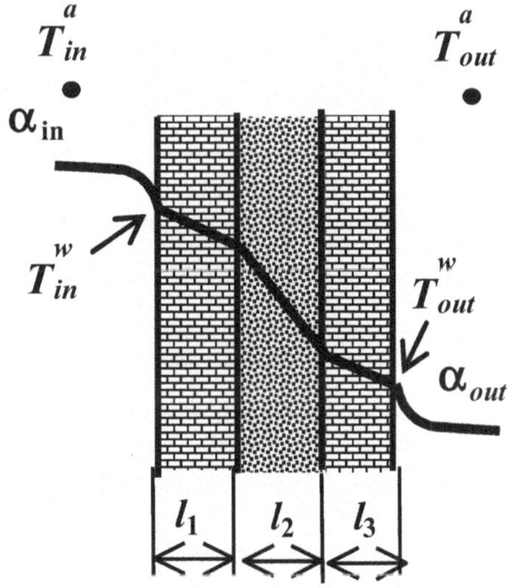

Fig. 11.2 Heat conduction through a multi-layer wall

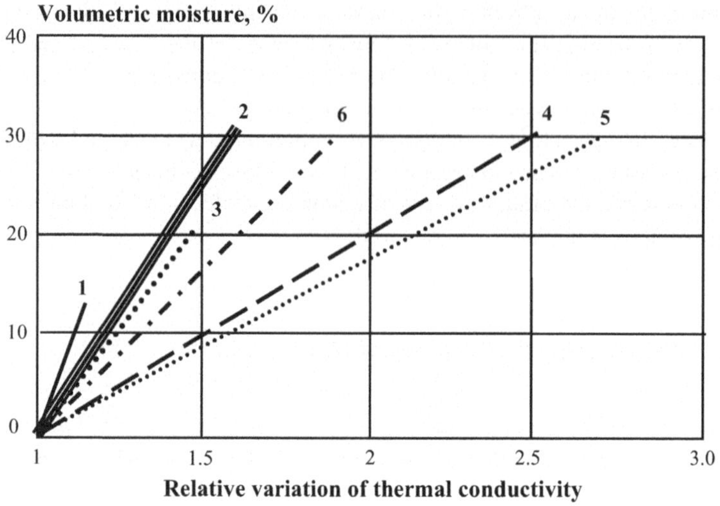

Fig. 11.3 Thermal conductivity of building materials *vs.* moisture
1. concrete, density 2000 kg/m^3,
2. expanded-clay concrete, density 1200 kg/m^3,
3. red brick, density 1800 kg/m^3,
4. slag (rock) wool, density 200 kg/m^3,
5. glass wool, density 200 kg/m^3,
6. slag, density 1200 kg/m^3.

Thermal properties of typical building materials were reported in Table 2.3. The influence of moisture on thermal conductivity of some building materials is shown in Fig. 11.3.

To characterize the components of wall thermal resistance, the following definitions have been introduced:

thermal resistances of boundary layers on both inside and outside surfaces:

$$R_{in} = \frac{1}{\alpha_{in}}, R_{out} = \frac{1}{\alpha_{out}} \tag{11.8}$$

and the thermal resistance of an N -layer wall:

$$R_t = \sum_{1}^{N} \frac{l_i}{\lambda_i}, \tag{11.9}$$

where l_i, λ_i are the thickness and thermal conductivity of the i-th layer.

The total *wall thermal resistance* is:

$$R = R_{in} + R_t + R_{out}. \tag{11.10}$$

We reiterate that the parameter $U = 1/R_t$ is called the *overall (total) coefficient of heat transmission*, or *U-value*.

In a building, the parameter of thermal resistance is normally used for areas with uniform temperature distributions; if temperature is essentially non-uniform across a wall, the *reduced* thermal resistance is introduced to specify a mean value which characterizes the thermal resistance of large and thermally-uneven areas. Obviously, the use of IR thermography makes the process of determining the average temperature easier and more accurate than when using contact temperature sensors.

In Russia, according to SNiP 23-02-03 "Thermal protection of buildings", the requirements for the thermal insulation of buildings was revised twice: on 1 September, 1995 and on 1 January, 2000. For instance, in the city of Tomsk the total wall thermal resistance in residential buildings, hospitals and schools was required to be 2.15 $W^{-1} \cdot m^{2} \cdot K$ at the first revision and 3.76 $W^{-1} \cdot m^{2} \cdot K$ at the second revision. In fact, the value approved by the local Tomsk Territorial Building Norms is even higher: 3.83 $W^{-1} \cdot m^{2} \cdot K$.

Steady-state regime

In the steady-state regime, the outside wall temperature can be calculated by the following formula (see Sect. 3.12):

$$T_{out}^{w} = T_{out}^{a} + \frac{R_{out}}{R}\left(T_{in}^{air} - T_{out}^{air}\right). \tag{11.11}$$

Defects appearing in thermal insulation change the wall thermal resistance R by the ΔR value, thus causing the corresponding temperature signal:

$$\Delta T = \frac{R_{out}}{R}\frac{\Delta R}{R - \Delta R}\left(T_{in}^{air} - T_{out}^{air}\right). \tag{11.12}$$

Eq. (11.12) means that by determining ΔT, e.g. with an IR camera, one could evaluate variations in thermal resistance. In particular, *decreasing* R by ΔR will lead to a positive ΔT, which creates an increased outside wall temperature. This would be a relatively warm area, which could be easily detected by an IR camera.

The heat transfer through building walls can be regarded as steady-state only to a certain approximation. The higher the diurnal temperature variations, the less accurate the Eqs. (11.5, 11.6, 11.7, 11.11 and 11.12) will be. The Russian national standard GOST 26254-84 "Buildings and constructions, methods for determining envelope wall thermal resistance" suggests the following formula for determining R, which is similar to Eq. (11.5):

$$R = R_{in} + R_{unif} + R_{ext} = \frac{t_{in} - \tau_{in}}{q} + \frac{\tau_{in} - \tau_{ext}}{q} + \frac{\tau_{ext} - t_{ext}}{q}, \tag{11.13}$$

where R_{in}, R_{ext} are the thermal resistances of the boundary layers on the inside and outside surfaces, R_{unif} is the thermal resistance of a uniform wall area, (see Eq. 11.9),

t_{in}, t_{ext} are the indoor and outdoor air temperatures *averaged* within a particular time period, τ_{in}, τ_{ext} are the indoor and outdoor wall temperatures also averaged within the same time period as t_{in}, t_{ext}, q is the averaged thermal flux density.

Heat exchange coefficients are typically determined by employing heat fluxmeters. The use of the values recommended by SNiP 23-02-03 ($\alpha_{in} = 8.7\ W/(m^2K)$; $\alpha_{out} = 23\ W(m^2K)$ is not recommended because actual values, particularly, when dealing with α_{out}, may deviate significantly from handbook recommendations. The heat exchange rate varies by vertical location along the building height and is influenced by the presence of architectural elements, such as windows, structural ribs, columns etc. This makes it difficult to determine R across the full field of view of an IR camera. In some cases, the thermograms of flat surfaces can be converted into maps of thermal resistance but the accuracy of such data cannot be better than 15–30%. The relevant standard requires the use of a large number of contact heat flux sensors, but this cannot guarantee reliable results due to the possible presence of local heat sources or heat sinks, curved surfaces, coating detachment etc. Therefore, it is advisable to perform a preliminary IR survey of building walls before attaching contact heat flux sensors so they can be attached to areas with uniform temperature distributions. It is worth noting that the contribution of the boundary thermal resistances (respectively $R_{in} = 0.115\,m^2K/W$ and $R_{out} = 0.0435\ (m^2K)/W$, recommended by SNiP 23-02-03) to the total thermal resistance is relatively small. Therefore, the procedure of determining wall thermal resistance is reduced to the formula:

$$R_t = \frac{T_{in}^{w} - {}''^{w}_{out}}{Q}. \tag{11.14}$$

In addition to heat exchange intensity, other factors that may affect the accuracy of determining R by IR are as follows: (1) incorrect emissivity settings, (2) difference between the air temperature at the inspection time and 8–12 hrs before, and (3) non-linear temperature distribution within the envelope walls.

A root-mean-square error in measuring R_t by Eq. (11.14) in a steady-state regime can be evaluated by assigning root-mean-square errors in measuring both thermal flux and wall temperatures. Let us limit ourselves to a simple graphical illustration by using a random number generator that allows modeling IR camera readings and heat flux meters at different times.

Consider an example described in Table 11.3 (by analyzing two thicknesses of brick walls).

Assume the wall temperature is measured with an IR radiometer having an accuracy of ±1°C, and the thermal flux is measured with an accuracy of ±1.5% (ITP-2 unit, Russia). Consider 20 measurement points with input values varying randomly within the specified interval of uncertainties. The plot of the total relative uncertainty in determining R_t is shown in Fig. 11.4. It is seen that, independent of wall thickness, its thermal resistance can be determined with an accuracy of ±6% of the true value.

Table 11.3 Determining the thermal resistance of a wall using steady-state procedures

$T_{in}^{a} = +20C^{\circ}$ $T_{out}^{a} = -10C^{\circ}$	
$\alpha_{in} = 8.7$ $W/(m^2K)$, $\alpha_{out} = 23$ $W/(m^2K)$	
Brick wall thickness 0.76 m	**Brick wall thickness 2.6 m**
$R = 1$ m²KW⁻¹, $R_t = 0.84$ m²KW⁻¹	$R = 3$ m²KW⁻¹, $R_t = 2.89$ m²KW⁻¹
$Q = 29.91$ Wm⁻²	$Q = 9.85$ Wm⁻²
$T_{in}^{w} = +16.56$ °C, $T_{out}^{w} = -8.70$ °C	$T_{in}^{w} = +18.87$ °C, $T_{out}^{w} = -9.57$ °C

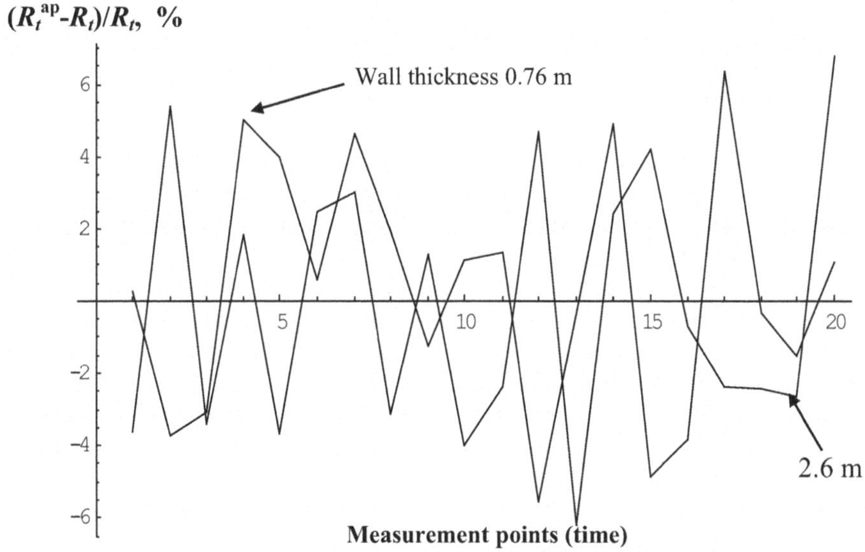

Fig. 11.4 Accuracy of measuring the thermal resistance of a brick wall using Eq. (11.14)

Transient regime

The necessity of averaging measurements in order to simulate a steady-state regime, as required by Eqs. (11.13 and 11.14), makes the determination of *R* a long procedure (up to 15 days). Presently, multi-channel digital flux meters and surface mounted thermometers are often used. The recorded data is analyzed on a computer to choose the time periods where the variations in the outdoor air temperature do not exceed ±1.5 C of the mean value (requirement of GOST 26254-84). Duration of such periods is to be from 1.5 to 3 days depending on wall thermal inertia. It is interesting that the dynamic character of heat transfer through a wall influences calculated *R* values less if a flux meter is placed in the centre of the wall; that is, of course, hard to do in practice. In Russia, there have been some attempts to develop a procedure for determining *R* by taking into account the temperature history of the

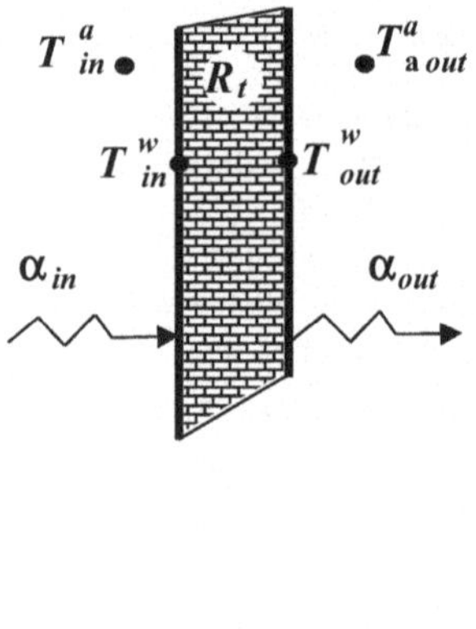

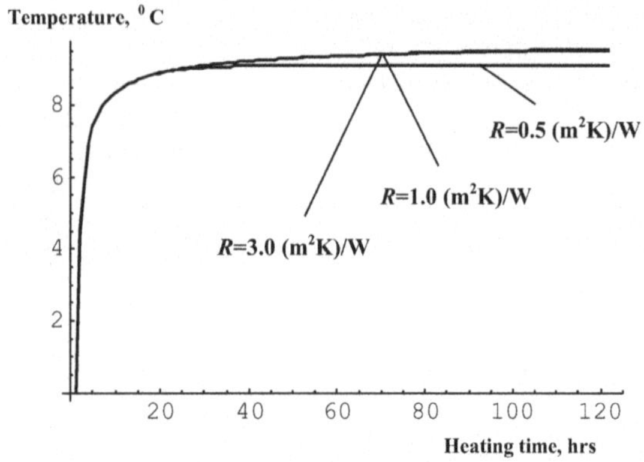

Fig. 11.5 Heat conduction through a uniform wall: (**a**) – heat exchange scheme, (**b**) – outside wall surface temperature *vs.* time (10 °C step increase in the ambient temperature)

outdoor temperature for several days preceding the inspection (see below); however, "metrology" of such an approach remains undefined. In summertime, wall thermal resistance can be determined by heating a building section for few days with an auxiliary heater.

Consider the heating of a homogeneous brick wall (Fig. 11.5a). The initial wall temperature has no affect due to the linear character of heat transfer. Let the outdoor air temperature sharply increase by 10 °C at $\tau = 0$. Determine the time necessary the surface temperature reaching a steady-state value.

Mathematically, we are dealing with the problem of heating a plate in a medium (or with a heat flux) where heat exchange coefficients are different on the front and rear surfaces. The latter fact makes an analytical solution cumbersome. Therefore, the data shown in Fig. 11.5b has been obtained by using the numerical ThermoCalc-2D software. Three accepted thicknesses of the wall 0.31, 0.76 and 2.6 m correspond to three values of thermal resistance ($R = 0.5$, 1 and 3 $W^{-1}m^2K$) if the heat exchange coefficients are chosen in accordance with SNiP 23-02-03, and the brick thermal conductivity is $\lambda = 0.9\ Wm^{-1}\ K^{-1}$.

It follows from Fig. 11.5b that, in the transient heating process, brick walls having thicknesses of 0.76 and 2.6 m behave similarly if one deals with the front surface of the wall. Let us introduce two characteristic parameters $\tau_{90\%}$ and $\tau_{99\%}$ which correspond to times when the surface temperature reaches 90% and 99% of its stationary value respectively. The corresponding results are presented in Table 11.4.

The results presented in Table 11.4 and Fig. 11.5 lead to the following conclusions.

- At the end of heating, the wall temperatures reach their steady-state values; the slightly different value in the case of wall thickness 2.6 m is due to the fact that the 10 day period is not long enough for the wall to transit into the steady-state regime.

Table 11.4 Step heating of a brick wall (outdoor temperature sharply increases by 10 C) ($\alpha_{in} = 8.7$ Wm^{-2} K^{-1}, $h_{out} = 23$ Wm^{-2} K^{-1}, $\lambda = 0.9$ Wm^{-1} K^{-1}, brick thermal diffusivity $a = 6.1 \cdot 10^{-7}$ m^{2}s^{-1})

Brick wall thickness, m	Steady-state surface temperature calculated by Eq. (11.11), C	Front surface temperature after 10 days heating calculated with ThermoCalc-2D	$\tau_{90\%}$, hrs (Fo)[a]	$\tau_{99\%}$, hrs (Fo)
0.31	9.14	9.14	7 (0.16)	26.5 (0.6)
0.76	9.57	9.57	11 (0.042)	82 (0.31)
2.6	9.86	9.70	13.5 (0.0044)	137 (0.045)

[a]In brackets is given the corresponding dimensionless time (Fourier number)

- Two brick walls of thicknesses 0.76 and 2.6 m, which, according to SNiP 23-02-03, approximately correspond to the "old" and "new" Russian standards of thermal resistance, behave very similarly in the transient process when dealing with the outside surfaces. The difference in the outside wall surface temperatures is 0.29 °C (in the steady-state regime) if the indoor-outdoor temperature difference is 10 °C; respectively, in the winter period (40 °C temperature difference) it will reach "only" 1.16 °C. This latter value is comparable to the accuracy of IR thermographic measurements. This example shows the difficulty in trying to determine the thermal resistance of thick building constructions. Note that energy saving in the latter case will be about 23 × 1.16 = 26.7 Wm^{-2}. Assuming that the outer (envelope) surface of a typical 5 story building is 2000 m^2, the total energy gain that would be achieved by replacing a 0.76 m-thick brick wall with a 2.6 m-thick one, would be 53,400 W. This would meet the new Russian thermal protection standard. The annual saving can be calculated by introducing a prescribed indoor-outdoor temperature difference, and the duration of a heating period (see Sect. 11.2.2).
- The front-surface response of the brick wall at 99% of the maximum level takes place during a rather long period of time, which is from 26 to 137 hours, depending on the wall thickness. The 90% temperature level is reached in from 7 to 13.5 hours.

In practice, flux meters are placed on internal walls to achieve the best accuracy. In this case, temperature fluctuations will penetrate through a wall for a much longer period of time. The thicker the wall, the longer the time $\tau*$ that is needed for correctly determining the "steady-state" wall thermal resistance R_t by Eq. (11.14). Let us introduce the relative uncertainty RU of determining R_t as: $RU = (R_{true} - R_{ap})/R_{true}$, where the subscripts "*true*" and "*ap*" specify the true and apparent values of R_t.

Figure 11.6 allows the approximate evaluation of the dimensionless time $Fo* = a\tau*/L^2$ as a function of the Biot number $Bi^i = \alpha^i L/\lambda$, which is related to the indoor ("*i*") environment where the heat flux is measured. For example, if one needs to evaluate R_t of a 0.76 m thick brick wall (see Table 11.3) after a step change of the outside air temperature with an accuracy of 5%, one has to wait for 140 hrs to

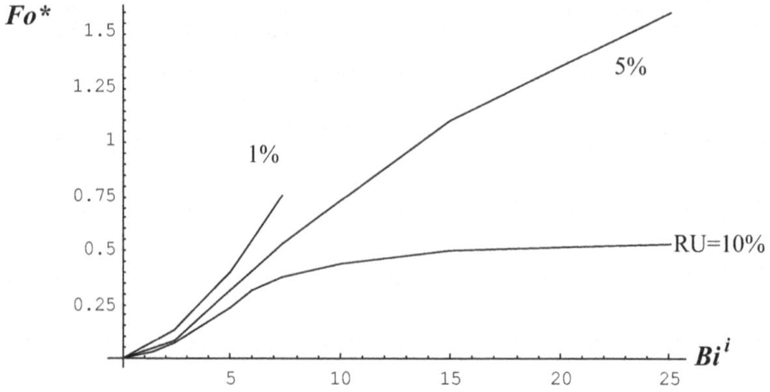

Fig. 11.6 Step-function heating of a "thick" wall (determining *R*)

measure both the wall temperature and the heat flux in order to correctly apply Eq. 11.14. Providing an accuracy of 1% on a very thick wall would require unrealistically long $\tau*$ times, thus making such the procedure impractical.

A comparison of Figs. 11.5 and 11.6 shows how different the time scales are when analyzing front and rear surface temperature responses of thick walls in view of sharp variations in the ambient temperature.

When using Eq. 11.14, the simplest way to shorten measurement time is simultaneously measuring the thermal flux on both the inside and outside wall surfaces. For example, applying the expression

$$R_t = \frac{T_{in}^{w} - T_{out}^{w}}{\left(Q_{out} + Q_{in}\right)/2} \tag{11.15}$$

allows the evaluation of the R_t value on a 2.6 m-thick brick wall in 12 days after the start of heating with an accuracy of 6% or less. However, it is worth reiterating that outdoor heat flux measurements are not very reliable because of solar radiation, wind, precipitations, etc.

Determining the thermal resistance of a wall by analyzing the temperature pre-history

Algorithms to improve the accuracy of determining wall thermal resistance in a transient regime are being developed worldwide. The international standard ISO 9869-1994 "Thermal insulation. Building elements. In-situ measurement of thermal resistance and thermal transmittance" allows the determination of *R in situ* by applying two methods: *averaging* and '*black box*'. The first method suggests the use of the "steady-state" formulas presented above by determining progressive mean values of the temperature and heat flux, particularly, by integrating values of $T(\tau)$ and $Q(\tau)$ as a function of time. These results become more reliable as the number of measurements increases. The "black box" method, which is based on a dynamic model, provides better results when the air temperature varies strongly in time. It is recommended that transient data, obtained over 3–5 days, should be processed by both techniques and the calculated R values be compared.

In this section, we will consider a simple illustration of how the accuracy of determining R can be improved by knowing the pre-history of the outdoor temperature.

Assume that initially a building wall is at a thermal equilibrium where R_t can be calculated by Eq. (11.14). Let the outdoor air temperature $T(\tau)$ start to change at a certain time and, for simplicity, we assume no solar irradiation. In practice, one can use meteorological records, but the accuracy of the meteorological data when surveying a particular building can always be questioned. Then, the wall temperature response at the time $\tau*$ from the stimulation $T(\tau)$ can be determined by using the Duhamel integral:

$$T_{out}^{w}\left(\tau *\right)=\int_{0}^{\tau *}T\left(\tau\right)T_{outD}^{w}\left(\tau *-\tau\right)d\tau, \tag{11.16}$$

where T_{outD}^{w} is the wall temperature response toward the Dirac pulse stimulation. In the same way, one can determine the magnitude of the heat flux at time $\tau*$. Afterwards, the calculated values of temperature and heat flux are subtracted from the corresponding experimental values measured at $\tau*$ in order to retrieve the values that existed before the outdoor temperature started to change. Finally, Eq. (11.14) is applied in order to determine R_t.

Example The ThermoCalc-2D software was used to calculate the temperature response of a 0.76 m-thick brick wall subjected to combined thermal stimulation. First, the outdoor air temperature "jumped" from 0 to 21.7 °C, and the temperature evolution was followed for 210 hrs. At this time, both the temperature and the thermal flux are measured, thus allowing the estimates of R_t by Eq. (11.14) with an error of 0.5% on both the outside and inside wall surfaces. Then, the ambient temperature increased linearly from 21.7 °C to 31.7 °C (Δ = 10 °C) in 10 hrs. At this time, 220 hours, another measurement was performed. The plot of the thermal stimulation function is shown in Fig. 11.7, and the calculated results are presented in Table 11.5. It is seen that the straightforward use of the "new" values of temperature and thermal flux in Eq. 11.14 leads to considerable errors in thermal resistance values on both surfaces. Introducing the corrections by Eq. (11.16) allows the prediction of the temperature and thermal flux values at 210 hrs, thus improving the estimates of R_t.

Unfortunately, the mathematical processes above are not accurate when temperature and heat flux are changing. This makes the determination of wall thermal resistance fairly difficult in outdoor applications, where ambient conditions are seldom constant.

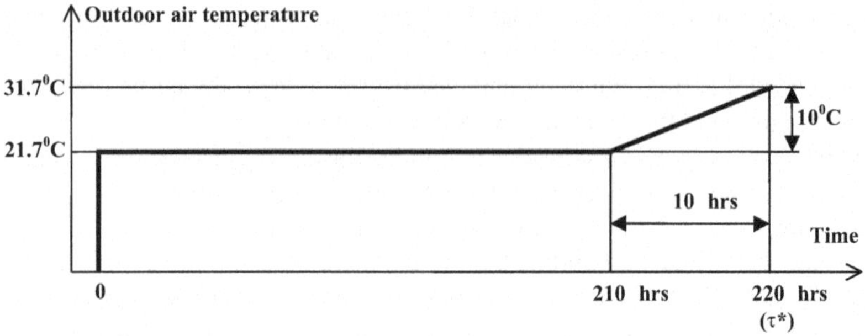

Fig. 11.7 Outdoor air temperature *vs.* time in determining the thermal resistance of a 0.76 m-thick brick wall (see Table 11.5)

Table 11.5 Applying the algorithm of Eq. (11.16) for determining the thermal resistance of a 0.76 m-thick brick wall if the outdoor air temperature changes according to Fig. 11.7 (ThermoCalc-2D)

Time 210 hrs (steady-state regime)	Time 220 hrs (in 10 hrs after ambient temperature started to change)
$T^{w}_{out} = 20.8\ °C$, $T^{w}_{in} = 2.5\ °C$ $Q_{out} = Q_{in} = 21.9\ Wm^{-2}$	$T^{w}_{out} = 29.6\ °C$, $T^{w}_{in} = 2.5\ °C$ $Q_{out} = 48.9\ Wm^{-2}$, $Q_{in} = 21.9\ Wm^{-2}$
Thermal resistance by Eq. (11.14): $R_t = \frac{(20.8-2.5)\,^oC}{21.9Wm^{-2}} = 0.836\ W^{-1}m^{2o}C$ True thermal resistance by Eq. (11.9): $R_t = \frac{0.76m}{0.9Wm^{-1}\,^oC^{-1}} = 0.84\ W^{-1}m^{2o}C$ Error 0.5%.	Thermal resistance by Eq. (11.14) when measuring Q_{out}: $R_t = \frac{(29.6-2.5)\,^oC}{48.9\ Wm^{-2}} = 0.554\ W^{-1}m^{2o}C$ Error 36%. Thermal resistance by Eq. (11.14) when measuring Q_{in}: $R_t = \frac{(29.6-2.5)\,^oC}{21.9\ Wm^{-2}} = 1.237\ W^{-1}m^{2o}C$ Error 47%.

Special algorithm of determining R_t

By Eq. (11.16), outside excess surface temperature is $\Delta T = 7.6$ °C and excess thermal flux $\Delta Q = 27.6\ Wm^{-2}$.

By subtracting the values above from those at 220 hours, the expected parameters at 210 hours can be found, e.g. on the outside wall surface:

$T^{w}_{out} = 29.6–7.6 = 22$ C, $Q_{out} = 48.9–27.6 = 21.3\ Wm^{-2}$.

R_t by Eq. (11.14):

$$R_t = \frac{(22-2.5)\,^oC}{21.3\ Wm^{-2}} = 0.92\ W^{-1}m^{2o}C.$$

Error 9.5%.

11.2.5 Inspecting Roofs for Trapped Moisture

An accumulation of water in a roof is a serious problem, as it causes roof damage and water leakage into buildings.

Anomalous surface moisture can be detected in a steady-state regime due to evaporation and the accompanying temperature decrease. However, the detection of trapped water requires using a transient heat exchange, for example, the diurnal cycle where water, due to its high heat capacity, remains (relatively) warm during the night time and cold during the day time.

Roof inspections can be performed while walking on a roof, but the test efficiency will be low, and obtaining a general thermal picture will be difficult. In the USA, IR inspections of flat roofs are often performed from a helicopter or an airplane. In the last decade, the use of unmanned flying vehicles, in particular, drones and quadrocopters, has increased. Aerial inspection provides a very efficient survey and allows easy coordinate mapping by GPS, and the identification of defect areas.

11.2.6 IR Imagers for Building Applications

In building diagnostics, both SW (2–5.5 μm) and LW (7–14 μm) IR imagers can be successfully employed if they meet the following technical requirements:

- measured temperature range, °C..................................from −20 to +100,
- sensitivity, °C...under 0.2,
- ambient temperature, °C..from −15 to +40,
- field of view (additional lenses are recommended), angle degree ....from 7 to 20,
- image format, pixel..128 × 128 or higher,
- frame rate...1 Hz or higher,
- measurement accuracy...not worse than ±1 °C; (higher accuracy may be required when determining wall thermal resistance),
- emissivity compensation option,
- digital data recording and IR image video output,
- small mass and dimensions,
- battery power supply.

The world market offers many IR cameras that meet or exceed the above-mentioned requirements (see Sect. 9.5 and Table 9.3). Determining heat losses and wall thermal resistance requires the use of more advanced IR thermographic systems that ensure accurate temperature readings at each pixel; however, even in this case, the reliability of quantitative estimates is typically low.

Auxiliary equipment

An IR camera is not the only tool for performing thermal diagnostics of buildings envelopes. The auxiliary tools are:

- IR thermometer,
- heat flux meter,
- anemometer,

- mercury or other type of thermometer,
- manometer,
- air pump,
- moisture gauge,
- psychrometer,
- anemometer.

Thermal fluxmeters allow direct measurement of heat losses and estimation of heat exchange coefficients. Thermometers should measure temperature in the range from −20 to +100 °C with the scale division 1 °C or better. Air permeability is evaluated by lowering air pressure with a pump while pressure is measured with a manometer ("air door" technique, see the standards ASTM E1827 "Standard test methods for determining air tightness of buildings using an orifice blower door" and/or ISO 6781 "Thermal Insulation - Qualitative Detection of Thermal Irregularities in Building Envelopes, Infrared Method".

Some combination gauges, such as KESTREL (USA) units, are available for measuring a set of the above-mentioned parameters: air temperature and humidity, dew point, wind velocity, etc. The most capricious are fluxmeters. They are very sensitive to magnitude and direction of heat flux, and personnel using them must have training and experiance. In the former U.S.S.R., some types of flux meters were developed at the Institute of Technical Physics (now Ukraine). These are currently manufactured by few Russian companies, e.g. Tekhnoas, Etalon, Stroypribor, Practic-NZ, and Karat. For example, a flux meter ITP-MG4.03 from Karat allows the measurement of temperature and thermal flux density according to the requirements of the appropriate Russian standards, such as GOST-25380-82 "Building and constructions. Method for determining density of thermal flux passing through building envelopes", GOST 26254-84 "Building and constructions. Method for determining wall thermal resistance" and GOST 26602.1-99 "Windows and doors. Method for determining thermal resistance". The main parameters of this unit are: thermal flux measurement range is from 2 to 999 W/m^2, temperature measurement range is from −30 to +100 °C, accuracy in heat flux measurement - ±7%, number of recorded values is up to 2000, observation period is from 1 to 400 hours, observation interval is from 1 to 180 minutes. Such units are often multi-channel, which allow data storage from many sensors (Terem-4.0 unit from Karat).

An example of a portable fluxmeter is the IPP-2 unit from Practic-NZ (measurement range from 0 to 2000 W/m^2, measurement time 3.5 minutes, accuracy 8%).

11.2.7 Practical Aspect of IR Building Thermography

In building IR thermography, the inspected area may vary from a few square meters to several tens of square meters depending on the distance and survey purposes. Respectively, the distance between the target and the IR imager is chosen depending on the camera optics and survey conditions.

Current IR imagers provide a temperature measurement accuracy as low as ±(1–2)% of measured values, or ±1–2 °C, but, as mentioned above, this accuracy may be insufficient when determining wall thermal resistance.

According to the international standard ISO 6781-83 "Thermal insulation, Qualitative detection of thermal irregularities in building envelopes. Infrared method". an inspection may be performed if the outdoor-indoor temperature difference exceeded 10 °C for the preceding 24 hours, In the severe Russian winter, this requirement is easily met. While conducting a survey, the outdoor-indoor temperature difference should not change more than by ±30%, the temperature variations in the building should be under ±2 °C, and the building envelope must not be exposed to solar radiation for the preceding 12 hours [10]. When determining wall thermal resistance in Russia, one must follow Russian standard, GOST 26254-84 [11]. One requirement is that temperature and thermal flux data be averaged over a time period from 1 to 3 days (depending on wall thermal inertia) while the mean temperature variations must be under ±1.5 °C. The total duration of this observation period can reach 15 days. The latter negates one important advantage of IR thermography, which is a high rate of inspection. In Russia, it is important for IR imagers to operate under low ambient temperatures. The performance of many radiometric units may be reduced at low temperatures, and some units cannot be used at very cold outdoor temperatures. In some cases, an inspection can be made with the IR camera operating inside a heated vehicle and looking through an open window.

Infrared indications should be compared to visual images obtained with a video or still camera. In recent years, the superposition (*image fusion*) of visual and IR images has become popular.

Emissivity problem

To detect hidden defects in building envelopes, the knowing the emissivity is desirable but not necessary, particularly, when defects are detected on thermally-uniform walls. Moreover, emissivity variations may serve as an additional indicator of wall quality. The emissivities of many building materials is presented in Tables 7.3 and 7.4.

When evaluating heat losses and wall thermal resistance, temperature readings must be corrected for emissivity; this is easily done in all modern IR cameras. The relationship between emissivity and true temperature measurement was discussed in Sect. 7.2.4.

Meteorological conditions

IR survey results depend on the geographical location of a building inspected. Inspection should not be performed during rain, fog and snow. Measurement errors are caused by the absorbtion of IR radiation by water in the air and these increase with distance. And snow, water or ice should not be present on a test surface, as this causes problems with emissivity and reflection.

Direct and scattered solar radiation, particularly, in the autumn-summer period, can warm up exposed wall sections, thus producing areas of elevated temperatures that are not caused by pure heat conduction through the wall. Solar heating is particularly important in roof inspections where it may also enhance the detection of water ingress sites. Smooth surfaces, such as window glass, ceramic tiles etc. reflect solar radiation well and can produce false temperature indications. In many cases, the presence of solar reflections can be confirmed if the operator simply moves the IR imager, as reflections also move while true temperature indications on test objects remain unchanged. Additionally, when inspecting windows, the IR camera may "see" reflections of cold radiation from the Sky, and this significantly decreases the apparent temperature of the window. Note that the direct or reflected IR monitoring of a cloudless Sky may show a temperature as low as −40 °C. So IR surveys should be performed in early morning or night time when the influence of the ambient is minimal. In day time, best results appear under a cloudy sky. In some cases (winter time and when comparing two similar areas), differential temperature signals may be evaluated even under solar irradiation. In such cases, the absolute temperature values will differ from true ones but only the temperature differences are of interest. In day time inspection, long-wave IR imagers are recommended to reduce the affect of atmospheric absorbtion. However, in all cases, the possibility of solar heating should be taken into account, as the intensity of solar irradiation can reach about 1 kW/m^2 at ground level on a clear day.

Strong wind significantly cools building facades and test objects down. It is recommended to conduct IR surveys when the wind velocity is less than 5–7 m/s.

Choosing an appropriate value of the heat exchange coefficient may be crucial when calculating wall thermal resistance. Experimentally, heat exchange coefficients can be determined by measuring thermal flux density; also, there are some relevant formulas recommended in the heat transfer theory. The results of an outdoor inspection can be corrected for wind velocity by multiplying the differential temperature signals by the coefficients given in Table 11.6.

Another important factor that influences the accuracy of determining heat losses and thermal resistance is the variation in outdoor temperature within a diurnal cycle. The relationship between temperatures of a particular building section may vary in time because of its different thermal inertia and this affects the calculated parameters of the building's thermal protection. Accounting for thermal inertia is described in the Russian standard GOST 26629-85 [12]. If possible, the analysis of building constructions and materials should be performed in laboratory conditions by placing a test object in a climatic chamber. It is worth stating that the determination of thermal resistance by using an IR camera under Russian standard GOST 26254-84

Table 11.6 Wind correction coefficients [6]

Wind velocity, m/s	Correction coefficient
1	1.00
2	1.36
3	1.64
4	1.86
5	2.06
6	2.23
7	2.40
8	2.54

requires that both temperature and heat flux be measured over a long period of time, so it has no obvious advantages over the use of standard temperature sensors (thermocouples, etc.).

Micrometeorology
The term "micrometeorology" (also "microclimate") refers to unusual environmental conditions that exist at a limited location, especially when these conditions are different from the area surrounding it.. For example, buildings that are located in lowland depressions may experience increased water condensation. Trees and bushes, as well as adjacent buildings, may affect solar radiation and wind.

Furnaces and water heaters in buildings may cause additional heating and reflections.

IR camera-object distance
With increasing distance between an IR camera and a test object, the test area widens, spatial resolution decreases and temperature measurements become less accurate due to atmosphere absorption. The latter phenomenon is negligible if the distance is less than 30–50 meters, which is typical of most practical applications. At greater distances, one should use correction formulas to account for the absorption of radiation in the atmosphere. Many contemporary IR imagers include built-in functions to allow corrections for distance and atmospheric absorbtion.

Indoor conditions
As a rule, most building defects, such as air leaks, cold spots, etc., can be better detected inside a building, except in situations where defects are located closer to the outside wall surface, such as detachments of external wall coatings or decorative elements. The main problems in performing indoor inspections are: (1) the use of fans and heaters, (2) shielding of test areas by furniture, carpets and other objects, (3) detachments of wall-paper and plaster, and (4) uneven painting of test areas.

Radiation from space heaters can be avoided by covering them with opaque shields or by carefully choosing both a test area and a viewing direction. Radiators and heaters should be turned off 12 hours prior to inspection and electric lamps should be turned off just before measurement. A thermographer should stand about 1 m away from a test object in order to avoid reflected radiation (Russian standard GOST 26629-85 requirement); electric lamps should be located at least at 2 m away from a test surface. While looking at windows, a thermographer may see himself and other heat sources reflected from the glass. But even more important is reflected solar radiation. In general, the inspection of semi-transparent objects, such as dielectric films and glasses, requires using a special methodology to discriminate between optical and thermal phenomena. For example, the temperature of window glass is usually measured by adhering a piece of adhesive tape to the glass and measuring the temperature of the tape; this method removes the problem of reflections, transparency and emissivity of the glass,. Objects that prevent direct surface viewing should be removed, and detached wallpaper and plaster can be erroneously interpreted as insulation defects. The influence of uneven emissivity can be evaluated by visual observation and/or by adding corrections into IR cameras. IR inspections are not advisable if the building's emissivity is less than 0.7 (Russian standard GOST 26629-85). Fortunately, most building materials, including oxidized metals, have a high emissivity and provide temperature measurements with reasonable accuracy (±1 °C). A simple method of estimating emissivity involves placing a piece of a high emissivity tape on a test surface. If the emissivity is unknown and cannot be measured directly, the operator can vary the emissivity setting in the IR camera to match temperature readings measured by contact sensors or other means.

Qualitative and quantitative data interpretation

Inspection results should be interpreted by a trained thermographer certified to Level II or III. Best results are achieved by analyzing black and white and/or color IR thermograms according to the thermographer's experience and established procedures or guidelines reported below. Black and white images may be easier to interpret if there are numerous sources of noise or if solar reflections are present. Color thermograms are illustrative and are often preferred in reports. The use of color is effective if thermal anomalies differ slightly in temperature from the background. The "Iron" palette may be regarded as a compromise between black and white and color palettes.

In qualitative data treatment, important factors are: knowledge of a building's thermal status, the thermographer's experience, and the surface characteristics of adjacent areas that serve as a reference when interpreting anomalies.

The following may be used to make decisions in building inspection [6].

- IR diagnostics should be complemented by visual inspection and/or photographs. Optical and IR images are typically very different in appearance. An efficient technique of diagnostics is the superposition of IR and visual images of the same area or adding defect outlines on the visual image after detecting them on the corresponding IR thermograms.
- Thermal patterns and anomalies should be evaluated by their temperature amplitude or by their size and shape with respect to a reference area. A reference area should be chosen similar to the analyzed one.
- Under identical conditions, an indoor inspection is more efficient than an outdoor one because defects produce stronger temperature anomalies on inner walls of buildings. Also some defects that are easily seen indoors, such as poor insulation of seams, often cannot be detected on the outer wall.
- The spatial resolution of IR inspection decreases with increasing distance to test objects. IR imagers with large array detectors may be necessary to detect small details.
- Surfaces viewed at large angles often seem colder than they really are due to the angular dependence of emissivity. The same phenomenon appears at long distances due to atmospheric absorption.
- When performing panoramic surveys of buildings, upper floors may look warmer due to convection and the peculiarities of the heating system in the building. However, upper floor windows may seem colder because of the cold sky reflections.
- Thermal anomalies appear on IR thermograms as areas of higher or lower apparent temperature and correspond to: (1) architectural details, (2) emissivity

variations, (3) uneven heat exchange between a test object and the ambient, and (4) defects of a different nature.

- When the indoor temperature is higher than outdoor, cold bridges look colder on inner surfaces and warmer outside.
- When the indoor temperature is higher than outdoor, air leaks look colder inside and warmer outside if the indoor pressure is higher than the atmosphere pressure. In the case of an inverse pressure gradient, air leaks look colder on both inner and outer surfaces.
- Indoor pressure depends on wind velocity and direction, as well as on the ventilation system design and the temperature gradient; it may vary in different building sections. In multistory buildings, the lower floor pressure may be higher and the upper floor pressure – lower than the outside pressure. When pressure varies along the height of a building, the sign of the temperature gradient will depend on the location along the building's height.
- Unpainted metallic components usually look colder, except in situations where adjacent warm objects produce spurious reflections.
- When analyzing the moisture content of a building, particularly in an attic, one must understand whether water penetrates the building from outside or comes from inside. Depending on the outdoor conditions, moisture on brick walls may look either warmer, because of a reduced thermal resistance, or colder, because of surface evaporation, than the surrounding area. Construction defects may lead to extensive water leakage into walls if they are accompanied by heavy rain. And these moist areas will persist for a long time even after construction ends and the building's heating and cooling systems are put into operation.
- Hot radiators may cause warm patterns to appear on the outside walls, typically, under windows, and these patterns may be erroneously interpreted as defects.
- Windows produce obvious thermal patterns on IR thermograms.
- Prominent thermal patterns having sharp boundaries are often created by surface (emissivity) phenomena, including solar reflection. Such cases are easily detected by visual observation. In general, the temperature indications caused by air leaks have sharper boundaries than those caused by thermal insulation defects.
- Windows may reflect the Sky at particular viewing angles. The temperature of the Sky can vary from ambient temperature (very cloudy conditions) to −40 C and lower (with clear sky).
- Solar reflections will" crawl" across an IR image while an operator is moving; alternately, real temperature phenomena do not change their patterns and positions under different viewing angles.
- When performing IR inspection of buildings, the variation of indoor temperature in different areas should be taken into account; temperature signals are proportional to the indoor-outdoor temperature difference.
- In the inspection of inter-panel seams in buildings, the IR image is affected not only by a pressure gradient but also by the defect type. It may be a variation in the thermal resistance of the seam or an air leak. In the latter case, whether the defect indication is relatively warm or cold depends on the pressure gradient in the building section.

Qualitative analysis of thermal anomalies is performed in order to evaluate defect severity in normal building operation. In Russia, the mean temperatures of floor, ceiling and walls must meet the requirements of SNiP 23-02-03. Defect severity is determined by:

- additional heat losses through defects
- discrepancy between experimental and standard values of wall thermal resistance
- danger of possible building damage caused by condensation and material corrosion. Moisture in the wall also leads to lower thermal resistance.
- Discomfort among building residents caused by indoor temperatures being out of comfort range

The relationships between the magnitude of temperature signals at defect sites and defect severity have not been thoroughly studied because of the necessity to collect a lot of statistical data. It is important that temperature signals should be normalized by the indoor-outdoor temperature difference. For example, a temperature signal of 1 °C in an inter-panel seam at an ambient temperature of 0 °C indicates a more serious defect than the same signal detected at an ambient temperature of −2 °C.

Russian building and health standards include the following requirements for an indoor temperature environment, e.g.: (1) air temperature from +18 to +22 °C depending on the type of building and the outdoor winter temperature, (2) the maximum difference between the indoor air and exterior wall temperatures is to be 4 °C, between the indoor air and ceiling temperatures – 3 °C and between the indoor air and floor temperatures −2 °C. In any case, the inner surface temperature should not be lower than a dew point temperature at a normal outdoor winter temperature. This requirement is related to thermally-conductive wall elements, such as panel junctions, reinforcement, through-seams filled by mortar, etc.

Outdoors, if defects are evaluated by temperature signals ΔT, defects should be considered serious if they produce relative signals $\Delta T / \left(T_{\text{int}}^{air} - T_{ext}^{air}\right)$ greater than 6% in the case of no air leaks and 12% if air leaks are present. In an indoor inspection, these criteria are recommended to be 15 and 25% respectively.

If measured values of thermal resistance are used for assessing the quality of thermal protection, serious defects are those leading to R reduction of more than 15%.

It is worth noting that the above-mentioned criteria are empirical and should be verified by further research.

11.2.8 Illustrations

Some features of IR building thermography are illustrated in Figs. 11.8, 11.9, 11.10, 11.12, 11.13, 11.14, 11.15, 11.16, 11.17 and 11.18. In most images the outdoor temperature was about −20 °C.

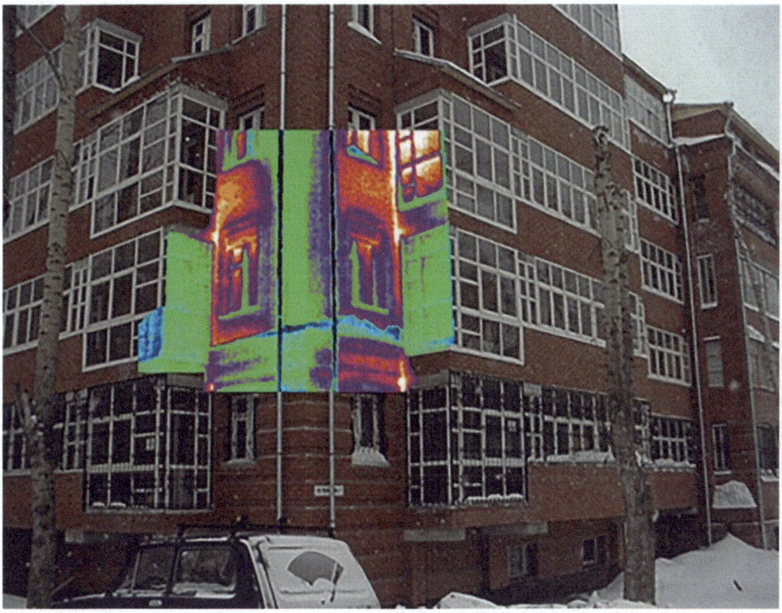

Fig. 11.8 IR inspection of a multi-storey residential building: superposition of the IR and visual images

Identification and locating of defect areas can be facilitated if the IR and visual images can be superimposed, as shown in Fig. 11.8; special computer programs for this are available but even simple graphical editing programs can be helpful.

IR inspection of small private buildings allows the evaluation of insulation quality and the identification of areas of major heat loss where further repair is recommended (Fig. 11.9).

The IR thermogram in Fig. 11.10a shows a defect in the vertical seam of a brick residential house with a differential temperature signal of about 3 °C. The outdoor inspection of seams is difficult due to low temperature signals. In some cases, such an inspection can be complemented by indoor inspection, as illustrated in Fig. 11.10b. The defect shown in Fig. 11.10a produced an indoor temperature signal of 8 °C (note that the kitchen furniture partially shields the analyzed seam).

The potential of IR monitoring compared to visual inspection are illustrated in Fig. 11.11. A reduced pressure in the ventilating shaft leads to air in-leaks along the total building height, producing temperature signals up to 5 °C.

In Fig. 11.12, two IR thermograms clearly show the difference between defect and non-defect seams observed in an indoor inspection.

The use of the active TNDT principle in building inspection is illustrated in Fig. 11.13 where the ingress of cold air between the support wall and the detached decorative plaster produces extended areas of a lower temperature (2 °C signal). In other cases, detached decoration may appear warm due to solar heating. IR surveys

Fig. 11.9 IR inspection of a private house

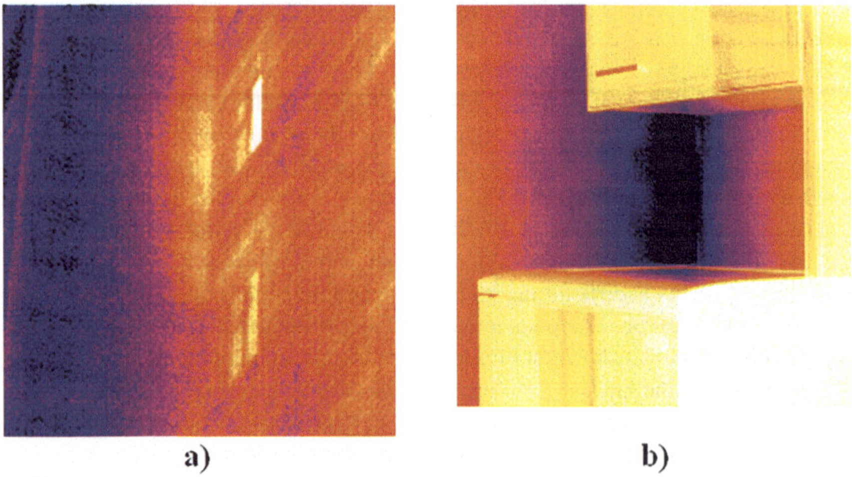

Fig. 11.10 Defect in a vertical seam in a building: the temperature gradient along the seam is 3 °C outdoor (**a**) and 8 °C indoor (**b**)

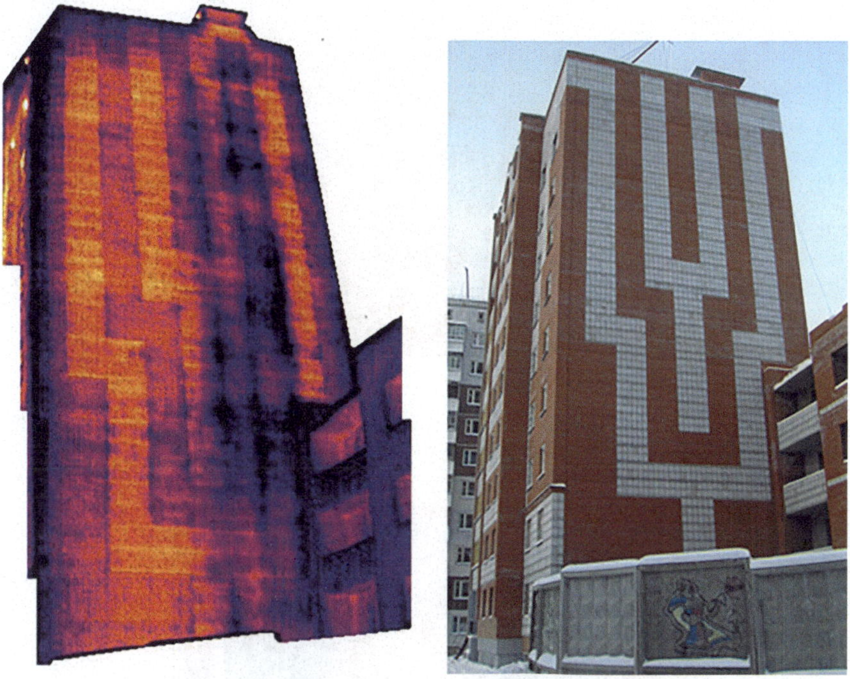

Fig. 11.11 Air leaks into a masonry building (dark blue areas) in the area of the ventilation shaft

allow the evaluation of defect type and size so that a decision can be made on the size and type of repair. The same principle of "mild" heating has been used in the inspection of wall frescos (see Sect. 11.3).

Figure 11.14 shows the facade of an administrative building where the fourth floor is clearly overheated (wall mean temperature is +4.7 °C) compared to the second floor temperature (−2.1 °C), while the outdoor air temperature is −4.0 °C. It was calculated that the optimization of the heating system in this building may save up to 20% of the heating energy.

Figure 11.15 shows serious through-defects in two vertical seams between the wall panels. As mentioned above, buildings made with panels are ubiquitous in Russia since Khrushchev's times, and these have proven to be a good target for IR diagnostics. Another example of numerous inter-panel seam defects is presented in Fig. 11.16a with a corresponding binary map of defects shown in Fig. 11.16b.

IR thermography is a very convenient technique to analyze the efficiency of heating systems, especially when checking systems in new buildings. For instance, the IR image in Fig. 11.17 demonstrates that one of two radiators is not fully operating. Sometimes, such an inspection can be accomplished by the use of an IR thermometer.

A roof inspection is illustrated in Figs. 11.18 and 11.19. The visual contrast of moist areas is low compared to adjacent areas even in the case of surface water, while the detection of hidden "water pockets" is generally impossible (Fig. 11.18a).

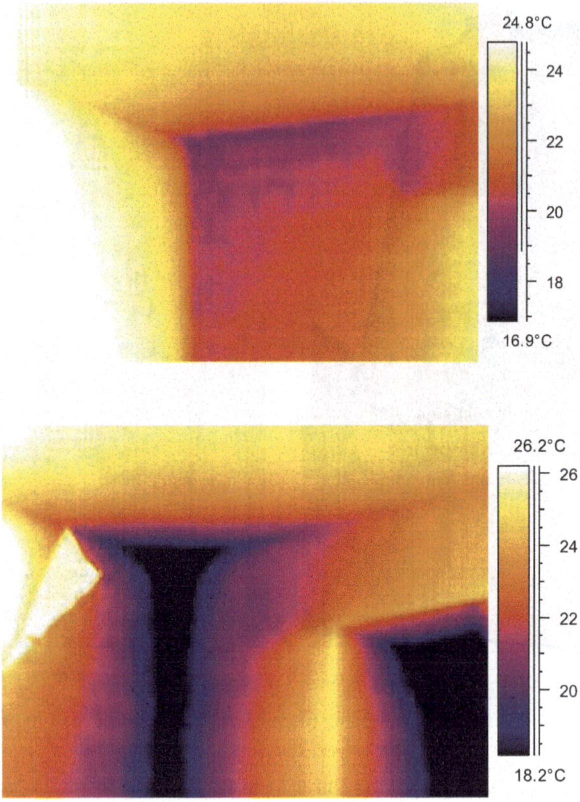

Fig. 11.12 Non-defective (above) and defective (below) vertical seams in a building (indoor inspection, defect temperature signal 11 °C)

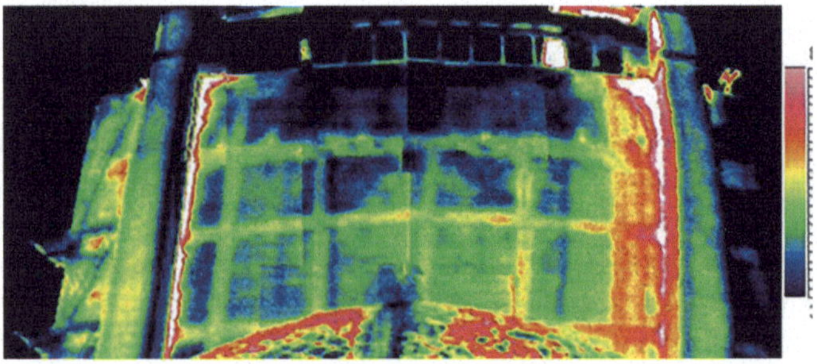

Fig. 11.13 Detecting detachments between plaster and support wall (temperature signal 2 °C)

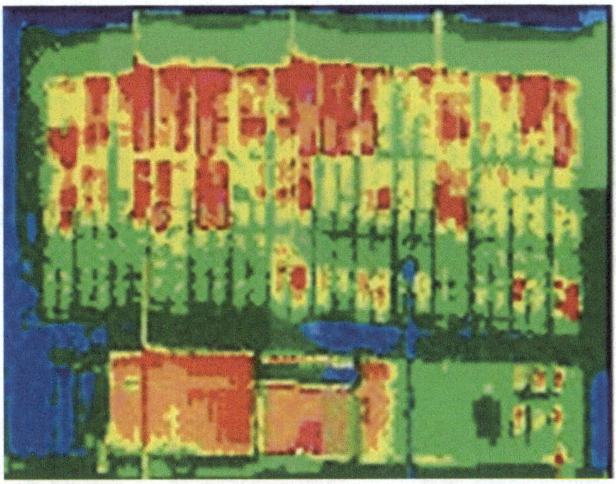

Fig. 11.14 Panoramic IR thermogram of the façade of an administrative building

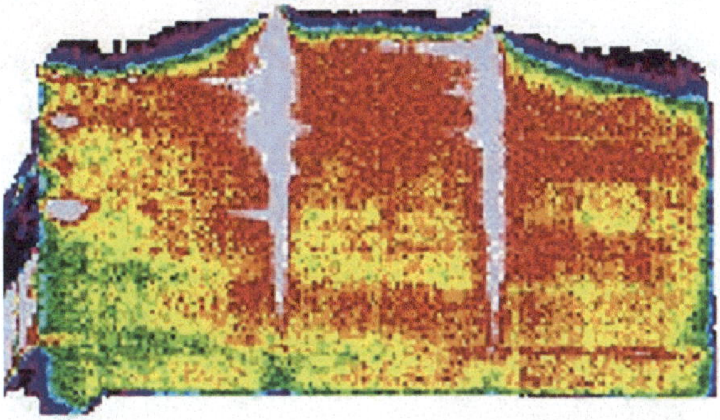

Fig. 11.15 Panoramic IR thermogram of the side wall of a power station main building

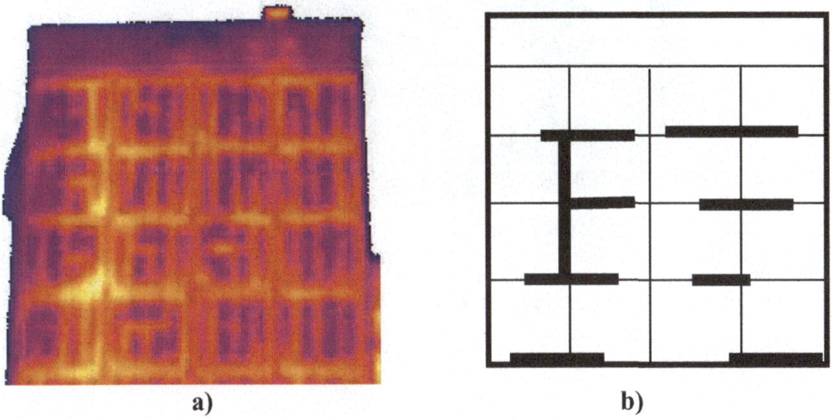

Fig. 11.16 Panoramic IR thermogram of the side wall of a residential panel building (**a**) and a binary map of defects (**b**)

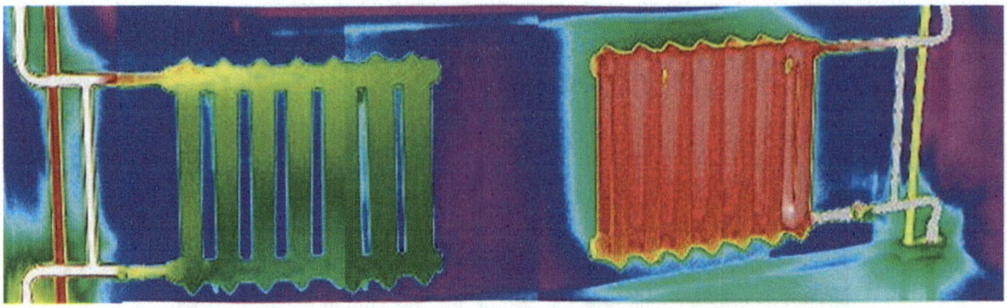

Fig. 11.17 Inspection of an in-house heating system (radiator temperature on the left +32 °C, on the right +60 °C)

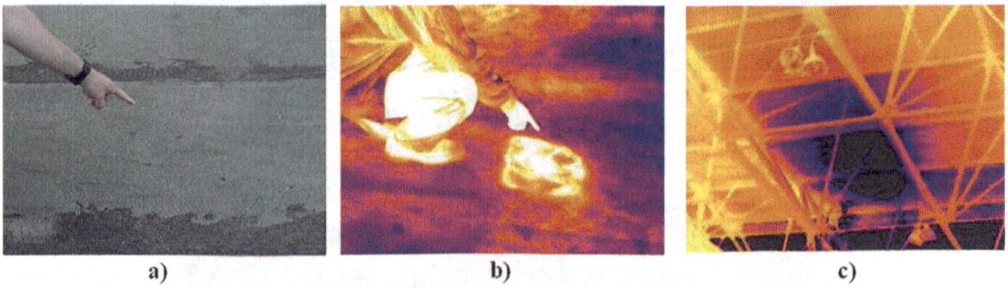

Fig. 11.18 IR inspection of moisture in a flat roof: (**a**) roof photo. (**b**) IR image, area a). (**c**) trapped water (power station building roof, indoor inspection)

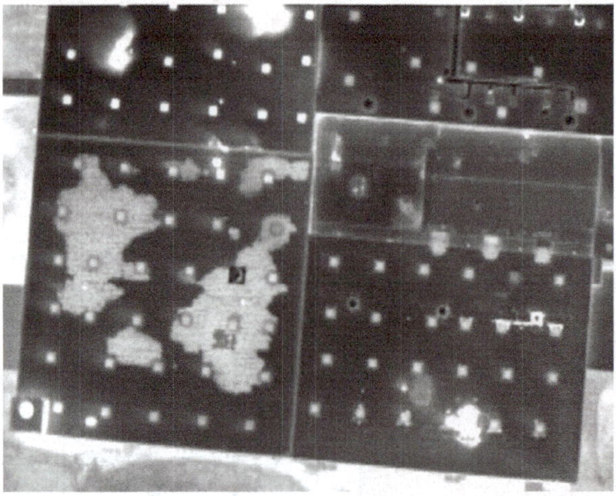

Fig. 11.19 Aerial IR survey of a flat roof: light spots correspond to moist areas (courtesy: G. Stockton, Stockton Infrared Services, USA)

If the observation time is properly chosen, abnormally moist areas look warmer in the evening and colder in the morning due to the high heat capacity of water (Fig. 11.18b). Indoors, such areas look colder (see the image in Fig. 11.18c, $\Delta T \sim 8$ °C). The panoramic image of a flat roof obtained from an airplane is shown in Fig. 11.19. Multiple warm spots in the roof insulation have been reliably identified as defects.

In the early morning of July 17, 1997, a dormitory of the Tomsk Higher Military Communication School collapsed causing the death of 12 students. An IR survey of the debris was performed on July, 25, i.e. 8 days after the catastrophe under favorable weather conditions (no wind, no rain and no direct solar radiation) [8]. Surprisingly, local temperature anomalies were detected in 5 of 24 joints between floor panels and vertical supporting columns. The amplitude of these signals was from 0.8 to 1.5 °C, while, in the areas of transition from plaster to brick, the noise fluctuations, were only 0.3 °C (Fig. 11.20a). It was suggested that the observed temperature signals were caused by either emissivity variations or the blackbody-like behavior of wall cavities but these suggestions were rejected after further study. Other explanations of the observed temperature anomalies included moisture in the joints of the building materials and the possible presence of mechanical stresses. The building debris survey was repeated one month later under very similar weather conditions (on August 26) to show the absence of the earlier detected temperature anomalies. However, the appearance of the high signals contradict predictions of the theory of thermomechanics, while the hypothesis of cyclic loads has not been confirmed in this test case.

Cornienko thermographically surveyed the power station of a hydroelectric dam of and found areas of abnormal thermal inertia in the concrete that he concluded were probably related to areas of abnormal mechanical stresses [14]. IR surveying also detected areas of moist concrete.

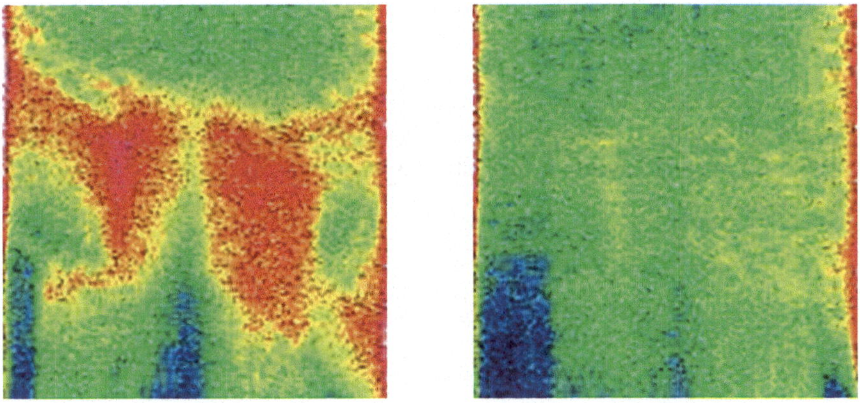

a) b)

Fig. 11.20 IR thermograms of the column-collar joints (Tomsk Higher Military Communication School dormitory that collapsed 17 July 1997): (**a**) survey 25 July 1997, temperature anomaly 0.8 °C, mean column temperature 13 °C), (**b**) survey 26 August 1997, temperature noise 0.3 °C, mean column temperature 12 °C)

Airport runways can be inspected when the sun provides a natural heating pulse. In a transient regime, cracks in the concrete and voids between the concrete and the soil, which are caused by natural ageing and mechanical impacts, can be detected [15]. Inspection must be performed in daytime and on dry runway surface. A disadvantage of this technique is the difficulty in evaluating the depth and thickness of subsurface defects. Therefore, ground-penetrating radar (radiation frequency 1–2.5 GHz) may be used in addition to IR thermography.

A similar inspection method has been used in the inspection of bridge decks where temperature gradients over deck delaminations reached 2–5 °C (US standard ASTM D4788 [16]). An IR imager was mounted on a vehicle moving at a speed of 16 km/hr. IR thermograms were video recorded concurrently with performing a visual inspection that identified surface clutter (patches, etc.). Bridge inspection services consider IR thermography as a very attractive method, even if its efficiency is affected by weather and the type of pavement. A drawback of the method is the inability to evaluate the depth of defects which can be located either between an asphalt top layer and a reinforced concrete base, or deep in the reinforced concrete. Surveying a bridge pavement with ground-penetrating radar complements IR thermography.

Astec Industries, USA, reported the use of an IR imager for evaluating the quality of asphalt pavement. When asphalt is applied, it is required to be within certain temperature limits to prevent premature aging and failure. US federal authorities spend from 300 to 400 million dollars a year for patch repair of federal highways, and these expenses are doubled when repairs of non-federal highway and streets are included. The use of IR thermography may help to reduce such expenses.

On Russian railways, an important inspection problem is the early detection of damage in contact system piers; it is particularly important while dealing with underground railway sections. Usually this problem is solved by using an acoustic emission technique. However, the Romanian power company RENEL reported the use of IR thermography to detect defects in reinforced concrete piers when exposed to solar radiation [17]. Such defects occur due to low quality construction and ageing processes. High quality concrete is characterized by an even temperature distribution while in defect areas the temperature may exhibit relative warm spots up to 3 C. According to the authors, the minimum detectable defect size was 150 × 150 mm, but the statistical reliability of this estimate, as well as the method itself, is still questionable.

11.3 Active TNDT of Objects of Art

11.3.1 Introduction

In many countries, there is increasing interest in the conservation of ancient art objects, including architectural monuments, sculptures and paintings. For many years, this activity has been supervised by UNICEF; national programs have been also existed in Italy, France, Spain and Germany. NDT plays an important role in these programs, especially visual/optical methods. However, IR thermography has found a niche in the inspection of architectural monuments and other objects of arts. This method is being used for a wide range of applications from the airborne monitoring of archaeological digs to detect underground structures, to evaluating sculptures to detect restoration or falsification. Both passive and active TNDT procedures can be applied to reveal temperature and emissivity differences in objects being tested.

IR thermographic inspection of wall frescos and other objects of arts was thoroughly described in [19–24].

Data processing algorithms have been borrowed from other active TNDT applications, including the NDT of composites, building surveys and landmine detection.

Inspection of wall frescos must be approached with caution. Ancient frescos by Giotto, Michelangelo, Fogolino and other artists are very valuable and irreplaceable, and only very non-destructive techniques may be used to test them. One must especially avoid overheating a fresco,

The depths at which detachments and inclusions occur may be from a few millimeters to a few centimeters. Therefore, the optimum observation times are long, and IR sequence acquisition time may be long.

Art objects, such as frescos, paintings and religious icons, usually have very uneven optical and thermal properties. Therefore, IR images exhibit a high level of noise. Specifically, if optical heating is applied, advanced data processing algorithms may be required to obtain worthwhile results.

11.3.2 Wall Frescos

A typical wall fresco is a multi-layer structure of (1) colored pigments mixed into plaster (2) a plaster layer (called an "arricio" in Italian), (3) a bonding mortar layer, and (4) a support wall (Fig. 11.21). The most critical defects are detachments between 0.5 and 1 cm-thick plaster and the support wall, because the detachments might lead to a loss of plaster or even fresco collapse. There is a fresco repair procedure in which a bonding material is injected into any detachments, but this requires knowing their locations precisely. Frescos are thermally stimulated by

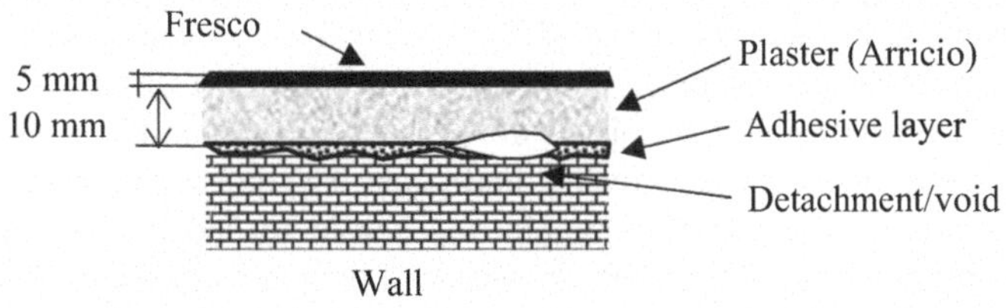

Fig. 11.21 Wall fresco scheme

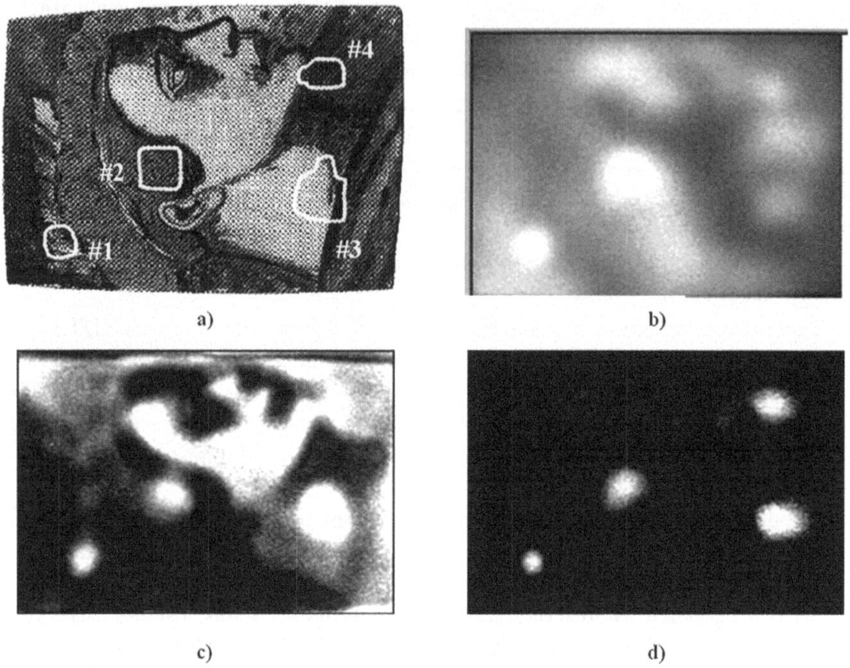

Fig. 11.22 TNDT of "Apollo head" fresco NDT reference standard [25]: (**a**) sample photo and defect location, (**b**) IR thermogram at 270 seconds (optimum observation time, $S = 0.8$), (**c**) maxigram (source sequence, $S = 1.5$); (**d**) maxigram (after 3D normalization, $S = 3.4$)

applying "mild" optical or air heating in order to increase their temperature by a few degrees Celsius [19–21]. In some cases, just the natural variations in indoor temperature may be sufficient to allow the detection of large defects. Heating can be uniform over a relatively large area or it can be localized and centered at a spot or on a line. In one example, line-scanning (or linear) heaters are moved at a constant speed across a surface. This allows the detection of cracks perpendicular to the heating line. As a rule, one should first capture an image sequence and then analyze dynamic heat conduction phenomena in order to enhance the signal-to-noise ratio in the detection of subsurface defects. Processing in the time (frequency) domain has proven to be very effective in the inspection of frescos.

Data processing algorithms have been developed by inspecting reference samples that were fabricated containing defects typical of ancient frescos [19]. An example of a fabricated reference specimen, named "Apollo's head", is shown in Fig. 11.22a. Defects 1 and 4 are pieces of sponge implanted in the plaster prior to hardening, and Defects 2 and 3 are air-filled voids. The defect parameters reported in Table 11.7 were evaluated by holographic interferometry and thermal tomography.

The specimen was heated for 30 seconds with optical radiation having a power density of about 1 kW/m^2, the acquisition time was 900 seconds, and up to 300 thermograms were captured. Preliminary experiments had shown that air-filled

Table 11.7 Parameters of defects in wall fresco NDT reference sample (Fig. 11.22a)

Defect	Lateral size, mm	Depth, mm	Thickness, mm
1	35 × 35	3	3
2	45 × 30	10	10
3	45 × 30	7	5
4	45 × 45	10	15

voids can be detected in plaster at depths up to 3 cm if the plaster surface is uniformly heated. The presence of fresco sections painted in different colors makes optical heating inefficient; for example, dark areas, which absorb more optical energy, look warmer in thermograms and can be wrongly identified as defects. Conversely, light colored areas do not absorb enough energy to create the required ΔT values. Moreover, the presence of sharp boundaries between areas of different colors causes significant thermal fluxes to flow in lateral directions, thus producing artifacts. Specifically, it was difficult to apply the algorithm of dynamic thermal tomography. The consideration above is illustrated by the IR thermogram in Fig. 11.22b, which was recorded at the optimum observation time 270 seconds. The area of the fresco "eye", as well as that between defects 3 and 4, may be thought to be a defect, although it simply absorbs more energy since the eye is darker. In its turn, defect 2 is reliably detected because it is located under the dark area and, hence, well-heated area. Figure 11.22c is a maxigram that shows temperature indications from both defects and surface clutter. .

The best data processing algorithm has proven to be the 3D filtration technique described in Chap. 6. A source image captured at 10 seconds was used as a normalizing image (mask), and this reduced surface clutter. The resulting maxigram is shown in Fig. 11.22d;

Another example of advanced data treatment is shown in Fig. 11.23 where the binary maps of defects are presented to evaluate the lateral size of defects. The image obtained by processing the timegram (no 3D normalization) shows artifacts that correspond to black (overheated) areas (Fig. 11.23a), while after filtration even "weak" defects can be detected (see two sequential thermal tomograms in Fig. 11.23b, c).

The advantages of 3D filtration can be expressed quantitatively by using the concept of signal-to-noise ratio S (see Fig. 11.23). All four defects have been considered to be a single defect area, while the rest of the sample is assumed to be defect free. Applying 3D filtration has doubled the S value.

Results of an inspection of a Fogolino fresco in Bergamo, Italy, are presented in Fig. 11.24a. After inspecting the whole fresco, the area shown in Fig. 11.24a was chosen for further analysis. The thermogram at the end of heating revealed a non-uniformity in the fresco (Fig. 11.24b). At the best observation time, the image became more uniform and revealed a weak indication of a defect (Fig. 11.24c). This defect became clearly visible in the thermal tomogram (Fig. 11.24d) that corresponded to the depth equal to the plaster thickness; in this case, the thermogram of Fig. 11.24b served as a mask for performing 3D filtering.

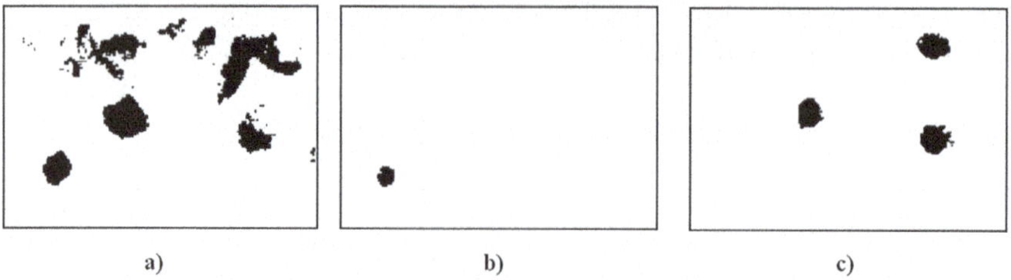

Fig. 11.23 Thermal tomography results in the inspection of "Apollo head" fresco [25]: (a) thermal tomogram of the layer with four defects, (b) tomogram of the layer with Defect #1 (after 3D normalization), (c) tomogram of the layer with Defects #1, 2, 3 (after 3D normalization)

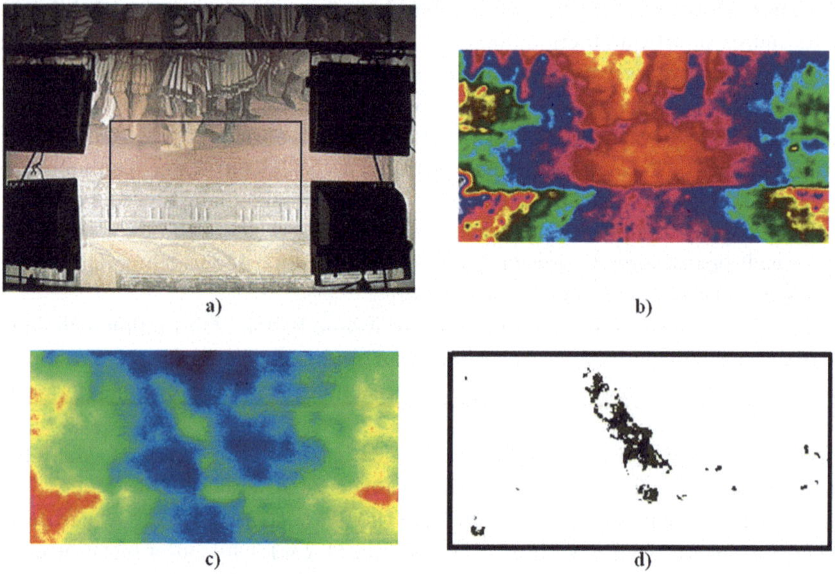

Fig. 11.24 TNDT of the Fogolino fresco in Bergamo, Italy. (Source images: courtesy E. Grinzato): (a) TNDT scheme, (b) IR thermogram at the end of heating, (c) IR thermogram at optimum observation time, (d) thermal tomogram of the 8.1–9.3 mm layer

11.3.3 Religious Icons Consisting of Paint on Wood

There are several techniques for making religious icons, but from the point of view of TNDT it is important that icons, similarly to frescos, represent multi-layer structures whose surfaces are visually inhomogeneous. For example, the icon reference sample shown in Fig. 11.25 was produced according to an ancient Italian recipe and it included 4 pieces of Teflon between different icon layers [26]. The defects were called: (1) "Bubble", (2) "Rod", (3) "Drop" and (4) "Tear". Some advanced data treatment algorithms mentioned in Sect. 6.10.1 including polynomial fitting were considered.

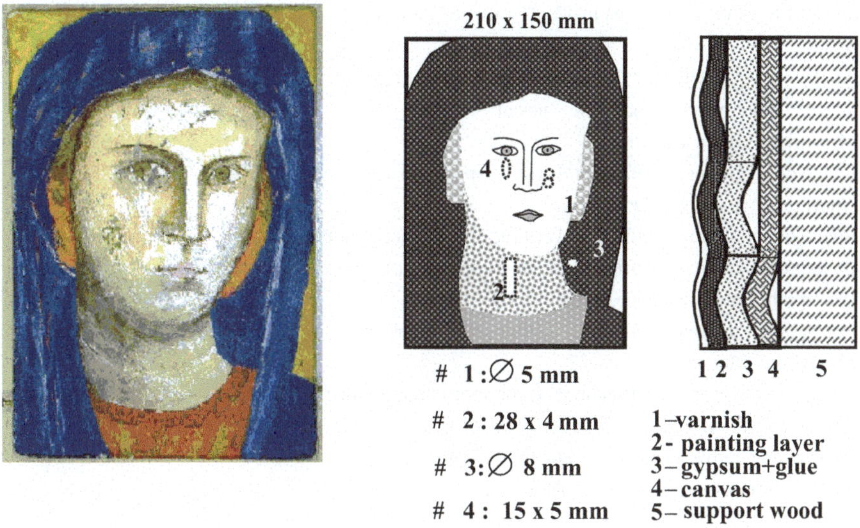

Fig. 11.25 Icon-on-wood reference sample. (Sample: courtesy E. Grinzato)

This sample was also inspected elsewhere by using laser shearography, speckle pattern interferometry, speckle de-correlation and IR thermography [26].

Regarding IR thermography, it was concluded that: (1) images of defects are of a reasonable quality, (2) the equipment cost is high, (3) the technique can be easily applied *in situ*. Although the size of the implants was known, there were some doubts about the real size of defects, as it is known that inserting pieces of Teflon between layers of CFRP may cause air gaps between the edges of the inserts and the host material. For this particular sample, the IR evaluation of "Rod" was 37 × 7 mm while the manufacturer reported that the "rod" was 28 × 4 mm.

The sample was heated for 3–5 seconds with linear quartz lamps, which provided about 300 W/m^2 of absorbed energy. IR images were captured at 10 to 50 Hz for 60 seconds by using a ThermaCam SC3000 IR imager; the number of images was 3000.

Inspection results were analyzed by using a number of methods used in pulsed TNDT: thermal tomography, pulse phase thermography, polynomial approximation and others. Statistical results are presented in Table 11.8; when calculating the signal-to-noise ratio, the whole icon surface was considered to be defect free while either one or three defect areas were analyzed.

In spite of the fact that in all cases $\Delta T > 0$, some processing algorithms produced defect indications both "warmer" and "colder" than the background; this changed the S values, which were calculated automatically by assuming that all defects produce temperature signals of the same sign. Therefore, TNDT optimization has been done only by the "Drop" defect, which was the deepest.

The "Tear" defect was not detected at all, probably, because its depth exceeded the detection limit for this particular sample.

Table 11.8 Statistical results in TNDT of wood icons (Fig. 11.25)

Image	S
Analyzing three defects: "Rod", "Bubbles" and "Drop"	
Source IR thermogram (at 2.5 seconds after heating)	1.29
Maxigram (source sequence)	3.91
Image of polynomial coefficient A_2 (three approximation members)[a]	3.25
Analyzing one defect: 'Drop'	
Image of polynomial coefficient A_3 (three approximation members)	5.79
Image of polynomial coefficient A_1 (four approximation members)	5.73
Phasegram (fitted sequence, frequency 0.084 Hz)	11.72
Image of polynomial coefficient A_0 (four approximation members)	11.66
Maxigram (fitted sequence)	11.37
Image of polynomial coefficient A_2 (four approximation members)	11.32
Phasegram (fitted sequence, frequency 0.042 Hz)	11.21
Phasegram (source sequence, frequency 0.084 Hz)	11.08

[a]Fitting coefficients are specified as: $A_0,\ A_1,\ A_2\ \ldots$

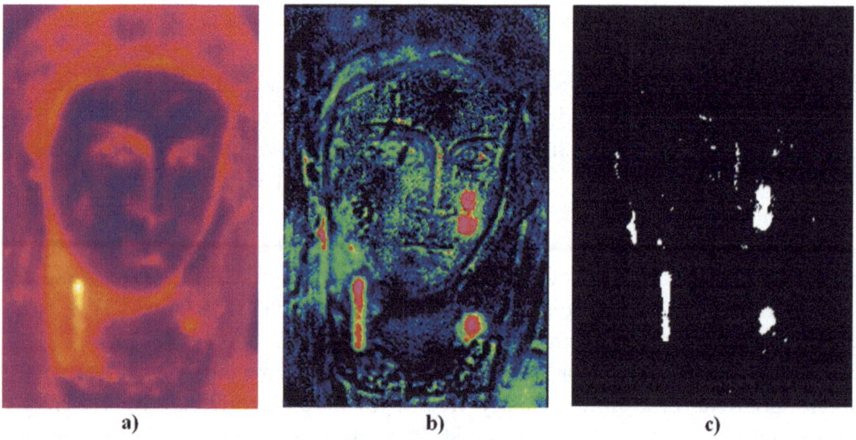

Fig. 11.26 TNDT of the icon-on-wood (optimization by defects "Rod", "Bubbles", "Drop"): (**a**) source image at 2.5 seconds after end of heating, $S = 1.29$ (same as Fig. 6.24a), (**b**) maxigram, after polynomial fitting and normalization, $S = 3.51$, (**c**) map of defects produced by image b), $P_{c.d.} = 80\%$, $P_{f.a.} = 3\%$

The optimum source image (Fig. 11.26a) was characterized by $S = 1.29$, with the most prominent defect being the "Rod", while the image of two other defects was fading. It was expected that all three defects would be optimally detected in the maxigram, as shown by the image in Fig. 11.26b ($S = 3.51$). The map of defects obtained by the maxigram showed defects with the respective probabilities of correct detection and false alarm $P_{c.\,d.} = 80\%$ and $P_{f.\,a.} = 3\%$ (Fig. 11.26c).

When optimizing results for the "Drop" defect, the highest value was $S = 5.79$ and it appeared in the image of the polynomial coefficient A_3 (Fig. 11.27b). The

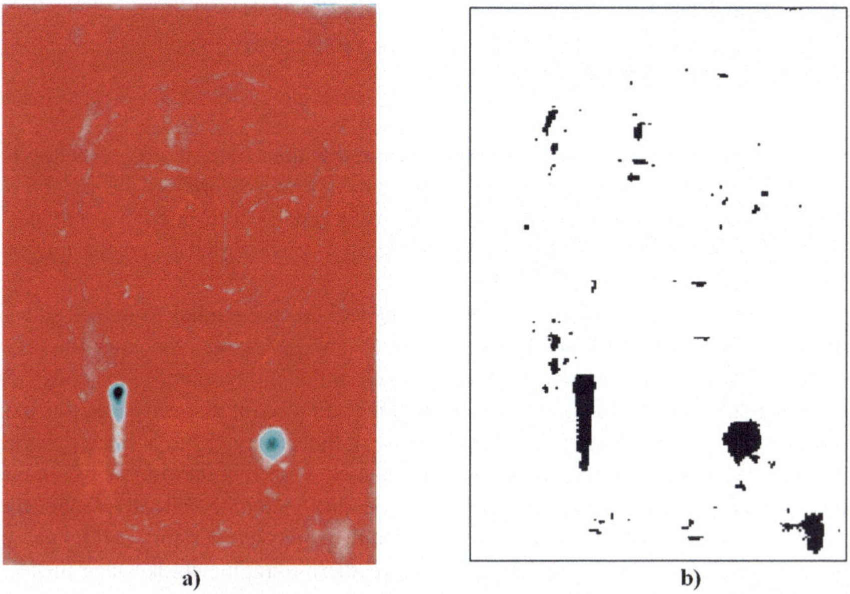

Fig. 11.27 TNDT of the icon-on-wood (optimization by defect "Drop"): (**a**) polynomial coefficient A_3, $S = 5.79$, (**b**) map of defects produced by image (**a**), $P_{c.\,d.} = 100\%$, $P_{f.\,a.} = 3\%$

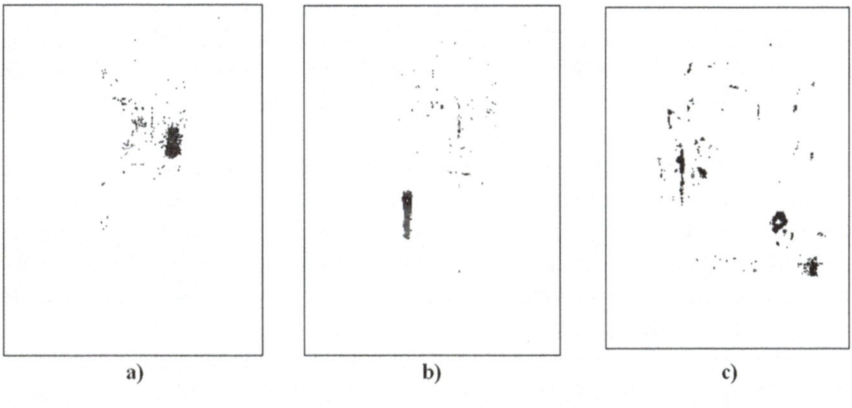

Fig. 11.28 Tomographic presentation of defects in the icon-on-wood: (**a**) layer 0.3–0.8 mm. (**b**) 0.81–0.91 mm. (**c**) 1.43–1.47 mm

corresponding binary defect map was characterized by $P_{c.\,d.} = 100\%$ and $P_{f.\,a.} = 3\%$ (Fig. 11.27b). It is worth noting that in the optimum source image this defect produced only $S = 1.2$.

Figure 11.28 includes thermal tomograms that show three of the above-mentioned defects in different sample layers. It is seen that the "Bubbles" defect is located within the 0.3–0.8 mm layer, and therefore its temperature indication appears during the heating period.

11.4 Smokestacks

11.4.1 Introduction

IR inspection of smokestacks is performed to assess their current structural condition, detect damage and defects, predict their lifetime, determine the extent of repairs required (if any) and evaluate the quality of repairs performed. Such inspections should be performed by properly accredited organizations, and thermographers should be certified by Level II.

The methods of performing smokestack inspections are similar to those used in buildings. However, smokestacks can be checked all year round, because the temperature difference between smokestack gas and ambient air generally exceeds 100 °C in operating smokestacks. However, the interpretation of inspection results can be even more difficult than in buildings because of the variety of smokestack designs and the lack of access to the inside of a smokestack. Also it is impossible to obtain, some operation parameters that would be necessary to mathematically model the smokestack thermal regime. These would include the gas temperature and its variation as a function of smokestack height. The IR thermographic smokestack inspection method is described in the guidelines reported in Chap. 12. Additional information can be found in [27–31]. In Russia, the high potential of IR thermography to evaluate operating smokestacks was discussed by Drozdov and Sukharev in the 1980s [18]. However, at that time, the IR thermographic method was not widely used because it was relatively new technology and the equipment was expensive.

Presently, the interest in IR inspections of smokestacks in Russia is increasing continuously for several reasons. First of all, many smokestacks built in the 1950–1970s have exceeded their expected lifetime and they require constant monitoring. Second, some smokestacks are "underloaded", i.e. they serve fewer boilers than expected during their design. Third, in the last decade, many power plants have converted from coal to natural gas fuel. In this case, smokestack operation parameters deviate from those before conversion; for example, a change in gas temperature may shift the dew point into a smokestack shaft. Nowadays, several Russian companies are active in the market of smokestack IR inspections. Some of these are ORGRES, TTM, Sibtekhenergo et al.

Smokestacks and gas flues at power plants should be externally inspected annually in the spring. Internal inspection should be performed within 5 years after bringing a new smokestack into operation and then at least every 15 years or as warranted by other inspection data. For smokestacks with brick and monolithic refractory, the internal inspection can be replaced with IR thermography and conducted at least each 5 years. According to the Russian document SP 13-101-99, all monolithic reinforced concrete smokestacks having a height over 100 meters, as well as brick and metallic smokestacks with height over 70 meters but operating under high temperatures (over 300 °C) or subjected to corrosive gasses, are to be thermographically inspected every 5 years. Visual inspection is often done from the ground by using binoculars. IR surveys should be performed at least from four points around the smokestack in order to cover total the smokestack perimeter. Prior to conducting a survey, it is recommended to collect and review all available technical documentation, which should include records of previous repairs, and select ground observation points. Inspection results are digitally recorded for archiving. It is important to stress that IR inspections of smokestacks should be accompanied by visual inspection, because many defects that may influence smokestack load-bearing capability cannot be detected thermographically.

Smokestack defects are usually caused by

- poor workmanship during construction,
- operation in a temperature regime above the temperature for which it was designed,
- lower gas temperature and reduced gas volume across the "boiler-gas flue-smokestack" path,
- general deterioration of thermal insulation and water condensation.

In brick-shaft smokestacks, the typical defects are:

- through-shaft and non-through-shaft vertical and horizontal cracks,
- masonry damage due to freeze-thaw cycles,
- poor cohesion between bricks and mortar because of sulfuric acid corrosion,
- masonry swelling between band rings,
- deformation (leaning) of the upper smokestack due to masonry deterioration.

The defects that are typical of reinforced concrete smokestacks are presented in Table 11.9. In smokestacks with forced ventilation, common defects are caused by blockage of the air channel or non-operating ventilation blowers.

Compared to visual inspection, IR thermography has the following advantages:

- smokestacks can be surveyed without ceasing boiler operation,
- visually invisible defects, e.g. defects in internal insulation layers and shaft cracks, can be detected;
- defect classification by severity is possible.

The limits of IR thermography are related to:

- problems of IR image identification and their classification by defect types;
- difficulties in surveying the tops of smokestacks due to a steep viewing angle and a long distance (this problem is solved by using telescopic lenses at larger distances),
- solar heating of smokestacks and the presence of reflected solar radiation. The latter is solved by surveying in early morning or at night.

11.4.2 IR Imagers for Smokestack Inspection

The same models of IR imagers that are used in building applications are generally recommended for inspecting smokestacks. In addition to technical performance, IR camera weight and size are important, as surveys may need to be performed in

Table 11.9 Characteristic defects in reinforced concrete smokestacks. (Courtesy: A. Alexandrov)

Defect	Description	Causes
Reinforced concrete smokestack cylinder		
1. Defective seam	Concrete seam characterized by large pores and voids	Poor quality workmanship, formation of large pores and voids containing water that penetrates seams. Seams experience damage along part or all of the perimeter.
2. Deteriorating seam	Concrete seam with visible signs of damage	Concrete delaminations, cavity formation, etc.
3. Peeling of concrete surface layer	Surface layer damage	Lower resistance to cold Soiling source materials with clay mixtures Using poor quality fillings Low strength of astringents and high mixture plasticity Violation of processing thermal regime
4. Cracks	Through or non-through cracks on the surface	Longitudinal bend of vertical reinforcement bars Crippling of concrete protection layers Impact of mechanical loads and temperature variations Locked out effect of reinforcement corrosion products Locked out effect of water in winter season
5. Deep through-cylinder damage	Concrete cylinder damage Exfoliation of large filled cavities and reinforcement from concrete through holes	Periodic moistening and drying, action of concrete freezing/thaw cycles, sulfuric acid and corrosion
6. Coarse grain concrete	Concrete damage in the form of small surface cavities Large surface cavities propagate in-depth up to 3–4 cm, deep ones – through whole shaft thickness	Inappropriate preparation of concrete Splitting of concrete mixture when falling down Poor concrete mixture vibrating Using hard or packed concrete mixture Splitting of concrete mixture during transportation Violation of concrete recipe Too dense reinforcement
7. Splits in concrete	Tearing off a piece of concrete with formation of crater-like cavity	Damage of concrete filling that leads to local concrete tearing off

(continued)

Table 11.9 (continued)

Defect	Description	Causes
8. Concrete exfoliation and damage	Surface concrete damage, including splitting with reinforcement exposure	Building spoilage, lower concrete strength, impact of wind and smokestack mass
9. Moist concrete	Moisture and ice accretion	Condensation of smoke gas that penetrates a cylinder. Concrete is being damaged because of lime leaching and freeze/thaw cycles.
Thermal and vapor/moisture insulation		
10. Thermal insulation breakdown including refractory bricks and rock-wool	Thermal insulation breakdown on refractory side (formation of white powdery mass)	Presence of large quantities of vapor and sulfur oxide in smoke gases which penetrate refractory and condense as acids
11. Vapor/moisture insulation (bitumen) breakdown	Fragile and crack-full vapor/moisture insulation	See above
Refractory		
12. Refractory deformation (swelling)	Swellings and barrel-like broadening of masonry	Refractory deformation because of sulfuric corrosion (sulfation and sulfate crystallization) Saturation of refractory with acids and sulfates which cause masonry to swell
13. Refractory breakdown	Total or partial refractory breakdown, bricks fall out forming holes	Chemical corrosion (sulfuric acid and sulfuric corrosion of refractory) Refractory swelling, up to failure
14. Refractory brick corrosion	Brick delamination	Sulfuric acid and sulfuric corrosion of refractory
15. Cracks	Vertical, horizontal or diagonal cracks on the refractory surface or penetrating through the stack	Stain forces, influence of temperature gradients, masonry damage and condensation
16. Damage of masonry mortar	Reduction of strength and cohesion between bricks and mortar, total or partial spalling and shedding of mortar	Chemical corrosion
17. Gap blocking	No gap between refractory and shaft No compensation gap	Chemical corrosion of refractory leading to its enlargement Blocking of compensation gap because of refractory swelling

places of difficult access, e.g. from roofs of neighboring buildings. If inspection is performed at 6–7 a.m., the affect of the radiation from the rising Sun becomes noticeable, therefore, long wave (LW) IR imagers are recommended. An IR survey of a smokestack typically consists of images taken at 4–6 observation points around the circumference and, at each of these observation points, 4–8 IR images taken at different heights along a vertical line on the smokestack. Using digital recording single IR thermograms may be combined to produce "stitched" (composite or panoramic) images. The use of both wide-angle and narrow-angle lenses (4, 7, 12, and 18°) is recommended for routine use, while a telescopic lens should be used for images of the tops of smokestacks. Voice comments, which can be attached to IR thermograms in some IR imagers, can be used to describe detected anomalies or inspection peculiarities.

The visual inspection which is typically performed in parallel with IR monitoring allows the evaluation of technical conditions, such as the efficiency of lightning guard and the possible corrosion of metallic constructions, as well as determining concrete quality by using a special hammer (Koshkarov's hammer) or more advanced electronic devices. A geodetic survey is a special type of smokestack inspection that is used to evaluate a smokestack base. In fact, evaluating how the temperature changes as a function of time in smokestack heels can lead to a prediction of the smokestack's lifetime.

11.4.3 Smokestack Description

Steady-state heat conduction through a three-layer cylindrical wall

The theory related to the inspection of smokestacks and gas flues involves the analysis of heat conduction through cylindrical and plane walls (see Sect. 11.4.1).

The thermal flux across a multi-layer cylindrical wall is defined in the following way (three-layer structure of Fig. 11.29):

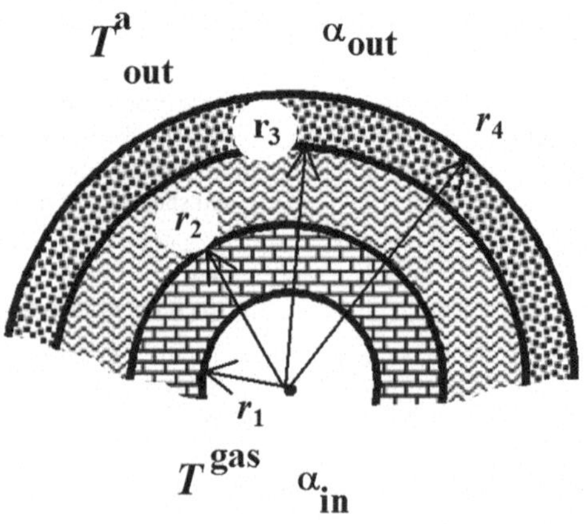

Fig. 11.29 Steady-state heat conduction through a multi-layer cylindrical wall

$$Q = \frac{T^{gas} - T^{a}_{out}}{R}; R = \frac{r_4}{r_1}\frac{1}{\alpha_{in}} + \frac{r_4}{\lambda_1} Ln\frac{r_2}{r_1} + \frac{r_4}{\lambda_2} Ln\frac{r_3}{r_2} + \frac{r_4}{\lambda_3} Ln\frac{r_4}{r_3} + \frac{1}{\alpha_{out}}; \tag{11.17}$$

$$Q = \alpha_{in} \left(T^{gas} - T^{w}_{in}\right); \tag{11.18}$$

$$Q = \alpha_{out} \left(T^{w}_{out} - T^{a}_{out}\right), \tag{11.19}$$

where T^{gas} is the smoke gas temperature, R is the shaft thermal resistance; all other specifications are same as in Sect. 11.4.1.

Note that the thermal flux in Eq. (11.17) is defined per unity area of external smokestack surface.

Smokestack design and characteristic defects

The design of a smokestack depends on the expected chemical composition of the smoke gas, the local climate and when the smokestack was built.

Temperature anomalies in smokestacks may look different depending on the operating conditions. The general characteristics of temperature signals caused by several smokestack defects are presented in Table 11.10 for both high and low gas pressure in the smokestack.

Gas flues made of brick, reinforced concrete panels or thermal-insulating blocks, may have the same types of defects as smokestacks (Table 11.10) but these defects tend to degrade much more quickly; for example, the lifetime of a gas flue might be less than 5–10 years if a sulfurous fuel is burned.

11.4.4 Performing Smokestack Surveys and Analyzing Results (Illustrations)

Prior to performing an IR inspection of a smokestack, a thermographer should formulate a survey plan to ensure as much coverage as possible. This is particularly important in the case of a cylindrical smokestack where side sections cannot be effectively monitored due to geometrical distortion and the Lambertian character of emitted thermal radiation. The minimum number of observation locations is four, but in practice, six to seven views may be necessary.

It is strongly recommended to analyze all available documentation on a smokestack to be inspected. This should include results and dates of previous inspections and repairs, drawings, technical certificates, safety requirements, entry procedures, etc.

One should also perform a visual inspection of a smokestack by using binoculars or other means. And one should take photos or videos from the same observation points as the IR imaging.

Smokestack and gas flue defects are recorded in a particular way. A map of defects found by IR imaging can be imposed on a map of defects that were detected

Table 11.10 Temperature anomalies and corresponding smokestack defects depending on smoke gas pressure[a]

		Character of temperature distribution		
#	Defect type	Smokestack under reduced pressure	Smokestack under increased pressure	Note
1.	Defective seam in concrete	Horizontal cold line	Horizontal warm line	
2.	Stack surface crack	Cold line	Warm line	To confirm visually
3.	Stack through-crack	Cold line	Warm line	To confirm visually
4.	Stack low resistance toward gas permeability	Cold area	Warm area	
11.	Stack destruction	Either cold or warm area depending on defect depth		To detect visually
6.	Refractory corrosion (up to 60%)	Warm area	Warm area	–
7.	Refractory breakdown	Warm area	Warm area	–
8.	Refractory low resistance toward gas permeability	Warm area	Warm area	–
9.	No thermal insulation between shaft and refractory	Warm area	Warm area	–
10.	Ash sediments	Cold area	Cold area	To confirm by documentation

[a]Table: courtesy A. Alexandrov

visually. It is important to note that the specification of surface defects adopted in the industry is not the same as IR defect findings, which are thermal (warm or cold) anomalies. Defects that are considered serious should be photographed separately and reported to the owner of the smokestack.

During inspection the following parameters should be measured:

- outdoor air temperature (at each observation point)
- wind direction
- wind speed

When writing a technical report, one generally uses both single image and stitched, or panoramic IR thermograms. Standard or customized graphic editing programs can be used,. Three examples of panoramic images of different types of smokestacks are shown in Fig. 11.30. Inspecting metallic smokestacks by IR thermography is usually inconvenient, except when they are refractory-lined. For example, Fig. 11.30a shows a metallic petrochemical plant smokestack and it is obvious that there is degeneration in the upper edge of the internal refractory layer. The typical defects in brick smokestacks, are air in-leaks; see the numerous defects in Fig. 11.30b. A trained operator is able to classify a smokestack status as "good" or "bad" by analyzing the temperature field uniformity. Such a simple classification can be applied to reinforced concrete smokestacks; examples of IR images of these are presented in Fig. 11.31. A uniform temperature distribution corresponds to a

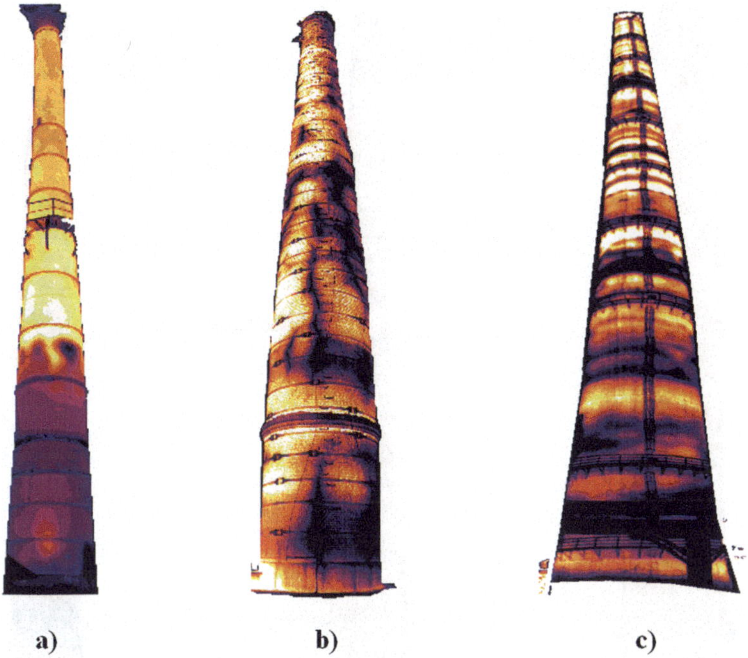

Fig. 11.30 Panoramic IR images of smokestacks: (**a**) 40 m-tall metallic (exterior) smokestack with a refractory liner in the bottom section (refractory damage at the top is clearly seen, evidenced by a temperature gradient up to 6 °C), (**b**) 60 m-tall brick smokestack with numerous cracks (temperature gradient up to 7 °C), (**c**) 150 m-tall reinforced concrete smokestack with deficiencies in thermal insulation (temperature gradients up to 3 °C)

defect-free smokestack (Fig. 11.31a); the single anomaly shown by an arrow in Fig. 11.31b was identified as a lack of mineral wool, while the numerous indications in Fig. 11.31c may be caused by numerous defects in the smokestack thermal insulation. Using IR diagnostics, one can evaluate the quality of a repair which may look fairly good when observed visually. But an IR survey can reveal hidden defects. Such a situation is presented in Fig. 11.32, where the IR image is imposed on the photo of a recently repaired brick smokestack. The IR thermogram clearly shows that air in-leaks are still present through a big crack that was detected during a previous inspection and then repaired. In reinforced concrete smokestacks, defects and their temperature patterns are quite variable. For instance, one should be able to identify a local deterioration of thermal insulation, which occurs due to an absence of, or shrinkage of, mineral wool mats. These may appear as circular warm areas which correspond to the "tear-off" zones, where the insulation thickness is reduced to accommodate concrete elements that collecting condensation (see Figs. 11.31c and 11.33). Concrete seams typically look as cold circles because of the enhanced porosity of concrete. Sometimes, such areas can be very extensive, indicating low quality concrete.

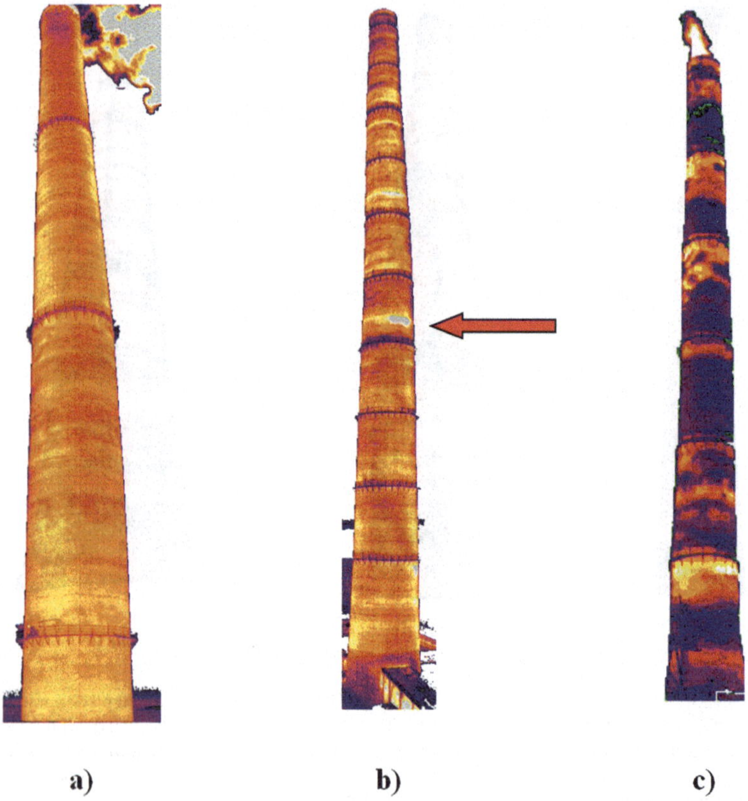

Fig. 11.31 Illustrations of the “thermal quality” of reinforced concrete smokestacks: (**a**) conventional “good” quality, with no defects. (**b**) “medium”, with some defects. (**c**) “low”, with a lot of defects

The thermogram of the joint between the reinforced concrete smokestack and the gas flue in Fig. 11.34 illustrates two common types of defects: thermal insulation deficiency and air in-leak. A thermogram of the zone where the gas coming out of the boiler hits the internal refractory, thus causing intensive wear, is shown in Fig. 11.35.

IR images recorded during smokestack surveys can be processed using software supplied with most IR cameras. Most software allows image recording, preliminary processing and producing IR thermograms in many palettes of color and black and white. On a map of a gas flue, one should indicate characteristic dimensions, panel sizes, etc. The vertical scaling of IR thermograms is done by light-colored signal foot plates, balconies and consoles, as well as by metal bands (rings) in the case of brick smokestacks. On reinforced concrete smokestacks, scaling and the assignment of coordinates may be accomplished by counting concrete seams if they are clearly seen in images.

Fig. 11.32 IR thermogram of a repaired vertical crack in a brick smokestack revealed the presence of an air in-leak with a temperature decrease of 3.7 °C

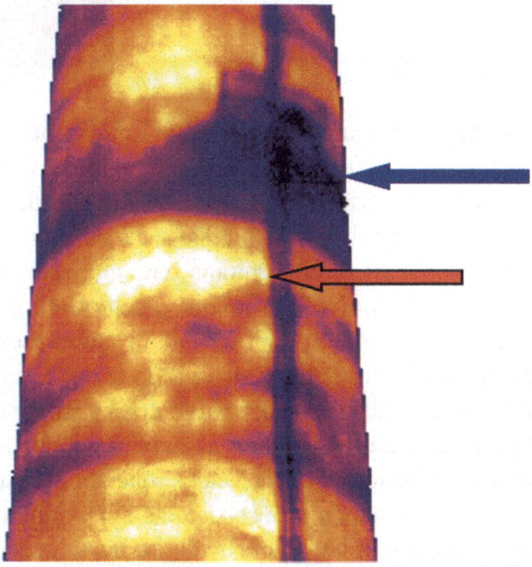

Fig. 11.33 IR thermogram of areas of poor thermal insulation (temperature increase 4 °C) and high concrete porosity (temperature decrease 5 °C) in a reinforced concrete smokestack

Defects may be identified by comparing experimental data to modeling results; if they match well, the hypothesis about the nature of a defect should be accepted. However, it is worth remembering that, as with the IR inspection of buildings, specific ΔT values may correspond to various combinations of construction features and defects. Mathematically, this is a reflection of the ill-posed character of the corresponding inverse heat conduction problem. Therefore, inspection conclusions are often subjective and they depend on the skill and experience of the thermographer.

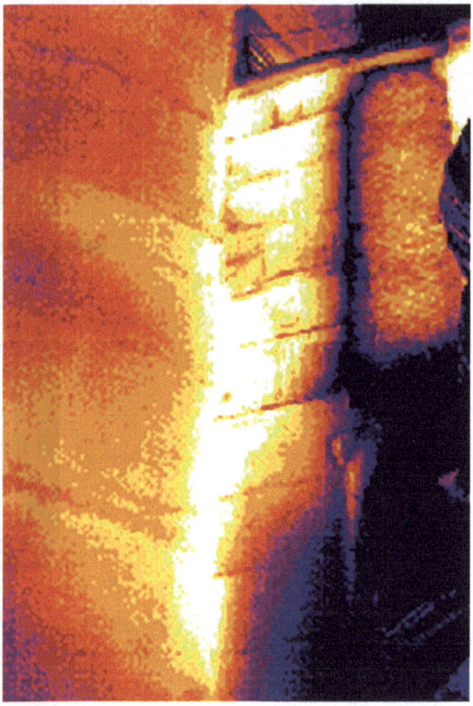

Fig. 11.34 IR thermogram of the joint between a brick gas flue and the reinforced concrete smokestack with poor thermal insulation (temperature increase 5 °C) and air in-leak (temperature decrease 6 °C)

The most uncertain parameters of smokestack thermal models are: (1) the smoke gas temperature and its vertical gradient, and (2) heat exchange coefficients on the internal and external surfaces of the smokestack.

After being detected, smokestack defects are shown on a map, which should include both IR thermographic data and also the results of visual inspection. The visual inspection, which involves determining the type of concrete, the status of lightning rods and grounding, and, in some cases, a geodetic survey, is an indispensable part of a smokestack survey. By using binoculars, one can detect exposed reinforcement bars ("rebar"), visible cracks, concrete degradation, delaminations, and the, appearance of leakage as evidenced by carbon smudges on the surface. On reinforced concrete smokestacks, one should carefully inspect concrete seams. An example of a map of defects detected in a reinforced concrete smokestack by applying IR thermography exclusively is shown in Fig. 11.36.

11.5 Electrical Installations

For many years, the inspection of electrical installations has been one of the most widely used and successful applications of IR thermography. In this application, temperature anomalies detected may reach tens of degrees Celsius, and this makes

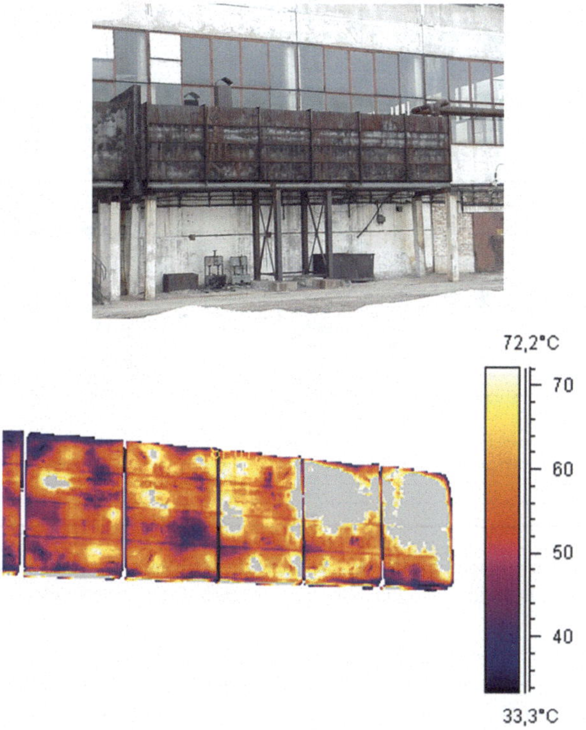

Fig. 11.35 IR thermogram of refractory wear in a metallic gas flue

the interpretation of results fairly straightforward. For several decades, major electrical power producers in Russia have been routinely using IR thermography to inspect electrical substations and indoor facilities. For example, in 2003, Mosgosenergonadzor, the company that supplies electricity to Moscow, conducted 170 IR thermographic surveys and concluded that about 50% of installations required urgent repair, another 40% should receive preventive maintenance at the next planned inspection, and 5% of problems should be fixed during the next scheduled shutdown.

Requirements for IR thermographic inspection in the electrical and power industry are described in a number of documents issued by the Russian Ministry of Energy and other organizations [32, 33]. The last edition of the document RD 153-34.0-20.363-99, "Basic statements on performing the IR thermographic diagnostics of electrical installations and high voltage transmission lines" summarizes the multi-year experience of some leading Russian organizations, such as Mosenergo, Lenenergo, Sverdlovenergo, Novgorodenergo, ORGRES, Mosisolator plant and some atomic power stations, and conforms with another basic document, RD 34.45-51.300-97 "Size and norms of the inspection of electrical installations". The methodical aspects of IR thermographic inspection were thoroughly

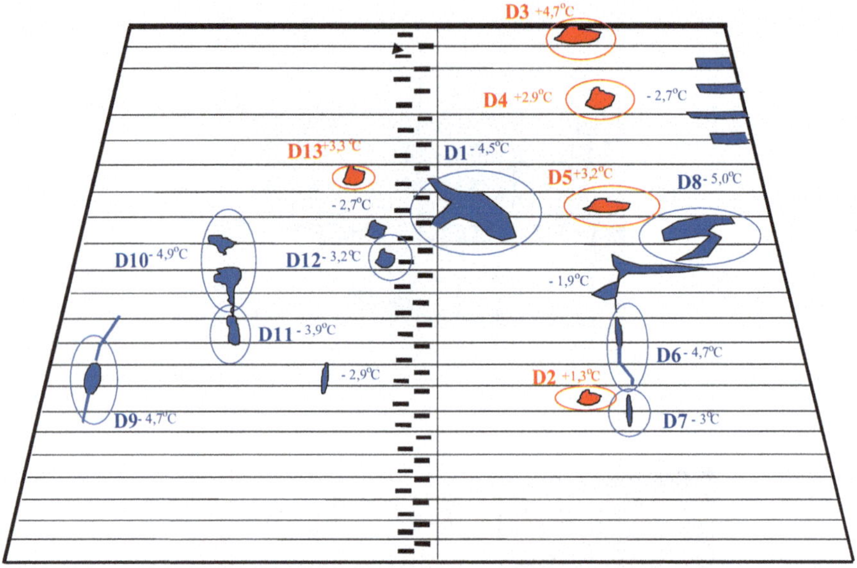

Fig. 11.36 Map of thermographically detected defects in a reinforced concrete smokestack (red areas designate poor thermal insulation, blue areas are air in-leaks, severe defects are marked with ellipses)

summarized by Bazhanov [34]. IR diagnostic applications of electrical equipment are described briefly in Table 11.1 and in detail in Table 11.11. Below we report some inspection peculiarities that appeared in [33, 34].

Joule's law states that the electric power dissipated in an operating conductor is equal to the current squared multiplied by the conductor resistance. The deterioration of electrical contacts caused by undersized conductors and corroded connections results in a higher resistance in the electrical joint and, therefore produces an abnormally high temperature. Defective electrical joints are common, and experience demonstrates that performing periodic IR thermography surveys makes it possible to minimize the number of defects.

Advantages of IR thermographic technique are as follows: (1) maintenance safety, (2) lower operating cost, (3) better and more economical maintenance, and (4) increased profit.

11.5.1 IR Thermographic Inspection of Electrical Equipment (Switchgear)

In the Russian document RD 153-34.0-20.363-99, the following frequency of IR thermographic inspections is recommended:

Table 11.11 IR thermography in the inspection of electrical equipment and installations [33, 34]

Electrical equipment at power plants and networks	Detected faults and correction work
Generators	Short circuits Poor quality of winding soldering Evaluation of brush gears Failures in operation of cooling system Checking excitation elements
Transformers	Focuses of magnetic leakage fields Formation of stagnant zones in transformer vessels because of dirt formation, winding swelling and displacement, failures of oil system Defects in connector lines Evaluation of cooling system efficiency
Commutation equipment	Bus and arc chutes overheating Evaluation of vessel insulation quality Defects of lead-in lines and dividing capacitors Cracks in bearing insulators
Oil-filled current transformers	Overheating of internal and external insulators Deterioration of internal winding insulation
Valve dischargers and overvoltage limiters	Encapsulation deterioration Diverter breaks Wrong component integration
Capacitors	Section breakdowns
Linear high-frequency obstructers	Overheating of contact joints
Current distributors	Overheating of switches, disconnects, current transformers, cables, buses etc.
Cable economy at power plants	Overheating of drive cables, evaluation of flammability risk of cables
Aerial transmission lines	Overheating of joints

- Electrical equipment in electrical distribution facilities having a voltage lower than 35 kV – once every 3 years, 110–220 kV – once each 2 years, and 300–750 kV – every year,
- Distribution facilities of any voltage if they are extremely dirty – every year,
- Extraordinary inspection of all kinds of equipment is to be done after natural and technical disasters, such as heavy wind, short circuits, ice storm, etc.,
- Overhead lines (inspecting all types of joints): (1) new lines should be inspected in the first year of operation, (2) lines that have been in operation over 25 years should be inspected every year if over 5% of joints are identified as defective, and once in three years if the number of defects detected is under 5%, (3) lines that operate with maximum current loads or supplying energy to important consumers, as well as operating in a contaminated atmosphere and/or under heavy wind and ice storm should be inspected every year, (4) other overhead lines should be inspected not less than once each 6 years.

As with other applications, one screening criterion is a differential temperature signal (rise) ΔT, which is the difference between the temperature of an analyzed element and either the air temperature or the temperature of the same conductor but far away from a test point. Inspection of electrical equipment is complicated by the influence of natural conditions, if it is performed outdoors, as well as by the relative small size of the inspected elements and their high reflectivity.

Document, RD 153-34.0-20.363-99, mentioned above, introduces the following criteria for making decisions: (1) an increase in temperature, defined as the difference between the measured temperature and the ambient temperature, (2) an excessive temperature which is defined as a temperature increase in an inspected element compared to similar phase elements operating under the same conditions, (3) a coefficient of defectiveness, which is defined as the ratio between a measured temperature increase and the temperature increase on the same conductor at least 1 m from a point of interest.

Maximum allowable temperatures of electrical equipment are shown in Table 11.12.

The status of conductors is evaluated by the coefficient of defectiveness according to Table 11.13.

Solar reflections are particularly noticeable in the 3–5 μm wavelength band, although solar heating of objects reduces test effectiveness in any spectral band. In order to exclude solar heating, it is recommended that IR surveys be performed in the dark, between sunset and sunrise, or under a cloudy Sky. Hot objects nearby may also provide spurious reflections, and this especially likely when inspecting components in a closed space or in the presence of many other objects in a complex system. This is typical of an electrical substation.

Wind becomes a significant factor when surveying components that are overhead and not screened by other equipment. A correction for wind velocity can be made using the following formula [34]:

$$\frac{\Delta T_2}{\Delta T_1} = \left(\frac{V_1}{V_2}\right)^{0.448}. \tag{11.20}$$

where the ΔT's are the temperature signals that are used for decision making. Equation (11.20) is valid for wind velocity from 1 to 7 m/s; if $V > 8$ m/s, temperature measurements are considered unreliable. A more correct approach taking into account a power level was suggested by Madding and Lyon [35].

Rain, fog and snow significantly reduce ΔT values and IR image quality; therefore, IR inspections may be conducted only in a slight rain or snow. When checking suspended insulators, the best test results are obtained in high humidity, as that causes a build-up of leakage currents.

Since temperature signals depend on the current load, measurement results are often reduced to 50% of a nominal load. The re-calculation formula is from Joule's law:

Table 11.12 Allowable temperatures of electrical equipment (according to RD 153-34.0-20.363-99, Russia)

Components	Maximum allowable value	
	Heating temperature, °C	Temperature increase, °C
1. Conductor lines (except contacts) and metallic non-conducting elements: Non-insulated and non-contacting insulated materials	120	80
Insulated and contacting insulated materials, which are characterized by the following classes of heat resistance, according to Russian standard GOST 8865-93:		
Y	90	50
A	100	60
E	120	80
B	130	90
F	155	115
H	180	140
2. Copper and copper alloy contacts:		
no coating, in air or oil,	75/80	80/50
with silver plates laid on, in air or oil,	120/90	80/50
coated with silver or nickel, in air or oil,	105/90	65/50
coated with silver with a thickness of at least 24 μm,	120	80
coated with tin, in air or oil	90/90	50/50
3. Ceramic metallic contact containing tungsten and molybdenum	85/90	45/50
4. Equipment outlets made of copper or aluminum and their alloys:		
no coating,	90	50
coated with tin, silver or nickel	105	65
11. Bolt joints made of copper or aluminum and their alloys:		
no coating, in air or oil,	90/100	50/60
coated with tin, in air or oil,	105/100	65/60
coated with silver or nickel, in air or oil,	115/100	75/60
6. AC 3 kV safety devices		
(connections made of copper or aluminum and their alloys without coating or coated with tin):		
plug-type spring joints	75/95	35/55
demountable bolt or screw joints, including outlets	90/105	50/65
metallic parts used as springs,		
made of copper,	75	35
made of bronze and similar alloys	105	65
7. Insulating oil in the upper layer of commutation devices	90	50
8. Built-in current transformers:		
winding	–	10
magnetic core	–	15

(continued)

Table 11.12 (continued)

Components	Maximum allowable value	
	Heating temperature, °C	Temperature increase, °C
9. Bolted joints of detachable outlets in oil or air	–	85/65
10. Connections in power transformers made of copper and copper alloys without silver coating in oil or air:		
with bolts or other elements ensuring rigid connection,	–	40/25
self-cleaning joints with springs,	–	35/20
non-self-cleaning joints with springs	–	20/10
11. Power cable conductors in presence of insulation (prolonged or breakdown regime) made of:		
polyvinylchloride or polyethylene,	70/80	–
vulcanized polyethylene,	90/130	–
rubber,	65/–	–
rubber of high heat resistance, treated insulating paper under the voltage source, in kV:	90/–	–
1 to 3	80/80	–
6	65/75	–
10	60/–	–
20	55/–	–
35	50/–	–
12. Collectors and contact rings, unprotected or protected by the insulation of the following class of heat resistance:		
A / E / B	–	60/70/80
F / H	–	90/100
13. Friction and frictionless bearings	80/100	–

Table 11.13 Evaluation of conductors by coefficient of defectiveness (according to RD 153-34.0-20.363-99)

Coefficient of defectiveness	Conclusion
Under 1.2	Beginning deterioration, that must be addressed
1.2–1.5	Existing defect that must be fixed in the next shutdown period
Over 1.5	Acute and serious defect requiring immediate repair

$$\frac{\Delta T_1}{\Delta T_2} = \left(\frac{I_1}{I_2}\right)^2. \tag{11.21}$$

However, a more correct analysis of load influence has shown that the power 2 in the latter formula may vary from 0.6 to 2 depending on target emissivity and target/environment temperature [36]. When correcting temperature signals for current load, the variation in thermal inertia among the test components must be taken into account. For instance, the time to reach thermal equilibrium for joints is about 20–30 minutes, while valve dischargers, because of their larger mass and thermal inertia, may take 6–8 hours to reach thermal equilibrium. Furthermore, brief, high current "spikes" should be monitored, as overheating may occur quickly and not be noticed. For example, thyristor panels at railway traction substations should be inspected when trains are passing by.

In the inspection of high-current facilities, such as generator buses, reactors etc., one should be aware of the presence of powerful magnetic fields. Such fields significantly distorted images on oscilloscope monitors that were typical in old models of IR cameras. LCD monitors that are currently used are less vulnerable to electromagnetic fields. Another phenomenon that may occur during surveys in the presence of powerful electrical currents is the existence of inductive currents, which may heat up bolts, bus holders and other elements, thus producing temperature signals that could be wrongly interpreted as other problems.

Corona discharge can be also interpreted as overheating if short wavelength IR imagers (~ 2 μm) are used, as the wavelength of corona discharge is around 1 μm.

Emissivity can always be a problem in the IR thermography of electrical facilities since components may be made of a wide variety of materials, painted or unpainted, shiny or oxidized. One way to validate test data is to compare test results on an installed component to a similar test on a reference component that is either identical to the test component or a similar component made of the same material. The true temperature may be measured by using a thermocouple or another contact temperature measuring device, and the emissivity in a IR measurement devide can be adjusted to provide a similar reading.

It is commonly believed that the inspection of materials with an emissivity lower than 0.2, which is typical of polished metals, is practically impossible due to reflections. Some publications state that the lower limit of emissivity for effective inspection is about 0.6–0.7. Low emissivity also results in a low level emitted signal, which may prevent an overheated component from being noticed.

Since emissivity varies as a function of angle, viewing angle is recommended to be between 0 and 40 degrees off perpendicular in the case of metals and 0–60 degrees in the case of dielectrics. Generally, the most accurate temperature measurement are obtained when viewing perpendicular to the surface, as long as there are no reflections caused by neighboring hot objects or the thermographer himself. When inspecting ribbed objects, such as the porcelain inlet insulators and transformers, there are some additional factors related to emissivity and viewing angle. Often the mean value across a certain area may serve as a good estimate of true temperature. It is worth noting that porcelain insulators are heated not only by the electric current passing through the winding but also by the dielectric losses in the insulation. These losses are characterized by a *dissipation factor* $tg\,\delta$. Its contribution can be evaluated by performing IR thermographic measurements without an electric load, i.e. when a the operating voltage is applied to a test object but no current is flowing.

11.5.2 Power Transformers, Autotransformers and Oil-Filled Reactors

The IR diagnostics of transformers has been partially investigated. IR is recommended as a complement to routine inspection methods, such as chromatography, oil analysis, etc. The temperature distribution on the transformer surface, shown by IR, is affected by internal defects that may be strongly dependent on heat transfer in the magnetic core and transformer winding, as well as on energy produced by eddy currents and Joule heating. The experience of the Russian company ORGRES has shown that IR thermography allows the detection of the following equipment failures [34]:

- Appearance of magnetic leakage fields caused by deterioration of magnetic core insulation,
- Disturbance in the operation of cooling systems (oil pumps, filters, ventilators etc.),
- Deterioration in the circulation of oil in vessels (stagnant areas), particularly, in transformers operating for a long time,
- Heating of inner contacts between windings and transformer outlets,
- Short circuits in coil windings.

Defectiveness criteria are specific for each piece of equipment [32, 33]. For instance, the temperature of the upper layers of oil under a nominal load must be from +70 to +95 °C and the maximum temperature gradient along the height should not exceed 20–35 °C.

All IR thermographic inspections should start with an investigation to understand the electrical functions, cooling systems, duty cycles and an analysis of failures of similar transformers. The surface of tanks, filters and cooling systems should be carefully cleaned of dirt, oil and rust. Inspections should be performed in the dark, between sunset and sunrise, with no wind and no rain, and in the no-load and maximum load conditions [34].

11.5.3 Oil-Filled Current Transformers

IR thermography also serves as an auxiliary inspection method for this type of transformer. Thermographic survey findings are to be identified and then checked by using other test techniques, including the determination of the insulation dissipation factor $tg\,\delta$, by chromatography, oil analysis or other means. It is recommended that inspections be performed at least once a year before the arrival of high summer temperatures. It is recommended that each transformer phase be examined from three points rotationally separated by 120 degrees from each other, to compare similar zones on each phase. If temperatures of each phase differ by more than 0.3 °C, a more careful analysis is advised under different weather conditions and different load. Also, the symmetry of phase currents should be evaluated. Even if the evaluation of the inner insulation is the primary purpose of the evaluation, one should also evaluate temperature in clip connections and inner switches. In the last case, surface temperatures in the compander of three phases should be compared (a compander is a processor that uses companding, which is a combination of a COMPressor and an exPANDer). It is proposed that an increase in the compander surface temperature that exceeds 10–15 °C probably indicates the presence of acute switch defects.

The peculiarities of the IR thermographic inspection of current transformers can be found in [33, 34].

11.5.4 Oil-Filled Voltage Transformers

Porcelain insulators of voltage transformers operating under load are heated up along their height. Therefore, the criteria in the previous section, a temperature increase of 0.3 °C is too low to be used as a criteria for screening insulators. Abnormally high temperatures in oil filled transformers might be caused by the following: (1) additional losses in the magnetic core, (2) moisture and contamination with oil accompanied by a higher dissipation factor $tg\,\delta$, (3) design or technical factors, (4) short circuits in the winding core. As in the case of current transformers, the establishment of a relationship between observable temperature findings and above-mentioned causes requires further research.

11.5.5 Oil, Air, Vacuum and SF_6 Switches

External joints and connections in switches are numerous by design and in most cases they can be evaluated by IR thermography. General information on this topic can be found in Table 11.14.

When inspecting oil breakers, one should view each phase on the surface of the object from 2 to 3 observation points. If local temperature anomalies are discovered, the inspection should be repeated 12–24 hours later, with the breaker under a different load. The normal operation of arc chute contacts causes no temperature anomalies; therefore, the oil temperatures in the vessel and on its surface are assumed to be similar. When contacts deteriorate, their resistance increases and they heat up, and warm the adjacent surface area relative to the surface over other phases. Extreme overheating of contact joints creates local warm areas due to convection or turbulence in the oil.

In air circuit breakers mounted in porcelain insulators, the joint temperature can be measured by the temperature of the metallic flanges that connect the insulators.

The failure statistics for vacuum and SF_6 breakers is poor, but it is believed that in this case the asymmetrical temperature of phase insulators may also serve as a defect indicator.

Table 11.14 TNDT of circuit breakers. (Adapted from [34])

Component	Test measurement
Oil breakers 6–10 kV: bus-outlet outlet with a flexible connection flexible connection-plug arc chute bus-tank low contact	Connection temperature
Oil breakers 110 kV and higher: bus-outlet current distributor of flange contact trolley current collection volatile and motionless contacts supporting insulators	Connection temperature Recording IR thermogram for visualizing defect location and type
Dead tank oil circuit breakers: bus-outlet arc chute	Same
Air circuit breakers: bus-outlet conducting connection of modules arc chute capacitive voltage divider isolating airways	Same
Vacuum and SF_6 breakers: bus-outlet switch contact system	Same

11.5.6 Disconnects and Eliminators

IR thermography is used to inspect contact joints and insulators in disconnects and eliminators. The joints can be overheated due to: (1) low reliability of clad copper contact outlets, (2) oxidation, (3) loose contact pressure, etc. The evaluation of the supporting porcelain insulators seems to be the most difficult task. Measurements are to be performed at positive ambient temperatures and high humidity, which increases leakage currents. The temperature of a good clean insulator exceeds the ambient temperature by about 0.5 °C, while overheated insulators may have an excess temperature up to 2 °C.

11.5.7 Oil-Filled Inlets

Oil filled high voltage inlets are used in power transformers and oil circuit breakers. In the first case, the bottom sections of the inlets are within the oil at a temperature of 95 °C and the upper sections correspond to the connecting insulator. In oil filled circuit breakers, the inlet temperature is fairly constant with height.

IR thermography is able to detect the following faults in oil filled inlets: (1) overheated connections between external conductors and inlet clips, (2) short circuit contours in expanders of hermetic inlets, (3) overheated internal contact joints, (4) low oil level (in this case, the temperature sharply changes between the transformer tank and inlet expander), (5) deterioration of oil quality (temperature grows up with higher $tg\,\delta$), (6) disturbed circulation of oil in inlets (porcelain insulators are locally overheated).

11.5.8 Connecting and Dividing Capacitors

Connecting and dividing capacitors of overhead (aerial) circuit breakers include porcelain insulators consisting of three packets, each packet consisting of 90 pieces. Failures in these devices are rare, and are mainly caused by oil contamination and, hence, an increase in the dissipation factor $tg\,\delta$. This causes general overheating of the insulators, which looks more severe in the areas with defective packets. When such a temperature anomaly is detected, one should: (1) compare the capacitor temperatures on each phase, (2) specify phase current loads, (3) repeat the measurement at a lower humidity, and (4) determine $tg\,\delta$.

11.5.9 Power Capacitors

Power capacitors consist of flat pressed sections assembled into stacked packets. The dielectric can be impregnated with synthetic or mineral oil. In the first case, the maximum allowable dielectric temperature must be under 95 °C, while in the second case it should be less than 65 °C.

Failure of power capacitors is common, due to degradation of the dielectric caused by internal overheating and breakdowns, loss of hermeticity, etc. An increase of $tg\,\delta$ by 15–20% increases the capacitor surface temperature by 3–5 °C. Using a capacitor of the wrong value may also cause overheating due to voltage re-distribution.

11.5.10 Valve Dischargers

Valve dischargers are widely used in outdoor electrical facilities and are manufactured for voltages from 15 to 500 kV. One design for dischargers for the voltages 110, 150 and 220 kV includes a porcelain cage with metallic flanges, an arc gap block and columns of resistors. And an arc gap block consists of a porcelain cylinder with four single arc gaps in which two electrodes and a washer are mounted.

Outside the porcelain cylinder, there are two ceramic horseshoe shaped resistors that align the voltage between the blocks coupled in series. IR testing helps to monitor the operation of shunt resistors and arc gaps, as well as to check the hermeticity of the component and the voltage distribution along discharger elements.

It is assumed that, on each discharger, the element voltage is linearly proportional to the element temperature. In general, the following thermal features of high-quality valve dischargers have been observed: (1) the upper elements located close to shunt resistors are heated uniformly on all phases, and (2) the temperature distribution across phase elements is similar for all.

The fusion of arc gaps leads to the overheating of non-shunt resistors. The break of a shunt resistor results in higher temperatures compared to other phase elements.

The following recommendations have been made (by ORGRES) for the IR thermographic inspection of valve dischargers with shunt resistors: (1) surveys at temperatures above 0 °C and high humidity should be performed no sooner than 6–8 hours after the discharger was put into operation, (2) temperature should be analyzed both by phases and on each independent phase, (3) the porcelain insulator temperature should be measured in the area of the arc gap blocks and shunt resistors, and (4) any IR radiometer used should have a temperature resolution of 0.1 °C or less, and (5) IR thermographic inspection must be complemented by the measurement of discharger conductance under operating voltage.

11.5.11 Overvoltage Limiters

A typical overvoltage limiter includes a block of varistors placed in a polymer (or silicone) shell and mounted in an insulating tube. The following failures are characteristic of these devices: (1) hermeticity deterioration, (2) moisture getting into quartz sand, (3) the displacement of some varistors, (4) varistor breakdown, and (5) the presence of leakage currents under the shell, etc.

IR thermographic evaluation overvoltage limiters is challenging due to the low amplitude of overheating. Both moist quartz sand and poor hermeticity may cause local hot spots on the insulator surface and these may be seen in an IR image. In the case of varistor breakdown, a temperature drop may be observed.

Overvoltage limiters may also be heated by a heat flux generated by transformer ventilation systems, as well as by solar radiation and other factors. This, plus the generally low amplitude of overheating, means that the inspection must be performed by experienced operators.

11.5.12 Switchgear Contact Points

Contact joints at indoor and outdoor switchyards are typical objects for IR thermographic inspection in the electrical power industry. Statistically, defects in these joints are distributed as follows: bolt joints 50%, disconnect defects 43%, compressed and stranded connections 5.1%, welded joints 1.3%, and defects in leads themselves 0.6%. Overheating may be classified in 3 or 4 grades. Some objects are screened according to the recommendations of the International Electrotechnical Commission. For example, one may find the following screening criteria for the IR thermographic inspection of joints operating at 50% of the rated load:

a temperature rise of less than 5 °C is an acceptable (non-defective) joint,
5–35 °C – joints may be repaired during scheduled maintenance,
35–85 °C – joints should be repaired in the next scheduled repair,
over 85 °C – unscheduled repair is needed within a 3 month period.

Similar to the above, in Russian standard RD 153-34.0-20.363-99, the screening of contacts, including bolted joints, is based on temperatures achieved at 50% of the rated load.

5–10 °C, it should be fixed during the next scheduled maintenance.
10–30 °C should be fixed at the earliest repair period.
over 30 °C identifies severe defects that should be repaired immediately.

The screening criteria above are similar to those used in the USA [37].

The inspection of a typical electrical substation takes a few hours for 1–2 thermographers who should be accompanied by an engineer familiar with the operation of the substation. The inspection may be facilitated by the use of a temperature alarm on the IR camera and a color palette or isotherm, that shows overheated objects as red on a black and white background. Survey results are recorded digitally or as a video, which can be augmented by voice comments. An inspection report should contain the description of detected defects, photos of the objects, and recommendations on their repair and maintenance. Many IR cameras now have built-in digital photo and video cameras that can record visual images.

In welded joints, defects may consist of air bubbles, caverns, undercuts, spills, cracks, slag and gas inclusions, wire burn and wire axis misalignment, choosing wrong heads, absence of protective coatings, etc.

Poor welding of anchor support loops in high-voltage power transmission lines may cause emergency switching. How long it takes for a defect to reach an critical status depends on the current load, wind velocity, vibration and a few other factors. IR inspection of transmission lines are typically inspected from a hovering helicopter, with an IR camera viewing through an open hatch or open window. In recent years the use of drones and quadrocopters for IR inspections has increased. Defective contact joints can be found at flight speeds up to 100 km/hr (60 mph).

It is recommended that such surveys be performed between sunset and sunrise, or, at least under a cloudy Sky because of possible solar reflections from shiny surfaces and soil.

It is believed that IR thermography is sensitive enough to detect defects that produce differential temperature signals over 5 °C.

In compressed connectors, defects appear due to a wrong choice of heads or insulator sleeves, imperfect insertion of wires into a head, insufficient compression, displacement of a steel core, etc. A compressed connector is considered good if its resistance exceeds an equivalent lead area by not more than a factor of 1.2.

Bus overheating at a hydro power station Overheated contact joint at an electrical substation

Fig. 11.37 IR thermographic diagnostics of electrical facilities. (**a**) Bus overheating at a hydro power station. (**b**) Overheated contact joint at an electrical substation

Typical defects in bolted joints are caused by an absence of washers when connecting copper wires to a flat outlet made of copper or aluminum, as well as the absence of plate-like springs and the insufficient tightening of bolts. The higher the current, the more frequently one should clean and service bolted joints. It is recommended that an IR thermographic inspection of bolt joints be performed at least once a year, as defects appear continuously, depending on the electrical load, the effect of chemical reagents, oxidation, level of tightening (torque), etc.

In twisted joints, assembly defects are common. For example, insufficient twisting and untrimmed wires can produce high contact resistance.

Two IR thermograms of defective contact joints are shown in Fig. 11.37.

11.5.13 High Frequency Blockers

High frequency blockers are typically mounted on insulator chains at 110 kV electrical substations. The objects of greatest interest to IR inspections are bolted joints, which should be checked 1–2 times a year.

11.5.14 Glass, Ceramic and Polymer Insulators

Suspended chains of plate insulators are characterized by a U shaped voltage distribution that can be distorted as a result of the breakdown of a single insulator. The highest temperature appears on the insulators that are adjacent to a phase lead. Broken insulators are at ambient temperature.. Temperature distributions are different on glass and porcelain insulators, and differential temperature signals are usually under 0.5 °C. Therefore, high sensitivity IR cameras are necessary for the reliable detection of such signals. Airborne IR surveying is possible from a low speed airplane or helicopter. The temperature changes on the surface of the upper insulators should not exceed 25–30%.

11.5.15 Power Cables

The IR thermographic inspection of power cables can be performed at various times. IR may detect areas with high $tg\,\delta$, evaluate the oil circulation in oil-filled cables, detect insulation deterioration and detect the rupture of aluminum jackets. The temperature of cable conductors is determined by using empirical values of the jacket temperature and calculating the temperature difference between the jacket and the conductor.

11.5.16 Storage Batteries

The IR evaluation of storage batteries detects significant temperature signals that appear because of the following battery defects: (1) short circuits; (2) defective soldered and welded connections, (3) non-uniform current delivery, (4) abnormal electrolyte temperature, and (5) formation of sludge.

11.5.17 Shielded Current Distributors

Current distributors with sectioned shields are mainly heated by inductive currents that are concentrated on section edges. The poor insulation of metallic elements or distributor shields may lead to the appearance of short-circuit contours with currents up to 3000 Amperes. Such powerful currents produce localized overheating and may cause power unit shutdowns.

In current distributors with continuous shields, insulation may rupture, thus causing overheating in the connection points between the shields and the generator/transformer bodies. The shield temperature should not exceed 110 °C at these points, and for safety reasons, the temperature of parts to which humans have access, should not exceed 60 °C.

11.5.18 Electrical Power Generators

Evaluating stator steel

The IR diagnostics of steel stators in turbo- and hydro-generators allows the detection of damaged areas in steel and the evaluation of their depth. Commonly, the inspection is performed with the rotor removed by energizing the magnetizing and control windings. Hydro-generators can be tested with the rotor installed. The stator surface is typically painted, so the emissivity is high; if necessary, remove dirt and rust. An IR thermogram should first be recorded before supplying voltage to the

Table 11.15 IR thermographic evaluation of stator steel. (Adapted from [33])

Hot spot "brightness"	Hot spot location	Time evolution	Description
1. Low	Gear tooth head	Quickly appears and quickly disappears	Relatively weak surface short circuit, or impact damage
2. High	Same	Quickly appears and survives for a long time	Strong surface damage Strong damage of sheets on a tooth neck
3. Low	Same	Quickly appears and quickly disappears	Relatively weak damage of sheets on tooth neck
4. Low	Same	Slowly appears and slowly disappears	Relatively weak damage of sheets in a notch
5. Low and high	Neighboring packet teeth	Quickly appears	Notch bottom damage
6. Low and high, intensified in some points	Neighboring notch teeth, frequently two parallel teeth	Quickly appears and disappears with various rate	Damage of tooth neck
7. Typically, low	Area covering several tooth heads	Slowly appears and slowly disappears	Deep areas of heating, also caused by uneven pressurization

Table 11.16 Maximal allowed differential temperature signals in stator steel. (Adapted from [38])

Steel	E41[a] (1511)	E42 (1512)	E43 (1513)	E44 (1514)	E31 (3411) Along rolling	E32 (3412) Across rolling	E330 (3413)
Maximum temperature elevation, °C	22		17		18	25	

[a]Russian national standard

magnetizing winding, then temperature distributions should be captured at 20 and 45 minutes during the stator heating and finally, in 15, 30 and 45 minutes during the cooling stage. The methodology for evaluating stator steel by IR thermographic findings is described in [38, 39]. Sequences of IR images are analyzed to look for areas of "brightness" that is indicative of overheated areas and to watch their evolution in time after switching off the thermal stimulation (see Table 11.15). Identification of defects is based on the location of overheated areas. Surface defects produce maximum temperature signals. The severity of defects is based on the difference between the maximum and minimum temperatures on the teeth, This difference should be under 10 °C. In addition, maximum temperature signals should not exceed the values given in Table 11.16.

Deep local heat sources produce weak surface signals that always meet the accepted standards for overheating, therefore, their analysis must be performed using a specific methodology such as that described in [38].

Tests for heating efficiency

Generators are tested for heating efficiency in order to detect temperature anomalies on the stator surface and to evaluate the operational quality of heat exchangers and gas cooling systems, as well as to check the bearing temperature according to the methodology described in [38, 39].

Checking the quality of soldered joints in the front sections of stator windings
IR can be used to inspect stator windings. One of the most important areas of inspection is the quality of the soldered (tin alloy) joints between the steel plates that make up a stator. The soldered joints are housed in protective boxes [39]. IR can detect poorly soldered joints and short circuits. In both cases the faulty soldered joints generate heat, the box temperature increases, and the overheated box can be detected by IR imaging.

In order to test the joints, the stator must be out of operation. A DC current of less than 50–70% of the nominal value should be applied to the stator windings with the side shields removed.

One specific box of soldered connections is typically equipped with a thermocouple or contact thermometer so it can be used as a reference. The IR thermograms of the boxes are typically accepted as the inspection results. Further evaluation would start with the boxes having the highest temperatures.

Electrical brushes
Electrical brushes can be inspected using either an IR imager or a spot radiometer equipped with a laser sight.. The inspection allows the adjustment of brush pressure, equalization of currents in parallel circuits equipped with brushes, verifiying that the fuses are the correct values, etc. The maximum allowed temperature of brush devices is 60 °C.

Thyristor excitation circuits
In these circuits, one can check the thermal status of electrical connections in conductors, thyristors, resistors, commutators etc. The following screening criteria are used: (1) the absolute temperature of contact joints which should not exceed 75 °C and the differential temperature signals should be under 35 °C, and (2) the thyristor temperatures in excitation circuits should not differ by more than 30%.

11.5.19 Electric Motors

Application areas for the IR thermographic inspection of electric motors are presented in Table 11.17.

The acceptable temperatures of stator windings are determined by the type of insulating materials (see the table in [33]). For collectors and contact rings, the excessive temperature should be under 60 °C. The temperature of friction bearings should not exceed 80 °C, while the temperature of frictionless bearings should not exceed 100 °C. In powerful electrical motors, one has to evaluate the operation of cooling systems; in this case, abnormal hot spots on the surface of the frame can be caused by coil short circuits, stator steel defects, local disturbances in the cooling air systems, etc.

Table 11.17 IR thermographic inspection of electric motors. (Adapted from [33])

Electric motor element	Detected defects
Stator frame	Detecting short circuits in stator windings and poor ventilation conditions
Collector and contact rings	Determining temperature of brushes and other components
Bearings	Determining temperature
Outlet boxes	Checking outlet connections of windings and cables
Cooling system	Checking thermal conditions and operation efficiency

11.5.20 Overhead Power Transmission Lines

The reliability of overhead power transmission lines depends on the quality of contact joints. These joints can be inspected by IR cameras, which can be mounted on flying or ground vehicles.

Breakdowns in high voltage overhead power lines can occur due to: (1) defective welding accompanied by breaks in the conductor, (2) insufficient compression of connectors, (3) conductor breaks, (4) conductor damage caused by the use of shims in the compressed joints,

Defective bolt joints are identified by excessive temperature (Russian national standard GOST 9024-90). Compressed, welded and twisted joints are evaluated either by applying a criterion of excessive temperature or the coefficient of defectiveness.

Airborne IR thermographic inspection is performed by a group of two thermographers and a technical expert who is familiar with the power line route. Typically, a helicopter will fly 10–30 m above and 30–60 m to the side of the transmission line with a speed of 60–100 km/hr. Highway power lines should be checked once a year and other lines should be checked every 3–6 years.

11.5.21 Electrical Facilities Operating at a Voltage of 0.4–10 kV

Electrical installations operating at a voltage of 0.4–10 kV typically include commutation devices, gauges, relays, power assemblies, cables and connectors. Thermographic inspection is preferably conducted through an open door or through a protective cage using the criterion of excessive or elevated temperature. In some cases, only the average temperature of the cell and temperature of the air coming out need to be recorded. When searching for a source of elevated temperature, the following factors are to be considered: (1) current load, (2) load type, (3) current distribution by phase, (4) location of a tested object, and (5) measurement method (directly or indirectly with respect to the defect point being analyzed).

It should be remembered that bus hot spots detected by thermography might be misleading due to some sources of noise, such as reflections from the sun and neighboring hot objects, inductive heating and variations in emissivity.

Electrical installations operating at a voltage of 0.4–10 kV should be thermographically inspected every 1–2 years, depending on the load.

11.6 Hot Water and Steam Generation Equipment

11.6.1 Steam Lines and Boilers

In the thermal engineering of steam systems and boilers, the requirements for thermal insulation, maximum allowed heat losses and normal outer surface temperatures are governed by various standards [40].

The insulation requirements are related not only to the allowed heat losses but also to the consumption of materials and cost. Actual heat losses might be different from those expected by design due to poor workmanship and the complex character of the surface heat exchange under varying ambient conditions. Moreover, in actual use, thermal insulation changes its properties, typically reducing its thermal resistance, and thus increasing the surface temperature and heat losses. This is particularly true in the case of high temperature and high-pressure equipment, such as boilers and piping where insulating materials experience harsh conditions. In power plants, insulation inspection and maintenance are economically important because a 1% excess heat loss requires the burning of about 300 kg of coal a year for each 1 MW of power generated. The use of standard contact heat fluxmeters is sometimes difficult due to of access difficulties, the presence of metallic structural components, high ambient temperature, etc. Therefore, applying IR thermography as a fast and remote technique can drastically reduce the scope of the inspection task.

The inspection of thermal insulation is intended to: (1) locate damaged areas and determine the size of repair needed, (2) evaluate the quality of the thermal insulation after its installation, repair or renovation, (3) analyze and document the current condition of the thermal insulation, and (4) determine the total heat losses and evaluate the operation efficiency of the primary equipment.

In all cases, inspection involves determining whether the thermal insulation meets the standard requirements for surface temperature and heat loss. Experimental inspection results are compared to design parameters or other standards to determine whether there are defective areas and, if necessary, outline a repair program.

The thermal insulation of primary and auxiliary equipment where the temperature of a heat-transfer medium exceeds 100 °C is subjected to mandatory inspection. According to Russian standards, at an ambient temperature of 25 °C, the thermal insulation temperature should not exceed 45 °C, if the temperature of the heat-transfer medium is under 500 °C, and 48 °C, if temperature of the heat-transfer medium is 500 to 600 °C. The surface temperature of thermal insulation on outdoor objects should not exceed 55 °C for metallic coverings and 60 °C in other cases.

Due to the complexity and variety of power station equipment, several classifications should be formulated.

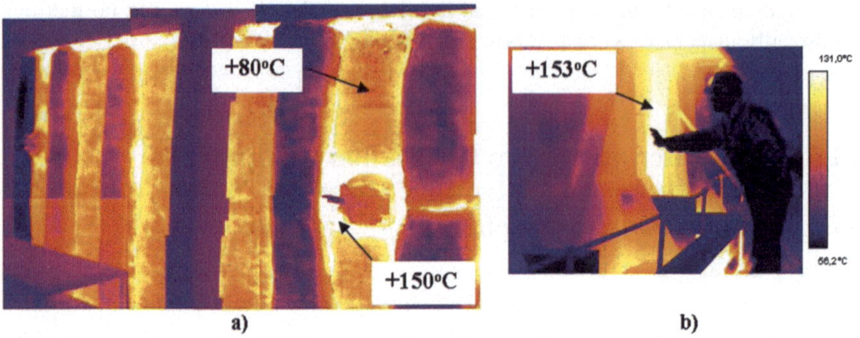

Fig. 11.38 IR thermographic inspection of the boiler brickwork envelope at a thermal power plant: (**a**) front surface, (**b**) thermal insulation collapse

In boiler houses, the following equipment is subjected to testing: water supply and vapor delivery pipes, drums, collectors, super-heaters, superheated steam pipes, condensate supply tubes, boiler brickwork envelopes, air and gas conduits, and coal-preparation plants. Figure 11.38 shows some examples of an IR thermographic inspection conducted at a Russian power plant. On the front surface of the boiler brickwork envelope (Fig. 11.38a), the mean temperature was about 80 °C, i.e. which is out of "spec" (45 or 55 °C), while around a recently repaired access hatch it reached 150 °C, even though the patch looked good visually. In a boiler section, where the air temperature was up to 60 °C, a large area of collapsed insulation with a temperature of 153 °C was found (Fig. 11.38b).

In turbine houses, IR inspection should be performed on the thermal insulation of turbines, steam and water supply pipes, heaters and condensers (deaerators), etc.

Thermal insulation can be surveyed with the same IR radiometers that are used in building applications but their temperature measurement range must be extended to 1500 °C. Also, equipment portability is more important.

Prior to performing IR diagnostics, it is important to review the thermal insulation design and related documentation. It is recommended that the following be specified: (1) equipment layout, (2) thermal insulation design for each system, (3) insulating materials used, and (4) design or normal thermal insulation temperature and heat losses. Inspection points must be chosen based on access, which includes the location of ladders and inspection galleries, etc. Visual inspection should always be performed first and should document, surface defects, such as insulation breaks, deflections etc., on the inspection plan or report. In areas of reduced insulation thickness, one should measure the temperature of the insulation temperature and ambient air temperature. It is highly recommended to match the visual and IR inspection data, so they can easily be compared. So all defects detected should be photographed, from the same point as the IR data, if possible. Many IR cameras include visual cameras for this purpose.

To compare with industry standards and requirements, the inspection results must be re-calculated using the formulas below.

Specific heat losses q (in W/m) over one meter of an isolated pipe are:

$$q = q_L L, \tag{11.22}$$

where q_L is the measured heat loss in W/m^2 from 1 m^2 of thermal insulation, and L is the thermal insulation length in meters.

Heat losses in W/m^2 should be based on an ambient temperature of 25 °C:

$$q_{25} = q_L \frac{t - 25}{t - t_a}, \tag{11.23}$$

where t is the temperature of the heat-transfer medium in °C, and t_a is the ambient temperature in °C.

The surface temperature of the thermal insulation at an ambient temperature 25 °C is:

$$t_s^{25} = \frac{q_{25}}{q_L}\left(t_s - t_a\right) + 25, \tag{11.24}$$

where t_s is the measured thermal insulation surface temperature in °C.

Thermal flux is calculated using known surface and ambient temperatures:

$$Q = 1.66\left(t_s - t_a\right) + \mathbf{A}\left[\left(t_s + 273\right)^4 - \left(t_a + 273\right)^4\right], \tag{11.25}$$

where c is a constant that takes into account the surface emissivity and the Stefan-Boltzmann constant. For painted, dirty and plastered surfaces c = 4.88 · 10^{-8} W/(m^2·K^4), and for aluminum covered structures c = 2.67 · 10^{-8} W/(m^2·K^4).

11.6.2 Boiler Surfaces

The surface of a boiler transfers heat energy to a fluid medium, either a liquid or air. Boiler tubes and collectors operate under severe conditions, so the materials must be carefully chosen to survive the operating environment. The heated surface should be inspected visually to locate surface defects, and thermographically to detect hidden defects. The visual inspection can be performed on a boiler when it's not operating. The goal is to compile an archive of visual and IR images that will be useful in subsequent inspections.

IR thermography is able to detect blocked boiler tubes by passing warm water (40–60 °C) through them. Anomalies (defects) detected visually should be marked on the surface. This will help to locate IR thermographic findings, which can then also be marked on the surface.

After performing an inspection, the possible causes for detected defects should be considered, and the severity of their effect on the efficiency of the boiler operation should be evaluated.

Visual inspection allows the detection of:

- metallurgical defects (cracks, skins, rolled-in scales, etc.),
- defects that affect thermal performance (corrosion, cinder, etc.).
- defects that occur when manufacturing, assembling or repairing a boiler (crimped tubes wall thinning, dents,, etc.),
- surface contamination (soot, sediment, etc.).

In an IR image, blocked or partially blocked tubes, appear as cooler areas compared to the warmer areas that indicate normally functioning (unblocked) tubes.

Tubes can be blocked because of: (1) formation of ferrous oxides, (2) presence of welding spits, sand, soil and foreign objects, and (3) sediment from ferric oxides and calcium compounds.

IR thermography cannot identify the sediment (laboratory analysis is needed) but it provides an overview of the general condition of a boiler, at least in the areas where direct viewing is possible. Due to low cost and efficiency of this inspection method, it is recommended that an inspection be performed whenever a boiler is taken out of operation. IR inspection provides information that can evaluate the condition of the boiler and that can be useful in planning future repairs.

11.6.3 Cooling Ponds

An IR thermographic survey of a power plant cooling pond provides the temperature distribution across the surface of the water, which reveals the pattern of the water flow in the pond, evaluates the efficiency of water cooling, and detects illegal water discharge. All these processes influence the thermal performance of a thermal power plant. For example, a relative increase of only 1 °C in the temperature of the water that is coming from the turbine condensers will cause an increase of about 25,000 tons of coal burned per year at a typical thermal power plant (about 4 millions kW power).

Airborne IR surveys are performed at two stages. First, a survey from an altitude of 200–300 meters provides a view of thermal details along the perimeter of the pond and objects along the shore. This survey shows areas of cooling water intake and discharge, as well as possible illegal water discharge. A second, IR survey from altitude above 300 meters, using a wide-angle lens provides an overall temperature distribution across the surface of the pond.

If it is not possibile to obtain a single IR thermogram of a whole pond, one should make a panoramic image that may include tens of single images that can be "stitched" together with appropriate software. Video recording is highly recommended and should include voice comments on location and anomalies observed.

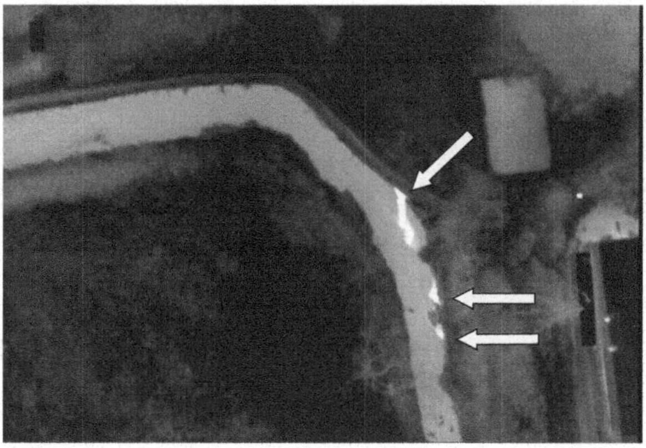

Fig. 11.39 Aerial IR monitoring of illegal wastes into a river. (Courtesy: G. Stockton, Stockton Infrared Services, USA)

When performing IR monitoring of a water surface, it is important to know that the emissivity of water is 0.96–0.98 at a viewing angle close to normal. However, at lower viewing angles, emissivity decreases and reflections may reduce the accuracy of temperature measurements.

Since IR surveying of cooling ponds is performed at long distances, factors such as atmosphere absorption, solar reflection, precipitation, fog, etc., may influence measurement accuracy. It is recommended to perform thermographic surveys on a cloudy day in the absence of wind, fog and rain.

Data analysis should provide an evaluation of water-cooling efficiency and any factors that may prevent cooling. Thermographic data can be used to modify the operation of a cooling pond. For example, decreasing the temperature of the cooling water may improve the vacuum in the turbine condenser and increase the power and efficiency of a turbine.

As an example of aerial IR monitoring used for ecological monitoring, Fig. 11.39 shows an elevated water temperature that indicates illegal waste being discharged into a cooling pond.

11.6.4 Community Heating Systems

Central underground heating systems generate hot water or steam at a central power plant and a piping system distributes the heat thoughout a community, an industrial complexes, or an entire city. These systems, are very common in cities in Russia, Scandinavia, and in some cities in Europe. In the US, such systems are still in use in parts of a few older cities, but they are common in US universities, private schools, and industrial complexes.

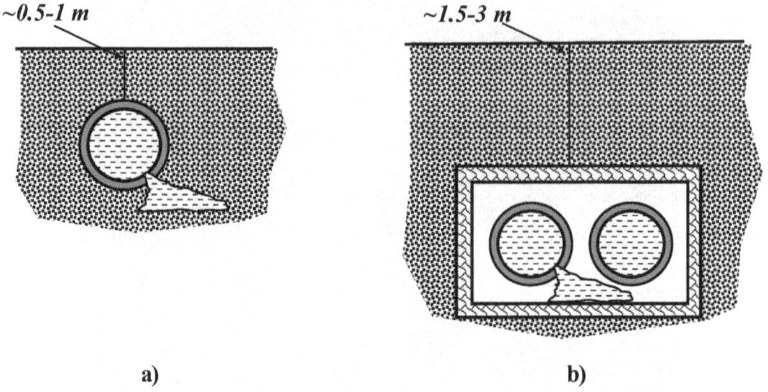

Fig. 11.40 Detecting water leakage from underground heating pipelines: (**a**) single pipe, (**b**) two pipes in chamber

These systems can be inspected thermographically for water and heat leaks by examining temperature distributions on the soil surface or, rarely, by comparing actual and calculated temperature elevations over buried pipes. Drawings of two types of pipe systems and leakage from them are shown in Fig. 11.40.

Scandinavian countries, have considerable experience with this inspection, especially Denmark and Finland, where small companies performing IR thermographic inspections of city heat networks are supported by local authorities. Surveys are typically performed at night using an IR imager mounted on a vehicle. It is worth noting that, in some countries, pipes are relatively shallow (0.5–1 m) and lack concrete enclosures (Fig. 11.40a); this makes IR thermographic inspection much easier. However, in Russia, the heating system pipes are located at a considerable depth (up to 3 m, Fig. 11.40b), and may be enclosed in a concrete enclosure. As a result, the surface temperature anomalies caused by a heat leak are greatly reduced due to heat diffusion into a great quantity of soil, and this IR leak detection process is not very effective; many leaks are not detected. Nevertheless, the use of IR thermography to compile maps of heat networks and to detect water leaks was attempted in Moscow, Saint-Petersburg and other Russian cities. Both Russia-made and foreign IR imagers were used in that research.

Note that, in Fig. 11.40b the pipes are enclosed in a concrete chamber or conduit. While this would seem to make it less likely that a leak would be detected, an ongoing leak may eventually fill the enclosure with hot water and make the leak more likely to detect.

Preparation to perform a survey of a heating network, should include studying drawings of the network, and learning about the types of insulation used, pipe routing, results of previous inspections, etc. Soil surface temperature should be measured if the temperature difference between the ambient air and the network heating water exceeds the minimum allowed value $\Delta T_{\min}$ defined by the formula:

$$\Delta T_{\min} = \Delta T_{res} \, R \frac{\alpha R_o}{1 - R_o}, \qquad (11.26)$$

where ΔT_{res} is the temperature resolution of the IR imager in °C, R is the designed thermal resistance in KW/m^2, α is the heat exchange coefficient on the surface of soil, and R_o is the relative thermal resistance in a leak site, which is defined as the ratio between the normal and designed values, but not more than 0.85. A value of h on the surface of soil is defined by the formula:

$$\alpha = \beta\sqrt{V}\left(6 + \frac{6.2}{V^2}\right), \tag{11.27}$$

where β is the constant (1.2 for soil, 1.4 for asphalt), V is the wind speed in m/s.

Soil temperature should be measured under (close to) steady state conditions and in the absence of precipitation, fog, and without direct or reflected solar radiation. The soil temperature should be measured 2–3 hours before sunrise. The soil surface must be free of vegetation, moisture, stones etc. The viewing angle should be less than 30°. The shape and amplitude of a temperature distribution over a buried pipeline depends on pipe depth, thermal resistance of pipe insulation, soil conductivity and the temperature difference between the soil and hot water. As mentioned above, Russian city heat networks are fairly deep, resulting in very weak temperature signatures on the soil surface. Moreover, thermal patterns over water leaks from a pipeline on a slope may appear far away from the leak. When evaluating thermal patterns, the existence of other heat sources in the area should be considered; these may include other pipes, heated building basements, thermal chambers, electrical power lines, pumps, etc. Inspection results should be recorded, and the thermal resistance of a reference heat network section should be calculated using experimental data or design documents.

The best way to perform an IR survey is with an airplane, helicopter, quadrocopter or drone. An IR image of a city heat network from an airplane is shown in Fig. 11.41. The site of a hot water leak is obvious (marked with an arrow) in thermal contrast with the surrounding buildings and streets.

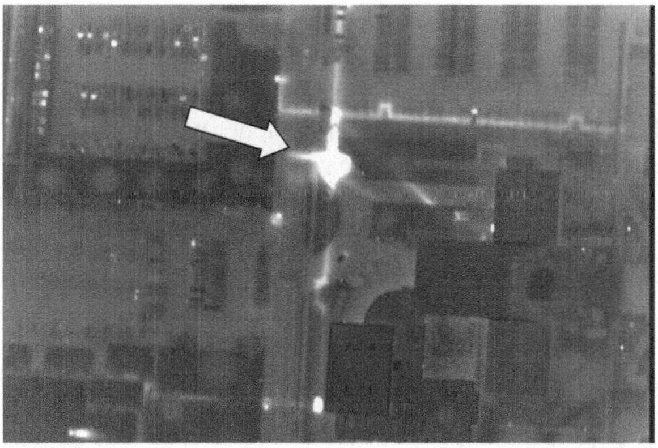

Fig. 11.41 Aerial IR monitoring of water leakages from underground pipelines. (Courtesy: G. Stockton, Stockton Infrared Services, USA)

When analyzing inspection results, it is typical to compare an experimentally measured maximum surface temperature to a "normal" value.

IR thermography is a convenient technique for detecting heat network pipe leaks. With reasonable accuracy, this technique is able to locate a leakage site, thus reducing the area of soil that must be excavated to make a repair, and this reduces the repair cost and time. In addition to IR thermography, other complemented leak detection methods may be used, including ultrasonic (Ultraprobe), devices. This improves the accuracy of the overall leak detection process.

11.6.5 Electrical Generator Turbine Vacuum Equipment

The fuel consumption of vacuum turbines depends on the level of vacuum. A routine technique for evaluating vacuum facilities involves the measurement of a pressure gradient, but while this will determine the existence of a leak, it cannot locate the leak site. The use of standard leak detectors to locate a leak in a turbine vacuum chamber is often difficult, since detector access may be impossible over some part of the chamber. Since IR is an optical method, there is no physical sensor required. IR thermography can detect air leaks into a vacuum due to a cooling effect that increases with higher equipment temperature and is proportional to the mass of the leaking air, i.e. with larger leak sites (holes, cracks, etc.) or higher vacuum.

In the USA, this method is routinely performed at the Peach Bottom nuclear power plant (PECO Energy Company). It is mainly applied to condensers that convert turbine vapor into water. The maximum efficiency of the operation of condensers is achieved at the highest possible vacuum. However, in practice, air leaks with a leak rate up to few cubic meters per minute can occur. Leaks are mainly found in: (1) turbine seals, (2) vacuum pumps (3) seals around inspection ports and manholes, (4) valves and flanges, (5) welded joints, etc. At the Peach Bottom power plant, a program for the IR thermographic inspection of the above-mentioned components has been implemented. However, it has been noted that the IR thermographic inspection can only detect (some) leaks and cannot measure leak rates or vacuum quality; thus it is qualitative rather than quantitative [41].

The temperature anomalies in some vacuum leak sites can be so significant that the leaks can be easily detected by a thermographer without much training or experience. However, there are more subtle cases, such as when inspecting flange connections, where the interpretation of data is not trivial and the leak detection process requires additional development and documentation. And such inspections would require inspectors with much more training and experience. The availability of previous IR thermograms of areas of interest can be useful when interpreting current inspection results.

Since the level of vacuum and the design of the corresponding facilities at thermal power plants are similar to those at nuclear power plants, it is expected that IR inspection would also be a viable technique at thermal power plants.

Three illustrations of detecting air in-leaks in the vacuum turbine path at one Russian thermal power station are shown in Fig. 11.42 (leaks marked with arrows).

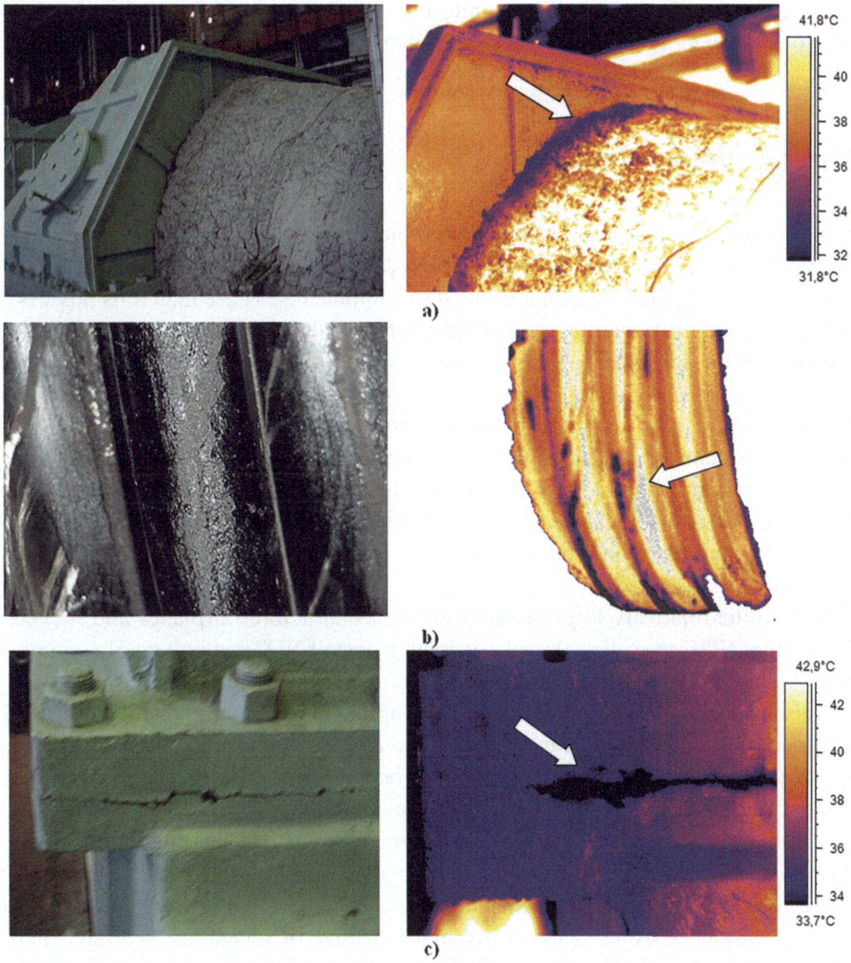

Fig. 11.42 Detecting vacuum deterioration in the turbine equipment at a thermal power plant: (**a**) joint between the turbine frame and the thermal insulation (mean temperature +41 °C, temperature decrease in defect area up to 8 °C), (**b**) welded compensator ribs (mean temperature +96 °C; temperature decrease in defect areas up to 6 °C), (**c**) air in-leak in the turbine frame

11.7 Nuclear Power Plants

At a typical nuclear power plant, a maintenance program involves up to 50 thousand test procedures, among which passive IR thermography is becoming increasingly important. In the USA, IR is used at all nuclear power plants. In Great Britain, the first applications of IR thermography at a power plant were at Heysham on, electric motors of 1.6 kW power. Subsequently this method was extended to commutators, valves and thermal insulation. In one example of the successful use of IR, a gas leak with a temperature 550 °C was detected, and this avoided a costly turbine shutdown.

IR thermography was used in monitoring the temperature of the collapsed reactor block at the Chernobyl nuclear power plant.

11.8 Aerospace Industry

In the aerospace industry, *novel materials* (composite, honeycomb, metal-non-metal structures and smart materials) are used increasingly for manufacturing frame elements, structural panels, thermal protection components. These materials possess a very good strength-to-weight ratio compared to metals. Structures for aircraft, space launch vehicles, missiles, helicopter blades, rocket motor cases and other numerous components are also made of composite materials. Their lifetime, even in harsh environments, can be fairly long, assuming they are free of defects when manufactured. It is important to realize that the defects that appear in composite materials are significantly different from the defects in metals because possible composite defects are typically located at lamina boundaries and include disbonds between composite laminates and honeycomb reinforcements, the presence of trapped water, etc.

Due to the relatively large areas of aerospace structure (airplanes and rockets) that require NDT, as well as the relatively high cost of NDT per unit area, a compromise between NDT sensitivity and inspection speed is particularly important. Because of this, the use of TNDT in aviation applications may be beneficial. The US aerospace industry is a major marketplace for TNDT systems. In the last decade, the Thermal Wave Imaging (TWI) company has become known for its successful development of TNDT systems, for example the ThermoScope and EchoTherm units (see Table 8.1). These units use the patented Synthetic Thermographic Signal Processing (STSP), or Thermographic Signal Reconstruction (TSR). The STSP (TSR) technology includes pulsed heating of a test object, along with the capture of an image sequence using a large array, high sensitive IR imager, presenting experimental data in the Log-Log format, performing polynomial data fitting, and finally, producing maps of characteristic heat transit times (see Sect. 6.10.3). Another achievement of this company is the development of the MOSAIQ software which allows the synthesis of large-scale IR images of large test objects, such as an airplane fuselage, by "stitching together" numerous separate IR thermograms.

11.8.1 Using TNDT in Aviation: The Repair Concept

The US company, NORDAM, chose TNDT as a method for inspecting aircraft panels [42]. The requirements for the inspection method included the following.

- Detect defects of different types, such as delaminations, corrosion, disbonds, impact damage, oil/water intrusion, etc.).

- Scan the surface of a part and produce an image.
- Should not require the disassembly or removal of a test object from an aircraft structure.
- Inspection data can be well documented and archived for future reference.
- Test equipment should be portable and easily set up.
- Inspection data should be evaluated on site.
- Test equipment should be capable of use by Level I operators.

In addition to traditional ultrasonic (UT) and X ray (RT) NDT, NORDAM began to use of thermal and NDT methods which, to a large extent, met the requirements above. A computerized TNDT system supplied by TWI is being used for inspecting aviation parts including:

1. CFRP laminates
2. GFRP laminates
3. Composite laminates reinforced with metallic or non-metallic honeycombs,
4. Composite panels with an embedded metallic lightning strike layer
5. Composite and metallic acoustic panels
6. Metal "bond" (Adhesively bonded) panels

A drawing of a reference sample that contains simulated defects representing those that may be found in air inlets of Boeing airplanes is shown in Fig. 11.43. The 1 × 0.6 m panel consisted of 25 mm-thick aluminum honeycombs joined to a massive aluminum plate with epoxy resin. The other surface, also bonded ot the honeycomb, is a perforated GFRP panel; with attached to a fine stainless steel grid that serves as an acoustic screen. This structure is designed to reduce noise from the jet engine. Artificial defects were placed in the two planes: first, between the honeycombs and GFRP, and second, between GFRP and the acoustic screen. The

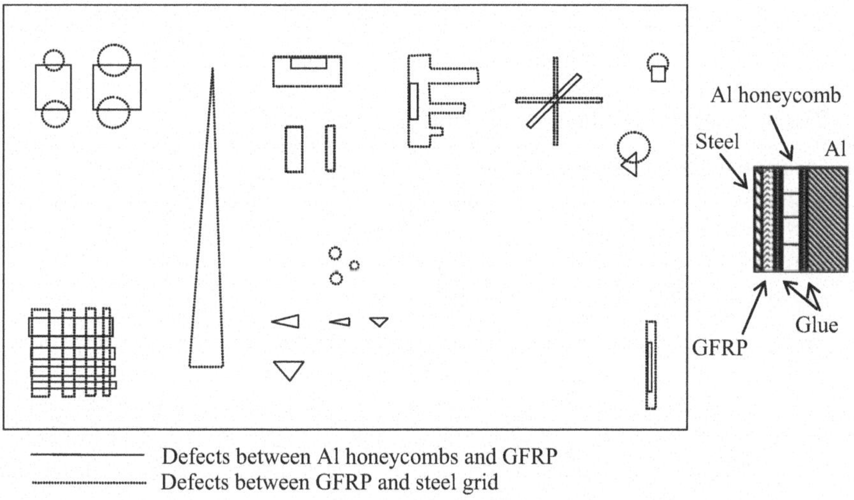

Fig. 11.43 Simulating defects in a Boeing aircraft air inlet (Adapted from [42])

simulated defects were made of an acrylic film and the adhesive was also removed on the surfaces of the film. Teflon and Grafoil inserts, which are typically used for ultrasonic (UT) testing, were not used in this case because of the low thermal resistance of these materials.

It is worth mentioning that one of the essential advantages of TNDT, compared to ultrasonics, is a high scan rate. For example, it was reported that the IR thermographic inspection of a Boeing 737 rudder whose area is 6.5 m^2, was performed in 3 hours and resulted in the detection of a defect that was caused by a lightning strike [42]. Note that an ultrasonic inspection of this part would have taken about 7 hours.

11.8.2 Detecting Water Ingress in Aviation Honeycomb Panels

Inspection speed is considered to be one of the most important characteristics of an inspection technique since the area of aerospace structures may consist of thousands of square meters. In the opinion of many NDT experts, TNDT may be less sensitive in detecting defects, but much faster in comparison to ultrasonic methods. Therefore, TNDT may serve as a good addition to UT methods, and in some cases it may be superior to UT. A good example of this is the use of the combination of passive TNDT and ultrasonic inspection for water ingress in aviation honeycomb panels.

Reinforced composite panels have two facesheets (skins) and a honeycomb layer between them. Numerous materials are used for the skins, including aluminum, GFRP, CFRP, etc. And honeycombs are generally metallic, usually aluminum, or non-metallic. In the Russian Ilyushin-86, 96 and some other airplanes, aluminum honeycomb panels are widely used. The aluminum skin thickness varies from 0.5 to 1 mm and cells are made of a 0.1 mm-thick aluminum foil. In the newer generation of airplanes, composite honeycombs are used.

In the space industry, honeycomb structures are widely used including facesheets made of GFRP or CFRP composites and honeycombs made of aluminum, Nomex, or other non-metallic materials.

The main defects in a honeycomb structure are:

1. Delamination in a facesheet (skin)
2. disbonds between the facesheet (skin) and the honeycomb. These are referred to as "skin-to-core" disbonds.
3. deformation of honeycomb cells such as crushed or compressed core.
4. presence of water in honeycomb cells

Water can get trapped in a honeycomb panel or in other cavities in an airplane as a result of condensation and/or direct water ingress through loose connections in wings, fuselage, keel and other airplane structures. This may occur from changes in temperature and atmospheric pressure during normal flight operations, or from rain, either while sitting outside on the ground, or when flying at high speed through a rainstorm. .

The standard techniques for detecting trapped water in aircraft parts are X-ray and ultrasonic.

X-ray is very efficient if access to both surfaces of a panel is possible. Water is detected due to the high radiation contrast between water and aluminum. A one-sided X-ray procedure is also possible, but this is rather expensive. However, even in this case the necessity of inspecting thousands of square meters by X ray film would be time-consuming. Plus, X-ray safety restrictions make this method undesirable to civilian airline companies.

Ultrasonic (UT) is quite sensitive for detecting water as long as the water is in contact with a skin, And in some cases, UT can evaluate the water height. The sensitivity of such a technique has been reported (in Russia) to be about 2 mm. The main disadvantage of ultrasonics for detecting trapped water, is that contact with the part is necessary and the inspection rate is slow. In fact, ultrasonic inspection is usually done at single a few centimeters apart. Spatial resolution is low, so relatively small quantities of water, such as the amount that may fill only few honeycomb cells, could easily be missed. Finally, the ultrasonic method is difficult to use on vertical surfaces, e.g. in the inspection of aircraft rudders. It is worth noting that there are a few companies in the world that are developing non-contact laser ultrasonic systems (laser excitation and detecting) but the equipment is very expensive (up to 1 million US dollars), and the equipment must still be relatively close to the part.

IR procedures for detecting water trapped in airplane panels have been routinely used by aircraft manufacturers such as Airbus and Boeing. Both companies have implemented thermally stimulated (active) TNDT of honeycomb aircraft panels either in a hangar or outside on pavement.

Trapped water can be detected passively (no heat applied) for a certain period of time after a plane lands, due to the large difference between the ground air temperature (from +20 to −20 °C) and the air temperature at cruising (10,000 meters) altitude (as low as −60 °C). Frozen water (ice) maintains a low temperature for a relatively long time and the corresponding temperature patterns are quite conclusive. An example of the thermograms obtained from the inspection of Toupolev-204 and Ilyushin-96 airplanes are presented in Figs. 11.44 and 11.45 (collaboration between Pergam Engineering and Innovation, Russia). It is interesting to compare TNDT images with UT findings which were obtained a few months before the thermographic survey. In some cases, the two tests produced very similar images of the trapped water (Fig. 11.44b), while in other cases the shape and location of areas of water were quite different, presumably because of water migration during continued flight use. Including the ground inspection of areas of a difficult access such as rudders and elevators (Fig. 11.44c), the total time of the IR inspection was about 1 hour The results presented in Fig. 11.45 demonstrate that passive TNDT was able to detect water trapped not in composite cavities, and also in aluminum honeycombs.

Airbus Industries uses an active TNDT procedure which implents a "heating blanket" applied to a test surface and heated by an electric current [43].

An active TNDT procedure for detecting trapped water in honeycomb aviation panels is under development in Russia [44]. The system shown in Fig. 11.46a uses an optical heat source consisting of 8 halogen lamps with a total power of 3 kW. An IR imager, such as the ThermaCAM 640, views an aircraft panel. In the test procedure

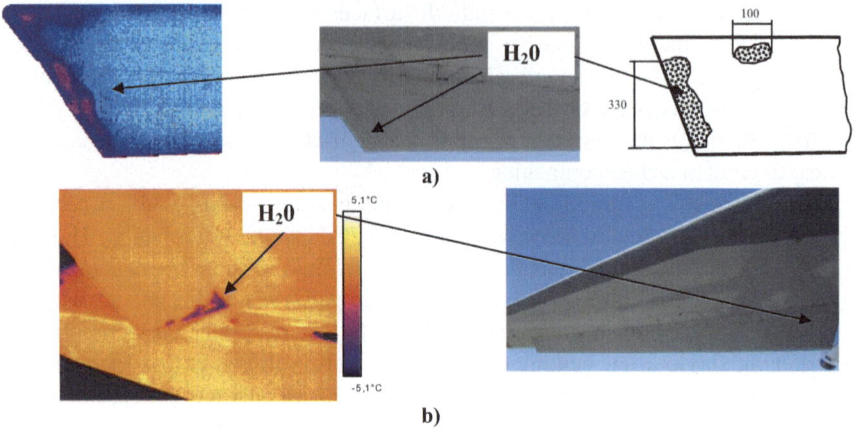

Fig. 11.44 Passive IR testing to find water trapped in honeycomb composite structures of a Russian Toupolev-204 airplane (1 hour after landing): (**a**) aileron (right: ultrasonic map of defects 2 months prior to the IR testing), (**b**) elevator

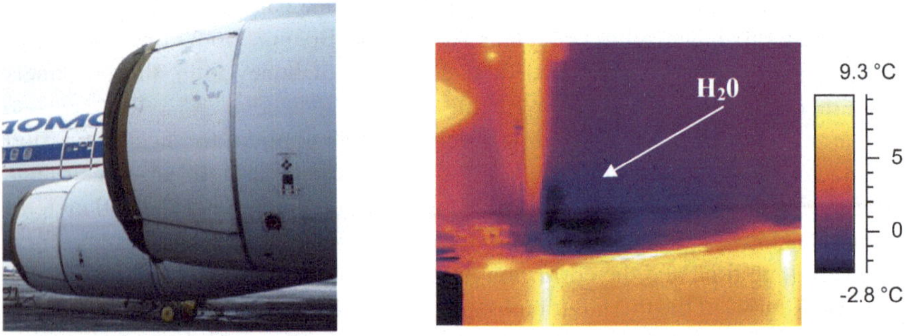

Fig. 11.45 Passive IR thermographic diagnostics of water ingress in the honeycomb aluminum air inlet of an Ilyushin-96 airplane (1 hr after landing)

a series of IR images of 0.5 × 0.5 m area are captured and they are pieced together to produce a panoramic image. While this active method is effective on painted panels, even in this case, the reflected radiation can be a problem. Areas of honeycomb panels that contain more than 3–6 g of water are seen as cold areas (Fig. 11.46b–f). The best results are obtained when inspecting composite skins, e.g. made of GFRP or CFRP reinforced with metallic honeycombs. The inspection of reinforced composite panels containing steel honeycomb is very effective due to the relatively low thermal conductivity of steel (Fig. 11.46d–f). In all the above-mentioned panels, temperature signals persist for a long time and areas of trapped water are easily observed, But in the inspection of aluminum honeycombs and aluminum skins, the temperature differentials survive for only 2–6 seconds (Fig. 11.46b, c). As predicted by TNDT theory, panels made of GFRP skins reinforced with aluminum honeycomb are optimal test objects, as the differentials survive for a relatively long time. And in this case, depending on skin thickness, it is possible to visualize the honeycomb

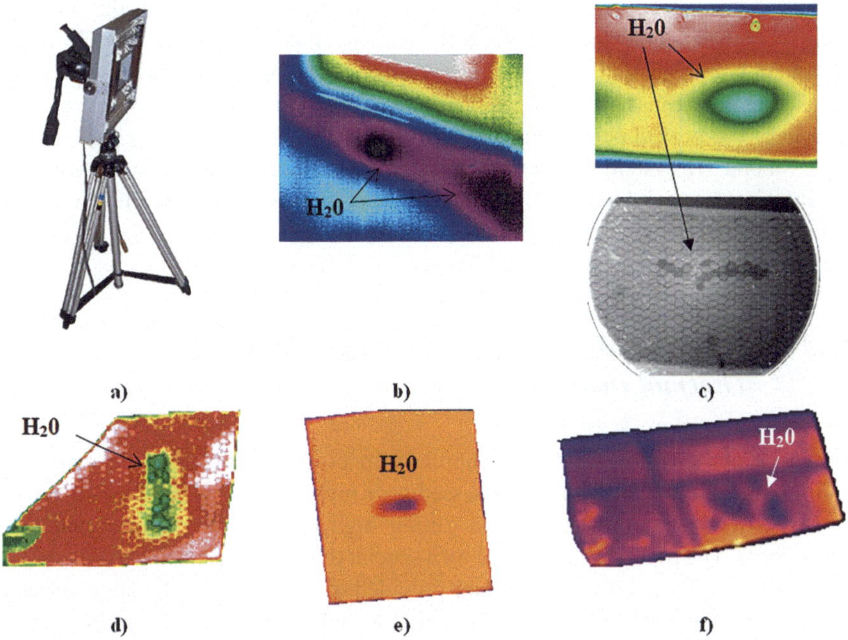

Fig. 11.46 Active TNDT of water ingress in aviation honeycomb panels: (**a**) optical heater combined with IR camera, (**b**) IR thermogram of aluminum panel with two water sites, (**c**) IR thermogram and X ray image of aluminum panel, (**d**) IR thermogram of "GFRP facesheet-aluminum core" panel, (**e**) IR thermogram of "CFRP facesheet-Nomex core" panel, (**f**) IR thermogram of steel honeycomb panel

cells and, hence, to detect not only trapped water but also disbonds between the individual honeycomb cell walls and the skin (Fig. 11.46d).

It is worth noting that trapped water, due to its high heat capacity, is generally considered to be easily detected by active (heated) TNDT. The test procedure is similar regardless of skin material. First, honeycomb panels are heated for a few seconds; afterwards, one records the surface temperature distribution. In the case of aluminum skins on aluminum honeycomb panels, advanced data treatment may be necessary, while, in the case of non-metallic honeycombs, capturing a single IR thermogram at the optimum observation time produces good results.

A drawback in applying TNDT (and UT) to the detection of trapped water is the necessity of having good contact between the water and the skin. Nevertheless, compared to the UT method, IR thermography is much less sensitive to the mass of water trapped in the cells. Therefore, a quantitative estimate is possible by using only the lateral size of defect areas and assuming that water fully fills the honeycomb cells.

As it was frequently mentioned above, optical heating sometimes causes reflected radiation. This can be reduced by using a shutter, which cuts off the radiation from the optical heater while the IR image "develops". Composite honeycombs can be effectively heated with less powerful heaters, such as air blowers (air guns, "hair dryers", or fans). Microwave heating is of a great interest as it enables the selective heating of water hidden under a skin while leaving other materials "cold". However, this can only be used with non-metallic materials. And commercial microwave heaters with horn antennas are not available. Plus, the use of microwave devices requires precautions to protect personnel.

A robotic IR system for detecting water in aviation panels is being used by Boeing Commercial Aircraft, USA. This system includes automated heating, data acquisition and processing.

11.8.3 Detecting Water Ingress in Space Shuttle Thermal Insulation

The lower surface of the US Space Shuttles were covered with black tiles made of a foam-like ceramic material that possessed very high thermal resistance and served as very effective thermal protection during high temperatures re-entry into the atmosphere. The tiles were protected by a special thin outer layer that prevented the penetration of water into the tiles and provided a smoother aerodynamic surface. However, this protection layer was quite fragile and cracks might occur during use, allowing water to penetrate the tiles and making them significantly heavier than dry tiles. The weight gain in the airframe could cause a reduction in the payload. And more importantly, the possible degradation or failure of the insulation caused by frozen or boiling water could be a serious problem.

The mechanism of water detection in thermal protection tiles is related to both the transient heat transfer processes that occur during landing and the natural water evaporation through coating cracks. In the USA, at the end of the twentieth century, a program called SILTS (Shuttle Infrared Leeside Temperature Sensing) involved the deployment of IR cameras on an airplane to record IR images of the Space Shuttle at high altitude, just after re-entry into the atmosphere. It has been assumed that moistened tiles can be detected during landing. Similar research was done in the former USSR for the Buran space shuttle program (Fig. 11.47). Russian

Fig. 11.47 Detecting water in Buran space shuttle thermal insulation due to natural evaporation (obtained with a TV-03 IR imager)

specialists approached the problem of quantitative water evaluation in both the passive (aerodynamic heating) and active (applying external heating) modes. Unfortunately, this research ended along with the Buran program.

11.8.4 Turbine Blades

Gas turbines are widely used in power generation and propulsion systems. The efficiency of a turbine depends mainly on the performance of turbine blades, which must withstand high temperatures in use. IR thermography has been used successfully for evaluating blade coatings and cooling channels. Research has been also performed on the detection of surface and subsurface fatigue cracks. In addition to laboratory research in this area, several companies have offered commercial turbine blade testing systems. For example, Thermosensorik, Germany, has developed a robotic pulsed TNDT system for the automated inspection and characterization of turbine blades (see Table 2.1). Siemens has also done a lot of development work in this area.

Inspecting turbine blade cooling channels

To minimize fuel consumption and operate efficiently, aviation turbines must operate at the highest possible temperature, pressure and rotational speed (RPM). Such extreme operating conditions impose stringent requirements on the quality of turbine blades.

The turbine blade manufacturing process is very complicated, and manufacturing defects include cooling channel blockage, displacement and size drift.

In industry, ultrasonic, X ray and neutron radiography NDT methods are used in the inspection of turbine blades. The ultrasonic technique is not very effective and is accompanied by noisy reflections from elements of the internal blade structure. This technique is also believed to be inappropriate for inspecting nickel alloys which are typically used in the production of turbine blades. The X ray technique produces results that are difficult to interpret, also due to internal structural elements, and only the method of computer tomography (CT) provides acceptable images of the blade cross-section and can evaluate the blade wall thickness with an accuracy of about 0.1 mm (0,004 inch). Unfortunately, CT equipment is expensive.

Maldague et al. demonstrated that a blockage in the inner cooling channels of a turbine blade can be effectively detected by observing anomalies in the surface temperature distribution when passing hot or cold air, or a liquid, through the channels [45].

In Russia, a similar test procedure was developed at Research Institute of Aviation Materials (VIAM) [46]. The main component of the experimental setup was a test chamber made of a vitreous material that is transparent in the IR spectral band. The chamber included a profiled frame that simulated inter-blade space, a blade mounting fixture, heat sources, and a PC.

With this system, over 10 types of cooled aviation turbine blades (D18, AL31F, soldered and Pratt & Whitney engine blades) have been tested. It has been concluded that TNDT is able to evaluate variations in the channel inner diameter with an accuracy of about 10%, as well as to detect ceramic debris in the channels. Blockage of a single channel caused the surface temperature signal to increase by about 1.6 °C at 50 °C).

A specialized TNDT system for inspecting turbine blades was also developed at the Salut enterprise, Russia, using a Russian IRTIS IR camera.

Depending on the blade construction and materials, the identification of blocked channels may require advanced data analysis, an example of which is shown in Fig. 11.48. While the difference between defective and non-defective blades is not visually apparent in the basic images, the use of the Fourier transform allowed the detection of a blocked channel.

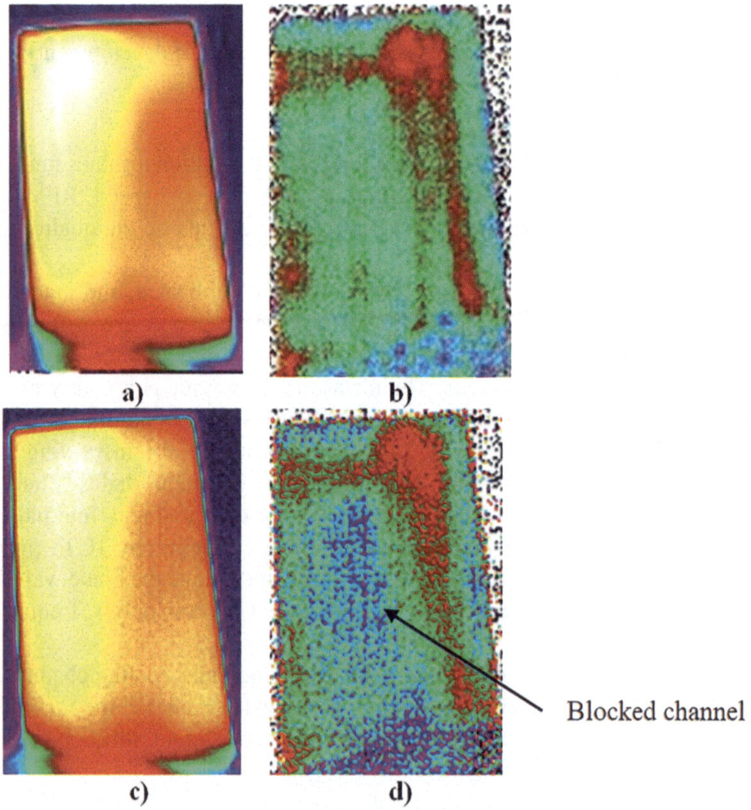

Fig. 11.48 TNDT of turbine blades heated with hot air: (**a**) IR thermogram of non-defective blade, (**b**) phasegram of non-defective blade, (**c**) IR thermogram of defective blade, (**d**) phasegram of defective blade

Inspecting turbine blade thermal coatings

Plasma-sprayed coatings are widely used in industry to protect critical structures against aggressive environments. A common feature of all thermal spray coatings is the lenticular or lamellar grain structure that results from the rapid solidification of small globules, which become flattened when they strike a cold surface at high speed. Bonding mechanisms at the coating/substrate interface and between the particles that make up the coating are still subject to speculation. Due to imperfect spraying technology, a variety of structural defects might appear either within the coatings or at the interface. Defects include delaminations between subsequent deposition layers, pores, non-melted particles and cracks. The ability to evaluate coating parameters and defects would be of benefit, for example, in the power generation industry where thermal barrier coatings are used to protect metals in gas turbine and diesel engine components. Coatings are typically made of ceramic materials and they may be overlayed with metallic layers. The blades of aircraft gas turbines are protected against oxidation and/or corrosion with platinum aluminides, gamma titanium aluminides, and MoCrAlY coatings.

Due to such factors as high sensitivity to defects in the relatively thin turbine blade coatings, and high test speed, active IR thermography is considered to be one of the best inspection techniques. Numerous publications describe where both pulsed and lock-in TNDT procedures have been used for this application [47–58].

The main benefits of using TNDT for this application are as follows.

- Thin coatings on substrates require high speed TNDT equipment for as a defect causes a thermal event that exists only for short observation time (microsecond to millisecond and millisecond to second range, respectively) depending on coating thickness and thermal properties. Very thin coatings are typically tested by using periodic thermal waves of high frequency (up to tens of kHz) or laser heat pulses. Thicker coatings allow either the use of longer thermal waves or, more often, pulsed TNDT.
- All types of coating structural imperfections affect both the temperature amplitudes and time evolution. The most reliable data is supplied by phase- (periodic thermal waves) or time-domain (pulsed thermal waves) analysis.
- Coating thickness variations and disbonds between the coating and the substrate create the strongest signal, while variations in coating thermal properties can decrease the reliability of the test data. A deeper mathematical analysis of the multilayer heat conduction problem shows that various phenomena related to material depth should be investigated by using thermal phase shifts or thermal transit times. However, the thermal resistance of interfacial discontinuities is more strongly related to the signal amplitude than to heat transit times.
- Since coatings typically cannot be painted black, emissivity/absorptivity fluctuations may pose a serious problem in the interpretation of test results. Uneven and residual heating can also be problems, but these are typically solved by applying data normalization.

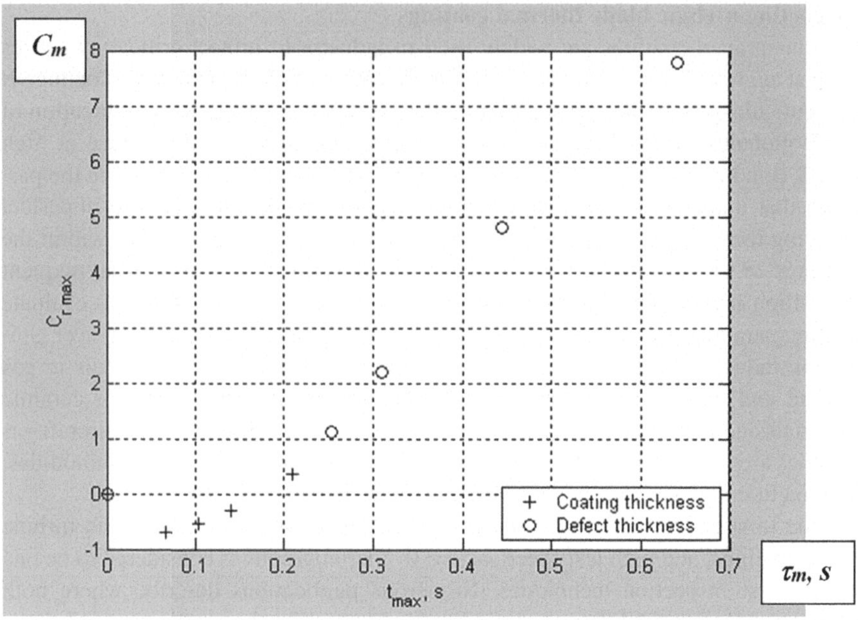

Fig. 11.49 Maximum running contrasts and their optimum observation times in the coating on the steel AISI 304 substrate (pulsed heating)

Coating defects, such as disbonds and thickness variations, can be easily modeled. In Fig. 11.49, the distribution of the maximum running contrast C_m (abscissa axis) and the time of its appearance τ_m (coordinate axis) are presented for a typical turbine blade coating used in the power generation industry. It is easily seen that the optimum conditions for detecting coating thickness variation takes place at about 0.03 second, and disbonds can be observed for 0.7 second and, in the latter case, the maximum signal may reach several tens of percent. It is also worth noting that C_m is always positive in the case of delaminations; in the case of thickness variation, the sign of τ_m depends on a reference thickness value. Thus, the thinning of a coating causes a negative value of C_m because the substrate material has a higher thermal conductivity.

The data in Fig. 11.49 has been approximated by the following inversion formulas:

$$d = 1.57024 C_m^{0.833941} \tau_m^{0.673748} l_c^{-0.361122} \tag{11.28}$$

$$l_c^{incr} / l_c = 359.603 C_m^{0.042124} \tau_m^{1.0322} l_c^{-2.03345}$$

Here d is the delamination thickness in μm, C_m is the maximum running contrast, τ_m is the time of C_m appearance in ms, l_c is the standard (reference) coating thickness in μm, l_c^{incr} is the increased coating thickness. Note that the second formula in

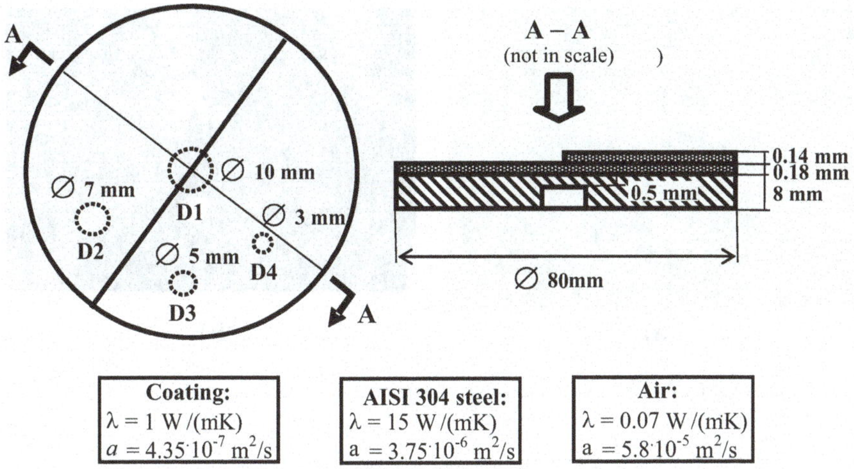

Fig. 11.50 Reference sample scheme (turbine blade coating, courtesy E. Grinzato)

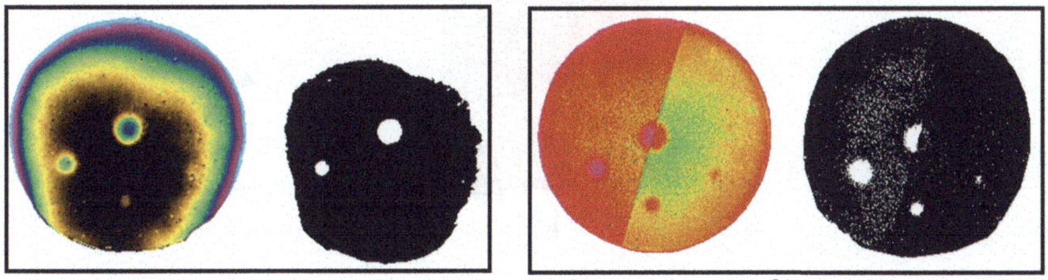

Fig. 11.51 Active TNDT of the turbine blade reference sample from Fig. 5.45. (Courtesy E. Grinzato): (**a**) raw IR thermogram at optimum observation time and map of defects ($S = 0.85$, $P_{c.d.} = 78\%$, $P_{f.a.} = 30\%$), (**b**) phasegram and map of defects ($S = 2.59$, $P_{c.d.} = 63\%$, $P_{f.a.} = 5\%$)

Eq. (11.28) assumes that a non-defect (reference) point always corresponds to a lower coating temperature, and the ratio l_c^{incr} / l_c is greater than one. Thus, both delaminations and thickness increases result in positive differential signals.

The estimated accuracy from Eq. (11.28) is from 5% to 30%.

TNDT tests were performed at CESI, Italy on a disk shaped reference sample that included a ceramic thermal coating on a stainless steel substrate There were two areas of different coating thickness (0.18 and 0.32 mm) and four delaminations of varying diameter (Fig. 11.50). The sample was heated with flash lamps which provided an energy pulse of 1500 J/m^2 in a 10 ms pulse. The recorded sequences consisted of up to 300 IR thermograms captured with a 20 ms interval. The results of these tests are shown in Figs. 11.50 and 11.51. The optimal image of the defects appeared 0.9 second after heating (Fig. 11.51a). However, the signal-to-noise ratio was less than one and the binary map of defects revealed only two defects with a high risk of a false positive. All four defects, as a well as areas of different coating

Fig. 11.52 TNDT of a 150 μm-thick turbine blade coating: a-raw image, b-map of defects ($P_{f.a.} = 5\%$, $P_{c.d.} = 96\%$)

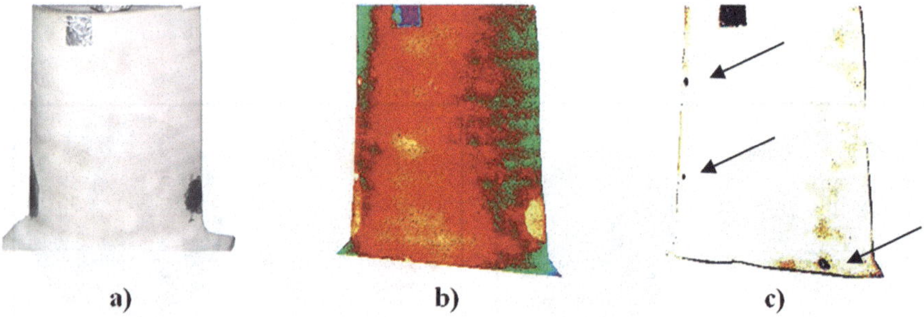

Fig. 11.53 TNDT a of turbine blade. (**a**) turbine blade photo, (**b**) optimal raw image, (**c**) correlogram. (Courtesy CESI, Italy)

thickness, became clearly visible after a Fourier transform was applied to the time evolution of the signals (see the phasegram in Fig. 11.51b). The corresponding map of defects shows all defects with a 5% risk.

Another example of TNDT of a reference sample is given in Fig. 11.52 where both the source image and the map of defects show two defect indications that have been identified as disbonds by analyzing the corresponding signals by Eqs. (11.28). Note the "salt and pepper" texture in the image.

The TNDT of a real turbine blade is illustrated in Fig. 11.53. Following the conclusions obtained on the reference sample, in this case the inspection results have been treated only by applying the Fourier transform and correlation techniques. Three small defects have been identified, as shown by arrows in Fig. 11.53, *c* (the square area on the top left corresponds to a piece of aluminum foil which was attached to the blade surface to allow the analysis of reflection phenomena).

Surface and subsurface cracks in turbine blades

The investigation of accidents and catastrophes in aviation sometimes requires the analysis of turbine blades made of heat-resistant nickel alloys. Surface cracks in metal structures are generally detected by applying liquid "dye" penetrants, but these may be useless in the case of subsurface cracks or surface cracks filled with mud. In the former USSR, Vavilov and Kasatkin analyzed the potential of active TNDT for the detection of both surface and subsurface cracks [59]. The theory of this method was described in Sect. 4.8.9. In the process of 'side' heating, the isotherm which propagates along the blade edge experiences distortion over a crack. In the case of surface cracks, the distortion is relatively strong and reliably located by IR cameras if a tested blade is placed into turbulent boiling water. The successful detection of subsurface cracks depends on their depth and an applied algorithm of advanced data treatment, such as the analysis of derivatives.

11.8.5 Detecting and Characterizing Corrosion in Metallic Aviation Structures

Aluminum panels

Aging aircraft may experience corrosion in aluminum panels whose thickness is from 0.8 to 2 mm (0.03 to 0.08 inch). The corrosion typically starts around rivets. The trend in the development of NDT techniques is to replace single point test procedures with those that provide images of relatively large areas. Some time ago, the US Federal Aviation Agency (FAA) performed a comparison study of NDT using a variety of equipment and methods, including TNDT [60]. As a result of this comparison study, it was recognized that the potential of TNDT has not yet been fully realized and deserves further investigation.

A typical scheme of applying TNDT in aviation is shown in Fig. 11.54a. A test area can be from 0.2 to 0.5 m^2 depending on the heater. Commercial TNDT equipment is manufactured by a few companies and typically includes a high-speed IR imager, a heater using either pulsed xenon or continuous halogen lamps, and a computerized data processing system. Some TNDT systems use robots, which perform the inspection automatically. The scientific principles related to such equipment are

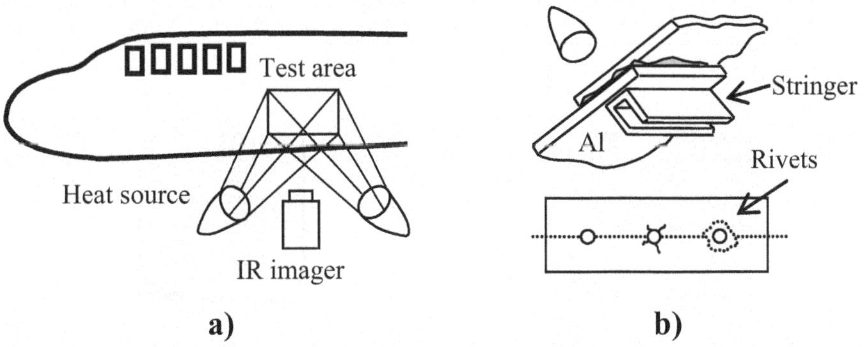

Fig. 11.54 TNDT of corrosion in aircraft panels. (**a**) test scheme, (**b**) testing rivets. (Adapted from [61])

in the software. Some developers prefer using a 1D Fourier transform (Thermosensorik), while Thermal Wave Imaging, uses thermographic signal reconstruction (TSR). Painted aviation panels do not typically require any additional treatment, while shiny aluminum surfaces have low emissivity (high reflectivity), so they are not efficiently heated and they produce very noisy reflections. Therefore, such panels are often coated with a black water-washable coating to increase the emissivity.

A typical section of an aviation panel is shown in Fig. 11.54b. Two lap-jointed aluminum sheets are attached to a stringer (rib) by epoxy resin and rivets (in Russian airplanes, sealants that prevent water ingress are used instead of resin). In use, the rivets become loose and around the rivets, separations in the epoxy bondline (interface) and subsurface cracks in the aluminum appear. Water from the atmosphere penetrates these joints, and causes corrosion in both aluminum sheets, starting around the rivets.

To perform flash TNDT on thin aluminum panels it is necessary to provide powerful flash heating and to capture thermal data at high frequency, because the optimum observation time of differential temperature signals is from a few tens to a few hundred milliseconds. The heating should be sufficiently powerful to produce strong enough thermal signals but not enough to damage the paint. Figure 11.55 illustrates TNDT of two aluminum sheets of thickness 1 and 2 mm containing four areas of manufactured corrosion. The heating was accomplished by 4 flash tubes with a total power of 20 kJ. The flash duration was only 1 ms, and the power density absorbed by the aluminum plates was estimated at 11.3 kJ/m^2, which was a loss of about 37% from the power provided by the heater. The values of temperature signals and contrasts reported in Fig. 11.55 were calculated for a 15 mm-diameter sound area chosen at the centre of the samples. The temperature over the corrosion sites was averaged over 20 mm diameter areas that covered visible defect indications. The noise standard deviation in the sound area was σ_n= 0.17 °C (C_n = 0.035) in the case of the 1 mm thick sample, and σ_n= 0.15 °C (C_n = 0.06) in the case of the 2 mm thick sample. The main detection problem in the experiment described above was the low level of temperature signals, ΔT_m.

The results in Fig. 11.55 validate the inversion formula (3.9), which enables the characterization of corrosion by experimental values of the maximum running contrast C_m. The best results are obtained for the largest defect in the 2 mm-thick plate (see Fig. 11.55). For the observed contrast C_m = 1.2, according to Eq. (3.9), we

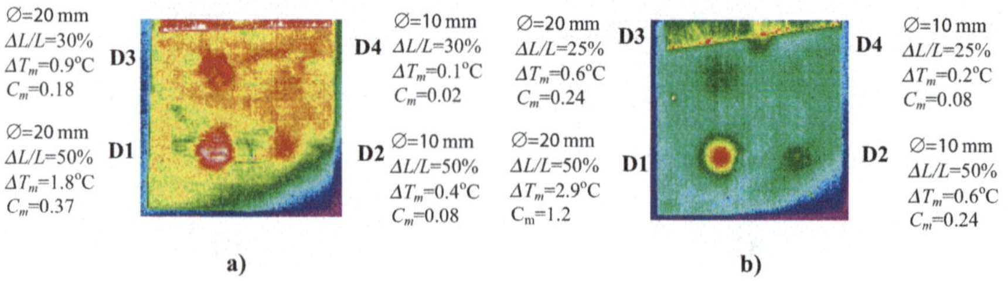

Fig. 11.55 TNDT of corrosion in aluminum: (**a**) aluminum thickness 1 mm (raw image at 150 ms), (**b**) aluminum thickness 2 mm (raw image at 300 ms)

obtain $\Delta L/L$~55%, which is close to the true value of 50%. Other areas tested produced a significant data spread, probably due to incorrect values of ΔT_m, that were caused by a low acquisition frequency, which was 6 Hz in this case. The test procedure allowed the detection of material loss from 25 to 30% in areas of about 20 mm (aluminum panels of thickness 1–2 mm). Corrosion detection limits can be improved by increasing the absorbed energy and increasing the acquisition frequency. However, we believe that the estimated detection limit of about 5–10% material loss, reported in [61], is overly optimistic for a practical inspection on an airplane in service. In laboratory conditions, NASA achieved a 2.5% detection level of material loss in a 1 mm-thick aluminum panel by using a 60 Hz IR imager [62, 63].

The same procedures as described above enabled the detection of disbonds in adhesively bonded aluminum panels.

Exfoliation corrosion in aircraft wing skins

Many aircraft (in the USA: E-3, B-52, C/KC-135) use countersunk steel fasteners to attach aluminum wing skins to aluminum spars and underlying structure as shown in Fig. 11.56. The fasteners are plated to minimize the electrochemical effects of dissimilar metal contact between the steel and aluminum. Over time, similar to the situation in Fig. 11.55, the plating wears away and exfoliation corrosion can develop (Fig. 11.56a). The most common method of detecting this kind of corrosion is visual inspection, which is able to find bulging or pillowing of the skin caused by a buildup of corrosion products. Since there are thousands of steel fasteners per aircraft, visual inspection is time-consuming and prone to false negative results, so the demand for a faster and more effective NDT method is high.

Experimental research was performed by ARINC, USA [64], to evaluate four different NDT methods. A commercial (TWI) EchoTherm TNDT system including a SW IR camera Amber Radiance HS (image format 256 × 256, frame rate 120 Hz) was used for temperature monitoring. Heating was provided by two linear flash lamps with an energy 6.4 kJ each. The tested area was 25 × 25 cm (10 × 10 inches),

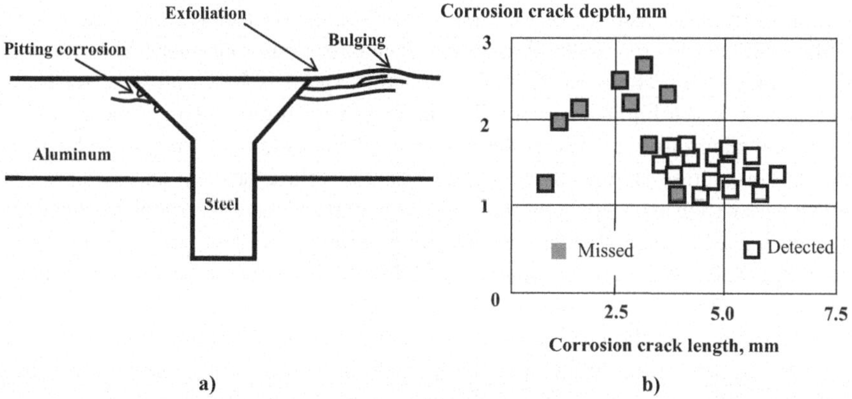

Fig. 11.56 TNDT of exfoliation corrosion in aircraft wing skin (**a**) corrosion diagram, (**b**) Data on detected and undetected (missed) cracks. (Adapted from [64])

which enabled the inspection of 18–24 steel fasteners at one time. Since the exterior paint was removed from the aircraft in the repair process, prior to performing TNDT, the test areas were painted with a black water washable tempera coating. Boundaries of the tested area were marked with chalk which was also visible in IR thermograms. It was found that, due to the high conductivity of aluminum, the "useful" information occurs within a 0.4 second time interval during which up to 50 IR images were captured. The data processing was performed in two ways: (1) by simply viewing source images at optimum observation times, and (2) looking at the change in slope of pixel values as a function of time, i.e. analyzing the first derivative of the temperature. Several samples that contained both corroded and non-corroded fastener holes were tested by TNDT. The extent of corrosion was confirmed by destructive metallographical analysis.

For the first 0.1 second after heating, corroded areas exhibited elevated temperature and round 'warm' spots. Optimum defect detection occurred at 0.058 second. Without advanced data treatment, the probability of correct detection was 73.5% and the probability of a false alarm was 1.5%. By the introduction of the analysis of temperature derivatives, these figures were improved to 81.6% and 1% respectively. The detection limit in this case was related to the length and depth of corrosion cracks. Cracks longer than 3 mm were reliably detected at depths less than 2 mm (see Fig. 11.56b).

11.8.6 Composite Materials

In the aerospace industry, composites such as carbon and glass fiber reinforced plastics, carbon-carbon composites and others, are widely used as primary materials. And composite laminates are used as facesheets in sandwich structures reinforced with non-metallic (Nomex or other) honeycomb, metallic (usually aluminum) honeycomb, plastic foam, balsa, and other core materials.

Due to the wide variety of composite materials and products, no single NDT method will be effective with all materials. The most commonly used NDT method is ultrasonic (UT). However, the UT inspection speed is low, and it is generally necessary to use a liquid couplant. The latter is very important in some cases, as composite materials may absorb water during UT inspection and they must be dried after the inspection. This is usually done in an autoclave or oven and this process can be very expensive, especially for very large parts. It has been reported that even after drying a honeycomb structure, IR thermography has been still able to detect remaining water. A general comparison of NDT methods for the inspection of large composite structures is presented in Table 11.18 [65].

Until recently, in the U.S. space industry, some large composite structures were inspected by tap testing (coin tap, sonic), which proved to be unreliable due to low spatial resolution and the subjective character of data interpretation. Tap testing could easily fail to find critical defects and result in many false negative findings. The tap method is still used in some (usually old and outdated) test procedures.

Table 11.18 Comparison of five NDT methods used for the inspection of large composite structures [65]

NDT method	Equipment cost	Portability	Scan rate	Spatial resolution	Special requirements
TNDT	Moderate	High	High	High	Non-reflective surface
Shearography	Moderate	High	High	High	Reflective surface[b]
Tap testing	Very low	Very high	Slow	Poor	Unreliable method
Ultrasonic	Moderate	Fair	Slow	High	Couplant[c]
X ray	Moderate	Fair	Slow	High	Safety concerns

[a]All methods produce images except tap testing
All methods require skill and training (less for tap testing)
[b]Test surfaces need not be highly reflective, but must reflect light to a certain extent
[c]Air-coupled ultrasonic systems exist and laser ultrasonic systems do not require a couplant, but these both have other limitations

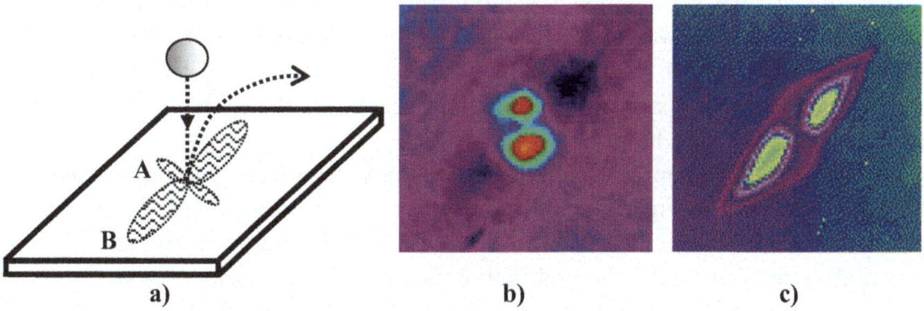

Fig. 11.57 Detecting mpact damage in CFRP: (**a**) producing reference impact damage, (**b**) front-surface IR thermogram under pulsed stimulation, (**c**) rear-surface IR thermogram under pulsed stimulation (early detection technique)

However, this method should never be used and it can usually be replaced with TNDT, shearography or other NDT methods [65].

Carbon reinforced plastic composites (CFRP)
CFRP (graphite epoxy) composites, due to their high strength and light weight, are ubiquitous in the aerospace industry, particularly in wings, control surfaces, and fuselages of military and commercial aircraft. A CFRP composite consists of layers of carbon fibers in a synthetic resin matrix.

TNDT can be used for detecting a conglomerate of micro-defects, such as porosity. However, most studies have focused on detecting delaminations, and these defects may be caused by impact damage during the use of a composite part. Figure 11.57a, shows the production of artificial impact damage by hitting the composite surface with a metallic ball of a known mass and diameter falling from a known height. Impact damage produces a defect pattern that looks like a butterfly. Except for a faint mark (dimple) at the impact site, there is usually no visible damage on the surface, but there might be an extensive delaminated area inside the composite. The area of delamination may increase with depth and their pattern is generally aligned along the fiber orientations. TNDT has been used to examine impact damage. Figures 11.57b, c show TNDT test results of both surfaces of impact damage to a CFRP panel, showing the typical "butterfly" pattern.

In 1998, a Round Robin test series was performed on a CFRP reference sample made by ONERA, France. This sample contained impact damage produced as shown in Fig. 11.57a. Under optical stimulation, defect **A**, located directly under the impact site, clearly exhibited two separate areas or "lobes" (Fig. 11.57b) of damage. The detectability of deeper delaminations depends on their size and depth. Delaminations can be also detected from an inspection of the rear surface, especially if they are located close to the rear surface. A rear-surface image is shown in Fig. 11.57c; this was obtained by Balageas et al. using the 'early detection' technique.

The use of the thermal tomography technique allowed the "slicing" of the sample into four layers to produce a synthetic image (depthgram) where pixel values correspond to the depth (Fig. 11.58, left). The sample was also checked by applying the C-scan ultrasonic method that produced the depthgram in Fig. 11.58, right. Both images match fairly well, although the ultrasonic depthgram has a better depth resolution. However, it is worth noting that the ultrasonic image took about 3 hrs by immersing the sample in water, which is not good for composite materials. However, the IR image sequence was obtained in a few seconds and the "manual" data treatment required a half an hour.

A schematic diagram of another 5 mm-thick reference panel fabricated by ONERA, is shown in Fig. 11.59a. Delaminations were simulated by Teflon inserts placed at depths of 1.3, 2.6 and 4 mm. These artificial defects are probably a more severe discontinuity in the thermal path than a real delamination, as there may be additional air gaps between the inserts and the host material. Building an artificial defect that exactly represents a real defect is impossible.

The results of applying tomographic data processing are presented in Fig. 11.59b, c. The advantages of thermal tomography are clearly seen in Fig. 11.59b, where the optimum source image shows only one defect at a depth of 1.3 mm, while the other defects are not seen because of uneven heating. The thermal tomogram of the 1.2–1.9 mm layer shows all three defects located at the depth of 1.3 mm. Three subsequent thermal tomograms presented in Fig. 11.59c show all six defects (note that the defect at the depth of 4 mm is partially shaded by the more shallow defect located at a depth of 1.3 mm).

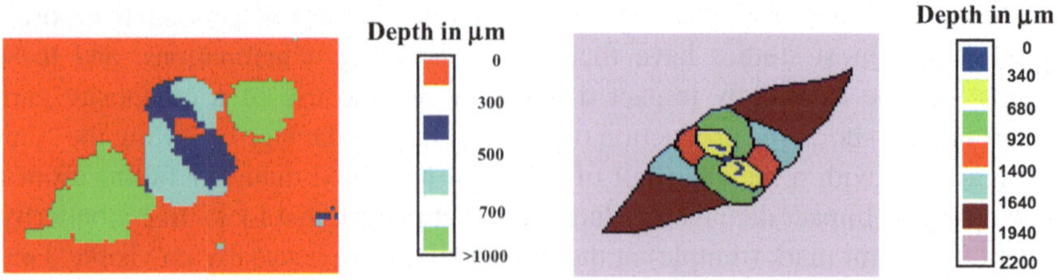

Fig. 11.58 Depthgram of impact damage in CFRP obtained by the thermal tomography method (left) and C-scan ultrasonic image (right)

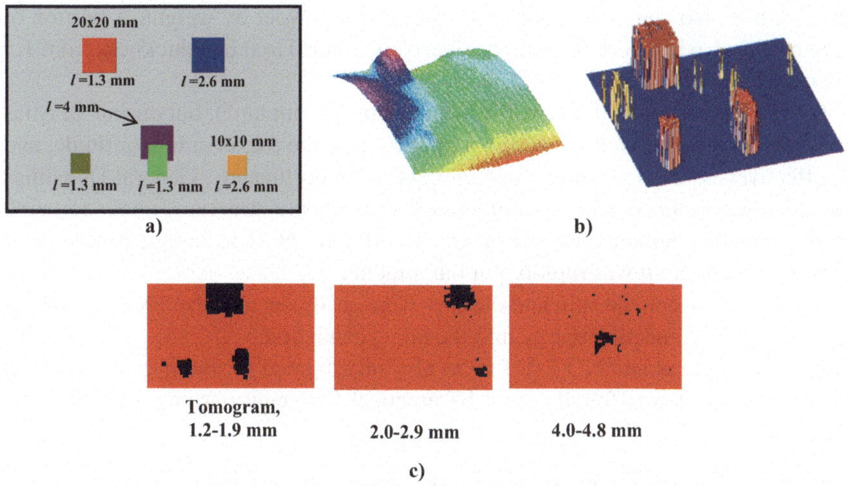

Fig. 11.59 Thermal tomography of 5 mm-thick CFRP sample containing Teflon inserts (l is the defect depth, ThermoFit Pro software): (**a**) defect location, (**b**) IR raw image (left) and tomogram of 1.2–1.9 mm layer (right), (**c**) thermal tomograms

Carbon-carbon composite materials

Carbon Carbon composite materials withstand a much higher temperature than CFRP. This material was used in the US space shuttle program (see also Sects. 11.8.7 and 11.8.8), as well as in the European Hermes space program. Since carbon fibers have a high thermal conductivity, Carbon Carbon composite materials are tested mainly by TNDT with flash heating. A study performed by Delpeche et Al. has shown that TNDT is a convenient technique for detecting delaminations and also characterizing their depth and thickness [66] in carbon-carbon materials. The evaluation of a 3 mm-thick Carbon Carbon reference sample has demonstrated that defect depth can de determined with an accuracy of about 10–14% compared to data obtained by micrography. The accuracy in defect thickness evaluation achieved by micrography was 95%, but it was suspected that the sample preparation for micrography, which entailed grinding and polishing, may have changed the defect thickness.

Polymeric composite materials

Starting in the 1950's the desire to reduce the takeoff weight of space launch vehicles has lead to the replacement of metal structures with lighter composite ones, including glass fiber reinforced plastic (GFRP) materials. In one example, the aft fairing of the US Atlas launch vehicle was made of 16 large, complex curved GFRP panels, including 4 doors [67]. The panels were made of GFRP facesheets with aluminum honeycomb attached by a layer of adhesive film, which contained a significant percentage of aluminum powder by weight. There were about 4 different densities of aluminum honeycomb used. The skin thickness varied from 1.5 to 9 mm. The gaps between adjacent honeycomb pieces were filled with a foaming adhesive that also contained about 50% aluminum powder by weight. A portion of some panels also had cork insulation adhesively bonded to it in a thickness from 1.5 to 15 mm.

The original NDT method used was tap testing ("coin tap"), but during an Atlas launch, radar data showed a metallic part falling off the rocket at high altitude, and the effectiveness of tap testing came into question. Fortunately, General Dynamics had already developed TNDT techniques for composite structures under research funding, and had demonstrated that it was an effective NDT technique for the Atlas composite parts. So it was rapidly put into practice.

Disbonds between the skin and the core were easily detected by TNDT, and they were repaired by injecting epoxy into the honeycomb core.

The use of ultrasonic NDT (UT) was also investigated for the inspection of the Atlas parts, but it was found to not be practical for many reasons, including the following

1. The large size and complex shapes of the Atlas parts would have required a very expensive ultrasonic C-scan system….millions of dollars.
2. UT would have been a very time-consuming process
3. The parts would have to be exposed to water and then dried, another expensive and time-consuming process
4. UT would have been inconvenient to apply on parts attached to a rocket.

As TNDT was implemented, tap testing continued for a while. This allowed an evaluation of tap testing results compared to TNDT and occasional destructive tests. It was found that tap testing falsely identified many areas as "defects", failed to find many defects, and undersized many defects that were found. It was discontinued.

Advantages of tap testing are:

1. Equipment is highly portable
2. Equipment is very inexpensive
3. Personnel training is minimal: anyone can do it! So management loves it.

Disadvantages of tap testing are:

1. No image and no means to document test results
2. Subjective, non-quantitative nature of inspection
3. Depends on human hearing;
   (a) hearing loss and background noise are problems,
   (b) Many companies that use tap testing do not require a hearing test or a quiet room! Some allow it next to riveting!
4. High level of false negatives
   (a) High probability of failing to detect defects.
5. High level of false positives
   (a) This causes unnecessary repairs and damage to the part.

6. Low spatial resolution
   (a) A typical instruction is to tap on ½ inch centers.
   (b) To do the above properly, a grid of ½ inch by ½ inch squares must be marked on the surface of the part.
7. Low inspection speed due to the need to mark an inspection grid on the part and remove it after the inspection
8. Tap testing is very unreliable. It may be worse than no inspection.

Because the GFRP facesheets were relatively thick and had low thermal conductivity, the IR temperature signals caused by typical defects in the Atlas panels would persist for as long as several minutes. Panels were heated with an industrial 1750 W hot air gun, or "hair dryer". During testing, the temperature increase did not exceed 10 °C. This inspection technique was suggested in 1990 (see Fig. 11.60). When developing the test procedure, artificial defects were produced by several methods, including inserting Teflon implants, inserting a metallic blade between the skin and the honeycomb core and removing it after cure, and also by impact damage.

TNDT allowed the detection of: (1) disbonds between an insulation coating and the CFRP skin, (2) delaminations between the skin and the honeycomb core, (3) core damage, (4) variations in the density of aluminum honeycombs, (5) injected epoxy resin glue, (6) embedded metallic parts, and (7) GFRP blocks under the skin (these blocks serve for mounting metallic parts).

Fig. 11.60 TNDT of the thrust structure of an Atlas launch vehicle by using a hot air gun while tap testing is being performed concurrently

TNDT proved to be very effective in the process of manufacturing the honeycomb structures because it allowed the detection of skin to core disbonds before the rear skin was attached, so the honeycomb cells remained open at the rear side, thus enabling the easy repair of defects by injecting epoxy into the open cells.

It was found that skin to core disbonds appeared most frequently in areas of higher curvature, and this was attributed to the "anticlastic" behavior of honeycomb materials.

Separate laboratory tests were performed to see if other heating methods were more effective than hot air. Xenon flash tubes with a total energy of 20 kJ were used, as was bank of halogen lamps. Cooling with liquid nitrogen was also tried. None of these methods provided better defect "detectability" compared to using hot air. It was reported that flash tubes provided good repeatability of results and enabled quantitative analysis but their use resulted in a much slower inspection [67].

Compared to the (original) tap testing NDT method, IR thermographic NDT demonstrated the following advantages [67]:

- quantitative test results,
- non-contact test method,
- area method, not a point technique,
- high spatial resolution,
- good documentation of test results using individual IR images and video methods,
- less dependent on human factors, such as hearing,
- separate defect types can be identified,
- defect boundaries can be accurately marked, facilitating accurate repairs.

The only advantage of coin tap was less expensive equipment.

According to [67], development of TNDT at General Dynamics (later merged with Lockheed Martin) resulted in the replacement of tap testing inspection by TNDT. Inspection procedures were written, reference samples were fabricated, and NDT personnel were trained and certified.

The reference samples included the following types of defects:

- structural defects,
- variation in honeycomb density and skin thickness,
- additional thermal insulation,
- delaminations in facesheets,
- skin to core (honeycomb) disbands,
- repair areas,
- damaged honeycomb cells,
- impact damage,
- abnormal and unacceptable adhesive thickness,
- variation in the number of composite plies.

IR testing using hot air heating was performed both indoors and outdoors, including when an Atlas was on a trailer inside buildings, as well as outside on a launch tower at Cape Canaveral. In outside testing, solar reflections occasionally appeared in the IR image and heated the test surfaces, thus creating false defect indications. However, trained operators could easily identify such areas as reflections as they did not change as a function of time. Strong wind, which is typical in open locations (launch pads, etc.), decreased the heating energy absorbed from the hot air source. Condensed water changed the surface emissivity and reduced temperatures during heating because of the high heat capacity of water. Outdoor inspection was least efficient (5–10 m^2/hr) because of changing weather conditions and the necessity to move test equipment. When inspecting Atlas parts in a factory environment, the scan rate reached 50 m^2/hr. It is worth noting that in the above-described test procedure, defects were marked manually during inspection, which made repair easy.

In the former USSR, solid rocket motor cases were made of filament wound composite material over a composite rib support structure [68]. In one variant, polyurethane foam filled in the space between ribs.

These rocket motor cases were subjected to a pressure proof load test after manufacture.

In these complicated structures, the following defects may occur:

- voids and delaminations in (a) filament wound shell, (b) interface between ribs and filament wound shell,
- excess adhesive,
- fiber breakage,
- anisotropy in wound fibers,
- porosity,
- density variations in the foam between ribs,
- low fiber content.

It was found experimentally that an increase in porosity from 0.6% to 6% reduced the inter-laminar shear strength by 45% and the compression strength by 37%.

Table 11.19 shows the evaluation of the NDT of solid rocket cases using a 7 point scale [68]. It is seen that the preferable techniques are thermal and optical holography (a form of laser shearography).

Russian researchers acquired extensive theoretical and practical experience on the TNDT of filament wound GFRP shells [68]. In these products, defects appear due to:

- poor adhesion between filament wound layers,
- mechanical malfunctions during the filament winding,
- violation of storage and usage requirements,
- contamination, with water, oil or other liquids, during manufacture,
- overpressurization during pressure testing the filament wound shell.

It is important to note that, because of several factors discussed above, air bubbles (inclusions) of specific sizes and configurations appear preferentially.

The most critical defects in filament wound structures are delaminations between subsequent filament wound layers. Statistical analysis of delaminations found by TNDT showed that the most common delamination thickness is 0.2 mm (0.008 inch). Analysis lead to the conclusion that, in order to ensure the detection of 97% of all delaminations, it is necessary to detect all defects of minimum thickness 0.23 ± 0.03 mm and a minimum lateral size of 25 × 25 ± 1 mm (1 × 1 inch) [68].

Table 11.19 Evaluation of NDT of rocket power shells made of polymeric composites. (Adapted from [68])

Requirements for NDT technique (by order of importance)	NDT methods						
	Ultrasonic (transmission)	Ultrasonic (impedance method)	Microwave	X ray	Computer Tomography (CT)	Thermal (TNDT)	Optical holography
Detecting delaminations between skin and ribs of reticular structure (air gaps present)	7	7	3	7	7	6	7
Detecting delaminations between skin and ribs of reticular structure (kissing defects)	5	5	4	6	6	6	7
Finding false defects	5	6	4	2	2	2	7
Evaluating defect size (contour)	6	6	5	7	7	7	5
Characterizing defect depth	1	5	3	6	7	7	2
Inspecting samples with foam filling	7	7	3	7	7	7	7
Safety	7	7	4	2	2	7	7
Inspection speed	5	5	2	5	6	6	7
Commercial equipment available	7	7	1	5	7	7	5
Automation possibility	7	3	2	6	7	7	5
Size of tested objects	7	4	1	5	2	7	7
Probability of correct detection	4	4	3	3	7	5	7
In total:	**48**	**69**	**35**	**61**	**67**	**74**	**80**

The optimization of TNDT of GFRP cylindrical shells was guided by solving a 2D heat conduction problem. Numerous studies devoted to the analysis of the relationship between GFRP defects and surface temperature signals were performed in the former USSR. Kush et al. developed a method of defect characterization that was based on the minimization of the function produced by the difference between experimental and theoretical data; this method was later called "non-linear fitting" (see p. 5.6) [69].

When this algorithm was tested in the 1980s on Russian EC computers, the computation time was tens of minutes. The long processing time was somewhat reduced by making an intelligent first guess, i.e. by knowing *a priori* the expected defect size. In modern mathematical programs, such as MatLab, Comsol Multiphysics, etc., non-linear fitting procedures involving many parameters are trivial. However, even on powerful PCs, the corresponding calculations may require a few minutes of processor time in order to evaluate a single pixel value, depending on the number of unknown parameters and the adequacy of the mathematical model used. In other words, the possibility of solving a characterization problem using full-frame images remains unrealistic even with contemporary computers.

Boron epoxy patches

The structural integrity of aircraft components is a primary concern of both airplane manufacturers and airline companies. Airline companies, commonly repair aluminum skins by applying non-metallic, boron fiber/epoxy patches. TNDT has been shown to be an appropriate technique for evaluating the bond integrity between these patches and an aluminum skin. One-sided TNDT is performed on the "patch" (non-metallic) side. In one example [70], the "lock-in" TNDT of a boron epoxy patch applied to a 4 mm-thick aluminum skin near a fuel tank was reported (Fig. 11.61a). In order to demonstrate the potential of TNDT, an artificial delamination was induced during the application of the patch. As was shown by later testing, this artificial delamination was also accompanied by a few unplanned, naturally occurring defects. The sample was heated with 4 cyclic heating cycles for 4.25 seconds and then cooled down for 4.25 seconds; afterwards, the first derivative image revealed all of the defects (Fig. 11.61b).

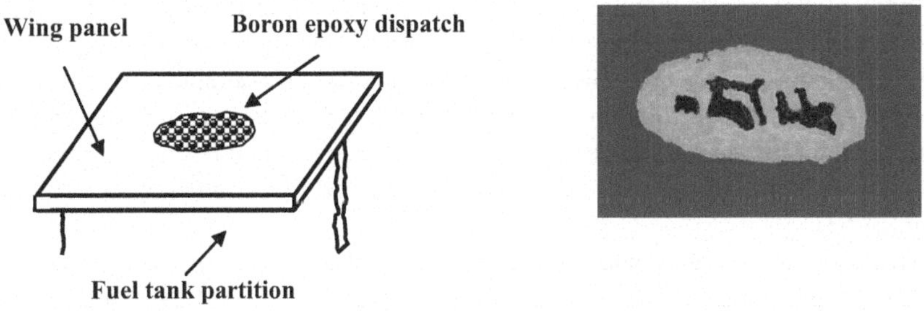

a) b)

Fig. 11.61 TNDT of a boron epoxy dispatch on an aircraft panel (**a**) dispatch location, (**b**) first derivative image (retouched). (Adapted from [70])

11.8.7 Components of Space Shuttles and Other Rockets

As mentioned above, aerospace structures include complex combinations of various materials joined by adhesive, rivets or welds. In this section, the use of TNDT for some specific inspection applications will be described.

In the USA, the Space Shuttle program required the development of novel NDT techniques; the same occurred in the former USSR in the Buran space shuttle program. NDT development continued as needed, and the unfortunate catastrophe of the Columbia Shuttle in 2003 led to a flurry of additional NDT development.

Mounting clamps for solid-fuel rocket boosters are covered with an anti-corrosion lubricant grease whose thickness is from 0.03 to 0.08 mm. A standard TNDT procedure was demonstrated to be able to detect grease thickness variations of about 0.003 mm, and it could also detect an absence of grease.

Another problem was that the faying (in contact) metal surfaces of the clamp and booster during vibration caused the premature wearing of the clamps. The problem was also addressed [71]. A routine TNDT test procedure detects an absence of lubricant. Inspection access to difficult areas was achieved with a (IR) mirror. When evaluating the lubricant thickness, it was found that scratches on the surface of the steel could also be easily detected thermographically, thus facilitating an overall inspection procedure.

Finally, the TNDT has proven to be effective in the inspection of thermal protection, in the area near the clamps. This thermal protection structure includes waterproof neoprene rubber on top of which is a cork-like material [71]. The integrity of the thermal insulation bond is important because pieces of insulation (cork or other) torn away during launch may damage the thermal protection tiles, as occurred in the Columbia catastrophe. In TNDT laboratory tests, air defects up to 3.3 mm thick were detected in the insulation.

Furthermore, in the field inspection of the thermal protection of Space Shuttles, defects missed by tap testing were found.

Space Shuttle solid-fuel booster rockets are manufactured by assembling a group of individual segments that are separated by a material called an inhibitor. NDT of such a structure examines the integrity of all bonded surfaces that are necessary for the functioning of the engine, but structures cannot be disassembled to perform the inspection [72]. The evaluation must be done *in situ* prior to covering the segments with the inhibitor and assembling them into a motor. Typical defects are disbonds between the inhibitor and the case and between the fuel segments and the case (Fig. 11.62).

By taking into account the thicknesses and the thermal properties of the materials involved (inhibitor: $a = 1.31{\cdot}10^{-7}$ m^2/s, lining: $a = 0.96{\cdot}10^{-7}$ m^2/s, fuel: $a = 1.78{\cdot}10^{-7}$ m^2/s), some US researchers proposed a concept of 'proxy heating', in which the heating of the inhibitor with a hot air gun leads to the uniform heating of the adjacent fuel layers [72]. Then the mass of fuel, in its turn, starts to act as a heat source. In this way, the authors believed that a one-sided TNDT procedure

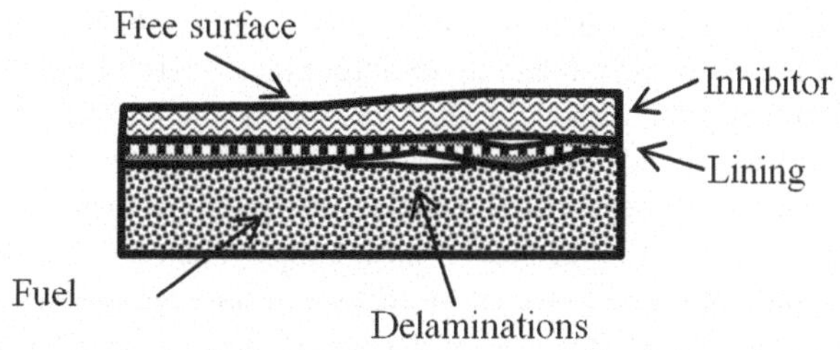

Fig. 11.62 Fuel-inhibitor interface. (Adapted from [72])

transitioned into a two-sided procedure. In any case, the presence of disbonds caused only faint defect indications in IR thermograms. Therefore, a special data processing procedure, the "Winfree filter", was created.

Choosing an optimal NDT method can be illustrated by an example of the evaluation of insulation on Centaur rocket cryogenic fuel tanks by General Dynamics, USA [73]. The Centaur rocket was a second stage on Atlas and Titan rockets. The company introduced a new type of thermal insulation for pressure stabilized stainless steel sheet (thin) metal fuel tanks that would contain liquid hydrogen and liquid oxygen [73]. This insulation consisted of pre-formed pieces of vinyl (PVC) foam adhesively bonded to the fuel tank. It was important to develop an NDT process that could detect areas where the foam was disbonded from the fuel tank. The optimal NDT method should provide: (1) non-contact inspection (to avoid damaging the foam), (2) portability (to operate on a launch pad), (3) one-sided access to the thermal insulation, (4) minimum detectable disbond size of 100 mm (4 inches), and (5) a scan rate from 1 to 60 m^2/hr.

General Dynamics conducted many tests comparing sonic, ultrasonic, thermal, laser shearography, X ray and mechanical (vacuum) methods of NDT. The thermal method could only detect defects by heating the metal tank from the inside and only if it was painted black, and neither of these were possible.

The optimal technique proved to be laser shearography. This method was used to inspect a number of Centaur tanks and no disbonds were ever found, either before or after cryogenic fueling. So NDT was discontinued.

The USA's experimental X-33 space launch vehicle program was sponsored by NASA; the prime contractor was Lockheed Martin. Goodrich Aerostructures (formerly Rohr) was a subcontrator responsible for the design, manufacture, and integration of the thermal protection system (TPS), which was the entire exterior of the vehicle. The X-33 TPS was to be built of a variety of advanced materials. The hottest areas, which were the nosecap and wing leading edges, were to be made of carbon-carbon composite, the same as the Space Shuttle. The upper surfaces of the vehicle were the coolest areas. The structural material was CFRP-aluminum honeycomb sandwich panels and insulation was high temperature ceramic fiber blankets or Nomex felt blankets, adhesively bonded to the exterior surface of the composite sandwich panels. These insulation materials were also used on the Space Shuttle.

The X-33 windward aeroshell was the area of intermediate temperature. The structure in this area was to be made up of approximately 1600 Inconel (high temperature metal) honeycomb panels. Inconel facesheets (thickness 0.15 mm or 0.006 inch) were metallically bonded (skin-to-core) to Inconel honeycomb.

The skin-to-core bondlines of these panels were tested by flash TNDT. The heating was performed with a 12.8 kJ Xenon flash lamp system and the IR imager was made by Santa Barbara Focal Plane (InSb IR detector, image format 640x512). All of the metallic parts were coated with water-based black paint prior to the inspection. The TNDT procedure allowed the detection of a 0.1 inch (2.5 mm) gap in the skin-to-core braze of a single honeycomb cell wall. No other NDT method was able to match this performance.

The original plan was to use TNDT for testing only the Inconel metallic honeycomb panels. It was planned that the graphite/epoxy aluminum honeycomb sandwich panels would be inspected by ultrasonic C-scan. However, there were problems in getting C-scan done on the very large parts. After a few preliminary tests, it was demonstrated that TNDT was a more efficient and effective method of inspecting the graphite/epoxy aluminum honeycomb panels also, and except for a small number of early parts, all of these parts were inspected by TNDT. TNDT was superior to ultrasonic for the following reasons [74]: (1) The TNDT scan rate was much higher than that of the ultrasonic method, (2) TNDT inspection cost was much less than that of ultrasonic, (3) TNDT could be performed in the factory area where the parts were fabricated, making it unnecessary to move the parts to the ultrasonic test laboratory, (4) ultrasonic inspection exposed the CFRP facesheet to water, which permeated the surface and became trapped inside the panel; TNDT was able to verify the removal (or presence) of the water, (5) location of the defects detected by TNDT could easily be marked on the surface of the parts, and (6) TNDT was able to find defects that had not been detected by ultrasonics. An additional benefit was that TNDT showed the locations of defects in relation to structural details. TNDT provided information that led to an understanding of how the defects were created and how to change the manufacturing process to prevent defects. As a result of this, the later panels manufactured had no defects. The general TNDT procedure used in these inspections was that the same as the one used for evaluating composite Atlas rocket components (see above).

It is worth adding that, in addition to being used for TNDT, IR thermography was also used to monitor test panels under simulated reentry conditions in Arcjet tunnels. These tests verified the thermal performance and analysis of X-33 panels.

Unfortunately, the X-33 program was cancelled before completion, due to problems with the composite cryogenic fuel tank. However the X-33 story demonstrates that TNDT and other uses of IR technology should be considered during the development of any new space vehicle program. This was also proven during the investigation of the Space Shuttle Columbia catastrophe where TNDT was widely applied (see Sect. 11.8.8 below).

11.8.8 TNDT of Space Shuttles on Launching Pad and in Outer Space

A new impetus for a wider use of TNDT in the US Space Shuttle program was created by the conclusions of the commission that investigated the Columbia catastrophe that occurred in 2003.

The Space Shuttle thermal protection system (TPS) was made of materials that could withstand temperatures up to 1650 °C (3000 F). Reinforced carbon-carbon (RCC) composite was used in the hottest areas, which are mainly the nosecap and wing leading edges, where the temperature may exceed 1510 °C (2750 F).

The failure investigation concluded that a piece of polymeric foam insulation tore off during the launch, hit one of the RCC wing leading edge panels, and broke a hole through it. When Columbia had finished its mission and was reentering the Earth's atmosphere, hot gasses entered the hole in the leading edge and melted the aluminum support structure beneath. The wing failed and the Shuttle broke apart.

Note that the Columbia catastrophe was not caused by a failure of NDT. It could be said that it was caused by a failure to use NDT on the polymeric foam, It was known that foam had come off in previous launches, as it had been seen falling in multiple launch videos. Unfortunately, it had not been thought that the falling foam could damage the Shuttle. This was an unfortunate underestimation.

As a result of the Columbia failure investigation, NASA launched a program to develop NDT methods, such as digital radiography, X ray computed tomography (CT), eddy current and others.,

TNDT was found to be the best method to inspect RCC panels. However, it could only inspect the integrity of the coating on the outside of the RCC panels; it could not inspect the bulk RCC material.TNDT had the following benefits: (1) non-contact, one-sided inspection, (2) reasonable sensitivity to critical coating defects, and (3) possibility of using it in outer space.

In the course of the NASA NDT development program, several concepts of TNDT were explored. To test Space Shuttle RCC panels on the launching pad at Cape Canaveral, a commercial TWI EchoTherm system including a Xenon flash tube heater (2 × 4.8 kJ), and a Phoenix IR imager (temperature resolution 25 mK, image format 320 × 256), was recommended [75]. A blind test was performed on an actual wing leading edge RCC panel that contained artificial defects. The critical defect requirement was:

$$D = 2.8l + 0.042, \quad (11.29)$$

where D is the defect diameter and l is defect depth in inches. It was found that TNDT was able to detect all defects that met the requirement above. The method of principle component analysis (PCA) was chosen for the processing of TNDT data (see p. 6.18).

The TNDT system described above was used for inspecting the RCC parts of the Discovery Space Shuttle after its mission in August 2005. A defect indication was found on a wing leading edge part and it was confirmed as deterioration of the RCC.

The failure investigation also concluded that a lack of in-flight inspection and repair technology, plus the underestimation of the severity of the incident led to the catastrophe during re-entry.

As a result, research was also performed on the possible use of IR thermography for performing active or passive TNDT on the Space Shuttle in outer space [76]. The strategy of orbital TNDT assumes that if an Orbiter Boom Sensor System (OBSS), which includes a video camera and two lasers, detects damage on the exterior surface of the Shuttle, the astronauts would perform a more detailed survey using a digital video camera of higher resolution, as well as an IR imager. The purpose of the survey would be to evaluate the size and severity of any suspected damage to the RCC. This data should result in a decision to either make a repair or continue the mission without it. For example, minor damage such as a narrow, shallow crack would be filled with a special paste, while more extensive damage would require the application of a patch. Two other possible options would be to dock the Shuttle at the International Space Station in order to perform a more major repair, or launch another Shuttle to perform a rescue mission.

Using flash TNDT in space was thought to be impossible. The EchoTherm system was relatively bulky, and the use of high voltage contradicted safety requirements. It was suggested to use the natural step heating that occurs when the Shuttle goes from shadow to being solar-illuminated as a means of thermal stimulation of defects during orbital flights. However, that heating pulse would have been much too slow to allow the detection of anything but gross defects.

The basic challenge within this project was the adaptation of a commercial IR imager (ThermaCAM S65 from FLIR Systems) to operate in outer space. Modifications were required for vacuum durability, radiation exposure, extreme temperatures, etc. For example, the control panel of the IR imager had to be modified so the astronauts could use it while wearing bulky gloves. The cost of this work was 6.4 million US dollars, including use on two orbital missions. It is interesting to note that it was one of very few NASA contracts which did not exceed the planned budget.

The modified IR imager was used during the Discovery mission in July 2005. After the conclusion of the Space Shuttle program this unit was delivered to the International Space Station to be used for inspecting electro-mechanical facilities. On July 12, 2006, during the mission of Discovery, Michael Fossum performed an IR thermographic inspection of the Space Shuttle at an altitude of 350 km. This was the first use of this technique in outer space.

11.9 Petrochemical Industry

The petrochemical industry is characterized by high power consumption and includes land-based and offshore petroleum exploration installations, production facilities, gas-oil separation plants, gas plants, refineries, petrochemical plants, pipelines and pumping stations. This industry involves continuous manufacturing processes and very expensive facilities. Petrochemical plants are major users of IR thermographic equipment and this is not difficult to justify, as the payback period is reported to be 0.5–2 years. There are five main application areas for IR thermography at refineries and petrochemical plants (see Table 11.20):

- surveys of electrical installations (power units and distribution systems),
- inspection of manufacturing processes and facilities, such as furnace tubes, reactors, catalyst lines, etc.,
- inspection of refractory and thermal insulation,
- measurement of smokestack temperature,
- evaluation of heat losses.

The IR thermographic inspection of electrical distribution systems is well developed and has been used for many years to assure the reliability of key equipment. IR thermography is used especially for evaluating rotating mechanical components such as shafts and bearings, or those that operate under pressure. TNDT can be used to inspect tanks and vessels of different types, piping, heat exchangers, coolers, etc. It is possible to detect the level of fluid inside a storage tank by a simple IR observation of the external surface. In fact, IR thermography is able to reveal the presence of water in lubrication oil tanks, contaminants in caustic tanks and hydrocarbon deposits in spent acid tanks. As in other IR applications, it is difficult to inspect shiny metal surfaces (steel, stainless steel, aluminum) and components that have reflective coatings to reduce radiant heat losses. IR thermographic diagnostics of compressors, electrical motors, pumps and other equipment containing rotating parts is possible but may be unreliable because of the low magnitude of temperature signals appearing in problem areas. The use of IR cameras for monitoring temperature patterns and conditions on smokestacks is believed to be superior to the use of multiple thermocouples or IR thermometers.

In recent years in Russia, various petrochemical plants, oil and gas producers, and transportation organizations have implemented IR thermographic inspections programs. Russian Federal industrial safety laws require companies to perform periodic inspections on both primary and auxiliary equipment, such as chemical reactors and smokestacks.

IR thermography is an effective method for the evaluation of chemical reactors, especially where nominal or critical temperatures have been established at various locations (Fig. 11.63a).

IR can detect various things in storage tanks, depending on the temperatures or thermal properties of the contents. The level of contents in a closed storage tank can be effectively determined, especially if the product temperature is very different from ambient. Boundaries between liquid, solid and gaseous phases can be identified due to cyclic temperature variations and the heat capacity phenomena. In some cases, it is possible to detect sediment, or sludge, which reduces the "payload" volume of tanks, reduces the production efficiency and causes financial loss. It is sometimes possible to detect a gas-foam boundary, as well as thin layers of sludge. The

Table 11.20 IR thermographic inspection in the petrochemical industry. (Courtesy: FLIR Systems)

Application area	Equipment	Undesirable features	Causes	Consequences
Power supply	Distribution systems	Elevated temperature	Resistance change due to corrosion, defective insulation or mechanical damage	Power outage
	Electrical unit bearings	Elevated temperature	Overloading, inadequate or incorrect lubrication	Bearing damage
	Chloride, fluoride, electrolytic elements	Elevated temperature	Inadequate bolt tightening	Ineffective production, power reduction
	Transformers, circuit breakers, dischargers	Elevated temperature	Various reasons	Outage
Process liquids	Safety valves	Elevated temperature of piping	Mechanical malfunctions, worn out catches	Product loss
	Steam traps	Elevated temperature	Mechanical malfunctions	Steam leakages (energy loss)
	Underground pipes	Elevated temperature	Leakage due to corrosion or poor welding	Product contamination and loss
	Vessels, tanks	Temperature difference between vaporized, liquid and solid phases	level of liquid or solid phase	Product contamination, evacuation fails, inaccurate information on contents
	Gas pipe lines, pipe lines	Warm/cold spots or areas	Creation of liquid/ solid phase	Reduction of liquid flow, pipe corking
	Heat exchangers	Temperature anomalies with respect to reference	Sediments, shell wear	Reduction of liquid flow, appearance of holes in tubes
	Air refrigerators	Lower temperature in some tubes	Sediments, corking	Reduction of liquid flow and deterioration in heat transfer
Refractory and thermal insulation	Reactor vessels, piping	Elevated temperature	Refractory wear, cracking	Appearance of holes, ruptures as result of creeping and stressing under impact of hydrogen and high temperature

(continued)

Table 11.20 (continued)

Application area	Equipment	Undesirable features	Causes	Consequences
	Smokestacks and gas flues for conducting furnace gas	Elevated temperature	Refractory defects, cracking	Burns and ruptures as result of creeping and stressing under impact of sulphuric acid
	Piping, vessels, furnaces, heat exchangers, reactors, building constructions	Mechanical malfunctions	Elevated temperature	Energy loss, productivity reduction and process fails
Furnaces	Piping	Elevated temperature of tubing	Excessive flame, poor flame torch, coking, external sediments И	Oxidation, rupture, downtimes, shorter tube lifetime
	Furnace pipe steaming	Cold spots on pipes Warm spot on pipes	Incomplete coke removal Incorrect chemical cleaning of scum	Lower temperature when setting furnace in operation, reduction of operation cycle Overheating, oxidation, malfunction due to rupture, reduction of operation cycle
Energy saving	Piping, vessels, furnaces	Temperature higher than optimal	Enhanced energy consumption	Energy losses

boundary between two liquids can be monitored much more easily, if the liquids, for example, oil and water, have different heat capacities. As mentioned above, IR thermography is hard to apply in the case of low-emissivity metals and also if there is a double wall, ventilated gap or a layer of insulation (foam or other) on a tank. In some cases, these difficulties can be overcome by placing a strip of high-emissivity material on the tank surface, or making a "window' in the insulation. Surface temperature gradients can be enhanced by heating the surface of the tank, using powerful hot air blowers or other methods. It is only necessary to heat a vertical strip on the surface of a tank and not the whole surface. An example of an IR thermogram of a tank with a clearly visible gas-product level is shown in Fig. 11.63b.

Promising results have been obtained in the evaluation of the thermal insulation systems on chemical tanker ships. These typically include a metal frame, polyurethane insulation and sheet metal protection. Because the area that must be inspected is vast for very large ships (up to 30 thousand tons), there is no alternative to IR thermography for the quick identification of anomalies and areas of interest.

In Russia, regular inspections of storage tanks containing liquid ammonia are mandatory. Such tanks are above-ground cylindrical vessels mounted on a reinforced concrete base and surrounded by a protective wall. In Fig. 11.64b the IR thermogram of a liquid ammonia tank reveals a faint temperature elevation of 0.3 °C

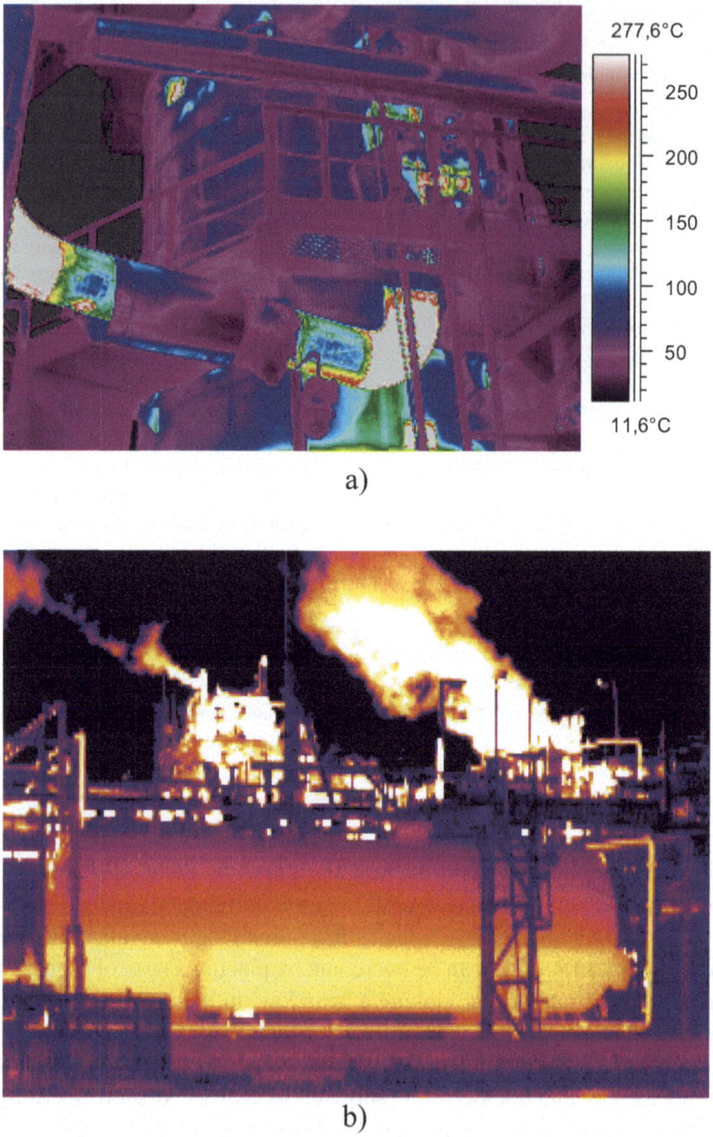

Fig. 11.63 IR thermograms of a chemical reactor (**a**) and a tank containing a liquid (**b**)

in the top section of the tank where the perlite insulation collapsed or was compressed. The inspection of such tanks should be done at night in order to avoid solar heating and reflections which may mask the weak temperature signals. The most difficult area to inspect is the tank dome, where IR camera readings often suffer from reflections caused by the cold Sky and nearby gas "flares", which are burning gas plumes, from adjacent facilities.

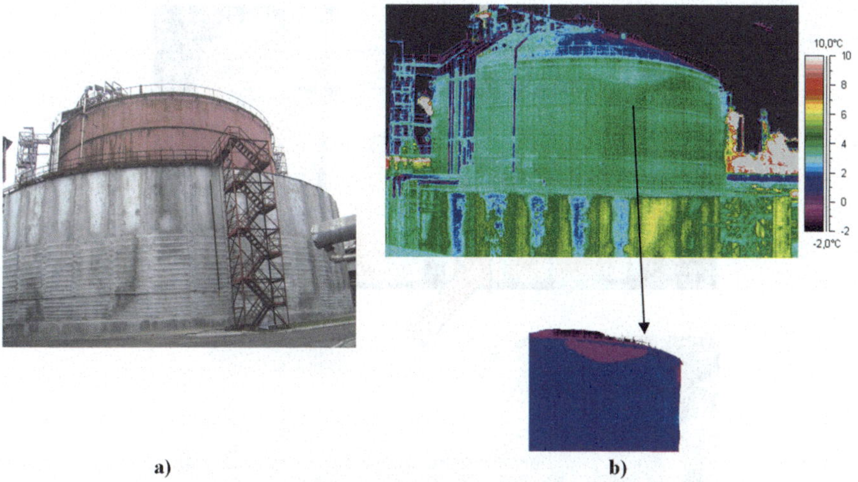

Fig. 11.64 Inspecting thermal insulation of a liquid ammonia storage tank: (**a**) photo of a liquid ammonia storage tank, (**b**) IR thermogram of the side surface with a lack of pearlite sand (below – processed image, temperature signal 0.3 °C)

Figure 11.65 shows a heat exchanger at a Russian chemical plant. A significant quantity of sediment on the inner surface of the heat exchanger was found by applying IR thermography. This probably prevented a costly breakdown.

11.10 Electronics

Analyzing temperature distributions in electronic components, Printed Circuit Boards (PCB's), Integrated Circuits (IC's) and especially micro-circuits, is complicated by the small size of components, the high density of components, low temperature signals and large variations in emissivity. IR equipment, including optical fiber units, are often more convenient to use than contact temperature sensors, due to the non-contact and non-invasive nature of the inspection, the practically unlimited temperature measurement range, and the ability to perform measurements at many points simultaneously. IR cameras can produce images with very high spatial resolution.

IBM in the USA has used TNDT for many years to detect faults in electronic boards and to predict their lifetime. Radio Systems Sweden optimized mobile telephonic communication systems by the thermographic analysis of electronic circuit boards; the previously used contact method proved to be ineffective.

It is important to understand that there is a direct relationship between component life and temperature in electronics. For several types of components (transistors, diodes, oxide cathodes, resistors, etc), statistical data on thermal performance has allowed the prediction of component life as a function of temperature. The

Fig. 11.65 Detecting internal sediments in a heat exchanger

effectiveness and speed of TNDT becomes especially important when inspecting a large number of identical components. After some preliminary study, the temperature distributions of such components are well known, including those that indicate parts that should be rejected. Therefore, it is not necessary to monitor 100% of the surface of each part to detect faulty components. Typically, one would determine temperatures at 10–100 characteristic points, determine a set of significant parameters, and finally apply an accepted identification, or pattern recognition, algorithm. IR thermograms can be analyzed by trained thermographers, or automatically by a computer. Decision making by operators is especially efficient when defects create a significant temperature pattern, as short circuits typically do. Ericsson, Sweden, found that integrated circuits that have no localized overheating serve for longer periods of time. Detection of microcircuit defects that produce weak temperature signals can be difficult and identification algorithms can be more complicated. Inspections of circuits are often performed with component cases (covers) removed.

The IR equipment used in microelectronics includes micro-pyrometers and microscopes. Systems intended for the inspection of microchips and other semiconductor devices may also include programmable scanning units and automated data processing devices. For example, the CompuTherm IR microscopic system from EDO possessed a spatial resolution of 15 μm and a temperature sensitivity of 0.1 °C up to +600 °C. The frame rate was relatively slow (2 seconds), but fast thermal events could be captured with a time response of about 60 μs. This unit allowed the inspection of coating quality, as well as detecting short circuits and breaks in microchip conductors.

Specialized software is used in electronic inspections and one example of this was the EQUAL program developed by FLIR Systems. This was intended to determine emissivity at all test points, provide corrected temperatures and produce "true temperature" maps. The use of the EQUAL program required placing a test object into a box with temperature from +40 to +50 °C in order to achieve a uniform temperature distribution and then to produce emissivity maps. Then, current inspection results were converted into true temperature by applying a special pixel-based correction algorithm. A similar program was also supplied with the CompuTherm equipment.

In the former USSR, within its domestic electronic industry, various aspects of TNDT were thoroughly investigated. For instance, at the Electronstandart institute, several specialized inspection units, such as 14EEP300-021, 09EI-1, ASTD and others were developed [77, 78]. The theoretical aspects of modeling IR inspection electronics and predicting component lifetime were formulated by Danilin et al. in 1974 [79]. Since then, Russian research on the IR thermographic inspection of electronics has dwindled, probably, due to overall use of foreign electronic components and equipment.

A serious technical problem, in using IR thermography in electronics, is the variation in component emissivity, which can exceed 15% (see Table 7.5). Some components may have metal cases or coatings. It is important to know the true temperatures of electrical components because it strongly affects component life time. To make the emissivity of all electronic components in a circuit uniform, IBM used a white powder coating made of aluminum oxide. This could also be accomplished by using a "conformal" coating. Using these methods to improve emissivity, TNDT was sufficiently sensitive to detect damage in single leads in integrated microcircuits.

For the inspection of electronic components in manufacturing TNDT should meet the following requirements: (1) high stability of temperature readings in time (achieved through periodical calibration of test equipment by using reference "blackbodies"), (2) coordinate identification (easily achieved with automated systems), (3) removal of emissivity noise, i.e. working with true temperatures, (4) high scan rate needed for inspecting components with low thermal inertia, such as films and microchips, and the ability to check 100% of identical components, (5) optimization of electrical/thermal stimulation which can either be simulated use profiles or by applying a certain combination of stimulation parameters; the latter improves the detection of particular defects, (6) choosing a limited number of test points where temperature is well-correlated with defects to be detected, and (7) implementation of automated TNDT systems.

TNDT application areas in electronics are presented in Table 11.21 (see also reference books [78–80]).

Figure 11.66 shows the temperature distribution on a PCB revealing some overheated components.

Table 11.21 TNDT of electronic components

Components	Detected defects	Notes
Semiconductors	Defects in p-n barriers (surface degradation, inter-metallic joints, meso-plasma), uneven current density, cracks, gas-filled bulbs between crystals and substrates, heat sink deterioration, defects in fusion welds, crystal damage, lead breaks and short circuits	In many cases, thermal resistance serves as an indicator of the presence of defects. Pulsed power supply is often useful allowing analysis of thermal inertia. Hidden defects are typically well-localized in IR thermograms.
Integral microcircuits	Heat sink defects, lead breaks and short circuits, poor metallization, resistive film splits, poor adhesion, capacitor breakdown, volumetric defects in semiconductor materials	Spatial resolution may reach 15–20 μm. Inspection is performed at 10–100 test points with the case (cover) removed.
Multilayer printed circuit boards (PCB)	Thinning and corrosion wearing of conductors, poor metallization, conductor delamination	Thermal stimulation by pulsed electric current is typically used (optical heating in some cases). Complicated topology of temperature distributions, references are required.
Equipment	Incorrect connections of components, poor assembling, inadequate displacement of components on PCB.	TNDT is efficient at designing, manufacturing and operation stages. It is recommended in mass production of identical components, in particular, if a reference is available. The test procedure should be optimized by properly choosing test points and suppressing emissivity phenomena.
Resistors	Local thinning, non-conducting inclusions, poor contacts, cracks	When inspecting cylindrical resistors, four temperature profiles are to be recorded and compared with the reference. The size of defects detected is about 15 × 15 μm.
Capacitors	Electrolytic capacitor breakdown Short circuit of capacitor layer in integrated circuits (ICs)	Inspection is difficult because of low signals and emissivity noise.
Wire	Thin areas, nicks, cracks	Both contact (by electric current) and non-contact (by microwaves) heating techniques are used. Scanning rate up to 4 m/min. Wire thinning up to 20 μm can be detected.
Cathodes	Uneven coating	Temperature increase of 50–60 °C reduces cathode life time by an order of magnitude. TNDT at lower voltage excludes X ray NDT and allows improving component life time.

(continued)

Table 11.21 (continued)

Components	Detected defects	Notes
High-temperature film coatings	Delaminations between the coating and the substrate, coating thickness variation	Transient TNDT is preferable. Observation time is a few tens of milliseconds
Testing the welds on ICs connections to PCB contacts	Faulty fusion, fusion of the gold coating in the gap between the electrodes, gold flowing onto contacts, melting of IC connections and appearance of high resistance areas, fusion of connections and contact areas	Under point thermal stimulation, low-quality welding joints are characterized by specific temperature responses. Presence of emissivity noise may be a problem.

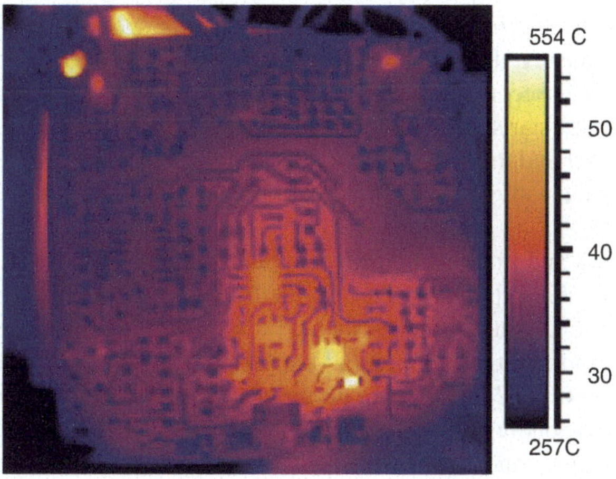

Fig. 11.66 IR thermogram of a PCB with overheated components

11.11 Rolled Metals

Typical defects in rolled metals are described in Table 11.22. These reduce the yield of a high-quality product. Inspection equipment for this application must be non-contact, highly efficient and resistant to, or protected from, high temperature. At metallurgical plants, ultrasonics is a typical NDT tool with ultrasonic waves being applied *via* a water jet. X ray methods have been used, but safety considerations limit the use of this technique.

In Russia, the VEMO Institute developed an automated system for the active TNDT of rolled metals [5]. A sheet of metal moving over a rolling bed is heated with gas burners or quartz tube lamps. The temperature distribution on the rear surface of the sheet is monitored with either an IR line scanner or IR imager. The number of necessary inspection units depends on the size of a rolled sheet and the distance between the sheet and the equipment. A pilot system implemented at a Russian metallurgical plant employed a Thermovision 550 IR imager that provided

Table 11.22 Typical defects in rolled metals [5]

Type of defects	Defect characteristic
Central porosity, bubbles	Diameter 10–30 mm.
Shrinkage	Length 2–20 mm, different depths.
Cracks	Diameter 1–30 mm. Cracks parallel to rolling direction.
Flakes	Length of hair-like defects 1–100 mm.
Hair line cracks	Defects parallel to the rolling direction, area over 100 mm^2.
Delaminations	Unwelded cracks with diameter 2–20 mm.
Voids	

a 2 × 2 m field of view at 6 m from the rolled sheets with a spatial resolution of 4 mm. An industrial system was designed involving several IR imagers to scan a complete rolled sheet. The heat power density was from 100 to 400 kW/m^2, and each point was heated for 0.5–1 second at a scanning speed of 100–2200 mm/s. With these parameters, the time needed for analyzing a single IR image was 1 second, including 0.04 second for data recording and 0.96 second for data processing.

As a metal sheet starts to move, a movement sensor supplies a start signal, which turns on the heater and the temperature monitoring unit. The heated areas are monitored thermographically at an optimum observation time that is defined by the operator.

The modeling and optimization procedures related to the inspection of rolled metals are detailed in [5]. In accordance with the TNDT theory discussed in Chaps. 3 and 4, some critical test parameters should be optimized, such as observation time, heating power and duration, differential temperature signals ΔT and running temperature contrast $C = \Delta T/T$. Software developed for this application provided solutions to the corresponding inverse problem, thus providing quantitative estimates of defect area, thickness and depth.

The main sources of noise are cinder (oxidation particles/surface contamination), rust, cracks, and shiny areas, as well as variations in metal sheet thickness and ambient temperature. Cinder and rust are thermal insulators; therefore, these areas are characterized by elevated temperature during external heating. However, this does not generally lead to significant temperature variations. Shiny zones behave similarly but they produce spurious reflections which may be identified by the TNDT automatic devices as defects (false positives). Occasional lateral displacements of rolled sheets may reach 50 mm, and this may change the absorbed energy by 10%, but this does not affect defect detectability much because decision making is based on the running contrast, which is independent of energy input. The influence of cracks can be neglected because the corresponding areas of temperature disturbance are small, while the ambient temperature represents an additive signal and does not affect defect identification.

The accuracy and validation of the TNDT system described above was ensured by testing reference samples in which defects were simulated by flat bottom holes sealed with metallic plugs. The system enabled the detection of defects with a diameter of 5 mm and a thickness of 0.05 mm with the mean error of estimating defect area being under 12%. These relatively high detection parameters can be achieved only by applying good data processing algorithms.

Table 11.23 Comparison between thermal and ultrasonic NDT of rolled metals. (Adapted from [5])

Technical feature	Thermal NDT	Ultrasonic NDT	Required (desirable) value
Minimum size of detected defects, mm	2	5	2
Tested material thickness, mm	0.5–30	3–30	0.5–30
Width of test area, mm	50–4000 (depends on heaters)	50–2500 (limited by ultrasonic sensors)	50–5000
Material maximum temperature, °C	500	100	800
Maximum speed of metal sheet movement, m/s	3	1	3
Possibility to inspect motionless samples	Yes	No	Yes
Allowed variations in sample thickness, %	80	10	100
Possibility to inspect samples of arbitrary shape	Yes	No (only sheets)	Yes
Allowed sample displacement in respect to NDT system, mm	100	10	100
Accuracy in evaluating defect coordinates, mm	10	15	5
Accuracy in evaluating defect depth, in % of sample thickness	10	30	10
Necessity of contact between sample and NDT system	Non-contact (3000 mm)	Contact (10 mm through an immersion medium)	Non-contact
Requirements to surface finish	Medium	High	Low
Requirements to the environment	Medium	High	Low
Approximate price, $	470 K	10 million	–

Table 11.23 presents a comparison between thermal and ultrasonic NDT in application to rolled metals. It is seen that TNDT meets the most of users' requirements of an optimal inspection method.

11.12 Welded and Soldered Joints

11.12.1 Temperature Control in Welding

The replacement of manual welding with robotic welding is a goal in modern industries. In automated welding processes, the quality of welded joints depends on how well robots are able to adapt to changes in materials, welding parameters, and overall welding geometry. Fluctuations of some parameters are random; therefore, their evaluation and control in real time may improve the quality of weld joints. The term "active" means either the control of welding parameters or the illumination of a welded area with an additional light source, e.g. a laser. Passive systems may analyze the thermal emission of material to be welded. Many active optical control systems are sensitive to the surface profile in front of a molten pool, and there may be a time delay between the welding and observation times. Pyrometric systems are free from this drawback because they measure the temperature of a molten pool. The geometry of a molten pool, the absolute temperature values and the temperature symmetry are directly related to the parameters that affect welding quality. Such parameters include the position of a welding area, welding root access, material thermal properties and thickness. Temperature measurement devices that operate in the visual or near-IR wavelength band are limited in their ability to measure temperatures adjacent to the molten pool. Besides, the welding arc, which has a temperature from 10,000 to 15,000 °C, is a strong source of optical noise. Therefore, IR temperature measurement systems that operate in the medium IR wavelength range are typically used for welding testing. Numerical analysis of a welding process has shown that IR signals necessary for controlling a welding process can be obtained by measuring the temperature profile along a line in front of the arc zone and perpendicular to the path of the weld [81]. This conclusion justifies the use of point (spot) radiometers or line-scanning IR thermometers and makes it unnecessary to use high-cost IR imaging radiometers. Such radiometers can use IR optical fibers that can withstand the aggressive welding environment and can transmit the IR radiation from the weld zone to the IR detector, which is located at a distance from the molten pool. Simple radiometric devices can measure the temperature of the metal at a particular distance in front of the molten pool, thus supplying information on the pool width and weld penetration depth. However, the most efficient are IR line scanning systems which measure discrete temperatures along a chosen line (see Fig. 11.67). The length of IR optical fibers may reach 5 meters. Scanning is achieved by using a bundle of single fibers, each of which measures the temperature at a particular point. In the system described in [81], eight PbS IR detectors were employed (spectral band 2–2.5 μm, measured temperature to about 500 C) to achieve 50 scanned lines per second with a spatial resolution of 1 mm. The noise was effectively suppressed when the scanning head, which was mounted in the same shell as the welding unit, monitored the surface perpendicularly rather than at an angle, as shown in Fig. 11.67. This system allowed an analysis of how the speed, as well as the arc current and gas flow, of oxyacetylene welding, influenced the temperature distribution and weld seam quality.

11.12.2 Spot Welding

Spot welding of sheet metal is performed by applying short pulses of electric current between electrodes, and this is widely used in the automotive industry. A typical car contains from 3 to 5 thousand spot welds, and their quality control and NDT are serious challenges for manufacturers. The use of ultrasonic and eddy current inspection techniques is marginal because of low inspection rates. The quality of

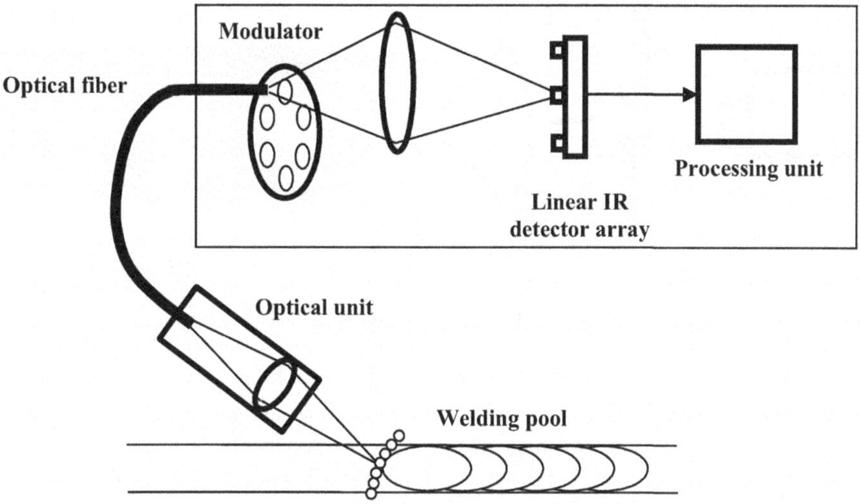

Fig. 11.67 Line-scanning IR pyrometer for monitoring welding temperature. (Adapted from [81])

welded spots is routinely tested by destructive methods on "sacrificial" cars that are destroyed in the process. The cost of these inspections to the Ford Motor company is over 100 million dollars a year, and the inspection time is 7–10 days per car.

The quality of spot welds is determined by the diameter and structure of the fusion penetration. Both parameters affect the heat conduction through the weld. In Russia, the first studies were performed by Bekeshko and Popov in the 1970s [82, 83]. The test procedure described in [82] employed a plasmotron for the short-time heating of weld joints for both one- and two-sided TNDT procedures. The dynamics of temperature distributions in the weld areas served as a quality criterion. In the later study, the same authors suggested the evaluation of joint quality by determining the apparent diameter of a circular isotherm that was evolving in time and was linearly proportional to the diameter of fusion penetration [83]. The experiments were performed on two 5 mm thick 10G2C steel samples heated with a 500 W halogen lamp. The theoretical analysis of the cooling process for spot welds with internal defects was performed by Vavilov et al. [84]. In spite of the simplicity of the analyzed model, which made it difficult to apply in practice, this study was one of the first in which an analytical expression for the dynamic temperature signal $\Delta T(\tau,\ r_d)$ *vs.* defect diameter r_d was derived. In further research, the thermal method of evaluating weld joint quality by the corresponding isotherm diameter has received no further development because the results could not be reproduced. This may be explained by the poor quality of IR imagers available at that time. However, this research topic recently gained a new impetus by the implementation of the principles of pulsed TNDT, which is shown in Fig. 11.68a [85, 86]. The heating is performed by using flash tubes, and the temperature distributions are captured by a high speed IR imager. When focusing the optical radiation of a 6 kJ flash tube onto a 2.5 cm-diameter area, the essential difference between defective and sound spot

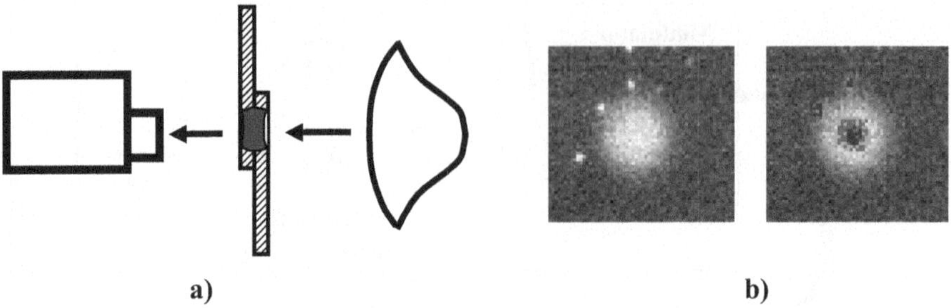

Fig. 11.68 Pulsed TNDT of spot welding (**a**) and IR thermograms (**b**) of a non-defect (left) and defect (right) weld joint

welds was observed (defective joints were produced by changing welding parameters while welding of 0.8 mm-thick sheets of low carbon steel). The heat pulse duration was 5 ms, while the maximum differential signals appeared over defective weld joints at times from 15 to 65 ms. Note that the temperature of defective weld joints was clearly lower than that in sound joints (Fig. 11.68b).

11.12.3 Diffusion Welding of Powerful Thyristor Billets

TNDT of powerful semiconductor rectifier (thyristor) billets was described in [87]. A billet consisted of a polished silicon layer (diameter 40 mm, thickness 390 μm) which was diffusion welded to a tungsten heat sink (diameter 40 mm, thickness 2.2 mm) using a 100 μm thick aluminum foil. Diffusion welding was accomplished by using a UDS-3 unit in the following conditions: temperature +540 °C, pressure 15 MPa, delay time 300 seconds, vacuum 66.5 mPa. Artificial defects were introduced into the aluminum foil by etching it in a weak solution of hydrochloric acid to a depth of 15–20 μm (see Fig. 11.69a). In real billets, defects can appear between the silicon and aluminum, as well as between aluminum and tungsten (Fig. 11.69a). At the final stage of manufacturing, aluminum electrodes are sprayed onto the surface of silicon. It was found that, in the absence of sprayed aluminum, internal defects are easily seen in a pre-heated billet directly through the silicon due to its transparency in the IR wavelength band. It is obvious that pure optical rather than thermal phenomena take place in this case. An example of introduced artificial defects and the corresponding IR image are presented in Fig. 11.70.

If the metallization of silicon is accomplished, the internal defects can be detected by using a two-sided procedure of pulsed TNDT. The temperature dynamics served as a decision-making parameter, in particular, Parker's technique was used. Some examples of billet temperature responses are shown in Fig. 11.71. The mean values of apparent thermal diffusivity that characterize the relative quality of billets are given in Table 11.24.

The statistical evaluation of TNDT results showed that the technique described above was able to detect defects that exceed 5% of the billet area.

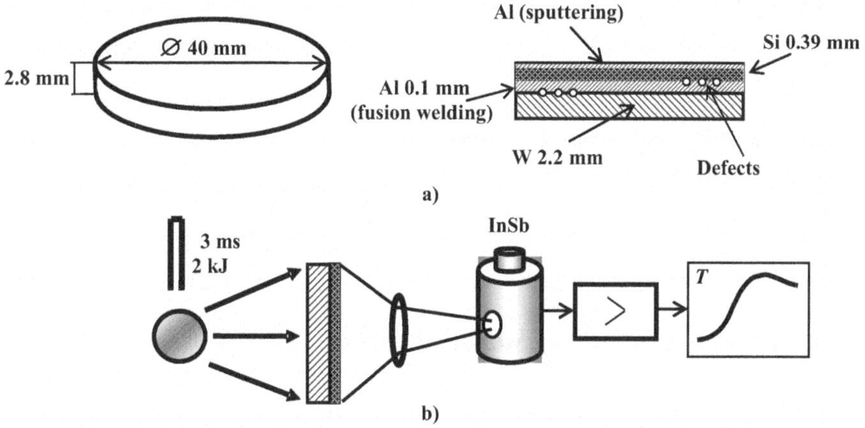

Fig. 11.69 Active TNDT of fusion welds in powerful thyristor billets: (**a**) thyristor scheme, (**b**) TNDT scheme

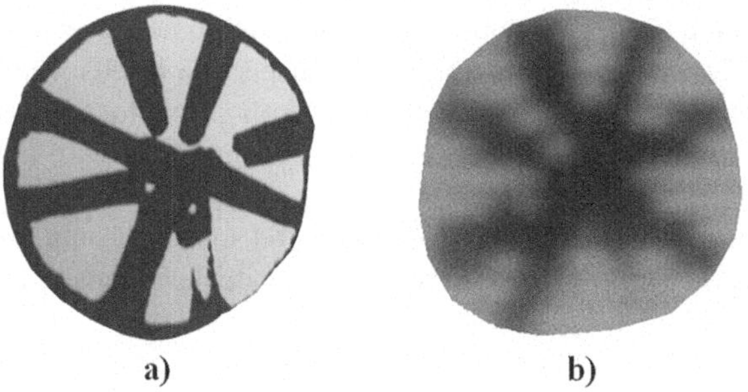

Fig. 11.70 TNDT of powerful thyristor billets without metallization: (**a**) artificial defects (metallographic section after sample destruction), (**b**) IR thermogram of defects (TV-03 IR imager)

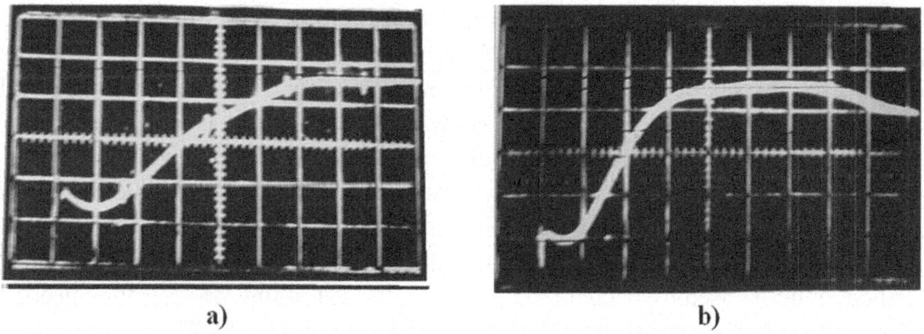

Fig. 11.71 Pulsed TNDT of powerful thyristor billets with metallization: (**a**) sound sample, (**b**) 20 × 20 mm defect in a fusion weld

Table 11.24 Defect detection in powerful thyristor billets by the $\tau_{1/2}$ parameter (Parker's technique)

Defect type	Apparent integral thermal diffusivity, 10^{-6} m^2/s
Non-defect billets (welding temperature +540 C)	59.5
Diffusion welding at lower temperature (+450 C)	60.5
Defect size 5 × 5 mm	61.6
10 × 10 mm	61.7
15 × 15 mm	62.0
20 × 20 mm	62.6

11.12.4 Active TNDT of Soldered Joints

In 1986, a photo-recording TNDT unit called FID-1 was developed at Tomsk Polytechnic University (former USSR) following a prototype system TD-21A developed by Storozhenko et al. at the Kharkov Institute of Radio Electronics [88, 89]. This unit was designed for the inspection of metallic (steel-brass) cylindrical samples that were joined by soldering (see the test scheme in Fig. 11.72a). Samples were heated with a linear heat source made of a nichrome bar with a cross-section of 8 × 2 mm. The line frequency was 10 Hz. Inspection results were recorded as a set of temperature profiles (Fig. 11.72b) on a roll of photographic film that was moving in front of an oscilloscope that was synchronized with the IR scanning (Fig. 11.72c).

Since the data recording was analog, the FID-1 unit is of only historic interest; but it illustrates the level of Russian TNDT hardware development in the 1980s.

11.13 Predicting the Lifetime of Cutting Tools by Determining Thermal Diffusivity

In mechanical engineering, the implementation of flexible automated machining lines has put the problem of predicting lifetime of the cutting instrument which is widely used by robotic machines. The instrument quality can be reliably evaluated by a mechanical test which is destructive by its very principle. Conventional NDT methods, such as eddy current, ultrasonic and others are characterized by the correlation coefficient between tool wear resistance and measured parameters lower than 0.2 that is unacceptable for practitioners. At the Tomsk Institute of Introscopy, it was proposed that measuring the thermal diffusivity of cutting tools by the Parker method could be used to evaluate their wear and to determine whether they needed to be replaced. The experimental setup is presented in Fig. 11.73. A cutting tool (insert) is placed into a cassette and heated with a pulse of optical radiation of 1 ms duration and 7.5 J energy (laser wavelength 1.06 μm). By the rear-surface temperature response (Fig. 11.74a), one can determine the apparent integral thermal diffusivity of a tested tool. In the calibration test, tool wear resistance was determined by

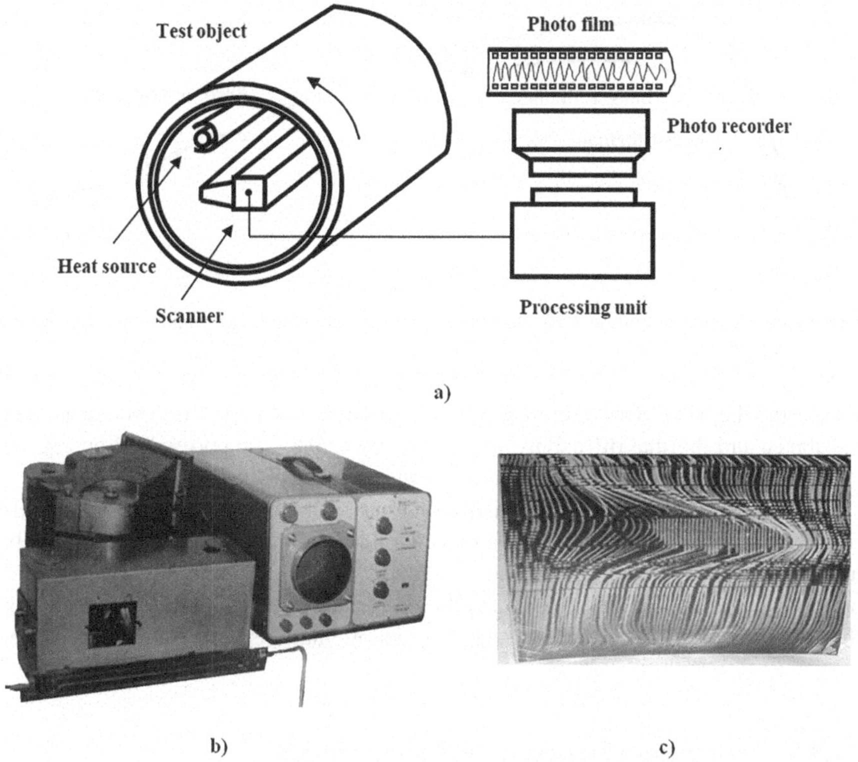

Fig. 11.72 Photo-recording unit FID-1 for TNDT of soldered cylindrical structures (Tomsk Institute of Introscopy, Russia,1986): (**a**) test scheme, (**b**) unit photo, (**c**) – test data recording example

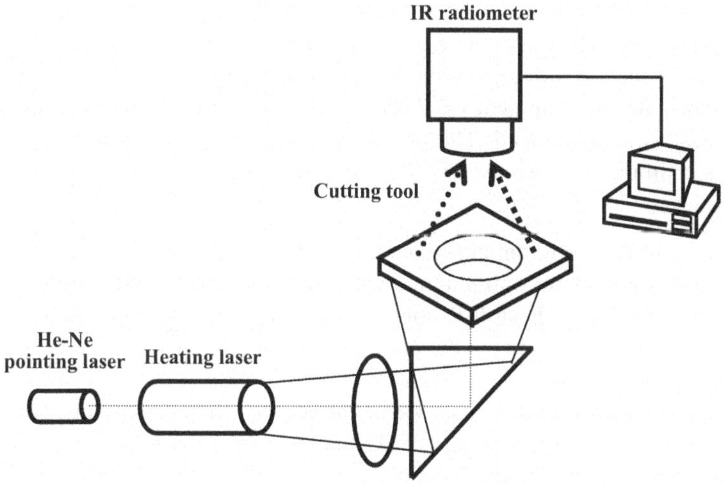

Fig. 11.73 Determining the apparent thermal diffusivity of a cutting tool by Parker's method

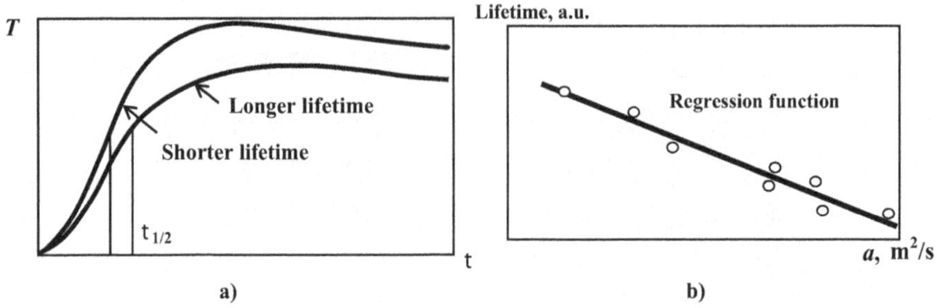

Fig. 11.74 Evaluating cutting tool lifetime by thermal diffusivity. (**a**) temperature signal, (**b**) regression function

measuring the wear of the cutting edge. The coefficient of correlation between wear resistance and thermal diffusivity was from 0.65 to 0.8. The corresponding regression field and the regression curve are shown in Fig. 11.74b.

The testing, which was accomplished on a series of cutting tools supplied by different manufacturers, showed that a considerable spread in the thermal diffusivity and, respectively, in the tool lifetime, appeared within each single series. The evaluation of the quality of single tools can be performed in a few minutes considering that 3–5 measurements are needed for different orientations of a tool in the cassette.

11.14 Detecting Corrosion in Thick Metals

There is a huge amount of massive metallic objects, such as tanks, vessels, support structures, pipes and boilers in developed countries. By some estimates, the financial loss from metallic wear and corrosion may reach up to 3% of the gross national product. Hence, the timely detection of corrosion in key parts of technical installations is important. The operating regulations of such installations typically include several NDT corrosion detection methods, starting with, ultrasonic and X ray. In recent years, the development of TNDT to detect corrosion in massive metallic objects has been encouraged. The principles of applying TNDT to thick metallic objects are similar to those used in the inspection of thin aluminum panels. Some peculiarities are listed below.

- In steel samples, differential signals over corrosion exist for longer times (up to several seconds) than in thin aluminum panels; however, with increasing metal thickness, the lateral heat diffusion also increases, and the problem of warming up a metal through its whole thickness requires applying a considerable amount of energy to a sample.
- A compromise between the factors that affect the efficiency of corrosion detection is expressed in the fact that TNDT may be very successful in the inspection of steel with a thickness up to 4 mm.

- Evaluating thicker metals is possible if the test object has high and even emissivity, a minor surface curvature, and free access to the test area. Powerful halogen lamps are considered to be the most efficient heat source for heating thick metals.

At NASA, an IR line scanning TNDT system was developed for detecting corrosion in power plant boiler tubes [90]. The boiler surface area may reach 9000 m^2 and access may be challenging. Therefore, NDT methods such as eddy current and ultrasonics, are difficult to use and relatively slow. The TNDT system included a 3 kW, 40 cm. long, linear heat lamp (a) and an IR camera with a temperature resolution of 0.025 °C. The system moved vertically along a boiler wall at a speed up to 5 cm/s. The sample excess temperature was under 10 °C. The IR thermographic system continuously captured IR thermograms, but the final diagnostic image was synthesized by a computer that selected the scanned lines at a certain time delay (observation time) in the range 0.5–1 second.

It is worth mentioning that in the former USSR, line-scanning TNDT systems were developed by Denisov et al. in the 1970s [89]. Those systems proved to be very efficient for the inspection of cylindrical GFRP structures.

When analyzing inspection results from these linear scanning systems, the material loss in the front surface of the boiler tubes can be calculated by using the following simplified solution for heating an adiabatic plate with a moving line source:

$$T(x,z) = \frac{q}{LVC\rho}\left[e^{-V\frac{(x+|x|)}{2a}} + 2V\sum_{n=1}^{\infty}\cos\left(\frac{n\pi z}{L}\right)\frac{e^{-\frac{Vx+|x|\sqrt{(2n\pi a/L)^2+V^2}}{2a}}}{\sqrt{(2n\pi a/L)^2+V^2}}\right], \quad (11.30)$$

where L is the sample thickness, V is the scanning speed and q is the heat source power in W/m.

Equation (11.30) is valid for the "far zone", where the initial coordinate reference point corresponds to the moving line [90]. At negative values of coordinate x, i.e. when an analyzed point is behind the heated line, the contribution of the infinite series can be neglected under certain conditions. If the test speed is low ($V < 2a/L$), this condition occurs if

$$x < \frac{2a}{V - \sqrt{(2\pi a/L)^2 + V^2}} \approx \frac{L}{\pi}. \quad (11.31)$$

At high speed ($V > 2a/L$), the contribution of the infinite series becomes small when the distance behind the heated line is larger than the sample thickness. In this case, the surface temperature is approximately:

$$T \approx \frac{q}{LVC\rho}, \quad (11.32)$$

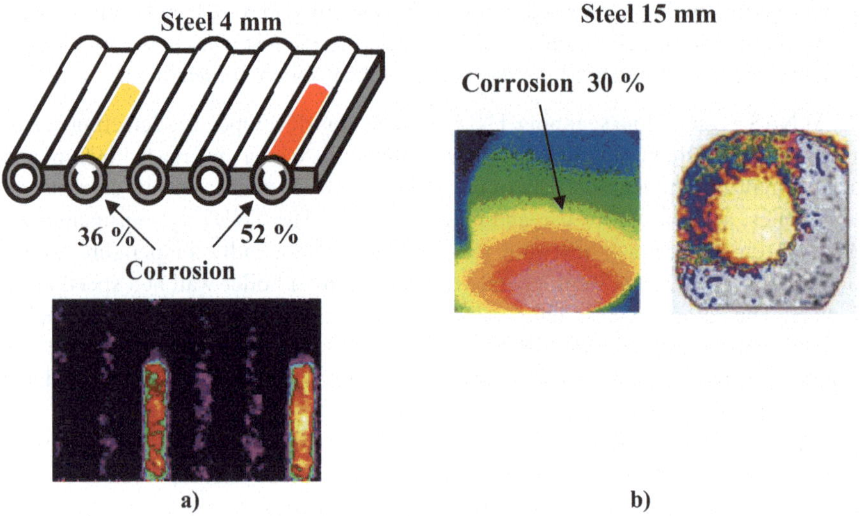

Fig. 11.75 Detecting corrosion in steel: (**a**) 4 mm-thick boiler section (flash heating 6 kJ, 10 ms), (**b**) 15 mm-thick steel sheet (heating with halogen lamps 1 kW, 300 s), left – raw image, right - thermal tomogram

Note that it is inversely proportional to the sample thickness. This last expression allows the evaluation of material loss.

Laboratory tests performed on samples cut from real boilers showed that TNDT was able to detect up to 18% material loss if the sample was first cleaned with water or sand and then coated with a water-washable black paint. The same samples were tested with ultrasonic methods. The agreement between thermal and ultrasonic test results was reasonable. The uniformity of heating was improved by developing a special scanning system (length 7.3 m, width 1.4 m and mass 230 kg). The system was mounted inside a boiler which had a 0.5 × 0.5 m hatch.

An example of this application is shown in Fig. 11.75a, which displays the thermogram of a section of boiler tubes heated with two Xenon flash tubes (ITC-CNR, Italy). A tube with a material loss of 36% is clearly seen. It is important to note that no special surface treatment (no paint) was applied prior to this inspection., The image was produced through advanced data treatment. The inspection of a thicker steel sample is illustrated in Fig. 11.75b. The application of a long duration (300 seconds) 1000 watt Halogen lamp heating pulse, followed by an unprocessed IR image (on the left) did not accurately reveal a 30% material loss. However, by applying an algorithm of dynamic thermal tomography, the indication of a round-shaped defect appeared (on the right).

Maldague described laboratory experiments on detecting corrosion in pipe bends [91]. A tube was alternately stimulated by alternating hot and cold water passing through the tube. IR thermograms were recorded in the transient stage at times from 1 to 3.5 s. The material loss reached 80% in a tube whose wall thickness was 5 mm.

The advantage of using this type of internal thermal stimulation was the absence of radiant noise which is typically produced by external optical heaters. The same defective areas were also detected by using a one-sided TNDT procedure where the tubes were heated with a hot air gun; however, differential temperature signals were significantly lower compared to the internal heating procedure.

11.15 Automotive Industry

IR thermography has been successfully applied to the inspection of race cars, although this technique can obviously be applied to the general car industry. For example, poor brake performance may be related to lower brake temperature. The evaluation of engines and exhaust systems allows the determination of the optimum operating temperature of critical components. In the same way, one can detect overheated bearings caused by insufficient lubrication. Note that in such test cases a reference point or temperature should be used. By inspecting tires one can detect relatively hot and cold areas, which can indicate patterns of unusual stress or areas of poor contact between the tire and the pavement. IR thermography has been also used in the analysis of window heating systems, cabin tightness, radiators, and even stitching of leather upholstery. Some laboratory research on diesel engines has demonstrated the ability of IR thermography to evaluate the degree of mixture reduction and leaks from cylinders [92]. Thermal methods are effective in the analysis of defects in wheels, windshields and headlights.

In the 1990s, AGEMA Infrared (now FLIR Systems) used a modified IR system to monitor temperature patterns on revolving tires under load. Internal defects, such as disbonds between the cord and the rubber, were detected dynamically, i.e. by analyzing thermal energy caused by roller friction. However, due a lack of recent publications on this topic, it seems that the research on IR thermographic inspection of tires has not been continued, probably, due to the appearance of other NDT techniques, such as holographic interferometry and laser shearography.

11.16 Pulp and Paper Industry

Historically, the pulp-and-paper industry was one of the first successful application areas for passive TNDT. For example, in a US poll on the use of IR thermography in industry, conducted in 1995, the largest number of users was reported in the pulp-and-paper industry. While a warm paper sheet is moving, moist areas are clearly seen due to evaporation. Since paper humidity is a critical quality parameter, IR thermographic monitoring is widely used in the pulp-and-paper industry of the USA, Great Britain, Scandinavia and other countries.

11.17 Boat Manufacturing (Composite Materials)

Hulls of large boats are generally made of welded steel or aluminum. And their inspection uses visual, X ray, ultrasonic, tap testing, and other methods. However, modern racing, cruising, fishing and luxury (yachts) boats are mostly made of composite materials, particularly GFRP and CFRP. The inspection of composite boats is historically performed by tap testing. However, this is an unreliable method and TNDT is used where the availability and cost are acceptable.

Jones and Lindgren reported the results of TNDT of a mine trawler MHC-51 of which the hull was made of GFRP [93]. Positive results were obtained by testing the hull frame, elements of the after deck, keel and webs. Burleigh used TNDT to inspect boats for the America's Cup races in San Diego and New Zealand.

11.18 Food Industry

In recent years, food producers and supermarkets have greatly increased their consumption of electrical energy. Simultaneously, an increasing number of food-related installations operate 24 hours a day. Frozen products are kept at a temperature of −18 °C or lower, and even a short temperature increase may compromise product quality. The use of IR thermography can result in the timely detection of defective components in electrical facilities that operate day and night. IR applications include the evaluation of energy losses in buildings and the evaluation of refrigerators, electrical motors, bearings, pumps, compressors, etc. For instance, the timely replacement of defective valves in refrigeration systems may save thousands of US dollars. Preventing the failure of walk-in freezers and large refrigerated storage rooms can prevent major financial loss. To illustrate this statement, the IR thermogram of a cold storage room in a Russian refrigerated industrial complex is shown in Fig. 11.76. IR thermographic monitoring was performed to evaluate the extent of repair required. Due to limited funding, it was advised to limit repair to only the sites of defective thermal insulation that are within the ellipses in Fig. 11.76.

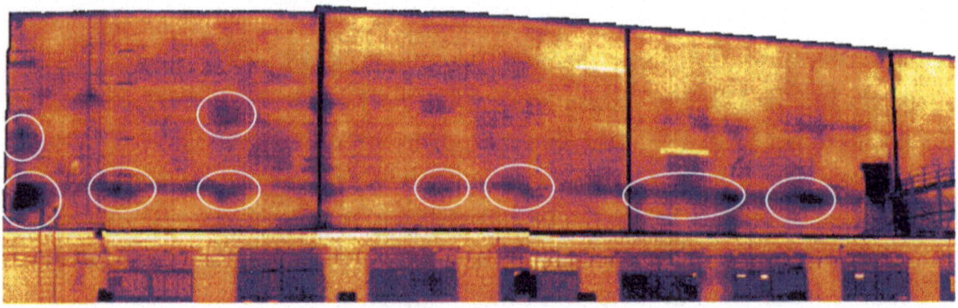

Fig. 11.76 IR thermogram of a refrigerator building wall (bridges of cold are marked by ellipses)

11.19 Detecting Gas and Oil Leaks from Pipes and Tanks

The detection of gas and oil leaks from industrial facilities and underground pipes, as well as from natural sources of these products, is a serious problem that is being addressed by using a variety of diagnostic techniques. The most successful implementation of IR thermography has been in the inspection of gas pipelines. This application is being widely discussed at the international level based on the experience of the USA, Sweden, Japan and Russia. In 2000, an international working group, which was formed in 1997, released a report entitled "State of the Art of Technologies for Remote Detection of Natural Gas" (Report SOC 110, ISSN 1102-7371, June 2000). During 2000–2003, the European Community financed a project called VOGUE (Visualization of Gas for Utilities and Environment).

A standard IR camera is able to detect gas leaks if they are accompanied by detectable temperature changes. However, the detection of some gas leaks cannot be accomplished by using a conventional IR camera; they may require the use of spectral filters or other special techniques.

In trunk pipelines, the gas is under high pressure, therefore, gas leaks can be detected passively by a local temperature decrease due to the *throttle effect*. For example, it has been reported that such leaks are reliably detected under snow, which fortunately has a high uniform emissivity. Russian experience in this application area was described by Aleev et al. [94]. If the ground temperature distribution is uneven, the identification of gas leaks becomes difficult even if leaks produce significant temperature anomalies.

If gas pressure is low, the temperature signature at the leak site is weak. However, it is possible to detect leaks in narrow spectral bands where a gas strongly *absorbs* IR radiation; for example, methane has significant absorption at 3.39 μm. Due to gas absorption phenomenon a warm gas can be seen on a cold background due to enhanced *emission* in the same wavelength band where it also possesses strong good absorption. Conversely, on a warm background, a gas is seen as a cold area due to *absorption*. This is the detection principle used in the Gas Finder IR equipment commercialized by FLIR Systems (see Chap. 9).

Some preliminary results on the passive detection of gas leaks from pipelines under relatively low gas pressure were reported by Ljungberg in the framework of the VOGUE project [95]. It has been confirmed that the detection efficiency depends on weather conditions, defect diameter, pipeline depth and soil properties. For example, leakage rates from 1 to 20 liters a minute from a 1 mm diameter hole were not reliably detected under certain conditions, while under other conditions; leakage rates from 1 to 5 liters a minute from a 1 mm diameter hole were reliably detected. Leakage site size and shape, as well as wind speed, influence the formation of gas clouds. Optimum detection conditions are when the gas is floating at few decimeters above the ground. In this case, temperature signals may reach 2 °C. If the wind speed is higher than 3 m/s, gas clouds rise to several meters and mix with air; therefore, their detection by purely thermographic indications becomes unlikely or impossible.

In an active TNDT method, the ground surface may be irradiated with a laser that operates at a gas absorption/emission wavelength. In practice, dual wavelength band systems are used where the second wavelength band is chosen outside but close to the absorption band. Such systems suppress the effect of emissivity variations. Whether a laser system can be used depends on the distance to a test object and the reflectance of the background (soil). In Russia, *combined* systems which include both active (laser) and passive (IR imager) channels are being developed by Pergam.

The detection of leaks from pipelines, especially from those buried in soil, represents a challenging technical problem. In the book [94], the IR surveying conditions are classified in three groups: (1) related to a pipeline (geographical, agro-physical and construction), (2) temporal and meteorological, and (3) related to flight conditions.

Northern pipeline sections are typically buried at depths from 0.5 to 1 m and often cross water. Performing IR thermography at night is most efficient due to a lower noise background. However, to meet safety requirements, aerial surveys are typically performed during the day, when considerable thermal noise appears from solar heating and the shadowing effect of vegetation. Obviously, performing IR thermography is preferable when the Sky is cloudy.

Since real gas and oil leaks generally occur suddenly and must be eliminated as soon as possible, the development of detection technologies may include the simulation of breakdown situations. For example, an "open" gas leak was simulated by discharging gas into the atmosphere through an extracting valve on a 6 mm diameter copper tube [95]. The tube length (30–80 m) was chosen to reach the area where the leakage was simulated. The open end of the tube was placed in soil so the gas plume came into the atmosphere under a 20-30° angle relative to the horizon. The leakage rate was such that the visible gas cloud size did not exceed 0.5 m.

"Closed" leakages were modeled by injecting gas into soil, snow or water.

Russian experimental results on the evaluation of integrity of pipelines by means of IR thermography in different seasons were summarized in [94].

11.20 Aerial IR Surveys and Monitoring

IR monitoring of the Earth in the middle IR band became possible after portable IR systems were developed and they were mounted on aircraft.

It is believed that IR images of the Earth in the middle and long wavelength bands are generally less informative than images in the visual and near-IR bands. However, one of the main advantages of aerial IR monitoring is its ability to operate at night time and to detect sources of heat emission which are invisible in "normal" visual photos. Heat sources include electrical power production installations, transmission lines pipelines, volcanoes and geothermal sources, forest and peat fires, animals and human beings, etc. In some cases, it is possible to detect hidden objects and even analyze their operational performance (pipelines, archeological objects, nuclear submarines, etc). By using the difference in thermal inertia between two objects, and by performing IR surveys in both day and night time, one may evaluate soil moisture and/or density. The status of agricultural crops can be also evaluated by some spectral features of IR radiation, while vegetation diseases can be diagnosed by its spectral features or elevated temperature. The monitoring of the surface of bodies of water allows the analysis of flow patterns in the ocean, lakes and fresh water streams. IR can also detect thermal and chemical pollution, (see Fig. 11.39).

An IR survey from a helicopter or aircraft (including drones and quadrocopters) helps to reveal forest fires and follow the propagation of underground fires, as well as detect hidden fissures in glaciers, etc. An IR survey is more reliable than the more common visual aerial survey because IR radiation transmits better through fog, smoke and ground clutter (leaves, etc.) than visible radiation. Night time IR thermographic monitoring has been used for searching and rescue of lost persons, counting wild life, etc.

Some space based IR systems have been developed for military purposes, as well as for surveying natural resources (in the USA: ERTS, SkyLab, HCMM, Spot, Irs, ADEOS, Landsat; in the former USSR and Russia: Meteor-Pogoda, Cosmos, Resource-0). For instance, orbital space stations can survey the Earth in 1–2 weeks over a rather narrow strip with a spatial resolution of about 10 m. Satellite systems are equipped with multi-channel systems for performing IR reconnaissance from the visual to the middle IR wavelength bands; typically, there is one channel operating in the range from 10 to 12 μm. A correlation between the data in the visual and IR bands has been found. Satellite thermal images are used for meteorology where the IR images provide important information about thermal processes in the atmosphere. And it is being used to collect data on forests and other agricultural areas.

Russian Resource series satellites developed by the Institute for Electro-Mechanics (Moscow) had a mass of 1900 kg including a 500 kg payload. The MSU-E scanning device operated in the 0.5–0.9 μm wavelength band and had a spatial resolution of 45 m across and 32 m along the flight direction with the scanned strip being 45 km. The IR scanning unit MSU-SK provided spatial resolution from 140 to 550 m in the following bands: 0.5–0.6, 0.6–0.7, 0.7–0.8, 0.8–1.1 and 10.4–12.6 μm. It is worth mentioning that satellites of super-high spatial resolution operating in the visible spectrum are able to detect objects on the Earth surface smaller than 1 m.

The USA's Landsat-7 satellite launched in 1998 had a mass of 2200 kg and was equipped with a digital 380 Gbit memory that allowed the storage of up to 100 images that represented an area of 185 × 185 km. The Earth's radiation is sensed in 8 bands from 0.45 to 12.5 μm; the spatial resolution is 15–30 m in the near-IR band and 60 m in the thermal band. The rate of data transfer is 150 Mbit/s to be captured by 21 receiving stations. In 2017–2018, the USA launched two weather satellites GOES-16 and GOES-17. The latter one possesses 16 channels: 2 visible, 2-near-IR and 10-IR. These satellites monitor aerosols, clouds, thunderstorms, hurricanes, rainfall, moisture, etc. Their spatial resolution varies from 500 m to 2 km depending on the wavelength band. The first Russian satellite equipped with an IR camera, Canopus-B-IR, was launched in 2017 to monitor forest fires.

The price list for satellite imaging is presented in Table 11.25.

Table 11.25 Satellite imaging price list [96]

Satellite/sensor	Number of channels	Image price, $	Scanned strip width, km	Spatial resolution, m	Price for imaging 100 km², $
Meteosat/MVIRI (USA)	3	113	hemisphere	2500	free
NOAA/AVHRR (USA)	5	115	2400	1000	free
Resource-0/MSU (Russia)	4	1300	600	150	0.36
IRS (India)	4	300	130	36	2
MOS (Japan)	4	316	100	50	3
Landsat/MCC (USA)	4	1233	185	80	4
Resource-0/MSU-E (Russia)	3	200	45	35	10
Landsat/TM (USA)	7	5180	185	30	16
SPOT/HRV-XS (France)	3	2487	60	20	69
SPOT/HRV-P (France)	1	31,337	60	10	87

11.21 Detecting Antipersonnel Landmines

According to the United Nations (UN), there are over 100 million (and 200 to 300 million according to other estimates) landmines deployed in more than 80 countries worldwide. It is important to note that the deployment rate for new mines is about 1 million a year, and that is 20 times higher than the current rate of mine removal. Up to 26 thousand people become victims of mine explosions every year, or 1 person every 20 minutes. Afghanistan, Cambodia, Kuwait, Iraq and several other countries are some of the most dangerous territories. Obviously, the problem of detecting landmines is important from a humanitarian point of view. This problem is being explored by military, political and social organizations, such as UN, UNICEF, etc. There are several international programs and projects being conducted on this topic worldwide. For example, the Coleman Research Corporation and US Army Centre for Environmental Studies are active in the USA. During the last few years, a conference entitled "Detection Technologies for Mines and Minelike Targets" has been held in the USA within the Defense & Security symposium (former AeroSense) by SPIE (Society of Photo-Optical Instrumentation Engineers). In 1996, the Mine Warfare Association, USA, organized a symposium devoted to the technologies and problems of mine use. Several international conferences on this topic have been held worldwide in the last two decades, see, for example, *m.technav.ieee.org*.

Numerous techniques have been developed for detecting surface and buried landmines. Due to the variety of mine types, geometric forms and sizes (up to 700 modifications), as well as the installation techniques, the most reliable detection

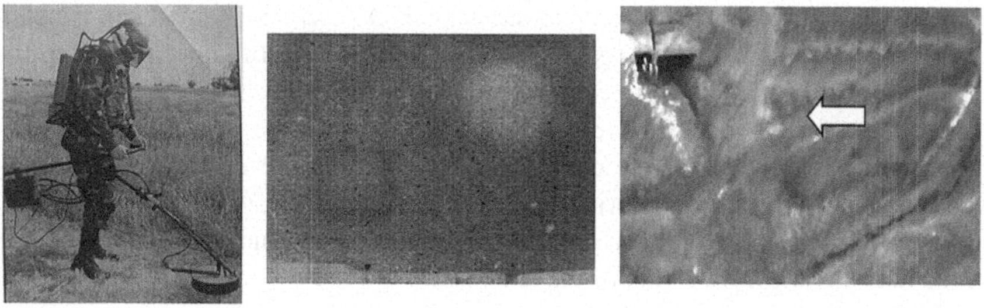

Fig. 11.77 Detecting buried antipersonnel landmines: (**a**) US Army officer uses a metal detector, helmet-mounted IR camera and ground penetrating radar (**b**) IR thermogram of landmines buried in sand at 3 cm depth: caseless mine (right) and aluminum case mine (left), solar irradiation power density ~250 W/m^2, (**c**) IR thermogram of two mines buried in a forest site (from an altitude of 30 m)

results are achieved by combining several different physical principles. In practice, visual inspection, as well as bomb-sniffing dogs and standard metal detectors, are widely used. However, modern caseless landmines may contain only very small metallic parts and this renders metal detectors useless. On the other hand, in populated areas, there may be numerous buried metal objects (trash, debris etc.) that cause a high number of false alarms. Currently, IR systems and ground penetrating radar are considered to be the most promising methods for landmine detection (Fig. 11.77a).

The use of various NDT techniques is based on the fact that buried mines possess various physical properties (density, mass, heat capacity, thermal conductivity, electrical conductivity, optical properties, etc.) that are different from the surrounding soil. Their presence also changes the local soil structure. Their installation is accompanied by the formation of air gaps around a mine, as well as by a change in the soil moisture and emissivity. Also, the vegetation (grass, leaves, bushes, etc.) will be disturbed. Both passive and active TNDT procedures can be implemented if the corresponding identification parameters (temperature or/and emissivity) produce thermal signals above the noise level.

IR imagers have been used to detect buried landmines for decades. In the well-known monograph by Hudson [97], it is mentioned that the detection limit may reach 40 cm (about 16 inches). That seems to be high now, based on tests performed in the last decade. The mechanism of pure thermal conductivity analyzed in Chap. 4 may guarantee the detection of landmines at depths up to 5–10 cm. The presence of other physical phenomena and noise factors leads to a complicated combination of identification parameters, and this can render decision making for detection unreliable.

The thermogram in Fig. 11.72b shows two landmines buried at the depth of 3 cm in sand. In spite of the uniform background, the surface temperature signals constitute only fractions of a degree Celsius due to weak solar radiation. It is interesting to note that, due to the low thermal conductivity of trinitrotoluene (TNT), caseless landmines produce "positive" temperature signals ΔT, while landmines in a metallic case produce by "negative" signals. Under more powerful solar radiation, the detectability of buried landmines is significantly improved; for example, the US Army successfully used IR imagers to detect mines during the "Desert Storm" operation in Kuwait. A much more complicated test case is presented in Fig. 11.77c, where the footprints of two buried landmines are nearly lost in clutter caused by vegetation. In this case, only the round shape of landmines and the operator's experience may allow the detection of the landmines.

11.22 IR Thermography in Surveillance, Security and Anti-Terrorism Activity

The first night vision units which appeared before WWII used active illumination of a scene by an IR source and were therefore easily detected by similar equipment used by the enemy. Modern passive night vision units use image intensifiers and are manufactured as goggles, mono-vision telescopes or binoculars. Such devices can be easily blinded by intensive light sources in the scene. The introduction of portable low-cost FPA IR imagers on the market increased their use for night vision (hunting, surveillance, military, etc.). The main advantages of IR imagers are their passive operation, relative independence from weather conditions, the absence of the blinding phenomenon and an increased operation distance. In the last decade, these advantages have been appreciated by the US coast guard, police and customs. Numerous applications of IR imagers in night time police operations have been reported. Often, criminals have been arrested in full darkness, hiding in bushes or under leaves, and sometimes being detected from a helicopter. Passive IR units can be installed on the perimeter of guarded areas, although their price is still high compared to conventional night vision devices. However, the prices of simple non-radiometric IR imagers used for night driving, hunting, surveillance etc., has decreased quickly and are now in the range of $2000 to $6000. Some years ago, FLIR adopted a version of its ThermaCAM E2 to a police version called the ThermoVision Scout. This unit has a built-in algorithm named "InstAlert", which allows the deliberate colorizing of certain objects, such as human beings, in red, thus making them easily visible on a gray background.

The HEAT company developed two types of commercial systems where signals from an image intensifier and an IR imager were combined (image fusion option). In the first of these systems, an optical image supplied by the image intensifier is imposed on the corresponding thermal image with a 1280 × 1024 format. The user can control the image brightness of each channel in order to adapt to particular inspection situations. The second type of unit works with digital images and involves several algorithms of preliminary data processing, thus facilitating decision making and detection. Such units utilize IR thermographic modules with an image format of 160 × 120, 320 × 240 and 640 × 480 and may include some additional devices, such as GPS.

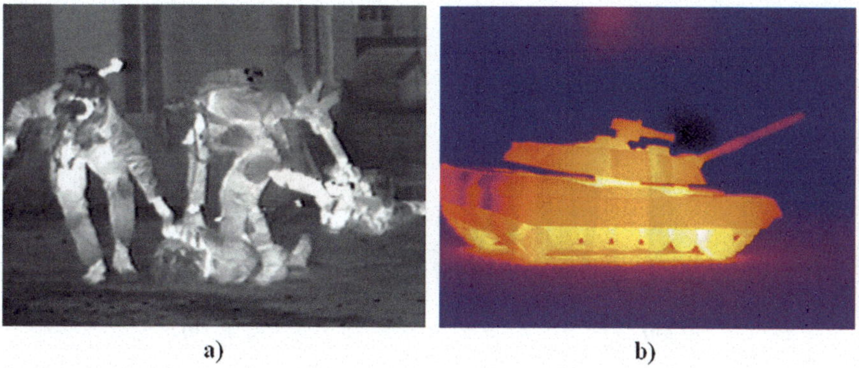

Fig. 11.78 Night vision by using IR thermography: (a) police operation (courtesy: Inframetrics), (b) IR thermogram of a tank

IR imagers are increasingly used in the detection of forest fires. Some simplified models have been developed specially for this application; for example, IR imagers from BAE Systems can operate under high ambient temperatures. IR equipment allows firemen to detect temperatures that are close to the combustion temperatures of paper, wood, etc. Also IR radiation is not obscured by smoke and water vapor, so the IR image provides an image of objects (bodies, furniture) that can not been seen visually. IR cameras make firefighting safer and more productive. .

US companies Raytheon and General Motors pioneered the use of IR thermography for night driving. It is expected that, in the near future, many car producers will be installing low-cost IR imagers on their vehicles.

IR has been used in Finland for many years (SETTLA Company) for search and rescue. This has been typically accomplished in the late evening by an IR imager mounted in a helicopter. It has been reported that IR devices have been used for rescuing people under building debris after earthquakes and other catastrophes.

IR thermographic surveillance, including night vision, is an important tool for anti-terrorism activity in the recent years.

Two illustrations of using IR imagers as night vision instruments are shown in Fig. 11.78.

Recently, passive and active systems in the Terahertz range appeared on the market. Figure 11.79, shows images obtained with a ThruVision system, (Great Britain) which operates at wavelengths in the range around 1 mm.

11.23 Military Applications

The review in Sect. 7.1 shows that, as with many other scientific endeavors, the development of IR technologies has been driven by military demands. Description of all the military applications of IR thermography is well beyond of the scope of

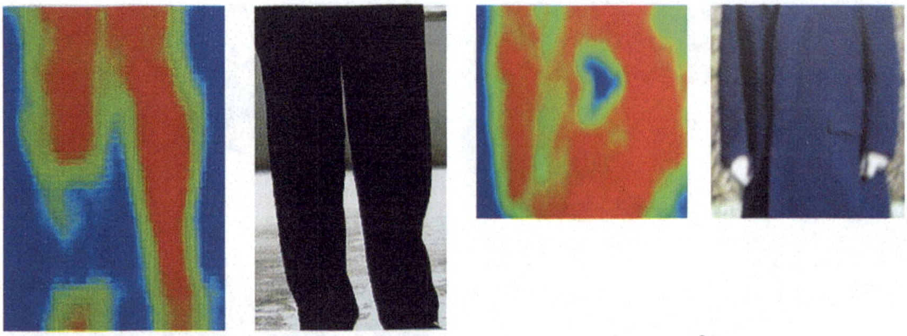

Fig. 11.79 Detecting hidden objects on a human body by using an imager operating in the Terahertz wave frequency range (T-4000 unit from ThruVision, Great Britain) (**a**) drugs under pants leg, (**b**) gun under a overcoat

this book so we will limit ourselves to a short list of major application segments. Note that IR detection of landmines was previously discussed since this application is closely related to active TNDT. The advantages of the IR technique for military applications are the non-contact, remote operation, the ability to avoid visual surveillance, the ability to detect heat sources, etc. The main military application areas are:

- detection and tracking of satellites,
- early detection of missile launches by their jet plumes
- detection of airplanes, warships, tanks and other military vehicles by the thermal radiation emitted by engines and power plants,
- detection of soldiers and armed forces,
- detection of buried objects (landmines) by their surface thermal contrast,
- search for nuclear submarines by their warm trace on the surface of the ocean,
- missiles guidance onto aerial, naval and land targets,
- navigation and control of flying vehicles by the detection of landmarks,
- IR reconnaissance of land and, in particular, military objects and thermal mapping,
- remote detonation of explosives,
- night vision for military vehicles,
- night targeting,
- detection of camouflage,
- counteraction of enemy IR methods,
- guarding, surveillance, and observation,
- IR communication,
- detection of flames in military objects,
- damage assessment.

The list above shows that the military applications of IR thermography are mainly related to the detection, reconnaissance and tracking of fighting forces and materiel.

The surveillance function of IR thermography was described in Sect. 11.22.

And there are IR, or thermal, position finders (locators) which detect of objects by their thermal radiation, and determine a path to objects. Automatic missile guidance is accomplished by using special modulators of IR target radiation. Thermal detectors were used during the WWII, mainly for detecting aerial targets. Presently, these devices are widely used in the naval and aerospace areas.

Thermal radiation of the scene is a complicated mixture of targets and background, as well as heat sources and atmospheric conditions. Before being sensed by an IR system, this radiation passes through the atmosphere where it experiences absorption and dissipation. Atmosphere turbulence may cause refraction of thermal radiation and, hence, its distortion at the IR detector(s). If an IR device produces an image, the operator makes a decision about target coordinates and characteristics. A set of parameters that are related to target radiation is called a *signature*. The sensing and evaluation of target radiation is the basis for the detection and identification of target signatures. Table 11.26 describes factors and subfactors influencing performance of target acquisition. Since there is no unique model of target identification, one should develop such models which should be related to particular IR systems and specific detection scenarios.

Military IR equipment was widely used by British troops during the Falkland Islands conflict with Argentina, as well as by the US Army in Vietnam, Kuwait and Iraq. There have been some fragmentary reports regarding the use of IR imagers by the Russian army in Chechnya.

The price range for military IR imagers is high. For example, the price of a US-made IR tank sight is about \$100 K, a gunsight is about \$30 K, and a helicopter-borne anti-tank system can reach \$600 K.

In recent years, increasing attention is being paid to protecting the key personnel of governments and large companies from kidnapping and terrorism. To provide road protection in dangerous areas, Ibis Tek (U.S.A.) developed a system called Cobra which can be mounted on an off-road car. The system is equipped with a 360° swiveling machine gun. For night vision and weapon aiming an EMX FPA 160 × 120 format IR camera is used. The Cobra is cheaper than many other military vehicles and requires fewer personnel.

11.24 Medicine

11.24.1 IR Thermography

In 1939, Hardy proposed measuring the temperature of a human body by IR radiation. In years past, medical diagnostics became an obvious and dynamic application area of IR thermography. In the former USSR, the peak interest in medical IR thermography, also known as IR screening, took place in the 1980s when thermal diagnostics was performed in over 100 cities. The organization of a diagnostic network across the country was under discussion, similar to X-ray diagnostics facilities. In

Table 11.26 Factors and subfactors influencing target acquisition performance [98]

Major factor	Subfactors	
Mission-related factors	Reconnaissance Surveillance Navigation Vectored attack	Attack targets of opportunity Attack predefined targets Damage assessment
Target and background	Type Size Shape Contrast Temperature Reflectance Emittance	Motion Shadow Masking Camouflage Clutter Cues
Environment	Visibility Cloud cover Sun angle Solar loading Diurnal variations Seasonal variations	Illumination level Attenuation Transmittance Scattering Patch radiance Turbulence
Platform	Crew size and interaction Vibration environment Target exposure time	Altitude Range Speed
Sensor parameters	Sensor type Spectral band Field of view Resolution Dynamic range Sensitivity	Sample spacing Interlace Frame rate Line of sight (LOS) panning capability Field of regard Search field size
System display	Screen viewing angle Aspect ratio Color Signal-to-noise ratio Resolution Phosphor persistence Gray-scale rendition	Interlace Contrast ratio Veiling glare Surround lighting Flicker Frame rate
Observer variables	Training Motivation Experience Prebriefing Stress Task load	Visual acuity Search pattern Fatigue Age IQ

spite of a decline in IR medical activity, there is still a Russian system for the qualification and certification of medical thermographers. Russian domestic IR imagers TKVr-IFP were developed at Institute of Semiconductor Physics, Russian Academy of Sciences, and were certified for medical diagnostics and this unit is being used by several hospitals across the country.

The initial euphoria that accompanied IR thermography as a universal non-invasive medical diagnostic tool, has given way to the understanding of a variety of relationships between pathological processes in a human body and their resulting temperature patterns on the surface of a body. After a brief period of stagnation medical IR is experiencing renewed interest due to the appearance of a new generation of IR imagers and greatly increased computer data processing. The world community of medical thermographers are represented by several national societies of thermology and other organizations (Institute for the Advancement of Medical Thermology, USA) that arrange conferences on the subject. In the USA, IR imagers are used as a diagnostic tool not only by large hospitals but also by private doctors, especially chiropractors. A great increase in the use of IR thermographic screening occurred during the SARS (Severe Acute Respiratory Syndrome) pandemic and bird flu concerns in the early 2000's. In South-East Asia, IR cameras are installed at airports to detect passengers with elevated temperature.

The surface of a human body emits in the following wavelength ranges: under 5 μm – less than 1%, 5–9 μm – 20%, 9–16 μm – 38%, over 16 μm – 41%. The peak radiation is at 9.5 μm. At wavelengths above 5 μm, the emissivity of human skin is independent of racial and individual peculiarities and is about 0.98.

Inflammation, infection, tumor growth, circulatory problems, bruising and mechanical damage to tissues can cause asymmetry in the IR signature of a human body. The human body uses convective heat transfer to react, to reflex areas which often coincide with Zakharyin-Head areas.

The temperature rise that accompanies pathological situations in a human body is from 0.1 to 5 °C. It is believed that, by using modern IR cameras and computer data treatment, it is possible to perform medical diagnostics based on temperature variations of 0.02 to 0.1 °C. Noise temperature fluctuations on the surface of human bodies are determined by a large number of factors that are related either to an individual person (constitution, age, sex, etc.) or to the environment (season, time of day, temperature, humidity, etc.). Moreover, the earlier research fulfilled at the Institute of Radiotechnics and Electronics, Russian Academy of Sciences, demonstrated that the temperature in particular body areas is subject to periodic variations, and the frequency of these variations may vary for various pathologies. A noticeable temperature dynamic also occurs after applying various functional and chemical stimulations that can improve diagnostic reliability. Dynamic IR image sequences can be processed by specialized software, such as SCANPRO (Institute of Radiotechnics and Electronics). Algorithms for performing IR thermographic medical diagnostics were developed at the Moscow Institute of Radiotechnics, Electronics and Automation. A specialized diagnostic computer program is supplied with TKVr-IFP IR imagers.

The manifold character of thermal implementations of pathological processes and a lack of reliable references (for instance, one classifies up to 4 types of face IR thermograms) decrease efficiency of medical IR thermographic diagnostics. Unfortunately, in many cases, detectability of pathologies is fairly low (30–85% in gastroenterology), and the number of false-positive conclusions is high. It is believed that IR thermography complements standard medical diagnostic test methods. Passive thermography may be combined with thermal and ultrasonic stimulation techniques. In some cases (breast cancer and blood circulation system diagnostics), it is sufficient to place a patient's hands into warm or cold water. It is worth noting that conclusions on the statistical efficiency of medical IR thermographic diagnostics were obtained in Russia in the 1970–1980s on the basis of relatively primitive IR cameras (Rubin, TV-03, Raduga). The increasing use of modern FPA IR cameras may change these conclusions. For example, in accordance with the signal detection theory, the efficiency of diagnostics is limited by a high level of false-positive conclusions (false alarm), because any pathology may change body surface temperature distributions, and the challenge is to detect weak temperature signals on a noisy background.

IR thermographic examinations are conducted at an air temperature of 19–21 °C, humidity of 40–70%, and air speed below 0.25 m/s. In an examination room, all heat sources, including water heaters, electric lights and direct sunshine should be removed or shielded. Human skin areas to be tested must be shaved and free of any cosmetics and all creams. Before being inspected thermographically, a person should be kept at a constant temperature for about 10–15 minutes.

The main application areas of IR thermography in medical diagnostics are as follows:

- oncology: detection of pre-cancerous growths and evaluation of various oncopathologies,
- gastroenterology: duodenal and stomach ulcers, pancreatitis, enterocolitis, hepatitis, abdominal abscesses,
- urology: nephritis, urinary calculosis, varicocele, dropsy, cystitis, prostate cancer, tumors of kidneys and testicles,
- neurology – nerve injury; peripheral neuropathy, atherosclerosis, venous insufficiency, thrombosis, Raynaud's syndrome,
- trauma medicine, orthopedics and rheumatology - psoriasis, fractures, scoliosis, arthritis, burns,
- pulmonology – ARD (Acute Respiratory Distress), tuberculosis,
- cardiology and endocrinology – arrhythmia, hypertension, myocardial infarction,
- plastic surgery and transplants - evaluation of post-surgery circulation and grafts, .
- IR thermography can also be used in gynecology, reflexology, pediatrics, etc.

IR thermography has also been used for lie detection (as a polygraph), see Sect. 11.25.

Some examples of medical IR thermograms are given in Fig. 11.80.

11.24.2 Microwave Thermography (Radio-Thermometry)

IR thermography allows the recording and subsequent evaluation of the surface temperature of biological objects that may be related to subsurface processes. However, it is also possible to detect *in-depth temperature profiles* by recording electromagnetic (microwave) radiation in the wavelength range from 1 to 150 mm (*radiothermometry* technique) due to the fact that biological tissues are partially transparent to radiation in this range.

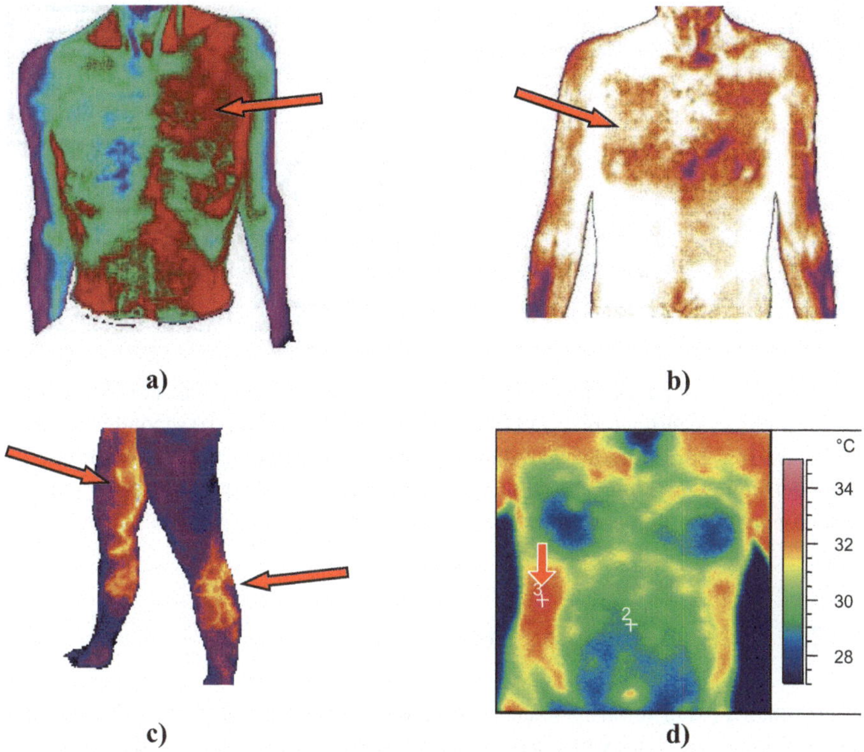

Fig. 11.80 Medical IR thermographic diagnostics: (**a**) cancer of the left lung. (**b**) lymphoma of thoracic lymph nodes. (**c**) varicose veins at an early stage. (**d**) liver cancer

There are two types of *microwave thermography*: *remote* (*radiothermography*) and *applicator-type*. The first technique is analogous to IR thermography enabling the non-contact determination of subsurface temperatures within a test body. Measurement results depend on emissivity and experimental conditions. However, the most common is the applicator-type microwave thermography which is achieved by using microwave detectors and horn antennas that are placed on the surface of test objects. Depending on frequency, calibration, and detector design, microwave thermography allows the determination of both absolute and relative temperatures within biological objects. For instance, at the Institute of Radiotechnics and Electronics, an RT-20 radiothermograph was developed. In combination with specialized software, this unit can be also used for performing thermal tomography (see Fig. 11.81).

Another implementation of the method for depth profiling biological objects is *passive acoustic thermotomography* developed by Guliayev, Pasechnik, Bograchev et al. [99]. Temperature profiles up to depth of few centimeters can be produced by solving the corresponding inverse problem. Some advanced mathematical techniques, including one by Tikhonov, have been used in this case.

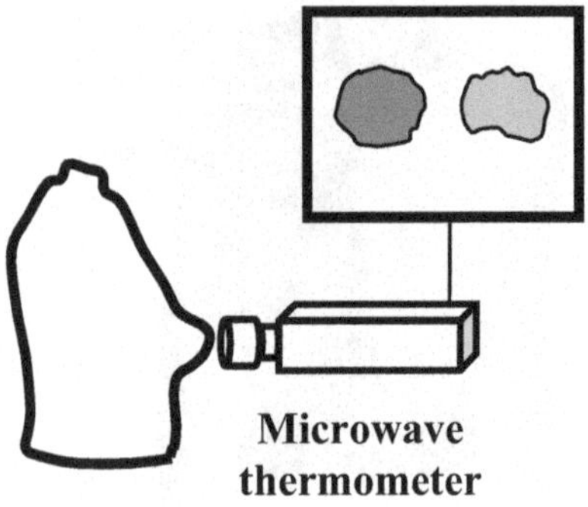

Fig. 11.81 Microwave thermometry in medicine

The main drawbacks of practical microwave thermometry are: (1) low spatial resolution (few centimeters), (2) large data storage time (over 10 seconds) because of the low power of microwave radiation, (3) necessity to use very sensitive microwave detectors which must be cooled to the temperature of liquid helium; typically, superconductive bolometers are used for this purpose.

11.25 Other Application Areas

In criminology and forensic science, there have been attempts to identify thermal traces of human activity which persist from few seconds to few hours depending on the type of impact and environmental conditions. In full darkness, a person can be detected at distances up to 2–3 km depending on the IR camera lens and weather. An IR camera is able to prove that a car recently parked in an area, and detect foot prints and recently fired shell casings. The attempts to find buried corpses have been also reported. Also, fake documents have been detected after examination in the near IR range,

Recently, the potential of IR thermography as a tool for lie detection was demonstrated. In the USA, polygraphs are routinely used for various reasons. It is estimated that the probability of correct conclusions is up to 90%. Polygraphs require contact with the test subject, and use several measurement "channels" (often three), such as blood velocity, respiration rate, intensity of perspiration, etc. The data is "scalar".

The advantages of IR thermography for lie detection are: (1) non-contact, (2) dependence between the squared temperature deviation (compared to the internal body temperature) on blood velocity, (3) vector nature of the data: if a dynamic temperature response is analyzed, (4) availability of an additional information channel that can generally enhance reliability of this method, (5) image-based documentation of test results that allows a comparison of data obtained over long intervals of time. The use of a Raytheon SW Radiance HS IR imager (25 mK temperature resolution), as a lie detector was reported in [100]. The most significant findings were found in the periorbital areas where temperature signals were small, but advanced data treatment resulted in a 78% correct result.

Some fragmentary reports have been published on using IR thermography at Customs inspection to detect smuggling, for example, inside a metallic box filled with an opaque liquid (oil). In Russia, it has been suggested to use IR thermographic monitoring at border guard stations to detect trucks smuggling goods in the night time. In laboratory experiments, the possibility of detecting explosive belts on a person has been demonstrated. However, it is obvious that the practical use of IR imagers for this purpose, for example, at airports, would be difficult, since using a low decision threshold, too many passengers will be stopped for checking (false positives), while a high decision threshold will lead to missing explosives (false negatives).

Kriksounov, a Russian expert in IR engineering in the 1980s, said that, "…possibilities of IR technologies are limited only by our imagination…". Diverse applications include analyzing temperature gradients in turbojet engines, investigations on Tokamak warming, controlling the operation of mixers that produce hot bitumen to bind radioactive wastes, research in fundamental hydro- and aerodynamics, analysis of beer quality in kegs, the study of billiard ball dynamics, and very many others. The Snell Group (USA) uses the slogan, "Think thermally".

References

1. Nondestructive Testing Handbook. Vol. 3 "Infrared and Thermal Testing", Techn. Ed. Xavier P.V. Maldague, Ed. Patrick O. Moore. A.S.N.T., Bellingham (2001)
2. Ljungberg, S.-A.: Infrared techniques in buildings and structures: operation and maintenance. In: Maldague, X. (ed.) Infrared Methodology and Technology. Nondestructive Testing Monographs and Tracts, pp. 211–252. Gordon & Breach Science Publish Amsterdam, Amsterdam (Netherlands) (1992)
3. Hart, J.M.: A practical guide to infrared thermography for building surveys. Building Research Establishment Report, Garston, Watford, U.S.A.-24 p
4. Evans, J.: Meteorology and infrared measurements. Proc. SPIE "Thermosense IV", Vol. 313. U.S.A., pp. 64–68 (1981)
5. Budadin, O.N., Potapov, A.I., Kolganov, V.I., et al.: Thermal Nondestructive Testing of Products, p. 476 (in Russian). "Nauka" Publish, Moscow (2002)
6. The diagnostics and determination of thermal parameters of building envelopes by using IR thermography. – Guidelines # 1305/442 from 10.01.2001 approved by Gosstandard, Russian Federation. 36 P. (in Russian)
7. Luong, M.P.: Infrared thermographic evaluation of fatigue behavior of concrete. Transact. 14th intern. conf. on structural mech. in reactor technol. (SMiRT 13), Lyon, France, pp. 155–162 (1997, Aug 17–22)
8. Vavilov, V.P.: Infrared thermography of mechanical stresses in building constructions (diagnostics of the TVVKUS casern in Tomsk after the catastrophe). Control. Diagn. **1**, 21–25 (in Russian) (1998)
9. Ivanov, G.S., Podolyan, L.A.: Energy saving in buildings. Heat Supply News. **7**, 8–13 (in Russian) (2001)
10. Standard ISO 6781-83. Thermal insulation, Qualitative detection of thermal irregularities in building envelopes. Infrared method
11. GOST 26254–84. Russian state standard. Buildings and constructions. Methods for determining envelope thermal resistance (in Russian)

12. GOST 26629–85. Russian state standard. The method of IR thermographic quality inspection. Thermal insulation of building envelopes (in Russian)
13. Edis, E., Flores-Colen, I., De Brito, J.: Time-dependent passive building thermography for detecting delamination of adhered ceramic cladding. JONE. **34**(3), 24–16 (2015)
14. Cornienko, S.G.: Detecting modes of deformation in monolithic concrete by using IR thermographic monitoring. Prikladnaya fizika (Appl. Phys). **1**, 134–138 (in Russian) (1997)
15. Moropoulou, A., Avdelidis, N.P., Koui, M.: Detection of defects of airport pavements using infrared thermography. CSNDT J. Vol 4020, 5–8 (2000)
16. ASTM Standard D-4788-88. Test method for detecting delaminations in bridge decks by using infrared thermography, 1997, U.S.A.
17. Frumuselu, D., Radu, C.: IR thermography applied to ground-level reinforced concrete constructions belonging to electricity networks. Insight. **40**(7), 501–504 (1998)
18. Drozdov, V.A., Sukharev, V.I.: Building thermography. Moscow, "Stroyizdat", p. 238 (1987). (in Russian)
19. Grinzato, E., Bison, P.G., Bressan, C. et al.: Active thermal testing of delaminations in frescoes' plaster. Proc. 4th intern. conf. on nondestr. testing of works of art, Berlin, Germany, pp. 357–366 (1994, 3–8 Oct)
20. Bison, P.G., Grinzato, E., Marinetti, S., Braggiotti, A.: Fresco thermographic inspection by convective heating technique. In: Thompson, D.O., Chimenti, D.E. (eds.) Review of Progress in Quantitative Non-destructive Evaluation, vol. 17, pp. 1769–1776. Plenum Press, New York (1998)
21. Grinzato, E., Bison, P.G., Bressan, C., Mazzoldi, A.: NDE of frescoes by infrared thermography and lateral heating. Proc. Eurotherm seminar #60, QIRT'98, Lodz (Poland), pp. 64–67 (1998)
22. Vavilov, V.P., Grinzato, E., Bison, P., Marinetti, S.: Thermal NDT of air-filled delaminations under wall frescos. Defectoskopiya (Rus. J. NDT). **7**, 73–83 (in Russian) (1994)
23. Grinzato, E., Bison, P.G., Marinetti, S., Vavilov, V.: Thermal NDE enhanced by 3D numerical modeling applied to works of art. In: Proc. 15th world conf. on NDT, Rome (Italy), (available only on CD). 9 p (2000, 15–21 Oct)
24. Spagnolo, S., Ambrosini, D., Paoletti, P.: Comparative study on the efficiency of some optical methods for artwork diagnostics. Proc. laser techniques and systems in art conservation. Munich, Germany. pp. 227–234 (2001, 18–19 June)
25. Grinzato, E., Bison, P.G., Marinetti, S., Vavilov, V.: Non-destructive Evaluation of Delaminations in Fresco Plaster Using Transient Infrared Thermography, vol. 5. No. 4, pp. 257–271. Res. in NDE, Springer-Verlag, New York (1994)
26. Grinzato, E., Marinetti, S., Vavilov, V., Bison, P.G.: Nondestructive testing of wooden painting by IR thermography. In: Proc. 8th Europ. Con. NDT, pp. 342–346, Barcelona (2002)
27. Enno, I.K., Duzhykh, F.P., Melentyev, N.N.: Nondestructive testing of chimneys by using IR thermography. Electricheskiye stanzii (Electrical stations). **6**, 23–26 (in Russian) (1988)
28. Vavilov, V., Demin, V., Shiryaev, V.: Inspecting smokestacks by IR thermographic smokestacks and heat conduction modeling. Proc. SPIE "Thermosense-XXIII", Orlando, U.S.A., Vol. 4360. pp. 324–332 (2001)
29. Volkov, E.P., Gavrilov, E.I., Duzhykh, F.P.: Gas-Collecting Chimneys of Thermal and Atomic Power Plants, p. 240. Energoatomizdat Publish, Moscow (1987). (in Russian)
30. Franchuk, S.V.: Tables of Thermal Properties of Building Materials, p. 264. Energoatomizdat Publish, Moscow (1969). (in Russian)
31. Kouznetsov, N.V. (ed.): Thermal Analysis of Boilers, p. 320. Energia Publish, Moscow (1973). (in Russian)
32. The Rules for Technical Exploitation of Electrical Stations in Russian Federation, 15th edn, p. 46 (in Russian). Ed. A. Alexandrov. ORGRES Publish, Moscow (1996)
33. RD 153–34.0-20.363-99. Management directive. The basics of IR diagnostics of electrical installations and high-voltage transmission lines, 2003 (in Russian)

34. Bazhanov, S.A.: Infrared Diagnostics of Electrical Facilities and Distribution Lines, p. 76. Electrotechnician Library.//Appendix to "Energetica" Journal, Energoprogress Publish, Moscow (2000). (in Russian)
35. Madding, R.P., Lyon, B.R.: Wind effects on electrical hot spots-some experimental IR data. Proc. SPIE "Thermosense-XXII", Vol. 4020, pp. 164–172 (2000)
36. Madding, R., Leonard, K., Orlove, G.L.: Important measurements that support infrared surveys in substations. InfraMation Proc. **3**, 19–25 (2002)
37. Kaplan, H.: Practical Applications of Infrared Thermal Sensing and Imaging Equipment.-Tutorial Texts in Optical Engineering, vol. TT34, p. 164. SPIE Press (1999)
38. OST 16.0.800.343–76 (Russian Industrial Standard). Turbo-generators. Testing stator cores
39. Methodical guidelines on exploitation of contact rings and brush gears of turbogenerators by power 165 MW and higher. # TI 34–70–024-84, Moscow, Russia, Soyuztechnenergo enterprise. 1984 (in Russian)
40. Khizhnyakov, S.V.: Practical Calculations of Thermal Insulation, 236 P. Russia, Gosenergoizdat Publ, Moscow (1976)
41. Lanius, M.A.: Infrared applications for steam turbine condenser systems. Proc. SPIE "Thermosense XXII", Vol. 4020, pp. 107–113 (2000)
42. Ducar, R.J.: Pulsed thermographic inspection and application in commercial aircraft repair. Proc. SPIE "Thermosense-XXI", Vol. 3700. pp. 77–83 (1999)
43. Airbus adopts infrared thermography for in-service inspection. Insight. **36**(10), 126 (1994)
44. Vavilov, V.P., Marinetti, S., Pan, Y., Chulkov, A.: Detecting water ingress in aviation honeycomb panels: qualitative and quantitative aspects. Polym. Test. **54**, 270–280 (2016)
45. Maldague, X., Cielo, P., Poussart, D., Emerson, P.: Thermographic nondestructive evaluation of turbine blades: methods and image processing. Ind. Metrol. **1**, 139–153 (1990)
46. Morozov, G.A.: Development of NDT techniques in aviation. Control Diagn. **7**, 3–8 (in Russian) (2002)
47. Cielo, P.: Pulsed photothermal evaluation of layered materials. J. Appl. Phys. **56**, 230–236 (1984)
48. Cielo, P., Dallaire, S.: Optothermal NDE of thermal-barrier coatings. Proc. '85 ASM's intern. conf. surface modification and coatings, Toronto, Canada, pp. 10–14 (1985, 14–16 Oct)
49. Maclachlan Spicer, J.W., Kerns, W.D., Aamodt, L.C., Murphy, J.C.: Determination of Degree of Thermal Barrier Coating Disbanding by Time-Resolved IR Radiometry, vol. 10B, pp. 1193–1200. Rev. of Progress in Quant. NDE, Plenum Press, New York (1991)
50. Maclachlan Spicer, J.W., Kerns, W.D., Aamodt, L.C., Murphy, J.C.: Measurement of coating physical properties and detection of coating disbonds by time-resolved radiometry. J. NDE. **8**(2), 107–120 (1989)
51. Osiander, R., Maclachlan Spicer, J.W., Amos, J.M.: Thermal inspection of SiC/SiC ceramic matrix composites. Proc. SPIE "Thermosense XX", Vol. 3361, pp. 339–348 (1998)
52. Lehtiniemi, R., Hartikainen, J., Rantala, J., et al.: Fast Photothermal Inspection of Plasma-Sprayed Coatings of Primary Circulation Seal Rings of a Nuclear Reactor. Part Two: after the Trial Run, vol. 12, pp. 1236–1240. Rev. of Progress in Quant. NDE, Plenum Press, New York (1993)
53. Varis, J., Hartikainen, J., Lehtiniemi, R., Luukkala, M.: A simple transportable imaging system for fast thermal nondestructive testing. Proc. Quant. IR Thermography-QIRT'92, Paris, France, pp. 235–238 (1992, July 7–9)
54. Netzelmann, U., Walle, G.: High-speed pulsed thermography of thin metallic coatings. Proc. Quant. IR Thermography-QIRT'2000, Reims, France, pp. 206–211 (2000, July 18–21)
55. Happlodt, P.G., Ellingson, W.A., Gardiner, T., Krueger, J.: Defect detection in multi-layered plasma sprayed zirconia by time-resolved IR radiometry: a comparison between analytical and experimental methods. Proc. SPIE "Thermosense XVI", Vol. 2245. pp. 210–219 (1994)
56. Shepard, S.M., Favro, L.D., Thomas, R.L.:Thermal wave NDT of ceramic coatings. Proc. SPIE "Thermosense XVII", Vol. 2473. pp. 190–193 (1995)
57. Bison, P., Marinetti, S., Grinzato, E., et al.: Inspecting thermal barrier coatings by IR thermography. Proc. SPIE "Thermosense XXV", Vol. 5073, pp. 165–173 (2003)

58. Vavilov, V.P., Marinetti, S., Cernuschi, F., Roba, D.: Thermal NDT of turbine blade thermal protection coatings. Defectoscopiya (Rus. J. NDT). **7**, 74–83 (in Russian) (2005)
59. Vavilov, V.P., Kasatkin, M.A.: Thermal NDT of heat-resistant nickel-based alloys. Defectoscopy (Rus. J. NDT). **4**, 51–54 (in Russian) (1990)
60. Alcott, J.: An investigation of nondestructive inspection equipment: detecting hidden corrosion on USAF aircraft. Mater. Eval. (5), 64–73 (1994)
61. DelGrande, N.K., Durbin, P.F.: Dual-band IR imaging to defect corrosion damage within airframes and concrete structures. Proc. SPIE "Thermosense XVI", Vol. 2245. pp. 202–209 (1994)
62. Syed, H.I., Winfree, W.P., Cramer, K.E., Howell, P.A.: Thermographic Detection of Corrosion in Aircraft Skin, vol. 12, pp. 724–729. Rev. of Progress in Quant. NDE, Plenum Press, New York (1993)
63. Zalameda, J., Rajic, N., Winfree, W.P.: A comparison of image processing algorithms for thermal nondestructive evaluation. Proc. SPIE "Thermosense XXV", Vol. 5073, pp. 374–385 (2003)
64. Pratt, J.: Detecting hidden exfoliation corrosion in aircraft wing skin using thermography. Proc. SPIE"Thermosense-XXII", Vol. 4020. pp. 200–209 (2000)
65. Burleigh, D.: A portable, combined thermography/shearography NDT system for inspecting large composite structures. Proc. SPIE "Thermosense-XXIV", Vol. 4710, pp. 578–587 (2002)
66. Delpeche, P., Boscher, D.M., Lepoutre, F., et al.: Quantitative Nondestructive Evaluation of Carbon-Carbon Composites by Pulsed IR Thermography, vol. 12, pp. 672–678. Rev. of Progress in Quant. NDE, Plenum Press, New York (1993)
67. Burleigh, D., Kuhns, D., Cowell, S., Engel, J.: Thermographic nondestructive testing of honeycomb composite structural parts of Atlas space launch vehicle. Proc. SPIE "Thermosense-", Vol. 2245, pp. 132–138 (1994)
68. Storozhenko, V.A., Vavilov, V.P., Volchek, A.D.: Nondestructive Testing of Industrial Products by Using Active Thermal NDT, p. 128. Tekhnika Publish, Kiev (1988). (in Russian)
69. Kush, D.V., Rapoport, D.A., Budadin, O.N.: Inverse problem of automated thermal NDT. Defectoscopiya (Soviet J. NDT). **5**, 64–68 (1988)
70. Cramer, K.E.: Quantitative thermal imaging of aircraft structures. Proc. SPIE "Thermosense-XVII", Vol. 2473. pp. 226–232 (1995)
71. Thompson, K.G., Crisman, E.M.: Thermographic inspection of solid-fuel rocket booster field joint components. Mater. Eval. (48), 1096–1099 (1990)
72. Welch, C., Eden, T.J.: Numerically Enhanced Thermal Inspection of Shuttle Solid Rocket Motor Inhibitor/Liner/Fuel Bondline, vol. 8B, pp. 2027–2034. Rev. of Progress in Quant. NDE, Plenum Publishing Corp, New York (1989)
73. Burleigh, D., Engel, J.: NDT of adhesively bonded fixed foam insulation for Atlas/Centaur cryogenic fuel tanks. Proc. 6th NASA NDE Conf., Johnson Space Center, Houston, Texas, pp. 86–94 (1989, May 2)
74. Burleigh, D.: Thermographic testing used on the X-33 space launch vehicle program by BF Goodrich Aerospace. Proc. SPIE "Thermosense-XXI", Vol. 3700. pp. 84–92 (1999)
75. Cramer, K., Winfree, W., Hodges, K. et al.: Status of thermal NDT of space shuttle materials at NASA. Proc. SPIE "Thermosense-XXVIII". Vol. 6205, pp. 62051B1–9 (2006)
76. Gazarik, M., Johnson, D., Kist, E., et al.: Development of an extra-vehicular (EVA) infrared (IR) camera inspection system. Proc. SPIE "Thermosense-XXVIII". Vol. 6205. pp. 62051C1–11 (2006)
77. Kontsevoy, Y.A., Kudin, V.D.: Inspection Methods in the Production of Semiconductor Devices, p. 140. Energia Publish, Moscow (1973). (in Russian)
78. Vavilov, V.P.: Thermal NDT of Composite Structures and Radio Electronic Components, p. 162. Radio i Svyaz Publish, Moscow (1984). (in Russian)
79. Danilin, N.S., Baklanov, O.D., Zagorovski, Y.I.: Theory and Methods of the Infrared Inspection of Radioelectronic Devices, p. 164. USSR Ministry of Defense Publish, Moscow (1974). (in Russian)

80. Use of Infrared thermography in electronics. https://www.ept.ca/features/use-infrared-thermography-electronics/
81. Boillot, J.P., Cielo, P., Begin, G., et al.: Adaptive welding by fiber optic thermographic sensing: an analysis of thermal and instrumental considerations. Weld. J. 209–217 (1985)
82. Bekeshko, N.A., Popov, Y.A.: Evaluation of weld spot joint quality by using IR thermography. Defectoscopiya (Rus. J. NDT). **3**, 123–127 (in Russian) (1971)
83. Bekeshko, N.A., Popov, Y.A.: Evaluation the diameter of weld spot joints by visualizing thermal distributions. Defectoscopiya (Rus. J. NDT). **6**, 86–90 (in Russian) (1972)
84. Vavilov, V.P., Gorbunov, V.I., Kouznetsov, V.B.: Theoretical analysis of thermal NDT of weld spot joints. Defectoscopiya (Rus. J. NDT). **1**, 21–27 (in Russian) (1973)
85. Shepard, S., Chaudry, B., Predmesky, R., Zaluzek, M.: Pulsed thermographic inspection of spot welds. Proc. SPIE "Thermosense-XX", Vol. 3361, pp. 320–323 (1998)
86. Jonietz, F., Myrach, P., Suwala, H., Ziegler, M.: Examination of spot welded joints with active thermography. JONE. **35**(1), 1–14 (2016)
87. Bragina, L.A., Vavilov, V.P., Ivanov, A.I., et al.: Evaluation of diffusion welding defects in semiconductor devices by using active thermal NDT. Defectoscopiya (Rus. J. NDT). **5**, 69–70 (in Russian) (1983)
88. Vavilov, V.P., Shiryaev, V.V., Tanaseychuk, S.Y.: Complex approach to designing a thermal NDT system for inspecting soldered joints. Defectoscopiya (Rus. J. NDT). **10**, 63–67 (in Russian) (1978)
89. Denisov, S.S., Volkov Ya, A., Storozhenko, V.A., et al.: A scanning optical head for active thermal NDT. Defectoscopiya (Rus. J. NDT). **6**, 116–118 (in Russian) (1975)
90. Cramer, K.E., Jacobstein, R., Reilly, T.: Boiler tube corrosion characterization with a scanning thermal line. Proc. SPIE "Thermosense XXIII", Vol. 4360, pp. 594–605 (2001)
91. Maldague, X.: Pipe inspection by IR thermography: NDT solution. Mater. Eval. **57**(9), 899–902 (1999)
92. Martinez, V.M., Martinez, B.T., Olmeda Gonzalez, P.O., Peralta Uria, R.W.: Fault detection in diesel engines using infrared thermography. Insight. **44**(4), 228–232 (2000)
93. Jones, T.S., Lindgren, E.A.: Thermographic inspection of marine composite structures. Proc. SPIE "Thermosense XVI", Vol. 2245, pp. 173–175 (1994)
94. Aleev, R.M., Ovsiannikov, V.A., Chepurski, V.N.: Aerial IR Thermographic Apparatus for the Inspection of Oil Transport Piping, p. 160. Nedra Publish, Moscow (1995). (in Russian)
95. Ljungberg, S.-A., Jonsson, O.: Passive gas imaging – preliminary results from gas leak simulations: a field study performed during real world conditions. Proc. SPIE "Thermosense XXIV", Vol. 4710, pp. 468–477 (2002)
96. Potseluyev, A.A., Arkhangelski, V.V.: Remote Sensing Methods for the Environmental Monitoring (School Book), p. 184. STT Publish, Tomsk (2001). (in Russian)
97. Hudson, R.D.: Infrared system engineering. Wiley-Interscience, New York (1969)
98. Accetta, J.S., Shumaker Exec, D.L. (eds.): The Infrared and Electro-Optical Systems Handbook. V. 3 "Electro-Optical Components", p. 668. ERIM and SPIE Optical Engineering Press, Bellingham/Washington (1993)
99. Pasechnik, V.I., Anosov, A.A., Barabanenkov, Yu.N. et al.: Determining in-depth temperature in biological objects by using passive acoustical thermotomography. - Proc. Nizhny Novgorod acoustical scientific session, NNGU, 2002, pp. 375–378 (in Russian)
100. Pavlidis I.T. Lie detection using thermal imaging. // Proc. SPIE "Thermosense-XXVI", 2004. V. 5405. P. 270–279

Chapter 12
Certification and Documents for TNDT

Abstract In this chapter, professional certifications for TNDT are summarized. An extensive list of certification documents, including national and international standards, inspection guidelines and other documents, is presented.

12.1 General Principles

NDT personnel in all NDT methods and in all industries should have the following:

1. Documented classroom training provided by a certified or designated instructor.
2. Written and practical ("hands on") examinations administered by a certified or designated instructor
3. Sufficient experience to properly perform NDT tasks.

The requirements for the above are defined under various NDT certification systems.

And requirements will vary for different NDT methods.

Requirements are higher for more complex NDT methods such as UT (ultrasonic), RT (radiography), and ET (Eddy current) and IR NDT.

And requirements are lower for simpler NDT methods such as MT (magnetic particle testing) and PT (Penetrant testing).

Personnel who have met the requirements are "qualified," and can be certified if they meet additional requirements.

The following terms are used in certification (from ASNT (American Society for Nondestructive Testing))

- Certification Standard: National or international documents describing the requirements for the qualification and certification of NDT personnel.
- Recommended Practice: A formal document that provides nationally or international guidelines, and describes the qualification and certification process for NDT personnel.

V. Vavilov, D. Burleigh, *Infrared Thermography and Thermal Nondestructive Testing*, https://doi.org/10.1007/978-3-030-48002-8_12

- Certification System: A standard or recommended practice governing the certification requirements, a third-party certification program (if applicable) or the employer's written practice, and additional employer documents used in the administration of their certification program.
- Certification Program: Documented procedures and processes for certification based on a recommended practice or other standard.

There are multiple NDT certification systems worldwide, but they can generally be divided into two main types: "employer-based" and "central" certification systems. In Central certification systems the qualification examinations are administered by an independent third party based on a central certification standard.

Most certification programs have three levels of NDT qualification: Level I, Level II and Level III. These certificates are intended to verify that a thermographer can: (1) operate an infrared camera, (2) collect good data, (3) produce professional infrared inspection reports, and (4) understand the techniques and limitations of infrared thermography for specific applications.

Certificates are issued by Level IIIs.

NDT certification requirements are provided by ASNT Recommended Practice SNT-TC-1A, NAS 410 and other documents.

An NDT Level I individual should be qualified to follow a written procedure, to set up and calibrate equipment and to inspect parts. A Level I may not be permitted to evaluate inspection results or to "buy off" or accept parts. An NDT Level I should receive necessary instruction and supervision from a Level II or III certified individual.

An NDT Level II individual should be thoroughly familiar with the scope and limitations of the method for which they are certified. In addition to the capabilities of a Level I, a Level II may evaluate test results according to the written procedures and "buy off" or accept parts. A Level II may write a test procedure, but the procedure must be approved by a Level III. A Level II may provide oversight to Level I personnel as they gain experience toward Level II.

An NDT Level III individual should be capable of developing, qualifying and approving inspection procedures and techniques, interpreting codes, standards and specifications, [1]. Level IIIs can perform inspections, "buy off" or accept parts and write and approve inspection practices and test procedures. A Level III may be responsible for an NDT operation. The NDT Level III may provide training and examinations to NDT Level I and II personnel for certification in the method(s) in which he/she is certified,

NDT training is based on Training Outlines.

An example of the ASNT outlines is given in Table 12.1.

Another International NDT certification system is implemented by The German Society for Non-Destructive Testing (Deutsche Gesellschaft für Zerstörungsfreie Prüfung – DGZfP) offering a comprehensive training program which is performed in its six regional training centers and several authorized training locations.

A tentative training program and some examples of examination questions used by ITT Technical Institute are given in Appendices 3 and 4.

Table 12.1 ASNT TNDT topical outlines*

1. Principles/Theory
2. Conduction
3. Convection
4. Radiation
5. The nature of heat and heat flow
6. Temperature measurement principles
7. Proper selection of Thermal/Infrared testing
8. Equipment/Materials
9. Temperature measurement equipment
10. Heat flux indicators
11. Performance parameters of non-contact devices
12. Techniques
13. Contact temperature indicators
14. Non-contact pyrometers
15. Infrared line scanners
16. Thermal/Infrared imaging
17. Heat flux indicators
18. Exothermic or endothermic investigations
19. Friction investigations
20. Fluid Flow investigations
21. Thermal resistance (steady state heat flow)
22. Thermal capacitance investigations
23. Interpretation/Evaluation
24. Procedures
25. Safety and health
26. Safety responsibility and authority
27. Safety for personnel
28. Safety for client and facilities
29. Safety for testing equipment

*This examination is 2 hours in length and has 90 questions of equal value

12.2 Basic Standards

This section lists several documents that are related to both active and passive ITT. Note that the standards, which describe calibration procedures for IR imagers and thermometers, are intentionally omitted.

ANSI/NETA MTS (American National Standards Institute/InterNational Electrical Testing Association) Standard for Maintenance Testing Specifications for Electrical Power Equipment and Systems, 2015 USA.

ANSI/ASHRAE 101-1981 (American Society of Heating, Refrigerating and Air-Conditioning Engineers) Application of infrared sensing devices to the assessment of building heat loss characteristics USA.

ASNT SNT-TC-1A. (American Society for Nondestuctive Testing) Personnel Qualification and Certification in Nondestructive Testing USA.

ASTM (American Society for Testing and Materials) Standard C1060-90. Thermographic inspection of insulation installations in envelope cavities of frame buildings, 1996, USA.

ASTM Standard C1153-90. Standard practice for the location of wet insulation in roofing systems using infrared imaging, 1990, USA.

ASTM Standard C 3259-95. Infrared determination of the temperature of applied coating on wood products during the coating cycle, 1995, USA.

ASTM Standard D 4788-88. Test method for detecting delaminations in bridge decks by using infrared thermography, 1997, USA.

ASTM Standard E 1149-90. Standard terminology relating to NDT by infrared thermography, 1990, USA.

ASTM Standard E 1186-87. Standard practices for air leakage site detection in building envelopes, 1987, USA.

ASTM Standard E1827. Standard test methods for determining air tightness of buildings using an orifice blower door.

ASTM Standard E 1316-93. Standard terminology for nondestructive examination, Section J, Terms, 1993, USA.

ASTM Standard E 1623-94. Standard test method for determination of fire and thermal parameters of materials, products and systems using an intermediate scale, 1994, USA.

ASTM Standard E 1862-97. Standard test methods for measuring and compensating for reflected temperature using infrared imaging radiometers, 1997, USA.

ASTM Standard E 1897-97. Standard test methods for measuring and compensating for transmittance of an attenuating medium using infrared imaging radiometers, 1997, USA.

ASTM Standard E-1934. Standard guide for examining electrical and mechanical equipment with infrared thermography, 1999 USA.

ASTM Standard E2582-19 Standard Practice for Infrared Flash Thermography of Composite Panels and Repair Patches Used in Aerospace Applications 2019 USA.

ASTM Standard C 1060-90. Thermographic Inspection of Insulation Installations in Envelope Cavities of Frame Buildings.

ASTM Standard C1046-95 (2007). Standard Practice for In-Situ Measurement of Heat Flux and Temperature on Building Envelope Components.

EPRI NR-6973. (Electric Power Research Institute) Guide for IR thermography, USA.

ISO (International Organization for Standardization) 9712:2012.

Non-destructive testing – Qualification and certification of NDT personnel.

The standard specifies requirements for principles for the qualification and certification of personnel who perform industrial non-destructive testing (NDT), including infrared thermographic testing.

ISO 10878. Non-destructive testing – Infrared thermography – Vocabulary.

The standard defines terms used in infrared thermography for nondestructive testing and forms a common basis for standard general use.

ISO 10880. Non-destructive testing – Infrared thermographic testing–General principles.

The standard provides general principles for infrared thermographic testing in the field of industrial NDT.

ISO 18251-1:2017. Non-destructive testing – Infrared thermography – Part 1: Characteristics of system and equipment.

This document describes the main components, and their characteristics, constituting an IR imaging system and related equipment used in NDT.

ISO 18434-1:2008. Condition monitoring and diagnostics of machines – Thermography – Part 1: General procedures.

This standard provides an introduction to the application of IR thermography to machinery condition monitoring and diagnostics, where "machinery" includes machine auxiliaries such as valves, fluid and electrically powered machines, and machinery-related heat exchanger equipment. In addition, IR applications pertaining to machinery performance assessment are addressed.

ISO 18436-7:2008. Condition monitoring and diagnostics of machines – Requirements for qualification and assessment of personnel – Part 7: Thermography.

This standard specifies the requirements for qualification and assessment of personnel who perform machinery condition monitoring and diagnostics using infrared thermography. A certificate or declaration of conformity to ISO 18436-7:2008 will provide recognition of the qualifications and competence of individuals to perform thermal measurements and analysis for machinery condition monitoring using portable thermal imaging equipment. This procedure may not apply to specialized equipment or other specific situations. The standard specifies a three-category classification program.

ISO 6781: 1983. Thermal Insulation – Qualitative Detection of Thermal Irregularities in Building Envelopes – Infrared Method.

This standard specifies a qualitative method, by thermographic examination (infrared method), for detecting thermal irregularities in building envelopes. The method is used to identify wide variations in the thermal properties, including air tightness, of the components constituting the external envelopes of buildings. This document does not apply to the determination of the degree of thermal insulation and air tightness of a structure.

ISO/TR 13154. Medical Electrical Equipment – Deployment, implementation and operational guidelines for identifying febrile humans using a screening thermograph – Laboratory accuracy of a screening thermograph – Calibration source – Example of relevant uncertainty terms for a screening thermograph, etc.

ISO 9869-1998. Thermal insulation – Building elements – In-situ measurement of thermal resistance and thermal transmittance.

ISO TC 163/SC 2 (ISO/FDIS 6946: 2007. Building components and building elements – Thermal resistance and thermal transmittance – Calculation method.

Military standard MIL-STD-185. Survey procedure of thermal image method for electrical components, USA.

Military standard MIL-HDBK-728, Vol. 1. Nondestructive testing, USA.

Military standard MIL-STD-2194. Infrared thermal imaging. Survey procedure for electrical equipment, 1988, USA.

NETA ATS-1999 (InterNational Electrical Testing Association) Acceptance Testing Specification for Electrical Power Distribution Equipment and Systems, 1999. USA.

CGSB 149-GP-2MP (Canadian General Standards Board) Manual for thermographic analysis of building enclosures. 1986, Canada.

French Standard A04-420. Essais nondestructifs. Thermographie infrarouge. Caracterisation de l'appareillage, 1993, France (in French).

French Standard A 09-400. Norme Française de thermographie infrarouge, France (in French).

French Standard X 10-023-82. Isolation thermique. detect infrarouge pour la detection qualitative d'irrégularites thermiques dans la structure externe des bâtiments, 1982, France (in French).

German Standard E DIN (Deutsches Institut für Normung) 54192: 2010. Zerstörungsfreie Prüfung – Aktive Thermographie, 2010, Germany (in German).

Japanese Standard JIS T 1141. Medical thermography (in Japanese).

Russian Standard GOST 18353-79. Nondestructive testing. Classification of methods (in Russian), 1979.

Russian standard GOST 23483-79. Nondestructive testing. Thermal methods. General requirements (in Russian), 1979.

Russian standard GOST 26629-85. IR thermographic inspection. Thermal insulation of building envelopes (in Russian), 1985.

Russian standard GOST 26782-85. Nondestructive testing. Optical and thermal test devices. General technical requirements (in Russian), 1985.

Russian standard GOST 25314-82. Thermal nondestructive testing. Terms and definitions (in Russian), 1982.

Russian standard GOST R 8.619-2006. Radiometric IR imagers. Calibration procedure (in Russian), 2006.

Swedish Standard SS024210. Thermal insulation: Thermography of buildings, Sweden.

Canadian Standard NMS Section 02 27 13. Thermographic Assessment – Building Envelope.

Canadian Standard GSB 149-GP-2MP. Manual for Thermographic Analysis of Building Enclosures.

12.3 Additional Standards and Guidelines

OIML (International Organization of Legal Metrology) R 141: Procedure for calibration and verification of the main characteristics of thermographic instruments (Confirmed 2008-10-31).

TTCTRAN.015 Guideline for Thermographic Inspection in Electrical Installations. Danish Technology Institute.

RESNET (Residential Energy Services Network) Interim Guidelines for Thermographic Inspections of Buildings (September 29, 2010).

Thermographic Inspection of Electrical Installations. – Publication. 556 556 776, AGEMA Infrared Systems Inc., 1985, Sweden.

Guidelines for Specifying and Performing Infrared Inspections. – Infraspection Institute., 1st Ed., Jan. 1988, USA.

Merkblatt über Thermografische Untersuchungen an Bauteilen and Bauwerken, DGZfP, Ausgabe October 1993, Berlin, Germany (in German).

Guidance Document RD 153-34.0-20.364-00. The method for IR diagnostics of thermo-mechanical equipment, 2003 (in Russian).

Guidance Document RD 153-34.0-20.363-99. The basics of IR diagnostics of electrical installations and high-voltage transmission lines, 2003 (in Russian).

Standard for Infrared Inspection of Electrical Systems and Rotating Equipment, Infraspection Institute, 2008, USA.

Standard for Infrared Inspection of Building Envelopes, Infraspection Institute, 2008, USA.

Standard for Measuring Distance/Target Size Values for Infrared Imaging Radiometers, Infraspection Institute, 2008, USA.

Standard for Measuring and Compensating for Emittance Using Infrared Imaging Radiometers, Infraspection Institute, 2008, USA.

Standard for Infrared Inspection of Insulated Roofs, Infraspection Institute, 2008, USA.

Standard for Infrared Inspection to Detect Pests and Pest Related Damage, Infraspection Institute, 2008, USA.

Standard for Infrared Inspection of Recreational Yachts & Small Craft Constructed of Fiberglass Reinforced Plastic and Composite Materials, Infraspection Institute, 2008, USA.

Standard for Measuring and Compensating for Reflected Temperature Using Infrared Imaging Radiometers, Infraspection Institute, 2008, USA.

Standard for Measuring and Compensating for Transmittance of an Attenuating Medium Using Infrared Imaging Radiometers, Infraspection Institute, 2008, USA.

Reference

1. ASNT: Level III Study Guide: Infrared and Thermal Testing Method (IR). ASNT, USA (2001)

Appendix 1: Sensitivity Functions

Table A1.1 Sensitivity functions for a semi-infinite body

Heating function	Surface heat exchange	$\frac{\partial T/T}{\partial \tau/\tau}$	$\frac{\partial T/T}{\partial \tau_h/\tau_h}$	$\frac{\partial T/T}{\partial Q/Q}$	$\frac{\partial T/T}{\partial \alpha/\alpha}$	$\frac{\partial T/T}{\partial \lambda/\lambda}$	$\frac{\partial T/T}{\partial a/a}$
	$\alpha^F = 0$	1/2 (1.1)[a]	–[b] (1.2)	1 (1.3)	– (1.4)	−1 (1.5)	1/2 (1.6)
	$\alpha^F \neq 0$	T. 2.4[c] (2.1)	– (2.2)	1 (2.3)	T. 2.4 (2.4)	T. 2.4 (2.5)	T. 2.4 (2.6)
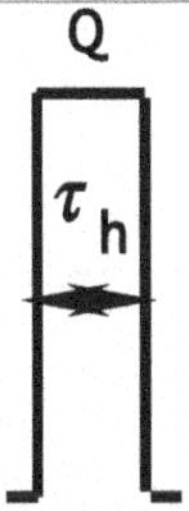	$\alpha^F = 0$	T.4 (3.1)	T.4 (3.2)	1 (3.3)	– (3.4)	−1 (3.5)	1/2 (3.6)
	$\alpha^F \neq 0$	T. 2.4 (4.1)	T. 2.4 (4.2)	1 (4.3)	T. 2.4 (4.4)	T. 2.4 (4.5)	N. 2.4 (4.6)
	$\alpha^F = 0$	−1/2 (5.1)	– (5.2)	1 (5.3)	– (5.4)	−1 (5.5)	1/2 (5.6)
	$\alpha^F \neq 0$	T. 2.4 (6.1)	– (6.2)	1 (6.3)	T. 2.4 (6.4)	T. 2.4 (6.5)	T. 2.4 (6.6)

[a]Solution number
[b]No solution
[c]See solution in Table 2.4

V. Vavilov, D. Burleigh, *Infrared Thermography and Thermal Nondestructive Testing*, https://doi.org/10.1007/978-3-030-48002-8

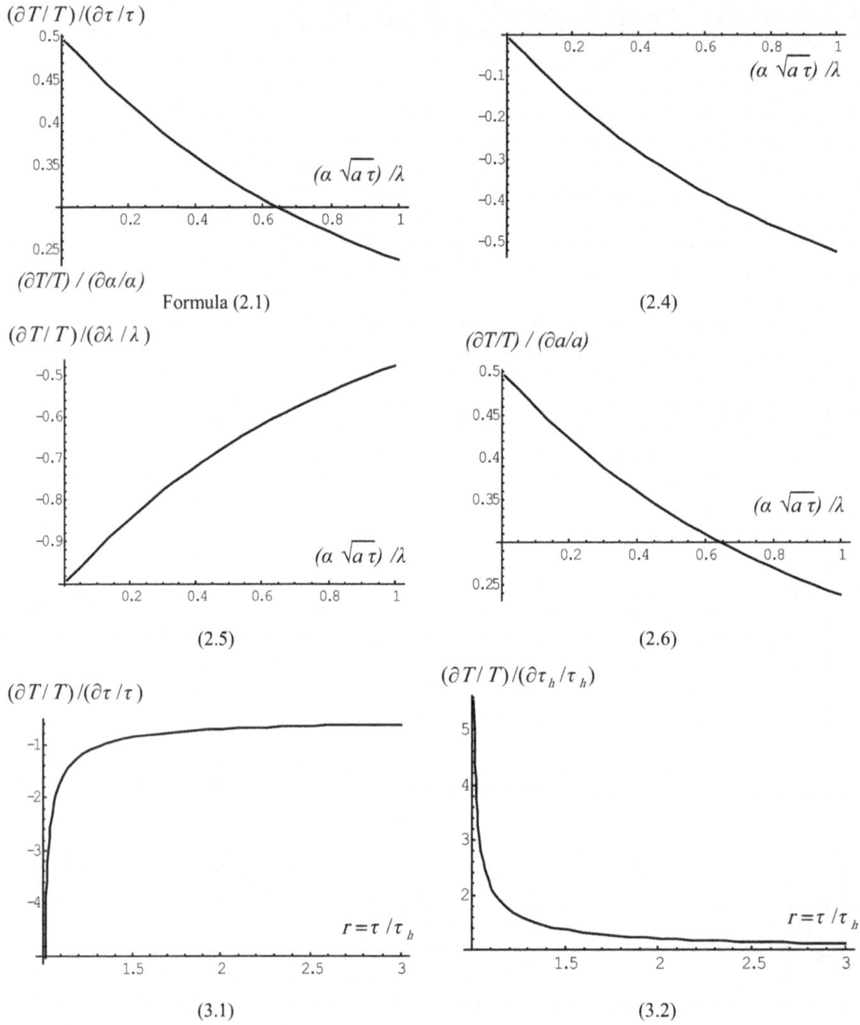

Fig. A1.1 Sensitivity functions for a non-adiabatic semi-infinite body (see formulas in Table A1.2). Sensitivity functions for a non-adiabatic semi-infinite body (see formulas in Table A1.2). Sensitivity functions for a non-adiabatic semi-infinite body (see formulas in Table A1.2)

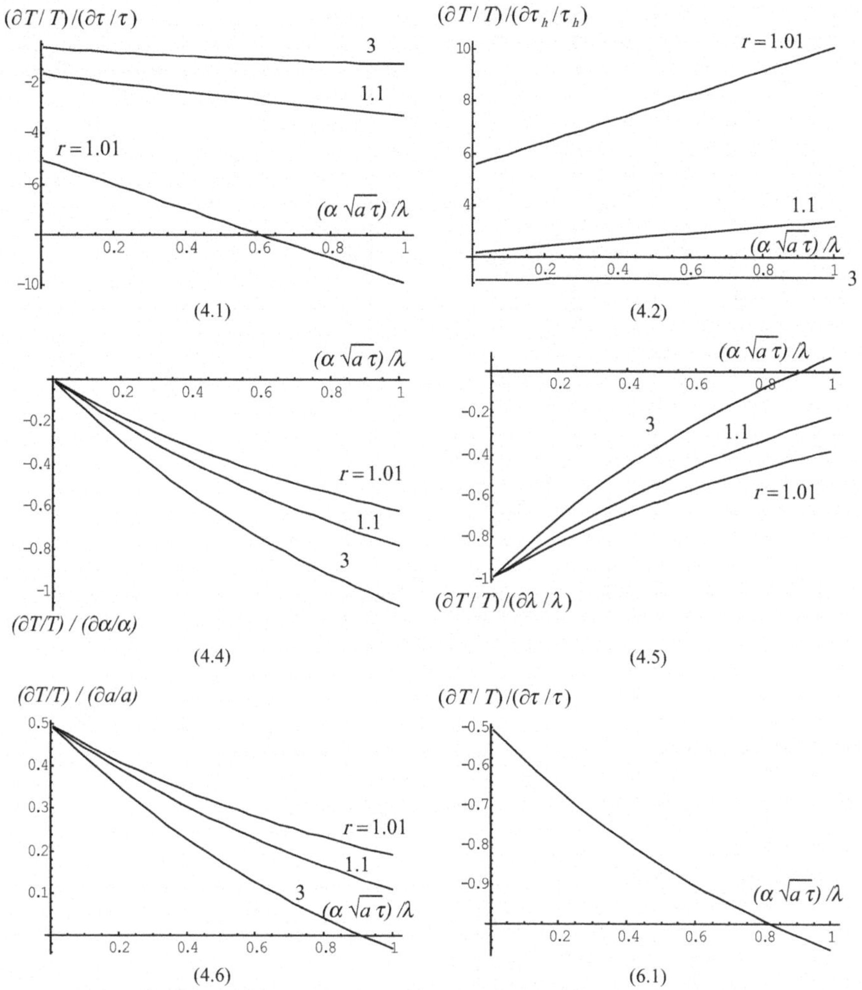

Fig. A1.1 (continued)

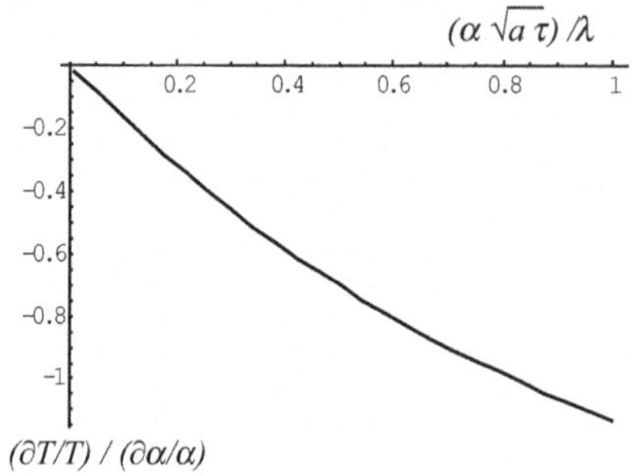

(6.4)

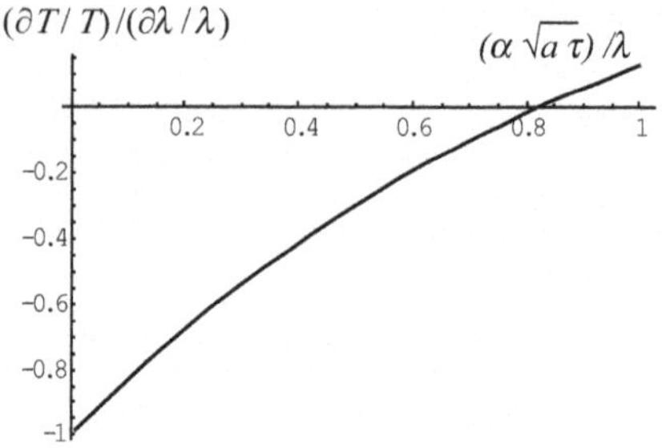

(6.5)

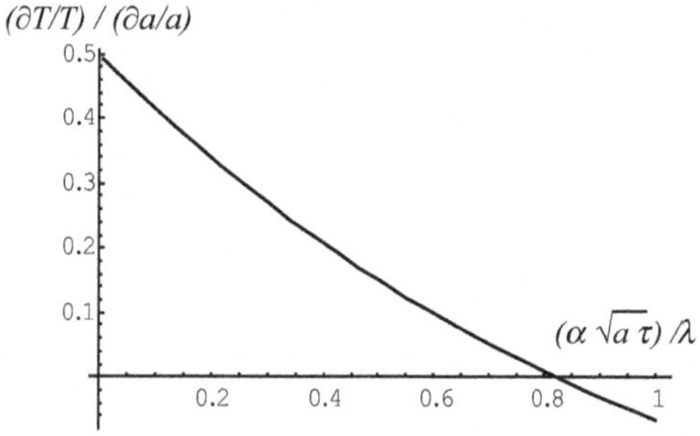

(6.6)

Fig. A1.1 (continued)

Table A1.2 Sensitivity functions for a semi-infinite body (see Table A1.1)

Formula number on Table A1.1	Formula
(2.1)	$\frac{\partial T/T}{\partial\tau/\tau}=\frac{\frac{x}{\sqrt{\pi}}-x^2e^{x^2}Erfc(x)}{1-e^{x^2}Erfc(x)}, \swarrow\!\!\!\nearrow\ x=\frac{\alpha\sqrt{a\tau}}{\lambda}$
(2.4)	$\frac{\partial T/T}{\partial h/h}=\frac{\left(1-\frac{2x}{\sqrt{\pi}}\right)-\left(1-2x^2\right)e^{x^2}Erfc(x)}{e^{x^2}Erfc(x)-1}, \swarrow\!\!\!\nearrow\ x=\frac{\alpha\sqrt{a\tau}}{\lambda}$
(2.5)	$\frac{\partial T/T}{\partial k/k}=\frac{\frac{2x}{\sqrt{\pi}}-2x^2e^{x^2}Erfc(x)}{e^{x^2}Erfc(x)-1}, \swarrow\!\!\!\nearrow\ x=\frac{\alpha\sqrt{a\tau}}{\lambda}$
(2.6)	$\frac{\partial T/T}{\partial\alpha/\alpha}=\frac{\frac{x}{\sqrt{\pi}}-x^2e^{x^2}Erfc(x)}{1-e^{x^2}Erfc(x)}, x=\frac{\alpha\sqrt{a\tau}}{\lambda}$
(3.1)	$\frac{\partial T/T}{\partial\tau/\tau}=-\frac{1}{2\sqrt{1-1/r}}, r=\tau/\tau_h$

Formula number on Table A1.1	Formula
(3.2)	$\frac{\partial T/T}{\partial \tau_h/\tau_h} = \frac{1}{2}\frac{1}{1+\sqrt{r}\sqrt{r-1}-r}, r=\tau/\tau_h$
(4.1)	$\frac{\partial T/T}{\partial \tau/\tau} = \frac{\frac{x}{\sqrt{\pi}}\left(1-\frac{1}{\sqrt{1-1/r}}\right)e^{-x^2} - x^2 Erfc(x) + x^2 e^{-x^2/r} Erfc\left(x\sqrt{1-1/r}\right)}{e^{-x^2/r} Erfc\left(x\sqrt{1-1/r}\right) - Erfc(x)}, \swarrow\nearrow$ $x = \frac{\alpha\sqrt{a\tau}}{\lambda}, r=\tau/\tau_h$
(4.2)	$\frac{\partial T/T}{\partial \tau_h/\tau_h} = \frac{x^2}{r}\frac{\frac{1}{\sqrt{\pi}x\sqrt{1-1/r}}e^{-x^2(1-1/r)} - Erfc\left(x\sqrt{1-1/r}\right)}{Erfc\left(x\sqrt{1-1/r}\right) - e^{x^2/r} Erfc(x)}, \swarrow\nearrow$ $x = \frac{\alpha\sqrt{a\tau}}{\lambda}, \swarrow\nearrow\ r=\tau/\tau_h$
(4.4)	$\frac{\partial T/T}{\partial \alpha/\alpha} =$ $= \frac{\frac{2x}{\sqrt{\pi}}\left(1-\sqrt{1-1/r}\right)e^{-x^2} + \left(1-2x^2\right)Erfc(x) - \left[1-2x^2(1-1/r)\right]e^{-x^2/r} Erfc\left(x\sqrt{1-1/r}\right)}{e^{-x^2/r} Erfc \rightleftarrows \left(x \rightleftarrows \sqrt{1-1/r}\right) - Erfc \rightleftarrows (x)}, \swarrow\nearrow$ $x = \frac{\alpha\sqrt{a\tau}}{\lambda}, \swarrow\nearrow\ r=\tau/\tau_h$
(4.5)	$\frac{\partial T/T}{\partial \lambda/\lambda} = 2x^2 \frac{\frac{1}{x\sqrt{\pi}}\left(1-\sqrt{1-1/r}\right)e^{-x^2} - Erfc(x) + (1-1/r)e^{-x^2/r} Erfc\left(x\sqrt{1-1/r}\right)}{Erfc(x) - e^{-x^2/r} Erfc\left(x\sqrt{1-1/r}\right)}, \swarrow\nearrow$ $x = \frac{\alpha\sqrt{a\tau}}{\lambda}, r=\tau/\tau_h$
(4.6)	$\frac{\partial T/T}{\partial a/a} = \frac{\frac{x}{\sqrt{\pi}}\left(1-\sqrt{1-1/r}\right)e^{-x^2} - x^2 Erfc(x) + x^2(1-1/r)e^{-x^2/r} Erfc\left(x\sqrt{1-1/r}\right)}{e^{-x^2/r} Erfc\left(x\sqrt{1-1/r}\right) - Erfc(x)}, \swarrow\nearrow$ $x = \frac{\alpha\sqrt{a\tau}}{\lambda}, \swarrow\nearrow\ r=\tau/\tau_h$
(6.1)	$\frac{\partial T/T}{\partial \tau/\tau} = \frac{1-2x^2+2\sqrt{\pi}x^3e^{x^2}Erfc(x)}{-2+2\sqrt{\pi}xe^{x^2}Erfc(x)}, \swarrow\nearrow\ x = \frac{\alpha\sqrt{a\tau}}{\lambda}$
(6.4)	$\frac{\partial T/T}{\partial \alpha/\alpha} = x\frac{-2x+\sqrt{\pi}\left(1+2x^2\right)e^{x^2}Erfc(x)}{-1+\sqrt{\pi}xe^{x^2}Erfc(x)}, \swarrow\nearrow\ x = \frac{\alpha\sqrt{a\tau}}{\lambda}$
(6.5)	$\frac{\partial T/T}{\partial \lambda/\lambda} = \frac{-1-2x^2+2\sqrt{\pi}x\left(1+x^2\right)e^{x^2}Erfc(x)}{1-\sqrt{\pi}xe^{x^2}Erfc(x)}, x = \frac{\alpha\sqrt{a\tau}}{\lambda}$
(6.6)	$\frac{\partial T/T}{\partial a/a} = \frac{1+2x^2-2\sqrt{\pi}x\left(1+x^2\right)e^{x^2}Erfc(x)}{2-2\sqrt{\pi}xe^{x^2}Erfc(x)}, x = \frac{\alpha\sqrt{a\tau}}{\lambda}$

Appendix 2: Calculating Surface Temperature of a Plate Heated by a Dirac Pulse (MatLab Platform, See Table 2.4)

```
%Program Heat Pulse *
%Themophysical properties, thickness
lambda=2;a=1e-6; %Themophysical properties, thickness
e=0.01;%Thickness (area S=1)
Q=1e5;%Energy density

% Coefficients of Stehfest's algorithm
v(1)=1/12;v(2)=-385/12;v(3)=1279;v(4)=-46871/3;v(5)=505465/
6;v(6)=-473915/2;
v(7)=1127735/3;v(8)=-1020215/3;v(9)=328125/2;v(10)=-65625/2;
%loop on time
tv=0;Trear=0;Tfront=0;Tmid=0;Fmid=0;
for t=1:1:1000,
      p=(1:10)*log(2)/t;k=sqrt(p/a);%calculation of the Laplace
parameters for a given time
    A=cosh(k*e);
    C=lambda*k.*sinh(k*e);
    Amid=cosh(k*e/2);
    Cmid=lambda*k.*sinh(k*e/2);

Tetarear=Q./C;Tetafront=Q*A./C;Tetamid=Amid.*Tetarear;Phimid=Cmid
.*Tetarear;
    tv=[tv t];
    rear=(log(2)/t).*(v*Tetarear');Trear=[Trear rear];
    front=(log(2)/t).*(v*Tetafront');Tfront=[Tfront front];
    midT=(log(2)/t).*(v*Tetamid');Tmid=[Tmid midT];
    midF=(log(2)/t).*(v*Phimid');Fmid=[Fmid midF];
end
%plot(tv,Trear,'k-',tv,Tmid,'k-',tv,Fmid/1000,'k-');
%axis([0 0.5  0 1]);
```


V. Vavilov, D. Burleigh, *Infrared Thermography and Thermal Nondestructive Testing*, https://doi.org/10.1007/978-3-030-48002-8

```
%hold on
%plot(tv,Tfront,'k-');
plot(tv,Trear,'k-',tv,Tmid,'k',tv,Fmid/1000,'k-');
axis([-10 100 0 10]);
hold on
%pause
Plot(tv,Tfront,'k-')
xlabel('Time[seconds]');
ylabel('Temperatures (Celsius),Flux(KW/m2)');

end
```

∗ lambda – thermal conductivity, a – thermal diffusivity,
e – plate thickness, Q – pulse energy density

Appendix 3: Psychrometric Table of Air Relative Humidity

Dry thermometer indication, °C	Difference between indications of dry and moist thermometers, °C										
	0	1	2	3	4	5	6	7	8	9	10
0	100	81	63	45	28	11	–	–	–	–	–
2	100	84	68	51	35	20	–	–	–	–	–
4	100	85	70	56	42	28	14	–	–	–	–
6	100	86	73	60	47	35	23	10	–	–	–
8	100	87	75	63	51	40	28	18	7	–	–
10	100	88	76	65	54	44	34	24	14	4	–
12	100	89	78	68	57	48	38	29	20	11	–
14	100	90	79	70	60	51	42	33	25	17	9
16	100	90	81	71	62	54	45	37	30	22	15
18	100	91	82	73	64	56	48	41	34	26	20
20	100	91	83	74	66	59	51	44	37	30	24
22	100	92	83	76	68	61	54	47	40	34	28
24	100	92	84	77	69	62	56	49	43	37	31
26	100	92	85	78	71	64	58	50	45	40	34
28	100	93	85	78	72	65	59	53	48	42	37
30	100	93	86	79	73	67	61	55	50	44	39

V. Vavilov, D. Burleigh, *Infrared Thermography and Thermal Nondestructive Testing*, https://doi.org/10.1007/978-3-030-48002-8

Appendix 4: TNDT Certification Syllabus

Level I (not less than 40 hrs)	Level II (not less than 80 hrs)	Level III
1. ITT classification		
1.1. Terms and definitions 1.2. Thermal ITT 1.3. Active ITT 1.4. Passive TITT 1.6. Typical defects detected by ITT	1.1. Terms and definitions 1.2. Thermal ITT 1.3. Active ITT 1.4. Passive ITT 1.5. Classification of ITT methods 1.6. Typical defects detected by ITT	1.1. Terms and definitions 1.2. Thermal ITT 1.3. Active ITT 1.4. Passive ITT 1.5. Classification of ITT methods 1.6. Typical defects detected by ITT
2. Physical principles of ITT		
2.1. Basics of the heat conduction theory: heat capacity, thermal conductivity and thermal diffusivity. Definitions, units. Law of heat conduction (Fourier) 2.4. Sources of thermal radiation. Temperature and its measurement. Temperature scales. 2.7. Blackbody radiation spectrum. Planck's law. Atmosphere transmission.	2.1. Basics of the heat conduction theory: heat capacity, thermal conductivity and thermal diffusivity. Definitions, units. 2.2. Law of heat conduction (Fourier). 2.4. Sources of thermal radiation. Temperature and its measurement. Temperature scales. Radiation and color temperature. 2.5. Emission, propagation and absorption of thermal radiation. Emissivity. The laws of Kirchhoff, Stefan-Boltzmann and Wien. Lambertian emitters. 2.7. Blackbody radiation spectrum. Planck's law. Atmosphere transmission. Radiation of blackbody cavities. Indicatrices of reflection and emission. Real body absorption. Absorption of solar radiation. Practical determination of emissivity.	2.1. Basics of the heat conduction theory: heat capacity, thermal conductivity and thermal diffusivity. Definitions, units. 2.2. Law of heat conduction (Fourier). 2.3. Determining material thermal properties (pulsed and harmonic thermal stimulation). Temperature waves. Measuring heat fluxes; fluxmeters. 2.4. Sources of thermal radiation. Temperature and its measurement. Temperature scales. Pyrometry. Radiation and color temperature. 2.5. Emission, propagation and absorption of thermal radiation. Emissivity. The laws of Kirchhoff, Stefan-Boltzmann and Wien. Lambertian emitters. 2.6. Blackbody radiation spectrum. Planck's law. Atmosphere transmission. Radiation of blackbody cavities. Indicatrices of reflection and emission. Real body absorption. Absorption of solar radiation. Practical determination of emissivity. 2.7. Thermal radiation in semi-transparent and selectively-absorbing media. Radiation spectrums. Background radiation. IR fibers. Polarization of radiation. Spectral filters.

V. Vavilov, D. Burleigh, *Infrared Thermography and Thermal Nondestructive Testing*, https://doi.org/10.1007/978-3-030-48002-8

Level I (not less than 40 hrs)	Level II (not less than 80 hrs)	Level III
3. Electro-optical systems		
3.2. Infrared imagers.	3.1. Optical systems for focusing, modulating and scanning of IR radiation. Optical materials. Lenses and their parameters. 3.2. Image quality. Image enhancement. Infrared imagers. Parameters of IR imagers (temperature resolution, frame frequency, image format etc.). IR line scanning systems and thermometers. 3.5. Influence of experimental conditions on the efficiency of IR thermography (background temperature, reflection and emissivity phenomena, atmosphere transmission etc.).	3.1. Optical systems for focusing, modulating and scanning of IR radiation. Optical materials. Lenses and their parameters. 3.2. Image quality. Image enhancement. Infrared imagers. Parameters of IR imagers (temperature resolution, frame frequency, image format etc.). IR line scanning systems and thermometers. 3.3. Spectral and spatial filtration of targets (defects). 3.4. Defect signatures and their identification. 3.5. Influence of experimental conditions on the efficiency of IR thermography (background temperature, reflection and emissivity phenomena, atmosphere transmission etc.). 3.6. Digital and analog data processing. Computer analysis of IR images.
4. ITT equipment		
4.1. Modern ITT equipment. Data acquisition. 4.2. Auxiliary equipment used in IR surveying. Metrology of IR devices.	4.1. Modern ITT equipment. Data acquisition. 4.2. Pyrometers, IR scanners, FPA IR imagers, auxiliary equipment used in IR surveying. Metrology of IR devices. 4.3. Heat sources in ITT (lasers, quartz halogen lamps, flash tubes, resistive heaters, etc.). 4.4. Reference (blackbody) sources. 4.5. IR endoscopes. 4.6. IR microscopes. 4.7. Presentation of test results (IR thermograms, spatial and temporal profiles, isotherms, image fusion, 3D images).	4.1. Modern ITT equipment. Data acquisition. 4.2. Pyrometers, IR scanners, FPA IR imagers, auxiliary equipment used in IR surveying. Metrology of IR devices. Computerized IR imaging systems. IR endoscopes. IR microscopes. Air borne IR systems. 4.3. Heat sources in ITT lasers, quartz halogen lamps, flash tubes, resistive heaters, etc.). 4.4. Reference (blackbody) sources. 4.5. IR endoscopes. 4.6. IR microscopes. 4.7. Presentation of test results (IR thermograms, spatial and temporal profiles, isotherms, image fusion, 3D images). 4.8. Spectrometric and dual-band systems. 4.9. IR thermometry at low temperatures.

Level I (not less than 40 hrs)	Level II (not less than 80 hrs)	Level III
5. ITT procedures		
5.1. Signal parameters in ITT. Choosing a ITT method and equipment depending on a test problem and a test sample. 11. Using IR thermography in civil engineering. 12. Using IR thermography in power production. 13. Other ITT applications.	5.1. Signal parameters in ITT. Choosing a ITT method and equipment depending on a test problem and a test sample. 5.2. Passive and active ITT. One-sided and two-sided ITT. IR thermographic measurement of temperature. Defect detection criteria. Basic relationships between defect parameters and temperature signals in ITT. Reference samples in ITT. 5.3. State-of-the-art of ITT and development trends. 5.4. Using IR thermography in civil engineering. 5.5. Using IR thermography in power production. 5.6. Material loss (corrosion) evaluation. 5.7. Defect detection, identification and characterization. Novel algorithms of data processing: pulse phase thermography, principal component analysis, thermal tomography, polynomial fitting, ultrasonic thermography, eddy current thermography. 5.9. Other ITT applications. 5.10. Other NDT techniques.	5.1. Signal parameters in ITT. Choosing a ITT method and equipment depending on a test problem and a test sample. 5.2. Passive and active ITT. One-sided and two-sided ITT. IR thermographic measurement of temperature. Defect detection criteria. Basic relationships between defect parameters and temperature signals in ITT. Reference samples in ITT. Statistics of defect detection. 5.3. State-of-the-art of ITT and development trends. 5.4. Using IR thermography in civil engineering. 5.5. Using IR thermography in power production. 5.6. Material loss (corrosion) evaluation. 5.7. Defect detection, identification and characterization. Novel algorithms of data processing: pulse phase thermography, principal component analysis, thermal tomography, polynomial fitting, ultrasonic thermography, eddy current thermography. 5.8. Laser thermographic systems (detecting gas leakages). 5.9. Other ITT applications. 5.10. Other NDT techniques.
6. Metrology and standardization		
–	6.1. Calibration of IR imagers. Equipment and test procedure calibration by reference samples.	6.1. Calibration of IR imagers. Equipment and test procedure calibration by reference samples.
7. Safety requirements. IR thermographic inspection at dangerous industrial objects		
	7.1. Technical installations used at dangerous industrial objects. Safety requirements. 7.2. Technical requirements to NDT equipment used at dangerous industrial objects.	7.1. Technical installations used at dangerous industrial objects. Safety requirements. 7.2. Technical requirements to NDT equipment used at dangerous industrial objects 7.3. Organization of NDT surveys at industrial enterprises.

Level I (not less than 40 hrs)	Level II (not less than 80 hrs)	Level III
8. Standardization and certification in ITT		
–	8.1. International standards on ITT. 8.2. Developing inspection guidelines. Protocol forms.	8.1. International standards on ITT. 8.2. Developing inspection guidelines. Protocol forms. 8.3. National certification system on ITT.

Appendix 5: Samples of Recommended Examination Questions in TNDT

Level 1

1. Heat energy is transferred from a body to a body by: A – conduction B – convection C – radiation D – all above-mentioned	2. In the scale of absolute temperatures, the initial point is: A – −273 °C B – 0 °C C – 1000 °C D – 6000 °C
3. At a particular temperature and a particular wavelength, the highest emissivity will adhere to: A – polished aluminum B – rough aluminum C – polished steel D – cast untreated steel	4. Surface reflection coefficient can be: A – from 0 to 1.0 B - from 0 to infinity C – from 0 to 10 D – variant A or B
5. Strong wind … outdoor IR thermographic inspection efficiency A – improves B – does not affect C – improves a little D – worsens	6. When a solid object is cooled down by wind, this phenomenon is related to the following heat transfer mechanism by: A – radiation B – conduction C – convection D – all above-mentioned
7. An operator should input the following correcting settings into an IR camera: A –emissivity B –ambient temperature C – object temperature D – A and B	8. The operation principle of IR imagers is based on using the following heat transfer mechanism: A – radiation B – conduction C – convection D – all above-mentioned
9. If all other parameters are the same, the more pixels are in an IR image, the … spatial resolution: A – worse B – is the same C – better D – depends on temperature sensitivity	10. IR thermography is typically well applicable to: A – non-metals B – unpainted metals C – painted metals D – A and C

V. Vavilov, D. Burleigh, *Infrared Thermography and Thermal Nondestructive Testing*, https://doi.org/10.1007/978-3-030-48002-8

11. The maximum of the Sun radiation is at about 0.5 µm. Therefore, an IR imager which is sensitive in the 7–14 µm wavelength band will … A – not sense it B – will still sense it because the Sun still emits some energy in the 7–14 µm wavelength band C – will still sense it because the Sun will heat up the sample surface and the sample will emit energy in the 7–14 µm wavelength band D – B and C	12. A blackbody is an ideal body which: A – reflects 100% of incident radiation B – absorbs 100% of incident radiation C – transmits 100% of incident radiation D – none of A, B or C
13. To compare to visible light, the wavelengths of IR radiation are: A – shorter B – about the same C – longer D – shorter or longer depending on temperature	14. Power density of thermal radiation is expressed in: A – J B – W C – J/m^2 D – W/m^2

15. The sum of the coefficients of reflectivity, absorptivity and transmittance is equal to: A – 0.1 B – 1.0 C – 1.5 D – 3	16. We measure the temperature of a polished metallic sample having some areas of rust. The value of emissivity is set up equal to 0.95. Temperature measurements will be more accurate in the areas of: A – rust B – polished metal C – depending on temperature D – depending on wavelength
17. Active thermal NDT means that: A – an external stimulating heat source is used B – a sample is mechanically treated C – no external heat source is necessary D – B and C	18. An object to be tested is characterized by temperatures in the range from 45 to 55 °C. In order to see all sample temperatures in color, the temperature range should be installed as follows: A – from 0 to 55 °C B – from 0 to 45 °C C – from 45 to 55 °C D – from 0 to 100 °C
19. Focal Plane Array (FPA) IR cameras use: A – opto-mechanical scanning B – electronic scanning C – both A and B depending on the detector D – built-in heaters	20. If an IR image is saved in the ∗.jpg format, will it keep temperature values? A – no, because the ∗.jpg format is used only for color pictures B – yes, if a special radiometric ∗.jpg format is used C – no, because only images with the ∗.img extension can keep true temperatures D – yes, because the ∗.jpg format was developed for temperature recording

Level II

1. In the case of a diffusive (Lambert) emitter, with increasing the observation angle and the reflection being negligible, the apparent temperature…: A – increases B – decreases C – oscillates D – can be unpredictable	2. Sources of noise in IR thermography are: A – uneven emissivity B – absorption in atmosphere C– external heat sources D – all above-mentioned
3. A wavelength of the maximum blackbody exitance (heat flux power density) depends on temperature: A – linearly B – inversely C – either A, or B D – is independent	4. Spectral behavior of blackbody exitance (heat flux power density)of is described by the: A – Planck's law B – Stephan-Boltzmann's law C – Wien's law D – energy conservation law
5. IR imager determines … temperature: A – absolute B – relative C – radiation (apparent) D – true	6. Typical IR imagers can operate in the following wavelength band: A – 3–5.5 μm B – 7–13 μm C – A and B D – 1–100 μm

7. If the user sets up an emissivity value higher than the true one, the IR imager temperature read-out will show temperature …to compare to the true one A – lower B – higher C – the same D – B or C	8. IR thermographic measurements become ineffective, if the emissivity is A – higher than 0.7 B – lower than 0,7 C – higher than 0.9 D – lower than 0.9
9. We perform the inspection of a multi-storey building from a ground position close to the building wall so the upper part of the building is seen under a large angle. Let us assume that the wall temperature is absolutely uniform. Then the temperature read-out on the upper part of the building will be …. than the temperature read-out on the bottom part (assume no reflections). A – higher B – lower C – the same D – will depend on the IR camera model	10. To suppress the influence of solar reflections in the outdoor inspection of electrical joints, it is recommended to: A – move a little with the IR camera B – re-calibrate the camera C – change camera sensitivity D – change the color palette

11. Rainy weather will … inspection of electrical joints: A – worsen B – improve C – not influence D – improve or worsen depending on distance	12. IR camera temperature resolution as a specified camera parameter…: A – depends on distance to object B – depends on atmosphere conditions C – A and B D – is independent on distance and atmosphere conditions
13. The serious problem in IR temperature measurement at short distances is: A – unknown emissivity B – unknown ambient temperature C – unknown distance D – A and B	14. If a blackbody (or graybody) absolute temperature increases by 2 times, its thermal radiation in the total spectrum will …: A – remain the same B – will decrease by 16 times C – will increase by 8 times D – will increase by 16 times
15. During an IR thermographic test, a sample emissivity is set up correctly in the IR camera but an inputted value of the ambient temperature is lower than in reality. The IR camera apparent temperature read-out will be … than a true value: A – lower B – equal to true C – lower or equal depending on a chosen sensitivity of the IR camera D – higher	16. The more humid is the atmosphere, the … is its influence on IR thermographic measurements A – less B – stronger C – no influence D – influence of the atmosphere conditions depends on IR camera settings
17. If at any wavelength a material absorbs thermal energy well, it will emit thermal radiation …: A – well B – badly C – depending on temperature D – depending on wavelength	18. An image obtained with an IR camera has 320 elements in the horizontal direction. The following formula which connects the Field of View (FOV) and the Instantaneous Field of View (IFOV) is correct: A – IFOV = 320∗FOV B – FOV = 240∗IFOV C – FOV=IFOV/320 D – FOV = 320∗IFOV
19. Focal Plane Array (FPA) IR cameras use: A – a single IR detector B – at least two IR detectors operating in different wavelength bands C – an array of IR detectors D – an array of external heaters	20. Results of evaluating temperatures of electrical joints should be typically reduced to: A – 50% of load B – 100% of load C – 10% of load D – A or B, depending on approved guidelines

Level III

1. What is the relationship between the Kelvin (T, K) and Celsius (t, ^oC) temperature scales? A – $T = t + 273$ B – $T = t - 273$ C – $T = 273/t$ D – $T = 273 - t$	2. What is the relationship between one degree interval in the Kelvin and Celsius temperature scales? A – 1 K = 1 °C B – 1 K = 10 °C C – 1 K = 0.1 °C D – 1 K = 273 °C
3. Can an IR imager be used as a fluxmeter when determining wall thermal resistance? A – No B – Yes C – Yes, if the IR imager is supplied with a laser pointer D – Yes, if the respective data processing program is available	4. Instantaneous Field of View (IFOV) of IR imagers is, as a rule, … than that of IR thermometers A – equal B – greater C – either A, or B D – less
5. For opaque bodies, absorptivity α, emissivity ε and reflectivity IR ρ are related as: A – $\alpha = \varepsilon = 1 - \rho$ B – $\alpha = \varepsilon = 1 + \rho$ C – $\alpha = \varepsilon = 1/\rho$ D– $\alpha = \varepsilon = \rho^2$	6. Increasing the roughness of a test sample surface will change its emissivity as follows: A – increases B – decreases C – does not change D – all above-mentioned
7. When monitoring the upper section of a chimney from a ground point, one should preferably use an IR imager with the following lens: A – short-focus B – any C – A or B D – long-focus	8. The following wavelength bands, which are typically used in IR thermography, correspond to the atmosphere transparency window: A – 1–3 and 5–10 μm B – 3–5 and 8–14 μm C – 1–10 and 2–20 μm D – A and B
9. Sensitivity of fotonic IR detectors is typically enhanced by cooling them with: A – liquid nitrogen B – thermoelectrical cooler C – Stirling cooler D – A, B or C	10. A blackbody-like object is heated up to 300 K. At what wavelength this body emits the maximum thermal energy? A – ~ 1 μm B – ~ 10 μm C – ~ 50 μm D – ~ 100 μm
11. Polymeric film absorbs IR radiation at 3.39 μm. What is the best operation wavelength of an IR thermometer to measure temperature of such film? A – 3.38–3.40 μm B – 3–5 μm C – 1–20 μm D – 8–14 μm	12. IR camera temperature resolution as a specified camera parameter…: A – depends on distance to object B – depends on atmosphere conditions C – A and B D – is independent on distance and atmosphere conditions

13. What spectral band can be recommended for the IR observation of remote objects under heavy meteorological conditions: A – 1–5 μm B – 8–14 μm C – 1–20 μm D – 0.3–100 μm	14. To measure temperature inside a vacuum-tight vessel by using a 3–5.5 μm, one needs to install a window made of: A – 'normal' glass B – quartz glass C – IR optical material transparent in the 3–5.5 μm band D – B or C
15. Spectral emissivity of a graybody… A – is constant at all wavelengths B – increases with an increasing wavelength C – is a harmonic function of a wavelength D – is a random function of a wavelength	16. Pulse phase thermography which uses the one-dimensional Fourier transform is applicable to: A – dynamic IR image sequences B – single IR images C – A and B D – steady-state processes only
17. Active one-sided TNDT is applied for the detection of air-filled defects in a solid material. In the heating stage, the temperature over defects …. than in a non-defect area A – is higher B – is lower C – oscillates D – is the same	18. Active two-sided TNDT is applied for the detection of air-filled defects in a solid material. In the heating stage, the temperature over defects …. than in a non-defect area A – is higher B – is lower C – oscillates D – is the same
19. How optimum observation times are related to defect depth in a one-sided active TNDT procedure? A – No particular relationship B – Increases approximately linearly with increasing depth C – Decays linearly with increasing depth D – Decays exponentially with increasing depth	20. Active TNDT is applied for detecting disbonds between materials with different thermal conductivity (metal-plastic). What is the optimum test procedure? A – One-sided procedure, heating metal B – One-sided procedure, heating plastic C – Two-sided procedure, heating plastic D – B or C

The manufacturer's authorised representative in the EU is Springer Nature Customer Service Centre GmbH, Europaplatz 3, 69115 Heidelberg, Germany. If you have any concerns regarding our products, please contact ProductSafety@springernature.com

Printed and bound by CPI Group (UK) Ltd, Croydon, CR0 4YY
05/10/2026
02228126-0017